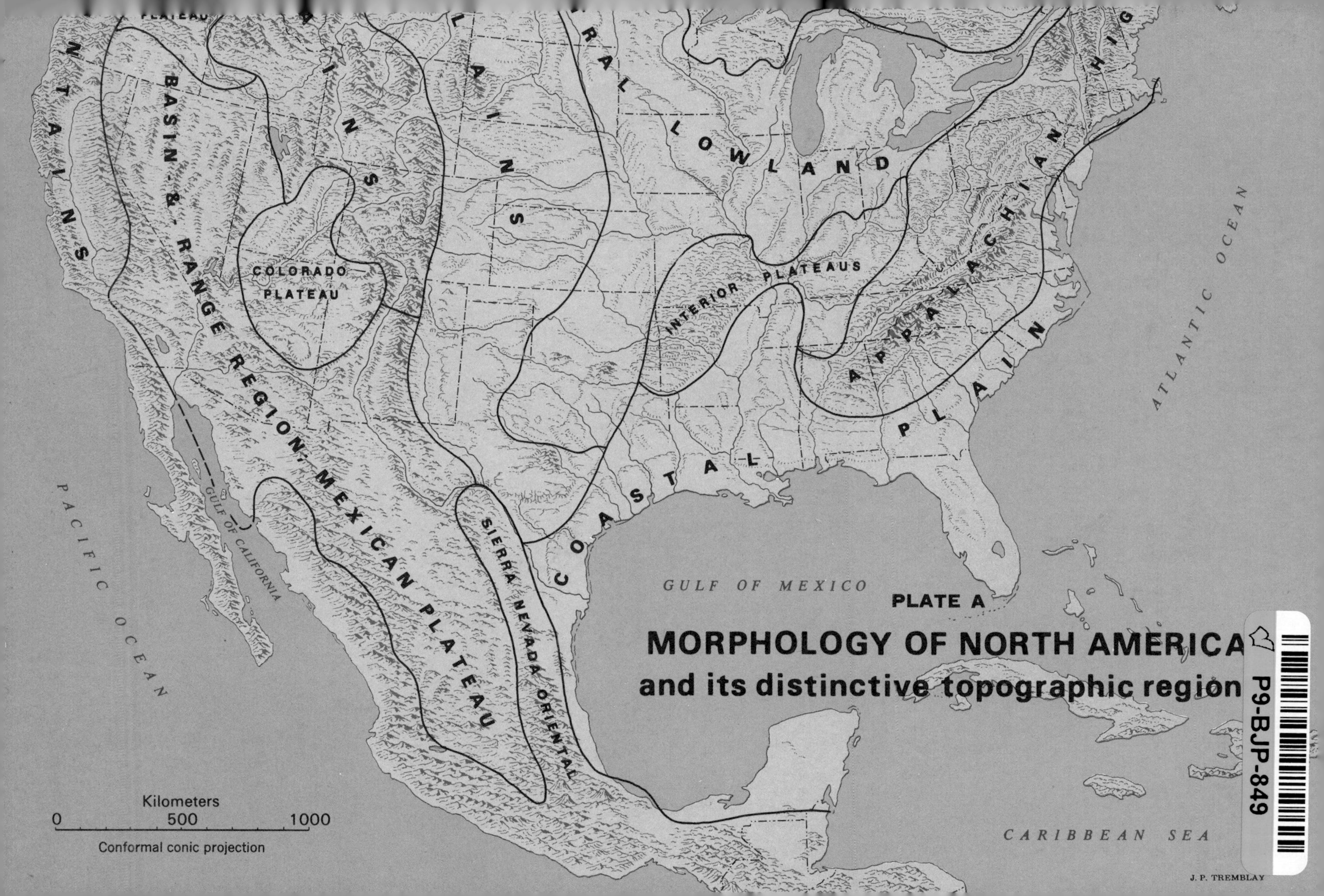
PLATE A
MORPHOLOGY OF NORTH AMERICA
and its distinctive topographic region
BASIN & RANGE REGION, MEXICAN PLATEAU
COLORADO PLATEAU
INTERIOR PLATEAUS
LOWLAND
APPALACHIAN
COASTAL PLAIN
SIERRA NEVADA ORIENTAL
GULF OF MEXICO
GULF OF CALIFORNIA
PACIFIC OCEAN
ATLANTIC OCEAN
CARIBBEAN SEA
Kilometers
0 500 1000
Conformal conic projection
J. P. TREMBLAY

# PHYSICAL GEOLOGY

# PHYSICAL GEOLOGY

SECOND EDITION

RICHARD FOSTER FLINT

BRIAN J. SKINNER

DEPARTMENT OF GEOLOGY AND GEOPHYSICS,
YALE UNIVERSITY

John Wiley & Sons

NEW YORK SANTA BARBARA LONDON SYDNEY TORONTO

STUDENT STUDY GUIDE
A student companion volume is available, which contains study problems, sample examination questions, summaries, and other study aids. It can be ordered from John Wiley & Sons, Inc. by giving this information:

Flint: STUDENT STUDY GUIDE to accompany
Flint and Skinner: PHYSICAL GEOLOGY,
Second Edition. By George D. Turner.

This book was printed by Murray Printing Co. and bound by Book Press.
It was set in Aster by Progressive Typographers.
Text design by Suzanne G. Bennett.
Cover design by E. A. Burke.
Cover photo by Terry Lennon.
The drawings were designed and executed by John Balbalis with the assistance of the Wiley Illustration Department.
Patricia Lawson supervised production.
Picture research was done by Marjorie Graham.



***Library of Congress Cataloging in Publication Data:***

Flint, Richard Foster, 1902–1976
Physical geology.

Includes bibliographies.
1. Physical geology. I. Skinner, Brian J., 1928– joint author. II. Title.
QE28.2.F55 1976 551 76-23206
ISBN 0-471-26442-3

Printed in the United States of America

10 9 8 7 6 5 4 3 2

# PREFACE

Our understanding of Earth with its multitude of materials and intertwined processes is a magnificent heritage from our predecessors. This understanding is built on centuries of patient observation, and brilliant insights. But only recently have we come to realize that each of us plays a small part in the changes that ceaselessly alter the face of the Earth. Though our individual contributions are small, the sum is large. We influence the atmosphere, streams, lakes, and oceans, we affect rates of erosion, we rely on Earth for our supplies, we cover its surface with cities; we are, in fact, a vital force in our own environment. Study of the Earth is a living, evolving, and tremendously exciting science. It is one to which each of us can contribute and in our turn bequeath increased understanding to those who will follow us.

Recent revolutionary advances have brought great changes in the depth and breadth of our knowledge of the Earth. At no time during the previous centuries have so many dramatic increases in understanding occurred within such a short period. Earth science is a field in a ferment, a subject laced with challenging excitement; new observations, new insights and new theories heighten the excitement every day.

In the previous edition of this book we pointed out that much of the excitement has arisen from discoveries related to three recent and revolutionary advances. The first involves an increasing awareness, particularly in industrial nations, of the effect of human activities on environments at the surface of the Earth, accompanied by widespread attempts to analyze such effects and to modify them where necessary. These attempts represent not only scientific concern, but public concern on a broad scale. They are, likewise, closely related to the "energy crisis," which has stimulated renewed appraisal of Earth's sources of energy and of the ways in which energy is used. We have finally understood that people are not just one of the minor forces of nature—they are a major force. What the Earth will be like in the future depends very much on how we act.

The second advance concerns the way Earth works, the dynamics of the lithosphere. Exploration of ocean floors and of Earth's crust and mantle has revealed crucial new facts that force the rethinking of Earth's dynamics. For the first time we can offer answers to such questions as the origin of ocean basins, why the continents are where they are, and how mountain ranges form where they do. Many processes are being reexamined; new concepts are replacing older assumptions. A fundamental group of these concepts has coagulated into the dramatic theory of plate tectonics.

The third advance springs from the start of systematic investigations of the Moon, Mars, Venus, Mercury, and other members of the Solar System. Earth and its fellow planets have a common birthright, and though each planet has evolved differently, common threads run through their histories. By unravelling those common threads we have been led to a deeper understanding of the history of Planet Earth. We are reaching a point where we can attempt answers to such questions as why Earth exists at all, why it is like it is, and are there likely to be other, hospitable, Earth-like planets in the Universe?

The present book, a successor to a long line of Yale textbooks of physical geology, has been written in sharp realization of our heritage of knowledge and of the need to integrate with the corpus of classical knowledge the many fruits of the current revolution. We have tried to do so in a cohesive form that is intelligible to students at a beginning level. Many of the chapters are entirely new and in one, where climate is discussed, a completely new topic is introduced; other chapters have been rewritten to whatever extent seemed desirable in light of new information. Inevitably the flavor of the book differs from its predecessors. More attention has been paid to the role of people, to their involvement with and reliance on resources from the Earth; plate tectonics is integrated throughout the book rather than being treated as an isolated topic and the total coverage has been greatly increased over the previous edition; the early history of continents is discussed more fully.

We have continued features of earlier editions that have proved to be well received. Among these are the review summaries at the ends of chapters and the unique scheme of defining technical terms within the body of the text. A term defined in the text appears in boldface italic type, and the defining phrase or clause appears in lightface italics. The page on which the term is defined is indicated by boldface type in the index. The definition thus occurs in context, in the presence of additional information, and perhaps also in one or more clarifying illustrations. Thus, in some instances, the reader can see how a definition is built and even why some definitions are difficult to frame. In addition, a glossary is included for those who may wish it.

We have used metric units throughout the book. For those wishing to convert from metric to English units, we have included an appendix of conversion factors.

For an outline, ideas, and phrases, besides help and guidance in many matters, our sincere thanks go to Margaret C. H. Flint.

As work on this edition of *Physical Geology* drew to a close, Richard Foster Flint died suddenly, tragically, and unexpectedly. With his passing we are all the poorer; he was a giant in his special area of geology, and he was one of those insightful scientists who contributed greatly to our heritage of knowledge and to the current revolution in Earth sciences. Richard Flint taught and trained many of the current leaders in geology. Equally importantly, he believed, and worked tirelessly to impart the belief, that an appreciation of the Earth and the way it works is as important for an educated person as an appreciation of literature or music. Richard Flint's eloquent teaching and concise, profound writing for the beginner enticed generations of students to see, know, and appreciate the fragile complexity of the world around them.

B.J.S.

# CONTENTS

**"Here are our neighbors."**

**"Let's invite them to lunch and get away from the crowd."** (*Luis Villota/Photo Researchers.*)

# PART ONE BASICS

# 1 EXPANDING HORIZONS FOR A SHRINKING EARTH

The dominant species
Agriculture and expanding populations
Energy consumption takes over
No more frontiers
Earth, small but intricate

You, who belong to the present generation of students, have inherited the fruits of countless thousands of inquiring minds. You are heirs to a vast legacy of knowledge about the Earth, our environment, our home. You can, if you will study this legacy, understand how the Earth works, how the natural activities that shape it are interwoven in complex ways, and how the environment of which you are a part was created. You can understand also a threat of frightening dimensions—how that environment can be changed, distorted, and even shattered, by the impact of human activity.

Being heirs to a body of knowledge is nothing new. Each generation enjoys that privilege. But, unhappily, yours is not merely another in a long line of generations; it is unique. It is the first to face the terrifying reality that man is now so powerful and so numerous that he rivals all the other forces of nature that change and shape our fragile environment. Therefore, you have a crisis on your hands, one that has come to a head so recently and so quickly that earlier generations, confused by long-accustomed ways of doing and thinking, have failed to join battle with it. But you have something that prior generations lacked; you have a point of view toward man's place in the natural world. You must now use your understanding of the Earth to impel your own generation, and future ones, to accept discipline, to make hard decisions, to sort out and choose a list of priorities by which the quality of life can be maintained and enhanced.

This responsibility, that now devolves on us all, has been enormously intensified by three great scientific discoveries. The first concerns life, and was finally stated by Charles Darwin in his theory of evolution. The second concerns matter and energy, and startled the world when physicists discovered the structure of the atom. The third concerns the Earth, and happened only when a new theory, the theory of plate tectonics, burst forth to transform our understanding of the Earth's dynamic movements.

These three discoveries have flooded the scientific world with new knowledge. Thanks to the scientists who are now analyzing that knowledge, you face the realization that Earth's capacities are finite. Earth's natural resources are limited, and its responsiveness to the demands of industrial mankind are weakening its ability to recover from thoughtless or deliberate exploitation. The demands of an increasing population have become so great that human ability to change and shape environments has come to equal that of natural forces.

At this point it helps us gain perspective if we look back into history and see how the crisis in Earth's development has come about. As we look back, we can see easily that the crisis had to happen some time, simply because of the presence of the species we call man. The only uncertainty was *when* it would happen. Now, just past the middle of the 20th Century A.D., in about the 4.6-billionth year since Planet Earth was formed, and in at least the 3-billionth year since life on the planet began, the crisis has arrived.

It came about in this way. Through billions of years of evolution, Earth's surface has been inhabited by various kinds of living things that gradually became more and more complex. The fossils found in Earth's layers of rock tell us that about 1 billion years ago, small but distinct animals as well as plants were living in the ocean. By 400 million years ago both plants and animals had invaded the lands, and animals with backbones had developed. By 150 million years ago reptiles (including dinosaurs) dominated the lands, although primitive kinds of mammals and birds had already appeared. By 60 million years ago, mammals had displaced reptiles as the dominant land animals. By 3 million years ago (and perhaps much earlier), man had evolved from the mammal stock and had begun to make tools from stone. The fossil human bones and the tools are here to prove it.

**The dominant species**

By half a million years ago *Homo sapiens,* our own species of man, had evolved and was making tools with increasing skill. His exceptionally large and complex brain enabled him both to make the tools and to use them in ever more sophisticated ways. Throughout his long existence, Stone-Age man was a skillful hunter of wild game. But as he increased in number, game became less abundant; the meat-eating population began to have trouble and, little by little, turned for its food to small mammals, birds, and fish, supplementing that diet with wild seeds and berries.

## Agriculture and expanding populations

Still, populations kept on increasing and so, between 12,000 and 9,000 years ago, a worldwide agricultural revolution got under way. Agriculture and the domestication of animals began to supplant hunting as the principal source of food. Agriculture made other things happen. Instead of living in caves as small bands of hunters, people began to build primitive dwellings, gathered into villages, and thus turned to a more settled life. Gradually, cities containing thousands of inhabitants grew up. Forests were cut down and were replaced by fields, for agriculture could provide much more food than hunting ever did. Gradually, streams were diverted, rivers were dammed, and hillslopes were terraced to provide more favorable growing conditions (Fig. 1.1). Such changes began early. Evidence in the country south of Baghdad in Iraq

**Figure 1.1**
**These mountain slopes in the Philippines were terraced 2,000 years ago for growing rice and are still being used today. (*WHO Monkmeyer.*)**

shows that in order to irrigate crops, people had begun to interfere with the Tigris and Euphrates Rivers as early as 7,000 years ago (Fig. 1.2).

The effect of this tremendous change was to replace natural groups of living things—wild animals and wild plants—with artificial groups consisting of domesticated plants and domesticated animals. But the change also had a side effect. It removed people one long step from direct participation in a wholly natural economy; it pushed them toward a life in a world of their own creation. Although the turn to agriculture increased the supply of food, the increase barely kept up with the steepening curve of human population.

**Energy consumption takes over**

Agriculture as the principal basis of human economy had lasted many thousands of years when, on both sides of the Atlantic Ocean, and beginning in the 18th Century, a marked change in the use of energy began to make itself felt. This was the start of the age of intensive use of energy, also called the industrial age because it is an age of high-energy industry.

Stone-Age people had an industry too: they made weapons and

Figure 1.2
**Irrigated vegetable fields in the Coachella Valley, southern California. The snaky line, center, is a canal bringing water from the Colorado River. The desert, right, is just as it was before irrigation. (*J. S. Shelton.*)**

tools of stone. But the energy needed to make stone tools was only the small amount supplied by human muscle. Other sources of energy remained to be tapped. Six thousand years ago people discovered metal ores and learned to smelt them and to work the metals into implements. The smelting required energy, most of which came from the burning of wood. Still, until the 18th Century, human industry drew its energy mostly from the muscles of men, horses, and cattle, and so production remained small.

The 18th Century brought with it the invention of steam engines and the ability to convert the energy locked up in wood, coal, and other fuels into other forms of energy. Machines could do work formerly done by men and by horses, so that, in theory, all individuals could have machine "servants" and machine "animals" working for them. Coal came into wide use as a fuel because it yields, on an equal-weight basis, far more energy than wood. Then, with the invention of internal-combustion engines, petroleum became a widely used fuel. From our vantage point in the last quarter of the 20th Century, it appears that nuclear energy could eventually become a successor to wood, coal, and petroleum. The use of high-energy fuels enabled a given number of people to exploit more plants, animals, and inorganic materials and to produce much more food and many more products than had been possible in the days of predominant agriculture. But high-energy fuels led also to huge industrial plants, giant cities, commuters and high-rise apartment buildings. Machines and cities are built with cement, metals, and other materials won from the Earth. Every year we seem to need more and more, until today we use Earth's materials in such vast amounts that it is hard to imagine where they all go (Fig. 1.3).

This huge change affected everyone who participated in a high-energy economy. It surrounded people with a whole range of artificial things. In doing so it removed them a further long step from direct participation in a natural economy. And it resulted in major changes in the environments in which energy use was highest. Stone-Age people had made little more change in their

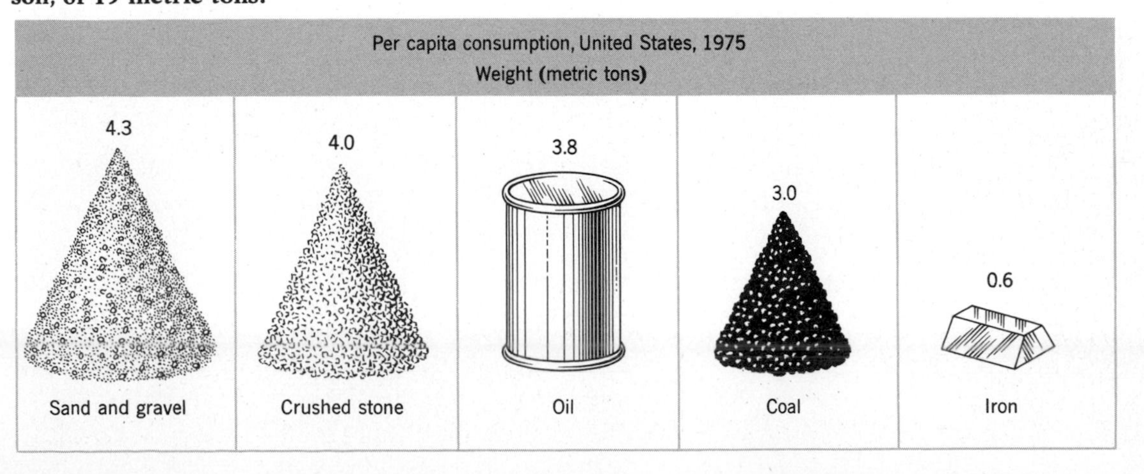


Figure 1.3
**The total amount of material mined and dug from the Earth increases each year. During 1975 the United States used nearly 3.4 billion metric tons of such material, mostly cement, sand, gravel, and crushed stone for building, coal and petroleum for fuel, iron and other metals for machines and a myriad other purposes. When we divide the total consumption of *all* minerals by the U.S. population we get an average yearly consumption, per person, of 19 metric tons.**

environments than did other species of animals then living; they fitted well into their natural surroundings. But with the turn to agriculture, things began to be different. The most obvious change was the destruction of forests, but in and near cities there was pollution of streams. With the age of intensive energy, the changes became far greater. There were many more people, many more cities, and new, powerful machines. The result was gradual destruction of the plant cover, devastation of whole landscapes, pollution of the air, of streams and lakes, and even of the ocean. Gradually, parts of Earth's lands have approached a state in which they will no longer be habitable. Yet the growth of Earth's human population (Fig. 1.4) continues unchecked. Despite slowdowns in some countries, including the United States, world population as a whole is still doubling every 33 years. Unless a miracle occurs, by the year 2000 the world will be swarming with seven billion human beings. The demand for food is already pushing forward ever more strongly the exploitation of plants and animals, and will tend to increase the existing intensities of pollution. That is to say, this demand will subject the delicately balanced natural economy to additional pressure. To put the case differently still, the human species is becoming less and less well-adapted to its proper place in nature. That this is true is shown by three basic indicators: (1) serious overcrowding, (2) decreases in per-capita consumption of food along with increases in per-capita consumption of energy, and (3) widespread destruction of terrain and pollution of air and water.

With a limited view confined to the natural community around us, we sometimes fail to realize that human activity creates tremendous changes in Earth's natural features. A major example is the group of changes being wrought by strip mining of coal in the Appalachian region of the United States. Destruction of terrain and pollution of water are clearly evident in the huge continuous gashes along hillsides. Broad horizontal shelves floored with

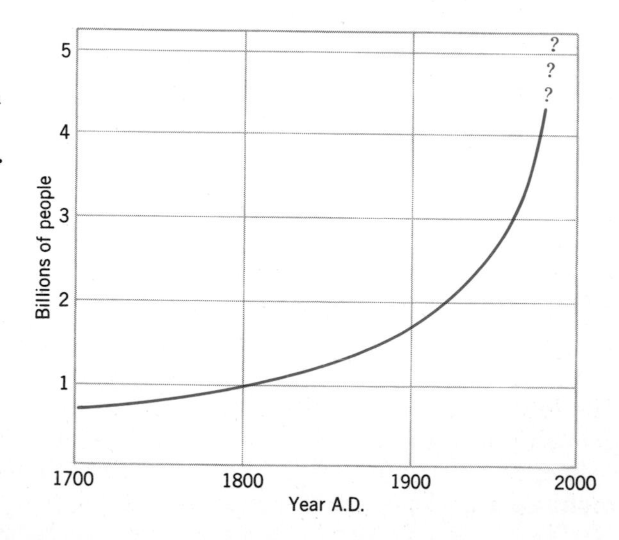


**Figure 1.4**
**From the time when man first appeared on Earth, his numbers have slowly increased. Beginning with the industrial age that commenced in the 18th Century, the world's population has approximately doubled every 33 years.**

broken rock waste and backed by sheer rock walls are the most conspicuous feature of many Appalachian landscapes. By 1975 such shelves aggregated more than 30,000km in length, a length that begins to approach the circumference of the Earth. As stripping proceeds, it destroys forests and soil, reduces the amount and quality of the water in the ground, pollutes streams by pouring into them quantities of debris and acid chemicals released from the coal, creates landslides, and causes erosion that leads to the piling-up of debris on valley floors. Slopes once forested become dangerous eyesores; entire stream systems become clogged and lifeless.

The bad effects of human interference with natural systems are evident also in the recent history of the Caspian Sea, the world's largest salt lake, about the size of the state of California. Since about 1930, its water surface has lowered by nearly 2.5 meters, leaving many port facilities literally high and dry. The water has increased in salinity and the commercial catch of fish has decreased by half. Also, because there were fewer fish to eat the mosquito larvae, there were epidemics of malaria.

The principal source of water for the Caspian is the huge Volga River. Because the lake has no outlet, the inflow of water is balanced by evaporation under a warm, dry climate. Since about 1930, contribution of water by the Volga has decreased markedly. The decrease resulted in part from decrease in rainfall and a warming trend in the climate, but also has been influenced to a considerable though unknown extent by withdrawal of water from the Volga for irrigation. A proposed plan to counteract the decline would divert a large Arctic river into the Volga. Although this would increase the flow into the Caspian, it might have other, unforeseen consequences.

We should not give the impression that man's interference with nature is always bad. Terrain can sometimes be rearranged on a very large scale without obvious unfavorable consequences. The conversion of shallow sea floor into usable land is an example (Fig. 1.5). Since about A.D. 1200, the Dutch people have been reclaiming land from the sea by building dikes and draining lakes and marshes. Today the reclamation process has become part of the high-energy economy, and the resulting dikes and other engineering works are huge. The reclaimed land adds up to more than 10 percent of the area of the Netherlands. But that much land—5,000 square kilometers—is less than the estimated area of Dutch terrain that has been lost to the sea, since the year 1200, by waves and currents along other parts of the coast. The engineers are not even holding their own!

Two vital lessons can be learned from the three examples described, and from the many other examples given in this book. The first is that any artificial change in a natural system will probably lead to unforeseen and undesirable side effects. The second is nicely illustrated by the Dutch reclamation example: the more a natural system is changed—the further it is pushed out of

**Figure 1.5**
**Land reclaimed from the sea in the Netherlands.** *(Adapted from a map published by Netherlands Ministry of Transport and Waterstaat.)*

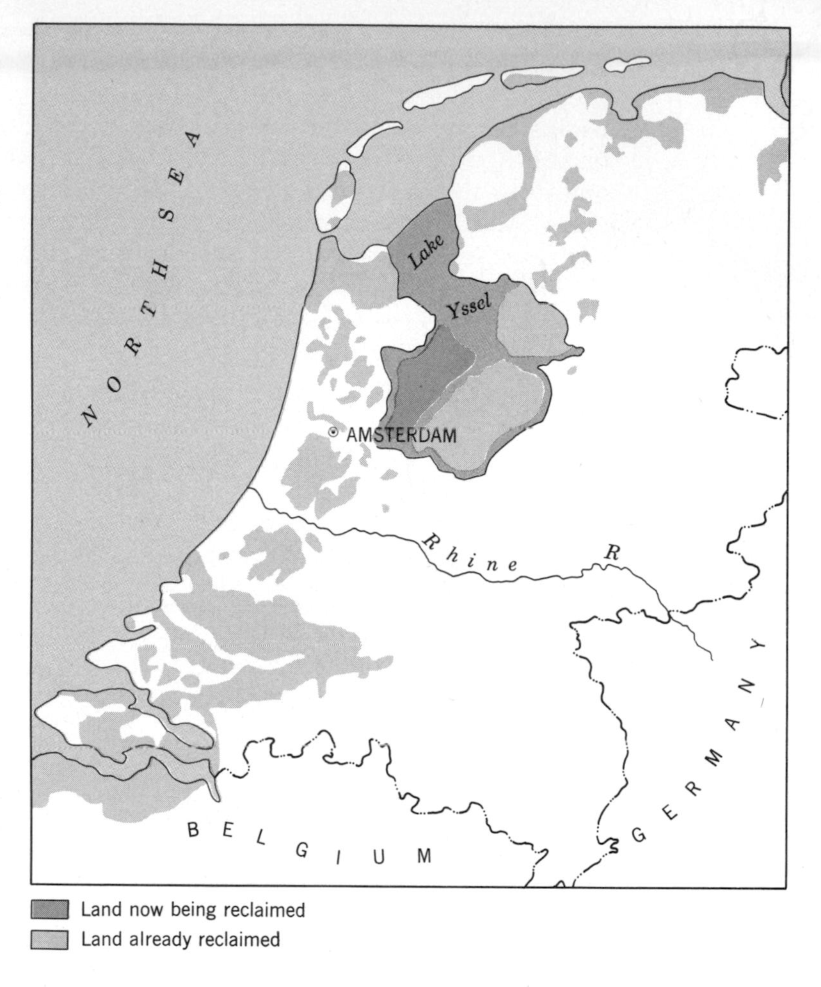


balance with nature—the greater becomes the effort that must be expended to maintain it.

**No more frontiers**

From this mini-review of the history of man in relation to his environment, we can return with more understanding to the crisis faced by the rising generation today. Why do you, today's young generation, have a responsibility that the older generation did not feel? Because you were born, or at least grew up, *after* two things had happened. One was that the use of nuclear energy was added to our technology. This demonstrated man's technical ability to set free forces that are destructive to a degree not previously dreamed of. The other was that exploration and scientific study dissolved the last frontier, the last boundary between our familiar, inhabited world and the unexplored, mysterious beyond.

While frontiers still existed, there existed also the sense of freedom to search the beyond for new lands and new resources. The restlessness of men took them on journeys and voyages of discovery. Among the many discoverers were Eric the Red and Leif Ericson, Marco Polo, Columbus, Sir Francis Drake, James Cook, Lewis and Clark, and Robert Falcon Scott. Those men were hardly responding to any personal need of land; their motives were various, including desire for adventure, curiosity, and greed for treasure or other exploitation. But their discoveries reflected a point of view, common to their contemporaries, that was rooted in the existence of more land, more of everything, beyond the horizon. This embodied the feeling that Earth has always provided its human population with more land whenever land was needed, more food and minerals when known supplies ran out, and that somehow, new land and new resources would be available through a future time of indefinite length. In short, there had developed a belief that a limitless cornucopia of riches would continue. That comfortable point of view made it easy to abandon land damaged by agriculture, by the cutting down of forests, or by mining, and to shift to another site. There was no obvious need for men to feel responsibility for Earth because Earth's habitable land had not been used up. As long as there were unoccupied lands, people could and did move into them. To the minds of some, the possibility of inhabiting other, nearby planets offered a way out. But the Space-Age examination of the Moon, Mars, Venus, and Mercury has made it abundantly clear that free habitation is impossible. Like it or not, we must live on Planet Earth, and we must accept the realization that all of Earth's frontiers have disappeared.

As the frontiers dissolved one by one, natural systems were damaged increasingly. The dissolution, in the 1950s and 1960s, of the last land frontier, the Antarctic Continent, brought into focus a huge problem. The human population continues to increase, and its mechanization continues to lay ever-increasing stresses on the environments in which people live. The present rates of increase are being maintained only by increased damage to natural systems, and this intensifies the damage already inflicted on them. Earth's supplies of soil, water, air, and even space itself are limited. There is not enough food for all the people living today, let alone for those who will be born during the next decade or the decade after that.

These facts explain why we chose the title for the first chapter of this book. The Earth *has* shrunk and *is* shrinking, in the sense that as its frontiers have disappeared, it has become smaller in terms of human expectations. This has created a problem entirely new in the history of people. To expand the horizons that have shrunk is the responsibility of the generation now young. These horizons must be expanded in new ways that will repair the damage done to Earth's natural systems and will make human activities work within the systems instead of against them. Solutions to the problem will be based on several kinds of knowledge, but of primary importance is the understanding that our Earth, as a planet, is small.

**Earth, small but intricate**

Although small, Planet Earth is made up of different materials, all of them involved in a wonderful array of interlocking movements. There is movement in the varied layers that form its interior, its outer, ever-changing crust that is continually being destroyed and renewed, and, at its surface, the huge mass and almost infinite variety of living things, interacting with each other and with Earth's nonliving parts. It is specially necessary to understand the repeating, cyclic pattern, at Earth's surface, of the activities on which environments depend, how the activities are energized, and the speeds with which they act.

To you, the reader, the facts of physical geology and related facts about the world of plants and animals offer a graphic explanation of these Earth systems for which your generation has suddenly become responsible. Earth sciences, of which physical geology is the foundation, are the study of our environment. To the average educated man or woman, regardless of occupation, this study and this knowledge is necessary as never before, to enable him or her to support or oppose with wisdom measures that will be proposed to deal with parts of the vast environmental problem. To the person who aims at a professional career in any of the many aspects of environmental studies, a broad view of Earth's natural activities is an essential beginning. The realization of Earth's finite limits is creating a social revolution. But an understanding of the limits can only come through scientific study of the Earth itself and how it works. This is the domain of ***geology,*** *the science of the Earth.*

# 2 THE EARTH, INSIDE AND OUT



## EXTERNAL PROCESSES

Living at Earth's solid surface, we are aware of two kinds of things, one tangible, the other intangible. The first consists of *materials*—substances such as rock, sand, clay, organic matter—that are common here. The second group consists of active *processes*—water flowing as a river and carrying mud or sand, surf pounding against a shore and making a beach, air flowing as wind and carrying sand and dust. These and other processes expend energy, derived from the Sun's heat, in working on the materials, which they continually move from place to place and in doing so, wear them down. The solid particles they are moving are mostly bits of rock; also moving, invisibly, are substances dissolved from rock. ***Rock*** is *any naturally formed, firm and coherent aggregate or mass of mineral matter that constitutes part of Earth's crust.*

The moving water and the solid particles in it are a very small sample of a great chain of processes that operate all over Earth's surface. The water that flows off through gutter and river comes from rain, which comes from clouds, which in turn form from moisture evaporated from the ocean. As rivers empty into the ocean, the traveling water, ocean to land to ocean, completes an endless circle and is ready to begin anew. The bits of rock carried in the flowing water are part of another circle, one that interlocks with the circle traveled by water. The particles of rock came originally from the firm rock that forms all continents and that is continually being broken into particles. Like the water itself, the rock particles are on their way to the ocean, where they will be spread out and deposited, and will eventually form new rock. *The complex group of related processes by which rock is broken down physically and chemically and its products removed is* ***erosion*** ("wearing away").

All the activities involved in erosion, and also in the deposition of the eroded materials, are together called *external processes* because they operate at or near Earth's solid surface. We cannot see, directly, the very different *internal processes* that are at work deep below the surface; but we can see some of their effects. Unexpected earthquakes, volcanic eruptions, and the sight of strange rocks at the surface remind us that Earth is far from quiet, and that down in the depths, things must be happening. The internal processes that work down there influence all living things, including people.

# EARTH'S INTERNAL ANATOMY

Although we cannot sample or examine Earth's depths directly, we can, by carefully examining its surface, infer a good deal about what goes on down below. Out greatest understanding of what is happening there comes, however, not from examining the surface, but from measuring things indirectly, things we cannot see. We measure the speed with which earthquake waves pass through Earth's body, we measure irregularities in the paths of orbiting spacecraft, which indicates that the pull of Earth's gravity varies slightly from place to place, and we measure variations in the force of magnetism at different places. From such indirect measurements we have learned that the solid Earth consists not of one single material, but of distinct layers, like the skins of an onion. Unlike the onion, however, each skin has a different composition.

### Core, mantle, and crust

(1) There are three such layers (Fig. 2.1): a massive *core* composed largely of metallic iron, part of it molten, (2) a *mantle* of rather dense rocky matter that encloses the core, and (3) a thin outer zone, the *crust,* consisting of somewhat lighter-weight rock. The crust is not uniform in thickness. That beneath the oceans, *the oceanic crust,* is about 10km thick, whereas the *continental crust,* which comprises the continents, ranges from 20km to 60km thick (Fig. 2.2).

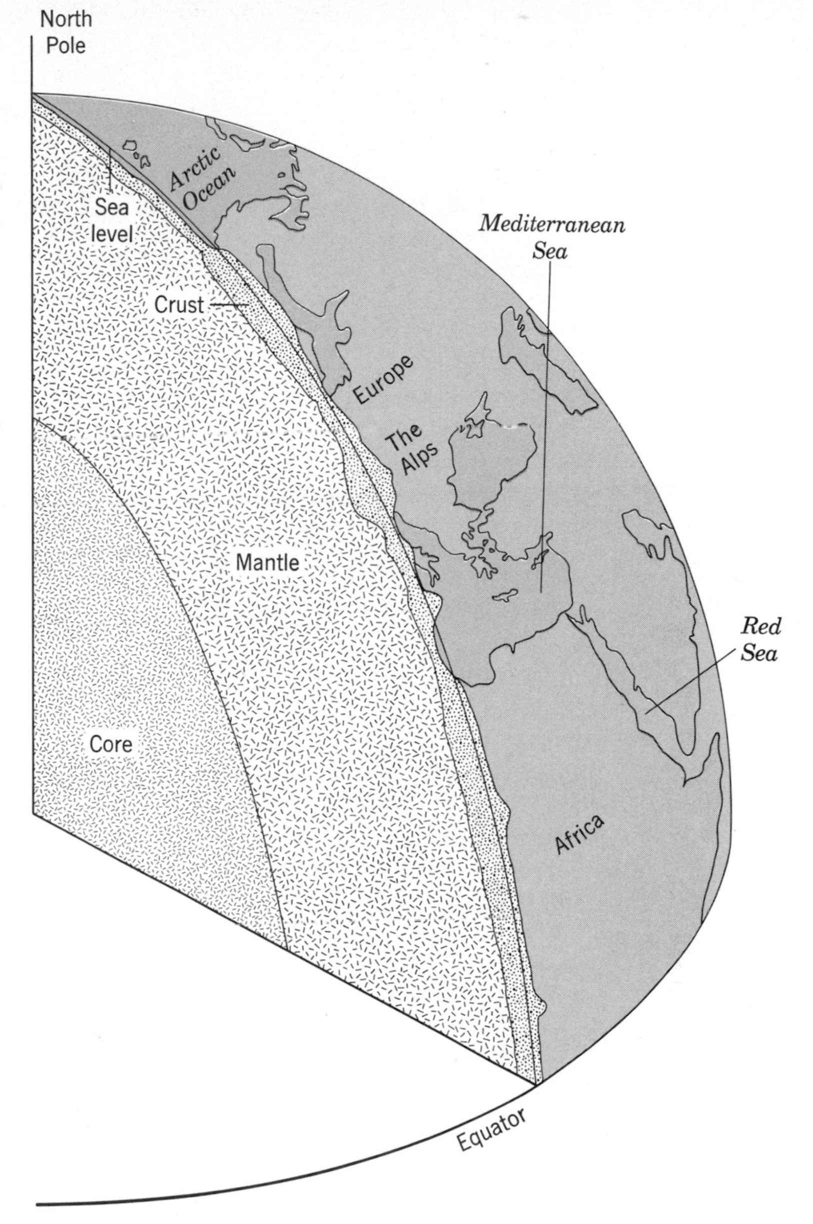


**Figure 2.1**
**A slice of the Earth, revealing layers with distinctly different composition. The core is composed largely of metallic iron and nickel, the mantle is a zone of dense rocky matter, and the crust (here shown with exaggerated thickness), is much less dense rocky matter with a composition distinctly different from that of the mantle. The slice cuts through the North Pole, central Europe, and Africa.**

The mantle and core have distinct compositions, and the boundary between them is sharp. The crust is more varied but its composition is distinct from the mantle below, and again, the boundary between them is sharp.

**Lithosphere and asthenosphere**

Just as water substance can exist, according to its temperature, in different states (solid ice, liquid water, gaseous water vapor), so do other materials in the Earth change their physical properties as their temperature and pressure change. The regions where physical properties change do not coincide exactly with the boundaries between core, mantle, and crust (Fig. 2.3). Within the core there is

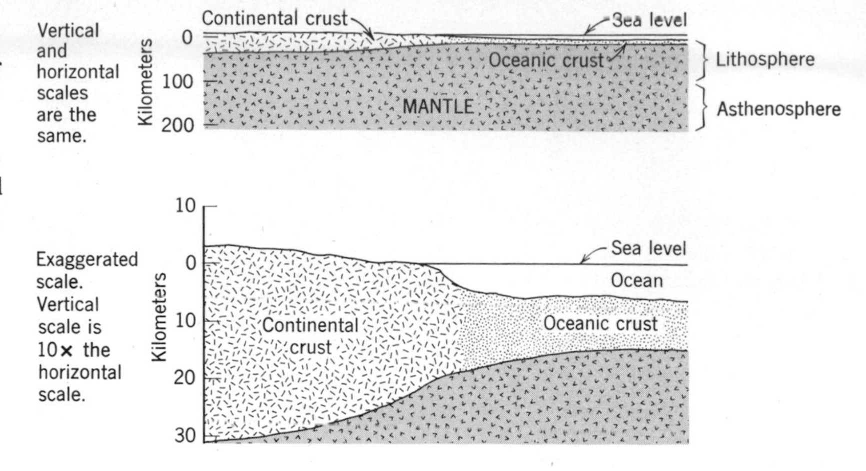


**Figure 2.2**
**Cross section of the crust and upper part of the mantle. The crust is of two different kinds. One kind underlies the continents and is 20km to 60km thick; the other kind underlies the oceans and is only about 10km thick.**

an inner region where pressures are so great that iron occurs as a solid. Surrounding the inner core is a zone where temperature and pressure are so balanced that iron melts and exists as a liquid. Analogous changes happen in the upper part of the mantle. Down to a depth of about 100km all rocks—crust plus an upper portion of the mantle—are rigid, hard, and brittle. This outer 100km is commonly called the *lithosphere,* meaning the "rock sphere." Below 100km, rocks in the mantle are still solid, but they are plastic somewhat like toffee or tar, and easily deformed, with many properties akin to very viscous liquids. This zone is called the *asthenosphere,* meaning the "weak sphere." The boundary between the lithosphere and the asthenosphere is distinct, but does not reflect a sudden change of composition. The compositions of the base of the lithosphere and the upper part of the asthenosphere are, as far as we can tell, identical. The boundary between them is merely one across which certain physical properties change rapidly, somewhat as they do at the boundary between water and a chunk of ice that floats in it.

Although the upper surface of the asthenosphere is distinct, there

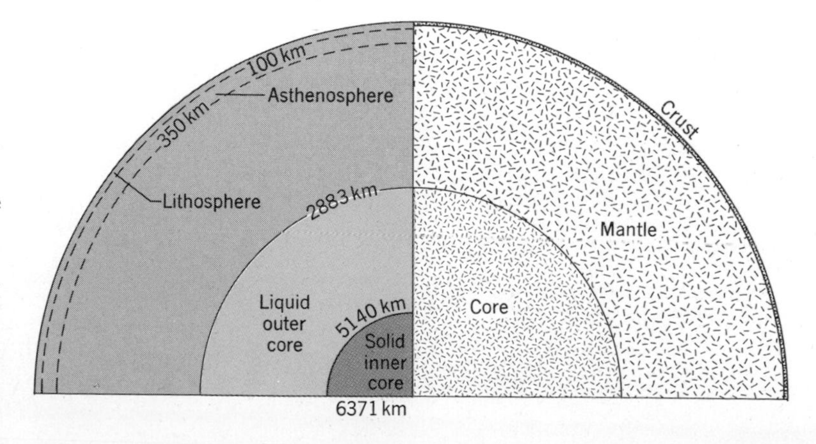


**Figure 2.3**
**Section through the Earth showing, on right, compositional layering of crust, mantle, and core. Left side of diagram depicts the way physical properties change: the *lithosphere* is brittle, rocky material; the *asthenosphere* is rocky too, but can be plastically deformed, like putty.**

does not seem to be a distinct lower boundary. Rock in the asthenosphere becomes gradually more rigid with increasing depth. Although it never quite reaches the brittle, rigid state of rock in the lithosphere, the plastic properties of the asthenosphere disappear by about 350km.

### How solid is the solid Earth?

Any rock we examine at the surface is obviously solid; so we think of the Earth as a solid. Most people therefore assume the rest of the Earth must be solid too, but this is not true. The familiar picture of a volcano that pours forth a stream of molten rock (*lava*) reveals that there must be at least some liquid below ground. An internal liquid of this kind, while still beneath the surface, is *magma,* although after it has come out upon the surface the same liquid is called *lava.* Magma forms because radioactive heat continually tends to raise the temperature of rock below ground, until eventually the rock melts. Yet only a small fraction of Earth's rocky material is molten at any time and the melting only happens in special, localized places. Also, we can easily see that rocks formed from cooling lavas differ from place to place; so we infer that the composition of magma varies from one place to another. Determining how these variations occur, and why and where they occur, allows us to infer a great deal about Earth's internal processes.

If a force of any kind disturbs it, magma, like other liquids, tends to move. Consequently, when subjected to uneven pressures, bodies of magma move slowly upward like toothpaste being squeezed through a tube, forcing their way through solid rock, gradually cooling, and in some places reaching the surface, where they emerge as lava.

Magma obviously exists; so the "solid Earth" is not as solid as it seems. Even in places where rock has not melted to form magma, there is evidence that other curious things have happened. Figure 2.4 shows rock that has been squeezed. It has actually flowed, as a pat of cold, stiff butter would flow out sidewise if you pressed down on it with your thumb. In contrast, Figure 2.5 shows layered rock that seems to have been bent. We infer that sidewise pressure must have contorted the layers without breaking them. Finally, in Figure 2.6 we see rock that has broken and shattered like glass. It was brittle, and so, when pressed, responded by shattering. These three photographs show that some of the forces at work within the Earth must be very strong. Such forces continually push bodies of magma around, and also slowly bend and fracture the solid Earth. We cannot see or easily measure the movement of material in the mantle, but after a good deal of study of evidence found in the continental crust, we can infer that great segments of the crust have been thrust up repeatedly to form mountains, and great blocks have been pulled down repeatedly to form basins. We can infer also that the rigid, rocky lithosphere is broken into a series of fragments, or plates, each about a hundred kilometers thick but as

**Figure 2.4**
**Rock showing evidence of extreme deformation,with evidence of flowing movement caused by squeezing. Layers that were originally flat and uniformly thick are now highly contorted, thickened in some places and thinned in others. The rock is gneiss, exposed in a road-cut in central Connecticut. The nearly vertical marks crossing the photograph are drill-holes used in blasting out the road-cut. Width of exposure, about 2 meters. (*B. J. Skinner.*)**

much as thousands of kilometers across, and that the plates slide at slow but measurable speeds over the top of the asthenosphere. This extraordinary and fascinating inference is the exciting and far-reaching theory of plate tectonics that was mentioned briefly in Chapter 1. We will discuss it further in Chapter 4 and later chapters.

# THE OUTER ENVELOPES

The outermost part of the solid Earth is surrounded by layers of water, air, and living things. The water, the air, and the living things are continually moving about and reacting with each other and with the crust, in many ways. It may be helpful to think of the outer part of Earth as consisting of four spheres: *hydrosphere, atmosphere, regolith,* and *biosphere.* Each is a shell or envelope, wrapping around and to some extent mixing with the sphere next within it. Each consists of characteristic materials. And in each, distinctive activities are at work.

### Hydrosphere

The hydrosphere is the "water sphere," embracing the world's oceans, lakes, streams, water underground, and all snow and ice, including glaciers (Fig. 2.7). A very small part exists in the atmosphere in the form of water vapor. Watching a steaming teakettle on the stove, we can see a bit of the hydrosphere entering the atmosphere around it.

Figure 2.5
**Strongly bent layers of sedimentary rock. Originally a stack of flat, horizontal layers of sediment, these spectacularly deformed rocks were bent and folded by the forces that formed the Canadian Rocky Mountains. Near Sullivan River, British Columbia. (*Courtesy, Geol. Survey of Canada.*)**

Figure 2.6
**Rock that has broken and shattered like glass. The shattered fragments were later cemented together by substances deposited from water slowly circulating through them. The fragments are igneous rock formed by cooling of lava. Specimen is about 20cm in diameter and is rounded because it is a boulder that has been tumbled around in a stream. (*Courtesy, Mackay School of Mines, Univ. of Nevada, Reno.*)**

## Atmosphere

Although the word *atmosphere* really means "vapor sphere," it would be more logical to label it "air sphere," because it consists of the mixture of gases that together we call air. It penetrates into the ground, filling the openings, small and large, that are not already filled with water. The atmosphere is never quiet. Some part of it is always moving, as we are well aware whenever we feel a wind blowing.

## Regolith

Where the atmosphere and hydrosphere are in contact with the surface of the land, reactions take place between them. For example, the atmosphere with its content of water vapor acts upon ***bedrock***, the *continuous solid rock of the continental crust,* breaking it up mechanically and causing it to decay chemically. We can observe the process at any ***exposure*** (sometimes also called *outcrop*), *a place where solid rock is exposed at Earth's surface.* As a result, the surface rock, although still a solid, is no longer a continuous solid. It has been subdivided into many pieces, most of them very small particles. This broken-up part of the crust has a separate name, ***regolith*** ("blanket rock"), because it lies like a blanket draped over (and in many places grading into) the continuous solid rock beneath. It is defined as *the blanket of loose, noncemented rock particles that commonly overlies bedrock.* We would usually have to use a hammer or even a high-speed drill to collect a sample of bedrock, but a shovel or pick would ordinarily be enough for sampling regolith.

Not all regolith has been broken up and left in place. Some of it has been moved and set down in a new site. It is on its way (with stopovers) to the ocean. *Regolith that has been transported* by any of the external processes is called ***sediment,*** a word that means "settling."

In some places, where bare bedrock is exposed at the surface, there is no regolith at all. In other places regolith is 100 meters or more in thickness. In most land areas the upper part of the regolith

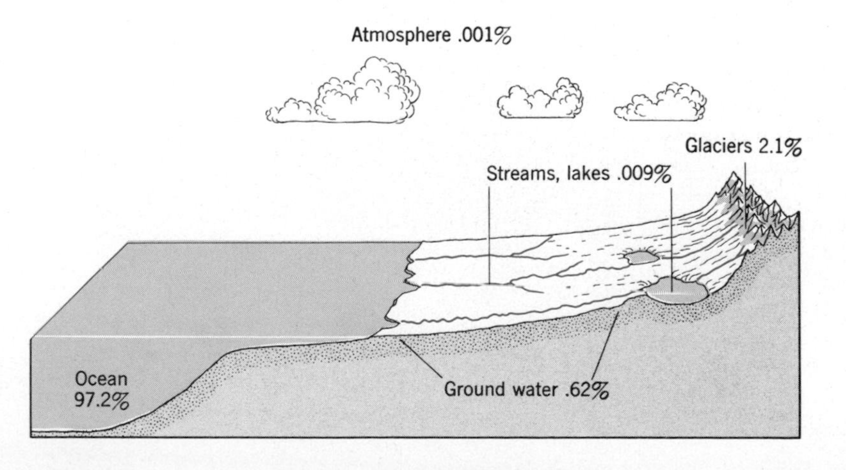


**Figure 2.7**
**Distribution of water in the hydrosphere.**

**The world ocean is the great reservoir; water, water vapor, and ice elsewhere make up less than 3 percent of the total.**

**The sum of the percentages is not 100% but 99.9%. The difference represents uncertainty in measurements and estimates. Although we do not like uncertainty, we do have to live with it. (*Data from U.S. Geol. Survey.*).**

forms soil, in which grow plants, including forests and the crops that are the principal food of human populations.

### Biosphere

The ***Biosphere*** (the "life sphere"), as its name implies, is *the totality of Earth's organisms,* and, in addition, organic matter that has not yet been completely decomposed, It embraces innumerable living things, large and small, grouped into millions of different kinds. We humans are one of the kinds, and we are nearly 4 billion in number. The composition of the biosphere is distinctive; its chief constituents are compounds of carbon, hydrogen, and oxygen, although it includes other chemical elements as well. All these chemical elements are drawn from the other spheres, but they are fixed in patterns peculiar to the biosphere. Whether they form parts of living organisms or dead ones, the organic compounds remain a part of the biosphere until they have been destroyed by chemical alteration.

The biosphere extends through, or at least into, each of the other three outer envelopes (Fig. 2.8). Although living things manage to exist through an enormous vertical range—nearly 20,000 meters—they are very sparse at great heights and great depths. Most of them are crowded into a narrow zone that extends from a little below sea level to a thousand meters or so above it.

Although very different from the air, water, and rock that make up

**Figure 2.8**
**Some data on the extent of the biosphere, shown in a schematic way.**

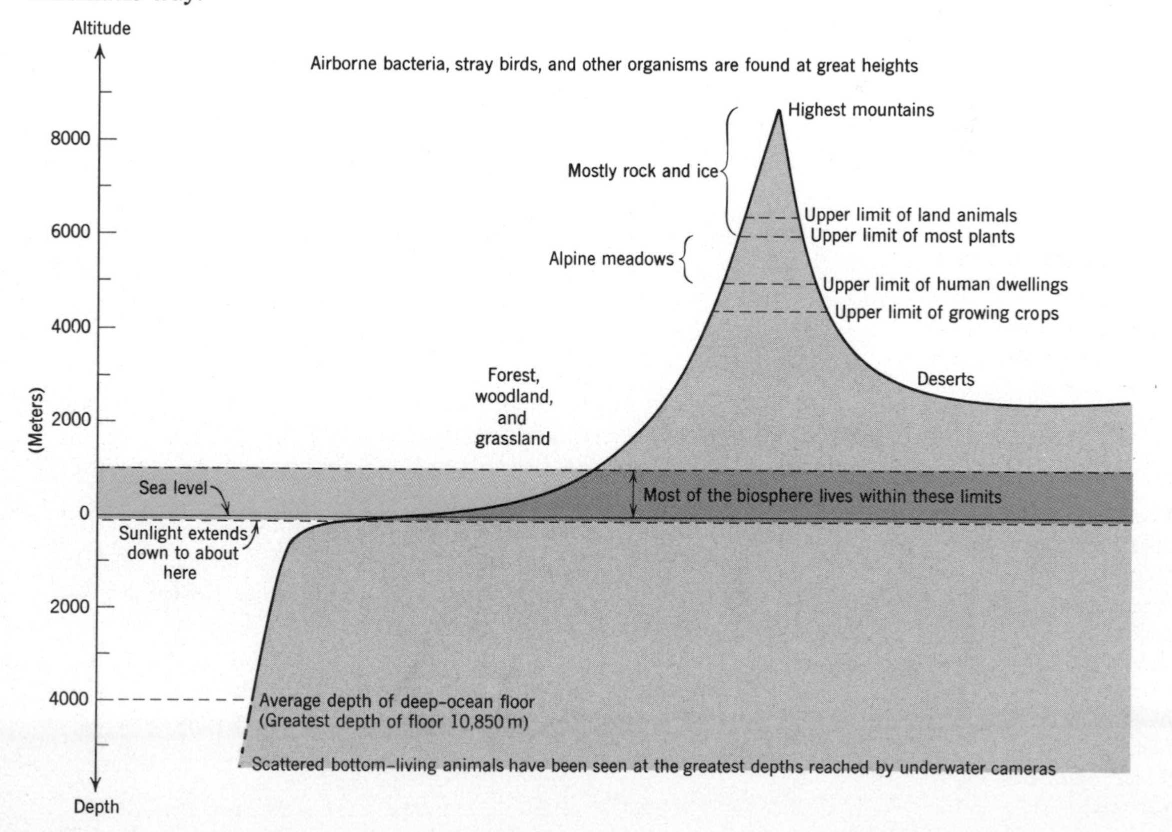

the rest of Earth's outer part, the biosphere is as much an integral part of Earth as are the other three outer envelopes. All organic matter, without exception, is derived from sources within the Earth. Eventually, in one form or another it returns to the inorganic Earth and ceases, for a time at least, to be part of the biosphere.

## THE BATTLE BETWEEN INTERNAL AND EXTERNAL PROCESSES

The crust and the outer envelopes interpenetrate each other to some extent, but seen from a distance, the surface or *interface* along which they are in contact looks sharp and distinct (Fig. 2.9). At this interface, exchanges and other activities take place. The busiest of the spheres are atmosphere, hydrosphere, and biosphere, because within them activities are going forward at higher rates than in the solid Earth.

Water and air penetrate the ground as far down as each is able to go. There they promote chemical reactions that cause rock to crumble slowly and become regolith. Driven by energy from the Sun, water and air move as flowing currents. In very cold places water in the air freezes and falls as snow, which in turn forms glaciers. Air currents, water currents, and glaciers (currents of ice) pick up solid particles and move them long distances. The biosphere penetrates air, water, and regolith so thoroughly that living things occupy every available nook and cranny within the other spheres, filling them all with additional activity. Partly through physical processes, but more importantly through biochemical reactions, organisms help to crumble rock and also to deposit solid substances on the ocean floor.

How fast do these things happen? At what rates do parts of the crust go up or down? These are fascinating questions to which we will return farther on. Calculation of the volume of sediment being carried by rivers today tells us that some high mountain ranges are losing by erosion, each year, rock material equivalent to a layer several centimeters thick. A basin in eastern Europe has been forming, deepening, and filling with sediment during the last four million years, at a rate of one centimeter every forty years—a rate slower than the rate of erosion quoted above, but still fast enough to have created a basin one kilometer deep. Planet Earth therefore is changing continually. As we shall see, the changes occur at different rates in different places, and we infer that rates may have changed from time to time in the past. There would, however, be no changes of any kind were it not for the existence of *energy*. Motions, processes, activities, changes of any kind demand the expenditure of energy. We turn next, therefore, to the sources of energy that drive Earth's internal and external processes.

**Figure 2.9**
**This satellite view of the Mississippi River delta reveals the apparent sharpness of the line that separates lithosphere from hydrosphere. Off the river's mouth are plumes of turbid water, loaded with suspended particles of rock derived from erosion of the land. Compare Figure 7.13.** ***(NASA photo, research by Grant Heilman Photography.)***

# KINDS OF ENERGY

Turn a page in this book. You are using energy. Whether walking outside, driving a car, or merely turning on the light—whatever the activity, you are using energy. Activities and energy are so intimately related that we may define ***energy*** as the *capacity to produce activity.* Energy is vital for our existence, and vital, too, for Earth's existence. Without energy, Earth would be a dead planet.

Energy appears in many forms, each of which produces characteristic activities. We speak of kinetic energy, meaning the energy of a moving body, and heat energy, meaning the energy of a hot body. Other forms are electrical-, chemical-, radiant-, and atomic energy. Each of these kinds of energy is important for some of Earth's activities (Table 2.1), but the four principal kinds are kinetic-, atomic-, heat-, and radiant energy.

## Kinetic energy

Every moving body has *kinetic* energy, named from the Greek word *kinetikos* (to move). The movements of Earth around the Sun and that of Earth spinning on its axis mean that Earth possesses kinetic energy. A ball moving in a tennis game, a boulder rolling downhill, a stream flowing down a valley, or tiny rock particles in a dust storm, all have kinetic energy. Many of Earth's processes involve kinetic energy. Besides running water and rolling boulders, winds, waves, icebergs drifting in the ocean, and moving glaciers are other common examples.

**Atomic energy**

Atom bombs and hydrogen bombs release vast amounts of the energy locked within atoms. Such energy is very destructive. But the same processes of release of energy from atoms go on continuously in nature (fortunately for us, in a controlled manner). To discuss these processes it is necessary, first, to talk about atoms. Appendix A contains a more detailed discussion; here we mention only the essential points.

If you asked a chemist to analyze a rock, he would report the kinds and amounts of the chemical elements present, because ***chemical elements*** are *the most fundamental substances into which matter can be separated by chemical means.* At present 103 of them are known. Each is separately named and identified by a symbol, such as H for hydrogen, Ag for silver, and E for einsteinium. The known elements and their symbols are listed in Appendix A.

If we asked what an element is made of, the chemist would reply that each element consists of a large number of identical particles called atoms. An ***atom*** is *the smallest individual particle that retains all the properties of a given chemical element.* We speak of an atom of hydrogen or an atom of lead, but we cannot see one because it is too small. When we handle a pure chemical element we are seeing instead an aggregation of a vast number of identical atoms. A cube of pure silver lcm on edge contains $58 \times 10^{21}$ atoms, a number so large that its significance is almost impossible to grasp. A faint idea of its magnitude may be conceived by imagining that we had somehow spread the cubic centimeter of silver thinly and evenly over the face of the entire Earth. Each square centimeter of Earth's surface would then be covered by approximately 10,000 atoms.

Everything on Earth is composed of atoms, but atoms in turn are built up from still smaller sub-atomic particles. The principal sub-atomic particles are *protons* (which have positive electrical

**Table 2.1**
**Activities Produced by Common Forms of Energy**

| Energy | Common Activity |
|---|---|
| Kinetic | Flowing water, wind, waves, land-slides |
| Heat | Volcanoes, hot springs, rainstorms |
| Chemical | Decaying vegetation, forest fires, rusting, burning coal |
| Electrical | Lightning, aurora |
| Radiant | Daylight, sunburn |
| Atomic | Heating Earth's interior |

charges), *neutrons* (which are electrically neutral), and *electrons* (which have negative electrical charges that balance exactly the positive charges of protons). Protons and neutrons are dense but very tiny particles and they join together to form the core or nucleus of an atom. Protons give a nucleus a positive charge, and we call *the number of protons in the nucleus of an atom* the ***atomic number.*** Electrons are even tinier particles; they move, like a distant and diffuse cloud, in orbits around the nucleus.

Elements are catalogued systematically, beginning with hydrogen, which has one proton and one electron, then helium, with two protons and two electrons, and so on. A new element is formed each time another proton is added to the nucleus. As a consequence, another electron is added to balance the electrical charge.

All atoms having the same atomic number are atoms of the same element and have the same chemical properties. The number of neutrons that accompany the protons in a nucleus can vary, within small limits, without affecting the chemical properties. Any chemical element may therefore have several ***isotopes,*** *atoms having the same atomic number but differing numbers of neutrons.* For some elements as many as ten isotopes have been discovered. But not all combinations of protons and neutrons are completely stable, so that some isotopes break up spontaneously, forming in the process, new isotopes and different elements. *The decay process by which an unstable atomic nucleus spontaneously disintegrates* is ***radioactivity.*** As we shall see in Chapter 5, the rate at which any radioactive isotope decays is constant, and can be used as a sort of clock.

The most common radioactive isotopes in the Earth are uranium-238, written $^{238}U$, (meaning an isotope with a total of 238 neutrons plus protons in the nucleus), $^{235}U$, $^{232}Th$, and $^{40}K$. When an atom breaks down by radioactive disintegration, a tiny amount of energy is released as heat. It is this heat energy that we refer to as atomic energy.

Radioactive isotopes of potassium, thorium, and uranium are widely distributed, in tiny amounts, through the crust and mantle, and their rates of disintegration are very slow. Nevertheless they generate sufficient heat to keep the Earth's interior exceedingly hot—at times hot enough to melt rock and form magma.

Atomic energy is released too by another and somewhat different process. When very light atoms, such as those of hydrogen and helium, are raised to temperatures of millions of degrees Celsius, the atomic nuclei combine, or *fuse,* to form heavier atoms, and as they do so, a great deal of atomic energy is released as heat. The combination of hydrogen atoms to form helium atoms is the process that produces heat in the Sun.

### Heat energy

We can think of heat as a special form of kinetic energy. It is the energy possessed by the motions of atoms. All atoms move constantly. The faster they move, the more heat energy they have, and the hotter a body feels. The atoms in a solid move within confined spaces—in a sense they rattle around their assigned places in the solid structure—but if they move fast enough (become hot enough) they break out of their fixed positions and the solid is said to have melted. With faster movement, which means more heat, atoms move with complete freedom, and the liquid is then said to have vaporized.

Hotness, or degree of heat, is a cumbersome term; so we use the word *temperature* instead. Temperature is an indication of the average speed of the moving atoms and is measured by arbitrary scales. A common one is the *Celsius scale,* in which we select as 100°C the temperature (or speed of atoms) just sufficient to boil water at sea level.

Heat energy is transmitted in two principal ways. The first is by *conduction,* which occurs when atoms pass on some of their motions to adjacent atoms. Conduction is the process by which heat is transmitted through the side of a cup of hot coffee; it is also the way by which most of Earth's internal heat energy reaches the surface. Conduction, therefore, is the means by which heat is transmitted through solids; it is a slow process because the transfer occurs atom by atom. The second way by which heat can be transmitted—by *convection*—is much faster. Convection occurs in liquids and gases in which the distribution of heat is uneven. When a liquid or gas is heated, it expands; so its density (mass per unit volume) decreases. The hot, less dense material floats up, with colder, more dense material sinking to replace it, thereby setting up a *convection cell* or *convection current.*

### Radiant energy

All the forms of energy discussed so far involve matter. How, then, is it possible for energy to pass through empty space? It is possible because energy can be transmitted by energy waves, commonly called *electromagnetic waves.* Radiant energy, therefore, *is* electromagnetic waves: cosmic rays, visible light, X rays, and radio waves are well-known forms of radiant energy.

The distance between the crests of two adjacent waves is a *wavelength.* The wavelengths of electromagnetic waves range from less than one thousand millionth of a centimeter (cosmic rays) to more than a million centimeters (radio waves). The names do not indicate any fundamental differences among electromagnetic waves. All such waves move with the same speed, the speed of light; the names that we apply are merely convenient ways of designating various wavelengths.

Radiant energy and its transmission through space by electromagnetic rays is vitally important for Earth. How else, for example, could we receive the atomic energy the Sun generates by nuclear fusion?

**Conversions between forms of energy**

When we burn gasoline in a car, we convert chemical energy to heat energy. When the car moves, the heat energy in turn is converted to kinetic energy. Conversion of heat energy to kinetic energy in a controlled manner was first accomplished in the 18th Century, when the steam engine was invented, thereby inaugurating the era of intensive energy use in which we live.

In all natural processes, energy is continually converted from one form to another. As radiation arrives from the Sun, it strikes Earth's surface, the rays are absorbed, and the ground is heated. Air in contact with the ground is warmed by conduction, expands, and rises; it is replaced by cooler, denser air from above and wind results (Fig. 2.10). The process of lifting warm air from Earth's surface to greater altitudes is an example of convection. Thus atomic energy that was generated as heat in the Sun was radiated to Earth by electromagnetic waves and eventually transformed to kinetic energy in a blowing wind. The vast number of Earth's internal and external processes result from the fact that many paths for such energy conversions are possible.

Before people realized that motion and heat were different expressions of energy, and that the different forms of energy could be converted back and forth, they had already begun to use different measurement units for each form. To avoid the confusion of multiple units, we will use only the calorie, the unit of heat energy. A ***calorie***, is *the amount of heat energy needed to raise the temperature of one gram of water by one degree Celsius.* A calorie is so small that in one tablespoon of sugar there are approximately 50,000 calories of chemical energy. To make us feel better about the calories we consume, nutritionists use a unit of 1,000 calories, which they label the Calorie. There are only 50 of these in a tablespoon of sugar. If we compare, in Table 2.2, the amounts of energy derived from different sources, it is apparent that an incomparably greater amount of energy comes from fusion and fission reactions than from burning.

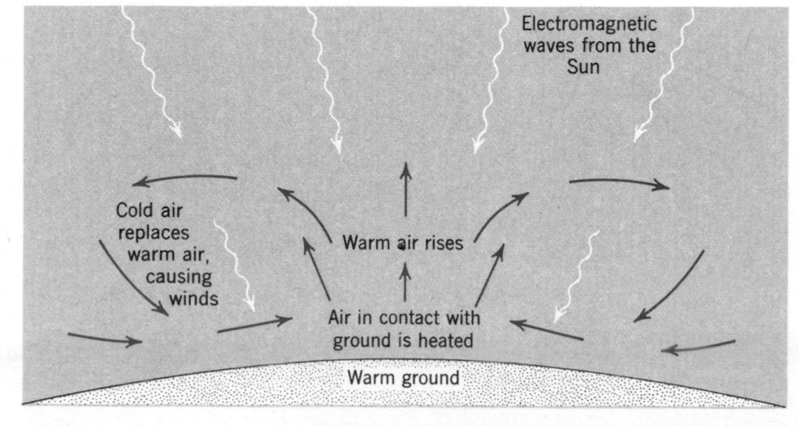


Figure 2.10
**Energy in the form of electromagnetic radiation from the Sun is transformed to kinetic energy of wind. The radiation heats Earth's surface. Air in contact with the surface is warmed in turn, expands as a consequence of the heating, and rises. Cold air flows in to replace the rising hot air; winds result. The complete chain of rising warm air and inflowing cold air is called a convection cell.**

**Table 2.2**
**Amounts of Energy Available from Different Sources. The Amount of Energy per Unit Mass from Nuclear Burning Is Vastly Greater Than the Chemical Energy Resulting from Burning**

| Reaction | Amount of Heat Energy |
|---|---|
| 1 gram of hydrogen fuses to form helium | $1.5 \times 10^{11}$ calories |
| 1 gram of U-238 decays by radioactivity | $0.2 \times 10^{11}$ calories |
| 1 gram of oil burns | $1.0 \times 10^{4}$ calories |
| 1 gram of coal burns | $0.8 \times 10^{4}$ calories |
| 1 gram of sugar is eaten | $4.2 \times 10^{3}$ calories |
| 1 gram mass moving at a velocity of 1000 cm/sec[a] | $1.2 \times 10^{-2}$ calories |

[a] 1000 cm/sec equals approximately 23 miles per hour.

# SOURCES OF ENERGY

Energy reaches Earth's surface from three principal sources. (1) Radiant energy arrives from the Sun; (2) kinetic energy arrives from the rotations of Moon, Earth, and Sun and appears as tides; and (3) energy reaches Earth's surface by continuous outflow of Earth's internal heat. Because the surface receives energy from three sources, we might reason that it is heating up. But this is not the case, because the average temperature of Earth's surface does not vary from year to year; so some sort of energy balance must be maintained. The source of the balance is not hard to find: Earth's surface does not merely receive energy; it also emits energy by radiating long-wavelength electromagnetic waves out into space. Earth's surface is said to be in a ***steady state*** or state of ***dynamic equilibrium,*** by which we mean *a condition in which the rate of arrival of some materials equals the rate of escape of other materials.* The rate at which energy reaches Earth's surface exactly balances the rate at which it escapes (Fig. 2.11).

### Energy from the Sun

Radiant energy reaches Earth from the stars, but in amounts that are tiny by comparison with the Sun's energy. When the Sun's rays reach Earth's atmosphere, approximately 40 percent are simply reflected back into space without change. It is this reflected radiation that the astronauts see when, standing on the Moon, they look at Earth (Fig. 2.12). The remaining 60 percent is absorbed, partly by the atmosphere (which becomes heated in the process) and partly by the land and by the sea. The energy absorbed by the sea warms the water and causes evaporation. The resulting water vapor forms clouds and eventually rain, snow, and all other forms

Figure 2.11
**Energy reaches Earth's surface from three sources; two are external, one internal. To maintain a heat balance, long-wavelength electromagnetic waves radiate energy into space.**

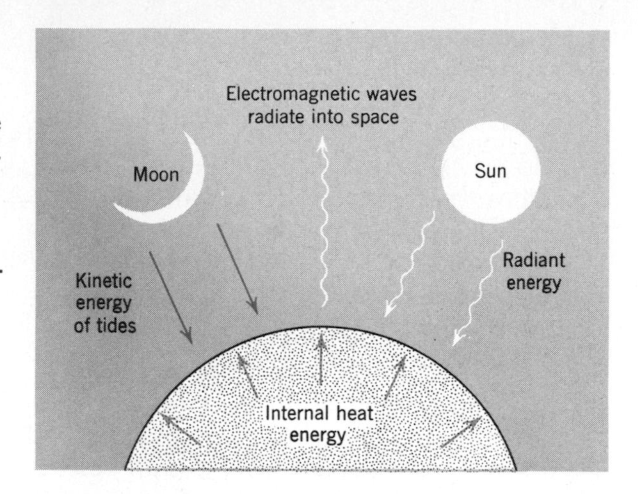


of precipitation. The energy absorbed by the land eventually warms the air, causes convection, and creates winds, which, blowing over the sea, create waves. Thus all the major agents of erosion that operate at Earth's surface—rain, ice, streams, winds, waves, and glaciers—are processes driven by the Sun's energy.

### Energy from tides

One does not always think of tides in the context of energy. Nevertheless, tides are the mechanism by which some of the kinetic energy from the motions of Moon, Earth, and Sun reach Earth's surface. The principal effect arises from interactions between Moon and Earth; so tides are principally lunar effects.

Gravitational attraction by Moon on Earth pulls seawater toward the Moon and creates tidal bulges in the ocean (Fig. 2.13). No

Figure 2.12
**Paths followed by Earth's incoming radiant energy from the Sun. The energy used in heating the atmosphere causes winds. Most of the energy absorbed by land and sea is used up in the evaporation of water to form clouds and cause rain, snow, and hail.**

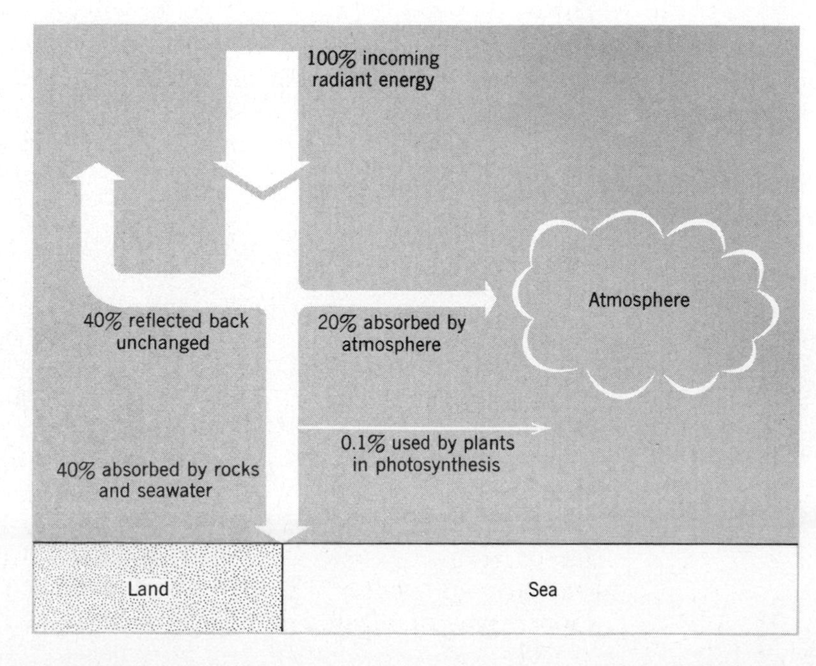

energy would be involved were it not for Earth's rotation about its axis and Moon's movement in orbit around Earth. As a consequence, the positions of the bulges move continuously, and at any spot on the sea we see two high tides and two low tides each day. However, as Figure 2.13 shows, the Sun also affects tides, sometimes aiding the Moon by pulling in the same direction, and sometimes opposing it by pulling at right angles. The Sun's effect is smaller than Moon's; so the two effects never cancel each other. The actual heights of high and low tides, therefore, vary on a cycle of approximately 14 days, matching the enhancement and opposition of tides by Sun and Moon.

Tidal bulges cannot move around Earth unhindered, because continents get in the way. Water therefore piles up against the continental margins whenever a tidal bulge arrives, and then flows back to the ocean basin as the bulge passes. The piling-up effect is the reason why tides are much higher along coasts than in the open ocean. Movement of water masses in coastal tides means kinetic energy is being used, and this energy must be taken from Earth's store of kinetic energy of rotation.

If the kinetic energy of rotation is being transferred to Earth's surface to cause tides, what is happening to Earth's rate of rotation? Earth's spinning motion, and therefore its energy, apparently remain from the days of its formation (Chapter 19); we know of no new kinetic energy that is being added. Therefore, removing kinetic energy can have only one effect. The tides are

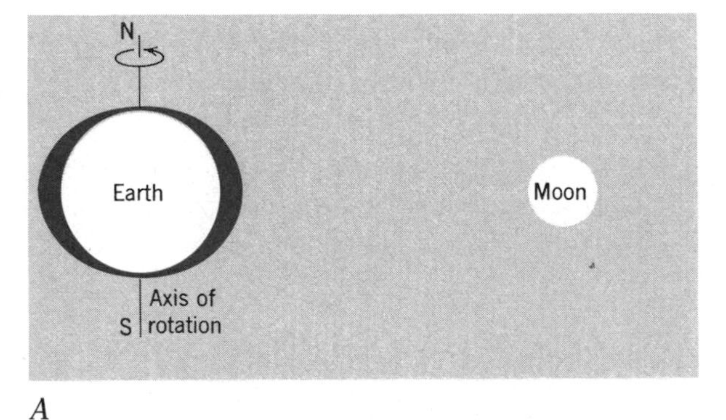


*A*

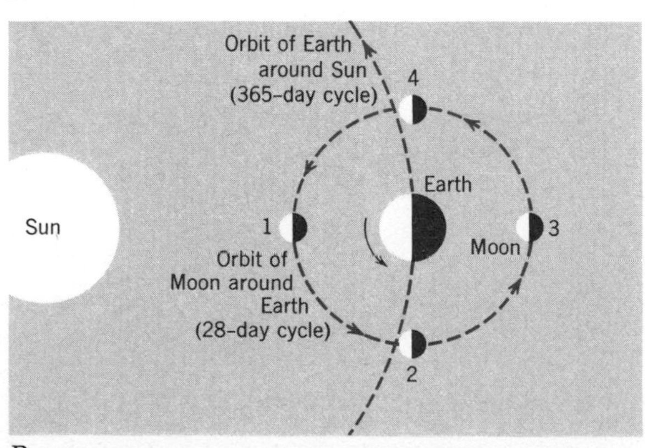


*B*

**Figure 2.13**
**The gravitational attractions of Moon and Sun on Earth raise tidal bulges in Earth's oceans.**

***A*. Idealized diagram of tidal bulges, relative to Earth's axis of rotation and to the position of the Moon.**

***B*. When Moon and Sun attract in the same direction, Moon positions 1 and 3, we experience highest tides. When Moon and Sun have opposing positions (Moon positions 2 and 4), we experience lowest tides.**

acting as weak but steady brakes, and Earth is gradually slowing down. The rate of slowing is, fortunately, not great because the rate of energy transfer is small. Astronomers have measured the exact length of the day over the past three centuries and find that it is increasing by 0.002 seconds each century. Over hundreds of millions of years the effect of this small increase can become very large. At a time some billions of years in the future, Earth could even stop rotating completely.

### Energy from Earth's interior

Anyone who has been down in a mine realizes that rock temperatures increase with depth. Measurements made in deep drillholes and mines show that the rate of temperature increase (the *geothermal gradient*) varies in different parts of the world from 15°C to 75°C per kilometer. We cannot, however, make direct temperature measurements beyond the deepest drillholes, which are only about 10km deep; so we have to use indirect means to estimate temperatures in Earth's interior. From physical properties that vary with temperature, such as the speeds of earthquake waves, we estimate that temperatures continue to increase toward the center and eventually reach values of 5,000°C or more in the core. When we consider Earth's size, obviously a vast amount of heat is stored within it, and as we have already seen, heat is being slowly but continually added by radioactive disintegration.

The average value of the heat flow that reaches Earth's surface is about $1.5 \times 10^{-6}$ calories/cm$^2$/second. To picture how small this amount of heat is, imagine that we have a gadget that can catch and use all the heat reaching one square meter of Earth's surface. We would have to gather heat for approximately four days and nights before we got enough heat energy to boil a cup of water. But the size of Earth's surface is enormous; so the total amount of heat that flows outward is also enormous. Just as the geothermal gradient varies from place to place, so does the heat flow, which is greatest near volcanoes and hot springs. Variations in surface heat flow tell us a great deal about activities down below and, as we shall see in later chapters, they allow us to infer that convection may be occurring in the seemingly solid rock of the mantle (Chapter 18).

### Comparison of Earth's energy sources

Table 2.3 shows the amounts of energy reaching Earth's surface every 24 hours from the three different energy sources. Because solar energy is vastly larger than the energy of either tides or internal heat flow, it is clear that the temperature at Earth's surface must be controlled by the Sun. Thus, if we fill our atmosphere with dust and other pollutant particles, thereby reflecting more of the Sun's incoming radiation, we might cause a reduction of temperature. Other energy sources are too small to offset man's effects on the environment.

Solar energy does not contribute to Earth's internal activities. Just

**Table 2.3**
**Comparison of the Total Amount of Energy Reaching Earth's Surface Every 24 Hours from the Three Principal Energy Sources**

| Energy Source | Energy Each 24 Hours |
|---|---|
| Solar radiation | $37{,}000 \times 10^{17}$ calories |
| Flow of internal heat | $6.6 \times 10^{17}$ calories |
| Tides | $0.6 \times 10^{17}$ calories |

as water cannot flow uphill, heat cannot flow from a cold body to a hot one; so the geothermal gradient prevents heat from flowing from the surface down into the interior. In a real sense, therefore, Earth's surface is a surface of conflict between different energy sources. Internal activities, driven by internal heat energy, raise mountains and cause irregularities on Earth's surface. External activities, driven by solar and tidal energy, continually erode and abrade the surface irregularities. Next, therefore, we must examine the consequences of the great energy conflict. To do this, we need to understand the materials from which Earth is constructed: minerals and rocks.

## SUMMARY

### EARTH LAYERS

1. The solid Earth is layered. It consists of a core, mantle, and crust, each differing in composition.

2. The crust consists of two parts, oceanic crust and continental crust.

3. The lithosphere, the outer 100km of the solid Earth, is hard brittle rock; beneath the lithosphere is the asthenosphere, a region of the mantle that is weak and easily deformed.

4. Great plates of lithosphere move slowly over the asthenosphere.

5. Outside the solid Earth are the envelopes we call hydrosphere, atmosphere, regolith, and biosphere.

6. Earth's active processes can be divided into two groups: internal and external. The two are in continual conflict.

### ENERGY

7. Atoms form 103 types of elements, and 88 of these occur naturally on Earth.

8. Each element has two or more isotopes. Isotopes are varieties of atoms having the same chemical properties but differing in their masses because of differing numbers of neutrons in their nuclei.

9. Some isotopes are not stable and disintegrate spontaneously by radioactivity. During radioactive decay, energy is emitted and new isotopes and elements are formed.

10. Energy is the capacity to produce activity. It appears in many forms: kinetic-, heat-, chemical-, electrical-, radiant-, and atomic energy.

11. Energy can be converted from one form to another. All of Earth's activities involve the conversion of energy.

12. Energy reaches Earth's surface from three sources. Radiant energy arrives from the Sun. Tidal energy comes from the kinetic energy of motions of Earth and Moon. Natural radioactive decay produces internal heat energy, which flows to the surface.

# SELECTED REFERENCES

Clark, S. P., 1971, Structure of the Earth: Englewood Cliffs, N. J., Prentice-Hall.

Scientific American, 1971, Energy and power, v. 224, no. 3, p. 36–200 (special issue consisting of 11 articles devoted to energy).

# 3 THE BUILDING BLOCKS: MINERALS AND ROCKS



## MINERALS

Walk outside and pick up a stone. If you can't find a stone, pick up some sand, gravel, or soil. Whatever you pick up, you will be holding a handful of minerals. Wherever you look you see minerals, and whatever you do you use products made from minerals. As bread is the staff of life, so are minerals the staff of civilization.

Most minerals are common, have little commercial use and little value. Some, like diamonds and rubies, are rare and are prized for their beauty. Others are the raw materials for industry and the basis of national wealth. Empires have been won and lost for minerals, and powerful countries have collapsed when deposits of valuable minerals became exhausted. The Romans conquered most of Europe and the Near East in their search for minerals containing copper, gold, tin, iron, silver, and lead. They built a remarkable empire by using the mineral wealth they found. When the mineral deposits became exhausted, or were captured by local tribes, Rome was deprived of its great sources of wealth and its empire slowly died. The United States, too, has prospered as a result of its abundant mineral wealth, and as some of its sources are now facing exhaustion, there are those who wonder if history is about to repeat itself. Others believe their fears will not be justified because, as some supplies run out, we may be able to find alternative and presently unknown sources, and in some cases it may be possible to substitute alternative materials.

The word *mineral* is often used loosely. We read advertisements for plant foods that provide "minerals" for plant growth, or for vitamin pills to provide "minerals" for healthy bodies. To avoid confusion we will use *mineral* in a specific, scientific way, and in order to do so we will give it an exact definition. Before attempting a definition, however, it is helpful to examine the two most important characteristics of minerals: composition and structure. And because most minerals are made up of more than one kind of atoms, it is helpful to start by discussing the way in which atoms combine.

### Compounds and ions

A few minerals, such as gold and platinum, are single chemical elements. Most minerals are *a combination of atoms of different elements bonded together* to form a ***compound.*** The reason atoms combine, or *bond,* is discussed more fully in Appendix A; the combination depends on how atoms transfer and share orbiting electrons. *An atom that has excess positive or negative charges caused by electron transfers* is called an ***ion.*** When the charge is positive (meaning that the atom gives away electrons), the ion is called a *cation;* when negative, an *anion.* The convenient way to indicate ionic charges is to record them as superscripts. For example, $Ca^{+2}$ is a cation, while $S^{-2}$ is an anion. Compounds contain one or more elements that are cations and one or more elements that are anions; for a compound to be stable, the sum of the positive charges on the cations and the negative charges on the anions must equal zero.

Sometimes two different atoms form such strong bonds that they seem to act as a single atom. A strongly bonded pair is said to form a *complex ion.* Complex ions act in the same way as single ions, forming compounds by bonding with other elements. For example, carbon and oxygen combine to form the very stable carbonate anion $(CO_3)^{-2}$. Other examples of important complex anions are the sulfate $(SO_4)^{-2}$, nitrate $(NO_3)^{-1}$, and silicate $(SiO_4)^{-4}$ groups.

Two broad classes of compounds are recognized. *Organic compounds* are made from carbon and hydrogen, with or without other elements such as nitrogen and oxygen. Organic compounds can form by direct combination of carbon and hydrogen, but most come directly or indirectly from the activities of living organisms. Mixtures of organic compounds are called organic matter. All other matter is said to be *inorganic* and its compounds are *inorganic compounds.* All minerals are inorganic. A few, such as gold and silver, are chemical elements, but all the rest are inorganic compounds. We have therefore defined one property of a mineral. But one property alone does not define a mineral. We must consider other properties as well.

Minerals: Inorganic compounds geometric, unchanging crystal structure Solids

### Structure

Minerals are solids. Whereas atoms in gases and liquids are randomly jumbled, in most solids the atoms are organized in regular geometric patterns, like eggs in a carton. *The geometric pattern that atoms assume in a mineral* is called ***crystal structure.*** The crystal structure of a mineral is a unique property, and all specimens of a given mineral have identical structures.

The packing of atoms in the mineral *galena*, PbS, the most common lead mineral, is shown in Figure 3.1. Notice that the sulfur atoms are larger than the lead atoms. Now anions tend to be large; cations tend to be small. The crystal structures of minerals, therefore, are largely determined by the packing arrangement of anions. The radii of some common ions are shown in Figure 3.2 and are given in *Angstroms* (abbreviated Å), a unit of length used for atomic measurement.

It is apparent from Figure 3.2 that some ions have the same electrical charge and are nearly alike in size; for example $Fe^{+2}$ and $Mg^{+2}$ have radii of 0.83Å and 0.78Å respectively. Because of their similarity in size and charge, ions of $Fe^{+2}$ are often found substituting for ions of $Mg^{+2}$ in magnesium-bearing minerals. The *structures* of the magnesium minerals are not changed as a result of

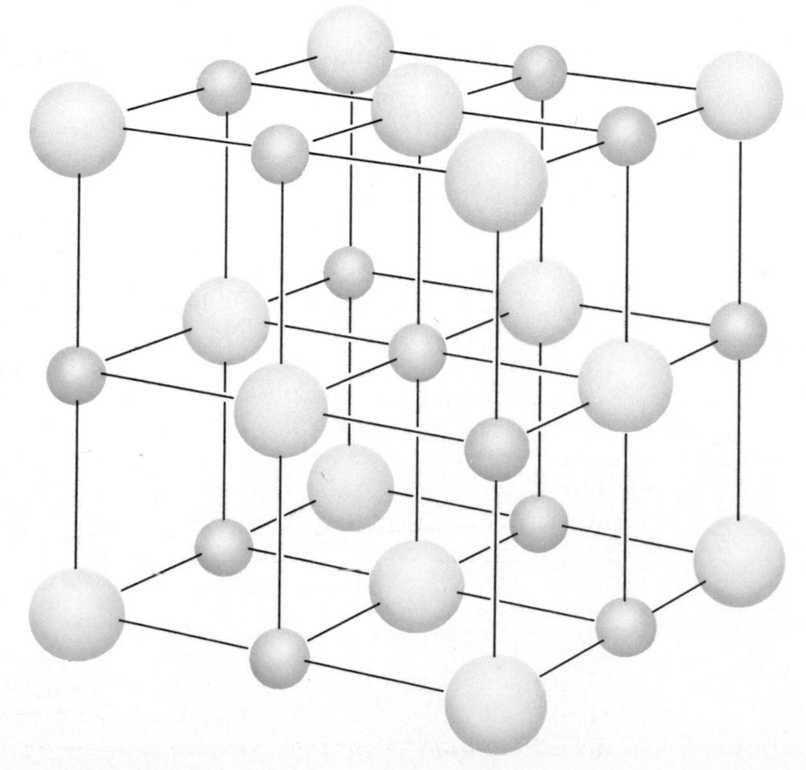

**Figure 3.1**
**Arrangement of atoms in galena, PbS, the common lead mineral. Pb is a cation with a charge of +2, S is an anion, charge −2. Therefore, for every Pb atom (dark) in the structure, there must be one S atom (light). The atoms are shown pulled apart along the blue lines, so that we can see how they fit together.**

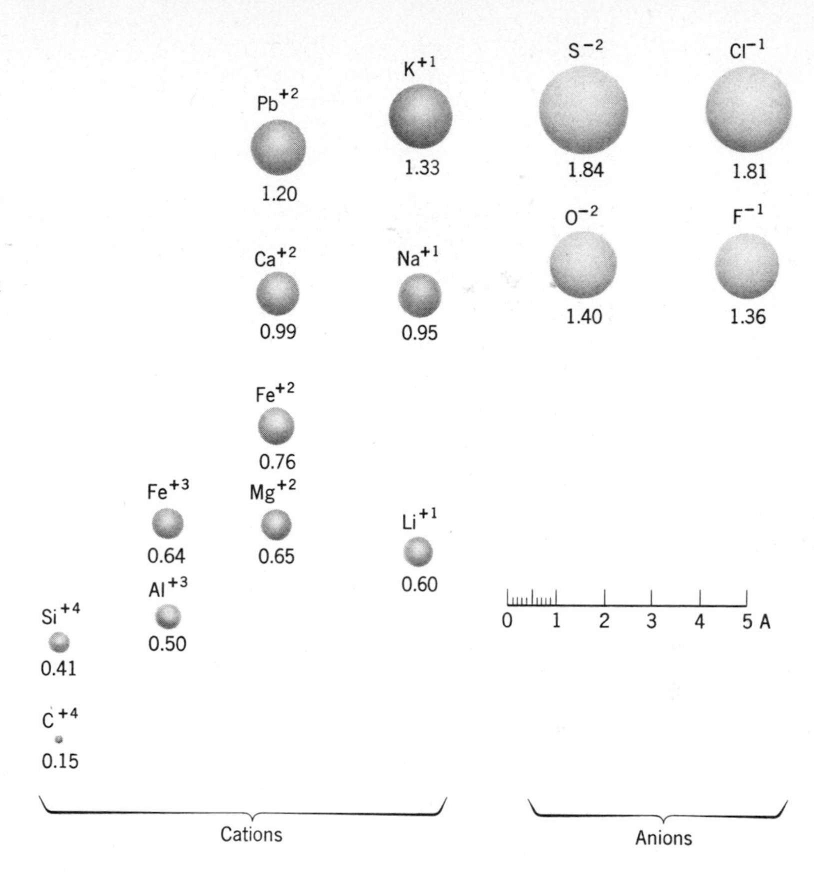


**Figure 3.2**
**Radii of ions of some common elements range from $C^{+4}$ at lower left to $Cl^{-1}$ at upper right. Ions are arranged in vertical groups based on charge, from $^{+}4$ at left to $^{-}1$ at right. Ions in each of the pairs $Si^{+4}$ and $Al^{+3}$, $Mg^{+2}$ and $Fe^{+2}$, and $Na^{+1}$ and $Ca^{+2}$ are about the same size and commonly substitute for each other in crystal lattices. Radii expressed in Angstroms (Å), where one Angstrom equals $10^{-8}$cm. (*Data from Fyfe, 1964.*)**

the substitution, but of course the *composition* of the mineral is affected. *The substitution of one atom for another in a random fashion throughout a crystal structure* is known as ***solid solution.*** There is a special way of depicting solid solutions in chemical formulas. When Fe substitutes for Mg in the mineral olivine, $Mg_2SiO_4$, for example, we simply write the formula $(Mg, Fe)_2SiO_4$, which indicates that the Fe substitutes for Mg but not for any other atoms in the structure. Variations in mineral composition caused by solid solution are often large and, as we shall see, important in the formation of common minerals and rocks.

Each mineral has a unique crystal structure, but some compounds are known to form two or more different minerals. The compound $CaCO_3$, for example, forms two different minerals. One is *calcite,* which is the mineral of which marble is composed; the other is *aragonite,* which is most commonly found in the shells of clams, oysters, snails, and other aquatic life. Calcite and aragonite have identical compositions, but entirely different crystal structures. *A compound that occurs in more than one crystal form* is called a ***polymorph.*** Some common polymorphs are listed in Table 3.1.

### Definition

Minerals are inorganic compounds, and each mineral has a unique crystal structure. We can now give an exact definition: ***minerals*** are *all naturally occurring, crystalline, inorganic materials*. The

**Table 3.1**
**Minerals with Identical Compositions but Different Crystal Structures Are Called Polymorphs. Some Well-known Polymorphs Are Listed Below, with the Most Common Variety Listed First**

| Compound | Mineral Name |
|---|---|
| C | Graphite<br>Diamond |
| $CaCO_3$ | Calcite<br>Aragonite |
| $FeS_2$ | Pyrite<br>Marcasite |
| $SiO_2$ | Quartz<br>Cristobalite<br>Tridymite<br>Coesite<br>Stishovite |

definition does not include all naturally occurring solid compounds. Some substances, such as glass, lack a systematic arrangement of atoms and do not have fixed compositions; they are called *mineraloids* and because they do not have crystal structures, they are said to be *amorphous* ("without form").

### Properties

The properties of minerals are determined by their compositions and their crystal structures. Once we know which properties are characteristic of which minerals, we can use the properties to identify the minerals. It is not necessary, therefore, to analyze a mineral chemically or to determine its crystal structure in order to identify most common ones. The characteristics most often used in identifying minerals are the obvious physical properties, such as color, shape of crystal, and hardness, plus some less obvious properties, such as the way the mineral breaks and its density. Each property is discussed in detail in Appendix B, which also contains a table of common minerals together with the most characteristic physical properties used in their identification.

### Composition

Approximately 2,200 minerals are known. Most have been found in the crust, because that is the only part of the Earth accessible to us. A few minerals have been identified only in meteorites, and two new ones were discovered in the Moon rocks brought back by astronauts. The total number of minerals may seem large, but it is tiny by comparison with the enormous number of ways in which the chemical elements can theoretically combine to form

**Table 3.2**
**The Most Abundant Elements in the Continental Crust** (*After K. K. Turekian, 1969*)

| Element | Weight % |
|---|---|
| Oxygen (O) | 45.2 |
| Silicon (Si) | 27.2 |
| Aluminum (Al) | 8.0 |
| Iron (Fe) | 5.8 |
| Calcium (Ca) | 5.1 |
| Magnesium (Mg) | 2.8 |
| Sodium (Na) | 2.3 |
| Potassium (K) | 1.7 |
| Titanium (Ti) | 0.9 |
| All other elements | 1.0 |
| Total | 100.0 |

compounds. The reason for the disparity between observation and theory is apparent when we consider the abundance of chemical elements—nine elements comprise the great bulk of Earth's crust (Table 3.2). The crust is constructed, therefore, of a limited number of minerals containing two or more of the nine abundant elements. Minerals containing other elements certainly do occur in small amounts, but most of the scarcer elements do not form minerals, or do so under very special and restricted circumstances. The less abundant elements occur, instead, in solid solution by replacing abundant elements in common minerals. We can estimate, for example, that only 0.01 percent of all the copper atoms in the crust reside in separate copper minerals. The rest of the copper atoms substitute by solid solution for atoms such as those of iron and magnesium in common minerals.

Looking at Table 3.2, we see that two elements, oxygen and silicon, make up more than 70 percent of the crust. Oxygen forms a simple anion, $O^{-2}$, and silicon and oxygen together form the complex *silicate anion* $(SiO_4)^{-4}$. In view of the abundances of silicon and oxygen, it is not surprising that silicate minerals are the most common naturally occurring inorganic compounds and that oxides are the next most abundant. Other natural compounds, although much less common than silicates and oxides, are sulfides, chlorides, carbonates, sulfates, and phosphates.

## THE ROCK-FORMING MINERALS

A few minerals—fewer than 20 kinds—are so common that they account for more than 95 percent of the continental and oceanic crust. These are the *rock-forming minerals*, so called because all rocks contain one or more of them. We have already seen that silicates and oxides are the most common minerals of the crust, but that silicates are much more abundant than oxides. Most rock-forming minerals, therefore, are silicates.

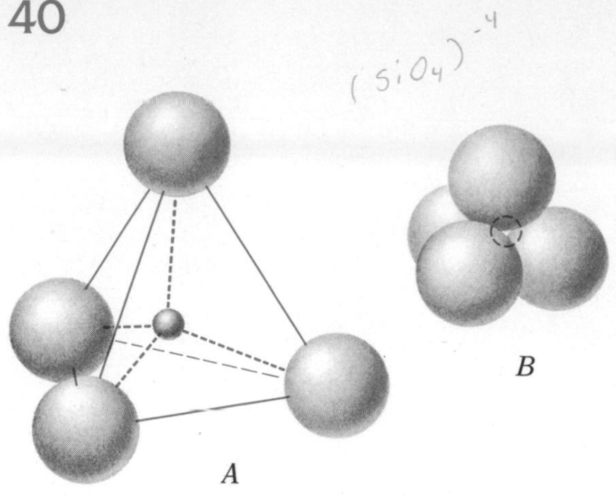


**Figure 3.3**
**Silica tetrahedron. *A*. Expanded view showing large oxygen ions at the four corners, equidistant from a small silicon ion. Dotted lines show bonds between silicon and oxygen ions; solid lines outline the tetrahedron. *B*. Tetrahedron with oxygen ions touching each other in natural positions. Silicon ion (dashed circle) occupies central space.**

## The silicates

Silicon and oxygen form the complex anion $(SiO_4)^{-4}$. The four oxygen atoms in the complex anion are tightly bound to the single silicon atom. Oxygen is a large ion (Fig. 3.2), while silicon is a small ion. The oxygens pack into the smallest packing space possible for four large spheres. As can be seen in Figure 3.3, the four oxygens sit at the corners of a tetrahedron—a pyramid—and the small silicon sits in the hole between the oxygens at the center of the tetrahedron. The shape of the complex silicate anion is therefore a tetrahedron and the structures and properties of silicate minerals are determined by the manner in which silica tetrahedra pack together.

Silica tetrahedra combine to form compounds in two ways. First, the oxygens of the tetrahedra can form bonds with cations so that the tetrahedra are acting like simple anions, isolated from each other in the structure and surrounded by cations. An example of this is found in olivine, in which two $Mg^{+2}$ ions satisfy the charges of each isolated $(SiO_4)^{-4}$ tetrahedron, yielding the formula $Mg_2SiO_4$. The second and entirely different way is for two adjacent tetrahedra to share an oxygen ion. As a consequence, two or more tetrahedra are joined to form an even larger anion unit, in the same way that beads are joined to form a necklace. *The process of linking silica tetrahedra into larger groups* is called **polymerization.** The simplest case of polymerization is that of two tetrahedra sharing a single oxygen atom. As Figure 3.4 shows, a large anion $(Si_2O_7)^{-6}$ results. There is one rare mineral that contains the $(Si_2O_7)^{-6}$ anion, but this simplest case of polymerization does not occur in any rock-forming minerals.

All the common silicate minerals have more complicated polymerizations. When a tetrahedron shares more than one oxygen with adjacent tetrahedra, large groupings, and even endless chains, sheets and networks of tetrahedra can be formed. The common polymerizations, together with the rock-forming minerals containing them, are shown in Figure 3.5, and are discussed below in order of increasing complexity of polymerization.

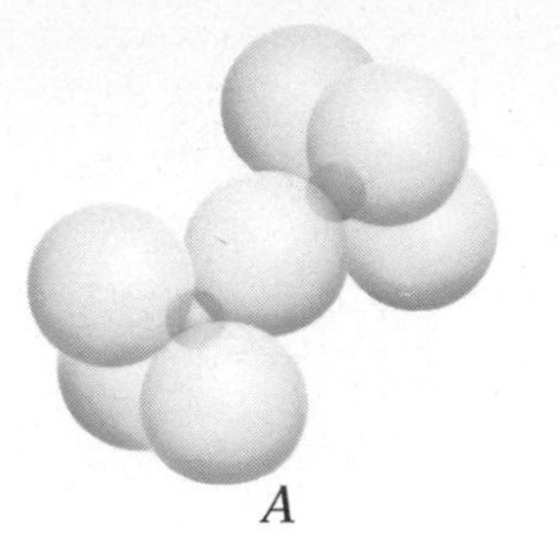


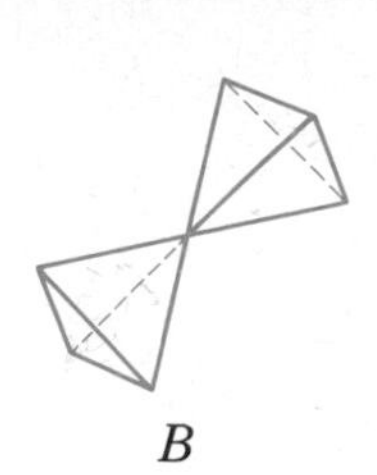


Figure 3.4
**Two silica tetrahedra share an oxygen, thereby satisfying some of the unbalanced electrical charges, and, in the process, forming a larger and more complex anion group. *A*. The arrangement of oxygen and silicon atoms in double tetrahedra, giving the anion $(Si_2O_7)^{-6}$. *B*. A geometrical representation of the silica tetrahedra, drawn so that an oxygen atom would sit at each apex and a silicon atom at the center of each tetrahedron.**

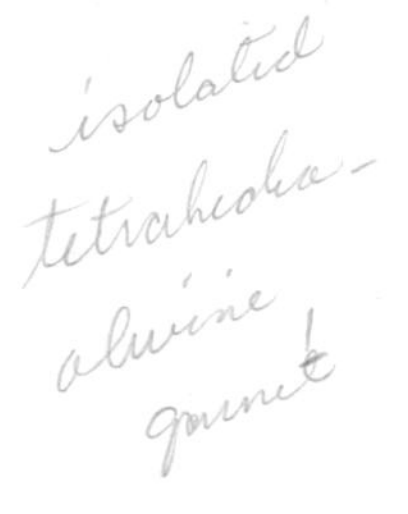

### Olivine

Two important minerals contain isolated silica tetrahedra. The first is a glassy-looking mineral, usually pale green in color, called *olivine.* Actually olivine is a family of minerals because Fe and Mg substitute for each other in solid solution, giving a general formula $(Mg, Fe)_2SiO_4$. Olivine sometimes occurs in such flawless and beautiful crystals that it can be used as a gem, *peridot,* and is a very common mineral in rocks of the oceanic crust.

### Garnet

The second important mineral with isolated silica tetrahedra is *garnet.* As with olivine, garnet is actually the name of a family of minerals. The garnets have the complex formula $A_3B_2(SiO_4)_3$ where A can be the cations $Mg^{+2}$, $Fe^{+2}$, $Ca^{+2}$, and $Mn^{+2}$, or any solid solution mixture of them, while B can be either of the triply charged cations $Al^{+3}$ or $Fe^{+3}$. Garnets are characteristically found in rocks of the continental crust. One of the most characteristic features of garnets is their tendency to form beautiful crystals. The iron-rich garnet, almandine, is deep red and is well known as a gemstone. Another important property of the garnets is their hardness, which makes them useful as abrasives for grinding and polishing.

### Pyroxene and amphibole

The *pyroxenes* and *amphiboles* are two families that have continuous chains of silica tetrahedra. They differ in that pyroxenes are built from a polymerized chain of single tetrahedra, each of which shares two oxygens, while the amphiboles are built from double chains of tetrahedra, equivalent to two pyroxene chains in which half the tetrahedra share two oxygens and the other half share three oxygens. These relations can be clearly seen in Figure 3.5.

The general formula for pyroxenes is A B $(SiO_3)_2$, where A and B can be any of the cations $Mg^{+2}$, $Fe^{+2}$, $Ca^{+2}$, and $Mn^{+2}$. The pyroxenes are most abundantly found in rock of the oceanic crust but are common also in many rocks of the continental crust. The most common pyroxene mineral is a shiny black species called *augite.*

Amphiboles are perhaps the most complicated family of all silicate minerals, with the general formula $A_2B_5(Si_4O_{11})_2(OH)_2$, where A can

| | Arrangement of silica tetrahedra | Formula of the complex anions | Typical mineral | |
|---|---|---|---|---|
| | | | Name | Composition |
| Isolated tetrahedra |  | $(SiO_4)^{-4}$ | The Olivine family | $(Mg, Fe)_2SiO_4$ |
| Isolated polymerized groups |  | $(Si_2O_7)^{-6}$ | Lawsonite | $CaAl_2Si_2O_7(OH)_2H_2O$ |
| |  | $(Si_3O_9)^{-6}$ | Wollastonite | $Ca_3Si_3O_9$ |
| | 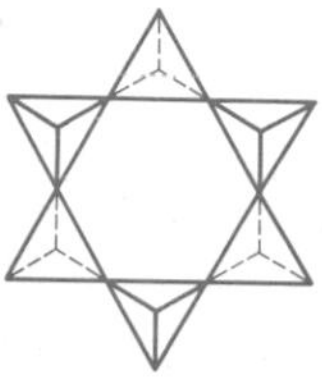 | $(Si_6O_{18})^{-12}$ | Beryl | $Be_3Al_2Si_6O_{18}$ |
| Continuous chains | 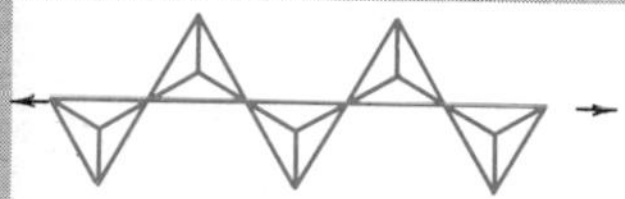 | $(SiO_3)_n^{-2}$ | The pyroxene family | $(Fe, Mg)SiO_3$ |
| | 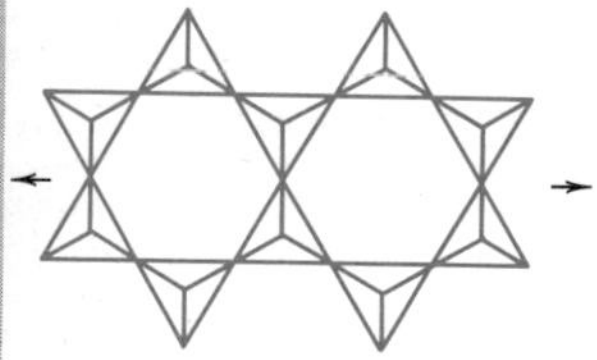 | $(Si_4O_{11})_n^{-6}$ | The amphibole family | $Ca_2Mg_5(Si_4O_{11})_2(OH)_2$ |
| Continuous sheets | 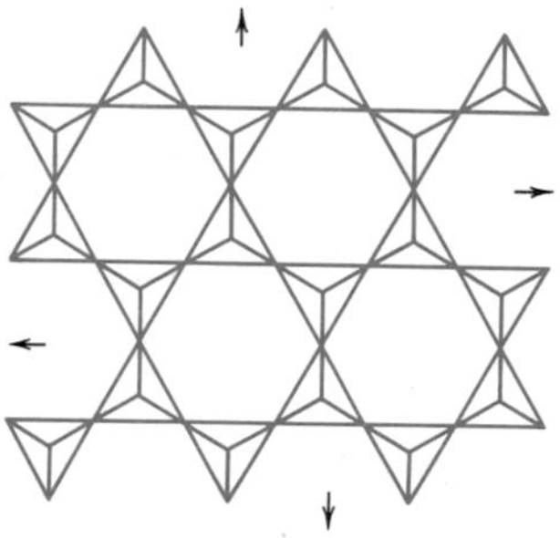 | $(Si_4O_{10})_n^{-4}$ | The mica family | $KAl_2(Si_3Al)O_{10}(OH)_2$ |
| Three-dimensional networks | Too complex to be shown by a simple two-dimensional drawing | $(SiO_2)$ | Quartz | $SiO_2$ |

**Figure 3.5**

**The way in which silica tetrahedra polymerize by sharing oxygens determines the structures and compositions of the rock-forming silicate minerals. Polymerizations other than those shown are theoretically possible, but have not yet been found in minerals.**

be either $Ca^{+2}$ or $Mg^{+2}$, and B can be $Mg^{+2}$ or $Fe^{+2}$. The amphiboles are characteristically found in the continental crust. The most abundant mineral species is hornblende, a dark green to black mineral that looks very like augite.

### Mica and clay

Members of the *mica* and *clay* families have as their basic building unit, a polymerized sheet of silica tetrahedra. The electrical charges in the sheets are balanced by $Al^{+3}$ cations in clays, leading to the formula $Al_4Si_4O_{10}(OH)_8$ for the clay mineral *kaolinite*.

With the micas a new principle must be mentioned to explain their compositions. $Al^{+3}$ cations are only a little larger than $Si^{+4}$ cations, and $Al^{+3}$ ions can replace the $Si^{+4}$ ions in silica tetrahedra by solid solution without affecting the polymerization. Because $Al^{+3}$ has a smaller charge than $Si^{+4}$, a substituted tetrahedron has an extra negative charge to be satisfied. This charge cannot be satisfied by polymerization so extra cations must be added. Approximately one-quarter of the silicon atoms in tetrahedra are substituted by aluminum atoms in the micas, and cations such as $K^{+1}$, $Mg^{+2}$, and even some extra $Al^{+3}$ must be added outside of the tetrahedra to balance the charges. The mica mineral *muscovite,* for example, has the formula $KAl_2(Si_3Al)O_{10}(OH)_2$.

Muscovite, the most common mica, is a clear and commonly colorless variety that derives its name from Muscovy, an old name for Russia, where it was widely used as a substitute for glass. *Biotite* is a dark variety rich in iron and magnesium. Micas and clays are most typically found in the continental crust.

### Quartz

The only common mineral composed exclusively of silicon and oxygen is *quartz*. It is the classic example of a crystal structure that has all its charges satisfied by polymerization of the tetrahedra into a three-dimensional network.

Quartz characteristically forms beautiful crystals, and is found in many beautiful colors. It is one of the most widely used gem and ornamentation minerals. Common names for some gemstone varieties of quartz are rock crystal (colorless), citrine (yellow), amethyst (violet), and agate (banded structure, a variety of colors). Quartz is a particularly abundant mineral in rocks of the continental crust. Indeed, it is so abundant that certain sedimentary rocks are composed entirely of quartz.

### Feldspar

The name given to the *feldspar* family is derived from two German words, *feld* (field), and *spar* (mineral). Early German miners were familiar with feldspar in their mines, and found the same mineral in the abundant rocks they had to pick, seemingly endlessly, from the fields around their homes. They were so struck by the abundance of feldspar that they chose a name to indicate that their fields seemed to be always growing new crops of crystals. Of course they were mistaken about fields growing new crystals, but they were not mistaken about the abundance of the feldspars: they are the most common group of minerals in the crust. The feldspars account for 60 percent of all minerals in the continental crust, and together with quartz comprise about 75 percent of the volume of the continental crust. Unlike quartz, which is rare in rocks of the oceanic crust, feldspars are also abundant in rocks of the sea floor.

Like quartz, the feldspars have structures formed by complete polymerization of all oxygen atoms in the tetrahedra. Unlike quartz, however, some of the tetrahedra contain $Al^{+3}$ substituting for $Si^{+4}$, so that other cations must be added to the structures to preserve the charge balances.

There are three principal feldspar minerals: potassium feldspar $K(AlSi_3)O_8$, albite $Na(AlSi_3)O_8$, and anorthite $Ca(Al_2Si_2)O_8$. Although $K^{+1}$ and $Na^{+1}$ substitute for each other to some extent, there are limits to the process of solid solution. The most important substitution is $Ca^{+2}$ for $Na^{+1}$ because, as can be seen in Figure 3.2, the two ions are much closer in size than either is to the size of $K^{+1}$. However, $Ca^{+2}$ and $Na^{+1}$ have different charges so the actual substitution involves two ions: ($Na^{+1}$ + $Si^{+4}$) for ($Ca^{+2}$ + $Al^{+3}$). The substitution is so effective that a continuous group of feldspar minerals, the *plagioclase* group, has compositions ranging from albite to anorthite.

### Other minerals

Although silicates are the most abundant minerals on Earth, a number of others—principally oxides, sulfides, carbonates, phosphates, and sulfates—are common enough to be called rock-forming minerals.

Some common oxide minerals are the compounds of iron, *magnetite* ($Fe_3O_4$) and *hematite* ($Fe_2O_3$); the oxide of titanium, *rutile* ($TiO_2$); and of course the solid form of $H_2O$, *ice.* Oxides are important ore minerals and the principal source of tin, iron, chromium, manganese, uranium, niobium, and tantalum.

The most common sulfide minerals are *pyrite* ($FeS_2$), *pyrrhotite* ($FeS$), *galena* ($PbS$), *sphalerite* ($ZnS$), and *chalcopyrite* ($CuFeS_2$). The sulfide minerals are exceedingly important as ore minerals, being the principal source of copper, lead, zinc, nickel, cobalt, mercury, molybdenum, silver, and many other elements.

The complex carbonate anion $(CO_3)^{-2}$ forms three important and common minerals: *calcite, aragonite,* and *dolomite.* We have already seen that calcite and aragonite have the same composition, $CaCO_3$, and are polymorphs. Calcite is much more common than aragonite. Aragonite, however, is widespread because it is commonly found in shells, coral reefs, and stalactites in caves. Dolomite has the formula $CaMg(CO_3)_2$.

One very important phosphate mineral contains the complex anion $(PO_4)^{-3}$. It is the mineral *apatite,* $Ca_5(PO_4)_3OH$, and it is the substance that our bones and teeth are made from. It is also a common mineral in many varieties of rocks and is the main source of phosphorus used for making plant fertilizers.

Sulfate minerals contain the complex anion $(SO_4)^{-2}$. Although many sulfates are known, only two are common, and both are calcium sulfate minerals: *anhydrite,* $CaSO_4$; and *gypsum,* $CaSO_4 \cdot 2H_2O$. Both form when sea water evaporates; they are the raw material used for making plaster of all kinds. The famous Plaster of Paris got its name from a quarry near Paris where a very desirable pure-white form of gypsum was mined centuries ago.

# SUMMARY OF MINERALS

We now know that minerals are naturally occurring elements or inorganic compounds, and that each has distinctive properties arising from its crystal structure and its composition. All rocks are made from minerals, but only a few minerals and mineral families form the mass of the crust. The common rock-forming minerals are listed in Table 3.3, which can be referred to as we proceed to discuss rocks and, in later chapters, how they are put together to form the Earth. The silicate minerals in Table 3.3 are divided into two categories: the *iron-and-magnesium-rich minerals* (those that contain iron and magnesium as their principal cations) and the *iron-and-magnesium-poor minerals.*

# MINERALS AS INDICATORS OF ENVIRONMENT

Minerals should not be regarded merely as objects of beauty or sources of economic materials. Contained within their makeup are

**Table 3.3**
**The Common Rock-Forming Minerals**

| Silicates | Oxides | Sulfides | Carbonates | Sulfates | Phosphates |
|---|---|---|---|---|---|
| *The iron- and magnesium-rich minerals* | | | | | |
| Olivines | Hematite | Pyrite | Calcite | Anhydrite | Apatite |
| Pyroxenes | Magnetite | Sphalerite | Aragonite | Gypsum | |
| Augite | Rutile | Galena | Dolomite | | |
| Amphiboles | Ice | Chalcopyrite | | | |
| Hornblende | | | | | |
| Garnets | | | | | |
| *The iron- and magnesium-poor minerals* | | | | | |
| Quartz | | | | | |
| Feldspars | | | | | |
| Potassium feldspar | | | | | |
| Plagioclase | | | | | |
| Micas | | | | | |
| Muscovite | | | | | |
| Biotite | | | | | |
| Clays | | | | | |
| Kaolinite | | | | | |

the keys to the conditions under which they formed. Study of minerals, therefore, can provide invaluable insight into chemical and physical conditions in regions of the Earth that are inaccessible to direct observation and measurement.

An understanding of the growth environments of minerals has come very largely from studying minerals in the laboratory. By suitable laboratory experiments, for example, scientists have been able to define the temperatures and pressures at which diamond grows, rather than its polymorph graphite (Fig. 3.6). Because it is possible to infer, by indirect means, the way temperature and pressure increase with depth in the Earth, we can state with certainty that rocks in which diamonds are found are samples of the mantle from at least 145km below Earth's surface.

The use of minerals to get information about environments is widely possible. Past climates, for example, can be deciphered from the kinds of minerals formed during erosion, and the composition of seawater in past ages can be determined by the minerals formed when seawater evaporated and deposited the salts it contained in solution. Rather than elaborate many examples, we will turn to an examination of rocks, for rocks are, after all, simply assemblages of minerals. The kinds of minerals that group to form a rock, and the ways in which the grouping occurs, is even more informative than are individual minerals.

## THE THREE KINDS OF ROCK

When one first looks at rocks they seem confusingly varied. Some appear platy or distinctly layered and display pronounced, flat

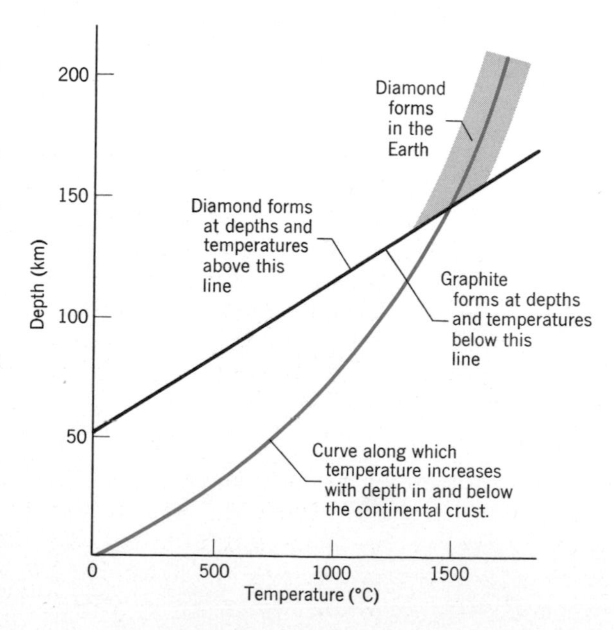


**Figure 3.6**
**Line separating conditions of temperatures and pressure of overlying rock (here plotted as depth) at which the two polymorphs of carbon, diamond and graphite, grow. At a pressure equal to that at depth 145km, the diamond-graphite line intersects the curve depicting the way Earth's temperature changes with depth. Diamond can form in rocks only at depths of 145km or more in the mantle.**

crystals of mica. Others are coarse and evenly-grained, and lack layering; yet they may still be made from the same kinds of minerals present in the platy, micaceous rock.

Despite obvious diversity, we can classify all rock into three families, according to the way the rock formed. The first family consists of ***igneous rock,*** named from the Latin word *ignis,* meaning fire. It is *rock formed by the cooling and solidification of magma; an interlocking aggregate of silicate minerals.* The second family consists of ***sedimentary rock,*** named from the Latin word *sedimentum,* meaning settling, and is defined as *rock formed from sediment by cementation or by other processes acting at ordinary temperatures at or near Earth's surface.* The third and final family is made up of ***metamorphic rock,*** named from the Greek words *meta,* meaning change, and *morphe,* meaning form; hence: change of form. Metamorphic rock is *rock formed within Earth's crust by transformation, in the solid state, of pre-existing igneous or sedimentary rocks as a result of high temperature, high pressure, or both.*

Most rock within the crust has formed, initially, from magma. It is estimated, for example, that 95 percent of all rock in the crust is igneous or metamorphic, derived from igneous rock. However, we can see in Figure 3.7 that most of the rock we actually *see* at Earth's surface is sedimentary. The difference arises because most igneous rock is formed as a result of internal processes that we cannot readily observe, whereas sedimentary rock is formed by the external processes that are happening around us at all times.

### Texture and mineral assemblage

Within each rock family, the various members share a common origin—by solidification of magma, by cementation of sediment, or by transformation of pre-existing rock. Yet the rocks within each family do not necessarily look alike. One igneous rock might contain only big mineral grains 20cm or 30cm in diameter; while another may consist entirely of grains as small as the head of a pin. One sedimentary rock has pronounced layered structure, while another is almost devoid of layering. The more closely we look at the kinds of rock within a family, the more differences we see. These differences are the clues we use to decipher how and where a rock formed. For example, large mineral grains in an igneous rock suggest slow cooling, while very small grains suggest rapid cooling. Slow cooling suggests, in turn, that an igneous rock formed deep underground, where the rock above it acts as a blanket that prevents rapid chilling. In contrast, rapid cooling suggests that the rock formed from lava poured out on the surface and cooled quickly.

Within a conveniently sized piece of rock held in the hand for close study, we can observe two kinds of small-scale features that provide many clues to the origin of the rock. The first feature is ***texture.*** This means *the sizes and shapes of the individual particles in a rock, and the mutual relationship between them.* For example,

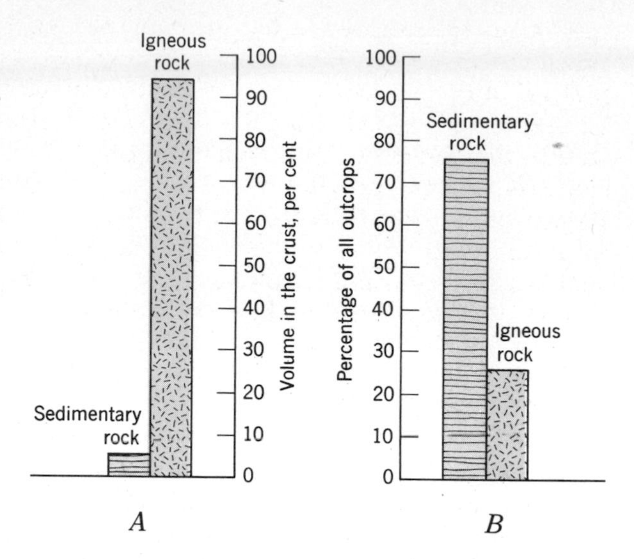


**Figure 3.7**
**Relative amounts of sedimentary and igneous rock in Earth's crust. Metamorphic rocks are included in the sedimentary or igneous category, depending on their origin. *A*. The great bulk of the crust consists of igneous rock (95 per cent), while sedimentary rock (5 per cent) forms a thin covering at and near the surface. *B*. The extent of sedimentary rock at the surface is much larger than that of igneous rock; so 75 per cent of all rocks seen at the surface are sedimentary and only 25 per cent are igneous. (*After Clarke and Washington, 1924.*)**

the mineral grains may be flat and parallel to each other, giving the rock a pronounced platy or flaky texture, like a pack of playing cards. In addition, the various minerals may be unevenly distributed and concentrated into specific layers. The rock texture is then both layered and platy. A texture both layered and platy is characteristic of metamorphic rock that was once a layered sedimentary rock.

Texture indicates much about the history of a rock. As we have noted, texture can tell us whether an igneous rock has cooled rapidly or slowly; but it can also tell us whether the grains in a sedimentary rock settled slowly down through still lake water, or were tumbled about by a rushing stream or by the surf along a coast. We shall discuss the most common textures later in this chapter, as we learn how rocks are classified.

The second feature we see in a hand specimen is the minerals the rock contains. A few kinds of rock contain only one mineral, but most contain two or more common rock-forming minerals. The minerals, and their percentages, immediately tell us the composition of the rock. *The varieties and abundances of the minerals present in rocks,* commonly called ***mineral assemblages,*** are important pieces of information in our reading of the rock record. They tell us, for example, whether an igneous rock was formed in the mantle or in the crust, because the compositions of rock from crust and mantle are quite different. Mineral assemblages can also aid us in distinguishing between different rock families. For example, calcite is a common mineral in sedimentary rock, but is rarely observed in igneous rock.

Sometimes the features in a hand specimen must be magnified so that they can be seen clearly. For this, a low-power magnifying glass can be a great help. Another convenient way to help with the examination of rock texture is to polish the surface of a hand specimen. Yet another technique—one we use for illustration many times in this book—is to make a *thin section.* This is prepared by first grinding a smooth, flat surface on a piece of rock. The flat surface is then glued to a glass slide and is ground down to a slice so thin that light passes through it easily. The appearance of the same rock on a polished surface and in a thin section is shown in Figure 3.8. Although we cannot expect everyone to have a chance to make or study thin sections, the photographs in Figure 3.8 should prove that the technique only makes more obvious the features already seen in a hand specimen.

# IGNEOUS ROCK

Now that we know what to look for in a rock, we can ask the meaning of the evidence we gather. Discussing rocks, family by family, we start with the igneous rocks because they are the most abundant family. The adjective *igneous,* as we have seen, pertains to fire. We use it for all the activities associated with the formation, movement, and cooling of magma in the Earth. Igneous events played an important part in shaping our ancestors' beliefs, including mythologies and religions. Even Stone-Age people were aware of igneous activity, because they saw lava pouring out of volcanoes. But they did not understand what caused these awesome eruptions, and they visualized volcanic "fires." The very word ***volcano,*** *the vent from which molten igneous matter, solid rock debris, and gases are erupted,* comes from *Vulcan,* the Roman god of fire. Let us, then, consider first the most obvious igneous rocks, those formed at Earth's surface where igneous matter from volcanoes cools and hardens.

## Volcanic or extrusive igneous rock

Igneous rock formed by volcanic eruption is called ***extrusive igneous rock,*** because it is *rock formed by the cooling of magma poured out onto Earth's surface.*

Not all magma comes out of a volcano as a smooth-flowing liquid. Often gases escape from a volcanic vent so violently that they splatter magma into small, hot fragments, and rip pieces of solid rock off the walls of the vent. Fragments from both sources form a pile or blanket around the vent. Dramatic photographs of erupting volcanoes show clouds of fine rock particles ("ash") thrown out by gases. Figure 3.9 shows the "ash" that was thrown out by an erupting volcano on Heimaey Island, Iceland, in April 1973. The "ash" was carried many kilometers away from the vent by the wind, covering and destroying roads and buildings. Volcanic ash has been found to travel many hundreds of kilometers from its source, and in some cases, such as the eruption of the volcano Krakatoa in 1883, "ash" was thrown so high in the air it circled completely around the Earth.

When applied to volcanoes, the word *ash* is misleading, because ash means, strictly, the solids that are left after something

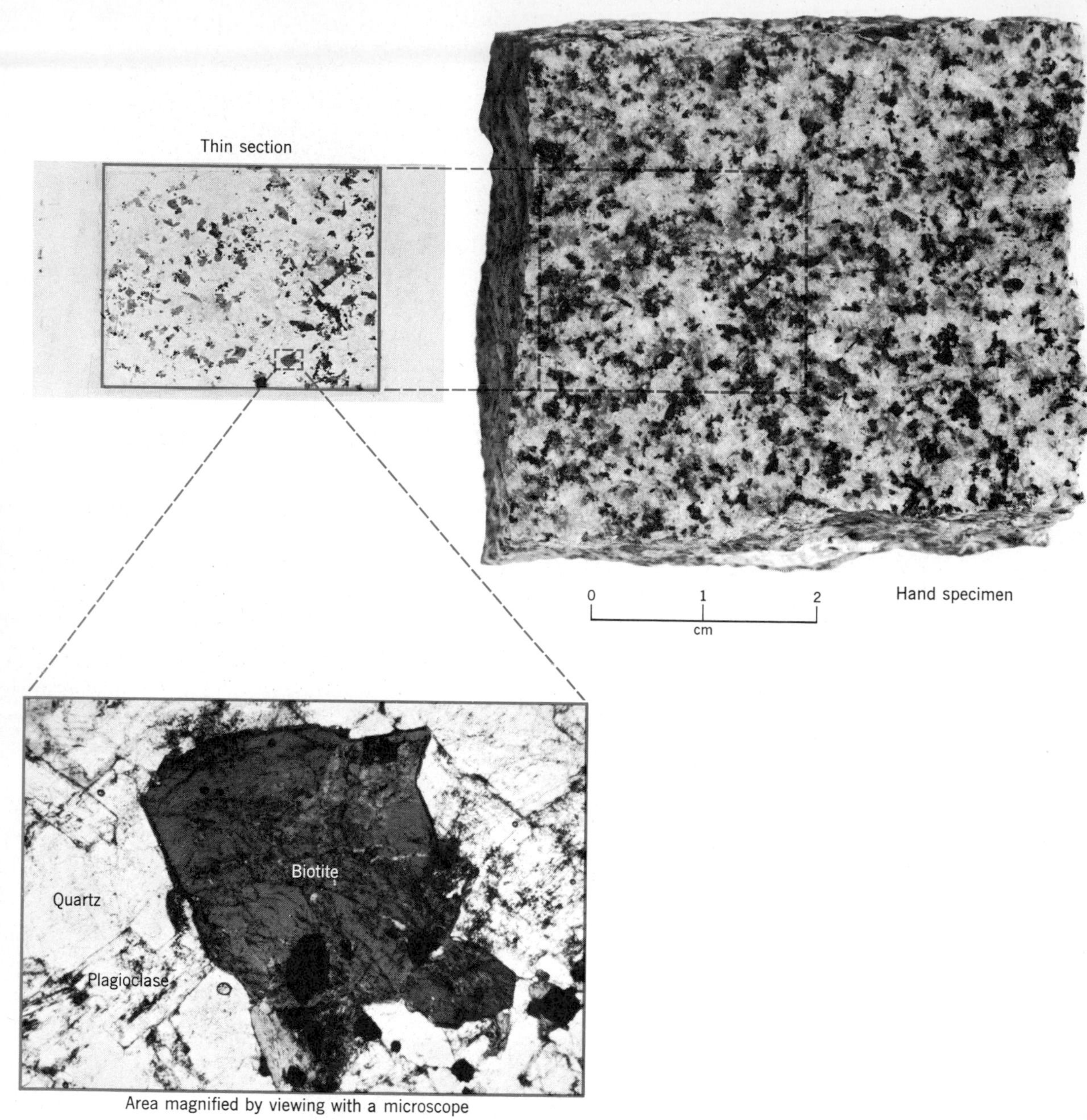


Figure 3.8
**In the study of rock, polished surfaces and thin slices of rock reveal texture and distribution of minerals to great advantage. The specimen here is igneous rock containing quartz, plagioclase, hornblende, and biotite. (*Yale Peabody Museum and B. J. Skinner.*)**

inflammable, such as wood, has burned. But the fine particles thrown out by volcanoes look so like true ash that it has become a convenient custom to use the word for them too. Volcanic particles, regardless of size, become packed together to form rock. *An extrusive igneous rock formed by the agglomeration of small volcanic particles* ("*ash*") is ***tuff*** (Fig. 3.10). Agglomeration means the gathering, or clustering together, of particles. It occurs in two ways. First, when ash is ejected from a volcano and settles on the ground, it may be so hot that its particles simply weld or fuse together. In many parts of western United States, welded tuff is common. The second way in which agglomeration occurs is when

**Figure 3.9**
**Volcanic ash, blasted out of a volcano on Heimaey Island, Iceland. The ash was distributed downwind from the vent, covering houses and roads in a small fishing village to a depth of nearly 5m. (*Klaus D. Franke/Peter Arnold.*)**

ash is too cold to weld; in this case the particles become tuff by later cementation. How this happens will be explained in the course of our examination of how the particles in sedimentary rock are held together, because tuff is in fact a sedimentary rock, all of whose particles are volcanic.

Many eruptions of volcanic ash have occurred within historic time; some have caused tragic disasters. One occurred in A.D. 79 during an eruption of Vesuvius, a volcano near Naples. A rain of ash from Vesuvius completely covered the wealthy Roman resort town of Pompeii at the southern edge of the volcano. The ash gradually became cemented to form tuff, and for nearly 1,800 years Pompeii lay buried and largely forgotten. When finally relocated and excavated by archeologists, the town revealed a vivid picture of Roman culture and the everyday activities of a Roman citizen.

Although eruption of ash is dramatic, the amount of rock formed from ash is small compared to that formed from liquid magma

**Figure 3.10**
**Hand specimen of tuff, an extrusive igneous rock formed by the packing of small rock- and mineral particles ejected violently from a volcano. In this specimen, fragments of igneous rock several centimeters across are trapped in a matrix of similar particles that range down to the size of dust grains. Clark County, Nevada. (*Yale Peabody Museum.*)**

A

B

Figure 3.11
**Two kinds of lava seen in Hawaii.** *A.* **Frozen lava from an eruption of Kilauea volcano in 1971. The highly fluid lava flowed 12km from its vent, then poured into the sea. The smooth, ropy surface of the frozen lava is locally called** *pahoehoe.* **The height of the cliff is 2 meters. (***D. Swanson, U.S. Geological Survey.***)** *B.* **Slowly moving, very viscous lava from an eruption of Mauna Loa volcano in 1935. As it flows the lava breaks up into hot, rubbly fragments. When flow ceases and the lava freezes, the resulting rock is a rough mass of fragments (locally called** *aa***), very different from the smooth rock in part A. (***H. T. Stearns, U.S. Geological Survey.***)**

that flows from volcanoes. *Magma that reaches Earth's surface through a volcanic vent, and flows out as hot streams or sheets,* is ***lava*** (Fig. 3.11). We have already noted that lava tends to cool rapidly by comparison with the magma that solidifies deep underground. Some lava cools and solidifies too quickly for its atoms to organize themselves into minerals; so, instead, the lava forms natural glass, the substance we call *obsidian.* Most lavas, however, cool slowly enough for crystals to form, taking months or even a few years for complete cooling. In such cases the resulting rock consists largely or entirely of small, crystalline mineral grains.

How might we distinguish between a tuff and a volcanic rock that consists of frozen lava? Many important characteristics are discussed in Appendix C, but two deserve mention here because they tell us much about how magma behaves. Nearly all the particles that constitute tuff are bits of igneous rock, although a few may be fragments of other rocks torn from the walls of the volcanic vents. The fragments, both small and large, are rough-edged and irregular in shape, two features indicating that the explosions were violent. On the other hand, solidified lava commonly contains features that reveal its former fluidity. The most unique feature is the presence of frozen bubbles, formed when gases bubbled out of the former liquid (Fig. 3.12). The process is akin to that of gas bubbling out of a bottle of soda water. The soda water effervesces when the bottle is opened because the carbon dioxide gas, formerly held in solution under pressure, forms bubbles and escapes. The same thing happens when magma rises to Earth's surface from deep within the crust. Because magma is sticky and viscous, gas bubbles cannot escape from it nearly as quickly as they can from soda water. Therefore, much of the lava cools before all the bubbles have escaped, leaving the upper parts of most former lava flows marked by "frozen" bubbles.

How common is volcanic rock? Anyone who has seen an active volcano such as Mount Etna in Sicily, Mauna Loa and Kilauea in Hawaii, and the volcanoes in Iceland and Central America, knows that these great mountain masses consist of volcanic rock, and that they must have been built up by numerous eruptions of volcanic fragments and lava. Furthermore, once we have recognized the features of active volcanoes, we can easily recognize old volcanic rock, even though the volcanoes that formed it have long since died. Almost everywhere we look, we find at least some evidence of former volcanic activity. In the Columbia Plateau in Washington,

**Figure 3.12**
**Frozen bubbles (*vesicles*) in lava are *small openings made by escaping gas originally held in solution under high pressure while the magma was underground.* The presence of vesicles in a rock is clear evidence that the rock is lava. Some lavas are so vesicular that they are frozen, rocky froth (called *pumice*). Mauna Loa Volcano, Hawaii. (*Yale Peabody Museum.*)**

in much of Ontario and Quebec, along the Blue Ridge in Virginia, in central Connecticut, southern Colorado, central Nevada, Texas, Oklahoma—in fact, in every one of the United States and in every province of Canada, we find volcanic rock. Even on the Moon, astronauts have found many former lava flows.

### Intrusive igneous rock

With so much magma poured out onto Earth's surface, how much magma fails to reach the surface and consequently solidifies down below? A single volcano such as Mount Etna has thrown out enough lava to build a cone-like mound whose volume is many cubic kilometers. Such persistent outpouring indicates that beneath the volcanic mound lies a large reservoir of magma. When a volcano becomes extinct, we infer that all the magma remaining in the reservoir has cooled and solidified. Although we cannot slice down through an active volcanic cone to see what lies beneath,

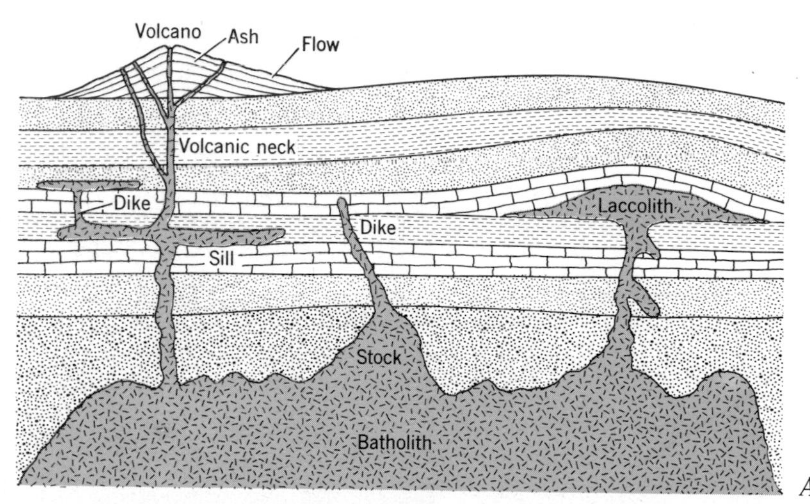

**Figure 3.13**
**(A) Diagrammatic section through part of the crust to show the various forms assumed by *plutons*, all of which are *intrusive igneous bodies, regardless of shape, size, or composition.* Many plutons were once connected with volcanoes, and indeed there is a close relationship between intrusive and extrusive igneous rocks. The common plutonic forms are: *dike, a sheet of intrusive igneous rock cutting across the layering of pre-existing rock; sill, a sheet that is parallel to the layering; laccolith, a lenticular intrusive igneous body above which the layers of invaded bedrock have been bent upward to form a dome; batholith, a very large intrusive igneous body that cuts across the layering of the intruded rocks and has an irregular shape;* and *stock, a small body with the same characteristics as a batholith.* (B) Shiprock, New Mexico, the eroded remains of a volcanic neck 400m high. The three prominent ridges radiating outward are made by dikes of intrusive igneous rock. (*J. S. Shelton.*)**

erosion has sliced and laid bare many former reservoirs full of frozen magma. Despite the enormous volumes of extrusive igneous rock, far more magma has cooled below the surface, where it has formed intrusive igneous rock. *Any igneous rock formed by cooling and solidification of magma below Earth's surface* is an ***intrusive igneous rock.***

Apparently, then, intrusive and extrusive igneous rock are closely related in composition; some of them even have similar textures. But intrusive and extrusive igneous rocks are commonly found in masses that have very different shapes. Extrusive rocks form great blankets, consisting of both fallen ash and former streams of flowing lava. Intrusive rock forms bodies with many complex shapes, because magma can flow and squeeze into innumerable openings in a rock. Although the shapes and sizes of intrusive masses are varied, in many cases the shapes were controlled by the viscosity—the stiffness—of the magma. The shapes of intrusive masses, then, are closely related to the kind of igneous rock that was formed (Fig. 3.13).

### Kinds of igneous rock

Examining a hand specimen, we note first its texture. We should, in general, be able to decide on the basis of texture alone whether we are seeing an intrusive or an extrusive igneous rock. The average size of mineral grains varies widely from one kind of igneous rock to another, but in all kinds of rock, the main control of grain size is the rate at which the magma cooled. Figure 3.14 illustrates that control and demonstrates that extrusive igneous rock is always fine-grained.

In most igneous rocks, all the grains are about the same size, but there is a special texture class called ***porphyry,*** *an igneous rock consisting of coarse mineral grains scattered through a mixture of fine mineral grains* (Fig. 3.15). The texture of a porphyry is a revealing clue. Porphyry forms when magma in a deep reservoir has partly crystallized to form a few large crystals, and is then rapidly moved upward. In its new cooling place it cools more rapidly, so that the new crystals are tiny. *The isolated large crystals in porphyry* are called ***phenocrysts.*** Because phenocrysts grow within a fluid and encounter no interference from crystals growing adjacent to them, most of them have perfect crystal forms. The mineral grains in most igneous rocks, however, are irregular because during the final stages of mineral growth and crystallization, all the mineral particles are crammed against each other, preventing the formation of smooth crystal faces.

Having determined by mineral texture whether an igneous rock is intrusive (coarse- or medium-grained) or extrusive (fine-grained), we should next determine the kinds of minerals present and roughly estimate their amounts. From this we get the information necessary to give a specific name to the rock by the use of Figure 3.16. In that figure, the classification is less complicated than it looks, and is so helpful that it is worth examining closely.

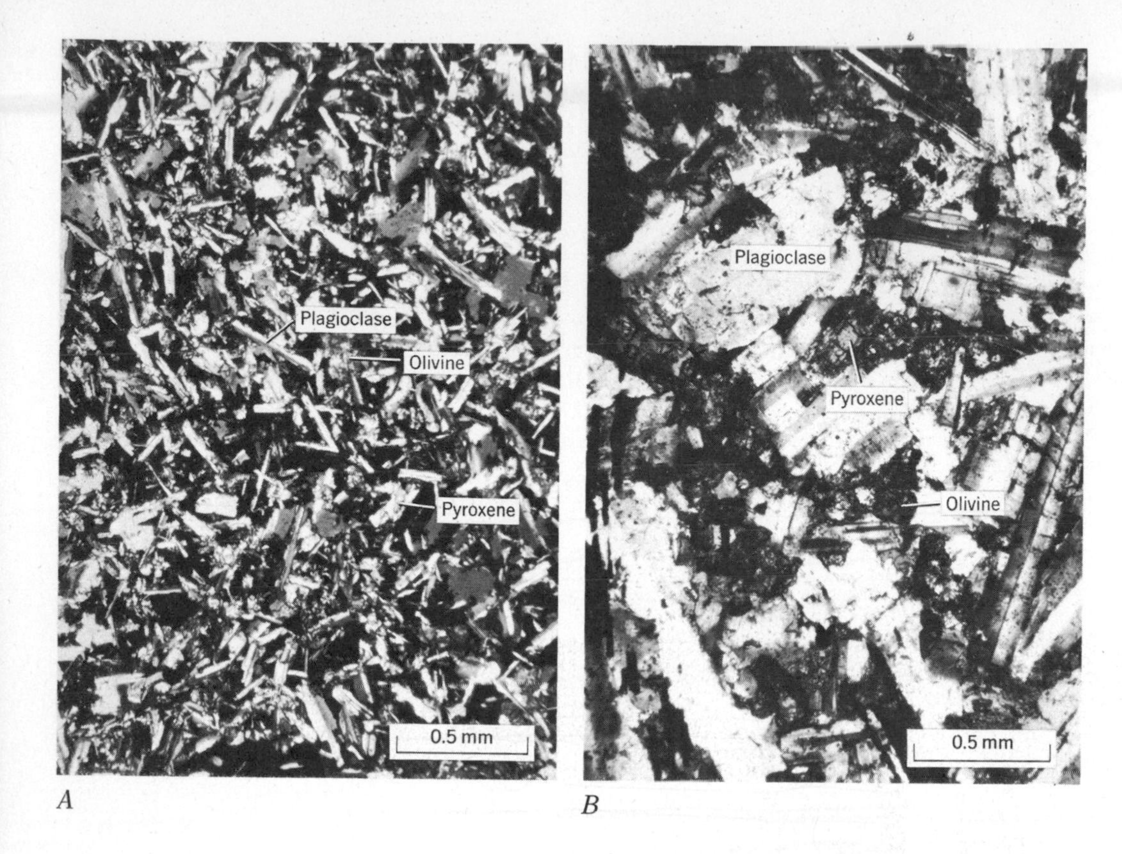


Figure 3.14
**The very different grain sizes that are visible in these thin sections of (*A*) basalt, (*B*) diabase, and (*C*) gabbro reflect the rate at which each magma cooled and thereby the control that depth plays on rate of cooling. The three rocks shown here have the same composition (plagioclase, pyroxene, and olivine). Gabbro is a coarse-grained, deep-seated intrusive igneous rock, diabase a medium-grained, shallow intrusive rock, and basalt a fine-grained, frozen lava. (*B. J. Skinner*.)**

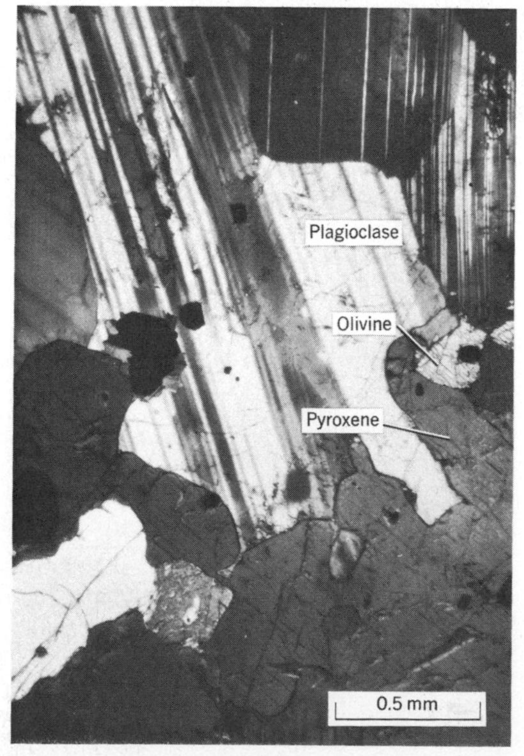

**Figure 3.15**
**Phenocrysts of plagioclase trapped in a matrix of sub-microscopic minerals reveal typical *porphyritic texture*. The grain size of minerals in the matrix is so small that individual grains cannot be seen with the microscope; the mass appears almost opaque. The porphyry seen in this thin section has the composition of basalt. It formed when the partly crystallized magma was extruded and suddenly chilled, preventing further growth of large crystals. If extrusion had not occurred and slow crystal growth had continued, the resulting rock would be a gabbro. (*B. J. Skinner*.)**

All the common igneous rocks are mixtures of one or more of six mineral families: quartz, feldspar, mica, amphibole, pyroxene, and olivine. When the percentages of each mineral in a hand specimen have been estimated, we find the correct place in Figure 3.16, and read the corresponding rock name, selecting the fine-grained or coarse-grained variety, whichever is appropriate.

The common igneous rocks, with suggestions as to how to identify them, are discussed more fully in Appendix C. Here we mention only the most common varieties.

First, consider granites and granodiorites, coarse-grained intrusive igneous rocks found only in the continental crust. These contain abundant feldspar and quartz, and are usually light-colored. Granites and granodiorites form huge batholiths in mountainous areas, and their origins seem to be associated with the origins of mountains. In Chapters 17 and 18 we shall see how this happens. Although bodies of granite and granodiorite are common, masses of extrusive igneous rock with the same composition—rhyolite—are less common. This fact provides an important clue that we shall use in Chapter 15.

Second, consider basalt, the fine-grained extrusive igneous rock that is characteristic of oceanic crust but is also found in the continents. The abundance of basalt on the floor of the ocean

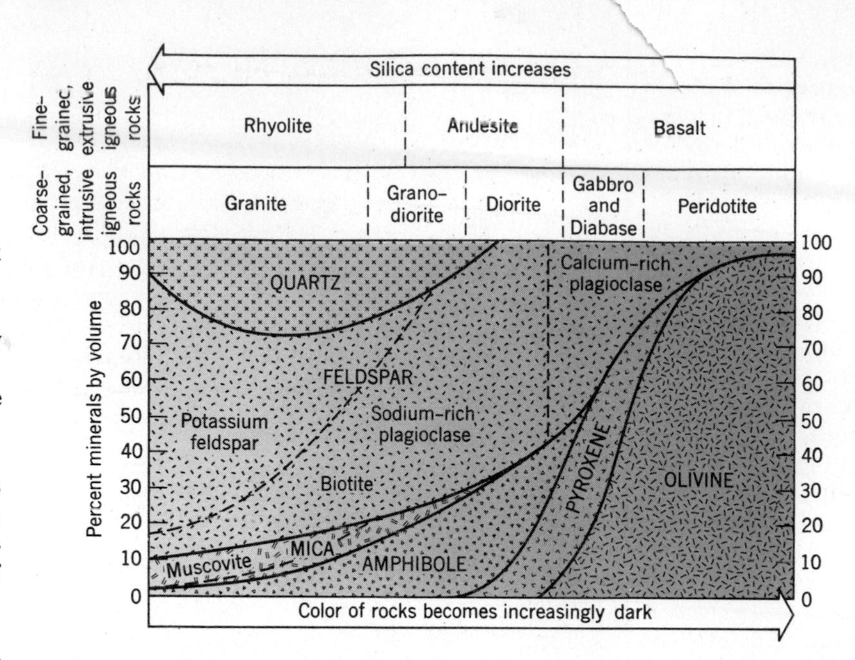


**Figure 3.16**
**Diagram showing the textures and proportions of the principal minerals in the common igneous rocks. Boundaries between kinds of adjacent rock in the table are not abrupt but gradational, as suggested by broken lines. In granites, for example, note the wide range in the proportions of minerals present: granites with nearly 75 per cent potassium feldspar belong at the left side of the diagram; others with only 20 per cent are near the boundary with granodiorite. To see the general range in composition for any granular rock, project the broken-line boundaries vertically downward; then estimate the percentage of a given mineral component by means of the figures at the right and left edges of the diagram. Only three kinds of fine-grained rock are included. Without considerable magnification it is not possible to estimate proportions of minerals in these rocks. (*Modified from R. V. Dietrich, Virginia Minerals and Rocks, Virginia Polytech. Inst.*)**

suggests its formation might have something to do with the origin of ocean basins. In Chapter 13 we shall see how this might happen. Basalt is dark colored because it contains dark-colored minerals such as pyroxenes and olivine. In complete contrast to the magma that forms granites, and which seems rarely to reach the surface, basaltic magma almost always gets up to Earth's surface. Basalt is therefore common, whereas deeply buried intrusive igneous rock with the same composition—gabbro—is rare. This fact too provides an important clue that we will use in Chapter 15.

The reasons why igneous rocks have the wide ranges of composition shown in Figure 3.16 are complex, but they can be reduced to two kinds of processes—how rock melts to form magmas, and how magmas crystallize to form rocks. These processes are discussed in Chapter 15, in connection with the origin of igneous rocks. Although igneous rocks may seem to be a confusingly large family of rocks to deal with, in fact they are not. In later chapters we shall see how the few magma compositions and the rocks discussed here, plus the basic properties that govern the formation of igneous rocks, provide clues to a great deal of Earth's history. Two important points about igneous rocks are these. First, rock formed from magma of granitic composition is confined to the continental crust, while rock formed from magma of basaltic composition occurs most commonly in oceanic crust. This distribution is a clue to some of Earth's most fundamental internal processes. The second point is that all igneous rocks are the products of internal processes. They are being formed continuously and brought to the surface, where they become the principal source of new sediment. Therefore, igneous rock is an important link through which internal and external processes interact.

# SEDIMENTARY ROCK

Like a perpetually restless housekeeper, nature is ceaselessly sweeping regolith off the nonweathered rock below, carrying the sweepings away, and depositing them as sediment in river valleys, lakes, seas, and innumerable other basins. We can see sediment being transported by every trickle of water after a rainfall and by every wind that carries dust. The mud on a lake bottom, the sand on the beach, even the dust on a windowsill is sediment. Because erosion and deposition of the products of erosion are continuous, we find deposited sediment almost everywhere.

When a thick pile of sediment accumulates, the particles near the base of the pile become compacted. Eventually they become cemented together to form a solid aggregate rock. Most commonly the cementation occurs when films of new mineral substance are deposited from water trapped in the spaces between particles of sediment. By compaction and cementing, then, sediment is transformed into sedimentary rock.

Sedimentary rock, as we saw in Figure 3.7, is the rock most commonly found at Earth's surface, because it forms a thin blanket over the igneous rock below. Its most obvious feature is sedimentary layering, strikingly exposed on mountainsides, in the walls of canyons, or in artificial cuts along many highways (Fig. 2.5). Such layers drew the attention of observant people even in ancient times. The most thoughtful observers realized that many sedimentary layers are composed of fragments of other rocks, spread out as loose sediment and eventually cemented to form new rock. Writings by Greek philosophers long before the time of Christ show that the meaning of sedimentary layers was understood even at that early time. Later, in 15th-Century Italy, Leonardo da Vinci wrote an explicit statement of the connection between erosion, sediment, and sedimentary rock. In his notebooks he recorded the close similarity between sedimentary rock high in the mountains of northern Italy and the sand and mud he saw along the seashore.

We see in sedimentary rock the same kinds of things that are visible in igneous rock: its constituent minerals and its texture. But because sedimentary rock does not resemble igneous rock, the questions we ask about it differ from those we ask about igneous rock. Some of these questions are: Where did the sediment come from? How was it transported? What led to its deposition? The terms we use in describing and classifying the kinds of sedimentary rock reflect such questions.

### Clastic sediment

Looking closely at a sediment, we see that its pebbles or sand grains are simply bits of rocks and minerals. A magnifying glass shows us that the finer sedimentary particles, too, are derived from broken-up rock but that generally the particles have undergone chemical changes; for instance, feldspars have been partly altered to clay. All sediment of this kind is known as ***detritus*** or ***clastic*** *sediment* from the Greek work *klastos* (broken), meaning *the accumulated particles of broken rock and of skeletal remains of dead organisms.* There is, of course, a continuous gradation of particle size, from the largest boulder down to sub-microscopic clay particles. This range of particle size is embodied in the classification of Table 3.4.

If a sedimentary rock is made up of mineral particles derived from the erosion of igneous rock, how can we tell it is sedimentary and not igneous? Besides obvious clues such as sedimentary layering, there are clues in texture as well. A typical clastic sedimentary rock contrasts strongly with igneous rock in the shapes and arrangements of the grains (Fig. 3.17). In igneous rock the grains are irregular and are interlocked. In sedimentary rock the particles are commonly rounded and show signs of the abrasion they received during transport. Clastic sedimentary rock also reveals cement that holds the particles together, whereas the grains in igneous rocks are held together by interlocking crystals. Another important feature of many sedimentary rocks is the presence of fossils. Life surely cannot tolerate the high temperatures under which igneous rocks form; so the presence of ancient shells or similar evidence of past life is an excellent clue to sedimentary origin. Similarly, the presence of features such as ancient ripple marks, marks of erosion by fast-flowing water, and mud cracks in the floors of ancient lakes tell us a rock was once a sediment.

Clastic sedimentary rock is classified mainly on the basis of size and shape of clastic particles as outlined in Table 3.4. When distinctive minerals are abundant in the rock, these too will aid in classification as explained in Appendix C.

### Chemical sediment

Certain kinds of rock contain fossils and other evidence of sedimentary origin, yet seem to be free of clastic sediment. The origin of such rock might be a puzzle were it not possible to find places like the Bahama Banks and the Persian Gulf, where similar

**Table 3.4**
**Definition of Clastic Particles, Together with the Sediments and Sedimentary Rocks Formed from Them** (*After C. K. Wentworth, Jour. Geol. Vol. 30 p. 377, 1922*)

| Name of Particle | Range of Limits of Diameter | | Name of Loose Sediment | Name of Consolidated Rock |
|---|---|---|---|---|
| | mm. | inches (approx.) | | |
| Boulder | More than 256 | More than 10 | Gravel | Conglomerate and sedimentary breccia |
| Cobble | 64 to 256 | 2.5 to 10 | Gravel | |
| Pebble | 2 to 64 | 0.09 to 2.5 | Gravel | |
| Sand | $1/16$ to 2 | 0.0025 to 0.09 | Sand | Sandstone |
| Silt | $1/256$ to $1/16$ | 0.00015 to 0.0025 | Silt | Siltstone |
| Clay[a] | Less than $1/256$ | Less than 0.00015 | Clay | Claystone, mudstone, and shale |

[a] Clay, used in the context of this table, refers to a particle size. The term should not be confused with clay minerals, which are definite mineral species.

**Figure 3.17**
**Examples of clastic sedimentary rocks.**

*A.* ***Conglomerate, a sedimentary rock that contains numerous rounded pebbles or larger particles,*** **is a coarse-grained example of many sedimentary rocks. Each fragment is itself a piece of some older rock or mineral, more or less rounded during transport. Canyon Range, Utah. (*R. L. Armstrong.*)**

*B.* **Thin section of sandstone, showing rounded sand grains that are cemented by calcite. (*B. J. Skinner.*)**

*C.* **Broken fragments of sea shells stuck together with calcite cement make a clastic sedimentary rock called *coquina.* St. Augustine, Florida. (*Yale Peabody Museum.*)**

*A*

*B*

*C*

rock is forming today. The rock is indeed sedimentary and the material deposited has indeed been transported. But the sediment is not clastic because its components were dissolved, transported in solution, and precipitated chemically instead of mechanically. *Sediment formed by precipitation of minerals from solution in waters of the Earth is* ***chemical sediment.*** It forms in two principal ways. One of these consists of biochemical reactions within the water.

Biochemical reactions result from the activities of plants and animals; for example, tiny plants living in seawater can influence the amount of carbon dioxide that is dissolved in the water and so cause calcium carbonate to precipitate.

The other way in which chemical sediment forms consists of inorganic reactions within the water. For instance, when the water of a hot spring cools, it may precipitate opal or calcite. Another common example is simple evaporation of seawater or lake water. After evaporation of the water, the salts that were formerly in solution remain as a residue of chemical sediment. All the table salt we eat comes from sedimentary rock formed in this way.

Most chemical sedimentary rocks contain only one important mineral and this is used as a basis for classification. The most common rocks formed by chemical precipitation are *limestone* and *dolostone,* which contain the minerals calcite and dolomite respectively.

**Transport and deposition of sediment**

Sediment is transported in many ways. It may simply slide down a hillside, or may be carried by wind, by a glacier, or by running water. In each case, when transport ceases, the sediment is deposited in a fashion characteristic of the transporting agency. When sediment is transported by sliding or rolling downhill, the result is a mixture of particles of all sizes. Much of the sediment carried by a glacier is deposited when and where the transporting ice melts. Such sediment also is a mixture of sedimentary particles of all sizes.

In the transport of sedimentary particles by wind or water, deposition happens when the flowing water or moving air slows down to a speed at which particles can no longer be moved. In a general way, therefore, the size of the grains in clastic sediment indicates the speed of the transporting medium. Coarse-grained sediment indicates deposition in fast-moving wind or water; fine-grained sediment indicates deposition by slow-moving wind or water.

The size of particles in sedimentary rock, and the way they are packed together, lets us draw inferences about the environment in which the original sediment was deposited. The existence of ancient oceans, coasts, lakes, streams, swamps, and all the other places where sediment accumulates today, can be demonstrated from clues in sedimentary rock. Clues to former climates, likewise, are found in the fossils within a sedimentary layer. Some animals and plants are restricted to warm, moist climates, others to hot, dry climates, and yet others to very cold climates. By using modern plants and animals as guides to the preferences of their ancestors, now fossils, we can draw inferences about the climate in which the latter lived.

# METAMORPHIC ROCK

Igneous rock is the principal product of internal processes, sedimentary rock of external processes. As sediment is piled up, layer on layer, sedimentary rock becomes so deeply buried and so greatly heated that it may melt and become magma. But long before the rock begins to melt, as temperature and pressure increase, new minerals start to grow in it, replacing some of the original minerals of the sedimentary rock. At the same time the new-mineral growth imprints new textures on the sedimentary rock. *The changes in mineral assemblage and rock texture, or both, that take place in the solid state within Earth's crust as a result of high temperatures and high pressures* are termed ***metamorphism.***

Metamorphism affects igneous rock also, but in a different way, because such rock responds differently to increased temperature and pressure. The explanation of the difference concerns water content. Sedimentary rock has, between its grains, innumerable open spaces that are saturated with water. When metamorphism takes hold, this intergranular "juice" serves to speed up chemical reactions in much the same way that water in a stew pot helps cook a tough piece of meat. In contrast, igneous rock has very little open space and contains little water. Before new minerals can grow within dry rock, the rock may have to be heated to much higher temperatures, and squeezed under higher pressures, than would be needed for mudstone or limestone.

### Mineral assemblages

In terms of the chemical elements present, the compositions of metamorphic rocks are essentially the same as those of igneous and sedimentary rock. Small differences exist because metamorphic heating eventually drives off the watery juices, which carry with them, in solution, small amounts of sodium, calcium, and other elements. But the chemical changes are small, and the principal differences between metamorphic rock on the one hand, and sedimentary and igneous rock on the other, are in mineral assemblages and textures.

An accumulating pile of sedimentary rock may eventually reach thicknesses greater than 20,000m. Under the influence of rising temperature and pressure, new minerals grow, and water is driven off. A new, water-scarce environment comes into existence, and in it, water-free minerals such as garnet begin to form. Finally, melting takes place, converting the metamorphosed rock into magma.

Metamorphism, therefore, creates progressive changes in mineral assemblages, and the changes depend on initial composition of the rock, temperature, and pressure. Each mineral assemblage is characteristic of a specific range of temperature and pressure.

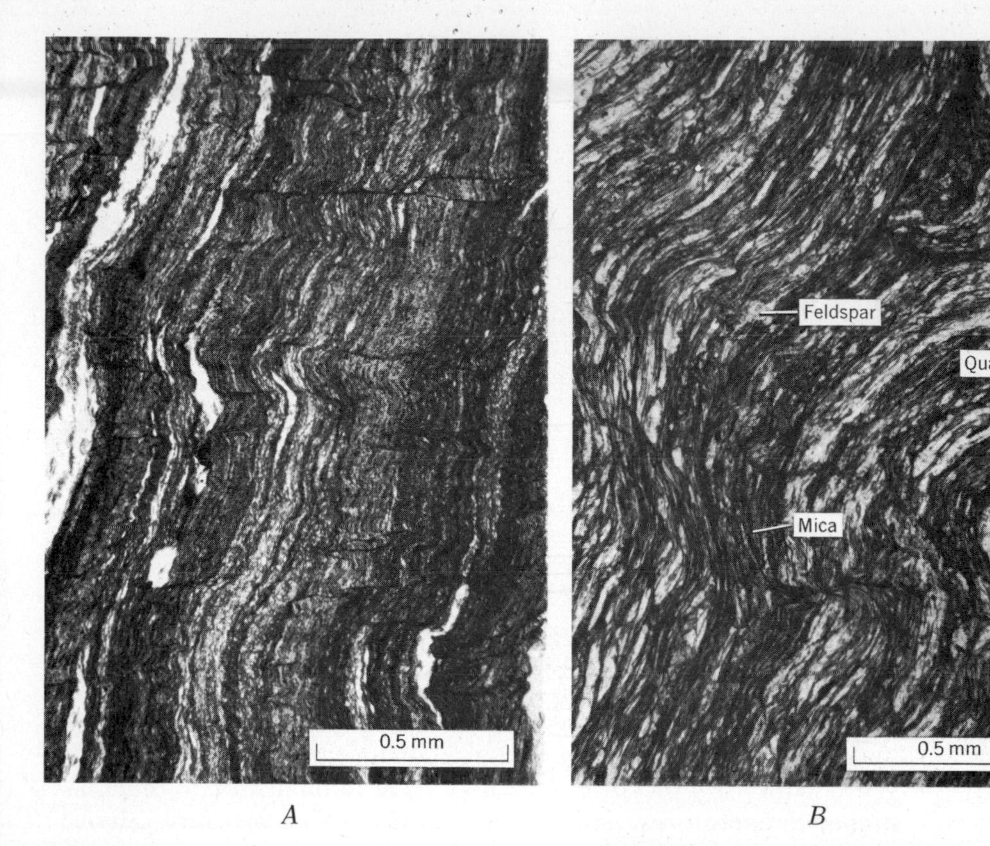


**Figure 3.18**
**The grain sizes in these thin sections of (*A*) *slate*, (*B*) *phyllite*, and (*C*) *mica schist* show how continued mineral growth occurs during metamorphism. The three rocks have the same composition. Mineral grains in the slate are barely visible. Grains in the phyllite are large enough to be seen easily, while those in the schist are large and obvious. (*B. J. Skinner*.)**

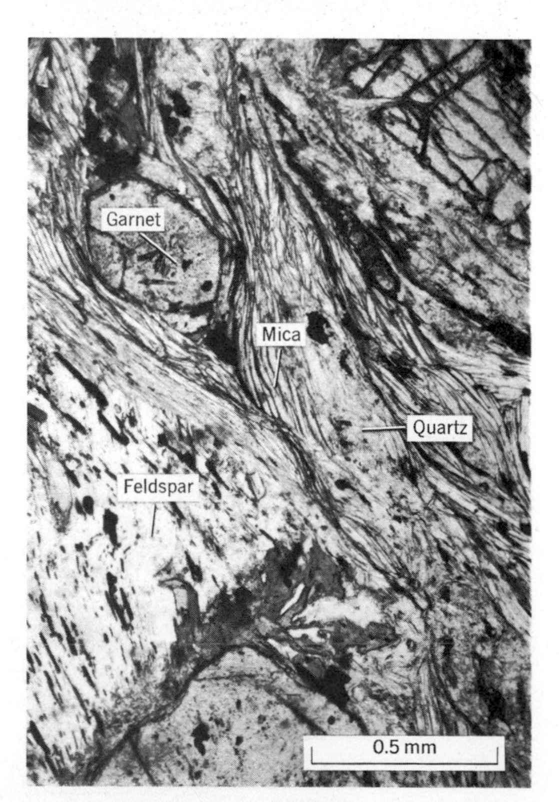

**Figure 3.19**
**Strongly foliated metamorphic rock. The light-colored bands are principally quartz, feldspar, and epidote; the dark bands quartz, hornblende, and biotite. The direction of foliation marks a direction of easy breakage. Because the layers are curved and contorted, flakes of rock that break along foliation surfaces also tend to be curved. Cornwall, England. (*Crown Copyright*.)**

Therefore, by studying mineral assemblages, we can draw many inferences about ancient rock temperatures, and sometimes about how deeply a rock body has been buried. But changes in temperature and pressure may occur otherwise than simply by burial, and metamorphic rock provides clues about this too. For instance, rock adjacent to an intrusive igneous rock becomes heated, and around the contact, a rim of metamorphic rock develops.

### Texture

One of the most obvious features of metamorphic rocks is a progressive increase in the size of mineral grains as rock is heated to higher temperature and squeezed to higher pressure (Fig. 3.18). The earliest minerals that grow within a metamorphic rock are silicates with layer structures: the micas. These minerals are distinctly platy because they grew in positions that minimize pressure; that is, with their flat sides perpendicular to the direction of maximum pressure. A similar thing happens when we press down on a heap of jumbled playing cards in a box. The more we press, the more the cards tend to line up parallel to the bottom of the box. So too, with metamorphic rock; the more it is squeezed, the more the platy minerals line up.

In the earliest stage of mineral growth, pressure is caused by the weight of the overlying rock; so the mineral plates grow parallel to the sedimentary layering. But with deeper burial, the sedimentary layers may become folded and contorted and the growing platy minerals no longer develop parallel to the layering. Eventually each sedimentary layer develops its own new mineral assemblage. Because the sedimentary layers differ slightly in composition, the new mineral assemblages differ but each displays the property of ***foliation,*** which is a *parallel or nearly parallel structure in metamorphic rock caused by a parallel arrangement* of platy *minerals* (Fig. 3.19).

### Kinds of metamorphic rock

The most obvious changes in metamorphic rock are textural. Although compositions vary as widely as the entire compositional range of sedimentary and igneous rock, textures of metamorphic rocks can be grouped into a few distinctive classes. Therefore, we use textures as the basis of describing metamorphic rocks, although where a particular mineral is very obvious, we modify the rock name by using the mineral name as a prefix.

The four commonest kinds of metamorphic rock are: *slate, phyllite, schist,* and *gneiss.* The sequence of names, as we have listed them, illustrates an important point about metamorphic rocks: it provides a clue to changes that happen systematically. For example, when mudstone is metamorphosed and tiny new grains of mica grow, the rock still looks like a mudstone because it is fine grained, but the new mica grains produce a pronounced ***rock cleavage,*** *the property by which a rock breaks into plate-like fragments along flat planes* (Fig. 3.20). The new metamorphic rock is *slate.*

Continuing metamorphism makes the mica grains grow larger, so that they can be seen with a magnifying glass; the rock is now *phyllite.* Eventually, the mica grains reach a size visible to the unaided eye, and new minerals such as garnet have formed; the rock has become *schist.* But as the new minerals grow, we observe that original differences in composition between layers of the mudstone are exerting a control. Some layers in the metamorphic rock are becoming rich in grains of feldspar, quartz, and other minerals. The rock becomes markedly layered, with alternating layers of coarse mica grains and coarse grains of quartz and feldspar; it is now *gneiss.* As we shall see in Chapter 18, these sequential changes are helpful clues to understanding how mountains are built.

## WHAT HOLDS ROCK TOGETHER?

We need not have had much experience in order to realize that some kinds of rock hold together with great tenacity, whereas

Figure 3.20
**Exposure of slate in northern Vermont, showing well-defined cleavage because of the growth of new micaceous minerals during metamorphism. Original sedimentary layering, somewhat folded, is still visible. The cleavage cuts across the layering from upper left to lower right. (*B. J. Skinner.*)**

other kinds are easily broken apart. The most tenacious rocks are igneous and metamorphic, because these possess intricately interlocked mineral grains. The growing minerals crowd against each other, filling all spaces and forming an intricate, three-dimensional jigsaw puzzle. Similar interlocking of grains holds together steel, ceramics, and bricks.

The forces that hold together the grains of sedimentary rock are less obvious. Sediment is transformed into sedimentary rock in four ways. (1) By pressure from overlying sediment or by vibrations of the ground arising from earthquakes, the irregular-shaped grains in a sediment can be packed into a tight, coherent mass. The interlocking of grains that results from this kind of packing is not strong, as in igneous rock, but under some circumstances it can hold sediment together. (2) Water that circulates slowly through the open spaces between grains deposits new minerals such as calcite, quartz, and iron oxide, which cement the grains together. (3) The weight of overlying deposits can squeeze water out of deeply buried sediment, compacting the sediment and reducing pore space. Compaction forces small grains close together and makes more effective the capillary forces exerted by remaining films of water between adjacent grains. (4) As sediment becomes deeply buried, its mineral grains begin to be recrystallized. Recrystallization resembles very low-grade metamorphism and has a somewhat similar effect on the texture of the rock. The newly growing minerals interlock and form strong aggregates, like those in igneous rock.

# WHAT BREAKS ROCK APART?

Studying an exposure of bedrock, we can see that the rock may be massive and difficult to break because no easy plane of fracture is present. Or we may observe that one or more easy directions of breakage are present. The breakage directions arise in three principal ways. (1) By original layering in a sedimentary rock. (2) By development of rock cleavage and foliation in metamorphic rocks. (3) By development of ***joints,*** *fractures on which movement has not occurred in a direction parallel to the plane of the fracture.* Rarely do joints occur singly. Most commonly they form *a widespread group of parallel joints,* called a ***joint set*** (Fig. 3.21).

As rock that has been deeply buried is slowly uncovered by erosion, it is relieved of the confining pressure exerted by overlying material, and so expands, and in the process develops joints. Joints do not let us infer much about the origins of rocks—they are found in all three families—but as we shall see in later chapters, they are extremely important in the control of weathering of rock, because they are passageways by which rainwater can enter it.

One special class of joints is restricted to certain igneous rocks, and in this one case they do afford a clue to origin. When a body of igneous rock cools, it contracts and sometimes fractures into small pieces, in the same way that a very hot glass bottle, plunged into

**Figure 3.21**
**This exposure of sedimentary rock in Alberta, Canada, displays three directions of easy breakage. The sedimentary layers, now tilted at a high angle, break the exposure into huge flat slabs. These in turn are broken by two sets of joints approximately perpendicular to each other. The joints in each set run at right angles to the sedimentary layering and tend to break the exposure into roughly cube-shaped fragments.** ***A combination of two or more intersecting sets of joints*** **is a** ***joint system.*** **(*B. J. Skinner.*)**

cold water, contracts and shatters. Cooling joints are found in igneous rock that cooled rapidly. When we see cooling joints, we infer that the rock cooled at or close to Earth's surface. Just as the hot bottle, if allowed to cool slowly, does not fracture, so a deeply buried body of igneous rock does not develop cooling joints because it must cool slowly. Unlike shattered glass, cooling fractures in igneous rock form regular patterns. For *joints that split igneous rocks into long prisms or columns,* we use the special term ***columnar joints*** (Fig. 3.22).

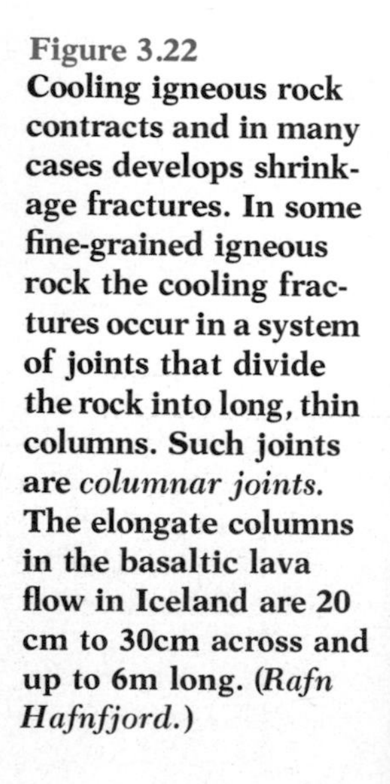

**Figure 3.22**
**Cooling igneous rock contracts and in many cases develops shrinkage fractures. In some fine-grained igneous rock the cooling fractures occur in a system of joints that divide the rock into long, thin columns. Such joints are** ***columnar joints.*** **The elongate columns in the basaltic lava flow in Iceland are 20 cm to 30cm across and up to 6m long. (*Rafn Hafnfjord.*)**

# CONCLUSION

The three rock families contain keys for understanding the internal and external activities of the Earth. Igneous rock results from internal processes, sedimentary rock from external. Metamorphic rock records intermediate steps by which both sedimentary and igneous rock are changed and influenced by internal processes. Our study of Earth's internal and external activities reveals that new rock is being made continuously. Extrusive igneous rock is forming at all active volcanoes, sedimentary rock along the margins of ocean basins, and metamorphic rock in the cores of high mountains (Himalaya) and adjacent to recently intruded masses of igneous rock (beneath the Imperial Valley in southern California).

There is, seemingly, a pattern to the ways in which igneous rock is weathered, and the weathered particles transported, deposited and turned, first into sedimentary rock, then into metamorphic rock, and finally again melted to form new igneous rock. Rock is the best evidence we have of the pattern, which consists of continuous motions and cyclic events taking place in and on the Earth. We turn next, therefore, to an examination of the great cycles of events that shape the Earth's surface.

# SUMMARY

1. Minerals are naturally occurring, crystalline, inorganic compounds. The atoms in each mineral are arranged in a definite geometric array, called a *crystal structure,* that is unique to the mineral.

2. Approximately 2,200 minerals are known, but of these about 20 make up more than 95 percent of the Earth's crust and are called the *rock-forming minerals.*

3. Silicate minerals are the most common minerals, followed by oxides, carbonates, sulfides, sulfates, and phosphates.

4. Beneath the regolith is a continuous solid body of bedrock. Where bedrock forms exposures that project through the regolith, we can study the kinds and structures of rocks.

5. The three rock families are igneous, sedimentary, and metamorphic. Igneous rocks form by Earth's internal processes, sedimentary rocks by external processes. Metamorphic rocks form when sedimentary or igneous rocks are buried and heated so that new minerals grow.

6. Igneous rock may be intrusive (meaning it formed within the crust) or extrusive (meaning it formed on the surface). The texture and grain size of igneous rock indicate how and where the rock cooled.

7. Igneous rock rich in quartz and feldspar, such as granite, granodiorite, and rhyolite, is characteristically found in continental crust. Basalt, rich in pyroxene and olivine derived from the mantle, is common beneath the ocean basins.

8. Sediment is transported by wind, streams, seawater, glaciers, and landslides. Then it is deposited, compacted, and cemented to form sedimentary rock.

9. Clastic sediment consists of broken, fragmental debris derived from weathering. Chemical sediment forms where materials carried in solution are precipitated.

10. By various processes of compaction and cementation, sediments become sedimentary rock. Common cementing agents are quartz, calcite, and limonite.

11. Mineral assemblages and textures in metamorphic rocks change continuously because of changing temperatures and pressures. Metamorphic temperatures may eventually become so high that rock begins to melt.

12. Exposures are commonly broken by one or more surfaces of easy breakage. Rock cleavage is breakage caused by the growth of new minerals in metamorphic rock; joints are breakages formed in rock by release of pressure, as the load of overlying rock is removed by erosion.

## SELECTED REFERENCES

Bayly, Brian, 1968, Introduction to petrology: Englewood Cliffs, N.J., Prentice-Hall.

Berry, L. G., and Mason, Brian, 1968, Elements of mineralogy: San Francisco, W. H. Freeman.

Ernst, W. G., 1969, Earth materials: Englewood Cliffs, N.J., Prentice-Hall.

Hurlbut, C. S., 1969, Minerals and man: New York, Random House.

Hurlbut, C. S., 1971, Dana's manual of mineralogy, 18th ed.: New York, John Wiley.

Pauling, Linus, 1964, College Chemistry: San Francisco, W. H. Freeman.

Pearl, R. M., 1965, How to know the minerals and rocks: New York, New American Library.

Pirsson, L. V., and Knopf, Adolph, 1947, Rocks and rock minerals, 3rd ed.: New York, John Wiley.

Tennisson, A. C., 1974, Nature of Earth materials: Englewood Cliffs, N.J., Prentice-Hall.

Vanders, I. and Kerr, P. F., 1967, Mineral recognition: New York, John Wiley.

# 4 EARTH'S CYCLES AND EARTH'S TECTONICS



## THE NETWORK OF CYCLES

Earth's crust and its enclosing atmosphere, hydrosphere, and biosphere are in constant turmoil. They are places of vast and continuing activities, forming a network of interconnected processes of different sizes and shapes, wheels within wheels, cogwheels turning other cogwheels. Some processes are large, others small; some are fast, others very slow. But all, no matter how intricately connected within themselves or with one another, are running; all are in progress, and all are constantly moving materials about. They did not all start at once, like the horses in a race. Some started billions of years ago, others more recently. The whole network evolved gradually, one process adjusting ever more closely to another. Today, the network has developed into an almost unbelievably interrelated whole, in which "everything is connected to everything else." The energy that keeps it all running comes, as we saw in Chapter 2, from three sources: heat energy from within the Earth, the kinetic energy of Earth's rotation, and radiant energy from the Sun.

"Wheels within wheels" is an apt description because the movements we are dealing with are cyclic. *Cycle* is a term that describes a sequence of recurring events. The rotation of Earth on its axis, the circling of Earth around Sun and the circling of Moon around Earth are all cycles. One result of the first cycle is the daily heating and cooling of Earth's surface. An important result of the first and third is the tide, the recurring rise and fall of the ocean surface. Cycles of one sort or another control almost everything that happens on Earth. Even the growth, development, and death of human beings is part of a cycle of events—the life cycle; and our human economy involves many kinds of cycles. For instance we cycle glass (but we call it *recycling*), a process in which bottles are ground up and melted down to form a liquid, from which new bottles are made. We "recycle" paper, which is pulped and fashioned into new paper. Any single particle of glass or paper substance thus may pass through the same series of states or forms again and again. The operation of Earth's natural processes, both internal and external, and the conflicts between the two groups of processes, involve distinct cycles. As we saw in Chapter 2, energy for the internal processes comes from radioactive heat in Earth's interior, whereas energy for the external ones comes largely from the Sun.

## WATER CYCLE

A cycle powered by energy from external sources is the ***water cycle****, the cyclic movement of water substance through evaporation, wind transport, stream flow, percolation, and related processes* (Fig. 4.1). Heat from the Sun evaporates ocean water and other surface water. The water vapor enters the atmosphere and moves with the flowing air (the winds). Some of it condenses and is precipitated, as rain or snow, back into the ocean; some is carried over the land

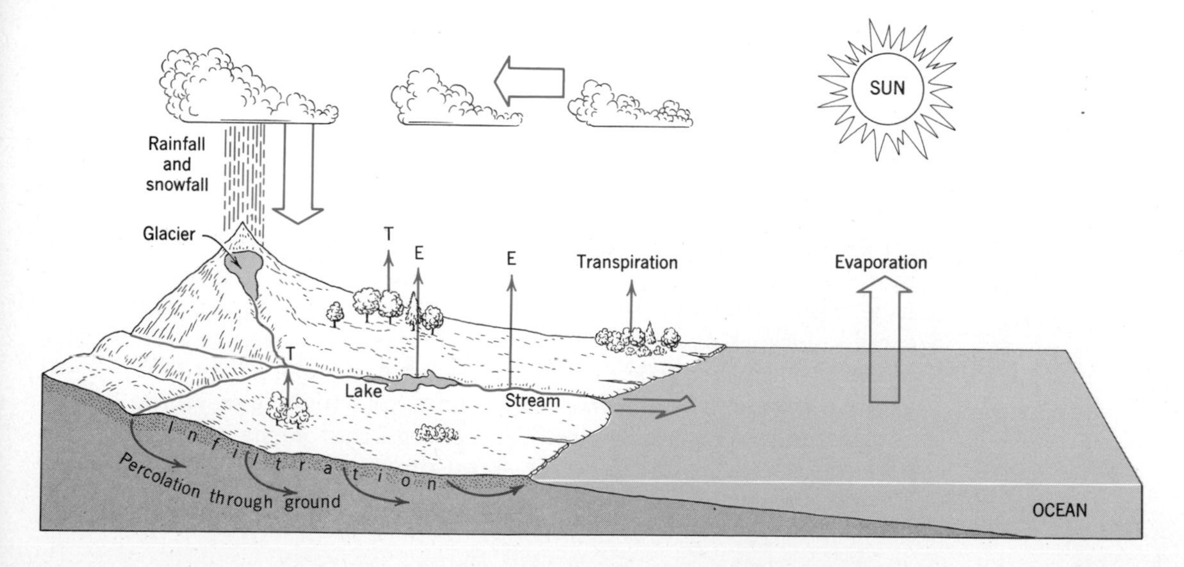


**Figure 4.1**
**The water cycle. E = evaporation; T = transpiration by plants.**

and is precipitated. The rain runs off the land as streams and so back into the ocean, completing the cycle. Snow lies on the ground, perhaps through a winter season, before melting and running off. In especially cold areas snow accumulates and forms glaciers. As glaciers it remains on the land longer, through many years or even thousands of years, but sooner or later it too melts or evaporates and so eventually returns to the ocean. The water cycle provides a good example of energy conversion from one form to another. Earth's lands, on the average, stand some 700m to 800m above the surface of the ocean. So when the Sun's radiant energy evaporates water and lifts air with its contained water vapor from ocean to atmosphere, and then precipitates water onto land at that altitude, the radiant energy has been converted to potential energy. The potential energy enables the water to flow downhill, back to the ocean. We make use of this principle when we build a dam across a stream to form a lake. The dammed-up water conserves and concentrates stream energy for use in generating electricity.

The quantity of water that participates each year in the water cycle

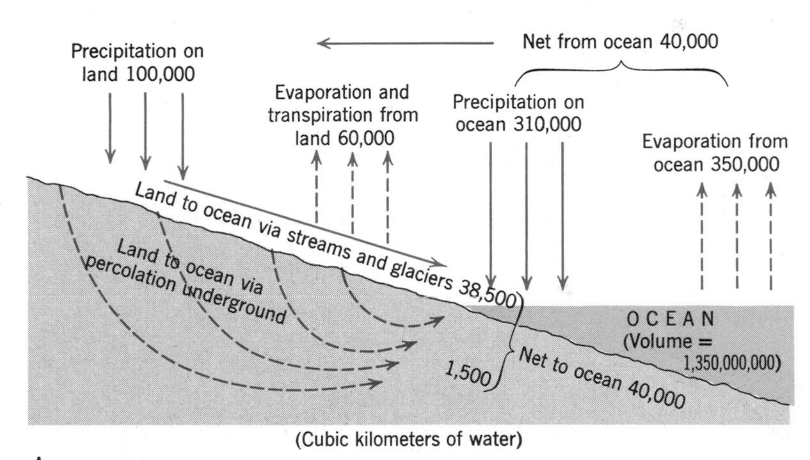


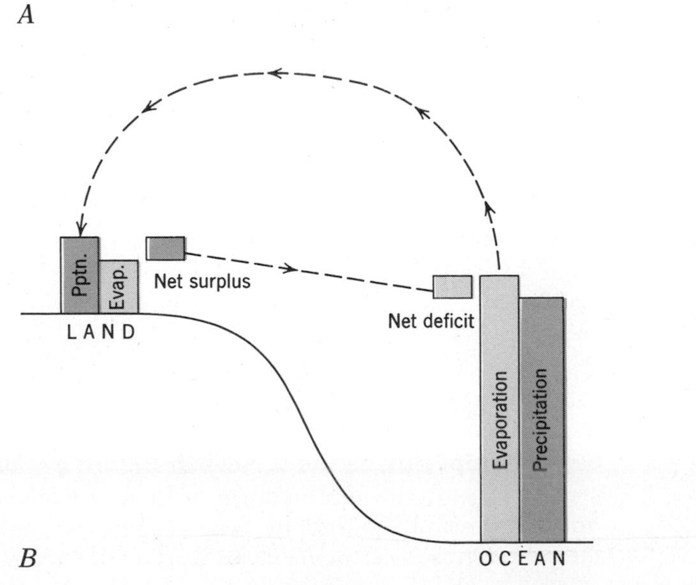


**Figure 4.2**
**Estimated annual world water budget, expressed in cubic kilometers of water.**

**A. The ocean loses by evaporation more than it gains by direct precipitation, but with the land it is the other way around. The surplus precipitation on the land goes back over or beneath the land surface, to the ocean and so balances the budget. The balance is shown by diagram in *B*, where the two little boxes are of equal size.**

**None of the numbers shown is precise, because all are estimates calculated from measured samples. (*Modified from data published by R. L. Nace.*)**

has been estimated for the world by comparing measurements of the amounts evaporated and precipitated at many places on land and at sea, as well as the amounts carried through streams, lakes, and glaciers. The estimated total quantity, called the *annual world water budget,* is 410,000 cubic kilometers, enough water to cover an area the size of the state of Illinois to a depth of more than 3km. What happens during a year's time is shown in Figure 4.2. From the numbers given in that figure we can calculate that an amount of water equal to the combined volumes of all the oceans goes through the water cycle once every 3,200 years.

The amount of water in the hydrosphere changes very little. Some water is lost for a time when chemical exchange locks it up as part of the composition of a mineral. Sooner or later the mineral becomes decomposed chemically. This frees the locked-up water from its chemical bonds and permits the water to re-enter the water cycle.

# ROCK CYCLE

Just as water substance travels through the water cycle, rock material travels, in its own ways, through a ***rock cycle,*** *the cyclic movement of rock material, in the course of which rock is created, destroyed, and altered through the operation of internal and external Earth processes.* Figure 4.3, shows that the rock cycle has two circuits, one that takes place within the continental crust, and one that involves the oceanic crust and the mantle. Within the continental circuit there are many possible paths of movement from one phase to another. In moving along these paths, a particle can travel thousands of kilometers laterally and tens of kilometers downward or upward.

### The crustal-rock circuit

Rock at the surface of a continent is attacked by the complex activities of erosion; it becomes subdivided into loose particles, which are carried away as sediment and deposited. The deposited sediment becomes cemented, usually by substances carried in ground water, and is thereby converted into new sedimentary rock. In places where such rock subsides, it can reach depths at which pressure and heat alter it to metamorphic rock, or depths at which internal heat is great enough to melt it, converting it to magma. The magma can then move upward through the crust, cool, and form a new body of igneous rock—most commonly granite. When uplift occurs, erosion gradually wears down the surface of the land to a depth so great that the top of the body of igneous rock is uncovered and itself begins to be eroded. Again igneous rock is attacked and broken up, its waste starting once more on its way to the sea. The crustal rock cycle has been completed and has begun again. But, as the arrows in Figure 4.3 show, the long crustal circuit can be interrupted by short circuits of various kinds. Some bodies of sedimentary rock are never melted, never metamorphosed, or never even buried deeply before they are

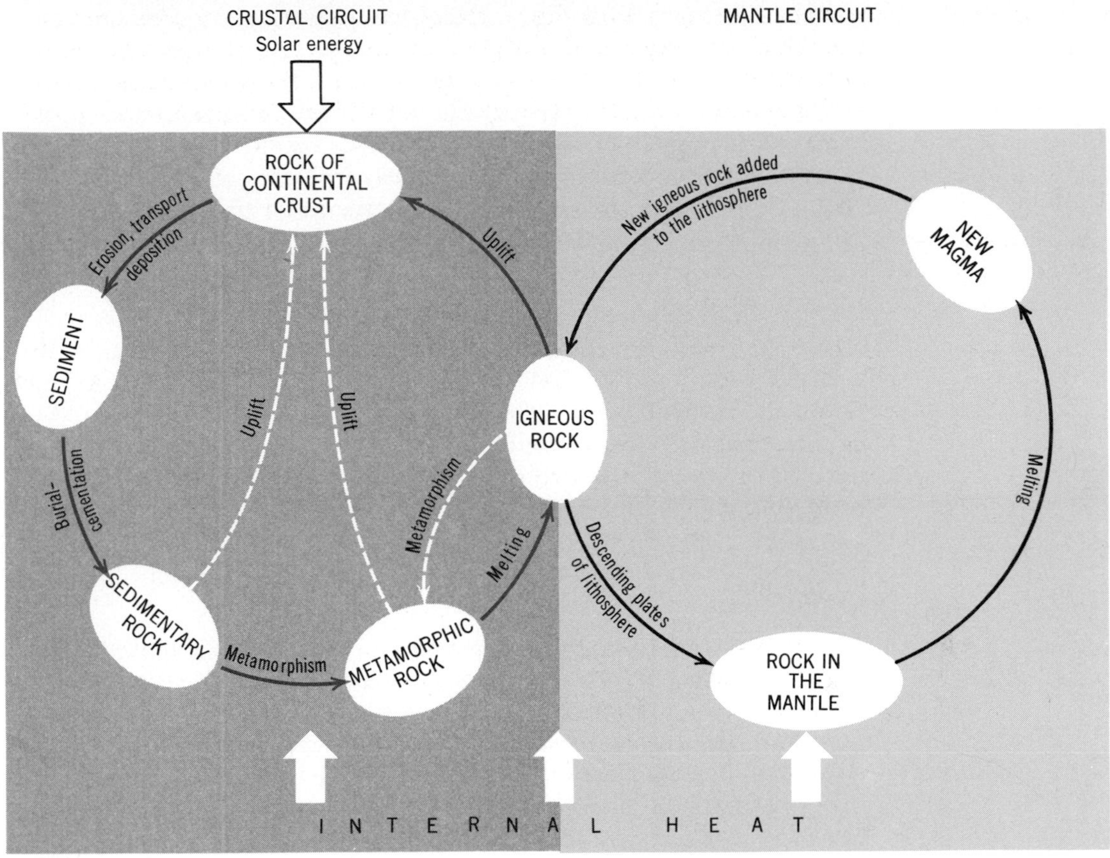

Figure 4.3
**The rock cycle, an interplay of internal and external processes. Rock material in the continental crust can follow any of the arrows from one phase to another. At one time or another it has followed all of them. Within the mantle circuit, magma rises from depth, forms new igneous rock in the lithosphere, and old lithosphere decends again to the mantle where it is eventually remixed.**

**Phases are labeled in capital letters; paths represent processes, labeled in lower-case letters.**

uplifted and eroded. But whether the circuits are long or short, sooner or later every rock body in the continental crust is exposed at the surface, where it is vulnerable to erosion and is ultimately destroyed.

As long as Earth's internal energy is supplied to the crust from below, and as long as solar energy reaches the crust from above, the rock cycle will continue to affect continental rock. Meanwhile, as erosion gradually breaks down rock and transports the waste as sediment to the sea or to basins on the land, it sculptures the land itself and greatly alters the landscape. Thus the form of the land goes through a cycle of its own, the erosion cycle, as shown in Chapter 7.

### The mantle circuit: plate tectonics

The circuit of the rock cycle that includes the mantle involves mysterious and little-understood processes. Nevertheless it has consequences for the Earth that are profound and far-reaching. This is so because the visible part of the circuit involves the lithosphere and its slow movement over the asthenosphere. This remarkable part of the mantle circuit, which has only recently been elucidated by the thought-provoking *theory of plate tectonics* mentioned in Chapters 1 and 2, is the principal shaping force of Earth's surface. The name of the theory recognizes these shaping effects. The word ***tectonics,*** derived from the Greek word *tekton,* meaning carpenter or builder, is the *study of Earth's broad structural features.* The theory proposes that the rigid, rocky lithosphere not only moves and is broken into a series of platelike pieces, but that it is because of movement by the plates that the major features on Earth—continents and ocean basins—are where they are and have the shapes they do. We must therefore outline what the theory of plate tectonics is all about if we are to understand the significance of the mantle circuit in the rock cycle.

We saw in Chapter 2 that the rigid lithosphere is about 100km thick, that it is capped in part with oceanic crust and in part with continental crust (Fig. 2.4), and that it overlies the weak, easily deformed asthenosphere. The plates into which the lithosphere is broken range from several hundred to several thousand kilometers across. Trying to visualize them, we can think of the skin of an orange. Imagine the orange to be the Earth, its skin the lithosphere. If we peeled the orange, then replaced all the pieces of peel, each piece, large or small, would be analogous to a plate of lithosphere.

Earth's lithosphere is broken into six large and several small plates, moving—at speeds of several centimeters a year—in the directions shown by the arrows in Figure 4.4. As the plates move, everything on the plate, including the continents, moves too. The discovery that the entire lithosphere moves is so recent that many questions remain unanswered. One concerns the shapes and sizes of plates in past ages. Plates change both in size and shape; this suggests that new plates can form by the breaking up, or welding together, of other plates. It is possible even that some plates disappear entirely and that at times during Earth's long history they have been stationary.

What makes the plates move? It seems likely that somehow, by means not yet understood but presumably caused by Earth's internal heat energy, slow but vast movements occur in the asthenosphere below the lithosphere. Like giant rafts on a slowly moving river, the plates of lithosphere are dragged along by the movements below. We do not know whether those movements affect the entire mantle or are confined to the asthenosphere. Nor do we understand yet how solid rock in the mantle can be deformed and can flow as readily as it seems to. All we can be sure of at present is what we observe: the lithosphere is broken into plates, the plates are moving, and where the plates abut against each other, a great many of the present day geological events are taking place—new mountains are growing, earthquakes are concentrated, and new mineral deposits are forming.

When we examine *how* plates move, we think of rafts, but a better analogy is conveyor belts. In a conveyor, the belt continually appears from below, moves along the length, then turns down, and passes temporarily from sight as it completes its circuit. Although broad and irregular rather than long and narrow, a plate of lithosphere acts like the top of a slowly moving conveyor belt.

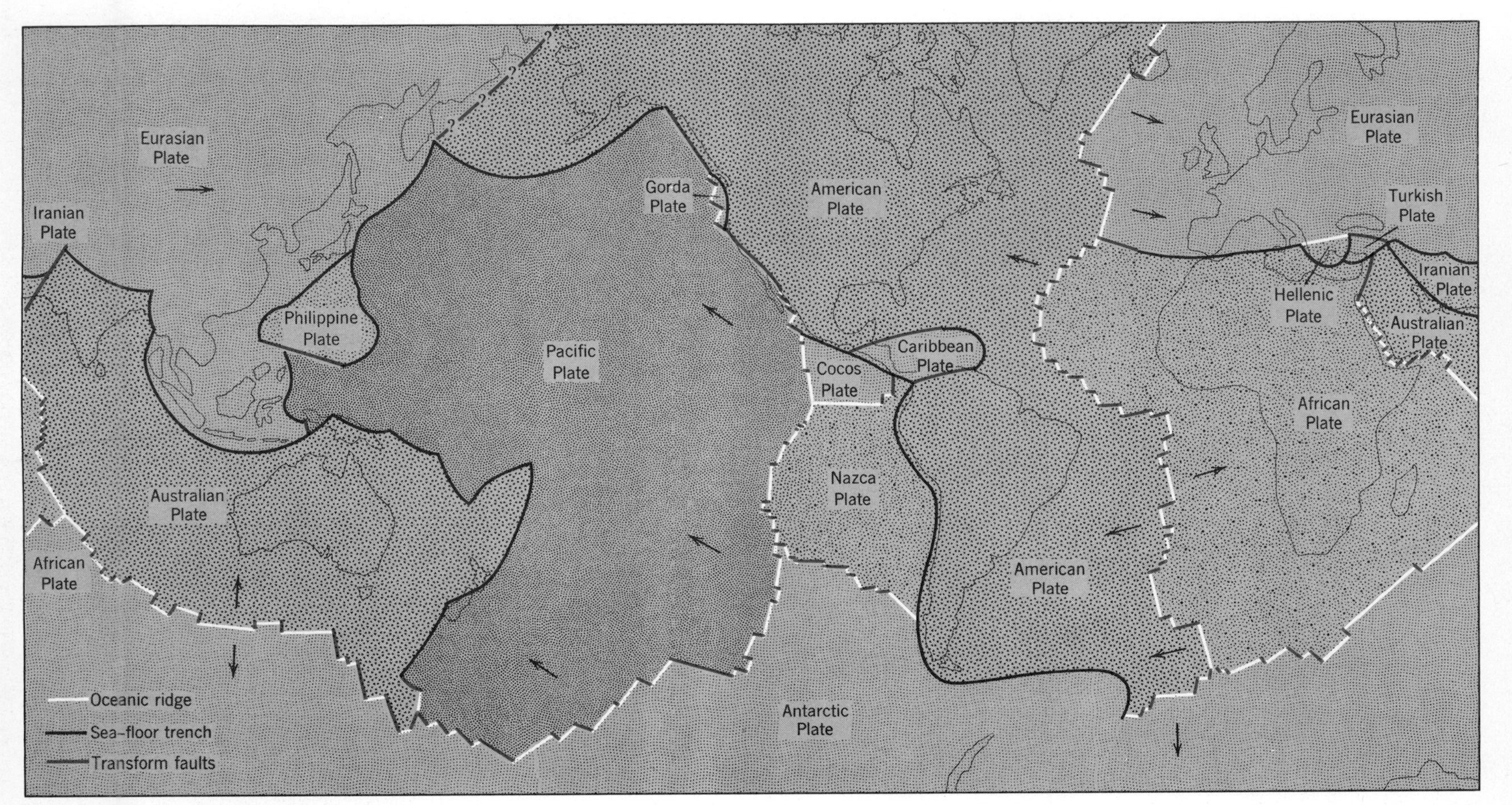


Figure 4.4
**Six large plates of lithosphere and several smaller ones cover Earth's surface and move continuously, in the directon shown by arrows. Plates have three kinds of margins: (1) growing or spreading margins, delineated by oceanic ridges, (2) margins of consumption, delineated by sea-floor trenches, and (3) transform faults. (*After Dewey, 1972.*)**

Along one *edge* (not the *side*) of each plate there is a long, clearly defined fracture in the oceanic crust. The plate moves directly away from the fracture, just as if it were a continuous belt rising up the fracture from the mantle below. This analogy is only partly correct, because the plate is not rising as a solid ribbon. It is being built, or rather added to, *as* it rises. New lithosphere is being created continually as magma wells up along the fracture and cools to form new igneous rock (Fig. 4.5). The sites of the newly growing edges of the plates are great ridges in the ocean floor, ridges that extend right around the globe. Along the center line of each ocean ridge is a deep trough. The trough is the surface expression of the fracture that marks the edge of the plate (Fig. 4.5). Because a plate moves—or spreads—outward, away from the ocean ridge, its newly growing edge is called a *spreading edge.*

The formation of new lithosphere is a continuous process. Movement of lithosphere away from the ocean ridge, like the movement of a conveyor belt, is a continuous process also. Finally, at a distance of a thousand or more kilometers from the spreading edge, the moving lithosphere bends downward, and, again like a conveyor belt, disappears back into the mantle. The edges along which plates of lithosphere turn down into the mantle, called *edges of consumption,* are marked by another, different set of deep trenches in the sea floor.

As the moving strip of lithosphere passes from view and slips slowly down into the depths of the mantle, we can no longer see it. Consequently what happens next is still largely conjecture. On one point we can be quite certain. The lithosphere plate does not turn under, as a conveyor belt does, and reappear at the spreading edge. Instead, it is reheated and slowly remixed in with the material of the mantle. Eventually some of the remixed material may again

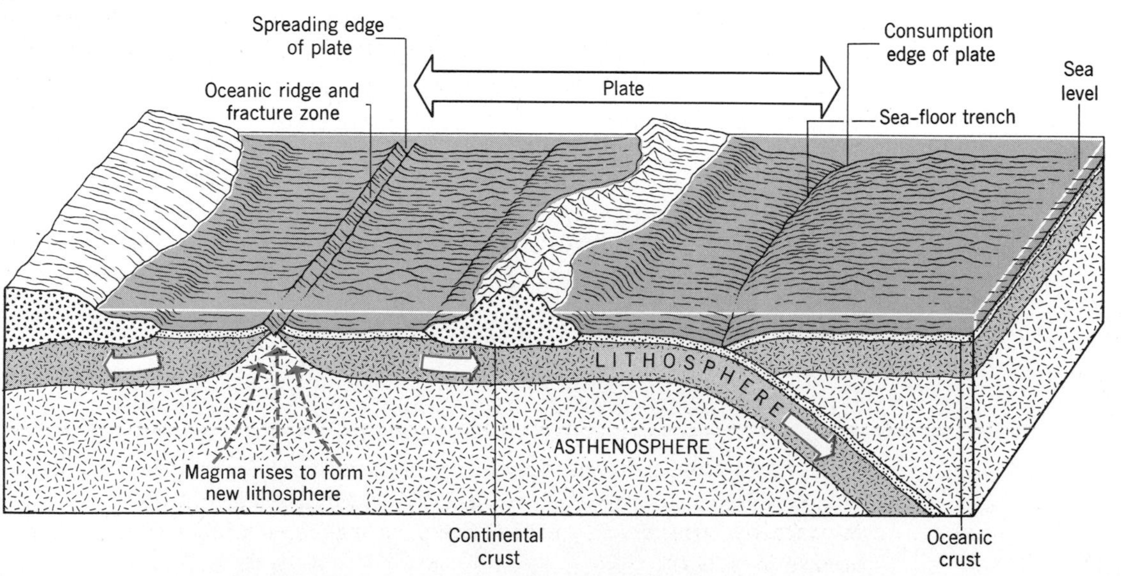


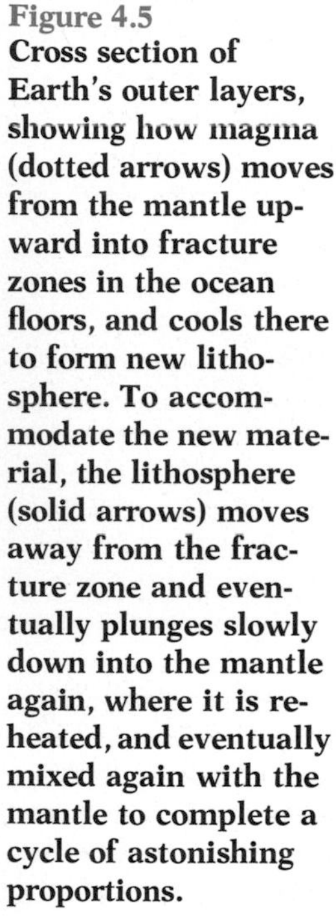
**Figure 4.5**
**Cross section of Earth's outer layers, showing how magma (dotted arrows) moves from the mantle upward into fracture zones in the ocean floors, and cools there to form new lithosphere. To accommodate the new material, the lithosphere (solid arrows) moves away from the fracture zone and eventually plunges slowly down into the mantle again, where it is reheated, and eventually mixed again with the mantle to complete a cycle of astonishing proportions.**

rise, as magma, at a spreading edge, thus completing a cycle through the mantle circuit depicting in Figure 4.5. How long this takes is not known, but the lithosphere trip, from spreading edge to consumption edge, may take as long as 200 million years. Because the volume of the whole mantle is huge, perhaps through all of Earth's long history, only a tiny fraction of the mantle material has been through the lithosphere trip.

The trip we have described is one for plates of lithosphere that consist of oceanic crust. Continental crust is not recycled by the mantle circuit; it takes a shorter trip that ends more suddenly. Because it is lighter and less dense than the rest of the lithosphere, continental crust "floats" in the lithosphere material: It is too buoyant to be dragged downward. So, in continent-size pieces, such crust moves around from place to place on Earth's surface, along with the plate of lithosphere on which it rides, much as a driftwood log, partly embedded in an ice floe, floats on a lake or river.

Such movement of continents causes remarkable things to happen. In parts of the Sahara Desert clear evidence shows that some 500 million years ago that region, hot and dry today, was covered by an enormous glacier, similar to the one that now covers the Antarctic Continent. A glacier in the Sahara was possible because 500 million years ago the Sahara was far from hot and dry. It then lay near the South Pole, and the plate of which it is a part has since been traveling slowly northward.

When a new spreading edge forms and passes beneath a large mass of continental crust, a different thing happens. The spreading edge does not, initially, split the continental crust. But as magma rises and as the newly formed plates begin to move away from the new spreading edge, the continental crust can be broken in two, much as a pane of glass, if pulled from two directions, will break into fragments. Africa, Europe, and the Americas are an example. At one time there was no Atlantic Ocean. Instead, the continents that now border it were joined together as a huge mass of continental crust. About 200 million years ago, for reasons that we do not fully understand, but that presumably involved changes in the mantle below, new spreading edges formed. They split the then-existing continental crust into the pieces we see today. These fragments then drifted slowly into their present positions. At first the Atlantic Ocean was a narrow body of water, but as the drifting continued, the ocean slowly widened to its present form and is still growing wider, by about 5 centimeters each year. There is abundant evidence to mark where the torn continental edges formerly fitted together. In one region, pieces of mountain ranges that once formed a long, narrow, mountain belt, like the Rocky Mountains of today, have been pulled apart so that now they lie on the two sides of the Atlantic. If these pieces are fitted back together, the mountains, now deeply eroded, fit like matched pieces of a jigsaw puzzle (Fig. 4.6).

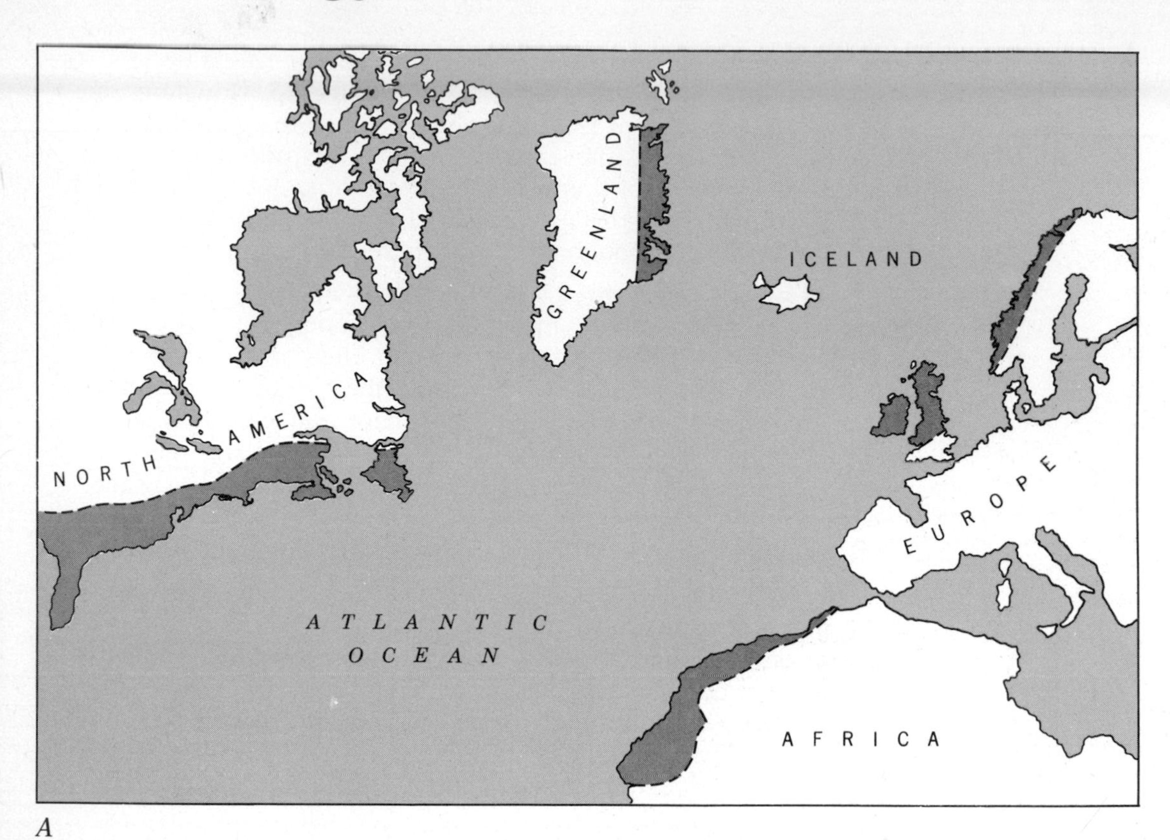


A

Another fact of moving plates of lithosphere is a necessary result of the geometry of a sphere. It is impossible to cover a spherical Earth with plates bounded only by edges of spreading and edges of consumption. There has to be a third kind of edge, analogous to the sides of a conveyor belt, along which plates simply slip past each other. These edges of slipping are great vertical fractures—or, to use a term defined in Chapter 14, *transform faults*—that cut right down through the lithosphere. One transform fault that is much in the public eye because it continually threatens earthquakes is the San Andreas Fault in California.

**Interactions between the two rock circuits**

We have now discussed both circuits of the rock cycle (Fig. 4.3). We need to remind ourselves again how these circuits interact and what a great influence the unseen mantle circuit has on the part of the rock cycle we can actually see in operation around us: the circuit in the continental crust. The influence of the mantle circuit is felt in two principal ways. First, some of the magma that rises from the mantle is intruded into the continental crust. There it forms igneous rock, which in time is lifted up and eroded, and thus becomes mixed with other sediment from the continental crust. Continual addition of mantle material to continental crust suggests that the volume of the crust should be increasing. Evidence gathered by scientists is not clear, but suggests that the volume of continental crust is constant. Probably this is because some of the sediment that mantles the oceanic crust is carried down to the

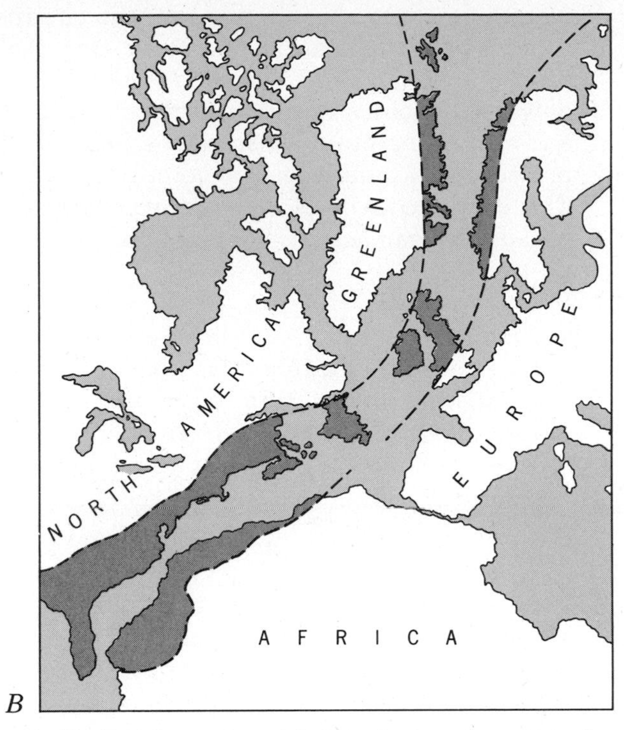


**Figure 4.6 (see opposite page) Opening of the Atlantic Ocean.**

***A.* Rock in eroded fragments of similar mountain belts (black)—each 350 to 470 million years old—is found on both sides of the Atlantic Ocean.**

***B.* When continents are moved and fitted together as they were 200 million years ago, the fragments are seen to form a continuous belt. The reconstruction provides evidence that the present continents were once part of a larger land mass broken up by moving lithosphere. Note that Iceland is not present in the reconstruction. It is a young land mass, and is a piece of the ocean ridge that marks the line along which the continental separation occurred. (*Adapted from P. M. Hurley, 1968.*)**

mantle by the moving lithosphere. The second way in which the influence of the mantle circuit is felt comes through uplift. Before rock can be eroded it must be lifted up as in the making of mountains and before sediment can be deposited, a basin to contain it must be formed. But new uplifts, and new down warpings to form basins, must proceed continually, for otherwise Earth would soon be eroded to a featureless, smooth landscape, and all basins would be filled with sediment. The forces that produce the needed uplift and down warping seem to result largely from forces in the mantle that in turn cause buckling and deformation in the moving plates of lithosphere.

# CHEMICAL CYCLES

When rock weathers to form sediment, some of the more soluble constituents from the rock dissolve and eventually concentrate in the ocean. This is the origin of many of the salts in seawater. When raindrops form they dissolve gases from the atmosphere and carry them down to Earth's surface where they react to form new minerals in the soil. Since material is continually transferred between Earth's spheres, why should the composition of the atmosphere be constant? Why doesn't the sea become saltier, or fresher? Why does rock two billion years old have the same composition as rock only two million years old?

The answers to these questions are the same. The chemical elements, like the bits of rock in the rock cycle and like the water substance in the water cycle, follow cyclic paths. These chemical cycles have many local circuits, and the circuits interact in many ways, but if we carefully measure the circuits we find that in any sphere, such as the atmosphere, the chemical elements added are just balanced by those removed. Thus, as the chemical cycles roll onward, the outer spheres—atmosphere, hydrosphere, biosphere, and crust—maintain constant composition. The cycle of carbon provides an example of how complicated a chemical cycle can be. This example involves the biosphere and is an interesting cycle because human activity may now be pushing it out of balance.

### The carbon cycle

Carbon occurs in four carbon reservoirs: (1) in the *atmosphere* it occurs in carbon dioxide ($CO_2$); (2) in the *biosphere* it occurs in organic compounds; (3) in the *hydrosphere* it occurs in carbon dioxide dissolved in lakes, rivers, and seawater; and (4) in the *crust* it occurs both in the calcium carbonate ($CaCO_3$) that forms limestone and in buried organic matter such as coal and petroleum. All four reservoirs are involved in the carbon cycle (Fig. 4.7).

The key to the carbon cycle is the biosphere, where plants continually extract $CO_2$ from the atmosphere, then break the $CO_2$ down by the process of photosynthesis to form organic compounds. When the plants die, or are consumed by animals, the organic compounds decay again by combining with oxygen from the atmosphere to re-form $CO_2$. This biosphere circuit of the carbon cycle is so rapid that the entire content of $CO_2$ in the atmosphere "turns over" every 4.5 years. But the biosphere and atmosphere circuits interact with circuits in hydrosphere and crust too.

Not all the dead plant matter in the biosphere decays immediately back to $CO_2$. A tiny fraction is transported and deposited as sediment, then is buried in sedimentary rock and is thus protected from the oxygen in the atmosphere. The buried organic matter has joined the slow rock cycle, and will only re-enter the atmosphere when uplift and erosion again have exposed the rock in which it is trapped.

Carbon dioxide derived from the atmosphere also dissolves in the waters of the hydrosphere. There it is used by aquatic plants in the same way that land plants use $CO_2$ from the atmosphere. Additionally, aquatic animals extract calcium and carbon dioxide from the water to make shells of $CaCO_3$. When the animals die, the shells accumulate on the sea floor, mixing with any $CaCO_3$ that may have been precipitated as a chemical sediment. When compacted and cemented, the $CaCO_3$ forms limestone. Some more carbon has joined the rock cycle. Eventually the rock cycle will bring the limestone back to the surface and erosion will break it down, returning the calcium, in solution, to the ocean and its carbon, as carbon dioxide, to the atmosphere.

All the circuits of the carbon cycle are in balance. Hence the amounts of carbon in crust, hydrosphere, atmosphere, and biosphere remain constant, despite the fact that carbon is exchanged among these spheres. The big question in the carbon

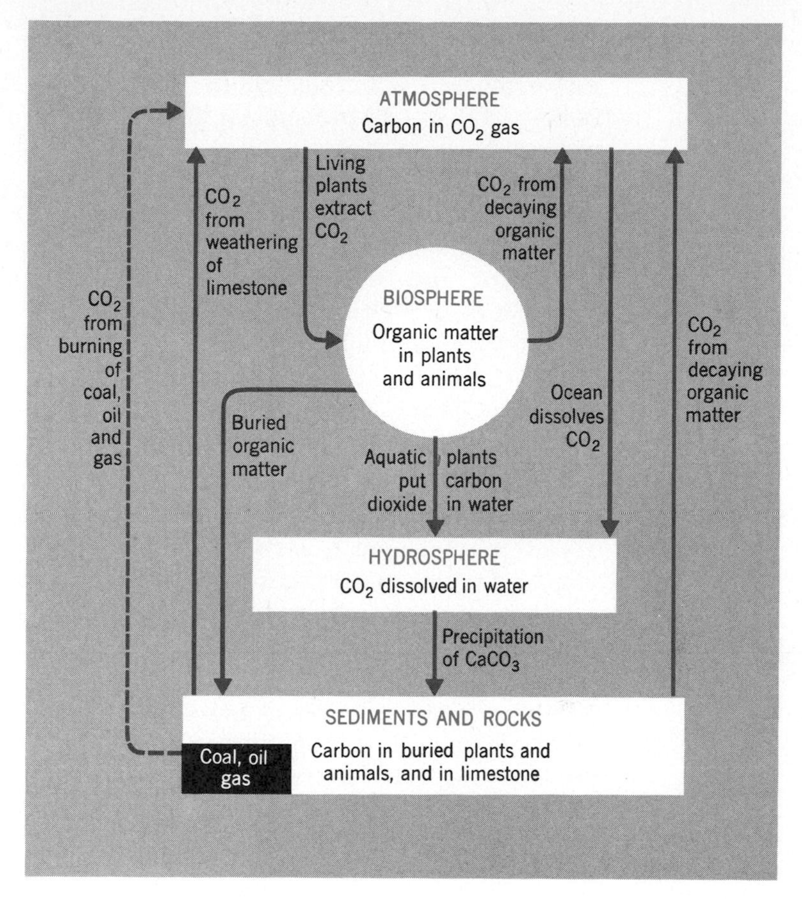


**Figure 4.7**
**The carbon cycle involves interlocking circuits in biosphere, hydrosphere, atmosphere, and crust. White boxes show main reservoirs of carbon; arrows denote the paths along which carbon moves. Circuits and reservoirs are in balance except for man's interference. As we burn coal and petroleum we speed up the circuit from sedimentary rock to atmosphere (dashed arrow). As a consequence, the $CO_2$ content of the atmosphere is rising.**

cycle is human activity. When we dig coal or pump oil from the crust, and burn them, we convert organic matter to $CO_2$ and we are tampering with the rate at which carbon goes through the rock cycle. The $CO_2$ accumulates in the atmosphere, and unless it can be dissolved in the hydrosphere and form more $CaCO_3$, the $CO_2$ content of the atmosphere must inevitably increase. The increase is now being observed (Fig. 4.8) and is leading to research, as well as debate among scientists as to its ultimate effect on the world's climate (Chapter 20).

The carbon cycle illustrates an important point—"everything is connected to everything else." The water cycle and rock cycle are fundamental to the biosphere. Without them, plants and animals could not get the nutrients they need for their life processes, nor the materials to build shells and skeletons.

# PRINCIPLE OF UNIFORMITY

How long have the cycles been rearranging the rock materials of the lithosphere, the water of the hydrosphere, and the chemical elements as they are doing today? Apparently for as long as the time that has elapsed since Earth's oldest known rocks were built, a time, as we shall see in Chapter 5, more than 3 billion years long. This seems a bold statement, yet we can state it with confidence. The many bodies of rock in the continents, regardless of their age, belong to the same recognized kinds as those we see being made today, whether by external or internal processes. All consist of the same range of materials, forming the same ranges of patterns. Each of these rock bodies fits into one phase or another of the rock cycle, in which materials are broken up, sorted, transported along the surface, deposited, cemented to form rock, buried deeply, bent and squeezed, melted and forced upward toward Earth's surface, only to be broken up once more. Because all the cycles interact with each other, we find in the rocks evidence of all the cycles.

The similarity of the old rocks to the recently made ones, and the way all of them fit into the many activities in the cycle, forces us to believe that the rock cycle, operating under the same physical and chemical laws we recognize today, has been going on and on, essentially unchanged, for more than three billion years. This belief is not new. It was first reasoned out in 1785, and early in the 19th Century was developed into a principle named the ***Principle of Uniformity.*** It says that *the external and internal processes we recognize today have been operating unchanged, and at the same set of rates, throughout most of Earth's history.* Recognizing this principle gives us a great ability. We can examine any rock, however old, and compare its characteristics with those of a similar rock that is being formed today in a particular environment. We then know that the old rock was formed in that same sort of environment. The Principle of Uniformity, therefore, provides a first and very long step in understanding Earth's history.

It is only fair to point out, however, that the Principle of

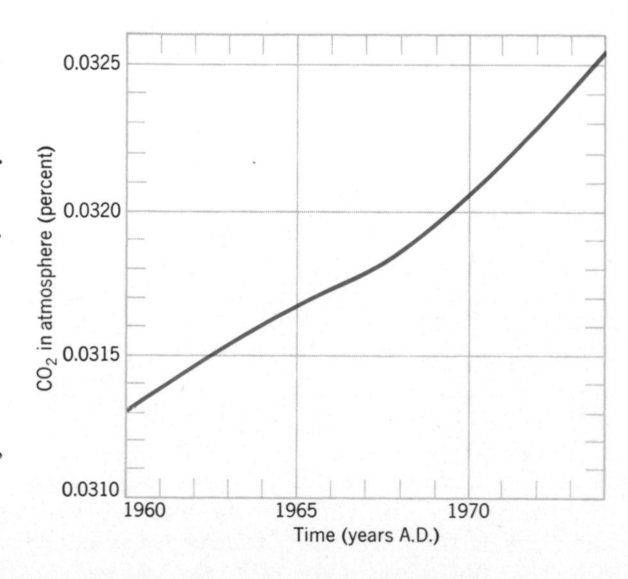


**Figure 4.8**
**Carbon dioxide in the atmosphere is increased through the burning of fossil fuels. Measurements are made daily at an observatory on the top of Marina Loa, a mountain in Hawaii. The reason for the change in slope of the curve about 1967 is not clearly understood. (*After Goldman, 1974, Jour. Geophys. Research, Vol. 79, p. 4550.*)**

Uniformity does contain a probable overstatement. The more we learn of Earth's history, the more we must question whether the rates of all cycles have always been the same as they are now. The evidence seems against constancy, and some rates may once have been more rapid, others much slower.

In order to evaluate the absorbing questions raised by an examination of Earth's history, we must turn next to evidence of past cycles. We must also explore the question of the age of the Earth, and thereby try to answer the question "How long have the cycles been operating and how, if at all, have they varied through time?"

# SUMMARY

1. The water cycle is driven by solar energy. In it, moisture evaporates (chiefly from the ocean), is precipitated, and returns to the sea directly or by longer paths.

2. In the water cycle, a quantity of water equal to the volume of the world ocean is recycled once every 3,200 years.

3. The rock cycle has two distinctly different, but interacting circuits; one of them builds continental crust, the other oceanic crust.

4. The rock cycle in the continental crust begins with magma, which solidifies and forms igneous rock. The rock is eroded, creating sediment, which is deposited in layers that become sedimentary rock. Deep burial leads to metamorphism and eventually, deep in the crust to temperatures and pressures so high that the rock melts and forms new magma.

5. Formation of lithosphere and oceanic crust is part of the rock cycle involving the mantle. Along an ocean ridge magma rises from the asthenosphere to form new lithosphere. The lithosphere, as a plate, moves slowly away from the ridge, and eventually reaches a point where the leading edge of the plate bends down and the plate slides into the mantle. There the plate is heated and remixed to complete the mantle circuit of the rock cycle.

6. The lithosphere is divided into six large plates and several small ones, each of them moving slowly over the weak asthenosphere below.

7. Moving plates of lithosphere can cause continents to move and ocean basins to open and close. The moving plates are the main force that shapes Earth's surface.

8. The moving plates of lithosphere are the surface expression of deep-seated, slow movements of mantle material, still little understood.

9. The chemical elements follow cyclic paths as they move back and forth among atmosphere, hydrosphere, biosphere, and crust.

10. The compositions of the spheres remain constant because the chemical cycles are in balance. Human activity is beginning to perturb some of the chemical cycles; in the carbon cycle human interference is increasing the carbon dioxide content of the atmosphere.

# SELECTED REFERENCES

Frisken, W. R., 1971, Extended industrial revolution and climate change: American Geophys. Union, Trans., v. 52, p. 500–508.

Garrels, R. M., and Hunt, Cynthia A., 1972, Water, the well of life: New York, W. W. Norton.

Garrels, R. M., Mackenzie, F. T., and Hunt, Cynthia, 1975, Chemical cycles and the global environment. Assessing human influences: Los Altos, Calif., Wm. Kaufmann, Inc.

Kuenen, Ph. H., 1963, Realms of water. Some aspects of its cycle in nature. Rev. ed., New York, John Wiley, Science Editions.

LePichon, X., Francheteau, J. and Bonnin, J., 1973, Plate tectonics: Amsterdam, Elsevier.

# 5 STRATA AND GEOLOGIC TIME



## THE GEOLOGIC COLUMN

**The rock cycle creates strata**

The rock cycle we discussed in Chapter 4, the great movements of material through a complex net of different paths, by which rock is made, broken down, carried off, and remade, is an *activity*. It is in progress today and every day. It is easy to see visible bits of this gigantic cycle in a heap of broken-up rock fragments at the base of a cliff, in a river brown with mud, or in a stream of lava flowing slowly away from a volcano. Not so easily appreciated but equally important is the evidence that the many-tracked rock cycle is not only moving and transforming materials today, but has been doing this same work, year in and year out, through a span of time so long that it amounts to billions of years. That evidence, too, is everywhere, but nowhere on Earth is it more clearly displayed than in the Grand Canyon. In Figure 5.1 we see that huge river valley, nearly two kilometers deep, one of the deepest cuts into the Earth's crust made by a river on any continent. This cut, the work of the Colorado River (visibly still on the job) reveals a thick pile of horizontal *strata*. Here we have a word that is used very frequently in geology. A ***stratum***, of which *strata* is the plural form, is *a*

**Figure 5.1**
**Colorado River flows past Desert View Point, Arizona, on its way through the Grand Canyon (*Joseph Muench.*)**

*distinct layer of sedimentary or igneous rock consisting of material that has been spread out upon the Earth's surface.* We will use the term many times in this book.

The strata exposed in the Grand Canyon are sedimentary, having been deposited one on top of the other as sediment transported by ancient rivers and spread out on the floor of a shallow sea. The physical characteristics of those particular strata, as well as the fossil animals they contain, indicate that the sedimentary layers were deposited in a basin occupied by an arm of the sea. As they lay buried for a long time after being deposited, the particles of sediment that composed each stratum became cemented to form sedimentary rock. Later the region was slowly lifted up, the Colorado River system formed by runoff from rainfall, and as the river started to cut into the crust, erosion began. The rock waste from the enlarging canyon, and from elsewhere along the river system, is the sediment that makes the river muddy today. Sediment has long been accumulating, spread out in flat layers, in the Gulf of California off the mouth of the river (Fig. 5.2).

The modern Gulf of California lies in an area far southwest of the position of that ancient arm of the sea in which the strata now

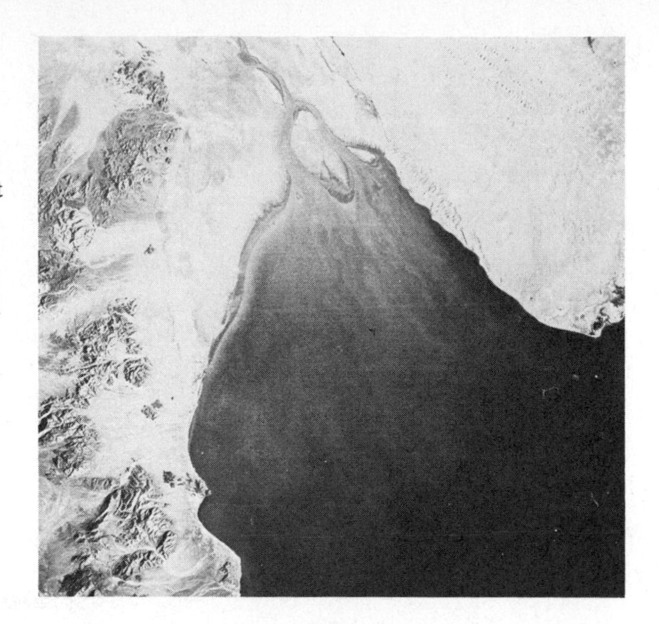

**Figure 5.2**
**Colorado River empties into the Gulf of California 250km southeast of San Diego, California. Out beyond the delta the water is turbid with silt and clay, brought from up the river and not yet deposited.** (*Apollo VI photo, NASA.*)

exposed in the walls of the Grand Canyon accumulated. But the sediment passing through the Canyon on its way to the Gulf consists of the same mineral particles and bits of rock that had been carried by ancient streams into the earlier sea (Fig. 5.3). This whole history, of course, is part of the rock cycle, and part of the

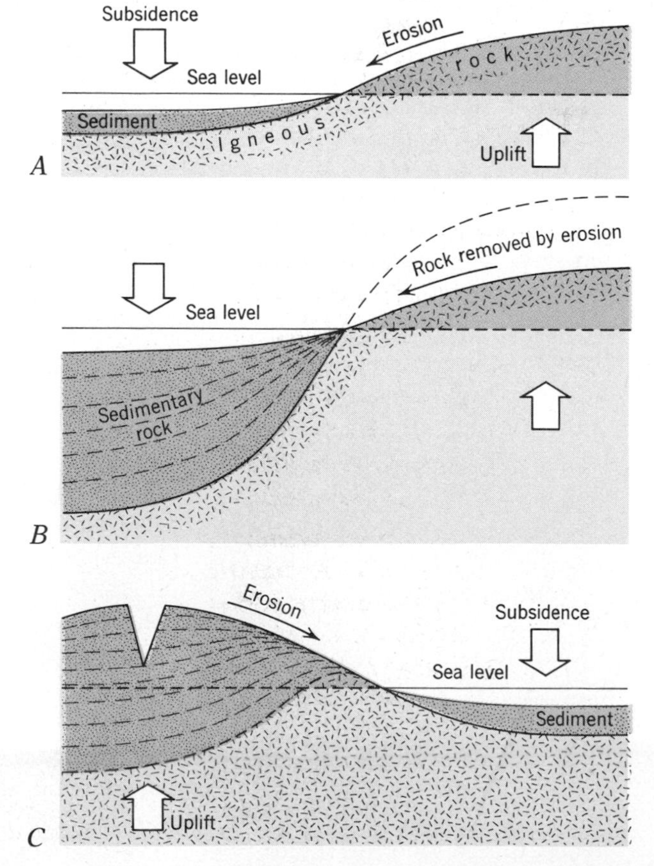


**Figure 5.3**
**The internal processes that cause uplift and subsidence determine the locations of land and shallow sea at any time, and guide the external processes of erosion of rock and deposition of sediment. The bite cut out of the uplifted area in *C* represents a situation like that of the Grand Canyon today.**

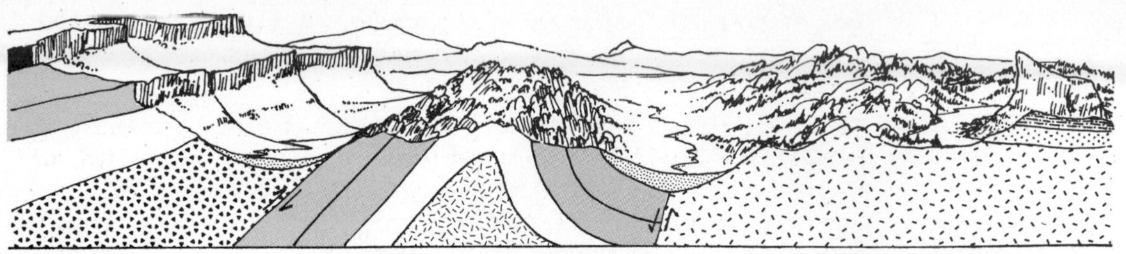

**Figure 5.4**
**A slice through continental crust shows three disconnected bodies of sedimentary rock separated in various ways by metamorphic and igneous rock.**

water cycle as well. Water has made many round trips between ocean and land while the rock particles are still in the midst of their second trip from land to ocean. How many still-earlier trips those particles made we do not know, but there must have been many.

Earth's continental crust is a patchwork of rock bodies large and small (Fig. 5.4). Some of these bodies consist of sedimentary strata, remnants of the fillings of former basins, that escaped erosion and dispersal to later basins. Other bodies, after slowly subsiding to depths far below the surface, have been squeezed and altered to form metamorphic rock. Still other bodies are igneous rock, made by cooling of some earlier rock carried downward and melted at great depth. Having been built piece by piece throughout a very long time, a time marked by many intervals of erosion that carried away a great deal of material, today they are disconnected.

In Chapter 3 it was shown that by examining rock of any kind in detail, we can learn much about its history. As records of Earth's history, sedimentary strata are useful in a special way, because although they occur today as separate, disconnected bodies (Fig. 5.4), they possess characteristics that make it possible to reconnect them on paper, and (still on paper) to arrange the individual layers in the order, or *sequence,* in which they were deposited originally. Why is it so useful to be able to do this with layers of sedimentary rock? To answer this question we must first say a little more about *sequence.*

### Sequence

Toward the end of winter it is often possible to see a layer of old snow that is compact and perhaps also dirty, overlain by fresh, looser, clean snow deposited during the latest snowstorm. Here are two layers, or strata, that were deposited in sequence, one above the other. The dirty layer underneath was deposited first and must therefore be the older of the two. The very simple principle involved here applies also to a whole succession of many snow layers. It applies equally well to layers of sediment and sedimentary rock. Known as ***the principle of stratigraphic superposition,*** it says that *in any sequence of strata, not later disturbed, the order in which they were deposited is from bottom to top.*

This principle implies a scale of relative time, by which the age of one stratum, *A*, can be fixed *in relation to* another layer, *B*, according to whether *A* lies beneath or above *B*. The principle does not enable us to fix the age of any stratum in years, as we reckon time with the use of a calendar; the ages derived from the principle are purely relative.

The principle of stratigraphic superposition was first forcefully presented and widely introduced to science by William Smith, an English civil engineer and land surveyor, shortly before the beginning of the 19th Century. His profession gave him an ideal opportunity to observe not only terrain but the rock that underlies it. While surveying for the construction of new canals in western England, he observed the sedimentary strata and soon realized that they lie, as he put it, "like slices of bread and butter" in a definite, unvarying sequence. He became familiar with the physical characteristics of each layer and, using the principle of stratigraphic superposition, with the sequence of the layers. By looking at a specimen of sedimentary rock collected from anywhere within a wide region, he could name the layer from which it had come and, of course, the position of the layer in the sequence.

### Fossils and their time significance

In the region where Smith worked, the strata contained abundant fossils of marine invertebrate animals. A ***fossil*** *is the naturally preserved remains or traces of an animal or a plant.* Preserved in accumulating sediment and later converted to rock, fossils form an ever-growing record of the kinds of plants and animals that lived at the times when the sediment that encloses them accumulated.

William Smith began to collect the fossils he saw, and soon realized that each layer contained distinctive kinds, enabling him to identify it by the sorts of fossil animals it held, without regard to its physical characteristics. In other words, he recognized that each assemblage of fossils was peculiar to the stratum in which it occurred and this constituted an identification tag for the stratum. In so doing, Smith discovered what we now call the ***law of faunal succession,*** which says that *fossil faunas and floras succeed one another in a definite, recognizable order.*

Today we know that this relationship between a stratum and its fossils is the effect of the evolution of living things through time. As successive generations of living things gradually change their form, the changes are carried from one part of the world to another through the migration or shifting of populations of organisms as they expand or otherwise change their living areas. The rates at which plants and animals spread or shift seem always to have been comparable with the rates at which organisms evolve. As a result the major evolutionary changes spread through large parts of the world within the generous time intervals represented by the major groups of strata. All this, however, was unknown in Smith's day and did not become clear until after 1859, when Charles Darwin put forth his famous theory of evolution.

### Correlation

Smith's discovery that strata containing similar assemblages of fossils are broadly similar in age, no matter where they occur, was not related by him to a scientific principle; it was purely practical. Nevertheless, it opened the door to the correlation of sedimentary strata through increasingly wide areas. By ***correlation*** we mean *determination of equivalence, in geologic age and position, of the sequences of strata found in two or more different areas.* Smith correlated strata, on the dual basis of physical similarity and fossil content, through distances at first measured in miles and later in tens of miles. But by means of fossils alone it became possible to correlate through hundreds and then thousands of miles (Fig. 5.5).

Figure 5.5 represents a comparatively simple situation, with flat-lying strata, good suites of fossils, and no complications. But returning for a moment to Figure 5.4, we can see in it a situation that is far more complex. Suppose that, on paper, we could unbend and smooth out all the layers of sedimentary rock in the figure, so that they would lie one on top of another in a single pile, in order of decreasing age from bottom to top. The smoothed-out pile would be a very useful standard reference for the strata in that region. If the known layers in all the continents were placed in one single pile, we should have a standard reference for Earth's entire continental crust as well as a vital link between present and past.

We do have such a worldwide pile—on paper. It is not complete, but it is being added to and refined continually. We call it the **geologic column** (Table 5.1), *a composite diagram combining in a single column the succession of all known strata, fitted together on the basis of their fossils or of other evidence of relative age.* Because the groups of fossils in each layer of the column record the gradual progress of evolution through time, the column represents not only a succession of layers but also the passage of time. It does not, however, tell us anything about how much time passed between the episodes of deposition of any two given strata. For the actual measurement of time we must look elsewhere.

# TIME

The question *"How much time?"* is as important as the geologic column itself. It must be answered if we are to attempt other questions such as, how old is the Earth?, how long has the rock cycle been at work?, how old is the ocean?, and how long has man lived on Earth?

## Indirect estimates of time

Many attempts, by indirect methods, have been made to subdivide

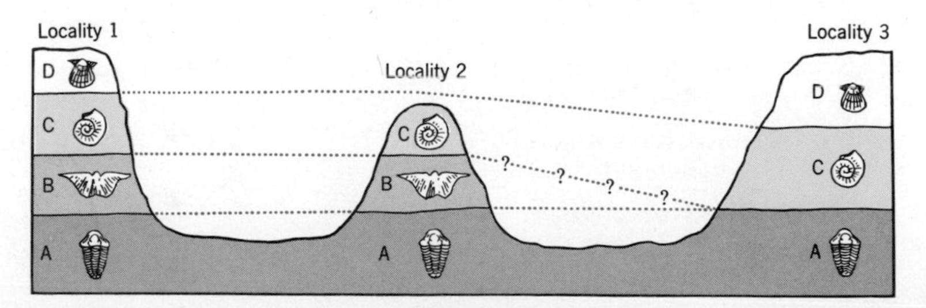


Figure 5.5
**Correlation of strata exposed at three localities, many kilometers apart, on a basis of similarity of the groups of fossils they contain.**

**The fossil groups show that at Locality 3, stratum B is missing because C directly overlies A. Was B never deposited there, or was it deposited but later destroyed by erosion, before the deposition of C?**

**Table 5.1**
**The Geologic Column, Major Worldwide Subdivisions, Selected Dates,[a] and Events in Evolution.**
**As the positions of the numbers show, the time divisions are not drawn to a uniform scale; the dates have merely been spotted onto the geologic column. If the scale were uniform, the table would have to be much longer.**

| Uniform Time Scale | Subdivisions Based on Strata/*Time* | | Systems/*Periods* | Series/*Epochs* | Radiometric Dates (millions of years ago) | Outstanding Events: In Physical History | In Evolution of Living Things |
|---|---|---|---|---|---|---|---|
| 0 | | | | | 0 | | |
| PHANEROZOIC | PHANEROZOIC | CENZOIC | Quaternary | Recent or Holocene; Pleistocene | | Several glacial ages; Making of the Great Lakes; Missouri and Ohio Rivers | *Homo sapiens* |
| | | | | | 2? | | |
| | | | Tertiary | Pliocene | | | Later hominids |
| | | | | | 6 | Beginning of Colorado River | Primitive hominids |
| | | | | Miocene | | Creation of mountain ranges and basins in Nevada | Grasses; grazing mammals |
| 575 | | | | | 22 | | |
| PRECAMBRIAN | | | | Oligocene | | | |
| | | | | | 36 | | |
| | | | | Eocene | | Beginning of volcanic activity at Yellowstone Park | Primitive horses |
| | | | | | 58 | | |
| | | | | Paleocene | | Beginning of making of Rocky Mountains | |
| | | | | | 65 | | Spreading of mammals; Dinosaurs extinct |
| | | MESOZOIC | Cretaceous | Many | | Beginning of lower Mississippi River | Flowering plants |
| | | | | | 145 | | Climax of dinosaurs |
| | | | Jurassic | | | | Birds |
| | | | | | 210 | | |
| | | | Triassic | | | Beginning of Atlantic Ocean | Conifers, cycads, primitive mammals; Dinosaurs |
| | | | | | 250 | Climax of making of Appalachian Mountains | Mammal-like reptiles |
| | | PALEOZOIC | Permian | | | | |
| | | | | | 290 | | |
| | | | Pennsylvanian (Upper Carboniferous) | | | | Coal forests, insects, amphibians, reptiles |
| | | | | | 340 | | |
| | | | Mississippian (Lower Carboniferous) | | | | |
| | | | | | 365 | | Amphibians |
| | | | Devonian | | | Earliest economic coal deposits | |
| | | | | | 415 | | |
| | | | Silurian | | | | Land plants and land animals |
| | | | | | 465 | | |
| | | | Ordovician | | | Beginning of making of Appalachian Mountains | Primitive fishes |
| | | | | | 510 | | |
| | | | Cambrian | | | Earliest oil and gas fields | Marine animals abundant |
| | | | | | 575 | | |
| | PRECAMBRIAN (Mainly igneous and metamorphic rocks; no worldwide subdivisions.) | | | | 1,000 | | Primitive marine animals; Green algae |
| | | | | | 2,000 | | |
| | | | | | 3,000 | | Bacteria, blue-green algae |
| | | | | | | Oldest dated rocks | |
| ~4,650 | Birth of Planet Earth | | | | 4,650 | | |

[a] Best estimates (after R. L. Armstrong, 1974, unpublished).

the geologic column by a scale of years. The earliest consisted of rough estimates of the time during which the rock cycle has been at work. If we assume that sediment has always been deposited at rates equal to today's rate, we can, at least in theory, say how much time has been involved in the deposition of all the sediment we now see preserved in sedimentary strata. But there are several difficulties. First, rates of deposition are probably not constant. Second, in every pile of sediment we find gaps representing intervals where deposition ceased, and we have no way of estimating the lengths of the gaps. Third, in rock older than Cambrian there are very few fossils to help put strata in their proper sequence. Because of these difficulties, early estimates of the duration of the rock cycle were unreliable and tended to be much too short.

A clever suggestion as to how to estimate the age of the ocean concerned the saltiness of seawater. Knowing that salt comes from erosion of common rock and reaches the sea dissolved in river water, why not measure the amount of salt in modern river water and calculate the time needed to have carried all the salt now in the sea? The answer gives an age for the ocean that is only a fraction of the age we calculate today for the ocean. The reason is that sea salt, like all chemical constituents, is cyclic, and the composition of the ocean is constant.

Perhaps the most interesting calculations were those made by physicists to estimate the time that Earth has been a solid body. Earth started as a very hot object, but once it had become solid, they argued, it could cool down only by the conduction of heat through the solid rock. By measuring the thermal properties of rock and the present temperature of Earth's interior, they calculated a time for Earth to cool to its present state. Their estimates of a few hundred million years were much too short because they did not take into account radioactivity, which continually supplies heat to Earth's interior. Instead of cooling, Earth now seems to enjoy a nearly constant temperature.

These indirect methods of measuring geologic events yielded ages that are too young. Such young ages could not be reconciled with the thick pile of sedimentary strata of the geologic column, nor did it seem possible for the vast evolutionary changes seen in the fossil record to have occurred within the few hundred million years estimated for Earth's age. To get around this dilemma what was needed was a way to measure geologic time by some process that runs continuously, that is not influenced by other processes and other cycles, and that leaves a continuous record without gaps in it. At the end of the 19th Century—in 1896—the discovery of radioactivity provided the needed method. That discovery opened the door to radiometric dating, a new and reliable means of measuring geologic time.

# RADIOMETRIC DATING

### Natural radioactivity

As we saw in Chapter 2, most chemical elements are stable and do not change, but a few are radioactive and are therefore unstable. These unstable elements are continually decaying. It is largely the heat created by this decay that makes Earth's interior very hot.

A radioactive element decays by throwing off—literally shooting out—particles from the nucleus of each of its atoms (App. A). By this process each atom becomes instantly converted into a different, "daughter" atom that has a mass number, an atomic number, or both, smaller than that of its parent. The daughter atom can be either an isotope of the same element as its parent or a different element altogether. An example is portrayed in Figure A.3, where we can see one atom of uranium-238 decaying and "bumping down" through a whole series of unstable daughters, finally ending up as one atom of an end product, lead-206, that is not radioactive.

This natural radioactivity of any unstable element follows a definite timetable. Think of a radioactive isotope that decays directly, in one step, to form a stable daughter end product. The number of decaying parent atoms continually decreases while the number of daughter atoms increases. The proportion, or percentage, of atoms that decay during one unit of time is constant. But although the proportion is constant, the actual number keeps decreasing because the parent atoms are being used up.

So, in a mineral sample, the overall radioactivity (the sum of the

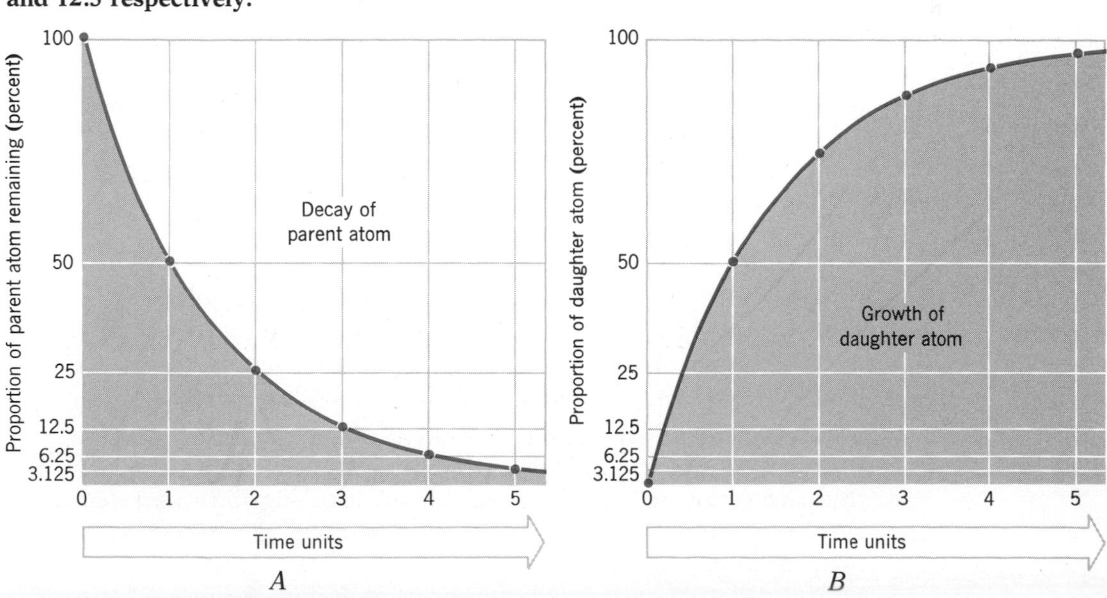


Figure 5.6
**Curves showing decay of radioactive atoms and growth of daughter products.**

***A*. At time zero, a sample consists of 100 per cent radioactive parent atoms. During each time unit, half the atoms remaining decay to daughter atoms.**

***B*. At time zero, no daughter atoms are present. After one time unit corresponding to a half-life of the parent atom, 50 per cent of the sample has been converted to daughter atoms. After two time units, 75 per cent of the sample is daughter atoms, 25 per cent parent atoms. After three time units, the percentages are 87.5 and 12.5 respectively.**

radioactivity of all the parent atoms remaining in the sample) decreases continually (Fig. 5.6). Because the rate of decrease is a percentage of the number of atoms left undecayed, we measure the rate in terms of the ***half-life*** of the parent, *the time required to reduce the number of parent atoms by one-half.* The time units marked in Figure 5.6 are half-lives. Of course they are of equal length, just as years are. But at the end of each one, the number of atoms that decay (and therefore the combined radioactivity of the sample) has decreased by exactly half.

While the proportion of parent atoms declines, the proportion of daughter atoms increases. Figure 5.6 shows that the growth of daughter atoms just matches the decline of parent atoms. That fact is the key to the use of radioactivity as a means of measuring time and determining ages.

### Radiometric mineral ages

Several elements possess naturally radioactive isotopes. We will select one of these isotopes, potassium-40 ($^{40}K$), to illustrate how minerals are dated. We can see from Table A.3 that potassium has three natural isotopes: $^{39}K$, $^{40}K$, and $^{41}K$. Only one, $^{40}K$, is radioactive and its half-life is 1.3 billion years. When $^{40}K$ disintegrates, it can do so in either of two ways. One way creates argon-40 ($^{40}Ar$) as a daughter atom; the other creates calcium-40 ($^{40}Ca$). Careful measurements have shown that 12 percent of the daughter atoms are always $^{40}Ar$ and 88 percent are always $^{40}Ca$.

When a potassium-bearing mineral crystallizes from a magma, or grows within a metamorphic rock, it traps a sample of $^{40}K$ in its crystal structure. The only way $^{40}Ar$ can be included in the structure is by solid-solution replacement of other atoms. Argon has unusual atomic properties, however, and does not readily substitute for other atoms. The newly growing mineral will therefore contain $^{40}K$, but be free of $^{40}Ar$. Once the mineral is formed, the $^{40}K$ becomes locked into the crystal structure and cannot escape. It steadily decays and produces $^{40}Ar$ as one of its daughter atoms. The $^{40}Ar$ that is produced is also locked into the mineral, albeit unwillingly, because the crystal structure makes a cage from which it cannot escape. All the $^{40}Ar$ atoms in a potassium-bearing mineral, therefore, must come from decay of $^{40}K$. All that now has to be done is to measure the amount of parent $^{40}K$ that remains, the amount of $^{40}Ar$ that has formed, and, the half-life of $^{40}K$ being known, calculate the *radiometric date*—the length of time the mineral has contained its built-in clock.

Dating by $^{40}K$ is not limited to minerals (such as muscovite) that contain potassium as a major element. Even minerals that contain small amounts of potassium that is substituting for other elements by solid solution will serve the purpose. Thus hornblende, a calcium-iron-magnesium silicate, can be used. Some rocks contain several different minerals that can be used for dating, and then it is possible to use the "whole rock" for dating.

Through the use of very sensitive instruments, radiometric ages can be measured by the use of any of several different radioactive isotopes. A particular isotope may be selected because it is present in one of the minerals in the rock we wish to date. Or it may be selected because its half-life is long (if we want to date a very ancient rock) or short (if we want to date something young). Table 5.2 lists some of the isotopes commonly used.

### Dating by carbon-14

Among the radiometric dating methods listed in Table 5.2, the one based on carbon-14 ($^{14}C$, also known as radiocarbon) is unique for two reasons. The half-life of $^{14}C$ is short, and the amount of daughter product cannot be measured.

Radiocarbon is continuously created in the atmosphere through bombardment of nitrogen-14 ($^{14}N$) by neutrons created by cosmic radiation. $^{14}C$, with a half-life of 5730 years, decays back to $^{14}N$. The radioactive carbon mixes with ordinary carbon $^{12}C$ and diffuses rapidly through the atmosphere, hydrosphere, and biosphere. Because the rates of mixing and exchange are rapid compared with the half-life, the proportion of $^{14}C$ is nearly constant throughout the atmosphere. As long as the production rate remains constant, the radioactivity of natural carbon remains constant because rate of production balances rate of decay.

**Table 5.2**
**Some of the Principal Isotopes Used in Radiometric Dating**

| Isotopes | | Half-Life of Parent (Years) | Effective Dating Range (Years) | Minerals and Other Materials That Can Be Dated |
|---|---|---|---|---|
| Parent | Daughter | | | |
| Uranium-238 | Lead-206 | 4.5 billion | 10 million to 4.6 billion | Zircon<br>Uraninite and pitchblende |
| Uranium-235 | Lead-207 | 710 million | 10 million to 4.6 billion | Zircon<br>Uraninite and pitchblende |
| Potassium-40 | Argon-40<br>Calcium-40 | 1.3 billion | 100,000 to 4.6 billion | Muscovite<br>Biotite<br>Hornblende<br>Whole volcanic rock |
| Rubidium-87 | Strontium-87 | 47 billion | 10 million to 4.6 billion | Muscovite<br>Biotite<br>Orthoclase<br>Whole metamorphic rock |
| Carbon-14 | Nitrogen-14 | 5,730 ± 30 | 100 to 50,000 | Wood, charcoal, peat, grain, and other plant material<br>Bone, tissue, and other animal material<br>Cloth<br>Shell<br>Stalactites<br>Ground water<br>Ocean water |

While an organism is alive and is taking in carbon from the atmosphere, it contains this balanced proportion of $^{14}C$. However, at death the balance is upset, because replenishment by life processes such as feeding, breathing, and photosynthesis ceases. The $^{14}C$ in the dead tissues continually decreases by radioactive decay, at a rate that decreases with time. The analysis for the radiocarbon date of a sample involves only a determination of the radioactivity level of the $^{14}C$ it contains. The daughter product, $^{14}N$, cannot be measured, and it is necessary to assume that the rate of production of $^{14}C$ has been constant throughout the last 50,000 years or so, the range of time to which the short half-life of $^{14}C$ limits the usefulness of the method.

Because the method is based partly on the assumption that $^{14}C$ in the atmosphere has been constant, the accuracy of carbon-14 dates has been checked against samples whose dates are known independently through historical information. Among these samples are grains of corn, wooden beams, prehistoric clothing, and furniture from ancient Egyptian tombs. In these samples, $^{14}C$ dates compare fairly well with historical dates. Although none of the historically dated samples is as much as 5,000 years old, the annual growth rings of long-lived trees provide a further check on $^{14}C$ dates of wood from the same trees that extends back more than 8,000 years. Many dates as old as 50,000 years have been calculated without the benefit of these independent checks. Although less accurate than the dates of younger samples, they are still very useful.

Because of its application to organisms (by dating fossil wood, charcoal, peat, bone, and shell material) and its short half-life, radiocarbon has proved to be enormously valuable in establishing dates for prehistoric races of man and for recently extinct animals, and in this way it is of extreme importance in archeology. It is of comparable value in dating the most recent part of geologic history, particularly the latest of the glacial ages. For example, the dates of many samples of wood taken from trees overrun by the advance of the latest of the great ice sheets and buried in the rock debris thus deposited, show that the ice reached its greatest extent in the Ohio–Indiana–Illinois region not more than about 18,000 years ago.

Similarly, radiocarbon dates afford the means for determining rates of movement, such as the rate of advance of the last ice sheet across Ohio, the rates of rise of the sea against the land while glaciers melted throughout the world, and the rates of local uplift of the crust that raised beaches above the sea.

#### Dating by fission tracks

When a radioactive isotope trapped in a crystal structure decays, the atomic particles given off will damage a small part of the structure as they travel through the crystal. If we had a way to count the number of damaged areas in a crystal, we would then have a measure of the number of radioactive disintegrations that had occurred, a measure independent of a count of the number of daughter atoms created.

Sensitive ways have been discovered to etch out the damage areas—called *fission tracks*—so they can be seen under high-powered microscopes. Then, if the concentration of the radioactive isotope (such as $^{238}U$) that caused the damage is known, it is a simple matter to calculate the age of the mineral from the abundance of tracks.

# OTHER DATING SCHEMES

Radiometric dates are reliable and widely used but there are other dating methods that can also be used under special circumstances, and which do not rely on radioactive decay.

We have already mentioned the annual growth rings in trees as a means of checking $^{14}C$ dates. Annual layering can also be found in the strata of certain lakes—particularly those in Arctic regions. The strata are successive layers of fine- and coarse-grained sediment, and each pair of layers is the deposit made during one year, the fine-grained being deposited in winter, the coarse in summer (Fig. 12.6). In carefully collected samples, scientists have been able to count back as far as ten thousand years and say when a given layer was deposited.

Annual layering, corresponding to a season's snow accumulation, can also be found in glaciers in Arctic and Antarctic regions. By drilling into the ice it is possible to collect and date samples of snow, now compacted into solid ice, that formed a thousand years ago.

An interesting method of dating the obsidian tools of Stone Age man deserves mention. When glass, such as that of a freshly chipped surface of obsidian, is buried in moist soil, the moisture starts to diffuse into the glass. The distance through which it diffuses is proportional to time, and the diffusion changes the properties of the glass. By measuring the depth of the changed, water-rich layer on the obsidian tool, one can estimate the length of time the sample has been buried.

A number of other possible dating methods are now being tested by scientists. Perhaps, within the lifetimes of those who read this book, it will be possible to date accurately most of the materials we find on Earth.

# GEOLOGIC TIME SCALE

Through the various methods of radiometric dating, the dates of solidification of many bodies of igneous rock have been determined. Many such bodies have identifiable positions in the geologic column, and because of this it becomes possible to date, approximately, a number of the sedimentary layers in the column.

We remember that the standard units of the geologic column consist of sedimentary strata containing characteristic fossils, but the typical rocks from which dates (other than radiocarbon dates) are determined are igneous rocks. It is necessary, therefore, to be sure of the time relations between an igneous body that is datable and a sedimentary layer whose fossils closely indicate its position in the column.

Figure 5.7 shows in an idealized manner how apparent ages of sedimentary strata are approximated from the apparent ages of igneous bodies. The age of a stratum is bracketed between bodies of igneous rock, the apparent ages of which are known.

In the figure, four series of sedimentary strata, whose geologic ages are known from their fossils, are separated by surfaces of erosion. Related to the strata are two intrusive bodies of igneous rock (A, B) and two sheets of extrusive igneous rock (C, D). From the apparent dates of the igneous bodies and the geologic relations shown, we can draw these inferences as to the ages of the sedimentary strata:

| Stratum | Age (millions of years) | Interpretation |
|---|---|---|
| 4 | $<34$ $<30$ $>20$ | age lies between 20 and 30 million years |
| 3 | $<60$ $>34$ $>30$ | age lies between 34 and 60 million years |
| 2 | $>60$ $>34$ | age of both is more |
| 1 | $>60$ $>34$ | than 60 million years |

To separate 1 from 2, dates from other localities are needed. Dates from igneous rocks elsewhere could also narrow the possible ages of 3 and 4.

Through this combination of geologic relations and radiometric dating, we are able to fit a scale of time to the geologic column. The scale is being continually refined.

It is a great tribute to the work of geologists during the first half of the 19th Century that the geologic column they established has

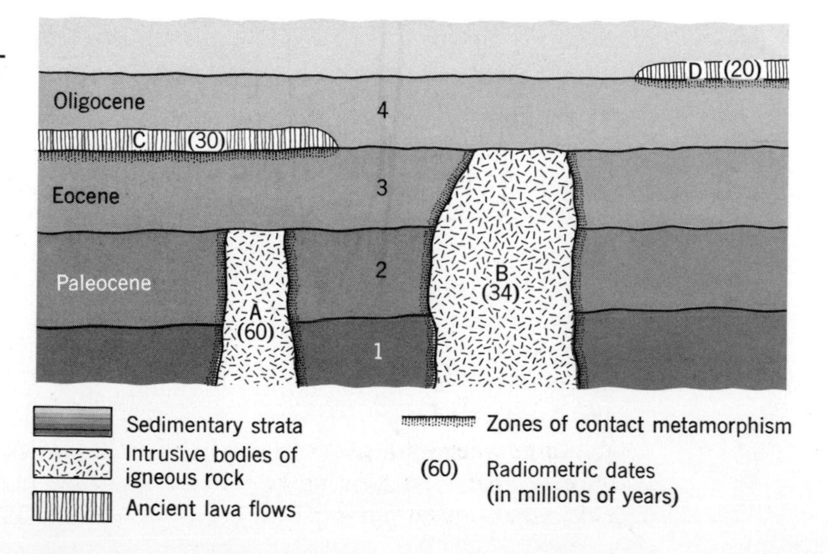


**Figure 5.7**
**Idealized section illustrating application of radiometric dating to the geologic column. For method see text.**

been fully confirmed by radiometric dating. Comparisons between the numbers column and the names column in Table 5.1 show this. They show also that the grouping of strata into the successively smaller subdivisions called *systems, series,* and *stages* is matched by the corresponding time units called *periods, epochs,* and *ages.* We can speak of the time units, of course, whether or not we know their dates. We could speak of events that occurred in the Devonian Period (or simply in Devonian time) even if we did not know that the dates of that period fall between 415 and 360 million years ago.

Of course the geologic column is not yet provided with a time scale that is complete. There are still plenty of gaps. One obvious gap near the top of the column occurs between the "old" limit of carbon-14 dating and the "young" limit of potassium/argon dating. To a great extent this gap is the result of scarcity of datable samples, and in time both it and the other gaps will surely be filled in.

# AGE OF PLANET EARTH

Table 5.1 shows us that the oldest rocks are the great assemblage of metamorphic and igneous kinds known as Precambrian rocks. Of the many radiometric dates obtained from them, the youngest are around 600 million years, the oldest around 3.8 billion years. The Precambrian unit of the geologic column, then, existed during a *minimum* time equal to 3.8 billion minus 600 million years, or 3.2 billion years—a span more than six times as long as the time elapsed since the Precambrian unit ended.

If some Precambrian rocks are 3.8 billion years old, the beginning of Planet Earth's history must be still farther back in time. This is confirmed by the oldest rock for which a radiometric date has been obtained. This is a boulder of granite from a layer of conglomerate in Greenland. The existence of granite proves both that continental crust was present and that the rock cycle was operating 3.8 billion years ago. Further confirmation comes from another of the very ancient Precambrian rocks, a body of granite in South Africa. Although itself an igneous rock, this ancient granite contains great chunks of quartzite, much as a pudding contains raisins. At an earlier time, before it became enveloped by the granite magma, the quartzite must have been part of a layer of sandstone. Even before that, the sandstone must have been loose sand. Clearly, then, the rock cycle must have been operating in its present manner well before the granite magma solidified. As far back as we can see through the geologic column therefore, we find evidence of the rock cycle and, because we see ancient sediment that must have been transported by water, we know that when that sediment was deposited, there must have been a hydrosphere.

We have been speaking of the oldest rock we can find. How much older might Planet Earth be? As we shall see in Chapter 19, good evidence suggests that Earth formed at the same time as the Moon, the other planets, and meteorites (small independent bodies that have "fallen" onto Planet Earth). Through various methods of radiometric dating, it has been possible to determine the ages of meteorites, and of "Moon dust" brought back by astronauts, as 4.6 billion years. By inference, the time of formation of Planet Earth, and indeed of all the other planets and meteorites in the Solar System, is 4.6 billion years ago.

### Changing length of the day

In its impact on our thinking about the age and history of the Earth, radiometric dating has been perhaps an even more important development for the 20th Century than the construction of the geologic column was for the 19th. The two great developments complement each other in a remarkable way. Radiometric dating and the geologic column can now be used to test that part of the Principle of Uniformity that we questioned in Chapter 4: the assumption that rates of the different cycles have been constant. Already there is compelling evidence to sustain the suspicion that rates have varied but perhaps no evidence is more compelling than the changing length of the day.

We read in Chapter 2 that tides in the ocean have a twice-daily period and a 14-day cycle through which the actual height of high and low tides increases and then decreases. The 14-day period results from Moon and Sun interacting to enhance or to counteract their gravitational pulls on the Earth (Fig. 2.13). The tides influence, in one way or another, most of the creatures living in the sea. Clams and other shellfish grow a microscopically thin layer of new shell material each day. The thickness of the layer depends on the depth of water covering the shellfish—thick layers at high tide, thin at low. We have already seen that highest tides occur approximately every 14 days—that is, twice each lunar month. When we examine with a microscope a section of a modern clam shell and measure the thickness of the daily growth layers, we find that there are variations, that the thickest layers form at times of highest tides, and that a repeating pattern of 14 thick and 14 thin layers emerges (Fig. 5.8). The shell of a modern clam is therefore an accurate recorder of tides. If tides have operated in the past, the same pattern should be seen in the shells of fossil clams. This turns out to be exactly what is observed, but as we look at older and older fossils, the number of thick and thin layers in a repeating pattern becomes larger. This means that Earth rotated more rapidly in the past because larger repeat numbers indicate there were more days in the lunar month than there are now (Fig. 5.9). The reason why Earth is slowing down is that its store of kinetic energy of rotation is being slowly used up. The tides, by friction against the sea floor, are acting as weak but steady brakes. The rate of slowing is small, but nevertheless it is sufficient to enable astronomers to confirm the evidence of the clam shells. Astronomers, measuring the exact length of the day through the past three centuries, find that it is increasing by 0.002 seconds each century. 600 million years ago, in Cambrian time, the rate of

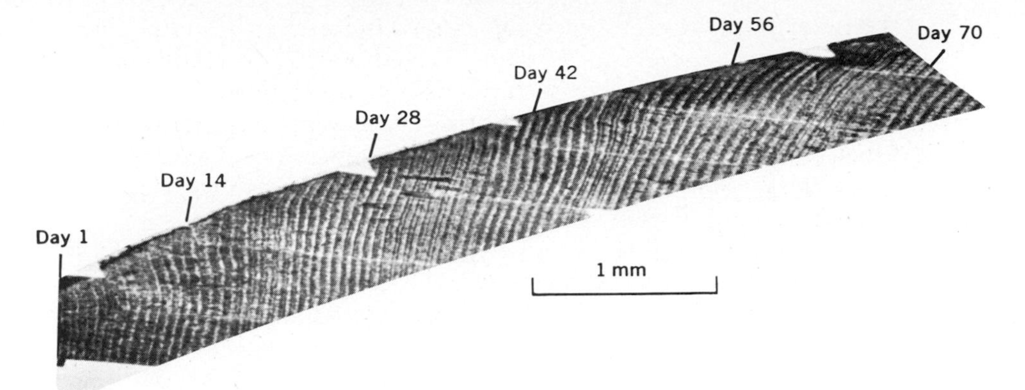


**Figure 5.8**
**Microscope photograph of the daily growth bands in a modern clam shell. The animal lays down a thick, new layer of shell each day, the thickness varying with water depth during high tide. A 14-day repeat pattern in the growth layers corresponds to the twice-monthly highest tides, when Moon and Sun exert their tidal attractions in the same direction.** (*After Pannella and MacClintock, 1968.*)

rotation was so rapid that a day—the time from dawn to dawn, corresponding to a single rotation—lasted only 21 hours. The length of the year was the same because it is fixed by Earth's movement around the Sun; so the Cambrian year contained 424 days each 21 hours long. As the length of the day changed, so must the rate of the water cycle have changed, because it is influenced by the length of time during which the Sun heats the ocean each day.

Before we discuss further how things may have differed in the past, however, we should examine how they operate today. We turn, therefore, to the great variety of processes that operate at Earth's surface—breaking down rock, transporting it, and changing the surface so as to create the familiar landscapes we see around us.

# SUMMARY

## GEOLOGIC COLUMN

1. Sedimentary strata are deposited in sequence, oldest at bottom and youngest at top.
2. Major groups of strata contain distinctive assemblages of fossils, by which they can be identified. The strata and their fossils are the basis for the geologic column.

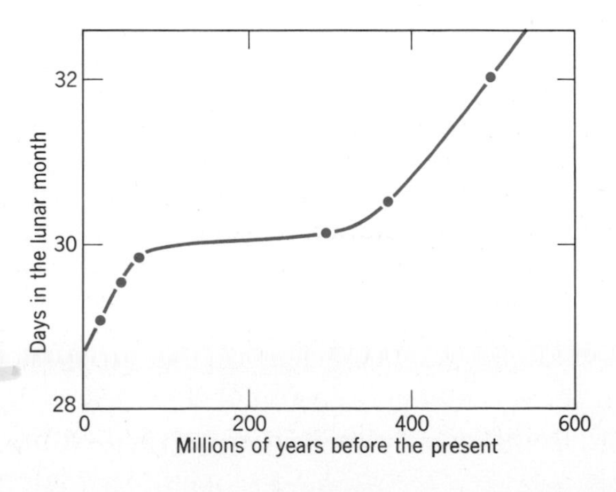


**Figure 5.9**
**Evidence from growth layers in fossil shells indicates that Earth's rotation is slowing down and the number of days in the lunar month is decreasing. Rate of slowdown appears to have varied with time, a factor attributed to the changing positions of continents and their effects on tides.** (*Pannella and others, 1968.*)

3. Decay of radioactive isotopes of various elements is the basis of radiometric dating.

4. A sedimentary layer is dated by being bracketed between two bodies of igneous rock that have been dated radiometrically.

5. The dates of the bracketed layers establish the correctness of the geologic column.

6. Measurement of carbon-14 activity, mainly in fossil organic matter, yields apparent ages through approximately the last 50,000 years.

7. The age of Planet Earth is obtained through radiometric dating of meteorites. It is about 4.6 million years.

## SELECTED REFERENCES

Berry, W. B. N., 1968, Growth of a prehistoric time scale: San Francisco, W. H. Freeman.

Eicher, D. L., 1968, Geologic time: Englewood Cliffs, N. J., Prentice-Hall.

Faul, Henry, 1966, Ages of rocks, planets, and stars: New York, McGraw-Hill.

Libby, W. F., 1961, Radiocarbon dating: Science, v. 133, p. 621–629.

Toulmin, Stephen, and Goodfield, June, 1965, The discovery of time: New York, Harper & Row.

Harland, W. B., and others, eds., 1964, The Phanerozoic time-scale: Geol. Soc. London Quart. Jour., v. 120S.

York, Derek, and Farquhar, R. M., 1972, The Earth's age and geochronology: Oxford, Pergamon.

**Colorado River eroding its way across Utah. (*William A. Garnett.*)**

# PART TWO
# EXTERNAL PROCESSES ON LAND

# 6 WEATHERING, SOIL, MASS-WASTING



## WEATHERING

We have now had a sweeping look at Earth's general character—its solid surface and its envelopes, its crust composed of minerals and rocks, the energy that drives its internal and external activities, the matter that moves gradually through the rock cycle, and lastly the steady march of time, so long that with its aid slow processes can accomplish results that are enormous. These processes are what we are going to examine next, beginning with the external ones because those are the most easily visible. Together they form a chain in which rock is broken up and moved, as sediment, which is eventually deposited, forming strata. The logical processes to start with are weathering and mass-wasting, by which rock is broken and the resulting pieces begin to move downslope on their journey toward the sea.

### ENVIRONMENT OF WEATHERING

Long ago, people began to select for their buildings, tombstones, and other structures stone that would be durable. Their success

A

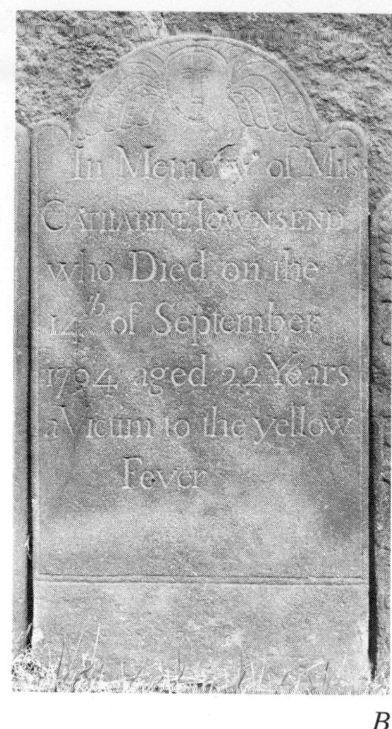

B

C

has been mixed. The durability of rock varies with climate, composition, and degree of exposure to weather. If gravestones (Fig. 6.1) made of firm rock have begun to crumble within a few centuries, what would have happened to rock exposed to the atmosphere through thousands or even millions of years? Fast or slow, mechanical and chemical alteration occurs everywhere at the interface between lithosphere and atmosphere.

The interface, as we noted in Chapter 2, is not sharply drawn. It is a zone rather than a surface, and extends down into the ground to whatever depth air and water penetrate. In this critical zone both hydrosphere and biosphere are involved also; in fact all the "spheres" are continually interacting. Within the interface, the rock constitutes a framework, full of joints, cracks, and other openings, some of which are tiny but all of which make the rock vulnerable. This framework is continually being attacked, chemically and physically, by water solutions. The result is conspicuous alteration of the rock.

The effects of such alteration are exposed commonly in cuts along highways and in other large excavations. Figure 6.2 shows fresh, unaltered bedrock (1) imperceptibly grading upward through rock (2) that has been altered but that still retains it organized appearance, into loose, unorganized, earthy regolith (3) in which the texture of the original rock is no longer present. Evidently alteration of the fresh rock is progressing from the surface downward. When exposed to the atmosphere no rock, whether bedrock or the stone in a man-made structure, escapes the effects

**Figure 6.1**
**(see opposite page)**
**Three dated gravestones in a Connecticut cemetery show the strong influence of rock composition on rate of chemical weathering. Climate is moist, with annual rainfall of more than 1,150mm.**

***A.*** **Marble, consisting of very soluble calcite, is greatly corroded; inscription is hard to read. Entire surface is rough.**

***B.*** **Medium-grained quartz sandstone containing feldspar and micas, much less soluble than marble, is slightly roughened overall and has flaked off in a patch near top.**

***C.*** **Fine-grained sandstone consisting almost wholly of quartz, very insoluble, is almost unaltered. Incised lettering is sharp and clear. Discoloration of lower part results from many years of partial burial or of lying on the ground.**

of ***weathering***, *the chemical alteration and mechanical breakdown of rock materials during exposure to air, moisture, and organic matter.*

The regolith we see in Figure 6.2 was formed *in place* (on the spot) by conversion of bedrock, and so we say it is *residual.* In many places, however, regolith is so different from the bedrock below that it cannot have resulted from chemical alteration. Instead, former regolith has been removed, and sediment transported from elsewhere has been deposited in its place. Both removal of the former residual material and deposition of the sediment could have been performed by a single agency such as a river, surf along a coast, or a glacier, or by two or more agencies working together.

# PROCESSES OF WEATHERING

If we could look closely at the bedrock in Figure 6.2 we would see that near the bottom of the exposure (1) the cleavage surfaces of the grains of feldspar flash brightly between the grains of quartz. Higher up (2) such surfaces are lusterless and stained. Near the top (3) the grains of quartz, although still distinguishable, are separated by soft, earthy material that in no way resembles the former feldspar, which has rotted away. Evidently the changes that have occurred are mainly chemical. However, some regolith consists of fragments identical with the adjacent bedrock. In them the minerals are fresh or only slightly altered. This relationship is commonly seen in the aprons of loose sliderock that mantle the lower parts of bedrock cliffs from which the sliderock must have been derived (Fig. 6.20). Because the coarse fragments of sliderock show little or no chemical change as compared with the bedrock, we conclude that bedrock can be broken down mechanically as well as decayed chemically. So we speak of *mechanical weathering* or *disintegration,* as distinct from *chemical weathering* or *decomposition,* even though the two processes work hand in hand, and even though their effects are inseparably blended.

# MECHANICAL WEATHERING

In many places regolith consists wholly of rock material that is identical in every way with the local bedrock. Hence no chemical changes like those just described can have occurred; the weathering processes that formed such regolith must have been mechanical rather than chemical. Mechanical breakdown is common and is brought about by freezing and thawing, by activities of plants and animals, and by the heat of fires.

### Frost wedging

In cold climates water freezes in both bedrock and regolith. Thermometers installed in such places show that in winter, at least, the water alternately freezes and thaws, in many places at least once daily. Laboratory experiments show that when water

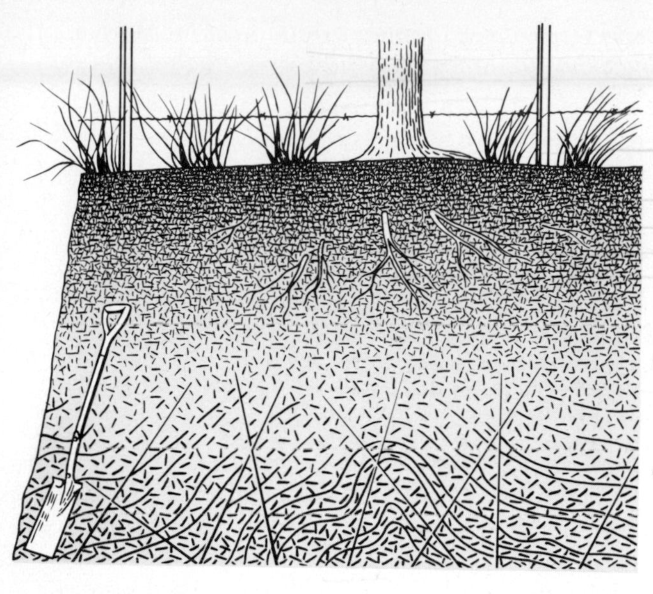


**Gradation upward from fresh bedrock (granite gneiss) to earthy regolith.**

crystallizes into ice, a volume increase of nine percent occurs. Consequently, when ice forms in an opening within rock, rock material is forcibly pushed up or pried apart—not only tiny particles but also blocks small and large, some of them weighing many tons. This mechanism is *frost wedging.* In Figure 6.3 we can see that movement of the largest block of rock has been sideways toward the left. This is the direction of easiest relief of the pressure exerted by water as it freezes in a former crack (now a wide gap)—the direction in which the least energy would be needed to move the block. This is why frost wedging is particularly effective at the faces of cliffs. But even with no cliffs present, it breaks up bare bedrock, littering the ground with broken pieces that in some places conceal the bedrock beneath.

You might think that because most cracks and cavities in rock are open to the atmosphere, freezing water could easily expand outward along the cracks. But water in the upper and outer parts of the openings tends to freeze first. This creates confined spaces, in which any further freezing can cause the rock to burst. As nearly all rock is cut by cracks of various kinds, frost wedging must be an effective rock-breaking process. Probably the wedging apart of tiny mineral grains loosens and, in the long run, moves more rock material than the more spectacular pushing of large blocks.

### Plants and animals

Seeds germinate in cracks in rock, and the growing plants extend their roots farther into the cracks. As trees grow, their roots wedge apart the adjoining blocks of rock, even those weighing a ton or more. Shrubs and smaller plants also send their rootlets into tiny openings, which they enlarge. Although it would be hard to measure, the total amount of rock breaking done by plants must be enormous. Much of it is obscured by chemical decay, which takes advantage of the new openings as soon as they are created.

**Figure 6.3**
**Mechanical weathering (chiefly frost wedging along joints) is disintegrating diabase bedrock in East Greenland, forming large, angular blocks. Curved surface (skyline of foreground) was created by glacial erosion during latest glacial age (compare glacially abraded rock, Fig. 10.10). The carbon-14 dates of samples collected nearby show that glaciers had melted back and that this locality was exposed to weathering about 8,000 years ago. Thus, the disintegration shown is the work of about 8,000 years of weathering. Mesters Vig, Greenland.**
**(*A. L. Washburn.*)**

Burrowing animals large and small (ants, for example) bring quantities of partly decayed rock particles to the surface, to be exposed more fully to chemical action. More than a hundred years ago, Charles Darwin made close observations in his English garden, and calculated that every year earthworms bring particles to the surface at a rate of more than ten tons per acre. After a study in the basin of the Amazon River, the geologist J. C. Branner wrote that the soil there "looks as if it had been literally turned inside out by the burrowing of ants and termites." The amount of rock material moved by burrowing organisms, cumulatively through hundreds of millions of years, must be huge. It illustrates again the cumulative effect of small forces acting through a very long time.

### Effects of heat

The heat of forest and brush fires breaks large flakes from exposed surfaces of bedrock. Because rock is a poor conductor of heat, fire heats only a thin outer shell, which expands and breaks away. Fires set by lightning must have been common during long ages before man began to disturb nature's economy, and fire has probably been an important factor in the mechanical breaking of rock.

## CHEMICAL WEATHERING

### Effects on potassium feldspar

The active agents of decomposition consist of chemically active water solutions and water vapor. As it falls through the atmosphere, rainwater dissolves small quantities of carbon dioxide, so that when it reaches the ground it has become a weak carbonic acid. As it moves on down through the soil, the acid solution is made stronger by addition of more carbon dioxide coming from the decay of vegetation.

The carbonic acid ionizes to form hydrogen ions and bicarbonate ions:

$$\underset{\text{(Water)}}{H_2O} + \underset{\text{(Carbon dioxide)}}{CO_2} \longrightarrow \underset{\text{(Carbonic acid)}}{H_2CO_3} \longrightarrow \underset{\text{(Hydrogen ion)}}{H^{+1}} + \underset{\text{(Bicarbonate ion)}}{(HCO_3)^-}$$

Hydrogen ions are extremely effective in decomposing minerals. They are so small (Fig. 3.2) that they can squeeze in between the atoms of a crystal and disrupt its structure. The effectiveness of $H^+$ ions is illustrated by the way in which potassium feldspar (orthoclase) is decomposed by hydrogen ions and water:

$$\underset{\text{(Potassium feldspar)}}{2KAlSi_3O_8} + \underset{\text{(Hydrogen ions)}}{2H^{+1}} + \underset{\text{(Water)}}{H_2O} \longrightarrow$$

$$\underset{\text{(Potassium ions)}}{2K^{+1}} + \underset{\text{(Kaolinite)}}{Al_2Si_2O_5(OH)_4} + \underset{\text{(Silica)}}{4SiO_2}$$

Here the $H^{+1}$ ions forcibly enter the potassium feldspar and displace potassium ions, which then leave the crystal and go into solution. Water combines with the remaining aluminum silicate radical to create the clay mineral called kaolinite. This combination of water with other molecules is *hydrolysis,* one of the chief processes in chemical weathering. The resulting kaolinite is a *secondary mineral,* because it was not present in the original rock. Kaolinite is the most conspicuous of the three products of the reaction. It is a common member of the group of very insoluble minerals that constitute clay, and as clay it accumulates and forms a substantial part of the regolith. Many of the potassium ions released during the decomposition of orthoclase are eventually taken up by plants; others enter into clay minerals other than kaolinite.

The silica, less insoluble than clay minerals, in part remains in the clay-rich regolith and in part moves away in solution. Many of the potassium ions likewise escape in water solution, and some of them, together with some of the dissolved silica, eventually find their way through streams to the sea. Some, however, are held in the clay-rich regolith and become food for plants.

We say that the matter carried away in solution has been *leached* from the parent rock. ***Leaching*** *is the continued removal, by water, of soluble matter from bedrock or regolith.*

### Effects on granite, basalt, and limestone

What happens in the weathering of potassium feldspar is a key to the weathering of rock, such as granite, that contains this mineral. Table 6.1 contrasts the chemical weathering of granite and basalt, showing stubborn minerals that persist, secondary minerals that form, and cations that are carried away in solution. A third kind

**Table 6.1**

**Chemical Weathering of Two Great Groups of Igneous Rocks, Represented by Granite and Basalt**

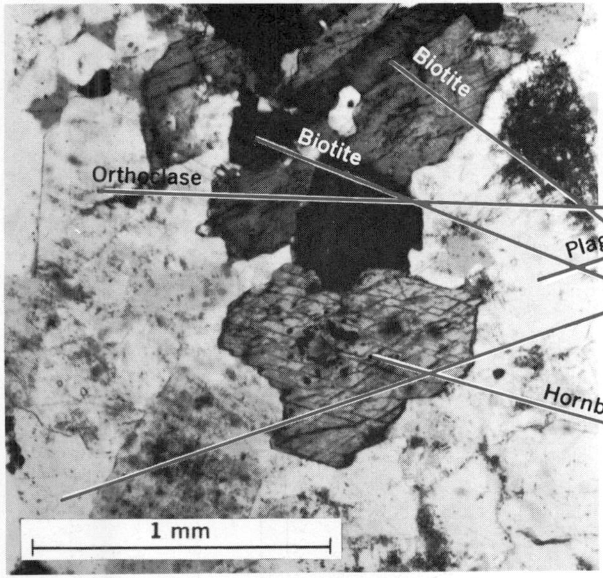

| Rock | Primary Constituents: Minerals | Primary Constituents: Cations | Weathering Products: Colloids | Weathering Products: Secondary minerals that form from colloids and ions | Weathering Products: Primary minerals that persist | Weathering Products: Soluble cations removed in solution |
|---|---|---|---|---|---|---|
| GRANITE | FELDSPARS | $K^{+1}$ $Na^{+1}$ | Silica, alumina | Clay minerals | | $Na^{+1}$ $K^{+1}$ |
| GRANITE | QUARTZ | | | | Quartz | |
| GRANITE | MICAS | $K^{+1}$ $Fe^{+2}$ $Mg^{+2}$ | Silica, alumina | Clay minerals | Some mica | |
| GRANITE | FERRO-MAGNESIAN MINERALS | $Mg^{+2}$ $Fe^{+2}$ | Silica, alumina | Clay minerals | | $Mg^{+2}$ |
| GRANITE | | | Iron oxides | Hematite, "limonite" | | |
| BASALT | FELDSPARS | $Ca^{+2}$ $Na^{+1}$ | Silica, alumina | Clay minerals | | $Na^{+1}$ $Ca^{+2}$ |
| BASALT | FERRO-MAGNESIAN MINERALS | $Mg^{+2}$ $Fe^{+2}$ | Silica, alumina | Clay minerals | | $Mg^{+2}$ |
| BASALT | MAGNETITE | $Fe^{+2}$ | Iron oxides | Hematite, "limonite" | | |

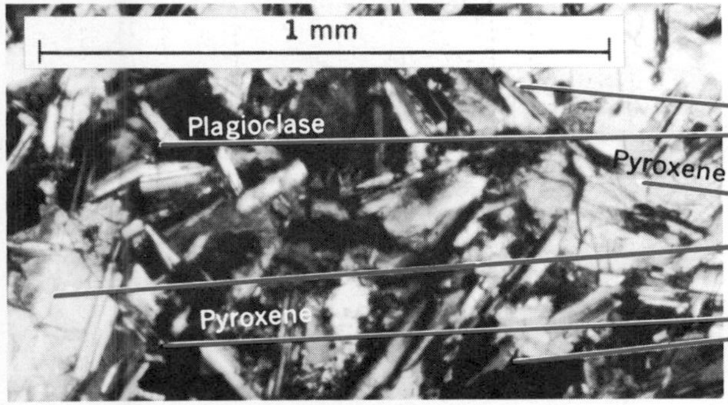

As the feldspars decay, the grains of quartz are loosened like bricks in a wall when the mortar between them crumbles.

Primary minerals are shown in capital letters; minerals constituting the product of complete weathering are in shaded boxes. Photographs (*left*) are thin sections of granite and basalt, with key minerals identified. Uncommon primary minerals (such as rutile) are not included, because they show up in the weathered product in very small quantities. (Upper photo G. M. Friedman; lower photo B. M. Shaub.)

Because basalt contains little or no quartz, its weathering products include clay minerals, and limonite.

of rock, limestone, is chemically weathered in a still different way. Limestone consists mainly of calcium carbonate, which is highly soluble. Carbonic acid in the ground readily dissolves the carbonate, leaving behind only the nearly insoluble impurities (chiefly clay and quartz) that are always present in small amounts. As limestone becomes weathered, then, the residual regolith that develops from it consists mainly of clay with particles of quartz.

Not only quartz but other minerals as well are stable in the environment at the Earth's surface, and so resist destruction by weathering. Like quartz, such minerals as gold, platinum, and diamond persist in weathered regolith and so are eroded and become sediment. Because these minerals are unusually heavy, they drop out of streams and become concentrated on the stream beds, forming *placers* (Fig. 22.13) from which they are mined. The widespread occurrence of these minerals in sedimentary strata of all ages indicates that weathering, as part of the rock cycle, has been going on through most if not all of Earth's history.

### Exfoliation and spheroidal weathering

A process that occurs commonly in rock that is cut by joints is ***exfoliation,*** *the separation, during weathering, of successive shells from rock,* like the "skins" of an onion (Figures 6.4 and 6.5). In some cases only a single shell is present, in others ten or more. The outermost shells tend to be flattish, whereas the innermost ones are spheroidal as corners become more and more rounded. Evidently exfoliation can take place not only at the surface but below ground as well, because we see it exposed in newly made cuts along roads (Fig. 6.4). The process is not restricted to a particular type of climate, although it is seen most commonly in dry climates. The spheroids created by exfoliation usually form a distinct pattern of nearly straight rows running in two or more directions. This is because the positions of the spheroids are controlled by joints (visible in Fig. 6.4) that existed in the rock long before weathering began. The joints were avenues of slow movement of water solutions that attacked the rock, causing chemical weathering.

Although the outer surface of exposed rock dries rapidly after a wetting, moisture that penetrates between mineral grains and into crevices remains long enough to cause some decomposition of feldspars and create clay. The process is accompanied by increased volume of the weathered rock. Probably the increase sets up small forces that cause shells to separate from the main body of the rock. Thus we have a *mechanical* effect that is a result of *chemical* weathering.

At this point we note two significant relationships. (1) The effectiveness of chemical reactions increases with increased surface area available for reaction. (2) Increased surface area results simply from subdivision of large blocks of rocks into smaller blocks. By merely subdividing a cube we can, while adding nothing to its volume, increase its surface area (Fig. 6.6). By subdividing it again and again we get a startling result. One cubic centimeter of rock subdivided into particles the size of the smallest clay minerals results in an aggregate surface area of nearly one acre. Weathering itself causes subdivision and so promotes further weathering. Although well represented in Figure 6.4, the weathering need not be spheroidal. The effects of decreased particle size are visible when we compare (1) and (3) in Figure 6.2.

## FACTORS THAT INFLUENCE WEATHERING

**Kind of Rock.** As Figure 6.1 shows, the minerals of which a rock is composed influence its decomposition. Quartz is so resistant to decomposition that rocks (such as granite) that are rich in quartz stoutly resist chemical weathering. In the eastern United States, from Maine to Georgia, hills and mountains that consist of granite are distinctly higher than surrounding areas underlain by less resistant rock. This relationship between mineral composition and the details of form of land surfaces is the result mainly of differences in rate of weathering.

However, rate of weathering of a rock is influenced not only by

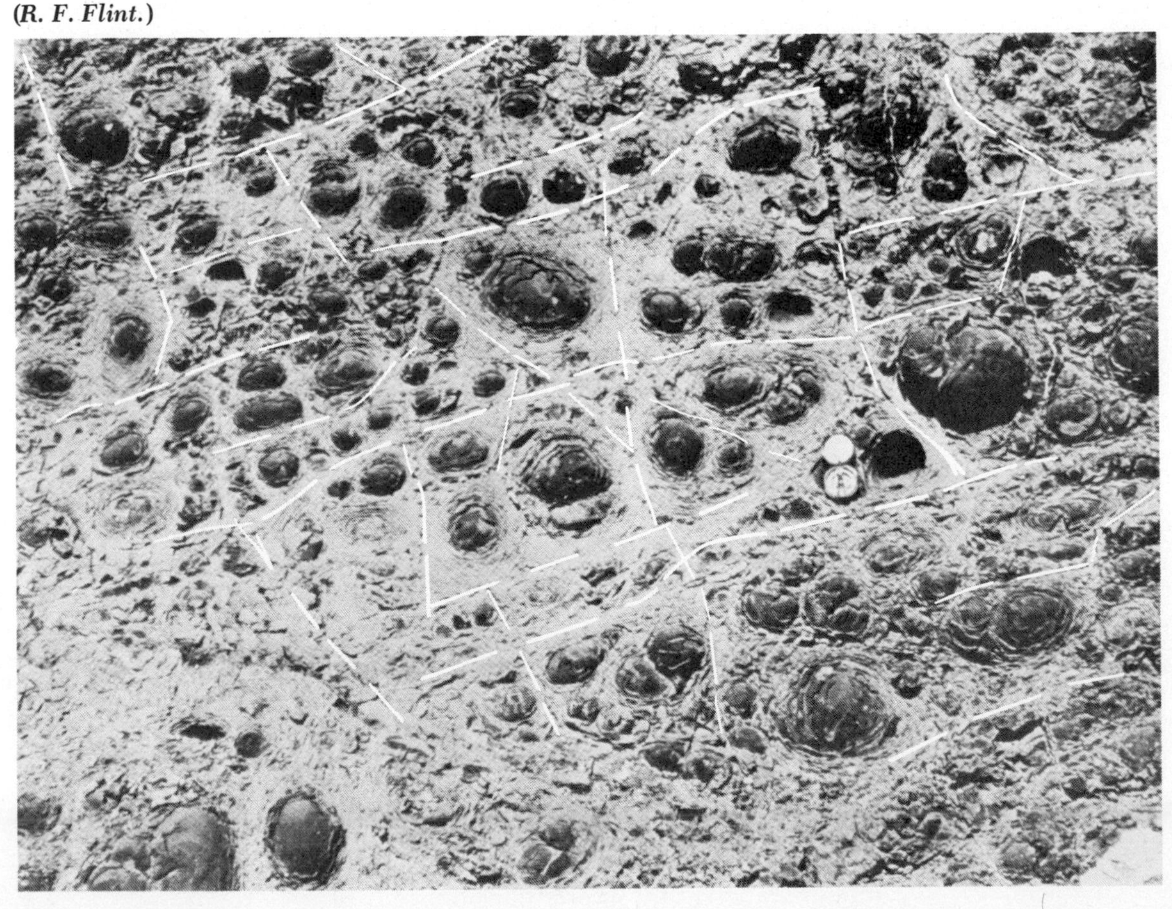

Figure 6.4
**Basalt, exposed along a road in western Argentina, is cut by three sets of joints, one nearly vertical, a second inclined gently from right to left, and a third (not visible) paralleling the plane of the photograph. Solutions penetrating inward from the joints have converted nearly cubic blocks of rock to spheroids, each with several successive shells. Diameter of mirror on the compass (*right center*) is 7.5cm. (*R. F. Flint.*)**

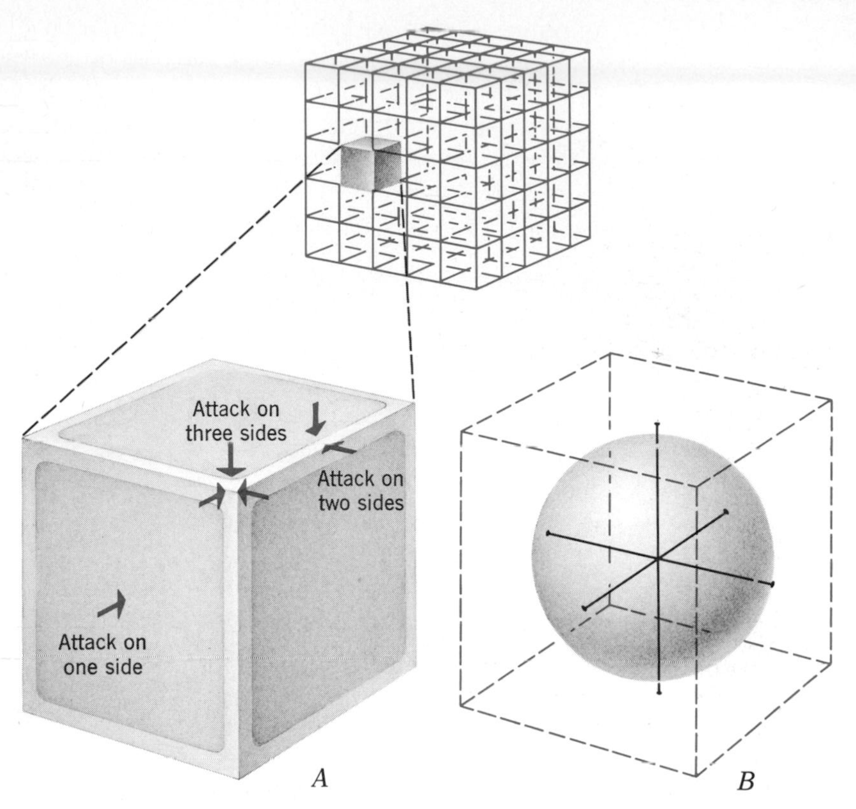


**Figure 6.5**
**Geometry of spheroidal weathering.**
***A*. Solutions that occupy joints separating nearly cubic blocks of rock attack corners, edges, and sides at rates that decline in that order, because the numbers of corresponding surfaces are 3, 2, and 1. Corners become rounded; eventually the blocks are reduced to spheres.**
***B*. Energy of attack has now become distributed uniformly over the whole surface, so that no further change of form can occur.**

minerals but also by structure. If a rock, even though it consists entirely of quartz (sandstone, quartzite), contains closely spaced joints or other partings, it can disintegrate very rapidly, especially when attacked by frost wedging.

**Slope.** When a mineral grain is loosened by decomposition on a steep slope, it is washed down the hill by the next rain. With the solid products of weathering moving quickly away, fresh bedrock is continually exposed to attack, so that weathered rock extends only to slight depth below the surface. In contrast, on gentle slopes weathering products are not readily washed away; so they accumulate to depths that in places reach 50m or more.

**Climate.** Both moisture and heat promote chemical reactions. So weathering generally goes deeper in a moist, warm climate than in a dry, cold one. In high mountains and in cold, high latitudes, wide areas are littered with frost-wedged pieces of rock, partly obscuring any effects of decomposition. Rocks such as limestone and marble, which consist almost entirely of (soluble) calcite, decompose quickly in a moist climate. In a dry climate, however, such rocks are very resistant, because with little rainfall, rock comes into contact with carbonic acid only rarely.

**Time.** Study of decomposition of the stone in ancient buildings and monuments readily shows that hundreds or even thousands of

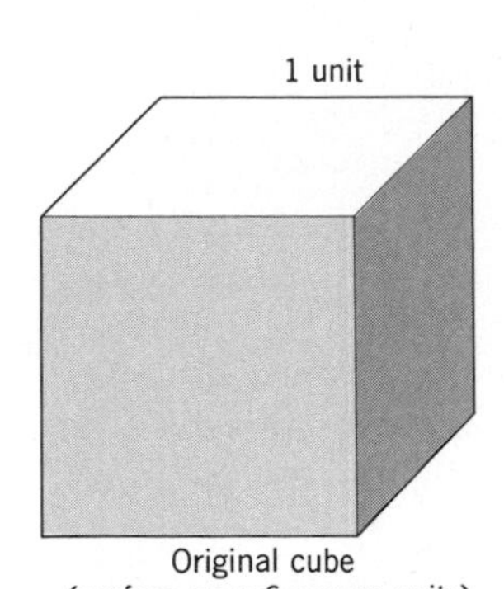


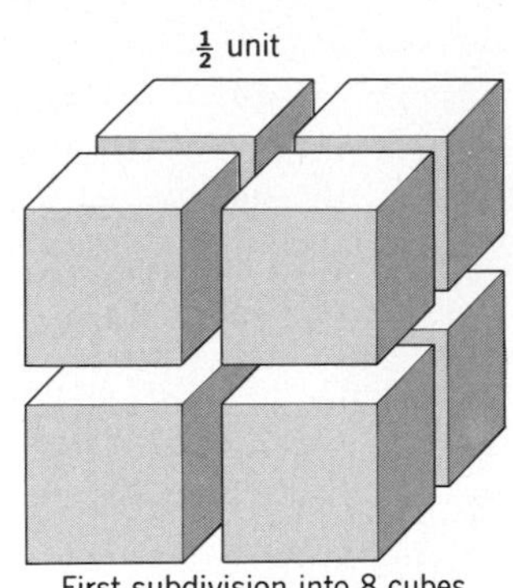


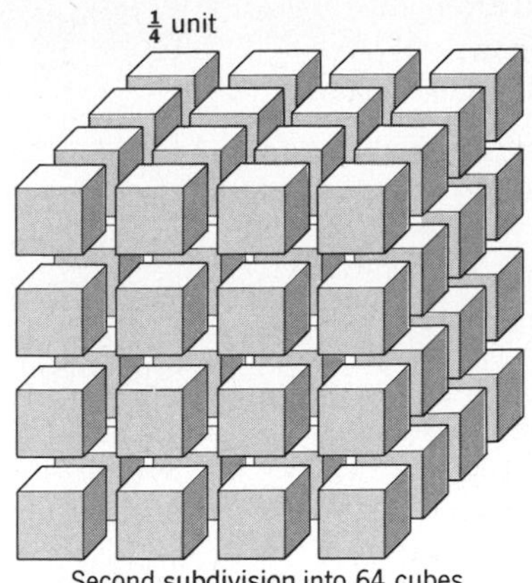


**Figure 6.6**
**Subdivision of a cube into smaller cubes. Each time a cube is subdivided by slicing it through the center of each of its edges, the aggregate surface area doubles. This greatly increases the speed of chemical reaction.**

years are required for hard rock to decompose to depths of only a few millimeters. Granite and other kinds of bedrock in the Sierra Nevada, New England, northern Europe, and elsewhere still exhibit polish and fine grooves made by glaciers during the latest glacial age, some 10,000 to 25,000 years ago (Fig. 10.9). We reason that in such rocks in such cool-temperate climates, it could take many tens of thousands of years, at the very least, to create weathered regolith like that shown in Figure 6.3.

But in some regions weathering extends far deeper and is far older. Mining operations have exposed bedrock that has been thoroughly weathered down to depths of 100m continuously through many millions of years.

The most promising way to measure the time involved in developing a zone of weathering is by radiometric dating. At present we have barely made a start at measuring rates of weathering.

# SOIL

### Origin

By breaking down rock to make regolith, weathering prepares the mineral framework for soil. But all soil contains at least a little, and commonly much, organic matter mixed with the framework. This organic part is essential to the usual definition of ***soil:*** *that part of the regolith which can support rooted plants.* It is the top part of the regolith, at the surface of the ground, although in places it includes the whole thickness of the regolith.

Partly through the activity of bacteria, the organic matter in soil is derived from the decay of plants. Living plants themselves are nourished by the decayed plant material and by decomposed mineral matter derived from the regolith by chemical weathering and pumped upward, in water solution, through roots. With the aid of the soil the plant world manufactures material from which it constructs its own fertilizer. These activities represent continual cycling of material between regolith and biosphere. The soil therefore is a busy, dynamic layer in which complex chemical, physical, and biological activities are happening continually. With its partly mineral, partly organic composition, soil is a great bridge between Earth's inorganic spheres and the teeming biosphere. To people it means food, and so it is the basic natural resource of every nation.

### Soil profile

As the weathering of bedrock and regolith proceeds (Fig. 6.7), the soil gradually ripens or matures. Normally it develops three layers, called *horizons*, that together constitute a ***soil profile***, *the succession of distinctive horizons in a soil, from the surface down to the unchanged parent material beneath it.* The soil profile is simply the upper part of the profile created by weathering.

The uppermost horizon, or *A* horizon, is commonly grayish or *blackish (at least at its top) because of the addition to it of* ***humus***, *the decomposed residue of plant and animal tissues.* The *A* horizon has lost part of its original substance through mechanical removal (by soaking downward through the ground) of clay particles and, more importantly, through the chemical removal of soluble minerals. Such removal, as we noted earlier, is *leaching.*

The *B* horizon, although poor in organic content, is a site of accumulation, for it has gained part of what has been leached from the *A*.

Although it is part of the soil profile, the *C* horizon is not part of the soil itself. It consists merely of the parent material (usually bedrock but in some places regolith) and therefore has no distinct lower limit. In much of the northern United States and Canada, the parent material of soils is regolith consisting of sediments deposited as recently as 10,000 years ago during the latest of the glacial ages (Chap. 10). Because of this rather recent origin, the soils of glaciated New England and the Great Lakes region are only partly developed. Soil scientists call them *immature.* The soils farther south, undisturbed by glaciation, are generally mature.

### Climatic soil groups

Parent materials differ widely, and strongly influence the character of soils, especially during the earlier part of soil development. But through a long period of time, the influence of climate is even stronger than that of bedrock in determining the character of soil. Under similar climatic conditions the profiles of mature soils developed on widely different kinds of rock become surprisingly alike.

Soils in the United States and southern Canada are divided into two groups on the basis of degree of leaching by rainwater. The most effective leaching occurs in areas with average annual rainfall of more than 635mm. An irregular line (Fig. 6.8) extending northward from central Texas to western Minnesota divides the more humid eastern half of the country from the drier western

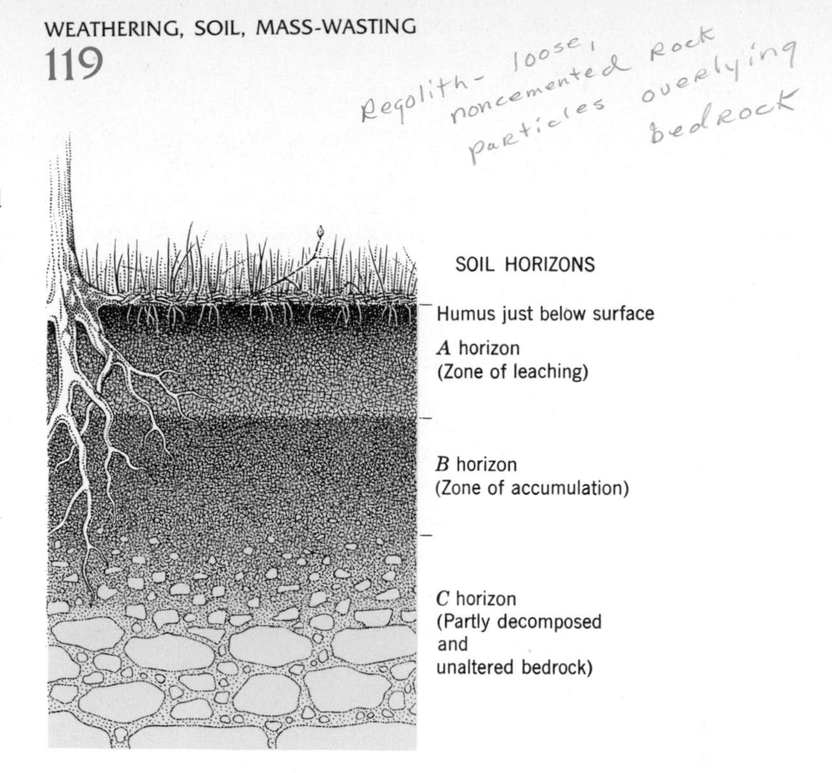


**Figure 6.7**
**The *A*, *B*, and *C* horizons of a typical soil profile. Each horizon grades down into the one beneath it.**

half. East of this line rainwater has leached from *A* horizons of mature soils large fractions of the calcium and magnesium carbonates that formed part of the parent material, and has deposited much clay and iron in the *B* horizons. West of the line, where rainfall is less abundant, carbonates accumulate in the upper part of the soil, making it strongly alkaline in contrast to the acidic soils of humid regions. An important part of the accumulation results from evaporation of water that rises through the ground, bringing dissolved salts from below. In large areas of Texas and adjoining states carbonates have, in this way, built up a solid, almost impervious layer. Such *a whitish accumulation of calcium carbonate developed in a soil profile* is generally known in western North America as ***caliche*** (kă lē′ chē).

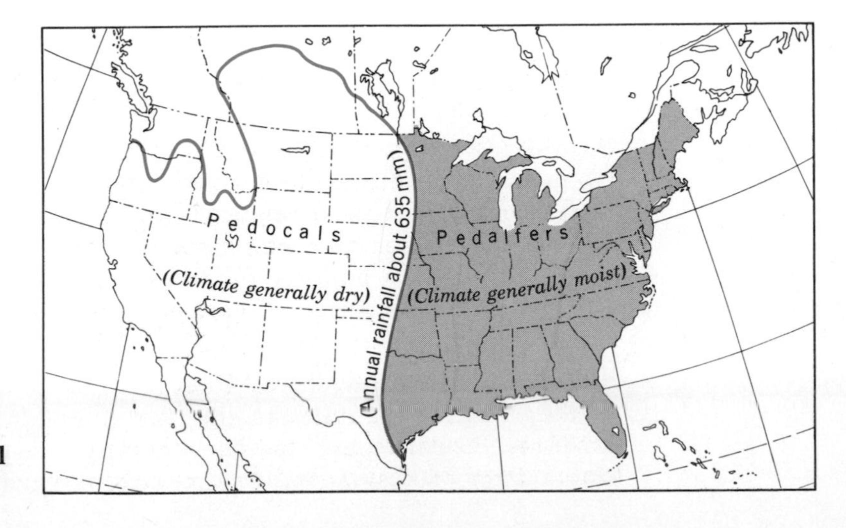


**Figure 6.8**
**General division of soils in the United States and southern Canada into two major classes, separated by a line along which annual rainfall is about 635mm.**

*A soil with calcium-rich upper horizons* is a **pedocal** (Greek *pedon,* "soil," and the first syllable of "calcium"). *A soil in which much clay and iron have been added to the B horizon* is a ***pedalfer*** (symbols for aluminum and iron—Al and Fe—added to the Greek root).

### Rate of soil formation

Although the making of soil constitutes part of the complex process of weathering, soil formation and weathering are not the same thing. The time factor in weathering chiefly concerns the decomposition of bedrock, which involves great lengths of time and is essentially a geologic process. The time required to form an orderly series of soil horizons in regolith involves far shorter periods. Soil scientists believe that on loose sandy material (not bedrock) having a forest cover, a soil might develop within a time as short as 100 to 200 years. Compared with Earth's long history, such a time is very short, but in terms of the world's need for foodstuffs, the rebuilding of soil lost as a result of human activities is a slow process.

## WEATHERING IS PART OF THE ROCK CYCLE

Weathering is an intrinsic part of erosion (Chapter 1) and so participates in the rock cycle. The place where weathering occurs is like the post office of origin in a worldwide sorting system, where outgoing pieces of mail get their first sorting and are sent off in various directions, only to be sorted again and again as they approach their destinations. In the sorting system begun by weathering, particles are detached or extracted from rock and are sent on their way down the nearest slope, into a stream, and so onward into a basin, where they are deposited. At every step along the way the particles are sorted according to their size, their weight, their shape, their durability, and other factors. Some are deposited at various points en route; others make it to a basin of some kind—the end of the particular system through which they are moving.

Weathering brings about a good deal of readjustment of minerals to environments at Earth's surface. Minerals that had formed at high temperatures as components of igneous rock, and others, in various kinds of metamorphic rock, that had formed under both high pressures and high temperatures, become unstable when exposed at Earth's surface, where both temperatures and pressures are much lower. Such minerals break down and their components form new, more stable minerals.

Of these new, stable minerals the most abundant are the clay minerals. Broadly speaking, then, chemical weathering is a huge clay-making process. It is also a process in which quartz is unlocked from granites and other quartz-bearing rocks; the quartz, being comparatively stable, persists in an unchanged state. The two most conspicuous products of weathering, new-made clay and old (but now free) quartz, go on into the next phase of the rock cycle, are transported and deposited, and eventually are transformed into strata of claystone and sandstone.

Meanwhile, other products of weathering are the cations $Na^{+1}$, $K^{+1}$, $Ca^{+2}$, $Mg^{+2}$. Most of the $K^{+1}$ goes immediately into plants and certain clay minerals. The other cations move slowly through the ground in solution (Chapter 9) and emerge at all points along the line into streams and so eventually reach the sea. There $Ca^{+2}$ and $Mg^{+2}$ are deposited, mostly as carbonates, eventually to form the limestone and dolostone that with claystone and sandstone constitute the common sedimentary rocks. The $Na^{+1}$ mostly remains in solution.

In the final analysis, therefore, weathering is one of two great links between all bedrock and the sediment derived from it. The other link consists of the processes that transport weathered detritus from the places of weathering to the places of deposition.

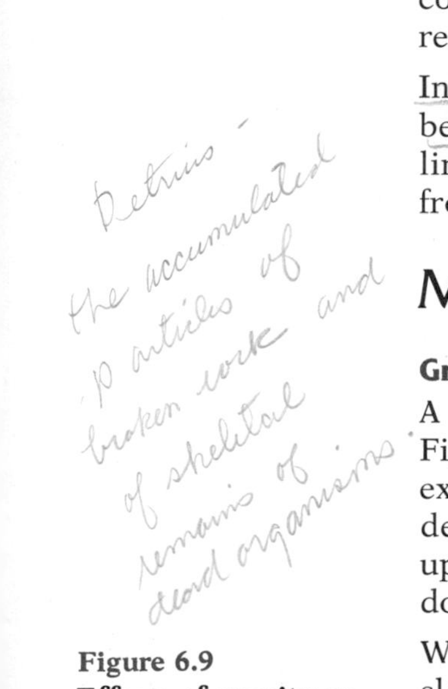


# MASS-WASTING

### Gravity

A smooth slope, covered with vegetation, like that on the right in Figure 7.3, shows little obvious evidence of activity. Yet if we examined the regolith beneath it, we would probably find particles derived from kinds of bedrock exposed only in areas that lie farther up the slope. We know, therefore, that the particles have moved downslope.

What makes them move is gravity, as it pulls persistently on all slopes. But, when we see a boulder lying on a hillslope, we realize that some force is holding it there. The holding force is likewise gravity. On a horizontal surface, gravity holds objects in place by pulling on them in a direction perpendicular to the surface. On any slope, gravity can be resolved into two component forces. The *perpendicular component of gravity* ($g_p$ in Fig. 6.9) acts at right angles to the slope and holds objects in place. The *tangential component of gravity* ($g_t$ in Fig. 6.9) acts along and down the slope. When $g_t$ exceeds $g_p$, objects move downhill, and we say the ***angle of repose*** has been exceeded. The angle of repose is *the steepest angle, measured from the horizontal, at which material remains stable*.

**Figure 6.9**
**Effects of gravity on objects lying on slopes.**

**Gravity can be resolved into two components, one perpendicular ($g_p$) and one parallel ($g_t$) to the surface. $g_p$ creates frictional resistance to sliding. When $g_t$ exceeds $g_p$ the object will move.**

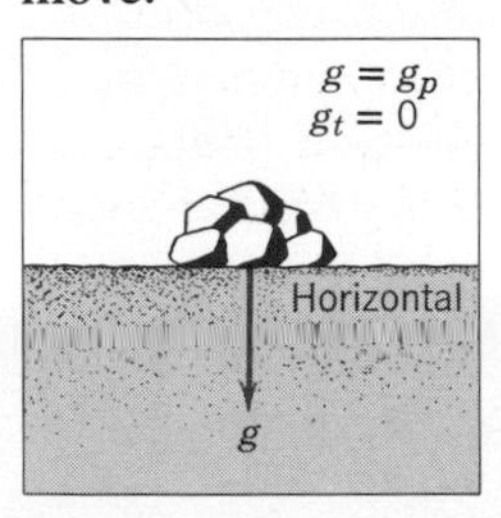


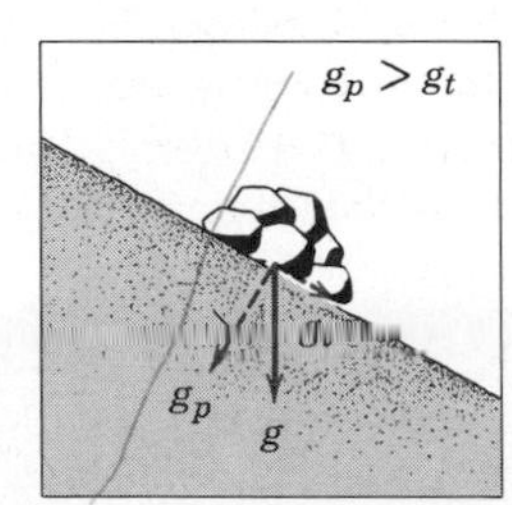


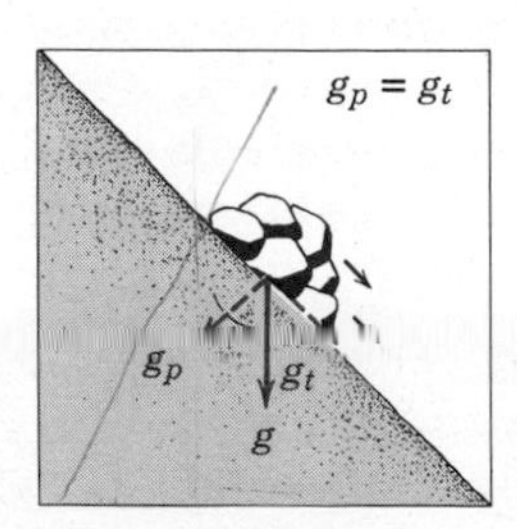


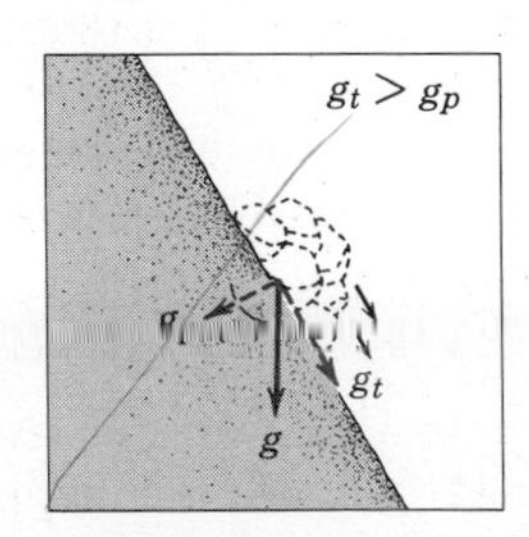

When the pull of gravity causes a body to move, the body acquires kinetic energy. Where does the energy come from? It is there all the time, waiting to be released. We call it ***potential energy,*** meaning *stored energy waiting to be used.* A rock on a hilltop reached its position because it was lifted, against the pull of gravity, when the hill was formed by one of Earth's internal activities. Similarly, the moisture in a raindrop was lifted to its position in a cloud by solar energy. When the raindrop falls, its potential energy appears as kinetic energy.

## Downslope movement

A particle loosened by weathering ceases to be bedrock and becomes regolith. Being loose and possessing potential energy, it will some time move downslope (Fig. 6.10). Sooner or later it reaches a stream or some other carrier, which transports it farther. Finally it ends up in the sea.

The beginning of the long journey, the trip down the nearest slope, can be very slow or very fast, but in either case it is controlled primarily by gravity. The movement is of several kinds, known collectively as ***mass-wasting,*** *the movement of regolith downslope by gravity without the aid of a stream, a glacier, or wind.* Although our definition excludes flowing water as a medium of transport, water nevertheless plays an important part in mass-wasting. Saturation of regolith with water makes movement easier. This is the main reason why some mass-wasting activities are especially common after long rains.

Mass-wasting affects whole bodies of regolith. Wide blankets and narrow streams of rock particles, large and small, move down hillsides, cliffs, and other slopes. But at times single particles, small boulders for instance, are detached from a cliff and fall, roll, or bounce downslope. This, too, is mass-wasting, because such movement depends on the force of gravity directly, rather than indirectly through a carrying medium.

Mass-wasting is not confined to the lands. Regolith, in the form of transported sediment, covers vast areas of the sea floor. Such material is seen moving down submarine slopes. Because such movements resemble mass-wasting, we think of mass movement as a universal process, submarine as well as subaerial, active wherever slopes exist.

Mass-wasting is nearly universal, because the Earth's land areas (and sea floors too) consist almost entirely of slopes, mostly arranged in systematic groups (Fig. 7.21). Natural flat surfaces that are truly horizontal are very rare. Furthermore, most land slopes are gentle; probably the majority lie at less than about 5° from the horizontal. Although steep slopes attract attention because they are conspicuous in terms of the amount of land surface they represent, they are rare.

The best way to visualize mass-wasting as a whole is to think of it as a natural system. Its input consists of the solid products of

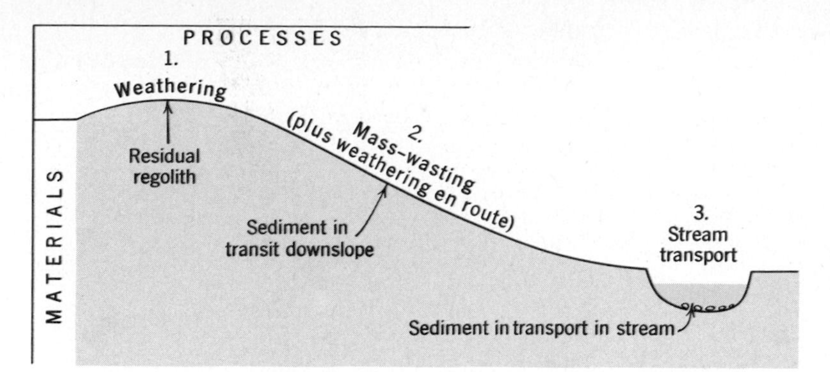


**Figure 6.10**
**Relation of mass-wasting (2) to weathering (1) and to stream transport (3).**

weathering contributed all the way down each slope; its output consists of sediments discharged, at the bases of the slopes, into carriers such as streams. As soon as regolith begins to move downslope it becomes sediment by definition. A steady state in the system is represented by a balance between input and output, and also by alteration of the slope itself to an angle that will just permit the quantity of regolith moving down it to maintain that balance.

# PRACTICAL IMPORTANCE

The broad complex of activities we call mass-wasting is subdivided into a varied group of separate processes. Although they differ in other respects, these processes share one characteristic: they all take place on slopes. Many of them were recognized even before the beginning of the present century, but their study was accelerated by the soil-conservation movement in North America, which gained impetus in the early 1930s. Scientists began to give close attention to the soil of farmlands and pastures, and hence to regolith as a whole. Their first objective was to help reduce rates of soil erosion by running water and wind. Such rates had been increasing alarmingly, partly through lack of understanding of the various factors that determine the stability of soil on slopes.

An additional reason for research on mass-wasting was the recognition that such research could help solve problems of other kinds: how stable are the foundations of proposed buildings, dams, and other structures built on or at the bases of slopes of various angles and in various materials? If they are unstable, what kind of engineering treatment will create stability, or should the proposed structure be built elsewhere?

# CLASSIFICATION OF PROCESSES

Studies made to help answer such questions led to attempts to define and classify the processes of mass-wasting. Some of the more obvious processes are illustrated in Table 6.2, in which the sketches are almost self-explanatory. It would be satisfying to be

**Table 6.2**
**Processes Involved in Some Kinds of Mass-Wasting**

| Process | Definition and Characteristics | Illustration |
|---|---|---|
| ***Rockfall* and *debris fall*** | *The rapid descent of a rock mass, vertically from a cliff or by leaps down a slope.* The chief means by which taluses are maintained. |  |
| ***Rockslide* and *debris slide*** | *The rapid, sliding descent of a rock mass down a slope.* Commonly forms heaps and confused, irregular masses of rubble. | 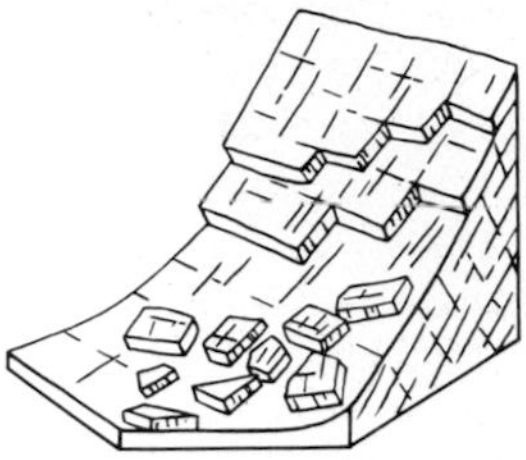 |
| ***Slump*** | *The downward slipping of a coherent body of rock or regolith along a curved surface of rupture.* The original surface of the slumped mass, and any flat-lying planes in it, become rotated as they slide downward. The movement creates a scarp facing downslope. |  |
| ***Debris flow*** | *The rapid downslope plastic flow of a mass of debris.* Commonly forms an apronlike or tonguelike area, with a very irregular surface. In some cases begins with slump at head, and develops concentric ridges and transverse furrows in surface of the tonguelike part. | 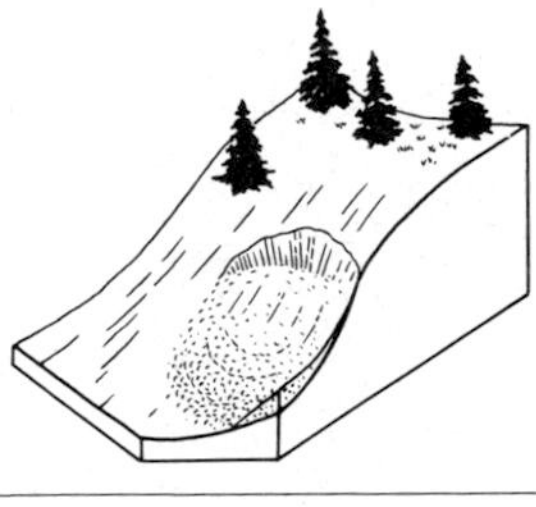 |
| **Variety: *Mudflow*** | *A debris flow in which the consistency of the substance is that of mud;* generally contains a large proportion of fine particles, and a large amount of water. |  |

able to classify the various processes entirely according to the kind of motion each displays. But this is not possible because some processes involve two or more distinct kinds of motion. Again, it would help if we could classify the processes according to their velocities. But we cannot, because the velocity of a single process at a single locality varies from one time to another.

The next few paragraphs describe the more rapid processes first, and then the slower ones.

### Falling and sliding

*Rockfall* and *debris fall, rockslide* and *debris slide,* and *slump* are generally small-scale processes, seen on cliffs and steep slopes where rock particles both small and large fall or slide downward. Slump is particularly common in places where slopes are kept steep and clifflike by erosion at their bases, as along stream banks (Fig. 6.11) and coastal cliffs (Fig. 6.12).

### Rock avalanche

Early in 1903 an unusual event destroyed much of the coal-mining town of Frank, Alberta, killing 70 people. With a great roar, a mass of rock having a volume of 30 million cubic meters rushed down the face of Turtle Mountain, more than 900m high, at a speed later estimated at nearly 100km per hour. Its momentum carried parts of the moving mass across a valley 3km wide, and 120m up the opposite side. Later investigation showed that the configuration of Turtle Mountain almost guaranteed an unstable situation (Fig. 6.13). Joints in the limestone strata were inclined toward the valley and were being subjected to dissolution and frost wedging. When

Figure 6.11
**Multiple slump in a cliff, 21m high, along the Genesee River near Avon, New York, April 1973. The cliff consists of till, a glacial deposit (Chap. 10), here rich in clay. Slump occurred after several days of rain. Flat area beyond road is a plowed field.** ***(Burr Lewis/Gannet Rochester Newspapers, courtesy R.A. Young.)***

Figure 6.12
**Slump on a large scale in coastal cliffs at Point Firmin, California.**

**Principal surface of slip is curved; it cuts cliff and displaces highway at two points. Other surfaces of slip, also curved, are shown by arrows. Erosion of foot of cliff by surf may have been the chief factor in removing support and causing movement.** (*Spence Air Photos.*)

the limestone becomes further weakened, the rock near the top of the mountain will be in serious danger of sudden movement, in another event similar to that of 1903. A somewhat similar event is depicted in Figure 23.4.

Formerly, the Frank event was referred to as a "landslide," a term we do not use here because it is ambiguous, implying that only sliding occurred. Rockslide and rockfall certainly were involved, but the dominant process is thought to have been *rock avalanche,* in which the dry, moving debris actually flows in a fluid-like manner.

### Debris flow

Many if not most of the conspicuous events in mass-wasting involve debris flow (Table 6.2). In many of them movement begins with slump and continues as debris flow. Figure 6.14 shows the idealized relationship between slump and debris flow, with well-defined masses, created by slump, breaking up into a single flowing mass that is ordinarily tonguelike in shape (see also Fig. 23.5). Many such features are 2 to 5km in length, although some are larger and others much smaller. Figure 6.15 shows the same features: slump blocks at head, tonguelike flowing mass downslope. Rates of flow range from less than one centimeter to about one kilometer per hour.

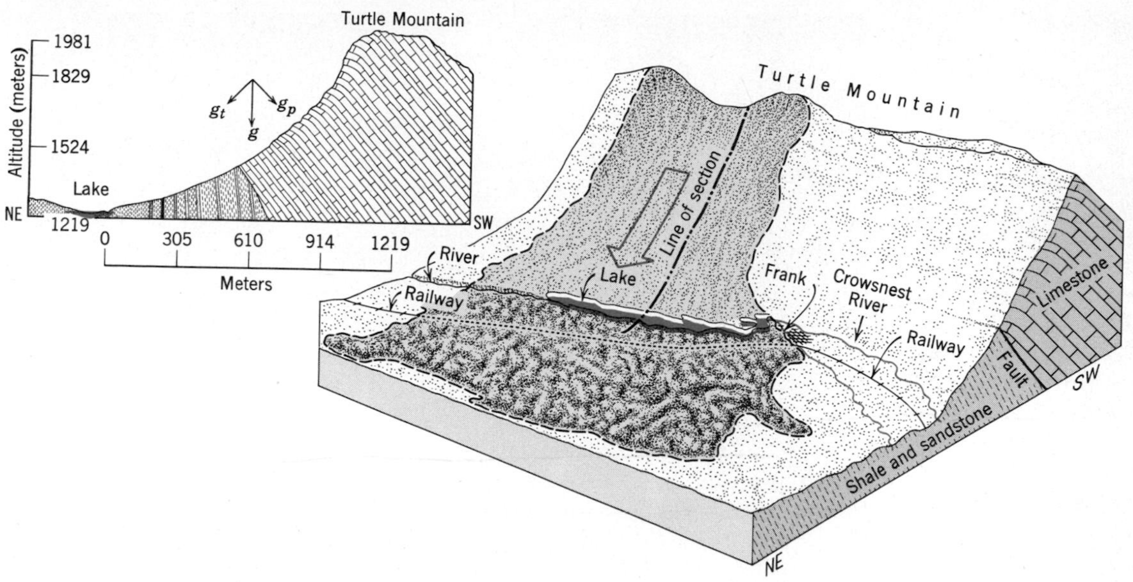


**Figure 6.13**
**The rock avalanche (large arrow) at Frank, Alberta, dammed a river to form a lake, destroyed 2,100 meters of railway line and part of the town of Frank, and spread a great apron of debris up the gentle slope beyond. Dotted line shows segment of railway that was destroyed. Section at left was made along a line running directly down the mountain slope.** ***(Data from Geol. Survey of Canada, modified by D. M. Kruden and J. Krahn.)***

### Mudflow

A fast variety of debris flow, mudflow is characterized by the presence of mostly fine particles and a water content that can amount to as much as 30 percent. It does not originate in slump blocks. As a result of its fine grain size and large water content, flowing mud tends to follow valleys as streams do.

Mudflow material grades from mud as stiff as freshly poured concrete to a souplike mixture not much thicker than very muddy water. In fact, after heavy rains in mountain canyons, mudflow can start as a muddy stream that continues to pick up loose material until its front portion becomes a moving dam of mud and rubble, extending to each steep wall of the canyon and urged along by the pent-up water behind it. On reaching open country at the mountain front, the moving dam collapses, floodwater pours around and over it, and mud mixed with boulders is spread out as a wide, thin sheet (Fig. 6.16) with destructive effects on farms and towns.

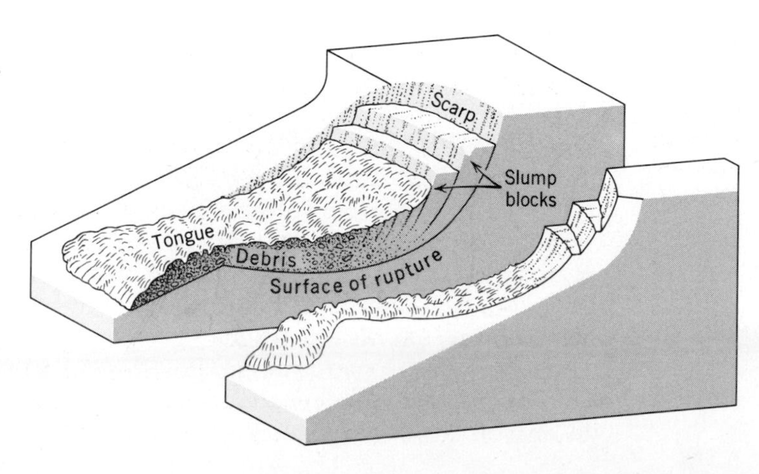


**Figure 6.14**
**Idealized sketch showing chief characteristics of many debris flows.**

**Figure 6.15**
**Slump and debris flow, near Oakland, California, December 9, 1950.**

**The slopes, although steep, are exaggerated somewhat by the angle of view. The whole feature is more than 200 meters long. (*Bill Young, from San Francisco Chronicle.*)**

Because of its great density, which enables it to move large, heavy objects, mudflow is destructive. Houses and barns in the paths of some mudflows have been carried from their foundations, and large blocks of rock are pushed along, rolling and sliding in the slimy mixture, some of them finally coming to rest on gentle slopes well out beyond the foot of a mountain range. The occurrence of huge, isolated boulders in such positions is not uncommon, and has led to the mistaken inference that the boulders were carried by former glaciers or by huge stream floods. Certainly some and probably many of them are the work of mudflow, the mud having been later removed by erosion (Fig. 6.17).

In regions of explosive volcanic activity, layers of volcanic ash are common. Noncemented fine-grained ash, mantling the slopes, is peculiarly susceptible to mudflow. A sudden heavy rain can bring this loose, easily transportable material into and down valleys in huge volume, as *volcanic mudflow*. Where downvalley slopes are gentle, the mushy material can flow long distances and can build up thick, flat-topped deposits of sediment that fill the containing valley from side to side. A valley fill of this kind, on the north side

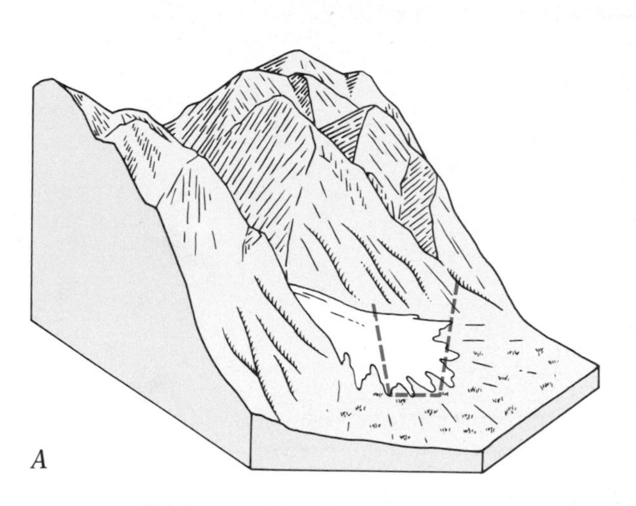

**Figure 6.16**
**Mudflow.**

*A.* **Typical mudflow setting, at base of mountains in an arid climate. Dashed line shows area like that in *B*.**

*B.* **Thin mudflow sediment at its downslope limit, forming small lobes on a slightly uneven surface of gravel. Mud has shrunk during drying and has been split by cracks. East base of Stillwater Range, Nevada. (*Eliot Blackwelder*.)**

of Mount Rainier in western Washington, is 72km long and as much as 100m thick; its estimated volume is well over a billion cubic meters. A similar though smaller volcanic mudflow buried and destroyed the Roman city of Herculaneum during the famous eruption of Vesuvius in A.D. 79 (Chapter 15).

### Creep

In contrast with those described so far, another group of mass-wasting processes moves at rates that are generally imperceptible. In this group the most widespread process is ***creep,*** *the imperceptibly slow downslope movement of regolith.* Its effects are evident in the consistent leaning of old fences, poles, and gravestones and in the derangement of roads (Fig. 6.18). Cuts that expose bedrock show that steeply inclined layers of rock curve strongly downslope just below the surface of the ground.

On sloping ground a cover of close-growing grass or other vegetation forms a protective armor against the cutting of gullies by running water and so keeps the slopes smooth. One might

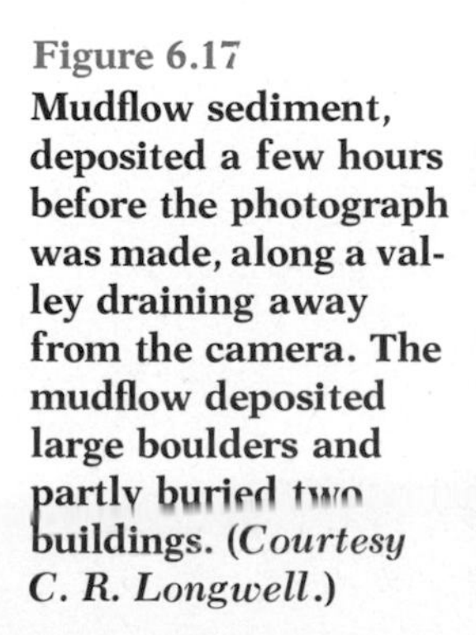

**Figure 6.17**
**Mudflow sediment, deposited a few hours before the photograph was made, along a valley draining away from the camera. The mudflow deposited large boulders and partly buried two buildings. (*Courtesy C. R. Longwell*.)**

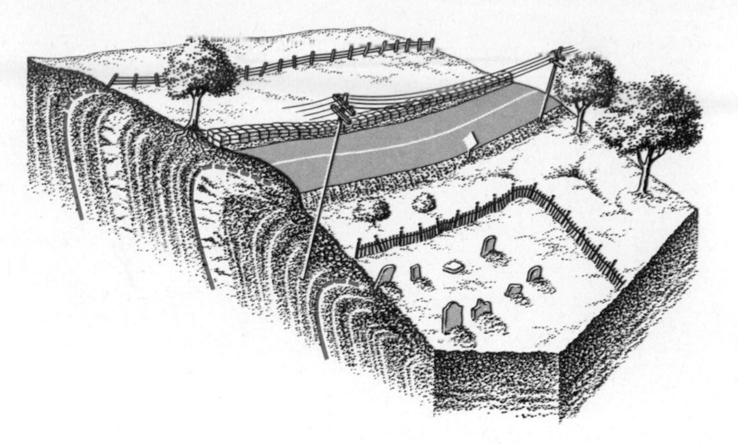

Figure 6.18
**Effects of creep on surface features and on bedrock. Blue lines emphasize strata bent over by drag of creeping regolith.** (*After C. F. S. Sharpe, 1938.*)

therefore suppose that ground so protected loses nothing to erosion except for mineral matter dissolved and carried off underground. But closer observation shows clearly that the regolith is creeping downslope, *carrying the vegetation with it.*

What are the causes of creep? There are several. One cause, common in regions with cold winters, is the freezing of water in regolith. The increase in volume pushes up the surface. This *lifting of regolith by freezing of contained water* is ***frost heaving.*** On a hillside the surface of the ground is lifted essentially at right angles to the slope; but when thawing occurs, each point tends to drop vertically and so moves downhill (Fig. 6.19). The movement consists of a complex series of zigzags.

The persistent activities that combine to make particles in the regolith creep downslope are listed in Table 6.3. Through countless repetitions on every slope, the effects of these activities add up to slow, persistent transport of regolith downhill.

# SEDIMENT DEPOSITED BY MASS-WASTING

### Colluvium

*Sediment deposited by any process of mass-wasting or by overland flow* (Chapter 7) is ***colluvium.*** Whether the depositing process is

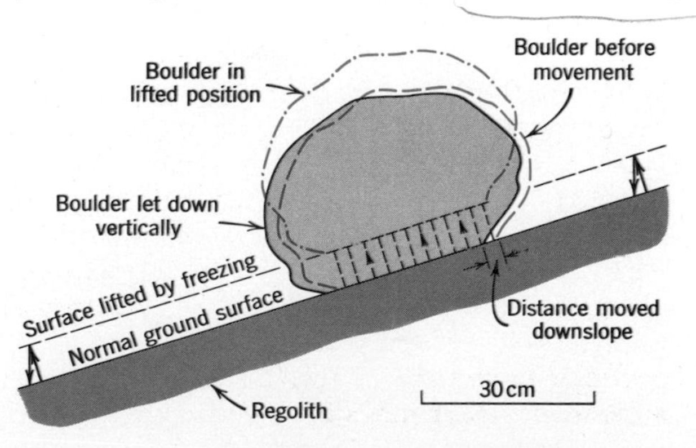


Figure 6.19
**A boulder moved downslope by alternate freezing and thawing. Diagram explains mechanism of one step in the movement.**

**Table 6.3**
**Causal Factors in Creep of Regolith**

| | |
|---|---|
| Frost heaving | Freezing and thawing, without necessarily saturating the regolith, cause lifting and subsidence of particles |
| Wetting and drying | Causes expansion and contraction of clay minerals; creation and disappearance of films of water on mineral particles causes volume changes |
| Heating and cooling without freezing | Causes volume changes in mineral particles |
| Growth and decay of plants | Causes wedging, moving particles downslope; cavities formed when roots decay are filled from upslope |
| Activities of animals | Worms, insects, and other burrowing animals, also animals trampling the surface, displace particles |
| Dissolution | Dissolution of mineral matter creates voids, which tend to be filled from upslope |
| Activity of snow | Where a seasonal snow cover is present, it tends to creep downward and drag with it particles from the underlying surface |

rapid or slow, the particles in a body of colluvium tend to lie in a chaotic jumble, because they were moved by falling, sliding, and similar activities. Also they tend to be angular in shape, because while moving, they did not collide frequently. These characteristics make it possible to distinguish colluvium from sediment deposited after transport in flowing fluids such as water (streams, surf) and air (wind). Such sediment tends to consist of rounded particles, sorted and deposited in layers.

**Sliderock; Taluses.** In areas where steep cliffs prevail and weathering is dominantly mechanical, accumulations of weathered particles usually mantle the bases of the cliffs. The particles, commonly angular and ranging in diameter from sand grains to boulders, are loosened from the bedrock of a cliff by mechanical weathering, accumulate, and move downward at various rates. *The apron of rock waste sloping outward from the cliff that supplies it* is a ***talus;*** *the material composing a talus* is ***sliderock*** (Fig. 6.20). From cliff to talus the movement is chiefly falling, sliding, and rolling; within the talus it consists of creep, generally so slow that the rock particles have time to become weathered en route. In such movement water plays little or no part. Weathering slowly

**Figure 6.20**
**Aprons of coarse sliderock form coalescing taluses against the base of a cliff in Banff National Park, Alberta. (*Jerome Wyckoff*.)**

converts the sliderock into fine-grained regolith which, with its pores of extremely small diameter, can hold much more moisture than sliderock and thus acquire both vegetation and soil (Fig. 6.21).

The activity in a talus constitutes a local system with input of large particles at its head, output of generally finer particles at its toe, and creep in progress from head to toe. The profile of the talus is concave-up, but with a radius of curvature so long that some profiles seem to be nearly straight. In each short segment down its length, the profile represents the angle of rest for the material in

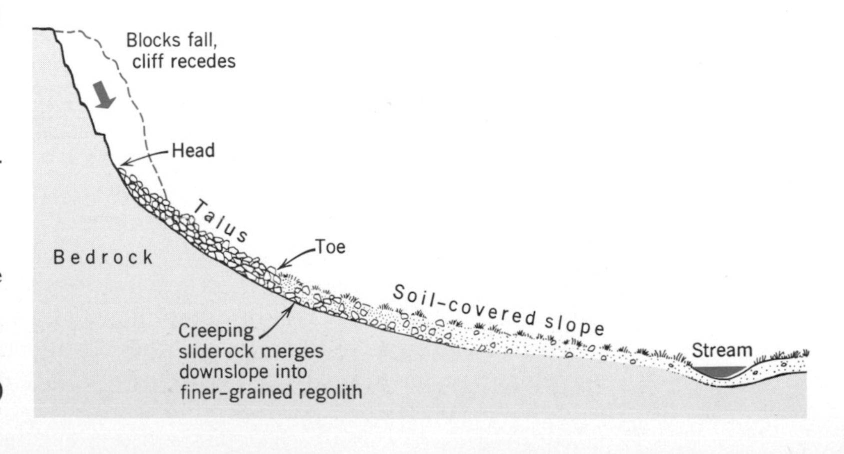


**Figure 6.21**
**Relationship of a talus to bedrock and to fine-grained regolith.**

**The sliderock that forms the talus grades downslope into finer-grained regolith that has been transformed from the sliderock by mechanical and chemical weathering. Also it has been added to by weathering of the bedrock beneath. Fine regolith is slowly fed to the stream, which carries it away. (*After C. F. S. Sharpe, 1938.*)**

that segment. Because the profile as a whole is the sum of all the short segments, in each of which the sliderock is stable, we call it an *equilibrium profile*, or stability profile. In it the force of gravity and other forces, mainly friction, are balanced nicely. The material added from upslope is just equaled by the material that moves on downslope. The same is true of the soil-covered slope downhill from the talus. That slope receives material from the talus and delivers an equal amount to the stream at its foot. Equilibrium profiles, based on a similar nice balance of forces, characterize other natural slopes such as the beds of streams and the beaches created by wave activity along coasts.

Despite the spectacularly rapid movements we have described, the part they play in the overall movement of material down the slopes of the lands is minor. Processes such as creep are slow, but they act continuously and they affect nearly all slopes. Hence they do much more work, in the aggregate, as agents of transport. With these almost universal activities in mind we can turn next to running water, the transporting agent that sooner or later receives and handles most of the sediments moved downhill by mass-wasting.

# SUMMARY

### ENVIRONMENT

1. The ground surface and a shallow zone beneath it represent an environment of low temperature and pressure, in which water and organic matter are present.

### PROCESSES

2. Chemical weathering and mechanical weathering, although very different, generally work in close cooperation.

3. Subdivision of large blocks into smaller particles increases surface area and thereby accelerates weathering.

### CHEMICAL WEATHERING

4. Carbonic acid is the prime agent of chemical weathering; heat and moisture speed chemical reactions.

5. Chemical weathering converts feldspars into clay. Grains of quartz, however, escape chemical decomposition, survive, and are carried away to be deposited as sand.

### MECHANICAL WEATHERING

6. The most obvious process of mechanical weathering is frost wedging, in which mineral composition is not altered.

7. Decomposition, however, is one of the causes of mechanical weathering.

### SOILS

8. A soil profile consists of horizons *A*, *B*, and *C*.

9. In moist climates, profiles of mature soils are similar even though they are developed in bedrock of different kinds.

10. Pedalfers form in areas of high rainfall, pedocals in areas of low rainfall.

### EFFECT OF BIOSPHERE

11. Animals and plants play a significant part in many processes of weathering.

### THE ROCK CYCLE

12. Weathering transforms bedrock into residual regolith, the chief source of sediment, which eventually becomes sedimentary rock.

13. Weathering is the source of sodium in the sea and of calcium in limestone.

### DOWNSLOPE MOVEMENT

14. Through mass-wasting, residual regolith created by weathering reaches the carrying agencies.

15. Because of the prevalence of slopes, mass-wasting is nearly universal.

16. Some mass-wasting processes are promoted by increased water content of the regolith.

### PROCESSES OF MASS-WASTING

17. Factors that cause creep include cycles of freezing and thawing, wetting and drying, and heating and cooling; also solution, and activities of plants and animals.

18. Although rock avalanche and debris flow are spectacular, imperceptible creep and solifluction, because they are widespread, move more material.

19. Mudflow is particularly common in arid climates where rainfall is sporadic and in areas covered with volcanic ash.

### SEDIMENT

20. Sediment deposited by mass-wasting is *colluvium*. It is generally nonsorted and either nonstratified or very poorly so. Although they may show wear, its particles are not rounded. These characteristics distinguish colluvium from sediment deposited from water.

21. Taluses develop at the bases of cliffs; commonly the material composing taluses grades outward into finer, soil-covered regolith.

## SELECTED REFERENCES

### WEATHERING; SOILS

Birkeland, P. W., 1974, Pedology, weathering, and geomorphological research: New York, Oxford Univ. Press.

Bradley, W. C., 1963, Large-scale exfoliation in massive sandstones of the Colorado Plateau: Geol. Soc. America Bull., v. 74, p. 519–528.

Carroll, Dorothy, 1970, Rock weathering: New York and London, Plenum Press.

Cruickshank, J. G., 1972, Soil geography: New York, Halsted Press Div., John Wiley.

Goldich, S. S., 1938, A study in rock weathering: Jour. Geol., v. 46 p. 17–58.

Hunt, C. B., 1972, Geology of soils. Their evolution, classification, and uses: San Francisco, W. H. Freeman.

Lyon, T. L., Buckman, H. O., and Brady, N. C., 1960, The nature and properties of soils, 6th ed.: New York, Macmillan.

Ollier, C. D., 1969, Weathering: New York, American Elsevier Publishing Co.

U.S. Department of Agriculture, 1957, Soil: Yearbook for 1957: Washington, U.S. Govt. Printing Office.

MASS-WASTING

Crandell, D. R., and Waldron, H. H., 1956, A recent volcanic mudflow of exceptional dimensions from Mt. Rainier, Washington: American Jour. Sci., v. 254, p. 349–362.

Eckel, E. G., ed., 1958, Landslides in engineering practice: Highway Research Board, Special Report 29, National Research Council, Washington, D.C.

Howe, Ernest, 1909, Landslides in the San Juan Mountains, Colorado: U.S. Geol. Survey Prof. Paper 67.

Kiersch, G. A., 1964, Vaiont Reservoir disaster: Civil Eng., v. 24, p. 32–39.

Legget, R. F., 1962, Geology and engineering, 2nd ed.: New York, McGraw-Hill, p. 106–128, 385–443.

Rapp, Anders, 1960, Recent development of mountain slopes in Kärkevagge and surroundings: Geografiska Annaler, v. 42, p. 1–200.

Sharp, R. P., and Nobles, L. H., 1953, Mudflow of 1941 at Wrightwood, southern California: Geol. Soc. America Bull., v. 64, p. 547–560.

Sharpe, C. F. S., 1938, Landslides and related phenomena: New York, Columbia Univ. Press. (Reprinted 1960 by Pageant Books, Paterson, N.J.)

Wahrhaftig, Clyde, and Cox, Allan, 1959, Rock glaciers in the Alaska Range: Geol. Soc. America Bull., v. 70, p. 383–436.

Washburn, A. L., 1966, Instrumental observations of mass-wasting in the Mesters Vig district, northeast Greenland: Meddelelser om Gronland, v. 166, no. 4.

# 7 STREAMS AND SCULPTURE OF THE LAND



## A FIRST LOOK AT A STREAM

A good way to understand the important part played by streams in transporting water and sediment down slopes is to sit beside a moderately rapid small stream a few yards wide and watch it. First you see the water flowing at different rates at various points. Out in midstream, flow is faster than near the banks, where eddies, little whirlpools, are usually in sight. This particular stream is clear, and its bed consists of sand and gravel. You can see pebbles rolling and sliding intermittently along the bed, whereas sand grains now and then make low jumps.

Walk along the stream, and note the channel winding from side to side in a rather smooth sequence of curves. At a place where the bank is undercut by the current at the outer side of a curve, chunks of the bank are slumping into the channel. If the bank is composed of sand, you can watch the current washing sand from the slumped mass and distributing it along the stream bed.

All this activity could have drawn your attention away from the darkening clouds, so that you have just time to take refuge under a large tree standing beside a plowed field, as an intense thundershower begins. For a time the tree acts as an umbrella; its leaves and branches hold back the raindrops, which do not reach the ground. But in the field beyond the tree, large drops strike the ground and splash up the loose soil, creating tiny craters as they do so. If you stepped out from under the tree, where it is still dry, raindrops would splash soil onto your shoes, a few centimeters above the ground surface.

At first the ground absorbs the water as it falls. But little by little the ground begins to glisten in patches, showing that it is saturated. The excess water starts to flow off over the surface toward the stream. Unless the field has been furrowed during plowing, the water may be flowing over it as wide, shallow sheets. It may be slightly turbid (muddied) with fine particles picked up from the soil. Where the water enters the stream, it makes turbid patches that soon blend and disappear. But from the grassy pasture beside the plowed field the water enters the stream through widely spaced grassy gullies, essentially as little, temporary streams of clear water.

The difference in behavior of the water, as it flows across plowed field and pasture, is the difference between overland flow and channel flow, a distinction we will describe in a few moments. Meanwhile, observe that the whole stream is swollen, is flowing faster than before, and has become turbid with silt and clay contributed from plowed fields, roads, and other bare areas unprotected by vegetation.

By now the leaves and branches of the tree can no longer hold back the rain, which is dripping onto the ground, though with less impact than on the unshielded ground outside. Since the tree is no longer an effective umbrella, you will have to make a dash for better shelter.

Observation and experiment by geologists and engineers indicate that all these stream activities are related to each other in a complex way. The stream you have been watching is one of many millions of drainageways by which water from rain and snow runs down over the land toward the sea (Fig. 4.1), enlarges valleys, and shapes most of the world's landscapes.

### Streams and people

Streams are useful for several reasons: (1) they are an increasingly important source of water for human and industrial consumption; (2) they are a comparatively small but essential source of energy; (3) many rivers are avenues of transportation and they have great recreational value. However, the floors of stream valleys are generally fertile and building on them is easy. Hence they tend to invite large populations which, then, must face the danger of damage by floods as well as the necessity of controlling pollution from the discharge of wastes into the streams.

## Streams as geologic agents

In addition to their immediate practical importance, streams are vital geologic agents because:

1: They move most of the water that goes from land to sea, and so are an essential part of the water cycle.

2: They carry eroded rock particles, as sediment, to the sea, and so are a prime link in the rock cycle. Sampling the loads of many rivers in many countries tells us that, every year, streams transport from land to ocean about 18 billion metric tons of rock waste mechanically, plus about 4 billion tons in solution. Thus about 82 percent of the average stream load consists of visible particles of rock; the rest is invisible, dissolved substances—the product of chemical weathering.

3: They shape the surface of the continental crust. The configuration of an enormous proportion of Earth's land area consists of stream valleys with hills in between, and is the work of streams and mass-wasting combined. Other agents, such as glaciers (flowing ice) and wind (flowing air), have shaped only minor parts of the lands, and even in many of those parts the principal shapes are the product of streams.

## Streams begin with raindrops

Erosion of the land by water begins even before a distinct stream has been formed. It occurs in two ways: by impact as raindrops hit the ground, and by sheets of water that result from heavy rains.

As raindrops strike bare ground they dislodge small particles of

Figure 7.1
**Apparatus for measuring sheet erosion, Marlboro, New Jersey.**
**Runoff, carrying soil particles, flows down a shallow slope; sediment is trapped in containers and measured by weight. A different kind of plant covers each plot. Results from a similar station are shown in Figure 7.2. (*U.S. Soil Conservation Service.*)**

loose soil, spattering them in all directions. On a slope the result is net displacement downhill. One raindrop is very little, but the number of raindrops is so great that altogether they accomplish a large amount of erosion.

What happens to rainwater after it reaches the ground? As Figure 4.1 shows, rainfall that is not returned to the atmosphere by evaporation or transpiration either infiltrates (sinks into) the ground or flows off over the surface as runoff. We can subdivide the runoff into ***overland flow***, *the movement of runoff in broad sheets or groups of small, interconnecting rills*, and ***stream flow***, *the flow of surface water between well-defined banks.* Stream flow is very obvious. Overland flow is not obvious; commonly it occurs only through short distances before ending in a stream valley. But it takes place wherever rain falls in excess of the amount that can be immediately absorbed by the ground. *The erosion performed by overland flow* is **sheet erosion.**

sheet erosion - performed by overland flow.

The great enemy of raindrops and overland flow as eroders is *vegetation.* If we examine carefully the world's natural land surface, setting aside areas covered with ice and snow and areas where people have stripped vegetation away or have covered it over with earth, concrete, or asphalt, we find that over a large proportion, the regolith is protected by a mantle of vegetation. The leaves and branches of trees break the force of falling raindrops and cushion the impacts upon the ground. More important, the intricate network of roots, especially grassroots, forms a tight mesh that holds soil in place, greatly reducing erosion compared with the erosion of bare ground. The mesh of roots is rarely broken, but if it should break it would be healed by new growth. Not only is the mesh a retainer of soil, but it also is a holder of water. Like a sponge, it absorbs rainwater, and retains it for a time, letting it percolate slowly down through the soil beneath, so that less water runs off over the surface.

Here we are dealing with natural vegetation. Where crops are

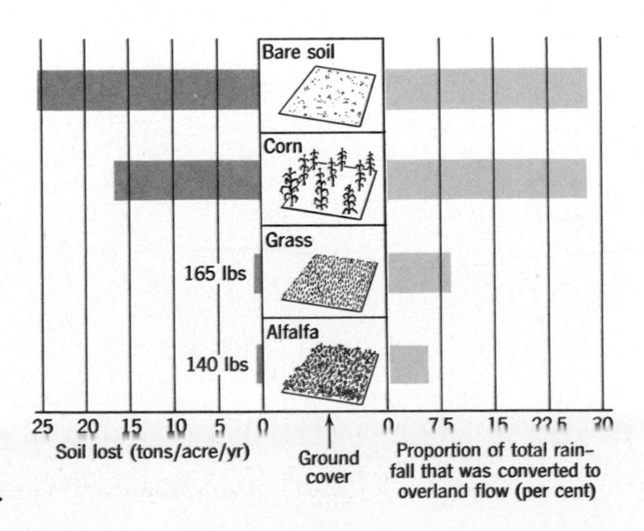


Figure 7.2
**Effect of plant cover on rate of sheet erosion, measured over 4 years at Bethany, Missouri, a station like that shown in Figure 7.1. Soil is silty, slope is 8 per cent, and annual rainfall is 1,000mm.**
**The measurements show that grass and alfalfa, with their continuous network of roots and stems, are nearly 300 times as effective as "row crops," such as corn, in holding soil in place. Erosional loss from bare soil shown here is at a rate of about 45cm/100 years.**

grown, during part of each year the surface is ordinarily bare. On bare, sloping fields, pastures that are too closely grazed, and areas planted with widely spaced crops such as corn, rates of erosion can be great. Because sheet erosion creates no obvious valleys, the damage it does to bare soil was not fully realized until accurate measurements began to be made. Now many experiment stations (Fig. 7.1) maintained by government agencies measure the erosion of soil (Fig. 7.2). The measurements have practical value not only to farmers, but also to everyone else with a stake in the economy because they show sheet erosion is a menace to soil left unprotected on slopes. In recognition of this fact, wise farmers reduce areas of bare soil to a minimum and prevent the grass cover on pastures from being weakened by overgrazing. When the protective plant cover is weakened, runoff and erosion are increased (Fig. 7.3). If crops such as corn, tobacco, and cotton must be planted on a slope, strips of such crops are often alternated with strips of grass or similar plants that resist sheet erosion.

Erosion of slopes, however, does not result solely from human activities. Under some natural conditions, without any agriculture at all, the splash of raindrops and the work of sheet erosion are so effective that they combine to remove large volumes of fine rock particles. For example, in some subtropical grasslands all the rainfall is concentrated within a single rainy season. During the long dry season, evaporation so depletes soil moisture that grass becomes sparse, covering no more than 40 to 60 percent of each square meter of ground. Although kept bare by natural causes, the soil is as vulnerable to erosion as soil laid bare by farming.

**Figure 7.3**
**Too many cattle were grazed on one of these two pastures in northern Kentucky, weakening the grass cover. Erosion did the rest. (*U.S. Soil Conservation Service.*)**

Although common on slopes, overland flow takes place only through short distances. The flowing water soon concentrates into channels and forms streams, to which we must now return in order to try to analyze their behavior.

# FACTORS IN STREAM FLOW

A ***stream*** is *a body of water that carries rock particles and dissolved substances, and flows down a slope along a definite path.* The path is the stream's channel, and the rock particles are an essential part of the stream itself. Five basic factors of a stream are:

1. ***Discharge*** (*The quantity of water that passes a given point in a unit of time*). Discharge is usually expressed in cubic feet per second (cfs) or in cubic meters per second ($m^3$/sec).
2. *Average velocity.*
3. *Size and shape of channel.*
4. ***Gradient*** (*the slope measured along a stream, on the water surface, or on the bottom*).
5. ***Load*** (of a stream) (*the material the stream carries*). The load consists of rock particles plus matter in solution. Unlike rock particles that constitute the mechanical load, dissolved matter generally makes little difference to the behavior of the stream.

# THE STREAM'S LOAD

### Turbulent flow

The particles of water in a stream do not move along straight or parallel paths. Their paths form bends and whirls (eddies, Fig. 7.4) so that the particles move in all directions and at different speeds. This kind of motion is *turbulent flow.* Turbulence in fast streams is much greater than in slow ones. Also it is greatest near the sides and bottom of the stream channel, where the flowing water drags against the solid banks and stream bed.

The way in which a stream moves its load of rock particles is easy

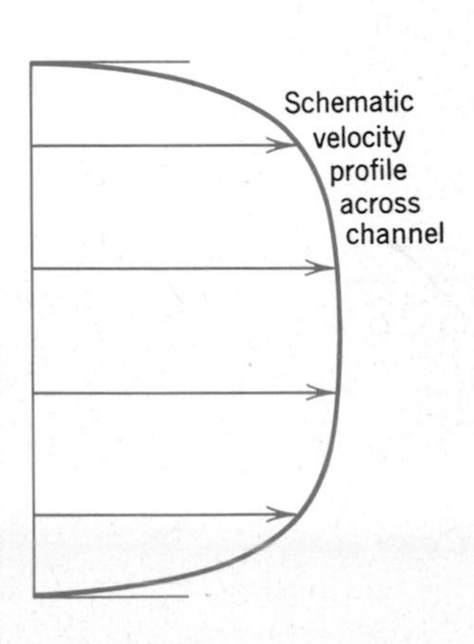


**Figure 7.4**
**Turbulent flow in an open channel, seen looking down onto surface of stream. Because of drag, turbulence at the surface is greatest near the channel sides. (*From a photograph in Prandtl and Tietjens, Applied Hydro- and Aeromechanics; courtesy Engineering Societies Monographs Committee.*)**

to observe in an artificial channel with glass sides, an apparatus found in many laboratories. As Figure 7.5 shows, the stream's load of solid particles consists ideally of *coarse particles that move along or close to the stream bed* (the **bed load**) and *fine particles suspended in the stream* (the ***suspended load***). In addition to these solid particles there is also *matter dissolved in stream water* (the ***dissolved load***), chiefly a product of chemical weathering.

### Bed load

As we watch water starting to flow through the channel, at first only a few particles move, by sliding or rolling. Pebbles start to move sooner than sand grains because they project higher into the current and thus into a zone of faster flow.

Now we increase the velocity a little. More particles on the bed start to move and to collide with one another. Some are propelled upward by the collisions and then are pushed forward by the current. Some are sucked upward by the faster current that flows above them, as the wing of an airplane is lifted by the flow of air above it. Gravity pulls the particles down to the bed again, but the result for each particle is a jump forward, downstream.

As we step up velocity just a bit more, particle movement becomes general. Seen from above, the particles travel parallel to the current.

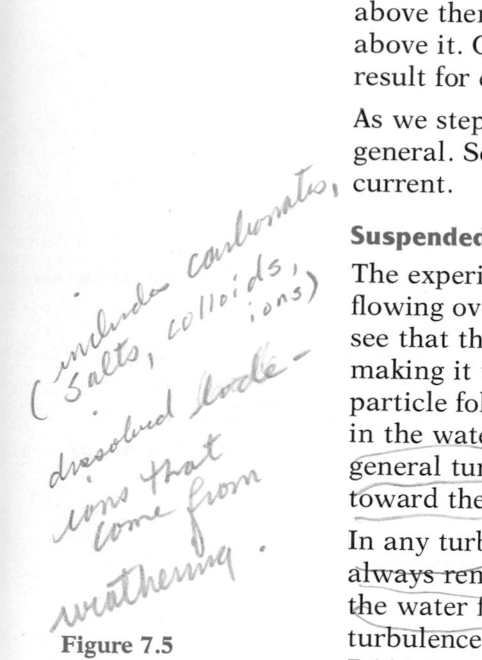


### Suspended load

The experiment we have been watching involved clear water flowing over sand. Now if we add silt and clay to the water we can see that these fine particles spread quickly throughout the water, making it turbid. If we look very closely we can see that each particle follows a very irregular path. The particles are *suspended* in the water, because upward-moving threads of current within the general turbulence exceed the speed at which each particle settles toward the bed under the pull of gravity.

In any turbulent stream, then, particles of silt size and clay size always remain in suspension; they move continuously, as fast as the water flows. Such particles settle and are deposited only where turbulence ceases; for example, on the floodplain shown in Figure 7.11*A*, in a lake, or in the sea. Because of these relationships, the transport of particles in the suspended load differs greatly from that of bed-load particles, which tend to move only intermittently.

Most of the suspended load in streams is derived from two sources:

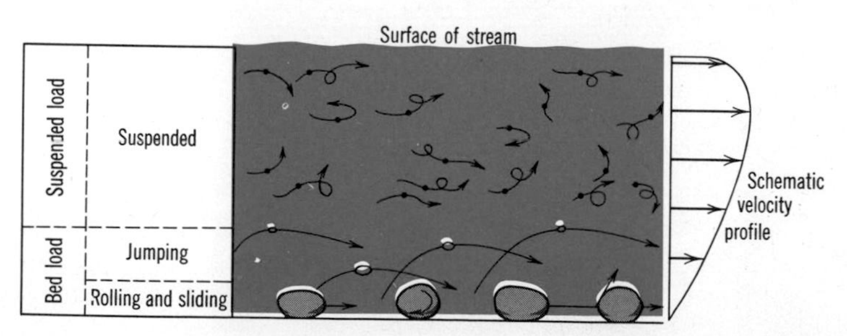


**Figure 7.5**

**Kinds of movement of rock particles carried in a stream; vertical distribution of bed load and suspended load. The stream flows from left to right.**

**The velocity profile (right) shows that the stream flows fastest (longest arrow) near the water surface and slowest at the bottom.**

(1) fine-grained regolith (mostly soil) washed from plowed fields and other areas unprotected by vegetation; and (2) sediment eroded (by washing and by slump) by the stream itself from its own banks.

# ECONOMY OF A STREAM

The channels of most natural streams consist partly or wholly of sediment. Streams move this material from place to place, and so their channels are continually being altered. Because stream and channel are closely related and are ever changing, we have to examine them together as an interrelated system. We can think of the ***economy*** of the stream and its channel as *the input and consumption of energy within a stream and the changes that result.*

The average annual rainfall on the area of the United States is equivalent to a layer of water 76cm thick. Of this layer, 45cm returns to the atmosphere by evaporation and transpiration (Fig. 4.2), 1cm infiltrates the ground, recharging the ground water (Chapter 8) and the remaining 30cm forms ***runoff,*** defined as *the water that flows over the lands.*

The average altitude of the United States is about 750m. Our 30-centimeter layer of runoff falls through that vertical distance as gravity pulls it down to sea level. The potential energy of this water, equal to the weight of the water times the height of the land, is converted into the kinetic energy of flow of all U.S. streams combined.

Part of that energy is spent in the geologic work of picking up and carrying rock particles. This kinetic energy makes streams the prime movers of rock waste from lands to ocean. This is why streams occupy a key position in the rock cycle, for they are among the chief agencies by which the sedimentary rock of the Earth's crust has accumulated.

Some of the sediment carried by streams is deposited on the land, at least temporarily. ***Alluvium,*** the general name for *sediment deposited by streams in land environments,* exists in great quantity, and still larger quantities of it have been converted into sedimentary rock. Much of the sediment, however, is carried into the sea and deposited there. Once it reaches the sea, it is no longer alluvium but becomes marine sediment.

### Factors in the economy

The economy of the system depends on a continual interplay among the five basic factors mentioned earlier: discharge, velocity, size and shape of channel, gradient, and load. The first factor is calculated from measurements made systematically at selected points along large and small streams for purposes of flood control, irrigation, water supply, and the like. The U.S. Geological Survey, for example, maintains nearly 6,500 measurement points, called gaging stations, in various parts of the United States.

The measurements show that as discharge changes, velocity and channel shape must change also. The relationship is defined and can be expressed by the formula:*

$$\underset{\substack{\text{Discharge} \\ \text{(cubic feet per second)}}}{Q} = \underset{\substack{\text{width} \\ \text{(feet)}}}{w} \times \underset{\substack{\text{depth} \\ \text{(feet)}}}{d} \times \underset{\substack{\text{velocity} \\ \text{(feet per second)}}}{v}$$

When discharge changes, as it does continually, the product of the other three terms must change accordingly.

With increased discharge, the stream erodes and enlarges its channel, instantaneously if it flows on alluvium, much more slowly if it flows on bedrock. The increased load is carried away. This continues until the increased discharge can be accommodated. In contrast, when discharge decreases, some of the load is dropped, making the channel less deep and less wide, and the velocity is reduced by increased friction. In this way width, depth, and velocity are continually readjusted to changing discharge.

Thus, a stream and its channel are related so intimately that we can think of them as a single system. The channel is so responsive to changes in discharge that the system, at any point along the stream, is continually close to a steady state.

### Floods

Seasonal distribution of rainfall causes many streams to rise in flood seasonally. As discharge increases during a flood, so does velocity. This has the double effect of enabling a stream to carry not only a greater load, but also larger particles. An extreme example resulted from collapse, in 1928, of the large St. Francis Dam in southern California. As the dam gave way, the water behind it thundered down the valley, moving as bed-load blocks of concrete weighing as much as 9,000 metric tons through distances of more than 750m. In most streams more geologic work is accomplished during regular seasonal floods than during intervals of low water. In addition to seasonal floods, rare, exceptional floods (Fig. 7.6), outside the stream's normal economy, occur perhaps only once in several decades or centuries. In such floods the geologic work done may be prodigious, but they happen so rarely that their long-term effects probably are less than the aggregate effect of normal activity throughout the very long periods intervening between them. As in the fable, this race, too, is won by the tortoise rather than the hare.

### Changes along a stream

If we traveled down a river from its head to its mouth, we would see that orderly adjustments occur along it. Among them are these: (1) discharge increases; (2) width and depth of channel increase; (3) velocity increases slightly; and (4) gradient decreases.

As tributary streams contribute additional water, discharge in the

* In U.S. Government practice these values are not yet quoted in terms of the metric system.

main stream increases in most cases, thus increasing velocity (Fig. 7.7). The demonstration that velocity increases downstream seems to contradict the common observation that water rushes down steep mountain slopes and flows smoothly over nearly flat lowlands. But the physical appearance of the water is not a true measure of its velocity, which increases downstream mainly because channels become deeper and wider in that direction.

**Long profile**

The gradient or slope of a mountain stream may be 60m per kilometer or even more, whereas that of the downstream part of a large river may be 10cm per kilometer or even less. Gradients of

Figure 7.6

**Flood of Quinebaug River at Putnam, Connecticut in August 1955, the result of extraordinary rainfall. View looking north, upstream.**

**River channel is at left, just out of view; its edge is seen as darker water in lower-left corner of picture. Stream shown is water that overflowed from main channel and is rejoining it at lower left. Although short-lived, the overflow eroded a large body of sand and gravel about 8m thick (remnants are near flooded railroad tracks) and deposited the resulting load as a series of bars and islands seen in foreground. (*Providence, R.I., Journal-Bulletin.*)**

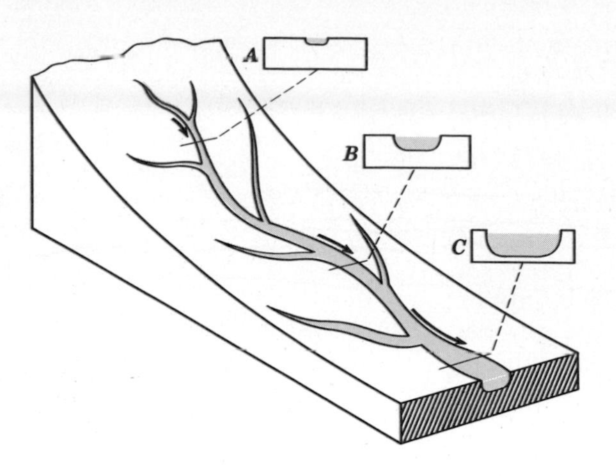


Figure 7.7
**Changes in the down-stream direction along a stream system.**
**Discharge is increased by entrance of successive tributaries. Width and depth of channel are shown by cross sections *A*, *B*, and *C*. Velocity is shown by relative lengths of three arrows. (*After Leopold and Maddock, 1953.*)**

most rivers, decrease downstream, and so the ***long profile*** of a stream (*a line connecting points on the stream surface*) is generally concave-upward (Fig. 7.20).

## Base level

The vertical position of the mouth of a stream is determined by its ***base level.*** This is *the limiting level below which a stream cannot erode the land* (Fig. 7.8). The ***ultimate base level,*** for streams in general, is *sea level* projected inland as an imaginary surface beneath all streams.* When a stream cuts down to that surface, its energy quickly approaches zero. For a stream ending in a lake, base level is the level of the lake (Fig. 7.8), for the stream cannot erode below it. But, if the lake were destroyed by erosion at its outlet, the base level represented by the lake surface would disappear, and the stream, having acquired additional potential energy, would deepen its channel. *The levels of lakes and all other base levels that stand above sea level* are ***local base levels.*** A common kind of local base level is the level of a belt of particularly hard rock lying across the stream's path. Even sea level itself changes slowly over long periods and this too affects the long profiles of streams.

## Fans

A common example of the effect of a local base level is the building of a fan. When a stream flows through a steep highland valley and comes out suddenly onto a nearly level valley floor or plain, it encounters an abrupt decrease of slope. It deposits that part of its load which cannot be carried on the gentler slope. The material it deposits takes the form of a ***fan,*** defined as *a fan-shaped body of alluvium built at the base of a steep slope* (Figs. 7.9, 9.5, and 9.9). The surface of the fan slopes outward in a wide arc from an apex at the mouth of the steep valley. The profile of the fan, from apex

* Although true in principle, this statement is not quite true in detail. A stream confined between channel banks can erode its bed slightly below sea level, as evident in Figure 7.12, where depth of base of the topset beds represents depth of stream-channel erosion.

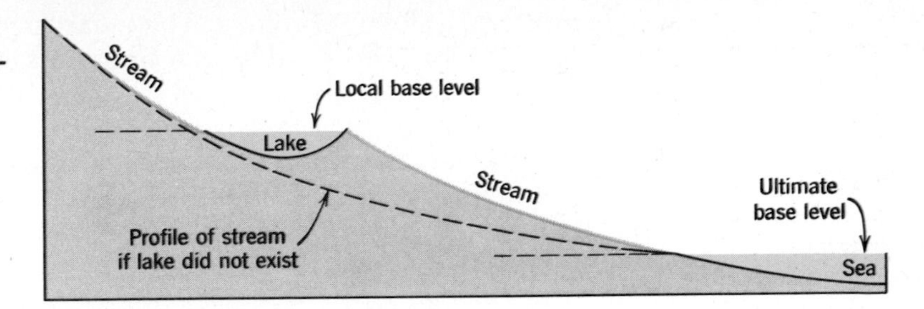


**Figure 7.8**
**Relation between ultimate base level (the sea) and local base levels such as a lake.**

to base in any direction, has the concave-up form characteristic of stream profiles. The exact form of the profile depends chiefly on discharge and on the diameters of particles in the bed load; hence no two fans are exactly alike. A small stream carrying a load of coarse particles builds a shorter, steeper fan than a larger stream carrying a load of finer particles.

Although a fan is originally localized by decrease of slope, as soon as its long profile has become smooth, the chief cause of further deposition on it is the spreading of water through a network of channels, with consequent loss of discharge and velocity in each channel and overall loss of water that percolates down into the underlying sediments.

Unless special circumstances preserve it, the fan will be destroyed by continuing erosion downward below the profile *cc′x* (Fig. 7.9). A fan, therefore, is likely to be a temporary deposit.

### Meanders

No stream is straight through more than a short distance. The pattern of most streams is a series of bends, and very commonly the bends are smooth, looplike, and similar in size. Such bends are termed ***meanders*** (from a Greek work meaning "a bend"), defined as *looplike bends of a stream channel* (Fig. 7.10). The meanders of the Mississippi and other large streams are under continual study by engineers and geologists concerned with problems such as floods and navigability. These people know that meanders are not accidental, that they occur most commonly in channels having gentle gradients in fine-grained alluvium, and that they occur even in streams having no load at all. Meanders represent the form by means of which, in making a turn, the river experiences least resistance to flow, does the least work, or dissipates energy most

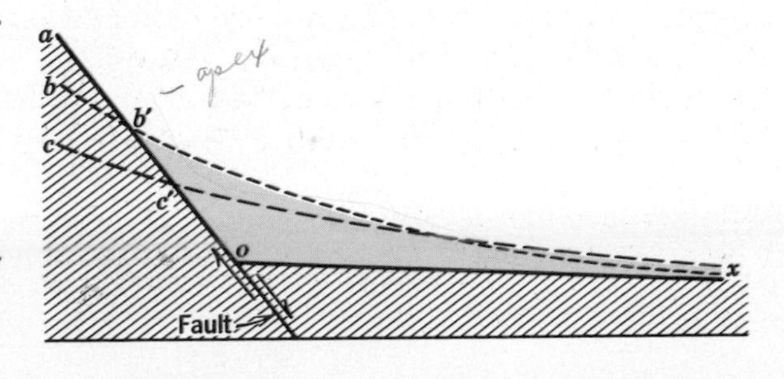


**Figure 7.9**
**Vertical section showing growth of a fan.**
**Bedrock is shaded; alluvium is blue. Line *aox*, profile of surface before deposition of fan. Line *bb′x*, long profile of stream at an early stage of fan building, with apex of fan at *b′*. Line *cc′x*, long profile at a later stage, after stream has cut away apex of fan *bb′x*, increasing fan radius while establishing a continuous, concave-up profile.**

nearly uniformly along its course. The meander form is therefore one of stability. Although stable as a form, a meander changes its position almost continually, as shown by year-to-year measurements of river channels, and by artificial streams. The shift or migration of a meander is accomplished by predominant deposition along one bank and predominant erosion of the opposite bank. Along the inner side of each meander loop, the place where speed of flow is lowest, bed load is deposited and accumulates as a distinctive *bar* (Fig. 7.10).

Meanders grow and bars enlarge rapidly, as can be shown easily with artificial laboratory streams. Also, because the valley floor as a whole slopes downward toward the mouth of the stream, slump (Fig. 6.11) is a little more rapid on concave banks that face upvalley than on other banks. As a result, meanders tend to migrate slowly down the valley, subtracting from and adding to various pieces of real estate along the banks, according to location, and causing legal disputes over property lines and even over the boundaries between counties and states.

The behavior of streams in laboratory channels shows that if the

**Figure 7.10 Meanders of Animas River above Durango, Colorado. View looking downstream. Both bars and oxbow lakes are conspicuous. (*J. S. Shelton.*)**

bank material is uniform, the meanders are symmetrical and migrate downvalley at the same rate. But, because the material of a bank is usually not uniform, the migration of the downstream limb of a meander can be slowed where it encounters resistant material. Meanwhile the upstream limb, migrating more rapidly, intersects and cuts into the "slow" limb. Thus the channel bypasses the loop between the two limbs and the cut-off loop is converted into a lake (*oxbow* lake, Fig. 7.10).

The aggregate length of Mississippi River channel abandoned through cutoffs since 1776 amounts to nearly 600km. But the river has not been shortened appreciably because the segments lost through cutoffs have been balanced by lengthening caused by enlargement of other meanders.

### Floodplain: natural levees

Every spring the Mississippi experiences a flood. Before the river began to be restrained by flood-control structures, it frequently overtopped its banks and inundated the lower parts of the valley floor. *That part of any stream valley which is inundated during floods is a **floodplain;*** the area of the natural floodplain of the Mississippi from Cairo to the delta is 78,000km$^2$, but more than half of that area is now protected against floods by dikes and other structures.

The channels of the lower Mississippi and many other meandering streams are bordered by ***natural levees**—broad, low ridges of fine alluvium built along both sides of a stream channel by water that spreads out of the channel during floods* (Fig. 7.11). Along the lower Mississippi, natural levees are 4m to 7m high. During the exceptional flood in late April 1973, the Mississippi River at St. Louis rose 13m above its low-water height, the highest flood since French fur traders began measurements in 1764. The water stood above the tops of levees and inundated low parts of the city.

The fine alluvium of which levees are chiefly built becomes still finer away from the river and grades into a thin cover of silt and clay over the rest of the floodplain. Natural levees were built, and are continually added to, only during floods so high that the

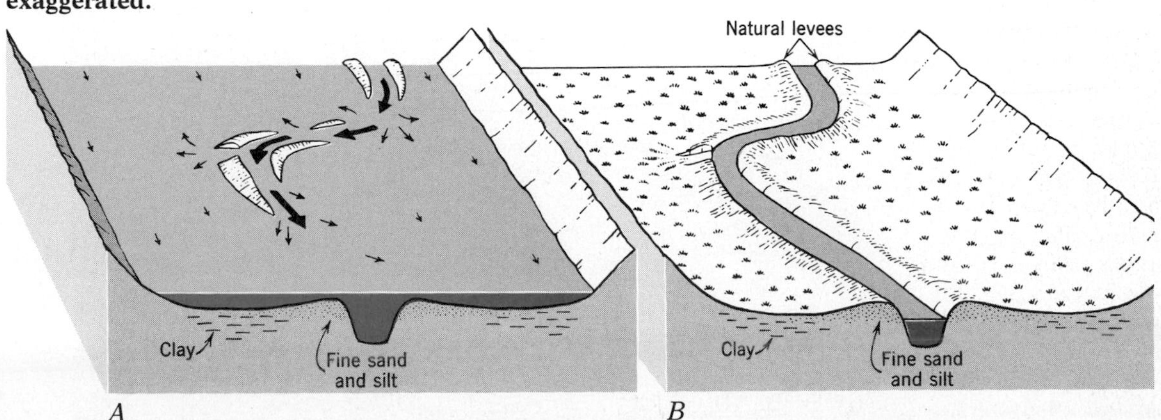


**Figure 7.11**
**Floodplain with natural levees.**
***A.* During a big seasonal flood much of the valley floor becomes a lake. Water remaining in the channel flows at high speed (large arrows). Water escaping from channel flows with diminishing velocity (small arrows) into adjacent broad, shallow areas. It deposits silt to form natural levees where it leaves the channel and blankets lower land with clay. Highest parts of levees, added to only during still higher floods, form islands.**
***B.* At times of low water, levees stand as low ridges along sides of channel; beyond them are swamp lands. Vertical scale exaggerated.**

floodplain is converted essentially into a lake deep enough to submerge the levees. In the water that flows laterally from the submerged channel over the submerged floodplain, depth, velocity, and turbulence decrease abruptly at the channel margins. The decrease results in sudden, rapid deposition of the coarser part of the suspended load (usually fine sand and silt) along the sides of the channel. Farther away from the channel, finer silt and clay settle out in the quiet water. In the vicinity of Kansas City, Missouri, during an exceptional flood in 1952, Missouri River water deposited a layer of silt as much as 15cm thick over wide areas of the floodplain. In some places fences and other obstacles caused silt and fine sand to accumulate to thicknesses as great as 1.5m. Under ordinary conditions flood-deposited silt is beneficial to agricultural lands, because it contains organic matter, washed from soils on the watershed, which acts as fertilizer.

### Flood control

The flood deposits of the Mississippi River, including the natural levees, have been built up during a long period, but the process has been interfered with by engineering works (mainly dikes built of earth and concrete) designed to prevent the river from generally inundating its floodplain. As described in Chapter 23, the natural levees have been heightened artificially by earth dikes to hold in ordinary floods, and at selected points spillways have been built to allow the water of the highest floods to escape harmlessly into natural channels that parallel the channel of the Mississippi.

### Deltas

As the water of the stream diffuses into the standing water of sea or lake, its speed is checked by friction, it loses energy, and deposits its load as a ***delta*** (Fig. 5.2), *a body of sediment deposited by a stream where it flows into standing water.* Although deltas are of several kinds, the type easiest to recognize and probably most common is shown in Figure 7.12. Although in form it somewhat resembles a fan, it differs from a fan because of two factors: (1) stream flow is checked by standing water, and (2) the level surface of sea or lake sets an approximate limit to upbuilding of the deposit, the top of which is flatter than the profile of a fan.

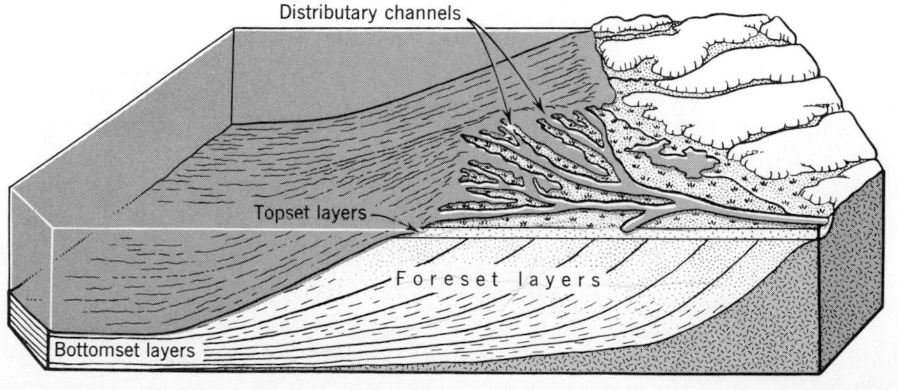


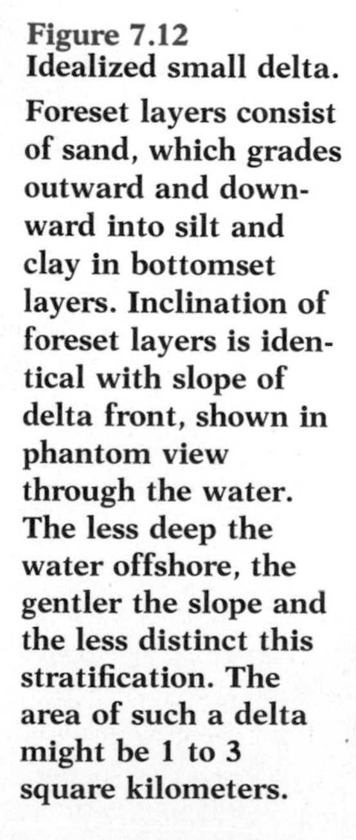

**Figure 7.12**
**Idealized small delta.**
**Foreset layers consist of sand, which grades outward and downward into silt and clay in bottomset layers. Inclination of foreset layers is identical with slope of delta front, shown in phantom view through the water. The less deep the water offshore, the gentler the slope and the less distinct this stratification. The area of such a delta might be 1 to 3 square kilometers.**

The particles in the bed load are deposited first, in order of decreasing weight; beyond this, the suspended sediments drop out. A layer deposited at any one time (as during a single flood) is sorted, grading from coarse at the stream mouth to fine offshore. The deposition of many successive layers creates an embankment that grows outward like a highway fill made by dumping. *The coarse, thick, steeply sloping part of each layer in a delta* is a ***foreset layer.*** Traced seaward, the same layer becomes rapidly thinner and finer, covering the bottom over a wide area. This *gently sloping, fine, thin part of each layer in a delta* is a ***bottomset layer.***

As successive layers are deposited, the coarse foreset layers, one by one, overlap the bottomset layers, producing the arrangement seen in Figure 7.12. The stream gradually extends seaward over the growing delta, erodes the tops of the foreset layers during floods, and at other times deposits part of its bed load in its channel and its suspended load in areas between channels during floods. The channel deposits and interchannel deposits form the ***topset layers*** of the delta. We define these deposits as *the layers of stream sediment that overlie the foreset layers in a delta.*

During floods the stream spills out of its channel and forms distributary channels, through which the water enters the sea independently, multiplying the topset deposits. The distributaries act in a way exactly opposite to that of the tributaries to a stream, illustrated in Figure 7.25*A*. Radiating distributary channels give the delta a crudely triangular shape like the Greek capital letter Δ, from which the deposit derives its name.

It may seem surprising that the suspended load, much of which has been carried hundreds of kilometers through the channel of a large river without being deposited, should drop out so abruptly to form part of a localized delta instead of remaining in suspension long enough to be carried far from land. But the salts dissolved in seawater act to coagulate, or flocculate, the suspended fine particles into aggregates so large that they settle to the bottom promptly.

Some of the world's greatest rivers, among them the Nile, the Hwang Ho, the Amazon, the Rhine, the Mackenzie, and the Colorado, have built massive deltas at their mouths. Each delta has its own peculiarities, and none is so simple as the small delta shown in Figure 7.12. The Mississippi delta, with an area of 31,000 square kilometers (not counting the submarine part), is really a complex of several coalescing subdeltas built successively during the last several thousand years (Fig. 7.13). Each subdelta was begun by a flood that created a new distributary.

### Artificial dams

The practice of building artificial dams across rivers, thus creating a reservoir upstream from each dam, has been widespread in many countries. In the United States, examples of whole series of dams along a single river and its tributaries include the Tennessee River system, the Missouri River system, and the system along the lower

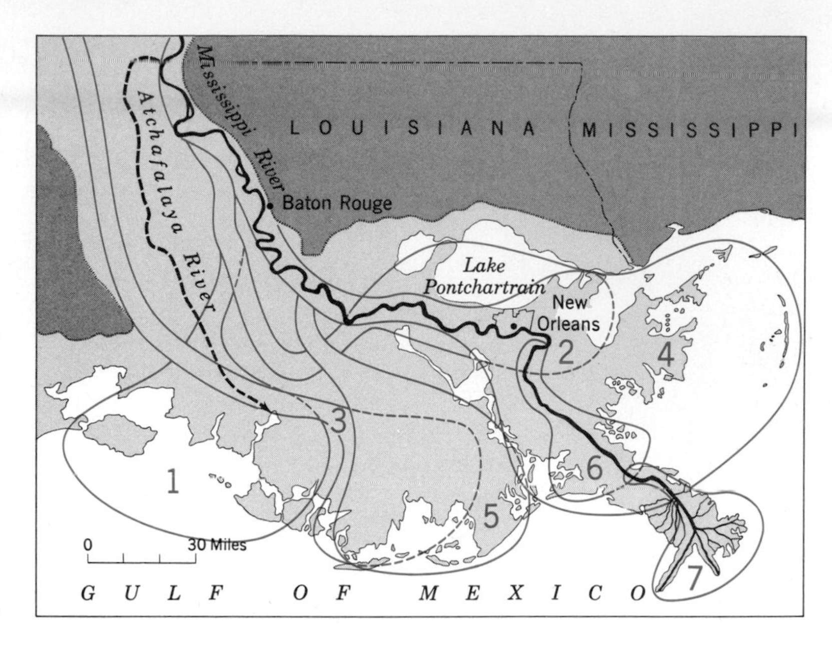


**Figure 7.13 Mississippi River delta.**
**The Mississippi River has built a series of overlapping subdeltas, while occupying successive distributary channels (numbered 1 to 7). Age of subdelta 1 is estimated at 3,000 years, of 3, at 1,500 years, and of 5, at 1,000 years.**
**Construction of subdelta 7 began more than 100 years ago. Discharge of the Mississippi has been gradually shifting to the Atchafalaya distributary. By 1958, 28 percent of the discharge was following this new route. Construction of a barrier to stop the diversion was begun in 1955. Had this not been done, by 1975, the percentage would have increased to 40. (*After Kolb and Van Lopik, 1958.*)**

Colorado. In the first two, the dams make possible the generation of hydroelectric power, and also reduce seasonal floods by allowing lowered reservoirs to fill during floods. The Missouri River system also provides water for irrigation in the comparatively dry climate of the Great Plains.

The Colorado River system generates power, provides irrigation water, and through aqueducts more than 400km long, furnishes much of the municipal water supply of Los Angeles and San Diego—that is, for nearly 10 million people. Figure 7.14 shows the essentials of the system of more than six dams and reservoirs. Parker Dam provides the municipal water supplies; the dams downstream from it (including the Morelos Dam in Mexico) furnish the irrigation water, while at the three dams upstream, Colorado River water generates electric power. Thus the water is used twice; first to fall steeply and generate power and then, via distribution systems, to supply cities and irrigate crops. So fully is it used that downstream from the Morelos Dam the river is sometimes hardly more than a trickle. Of course the reservoirs, which are artificial lakes, are traps for nearly all the sediment that the river formerly carried uninterruptedly to the Pacific Ocean. The accumulating sediment will eventually fill the reservoirs, making them useless, but the filling process is slow, so that complete filling is far in the future. Likewise there is now increased loss of water through evaporation, because evaporation is promoted by the large surfaces of the reservoirs, exposed to dry air.

Thus far, the Colorado River dams, although they have interfered with the natural steady-state condition of the river, appear to have done more good than harm to the environments through which the river flows.

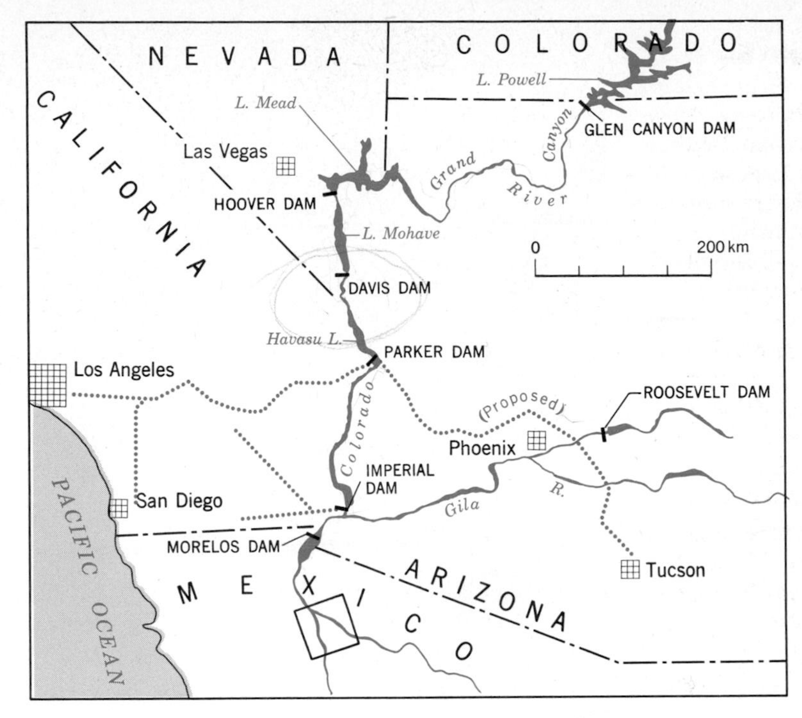


**Figure 7.14**
**Dams and reservoirs in the lower Colorado River region, with existing and proposed aqueducts and canals (dotted lines). The dams were built between 1938 and 1964. Square near base of map is area of the photograph, Figure 5.2.**

## Features of steep, rocky channels

Hitherto we have been discussing streams whose beds consist of sediment. Now let us turn to the erosion performed by turbulent mountain streams whose channels are steep and consist of bedrock. In streams, ***abrasion**, the mechanical wear of rock on rock,* is caused by friction, collisions between rock particles in the load and bedrock in the channel. A stream uses its mechanical load as tools. By rubbing, scraping, bumping, and crushing, it erodes bedrock and at the same time smooths and rounds the tools. Even where no bedrock is exposed in the channel, rounding of sand and coarser particles goes on, although rounding decreases rapidly as particle diameter decreases.

Under the special conditions in which a stream flows over a rock ledge or cliff, as at Niagara Falls (Fig. 7.15), the increased velocity of the falling water sets up strong turbulence at the base of the falls, and the stream bed is deepened. The cliff is gradually undermined and the falls retreats upstream. The retreat of the Canadian Falls at Niagara Falls has been rapid. Measurements by survey show that between 1850 and 1950 the rate of retreat averaged about 1.2m per year. This rapid rate is favored by the fact that the lip of the falls consists of strong, resistant dolostone beneath which is weak, easily eroded shale. As the shale cliff is eroded back, the dolostone lip is undermined and caves in piecemeal (Fig. 7.16). The long, deep gorge downstream from the falls was created by retreat of the falls, through successive positions, during periods of many thousands of years.

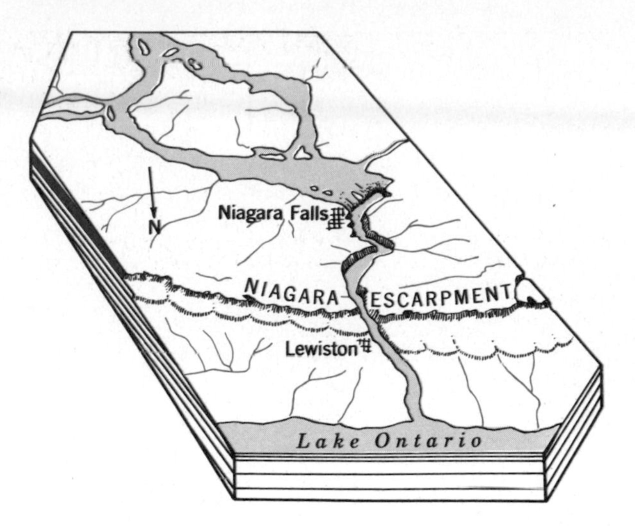


Figure 7.15
**Bird's-eye view of Niagara Falls, looking south, showing south-dipping strata and Niagara Escarpment, a great ledge formed by a resistant stratum, from which the Falls originated. By erosion the Falls has retreated through an aggregate distance of about 100km. Greatest length of block, 55km. Vertical exaggeration 2x.**

Whether a stream flows over a wide, gently sloping floor or in a steep, rocky channel, it is normally in a condition of steady state. We shall now see how this strong tendency toward the steady state in streams is reflected in the sculpture of the land.

# DRAINAGE SYSTEMS

### Relation of valleys to streams

Weathering, mass-wasting, sheet erosion, and stream activities not only create, transport, and deposit sediment; they also shape the land into systems of valleys. Groups or families of valleys are the commonest features of the land. Throughout wide regions the surface consists of little more than a complex of valleys created by erosion, and separated by higher areas that erosion has not yet consumed. Valleys exist in such great numbers that they have never been counted except in sample areas. The enormous number of valleys is commensurate with the huge volume of water that runs off over the land. Before the middle of the 18th Century, valleys were generally thought to be the result of catastrophes that somehow broke the Earth's crust and pulled it apart, creating

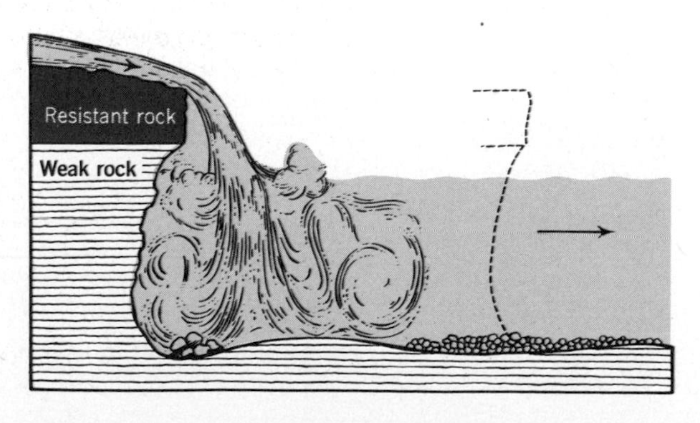


Figure 7.16
**A falls similar to Niagara Falls.**
**The stream bed downstream from the falls was created by successive positions of the retreating cliff; one former position is shown by the dotted line. Turbulence at base of falls keeps bedrock scoured nearly clean of sediment, but in less turbulent water downstream coarse gravel is deposited.**

paths for running water to follow. Today we know that with few exceptions, streams themselves make their own valleys, and, as we shall see in a moment, they can start these valleys very quickly (Fig. 7.17).

### Drainage basins and divides

Every stream or segment of a stream has its ***drainage basin,*** consisting of *the total area that contributes water to the stream. The line that separates adjacent drainage basins* is a ***divide*** (Fig. 7.25). On a map (Front end paper) we can trace the divide that encloses the huge drainage basin of the Mississippi River, an area that exceeds 40 percent of that of the conterminous United States.

Spacing of the streams in a drainage basin is orderly. When the streams in a large basin are measured accurately on a map, it appears that the distances between the mouths of tributaries are spaced in an orderly way, and that there is a mathematical relationship between the length of a stream and the area of its drainage basin. This orderliness measured on maps is analogous to the orderliness inherent in a stream's long profile, in which gradient decreases systematically from head to mouth, while discharge, velocity, and channel dimensions increase. All these relationships imply that in response to a given quantity of runoff, stream systems develop with just the size and spacing required to move the water off each part of the land with maximum efficiency.

Surface slopes converge toward the heads of small valleys, and runoff, moving at first as overland flow, soon concentrates in valleys (Fig. 7.18), forming streams. A system of streams does not necessarily require much time to develop, as indicated by the following example. In August 1959, an earthquake occurred at Hebgen Lake, near West Yellowstone, Montana. The movement

Figure 7.17
**This gully in northern Missouri was started in the 19th Century by uncontrolled runoff over farmland. After 1940, it was reclaimed by building an earth dam, which created a pond that trapped sediment and promoted vegetation. (*U.S. Soil Conservation Service.*)**

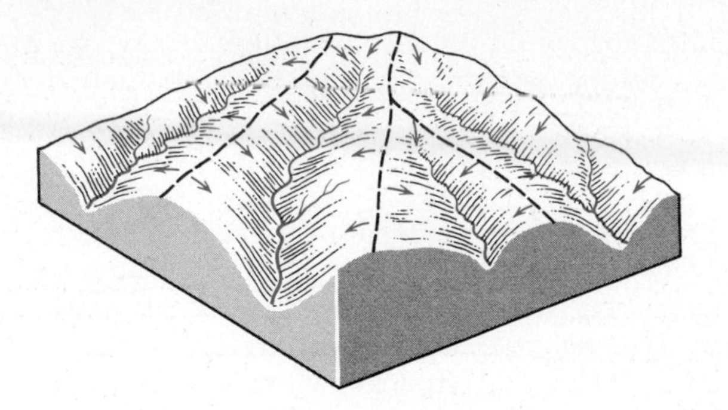

**Figure 7.18**
**Runoff moves over hillslopes as overland flow (suggested by arrows) and soon concentrates in small valleys. Broken lines indicate divides.**

tilted the country in such a way that a large area of silt and sand, formerly part of the lake bed, emerged and was subject to runoff. Small-size drainage systems began to develop immediately. Sample areas were surveyed and mapped one year and two years after the earthquake occurred. The results showed the same basic geometry that characterizes much larger and older systems. The small, newly formed valleys, together with the areas between them, were disposing of the available runoff in a highly systematic way, all within a period of two years after the surface had emerged from beneath the lake.

### Valleys, sheet erosion, and mass-wasting

If the sole agency involved in cutting a valley were the stream that flows through it, the valley should be as narrow and steep-sided as the one shown in Figure 7.19, resembling a cut made by a saw through a block of wood. The shaping of most of the land surface, including the valley sides themselves, is mainly the work of sheet erosion and mass-wasting of weathered rock material.

## SCULPTURAL EVOLUTION OF THE LAND

### Cycle of erosion

As long as rain falls upon it, any region that is above sea level is *sculptured* into a series of valleys, with hills between. In the

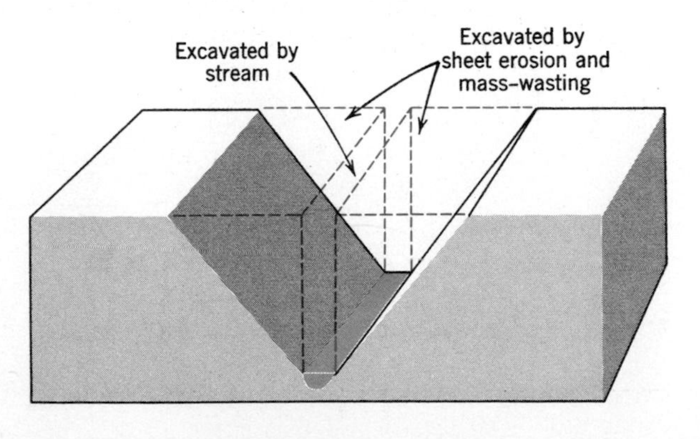


**Figure 7.19**
**In this idealized valley segment, the volume of rock excavated by the stream is compared with the much greater volume excavated by sheet erosion and mass-wasting. However, all the waste from the slopes, while being transported out of the area, had to pass through the stream.**

process, not only streams but mass-wasting and weathering all play their parts. As sculpture progresses, the land surface undergoes a series of gradual changes somewhat like those which affect an individual organism of any kind, from birth to death, including stages of youth, maturity, and old age. In the stage of youth, stream gradients tend to be steep (Fig. 7.20*A,a*) and erosion rapid. Valleys are actively deepened and make sharp cuts into the land. But they have not reached their full lengths, so that broad areas of the original surface remain uncut.

In the stage of maturity, valleys have lengthened so that the entire surface is cut up, but stream gradients have become gentler and streams are eroding more slowly. Slopes have developed smoothly curved profiles.

The mature surface is lowered toward base level very slowly, with the rate of erosion becoming increasingly slow as stream gradients and hillslopes become ever move gentle. A surface in a late phase of erosion, lying close to base level, is in the stage of old age. A land surface in a very late phase of old age is a ***peneplain*** ("almost a plain"), defined as *a land surface worn down to very low relief by streams and mass-wasting.* With ever-decreasing energy, erosion of a mature surface becomes so slow that many millions of years are necessary for the creation of an extensive peneplain. Despite its generally low relief, high steep hills can form part of it wherever the rocks are especially resistant to erosion.

We know of few if any peneplains that exist today in an undisturbed state. Numerous peneplains have been submerged beneath the sea, deeply buried by thick layers of sediment and later partly exposed by erosion of their covers of sedimentary rock so that we can see small parts of them. Others have been uplifted, and trenched by streams to form valleys and hills so that, again, only small parts remain. An example is shown in Figure 7.21. No peneplain, therefore, is seen today lying unaltered in the position in which it was made. This fact implies considerable instability of the Earth's crust during at least the last few million years, for if the crust had remained quiet, sculptural evolution of the lands should be more advanced than it is.

## Rates of erosion

How fast is the process of sculptural evolution? It is calculated that the surface of the conterminous United States is being

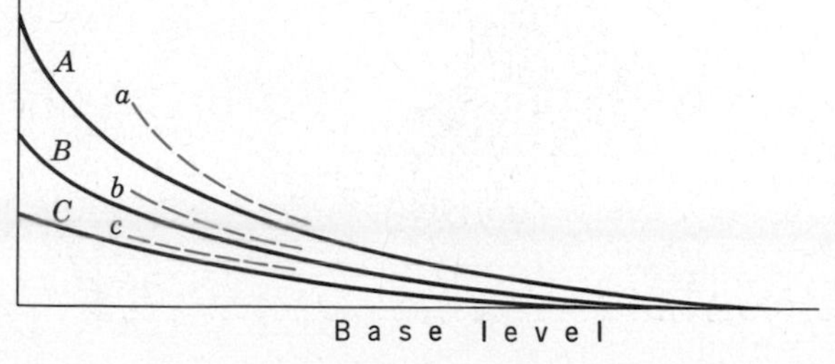


**Figure 7.20**
**Successive long profiles of a main stream in youth, maturity, and old age, (*A*, *B*, *C*) and one of its tributaries (*a*, *b*, *c*), changing with the passage of time. Although gradients progressively decrease, at all times the tributaries are nicely adjusted to the main stream, which in turn is adjusted to its base level.**
***Aa*, *Bb*, and *Cc* are profiles we would expect to find in stages of youth, maturity, and old age, respectively.**

**Figure 7.21**
**Result of a sudden increase in rate of erosion. The smooth surface of the area beyond the dotted line is the remnant of a formerly continuous surface, essentially a peneplain.**
**Following a sudden and considerable increase in stream energy, streams are deepening their valleys and are extending them headward; the surface is being rejuvenated. Probably the energy increase resulted from faulting, which lowered the extreme foreground relative to the land next behind, and so increased the fall from stream heads to stream mouths.**
**Streams are creating a new steady-state condition by building fans that help smooth stream profiles. View west, across south end of Inyo Range, toward Sierra Nevada, California.**
**(*John H. Maxson.*)**

stripped away at an average rate of about 6cm per thousand years. Erosion of 1cm every 166 years may seem a slow rate, but it involves the yearly removal to the sea of 1.3 billion tons of rock material from the area of the United States alone. Since prehistoric, Stone-Age people hunted big game in the United States 10,000 years ago, rock material equivalent to a layer 60cm thick must have been removed to the sea. The reduction of a broad region to a peneplain may take 15 to 100 million years, depending on height at the start, kinds of bedrock, amount of rainfall, and other factors.

### Interruptions in sculptural evolution

Our description of a land progressing slowly through a cycle of erosion assumed that the progress of erosion was smooth and uninterrupted by any outside influences. However, when we look closely at valleys we can often see evidence that the stable steady-state system has been interrupted. An example of local interruption is seen in Figure 7.17; a larger and less local one is that in Figure 7.21. Land areas in such condition are said to have been *rejuvenated* because, after reaching maturity or old age, they have taken on anew the characteristics of youth.

***Rejuvenation,*** then, is *the development of youthful topographic features in a land mass further advanced in the cycle of erosion.*

On a smaller scale, rejuvenation can result in the creation of stream terraces. A ***stream terrace*** is *a bench along the side of a valley, the upper surface of which was formerly the alluvial floor of the valley.* In a stream flowing on a broad valley floor, sudden increase in rate of erosion results in the cutting of a new valley within the older one. The floor of the older valley is left as a pair of stream terraces (Fig. 7.22), which, of course, will in time be entirely destroyed by continued erosion.

If an area subsides or is tilted so as to decrease the gradient of a stream that flows across it, the stream is likely to have to drop some of its load. The deposited load builds up an ***alluvial fill*** (*a body of alluvium, occupying a stream valley, and conspicuously thicker than the depth of the stream*), which gradually steepens the gradient to a point at which all the load can be carried (Fig. 7.22).

### Causes of interruptions

Interruptions in the stability of streams, valleys, and hillslopes are traceable to four chief causes.

**1. Movements of the Crust.** The principal cause is probably movement of Earth's crust. If the upstream part of a drainage basin is elevated in relation to the downstream part, energy increases because stream gradients increase and rejuvenation results. The streams flowing down the western slope of the Sierra Nevada in eastern California have been rejuvenated repeatedly by successive uplifts of the mountain range. Conversely, the western part of the same land mass, in central California, has been bent down and buried beneath accumulating sediments.

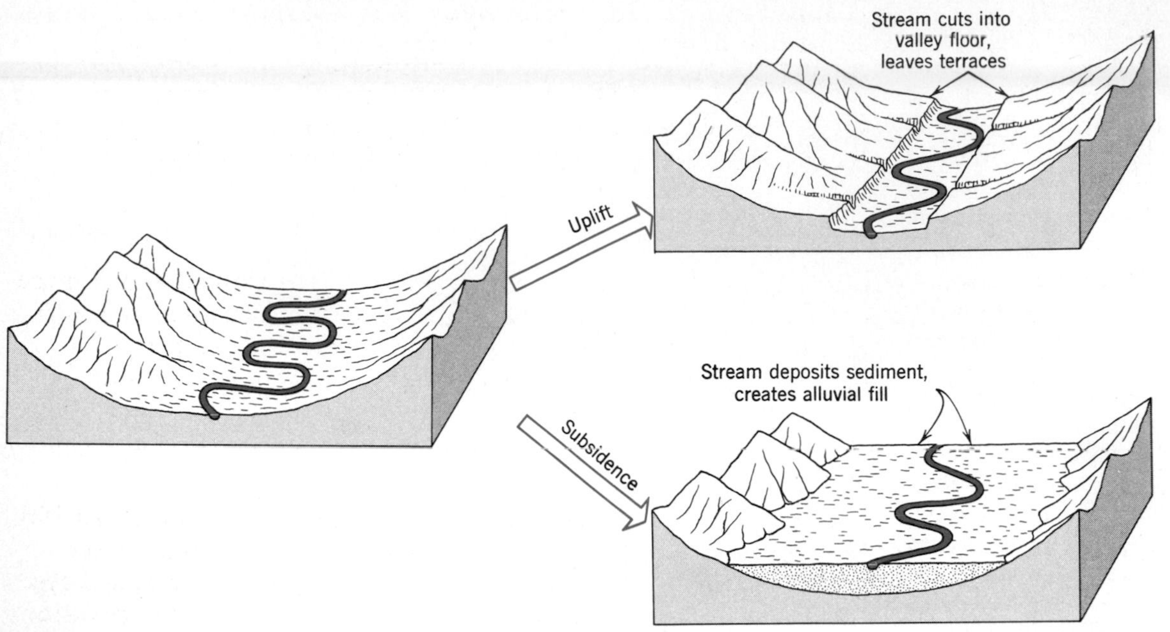


**Figure 7.22**
**Effects of suddenly increased or decreased stream erosion on a valley.**

**2. Change of Base Level.** Rise and fall of sea level (Chapter 11) change the base level of streams and can cause filling and erosion, respectively, in the segments of valleys that are near the sea. The making of a dam across a valley by a fan, a landslide, a glacier, a lava flow, or even by human activity, creates a local base level and can cause alluvial filling in the valley that is dammed. Erosion of the dam then can result in erosion of the fill.

A widespread effect of both (1) and (2) above is partial submergence of a coast, which "drowns" river valleys. Chesapeake Bay, nearly 300km long, was created by drowning of the Susquehanna, Potomac, James, and other river valleys. The St. Lawrence River is drowned well upstream from Montreal. San Francisco Bay, likewise, resulted from drowning of the Sacramento River/San Joaquin River system.

**3. Glacial Sediments** A melting glacier commonly delivers so large a load of sediment to a stream that the stream cannot carry it away and consequently deposits it as a fill. The lower Mississippi valley contains an alluvial fill more than 60m thick, believed to have resulted partly from the deposition of a copious load of glacial sediment and partly from rise of sea level.

**4. Change of Climate.** In dry southwestern United States, deepening and headward extension of innumerable small valleys have been going on since about 1880. The valleys are bare, steep-walled canyons cut into soft rock and loose sediment. Some of the canyons have grown headward faster than 1km per year and have been eroded to depths as great as 25m. Yet before the accelerated erosion began, Spanish settlers found the land surface stable and covered with vegetation.

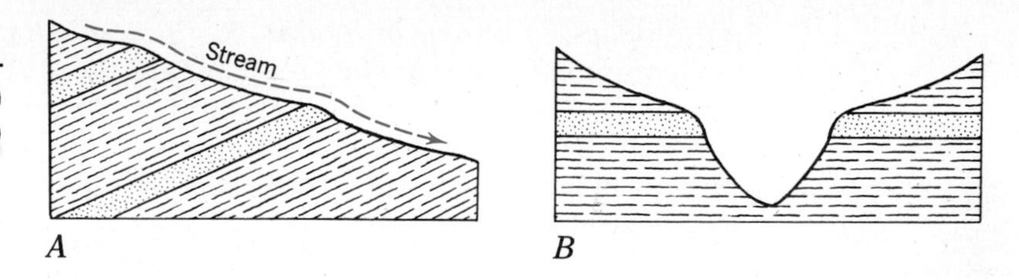


**Figure 7.23**
**Effect of outcrop of resistant strata (stipple) on the long profile (*A*) of a stream and on the cross profile (*B*) of a valley.**

Study of the valleys and of weather records suggests that the cause lies mainly in very slight changes of climate. During long periods of drought, with few but heavy rainstorms, the grass cover deteriorates and lays the ground bare to erosion. During long periods with more frequent but lighter rains, the grass cover improves and protects the ground; valleys tend to fill with alluvium.

A period of few, though heavy, rains during the last half of the 19th Century is believed to have caused the erosion now in progress, and overgrazing by cattle and sheep is thought to have accelerated the process. Some of the valleys contain clear evidence of repeated erosion and refilling (Fig. 9.6). Fragments of ancient Indian pottery buried in the alluvium, coupled with records of ancient Indian migrations, give approximate dates of an earlier erosion and filling. A still earlier erosion is probably prehistoric, dating back several thousand years.

# VARYING ERODIBILITY OF ROCK

### Profiles of streams and valleys

Some kinds of rock are weak and are easily eroded, whereas other kinds are strong and stubborn. These differences affect streams and valleys in several ways. When a stream flows over a layer of resistant rock, its long profile is steepened (Fig. 7.23*A*). A resistant layer exposed along the sides of a valley gives the cross-profile of the valley a steplike form (Fig. 7.23*B*). A valley is likely to be narrower where it cuts resistant rock than where it cuts weak rock. A ***water***

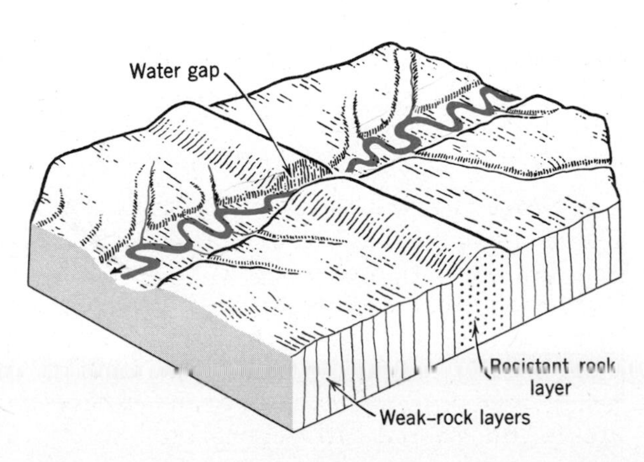


**Figure 7.24**
**Water gap formed where stream has cut through layer of resistant rock. In such rock, a valley is narrower and has steeper sides and gradient than in weak rock. Water gaps are common in the Appalachian region (Fig. 18.6).**

***gap*** (*a pass, in a ridge or mountain, through which a stream flows*) is a common feature at such a place (Fig. 7.24). The figure illustrates also that mass-wasting, sheet erosion, and stream erosion have lowered the surface underlain by weak rocks so effectively that the narrow belt of resistant rock has been left standing as a ridge above the general surface.

### Stream patterns

Not only the profiles of streams and valleys but also their patterns, as seen on a map, are affected by the kinds of rock on which they are developed. Patterns are affected also by the history of the areas in which they occur. Three common kinds of stream patterns are shown in Figure 7.25.

The ***dendritic*** ("treelike") ***pattern*** is *a stream pattern characterized by irregular branching in many directions.* This pattern is common in massive rock and in flat-lying strata. In such situations, differences in rock resistance are so slight that their control of the directions in which valleys grow headward is negligible.

The ***rectangular pattern*** is *a stream pattern characterized by right-angle bends in the streams.* Generally it results from the presence of joints (Figures 3.21 and 3.22) and faults (Fig. 14.11) in massive rocks or from foliation (Fig. 3.19) in metamorphic rocks. Structures such as these, with their geometrical patterns, have guided the directions of valleys.

The ***trellis pattern*** is *a rectangular stream pattern in which tributary*

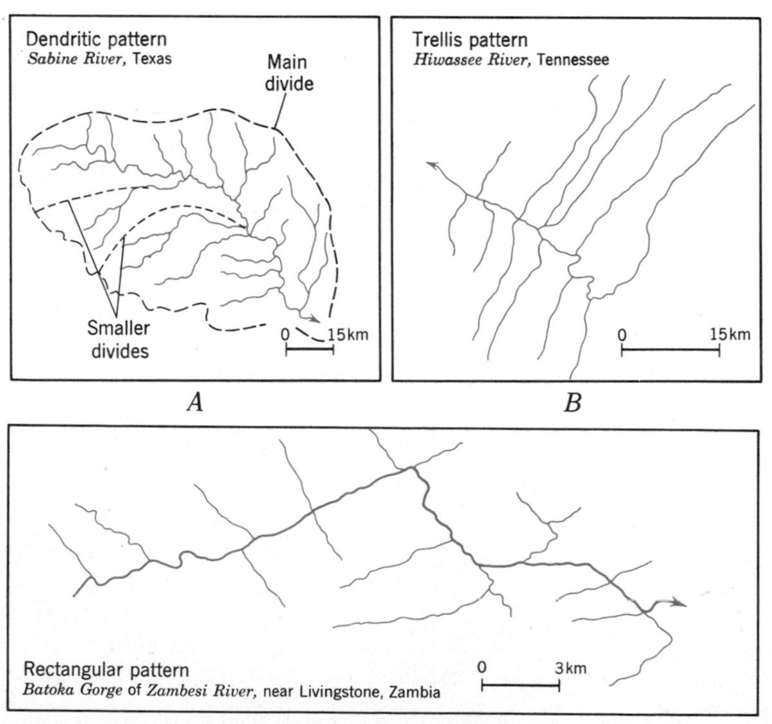


**Figure 7.25**
**Three kinds of stream patterns.**
***A* shows a drainage basin, enclosed by a main divide. Within the basin a smaller drainage basin, enclosed between two smaller divides, defines a tributary area. Other small divides are not shown. Patterns made by streams are like branching trees.**

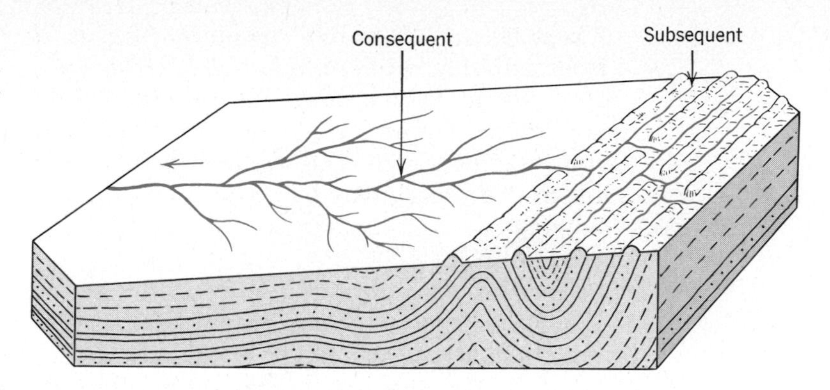


**Figure 7.26**
**Consequent streams contrasted with subsequent streams.**
**On the left the land surface is underlain by flat-lying strata. Drainage has developed under control of the slope of the land (shown by arrow) and is therefore consequent. The stream pattern is dendritic. To the right the same strata are folded. On them tributaries developed most readily along parallel belts of weak rock that determined stream locations. These tributaries are therefore subsequent streams, and in this case have a trellis pattern. The main stream crosses ridges of resistant rock through water gaps.**

*streams are parallel and very long,* like vines or tree branches trained on a trellis. This pattern is common in areas like the Appalachian region, where the outcropping edges of folded sedimentary rocks, both weak and resistant, form long, nearly parallel belts.

# CLASSIFICATION AND HISTORY OF STREAMS

### Kinds of streams

On the basis of their patterns and other characteristics, streams are classified into four groups, labeled consequent, subsequent, antecedent, and superposed. The streams in each group have distinctive origins and histories.

A ***consequent stream*** is *a stream whose pattern is determined solely by the direction of slope of the land.* Therefore, consequent streams generally occur in massive or flat-lying rocks and commonly have dendritic patterns.

A ***subsequent stream*** is *a stream whose course has become adjusted so that it occupies belts of weak rock.* When such belts are long and straight, subsequent streams constitute the long straight tributaries characteristic of trellis drainage patterns. Figure 7.26 illustrates the difference between consequent and subsequent streams.

An ***antecedent stream*** is *a stream that has maintained its course across an area of the crust that was raised across its path by folding or faulting* (Fig. 7.27). The name comes from the fact that the stream is antecedent to (older than) the uplifting.

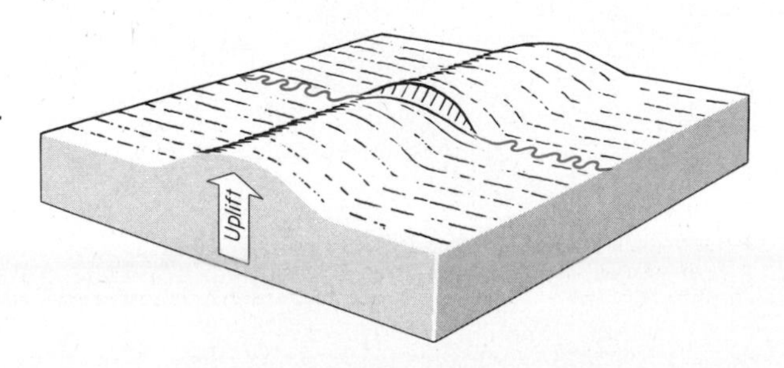


**Figure 7.27**
**This stream has an antecedent relationship to the present surface because of local uplift across its course. Stream has cut a deep gorge across the uplifted belt.**

A ***superposed stream*** is *a stream that was let down, or superposed, from overlying strata onto buried bedrock having composition or structure unlike that of the covering strata* (Fig. 7.28). Most superposed streams began as consequent streams on the surface of the covering strata. The streams' paths, therefore, were not controlled in any way by the surfaces on which they are now flowing.

### Stream capture

When the gradient of one of two streams, flowing in opposite directions from a single divide, is much steeper than that of the other, the steeper stream, having more energy, can extend its valley headward, shifting the divide against the other stream. In this way the steeper stream can capture the other one little by little. Alternatively, it can capture at one stroke a long tributary of the other stream by intersecting the tributary at its mouth. This process of ***stream capture*** (or *piracy*), *the diversion of a stream by the headward growth of another stream,* is illustrated in Figure 7.29. The Provo River shifted the divide at its head northward and eastward a distance of several kilometers until the divide intersected and diverted a principal tributary of the Weber River. Evidence of capture is of two kinds: (1) an abandoned segment of the valley of the diverted stream, and (2) tributary streams that are barbed with respect to the new (main) stream they have joined.

In this chapter and the one preceding it, we have followed the creation of rock particles by weathering, their movement down hillslopes into streams, and their transport as sediment toward the sea. These events complete three closely related phases of the rock cycle. Also, we have followed the sculpture of the land through the cycle of erosion. While a land mass passes through a slowly changing sequence of forms and is gradually eroded to form a peneplain, sediments resulting from erosion are transported and spread out to form new strata, on lower land as basin fills or on the floor of a sea. Thus the cycle of erosion and the rock cycle are complementary. Of course some peneplains are lifted up and destroyed by renewed erosion. But other peneplains sink down and are buried by layers of sediment. The buried surface is preserved until uplift occurs. The uplift renews erosion, which cuts away the cover and exposes the buried surface once more. Ancient

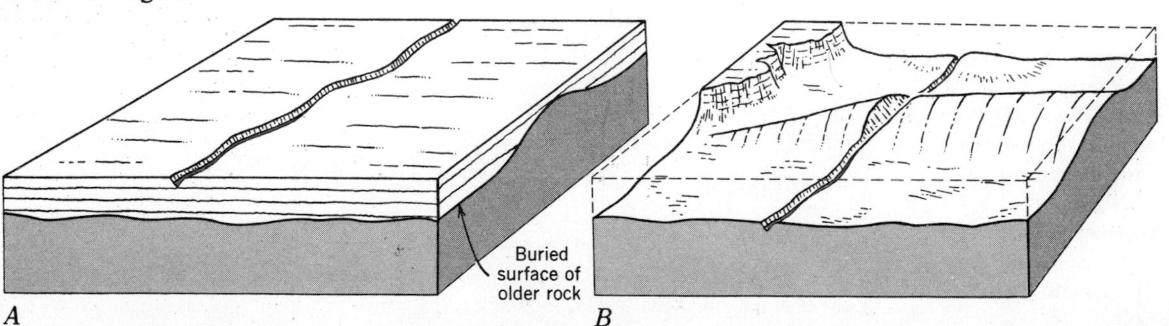


**Figure 7.28**
**Development of a superposed stream.**
***A.* Stream consequent on strata that bury a former land surface.**
***B.* After long-continued erosion the stream has become superposed and has cut a water gap through a hill that formed part of the older surface. Overlying strata have been removed by erosion, except for remnant in upper left. Compare the similar relationship, but with a different history, shown in Figure 7.27.**

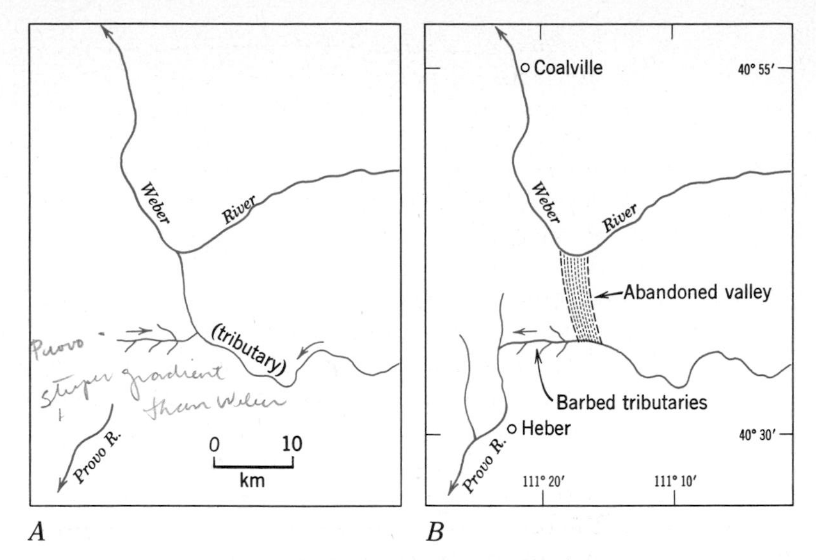


**Figure 7.29 Capture of tributary to Weber River by Provo River, upstream from Coalville, Utah.**
***A*. Reconstructed drainage pattern of an earlier time.**
***B*. Present drainage pattern. Provo River has extended its valley headward, capturing several small tributaries to Weber River and also the large tributary now part of Provo River. Abandoned segment of valley of former north-flowing stream is floored with stream gravel derived from the territory of the diverted stream. Small tributaries have barbed pattern, showing former flow toward the east. Probable cause of capture: Provo River, shorter and with a steeper gradient than Weber River, had the greater potential for erosion. (*After G. E. Anderson.*)**

peneplains created as long ago as Precambrian time, more than 600 million years ago, are now, after all that time, being again exposed to view.

# SUMMARY

## IMPORTANCE OF RUNNING WATER

1. As part of the water cycle, streams are the chief means of returning water from land to sea. As geologic agents, stream erosion and mass-wasting are foremost among the processes that erode the land and transport sediments from land to sea.

2. Splash of raindrops and sheet erosion effectively erode regolith on bare, unprotected slopes.

## STREAM FLOW AND BEHAVIOR OF ROCK PARTICLES

3. Turbulence characterizes the flow of nearly all streams, and is a prime factor in picking up and transporting sediment.

4. A stream's load consists of bed load, suspended load, and dissolved load.

## ECONOMY OF A STREAM

5. Long profiles of streams are concave-up curves. The profiles become gentler with time but are limited downward by base level.

6. Stream discharge, width and depth of channel, and velocity are intimately related, and continually adjust to eachother.

7. Because of increased discharge and velocity, a stream can carry a load both coarser and greater in amount during floods than it can transport at times of low water. Streams do most of their geologic work during seasonal floods.

## GEOLOGIC WORK OF STREAMS

8. Features of streams in alluvial valleys include meanders, oxbow lakes, bars, natural levees, and floodplains.

9. Fans are built at the toes of steep slopes; deltas are built at the mouths of streams. A common kind of delta consists of foreset, bottomset, and topset layers.

10. Streams tend to maintain a steady-state condition. If that condition is interrupted, the stream will return to it.

### SCULPTURAL EVOLUTION OF THE LAND

11. Most land surfaces consist of complexes of valleys, cut by the streams that flow through them.

12. Uninterrupted sculpture of a land mass follows a broadly predictable cycle of erosion, ending in a peneplain.

13. Weathering, mass-wasting, and sheet erosion together erode more rock material than streams do. The main work of streams is to carry away the material fed to them from slopes.

### INTERRUPTIONS IN SCULPTURAL EVOLUTION

14. The orderly progress of land sculpture is commonly interrupted. Among the interruptions are movements of the crust and changes in the position of base level. These can cause rejuvenation or deposition.

### EFFECTS OF VARYING ERODIBILITY OF ROCK

15. Streams tend to occupy belts of weak rock. Therefore the pattern of rocks exposed at the Earth's surface influences the pattern of streams.

## SELECTED REFERENCES

### STREAMS

Colby, B. R., 1963, Fluvial sediments—a summary of source, transportation, deposition, and measurement of discharge: U.S. Geol. Survey Bull. 1181, p. A1–A47.

Davis, S. N., and de Wiest, R. J. M., 1966, Hydrogeology: New York, John Wiley.

Davis, W. M., 1899, The geographical cycle: Geogr. Jour., v. 14, p. 481–504.

———, 1902, Base level, grade, and peneplain: Jour. Geol., v. 10, p. 77–111.

Denny, C. S., 1965, Alluvial fans in the Death Valley region, California and Nevada: U.S. Geol. Survey Prof. Paper 466.

Fisk, H. N., 1952, Mississippi River valley geology in relation to river regime: American Soc. Civil Engrs. Trans., v. 117, p. 667–682.

Hoyt, W. G., and Langbein, W. B., 1955, Floods: Princeton, N.J., Princeton Univ. Press.

Leopold, L. B., 1974, Water. A primer: San Francisco, W. H. Freeman, p. 34–162.

Leopold, L. B., and Langbein, W. B., 1966, River meanders: Scientific American, v. 214, p. 60–70.

Leopold, L. B., and Maddock, T., 1953, The hydraulic geometry of stream channels and some physiographic implications: U.S. Geol. Survey Prof. Paper 252.

Leopold, L. B., Wolman, M. G., and Miller, J. P., 1964, Fluvial process in geomorphology: San Francisco, W. H. Freeman.

Livingstone, D. A., 1963, Data of geochemistry, 6th ed., Chemical composition of rivers and lakes: U.S. Geol. Survey Prof. Paper 440, p. G1–G64.

Shirley, M. L., ed., 1966, Deltas in their geologic framework: Houston Geol. Soc., p. 233–251, maps of existing deltas assembled by A. E. Smith, Jr.

Sundborg, Åke, 1956, The River Klarälven. A study of fluvial processes: Geograf. Annaler, v. 38, p. 125–316.

Motion-picture film: Flow in alluvial channels (16mm color with sound). Shows stream flow and examples of ripples and sand waves formed in a laboratory channel. Available for free loan on application to Map Information Office, U.S. Geological Survey, Washington, D.C.

SCULPTURE OF THE LAND

Birot, Pierre, 1968, The cycles of erosion in different climates: London, Batsford.

Bloom, A. L., 1969, The surface of the Earth: Englewood Cliffs, N.J., Prentice-Hall, p. 81–102.

Cotton, C. A., 1952, Geomorphology, an introduction to the study of landforms, 6th ed.: New York, John Wiley.

Leopold, L. B., Wolman, M. G., and Miller, J. P., 1964, Fluvial processes in geomorphology: San Francisco, W. H. Freeman.

Morisawa, M. E., 1964, Development of drainage systems on an upraised lake floor: American Jour. Sci., v. 262, p. 340–354.

Schumm, S. A., and Lichty, R. W., 1965, Time space and causality in geomorphology: American Jour. Sci., v. 263, p. 110–119.

# 8 GROUND WATER; WATER SUPPLY



We need both food and water, but we can survive longer without food than we can without water, because the biochemical activities within our bodies are based on water solutions. We must, in some way, have access to water, whether from streams, lakes, or springs. Our ancestors the Stone-Age hunters had to stay fairly close to water that flowed or stood on the surface. Then at some time—when, we do not know—someone discovered that water exists below the surface and that one can find it by digging a hole.

When hunting had given way to agriculture and to a more settled way of life with permanent dwellings, many houses had their own laboriously dug wells. Then, as cities grew—beginning long before the Christian Era—more and more people lived crowded into limited areas, their dwellings commonly touching eachother. Individual, domestic wells became impractical because their water had become polluted with human and other wastes. The wells had to be replaced by community systems that brought water into the city from outside it. In nearly every city, therefore, water has become a critical resource. Each large community must ask itself: have we enough water? Is its quality adequate for the uses to which we put it? And are we using our water in an efficient way, with a minimum of waste?

The answers to these questions demand that we know the basic things about water supply: the quantity of Earth's water, where it is, how it is cycled, and the techniques of using it efficiently.

## GROUND WATER IN EARTH'S WATER INVENTORY

Table 8.1 is an inventory of Earth's water, carefully calculated or estimated from measurements. It tells us that an estimated 97.6 percent of the world's water is in the ocean, the main reservoir, and that the next largest amount consists of ice, most of it on the Antarctic Continent. Much less than 1 percent of the world's water is ***ground water,*** defined simply as *all the water contained in spaces within bedrock and regolith.* Small though the total volume of ground water is, it is about 35 times greater than the volume of water lying in lakes or flowing in streams on Earth's surface.

### Origin of ground water

Looking again at Figures 4.1 and 4.2, we can see that nearly all Earth's ground water (except for a tiny proportion that comes from magma) has its origin in rainfall, and is always on its way, however slowly, back to the ocean, either directly, or indirectly by coming out onto the surface and joining streams. This is how ground water plays its part in the water cycle.

In ancient Greece, 2,500 years ago, it was thought that the water in the ground originated as seawater, driven into the rocks by the winds and somehow desalted, or that it was created in some manner from rocks and air deep below the surface. Later it was recognized that rivers are fed, at least in part, by springs emerging from the ground and also that the discharge of rivers does not raise the surface of the sea appreciably. The truth that ground water is derived mainly from rain and snow was recognized by Marcus Vitruvius, a Roman architect of the time of Christ, who wrote a treatise on aqueducts and water supply, a matter of great practical importance to Romans.

**Table 8.1**
**Distribution of World's Water. (*From data in R. L. Nace, 1967, U.S. Geol. Survey Circ. 536, table 1, and other sources*)**

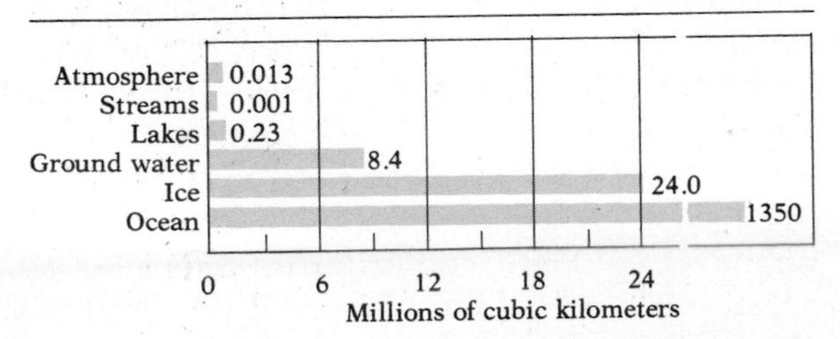

Although true, Vitruvius's statement that ground water comes from rain was not established on a quantitative basis until the 17th Century. Then Pierre Perrault, a French physicist, measured the mean annual rainfall on a part of the drainage basin of the River Seine in eastern France and also the mean annual runoff from it in terms of river discharge. After allowing for loss by evaporation, he concluded that the difference between the amounts of rainfall and runoff was ample enough, over a period of years, to account for the amount of water in the ground. Today we accept rainfall as the source of all ground water except for a tiny proportion that comes from magma.

### Depth

One of the reasons why people have been able to establish permanent settlements, not only in well-watered country but also in desert lands, is that few areas exist in which holes, intelligently located and sunk far enough into the ground, do not find at least some water. In a moist country the depth of an adequate well may have to be only a few meters; in a desert it may have to be hundreds. These facts have been learned by experience. Water is present beneath the land nearly everywhere, but whether it is present in usable quantity depends on depth of occurrence, kinds of rock present, kinds and amounts of substances dissolved in the water, and other factors. For this reason some places are much more favorable than others for obtaining ground water.

More than half of all ground water, including most of the water that is usable, occurs within about 750m of Earth's surface. This is estimated to be equivalent in amount to a layer of water some 55m thick, spread over the world's land area. Below the depth of about 750m, water decreases in amount, gradually though irregularly. Holes drilled for oil have found water lying as deep as 9.4km; but everywhere, at some depth, water ceases to be present. The chief reason is that the weight of overlying rock exerts such pressure on the rock below that the openings in it become so tiny that the water they contain is unable to move through them.

### Water table

Much of our knowledge of ground-water occurrence has been learned the slow way, from the accumulated experience of many generations of people who have dug or drilled millions of wells. This experience (Fig. 8.1) tells us that a hole penetrating the ground ordinarily passes first into a ***zone of aeration,*** (or *unsaturated zone) the zone in which open spaces in regolith or bedrock are normally filled mainly with air.* The hole then enters the ***saturated zone,*** *the subsurface zone in which all openings are filled with water. The upper surface of the saturated zone* is the ***water table,*** which, at any place, normally slopes toward the nearest stream. Ordinarily the water table lies within a few meters of the surface. Whatever its depth, the water table is a very significant surface, because it represents the upper limit of all readily usable ground water. We shall return to it shortly.

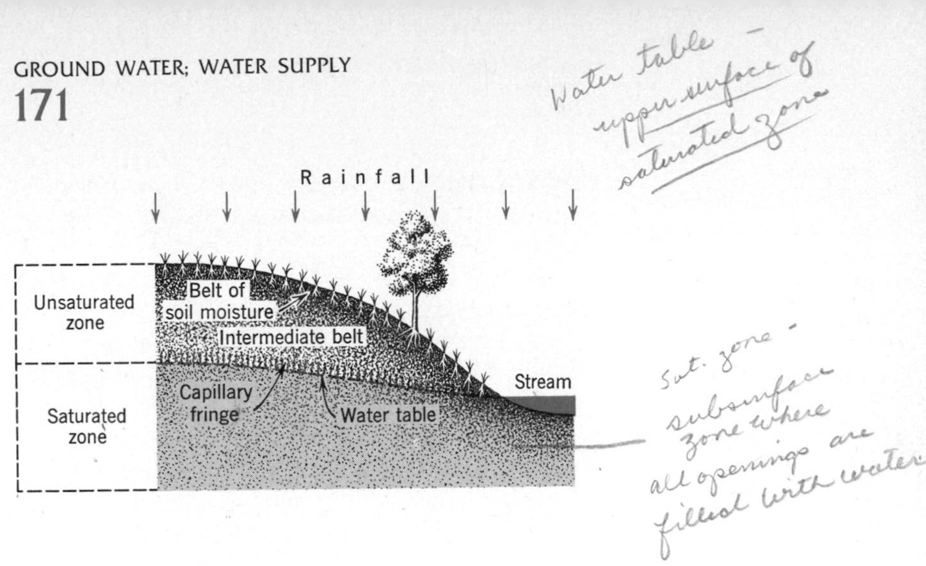


**Figure 8.1**
**Positions of saturated zone, water table, and unsaturated zone.** ***(After W. C. Ackermann and others, U.S. Dept. Agriculture.)***

# MOVEMENT

Most of the ground water within a few hundred meters of Earth's surface does not just lie there; it moves. But its movement is unlike the turbulent flow of rivers, measurable in kilometers per hour. It is so slow that velocities are expressed in centimeters per year. To understand why it is slow, we must understand the porosity and permeability of rocks.

## Porosity and permeability

The limiting amount of water that can be contained within a given volume of rock material depends on the ***porosity*** of the material; that is, *the proportion (in percent) of the total volume of a given body of bedrock or regolith that consists of pore spaces* (i.e., open spaces). So a very porous rock is a rock containing a comparatively large proportion of open space, regardless of the size of the spaces. Sediment is ordinarily very porous, ranging from 20 percent or so in some sands and gravels to as much as 50 percent in some clays. The sizes and shapes of the constituent particles and the compactness of their arrangement affect porosity, as does the degree, in a sedimentary rock, to which pores have become filled with cementing substances. In contrast, igneous and metamorphic rocks generally have low porosity, except where joints and cracks have developed in them.

***Permeability*** is *capacity for transmitting fluids.* A rock of very low porosity is likely also to have low permeability. However, high porosity values do not necessarily mean high permeability values, because size and continuity of the openings influence permeability in an important way. The relationship between size of openings and the molecular attraction of rock surfaces plays a large part. Molecular attraction is the force that makes a thin film of water adhere to a rock surface despite the force of gravity; an example is the wet film on a pebble that has been dipped in water. If the open space between two adjacent particles in a rock is small enough, the films of water that adhere to the two particles will come into contact with eachother. This means that the force of molecular attraction is extending right across the open space, as shown on the left side of Figure 8.2. At ordinary pressure, therefore, the water is held firmly in place; hence permeability is low. That is

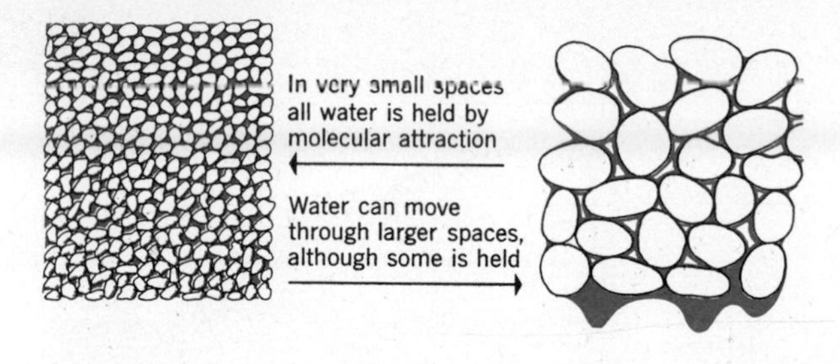


**Figure 8.2**
**Effect of molecular attraction in the intergranular spaces in fine sediment (*left*) and in coarser sediment (*right*). Scale is much larger than natural size.**

what happens in a wet sponge before it is squeezed. The same thing happens in clay, whose particles are so tiny that their diameters are less than 0.005mm (Table 3.4).

By contrast, in a sediment with grains at least as large as sand grains (0.06mm to 2mm) the open spaces are wider than the films of water adhering to the grains. Therefore the force of molecular attraction does not extend across them effectively, and the water in the centers of the openings is free to move in response to gravity or other forces, as shown at the right in Figure 8.2. This sediment therefore is permeable. As the diameters of the openings increase, permeability increases. With its very large openings, gravel is more permeable than sand and yields large volumes of water to wells.

### Movement in the unsaturated zone

Let us return for a moment to Figure 8.1. Water from a rain shower soaks into the soil, which usually contains clay resulting from chemical weathering of bedrock. Because of its content of extremely fine clay particles, the soil is generally less permeable than underlying materials. Part of the water, therefore, is retained in the soil by forces of molecular attraction. This is the belt of soil moisture in Figure 8.1. Some of this moisture evaporates directly and much is taken up by plants and is transpired (Fig. 4.1).

Water that molecular attraction cannot hold in the soil seeps downward through the intermediate belt (Fig. 8.1) until it reaches the water table. In fine-grained material a narrow fringe as much as 60cm thick, immediately above the water table, is kept wet by capillary attraction, the force that draws ink through blotting paper and kerosene through the wick of an old-fashioned lamp.

With every rainfall, more water is supplied from above, but apart from the belt of soil moisture and the capillary fringe, the unsaturated zone is likely to be nearly dry between rains.

### Movement in the saturated zone: percolation

The flow of ground water in the saturated zone is like what occurs when a saturated sponge is squeezed gently. Such movement is called *percolation*. In it, water particles move slowly through the very small open spaces along parallel, threadlike paths. Movement is easiest through the central parts of the spaces but diminishes to zero immediately adjacent to the sides of each space, because there molecular attraction holds the water in place.

The force of gravity supplies the energy for percolation of ground water. Responding to that force, the water "tends to seek its own level," percolating from areas where the water table is high toward areas where it is lowest; in other words, toward surface streams (Fig. 8.3). Only part of the water travels by the most direct route right down the slope of the water table. Other parts follow innumerable long, curving paths that go deeper through the ground. Some of the deeper paths turn upward against the force of gravity and enter the stream from beneath. This happens because, in the saturated zone, the water at any given height, such as $h_1$ in Figure 8.3, is under greater pressure beneath a hill than beneath a stream. The water therefore tends to move toward points where pressure is least.

Laboratory models, in which dye is injected into the percolating ground water at various depths, have been made. Paths followed by the dye resemble those in Figure 8.3, where they turn upward beneath a model stream at the base of a hill. The rate at which the water moves along these paths decreases sharply with increasing depth. Therefore most of the ground water that enters a stream has traveled to it via shallow paths not far beneath the water table.

# ECONOMY OF THE GROUND-WATER SYSTEM

The ground-water system operates continually, as a small part of the hydrologic cycle. Water moves through the ground and escapes into a stream valley on its way toward the ocean. Meanwhile the ground is continually *recharged* by rainwater entering the system from above. ***Recharge*** is *the addition of water to the saturated zone.*

### Fluctuation of the water table

As can be seen in Figure 8.3, the water table is a surface consisting entirely of slopes; its form is a subdued imitation of the ground surface above it. It is high beneath hills and low at valleys because, as we noted just above, the water tends to move toward valleys, where the pressure on it is least. In each valley the water seeps out into a

**Figure 8.3**
**Movement of ground water in uniformly permeable rock material.**
**Long curved arrows represent only a few of many possible paths. At any point such as $X$, slope of water table is determined by $\frac{h_2 - h_1}{l}$ where $h_2 - h_1$ is height of $X$ above point of emergence in surface stream and $l$ is distance from $X$ to point of emergence. (*After M. K. Hubbert.*)**

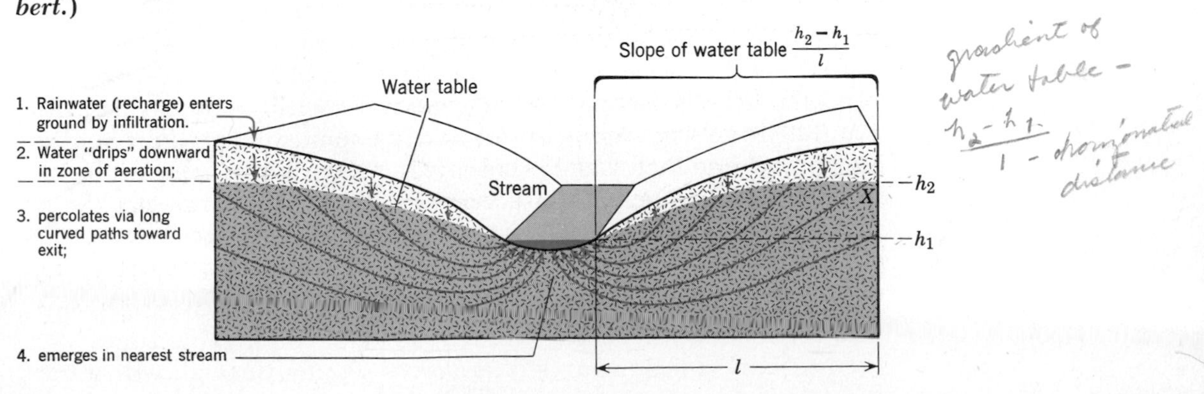

surface stream. If all rainfall could somehow be stopped permanently, the "hills" formed by the water table would slowly flatten, and would gradually approach the levels of the valleys. Percolation would gradually cease, and the streams in the valleys would dry up. Of course this could not happen, but in times of drought, when rain does not fall for perhaps several weeks, we sense the flattening of the water table in the drying-up of ordinary wells. When that occurs we know the water table has subsided below the bottoms of the wells. It is repeated rainfall, dousing the ground with new supplies of water from above, that prevents the water table from flattening very much.

### Velocity of flow

What, then, determines the steepness of the slopes of a water table and the rate of flow of the percolating ground water? Look again at Figure 8.3. Like the gradient of a stream, the slope of a water table between any point such as $X$ at height $h_2$ and the point where it emerges at height $h_1$ is measured by the difference in height $(h_2 - h_1)$ divided by the horizontal distance $l$.

It was discovered by experiment that velocity of flow of ground water increases with increasing slope of the water table, as long as permeability of the material remains uniform.

Therefore we can speak of the *economy* of a ground-water system just as we speak of the economy of a stream system. In the ground-water system, however, the terms are simpler because there is no alluvial channel that changes dimensions, as in a stream system. The water is percolating through the openings in a fixed framework of bedrock or regolith, although as the water passes from one kind of rock into another, permeability changes, and with it velocity changes too. Apart from changes of permeability, the important variable factor is the slope of the water table, which changes with rainfall as does the discharge of a stream.

### Velocity and discharge

Because of the large amount of friction involved in percolation, velocities are slow, commonly ranging between about 1.5m/day and 1.5m/month. The largest rate yet measured within the United States, in exceptionally permeable material, is only about 250m/year.

Velocity of percolation is measured between pairs of wells by various methods. In one method, two wells with metal casings are connected to form an electric circuit. A chemical compound that is an efficient conductor and is soluble in water is poured into the upslope well and percolates downslope. On its arrival at the downslope well it creates a short circuit between well casing and electrode; this is recorded on an ammeter. Distance between wells divided by elapsed time gives the velocity.

A commonly used formula says that slope of water table times permeability times the area of the cross section through which the water is percolating equals discharge. With this formula one can estimate the amount of water that a given well can be expected to deliver.

# ORDINARY WELLS AND SPRINGS

Wells and springs are divided into two classes, determined by the geometry of the rock bodies that supply them. In one class, water movement is unconfined and is therefore comparatively simple. In the other class, movement is confined to a particular part of the rock and in consequence resembles the flow of water through the pipes of a system of plumbing. The wells and springs of the first class we call *ordinary;* the others we call *artesian.*

### Aquifers

When we look for a good supply of ground water we search for an ***aquifer*** (from the Latin, "water carrier"), *a body of permeable rock or regolith through which ground water moves.* Bodies of gravel and sand are commonly good aquifers and so are many sandstones. But the presence of cement between the grains of sandstone reduces the diameter of the openings and so reduces the effectiveness of these rocks as aquifers.

It might seem that claystones, igneous rocks, and metamorphic

**Figure 8.4**
**The shaded areas are underlain by one or more aquifers that can yield, in individual wells, at least 50 gallons per minute of water containing no more than 0.2 percent dissolved solids. (*H. E. Thomas, U.S. Dept. Agriculture.*)**

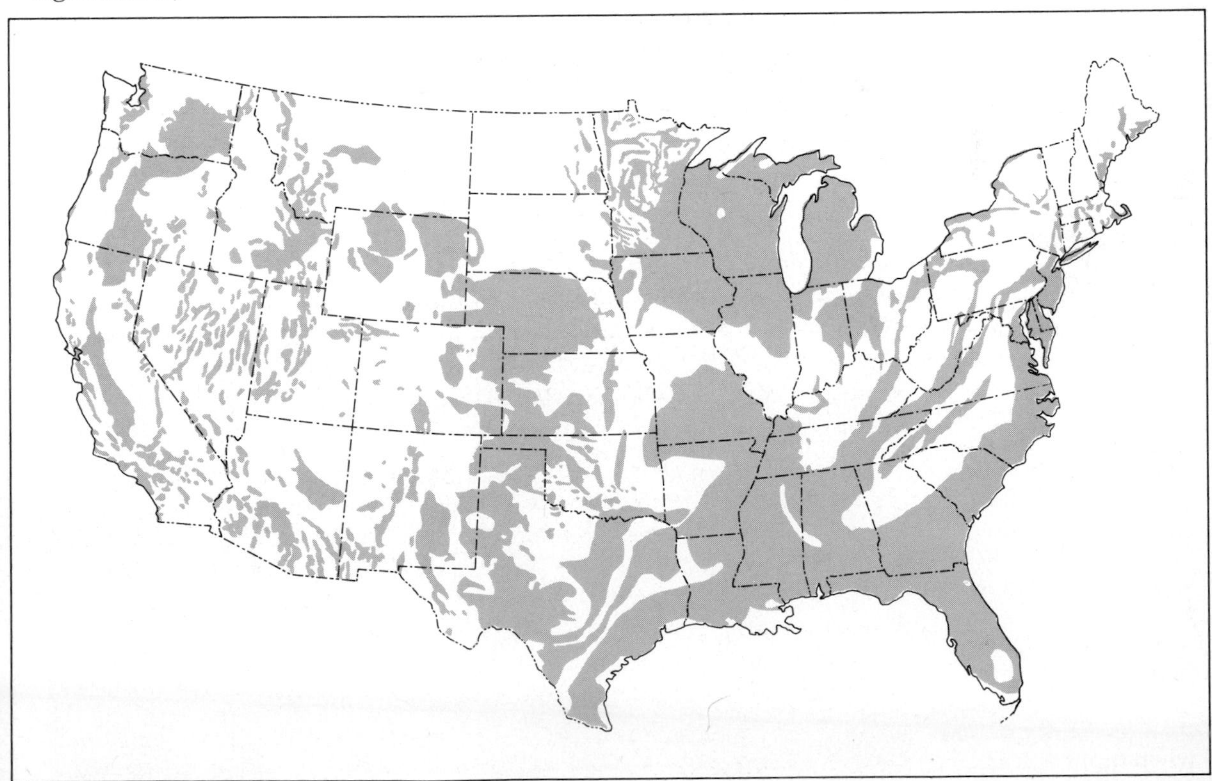

rocks should not be aquifers, because in them the spaces between grains are extremely small, and because samples of them, measured in the laboratory, are impermeable. What is true for laboratory samples, however, does not necessarily apply to large bodies of the same material. Many such bodies contain fissures, spaces between layers, and other openings such as joints that are too large for water flow to be controlled entirely by molecular attraction; these bodies can be aquifers. Even so, they are less effective as aquifers than are sandstones or conglomerates that are not completely cemented. Whatever their effectiveness, it is in aquifers that we find wells and springs. More than half the area of the conterminous United States is underlain by one or more aquifers (Fig. 8.4).

### Ordinary wells

An ordinary well fills with water simply because it intersects the water table (Fig. 8.5). Lifting water from the well lowers the water level and so creates a ***cone of depression,*** *a conical depression in the water table immediately surrounding a well* (Fig. 8.11). In most small domestic wells the cone of depression is hardly appreciable. Wells pumped for irrigation and industrial uses, however, withdraw so much water that the cone can become very wide and steep and can lower the water table in all the wells of a district. Figure 8.5 shows that a shallow well can become dry at times, whereas a deeper well in the vicinity may yield water throughout the year.

If rocks are not homogeneous, the yields of wells are likely to vary considerably within short distances. Massive igneous and metamorphic rocks, (Fig. 8.6*A*), for example, are not likely to be very permeable except where they are cut by fractures, so that a

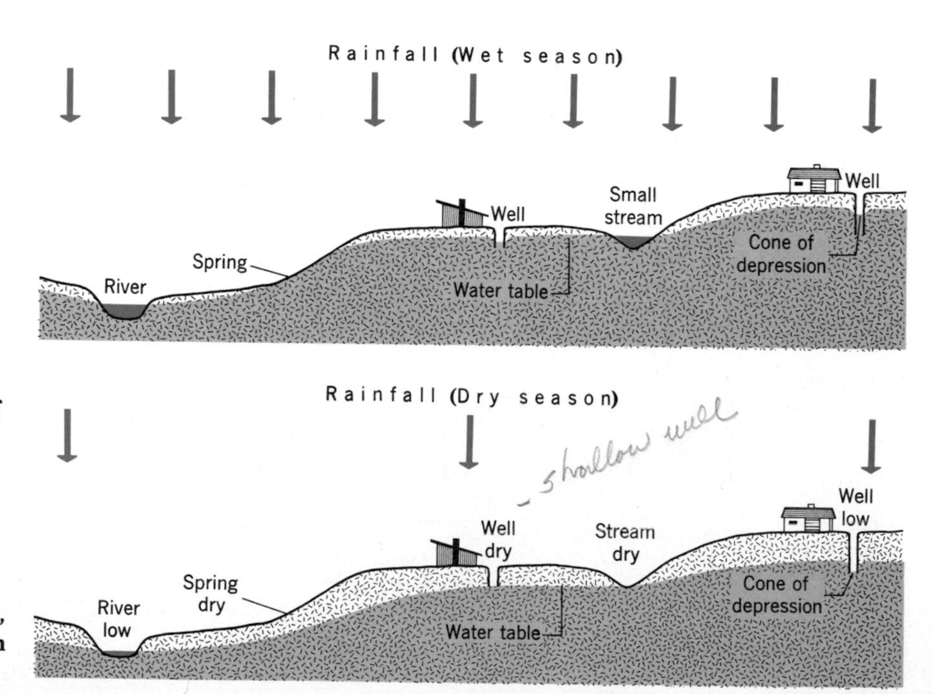


**Figure 8.5**
**Wells and a spring in homogeneous rock, showing cones of depression and effect of seasonal fluctuation of water table. The slopes of the water table are steeper in the wet season, when input of water into the system is greatest, than in the dry season when input is least.**

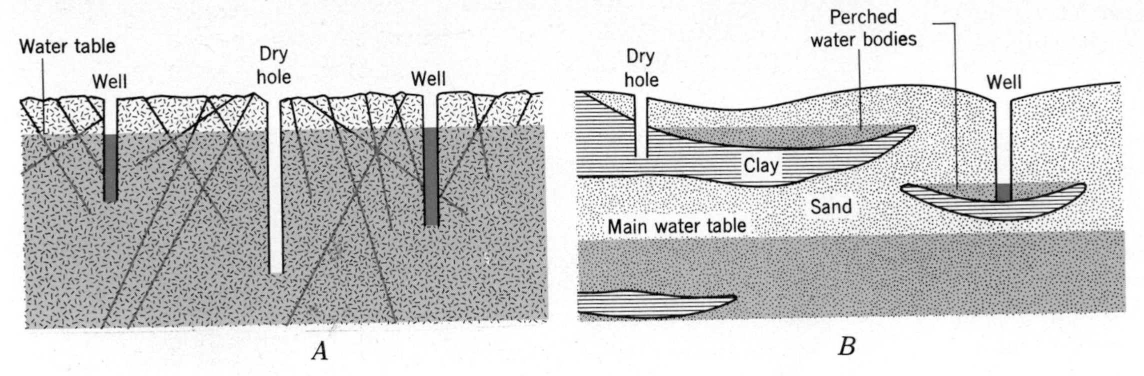


Figure 8.6
**Ordinary wells and adjacent dry holes in rock that is not homogeneous.**
***A*. In fractured massive rocks such as granite.**
***B*. In bodies of permeable sand containing discontinuous bodies of impermeable clay. Two perched water bodies are shown.**

hole that does not intersect fractures is likely to be dry. Because fractures generally die out downward, the yield of water to a shallow well can be greater than to a deep one. Again, discontinuous bodies of permeable and impermeable material (Fig. 8.6*B*) result in very different yields to wells. They also create ***perched water bodies*** (*water bodies that occupy basins in impermeable material, perched in positions higher than the main water table*). The impermeable layer catches and holds the water reaching it from above.

### Gravity springs

A ***spring*** is *a flow of ground water emerging naturally at the ground surface.* The simplest spring is an ordinary or *gravity spring.* Three common kinds of gravity springs are illustrated in Figure 8.7.

## ARTESIAN CIRCULATION

### Confined percolation

In some regions the geometry of inclined rock layers makes possible a special pattern of circulation of ground water. Three essentials of the pattern are shown in Figure 8.8.

1. A series of inclined strata that include a permeable layer sandwiched between impermeable ones.

2. Rainfall, to feed water into the permeable layer where that layer is cut by the ground surface.

Figure 8.7
**Three common occurrences of gravity springs.**

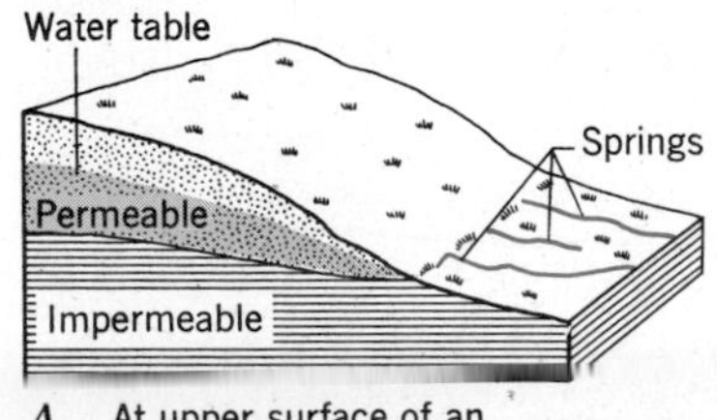


*A* At upper surface of an impermeable layer

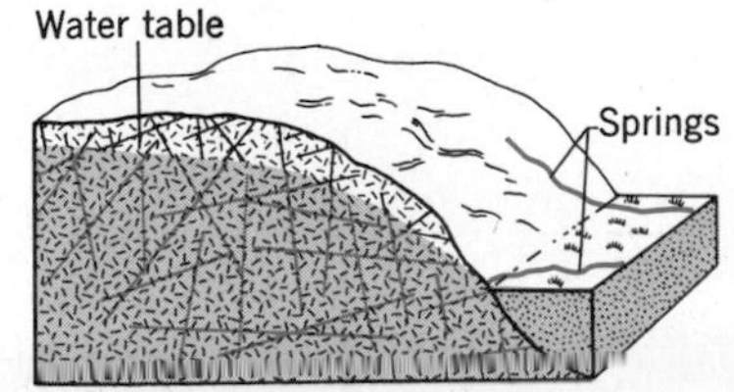


*B* In fractured massive rock such as granite

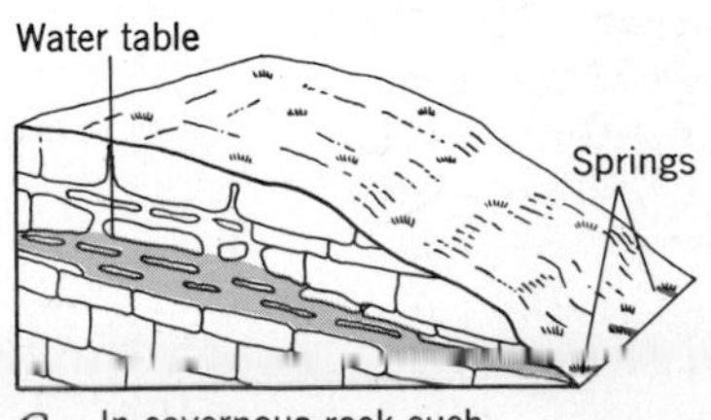


*C* In cavernous rock such as limestone

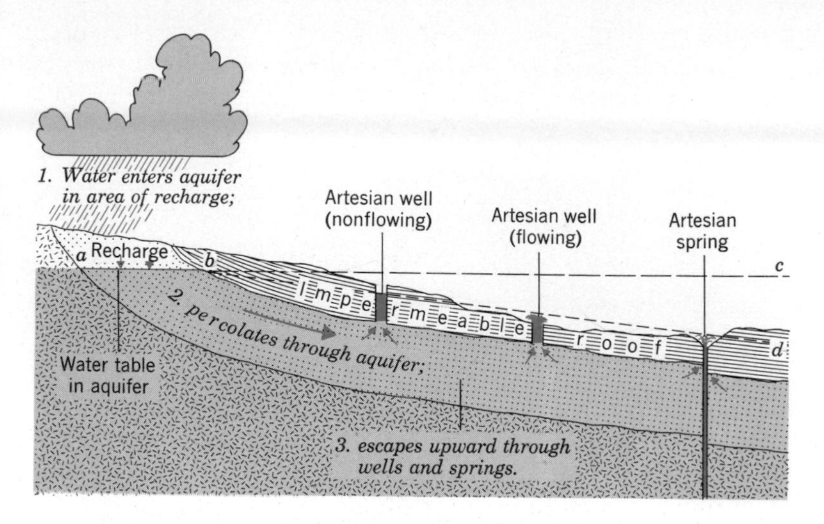


**Figure 8.8**
**Artesian wells and springs.**
**Three essential conditions are an aquifer, an impermeable roof, and water pressure sufficient to make the water in any well rise above the aquifer.**
**The water rises, in any well, to the height (*bc*) of the water table in the recharge area (*ab*), minus an amount determined by the loss of energy in friction of percolation. Thus the water rises only to the line *bd*, with slopes downward away from the recharge area.**

3. A fissure or a well so situated that water from the sandstone can escape upward through the impermeable roof.

When these essentials are present, we have a special kind of system. The input consists of rainwater, which enters the permeable layer (now an aquifer by definition) and percolates through it. The output consists of water forced upward through fissures or wells that perforate the roof.

Note, in Figure 8.8, the position of the water table. Except for a thin zone close to the surface, the whole series of strata is saturated with water. In the impermeable roof rock the water is motionless, because it is held in place by capillary attraction in the tiny spaces between the particles of rock. But in the aquifer it moves, provided only that water can escape through fissures or wells. Percolation in the aquifer, however, is *confined* between the impermeable strata above and below; it moves past the water that is held motionless in those strata. The aquifer is like a broad, flat, sand-filled pipe or conduit, holding its ground water confined under pressure of the column of water that extends up to the water table at its upper end (Fig. 8.8).

If, in the area of recharge shown in the illustration, rainfall reaches the ground in greater volume than that of water output through fissures or wells, only enough water to balance output can enter the system; the excess flows away over the surface. On the other hand, if wells draw out of the system more water than can enter it from the available rainfall on the area of recharge, the yield of the wells will diminish to a quantity small enough to be balanced by the recharge.

A well of this kind is an ***artesian well,*** *a well in which water rises above the aquifer.* The name comes from the French province of Artois, in which, near Calais, the first well of this sort in Europe was bored. When the factors listed above are unusually favorable, pressure can be great enough to lift the water above ground, creating fountains as much as 60m high.

Deep wells drilled through bedrock to intersect the water table are popularly called "artesian wells," but this is an incorrect use of the term. Such wells are ordinary wells, like those in Figures 8.5 and 8.6.

**Ocala Limestone artesian system**
The Ocala Limestone, a principal artesian aquifer in Florida, is an aquifer because it is full of caverns and smaller openings, intricately interconnected, created by dissolution. In the central and northwestern part of the peninsula this layer of limestone is exposed at the surface, but eastward and westward it becomes covered by overlying strata as it slopes downward toward both coasts. Impermeable roofs are provided by clayey layers within the limestone; so the Ocala is not merely *an* aquifer; it is a *series* of aquifers one on top of the other.

The age of the water at various places within the system has been determined. This was done by measuring the age of $^{14}C$ in $(HCO_3)^-$ dissolved in the water, at a series of wells along a line 133km long, generally parallel to the dip of the aquifer. Most of the $^{14}C$ entered the ground in rainfall on the recharge area and moved through the aquifer in the ground water. The age was found to increase systematically away from the recharge area. From the sum of the differences in age between samples from pairs of wells in the series is calculated an average velocity of percolation (through the whole distance) of 7m per year. According to this velocity, water in the well farthest from the recharge area has been in the ground for nearly 19,000 $^{14}C$ years.

There is a small error in these calculations. "Young" ground water has been percolating down into the aquifer, and is still doing so, along the entire distance of 133km. This "young" water dilutes the older water already in the aquifer, and so the average age of the water sampled in each well is a little younger than it would be if no younger water had been added from above. Hence the travel time calculated at 19,000 years is a little too small.

Tapping an artesian system is a very ancient art. Four thousand years ago many artesian wells, some of them as much as 100m deep, were in existence. The well near Calais, in France, was bored in A.D. 1126 and is still flowing today. In that area, at any rate, withdrawal has not seriously exceeded supply.

Artesian systems are not confined to wells. Some are *artesian springs,* in which ground water rises to the surface through a natural fissure rather than through a man-made hole (Fig. 8.8).

# WATER AND PEOPLE

We have been dealing with the way ground water moves and the places where it occurs. Now we turn to economic aspects of ground water: finding hidden water, quality of water, and problems of balancing need and supply. Our brief discussion treats surface water together with ground water wherever it seems easier to do so, for the two resources are very closely related.

### Finding ground water

In the days when the population of North America was mostly rural, in any region of fairly abundant rainfall a well could be dug to a few meters' depth with a good chance that it would yield enough water for the use of a family. Sometimes the sites for such wells were located by persons who used forked twigs and other kinds of "divining rods" and who claimed to possess supernatural powers. The search for water by this means, often called "dowsing," dates back at least to the time of Moses and is still widespread. Although no scientific basis for this kind of claim is known to exist, use of the divining rod persists, partly because in many areas shallow supplies of ground water are so widespread that successful results would be numerous even though sites were located at random. If the diviner were asked to indicate where water is *not* present below the ground and if his predictions were then tested by boring holes, the statistical results would soon reveal the unsoundness of his claims. But, because little money is spent on boring holes in attempts to avoid water, this has never been done.

Today the average depth of wells is deeper, and the rock drill has replaced spade and pickax to a large extent. Ground water is being found in aquifers that are well hidden, some of them buried deep beneath regions which, at the surface, are dry. Before a well is drilled, the rock or regolith is examined in detail, in an effort to picture conditions well below the surface, such for example, as those indicated in Figures 8.6, 8.7, and 8.8.

### Recharge

In moist regions recharge is accomplished by the natural infiltration of rainfall. But in dry regions it can occur by infiltration from the bed of a river that is leaking. The Platte River in Nebraska and the Nile in Egypt are examples of rivers that flow from mountains having good rainfall, into a much drier region in which the water table lies deep beneath the surface. Water from these rivers leaks downward and recharges the ground water below (Fig. 8.9).

Although most recharge is supplied directly by rainfall, the intense demand for water in some areas has led to artificial recharging of the ground. One example is the practice of *water spreading* in dry parts of the west. A common way to spread water for recharge is to build a low dam across a stream valley. This holds back water that would otherwise run to waste and allows it to seep downward and recharge aquifers beneath the stream bed. The water thereby stored underground is withdrawn through wells as needed.

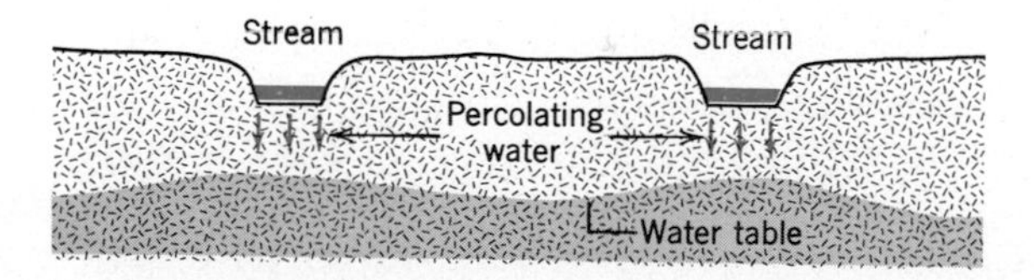


**Figure 8.9**
**Recharge of ground water in a dry region, by leakage from streams having their sources in mountains with abundant precipitation. Relation of water table to streams is the reverse of that shown in Figure 8.1, where leaking is impossible.**

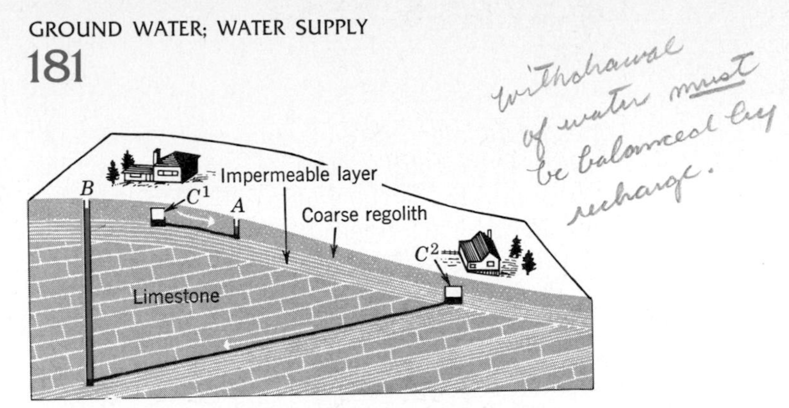


**Figure 8.10**
**Pollution of wells. The shallow dug well, *A*, was unwisely located a short distance downslope from a septic tank, $C^1$, and received polluted drainage (black) from it. The owner then drilled a deeper well, *B*. This well tapped layers of cavernous limestone inclined toward it from the lower septic tank, $C^2$. The water flowed through openings in the limestone, and reached the bottom of well *B* unpurified by percolation. The well owner must relocate his septic tank or else dig a shallow well located upslope from $C^1$.**

A chemical-industrial plant in the Ohio River valley near Louisville, Kentucky, was located on a valley fill of sandy alluvium about 50m thick. Water was needed for cooling. The plant bought city water (purified river water) in winter when the water was cold and fed it into wells in the valley fill, thus recharging the sandy aquifer. In summer the cold water stored in the ground was pumped out for industrial use.

In some districts an aquifer is recharged with used water. This practice has increased with the increased use of air conditioning, which requires a large volume of water. Some cities have laws requiring that water used for air conditioning be returned to the ground, where it successfully builds up the water table. This illustrates the basic principle of ground-water conservation: that withdrawal of water must, in the long run, be balanced by recharge; if it is not, then either recharge must be increased or withdrawal curtailed.

### Water quality

The *quality* of a body of water refers to its temperature and the amount and character of its content of mineral particles, solutes, and organic matter (chiefly bacteria) in relation to its intended use. The most common source of pollution of water from wells and springs is sewage, and the infection most commonly communicated by polluted water is typhoid. Drainage from septic tanks, broken sewers, privies, and barnyards contaminates ground water. If the water contaminated with sewage bacteria passes through material with large openings such as very coarse gravel or the cavernous limestone shown in Figure 8.10, it can travel long distances without much change. If, on the other hand, it percolates through sand or permeable sandstone, it can become purified within short distances, in some cases less than 30m (Fig. 8.11). The difference

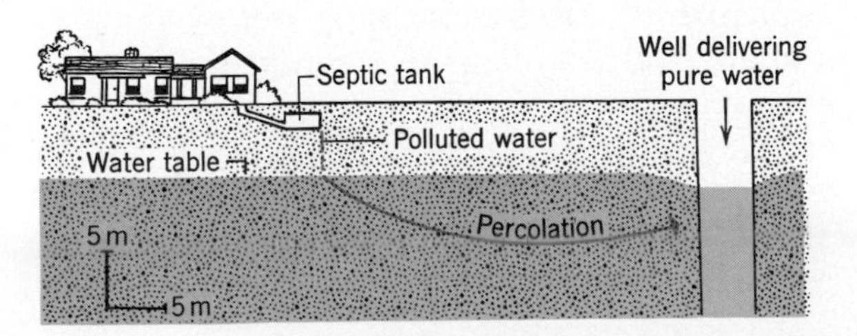


**Figure 8.11**
**Purification of contaminated ground water in sand and gravel during percolation through a short distance.**

lies in the aggregate internal surface area of the material through which it percolates. The large aggregate force of molecular attraction holds the water and promotes its purification by (1) mechanical filtering-out of bacteria (water gets through but most of the bacteria do not), (2) destruction of bacteria by chemical oxidation, and (3) destruction of bacteria by other organisms, which consume and oxidize them. Purification goes on in the zone of aeration as well as in the zone of saturation. Because clay particles are much smaller than sand particles, it might be thought that clay, with its much larger internal surface area, would be the ideal medium for purification. But it is not, because, as we have seen, it is almost impermeable. Particles of sand are large enough to permit rapid percolation, yet small enough to permit purification within short distances. For this reason treatment plants for purification of municipal water supplies and processing of sewage percolate these fluids through sand.

A substantial proportion of domestic sewage passes through septic tanks (Fig. 8.11) and then mingles with ground water. The mixture gradually becomes purified as it percolates toward the nearest streams. In contrast, the domestic waste from many areas, as well as much industrial waste, is dumped unaltered into surface streams. Although purification can be accomplished during stream transport, the distances involved are much greater than those required for the purification of water in the ground, and the amounts of sewage in many rivers are far too great to be dealt with by natural processes. In densely populated industrial countries, these facts constitute serious public-health problems.

A dramatic illustration of the difference between surface flow and underground percolation of contaminated water is this. In some communities much-polluted water that has traveled 30km or more through a river is pumped into the ground, where it becomes a part of the usable ground-water supply. In one city, percolation through a horizontal distance of 150m removed impurities from the sewage and made the water fit to drink.

**Balance sheet of water supply**

Ground water and surface water together are a resource that is an absolute necessity for human use. Table 8.1 shows how much exists, but part of the ground water is not available because it lies too deep, because it lies in rock through which it cannot flow, or because its quality is poor. Table 8.2 shows the four chief uses of ground water in the United States.

Whatever their actual amount, ground water and surface water replenish themselves as the water cycle rolls on and on. Normally, too, the water cycle is in a condition of steady state, so that the supply available to people is nearly constant. In parts of the world where people live in a simple agricultural economy, their withdrawal of water from streams, lakes, and the ground affects the steady state hardly at all because they use little water—possibly less than 10 gallons per day per capita. In

**Table 8.2**
**Use of Ground Water in the United States, 1970.**
**(*U.S. Geol. Survey Circ. 676, 1972, p. 12*)**

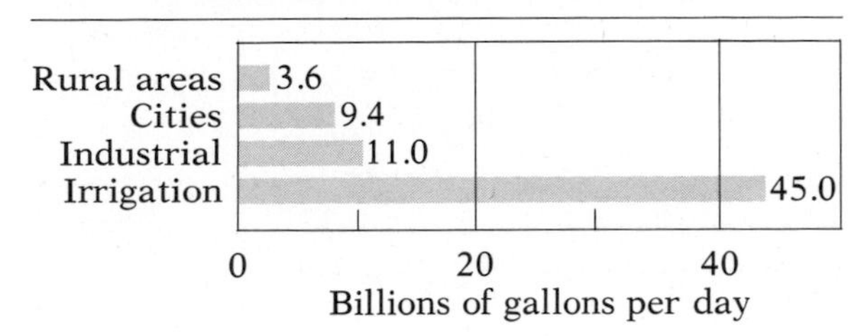


(1 gal. = 3.785 liters = .0378 cubic meters.)

contrast, in 1970 the population of the industrially advanced United States consumed 165 gallons per person per day in its houses alone. But when we add the per-capita share of the water used industrially, the total jumps from 165 to the enormous figure of 1,800 gallons per day—180 times our estimate of consumption by the simplest nonindustrial people. Of this total, about 1,500 gallons are used for irrigation alone. Such prodigal use of water *does* affect the natural, steady-state condition, particularly in dry regions where the demand for irrigation water is greatest. In parts of Arizona and of Israel, for example, it has been found that the $^{14}C$ ages of irrigation water drawn from deep wells are more—perhaps much more—than 10,000 years. We noted similarly "old" water in our discussion of the Ocala Limestone aquifer in Florida. But under Florida's moist climate, with abundant rainfall, recharge from rainfall keeps pace with withdrawal from wells. In dry Arizona, however, "old" water is relict from much earlier times. It represents rain that fell during the latest glacial age, when climates were cooler and moister. Under today's drier climates, rates of recharge are much less than the present high rates of withdrawal. Hence the water used today in many dry areas is being "mined"—that is, used up at a nonreplaceable rate—so that the supply now in the ground will become exhausted. In some irrigation areas, even if all pumping of wells were stopped, it would take more than 100 years to recharge the aquifers.

One of the obvious characteristics of industrialization is the concentration of people in cities. Today more than two-thirds of the people of the United States live in or around cities; only one-third live in the country. Not only do city populations use more water for all purposes (including air conditioning) in their buildings, but also the buildings themselves, as well as the streets and roads that give access to them, reduce local additions from rainfall. In cities the ground is covered largely with buildings, concrete, and asphalt, all of which send runoff along the surface and through sewers, rather than allowing it to soak into the regolith, as happens under natural conditions. It is urbanization, more than growth of population overall, that has created a demand which in many cities threatens to exceed supply. Furthermore, the

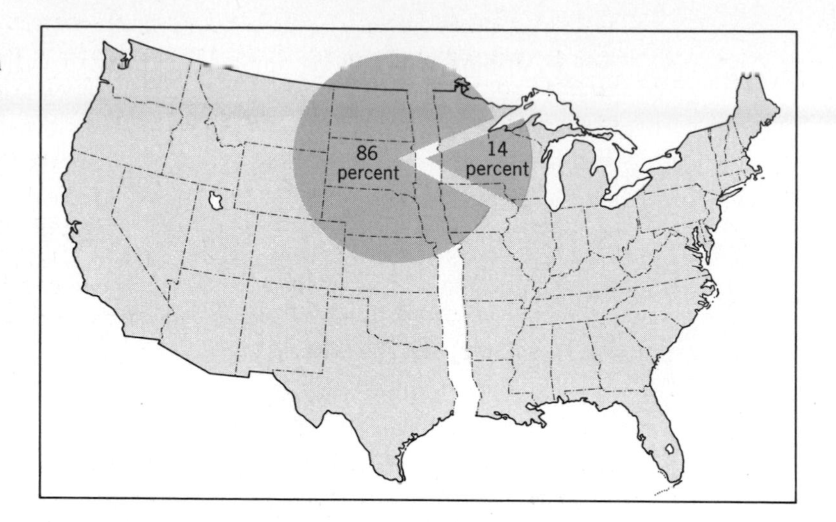


**Figure 8.12**
**Consumption of fresh water in the 17 Western States and the 31 Eastern States, 1970.** *(Murray and Reeves, U.S. Geol. Survey Circ. 676, 1972.)*

available supply is less than it would be were it not for pollution (chemical, physical, bacterial, or thermal) of many streams, some lakes, and some ground water by industrial wastes, agricultural poisons, and sewage.

Figure 8.12 shows that the western part of the United States consumes far more water than does the eastern part, although the eastern part contains more people—and more aquifers too (Fig. 8.4). The difference is mainly a consequence of the far greater use of irrigation (Table 8.2) in the west, where climates are drier and intensive agriculture is widespread.

Much of the water withdrawn from the ground and from surface sources and used for a variety of human purposes is returned to these natural reservoirs to be re-used. As one example, irrigation water may be withdrawn from a river, and the part not evaporated or used by plants is returned through the ground to the stream. This can happen again and again to a parcel of water during its long journey from the middle of a continent back to the ocean.

Many authorities believe that shortage of water in some large cities, although real, is less a matter of inadequate quantity than of inefficient planning and development. They believe that except for the dry southwestern part of the country, the potential supply of water, *if properly managed,* need not necessarily be overtaxed for some time to come.

# GROUND WATER IN THE ROCK CYCLE

As soon as rainwater infiltrates the ground, it begins to dissolve minerals in regolith and bedrock, as part of the general process of weathering. The dissolved matter contained in the ground water reappears in streams as dissolved load and is carried to the sea, where it joins other salts in solution and eventually enters into the building of limestone and other marine sedimentary rock. This chain of activities forms an important part of the rock cycle, helping to erode land and to build rock strata in the ocean.

By measurement of dissolved loads and discharges of streams in sample areas, it is possible to calculate the average rate at which lands are being lowered by chemical processes alone. In southeastern United States, with high rainfall, high water table, and extensive vegetation, lands are being thus lowered at rates of about 1cm/1,000 years. In dry parts of western United States, with low rainfall and water tables and with discontinuous vegetation, rates are very much smaller.

### Chemical composition

Analyses of many wells and springs show that the solutes in ground water consist mainly of chlorides, sulfates, and bicarbonates of calcium, magnesium, sodium, potassium, and iron. We can trace these substances back to the common minerals in the rocks from which they were derived by weathering. As might be expected, the composition of ground water varies from place to place according to the kind of rock in which it occurs. In much of the central United States the water is "hard," that is, rich in calcium and magnesium bicarbonates, because the bedrock includes abundant limestones and dolostones that consist of those carbonates. In some places within arid regions the concentration of dissolved substances, notably sulfates and chlorides, is so great that the ground water is unfit for human consumption. Furthermore, evaporation of water in the unsaturated zone deposits not only calcium carbonate but, in particularly dry regions, sodium sulfate, sodium carbonate, and sodium chloride. Soils containing these precipitates are loosely termed "alkali soils." They are unsuitable for agriculture because crops will not grow in them.

Some ground water is salty. Three causes of saltiness in ground water are:

1. Seawater, trapped in sediment on ancient sea floors, has persisted through the conversion of the sediment to rock.
2. Ground water, circulating slowly through sedimentary rock, has dissolved salts from the rocks themselves and so has become salty.
3. Along coasts, ground water has become mixed with seawater that has percolated into it.

### Deposition

The conversion of sediment into sedimentary rock, one of the great transformations that occur as parts of the rock cycle, is primarily the work of ground water. A body of sediment lying beneath the sea is generally saturated with water, as is a sediment lying in the saturated zone beneath the land. Substances in solution in the water are precipitated as a cement in the spaces between the rock particles that form the sediment. This activity transforms the loose sediment into

**Figure 8.13**
**Large cavern in limestone, partly refilled with dripstone and flowstone in the form of stalacites, stalagmites, and columns. Length of view along base of photo, about 6m. (*Luray Caverns, Virginia.*)**

firm rock. Calcite, silica, and iron compounds (mainly oxides), are, in that order, the chief cementing substances.

Less common than the deposition of cement between the grains in a sediment is ***replacement***, *the process by which a fluid dissolves matter already present and at the same time deposits from solution an equal volume of a different substance.* Evidently replacement takes place on a volume-for-volume basis because the new material preserves the most minute textures of the material replaced. Petrified wood is a common example. But replacement is not confined to wood and other organic matter; it occurs in mineral matter as well.

### Caverns

Limestone, dolostone, and marble are carbonate rocks that consist of the minerals calcite and dolomite in various proportions. These rocks underlie millions of square kilometers of Earth's surface. Although carbonate minerals are nearly insoluble in pure water,

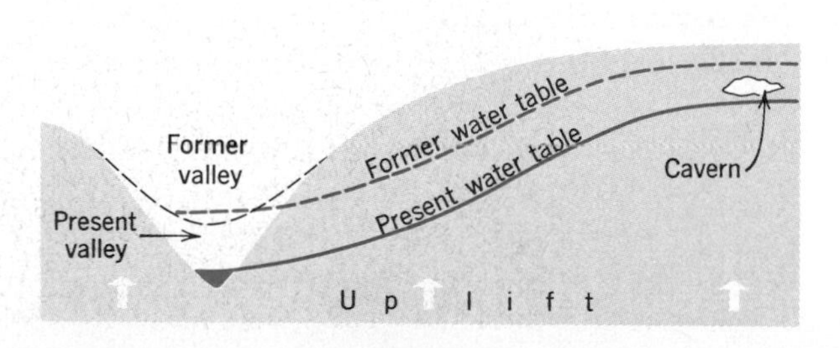


**Figure 8.14**
**Possible history of a cavern containing dripstone. Cavern was excavated below water table. When streams deepened their valleys, water table was lowered as it adjusted to the deepened valleys. This left cavern above water table.**

they are readily dissolved by the carbonic acid in ground water. As a result the ground water becomes charged with calcium bicarbonate, as shown by these reactions, the first of which we saw in our analysis of chemical weathering in Chapter 6:

$$\underset{\text{(Carbon dioxide)}}{CO_2} + \underset{\text{(Water)}}{H_2O} \longrightarrow \underset{\text{(Carbonic acid)}}{H_2CO_3} \text{ which ionizes to } \underset{\text{(Hydrogen ion)}}{H^{+1}} + \underset{\text{(Bicarbonate ion)}}{(HCO_3)^-}$$

The hydrogen ions attack the calcite and dissolve it.

$$\underset{\text{(Calcite)}}{CaCO_3} + \underset{\text{(Hydrogen ion)}}{2H^{+1}} \longrightarrow \underset{\text{(Water)}}{H_2O} + \underset{\text{(Carbon dioxide)}}{CO_2} + \underset{\text{(Calcium ion)}}{Ca^{2+}}$$

Such dissolution is a form of chemical weathering just as much as is the decomposition of igneous rock that contains feldspar and ferromagnesian minerals. In both cases the weathering attack occurs along joints and other partings in the bedrock. But whereas in granite the quartz and other insoluble minerals remain, nearly all the substance of a body of pure limestone can be carried away in solution in the slowly moving ground water. The process of dissolution creates cavities of many sizes and shapes. *A large, roofed-over cavity in any kind of rock* is a ***cavern.***

Although most caverns are small, some are of exceptional size. Carlsbad Caverns in southeastern New Mexico include one chamber 1,200m long, 190m wide, and 100m high. Mammoth Cave, Kentucky, consists of interconnected caverns with an aggregate length of at least 48km.

Some caverns have been partly filled with insoluble clay and silt, originally present as impurities in the limestone and gradually released by dissolution. Other caverns contain partial fillings of ***dripstone*** (*material chemically precipitated from dripping water in an air-filled cavity*) and ***flowstone*** (*material chemically precipitated from flowing water in the open air or in an air-filled cavity*). The "stones" take on many curious forms, which are among the chief attractions to cavern visitors. The most common shapes are ***stalactites*** (*icicle-like forms of dripstone and flowstone, hanging from ceilings*), ***stalagmites*** (*blunt "icicles" of flowstone projecting upward from floors*), and ***columns*** (*stalactites joined with stalagmites, forming connections between the floor and roof of a cavern*) (Fig. 8.13).

As its name implies, dripstone is deposited by successive drops of water. As each drop of water forms on the ceiling of a cavern, it loses a tiny amount of carbon dioxide gas and precipitates a particle of calcium carbonate. This chemical reaction is simply the reverse of the one by which calcium carbonate is dissolved by carbonic acid.

Dripstone can be deposited only in caverns that are already filled with air and are therefore above the water table. Yet many, perhaps most, caverns are believed to have formed below the water table, as is suggested by their shapes and by the fact that some caverns are lined with crystals, which can form only in a water environment. How can we reconcile the two conflicting environments? Probably the answer lies in repeated, long-continuing uplift and erosion of continents, which has been going on for hundreds of millions of years, responding, at least in part, to movements of plates of lithosphere. In the process, caverns, one by one, would be lifted above the water table and filled with air, after which they could begin to be filled with dripstone (Fig. 8.14).

### Sinks

In contrast to a cavern, a *sink* is *a large solution cavity open to the sky.* Some sinks are caverns whose roofs have collapsed. Others are formed at the surface, where rainwater is freshly charged with carbon dioxide and is at its most effective as a solvent. Many sinks, located at the intersections of joints where movement of water downward is most rapid, have funnel-like shapes.

Figure 8.15 shows a large, very young sink formed recently by collapse. It is one of more than 1,000 collapses that have occurred in recent years within an area of about 25 square kilometers. The cause may be lowering of the water table by drought and by excessive pumping of local wells. Some sinks are much larger than

Figure 8.15
**Sink, nearly 130m long and 45m deep, near Montevallo, Alabama, formed at 2 P.M. on December 2, 1973. Debris of the roof form a talus that conceals the bedrock. The bottom appears to contain a pond of rainwater. (*U.S. Geol. Survey.*)**

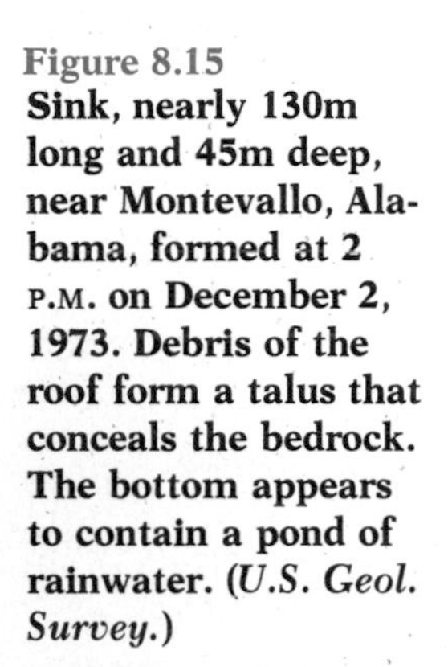

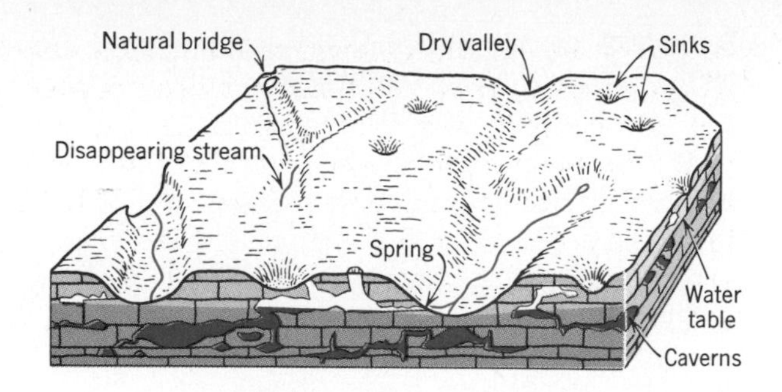


**Figure 8.16 Karst topography. Area of block is between 2 and 3 square kilometers.**

this one. A far older sink near Mammoth Cave, Kentucky, is 13 square kilometers in area.

### Karst topography

In some regions of exceptionally soluble rocks, sinks and caverns are so numerous that they combine to form a peculiar topography characterized by many small basins. In this kind of topography the drainage pattern is irregular; streams disappear abruptly into the ground, leaving their valleys dry and then reappear elsewhere as large springs. This has been termed ***karst topography*** (Fig. 8.16) because it is strikingly developed in the Karst region of Yugoslavia, inland from Trieste. It is defined as *an assemblage of topographic forms consisting primarily of closely spaced sinks.* Karst topography is developed through wide areas in Kentucky, Tennessee, southern Indiana, northern Florida, and Puerto Rico.

Sinks and caverns record the destruction of a very large volume of carbonate rock. It is calculated from measured amounts of carbon dioxide in ground water and from the solubilities of limestones that the amount of precipitation on northern Kentucky is capable, as ground water, of dissolving a layer of limestone 1cm thick every 66 years. This potential is far greater than the average erosional reduction of the surface of the United States by mass-wasting, sheet erosion, and streams. It depends on the presence of exceptionally soluble rocks.

# SUMMARY

### GEOLOGIC SIGNIFICANCE

1. In the water cycle, ground water plays a part in the return of water from land to sea.

### DISTRIBUTION AND ORIGIN

2. Ground water, derived almost entirely from rainfall, occurs nearly universally.

3. The water table is the top of the saturated zone. Its form is a subdued imitation of the ground surface above it.

### MOVEMENT

4. Ground water flows chiefly by percolation, at rates far slower than those of surface streams.

5. With constant permeability, velocity of flow of ground water increases as slope of the water table increases.

6. In moist regions ground water percolates away from hills and emerges in valleys. In dry regions it is likely to percolate away from beneath large surface streams.

ECONOMY

7. A ground-water system is an open system in each segment of which a steady state is approached.

WELLS AND SPRINGS

8. Ground water flows into most wells directly by gravity, but into artesian wells under hydrostatic pressure.

9. Withdrawal of water through wells creates cones of depression in the water table.

10. In a dry region, recharge can be supplied by a river that leaks into the ground.

GROUND WATER AND PEOPLE

11. A basic principle of conservation is that withdrawal of ground water must not exceed recharge.

GROUND WATER IN THE ROCK CYCLE

12. Ground water dissolves mineral matter from rock; much of the dissolved product eventually gets into the sea.

13. Ground water deposits substances as cement between grains and so reduces porosity and converts sediment into rock.

14. In carbonate rocks ground water not only creates caverns and sinks by dissolution but also, in some caverns, deposits mineral matter.

# SELECTED REFERENCES

Bretz, J H., 1956, Caves of Missouri: Missouri Geol. Survey, v. 39.

Davis, S. N., and DeWiest, R. J. M., 1966, Hydrogeology: New York, John Wiley.

Ellis, A. J., 1917, The divining rod, a history of water witching: U.S. Geol. Survey Water-Supply Paper 416.

Hubbert, M. K., 1940, The theory of ground-water motion: Jour. Geology, v. 48, p. 785–944.

Leopold, L. B., 1974, Water, a primer: San Francisco, W. H. Freeman.

McGuinness, C. L., 1963, The role of ground water in the national water situation: U.S. Geol. Survey Water-Supply Paper 1800.

Meinzer, O. E., 1923, The occurrence of ground water in the United States, with a discussion of principles: U.S. Geol. Survey Water-Supply Paper 489.

Moore, G. W., and Nicholas, G., 1964, Speleology. The study of caves: New York, D. C. Heath.

Murray, C. R., and Reeves, E. B., 1972, Estimated use of water in the United States in 1970: U.S. Geol. Survey Circ. 676.

Todd, D. K., 1959, Ground-water hydrology: New York, John Wiley,

U.S. Dept. Agriculture, 1955, Water: Yearbook for 1955: Washington, U.S. Govt. Printing Office.

# 9 DESERTS AND WINDS



# DESERTS

## WORLD DISTRIBUTION OF DESERTS

Although the word *desert* means literally a deserted, unoccupied, or uncultivated area, the modern development of artificial water supplies has changed the original meaning of the word by making many dry countries habitable. *Desert* has become a synonym for *arid land,* whether "deserted" or not. But aridity remains the chief characteristic of any desert.

Arid lands of various kinds together add up to 25 percent of the total land area of the world. In addition there is a smaller though still large percentage of semiarid land, in which annual rainfall ranges between about 250 and 500mm (10 and 20 inches). These dry and semidry areas are seen on a world map in Figure 9.1, forming a distinctive pattern. The meaning of the pattern becomes clear as soon as we grasp the general plan of circulation of the atmosphere.

#### Circulation of Earth's atmosphere

The atmosphere is continually in motion. It circulates in a definite pattern. The basic cause is that more of the Sun's heat is received

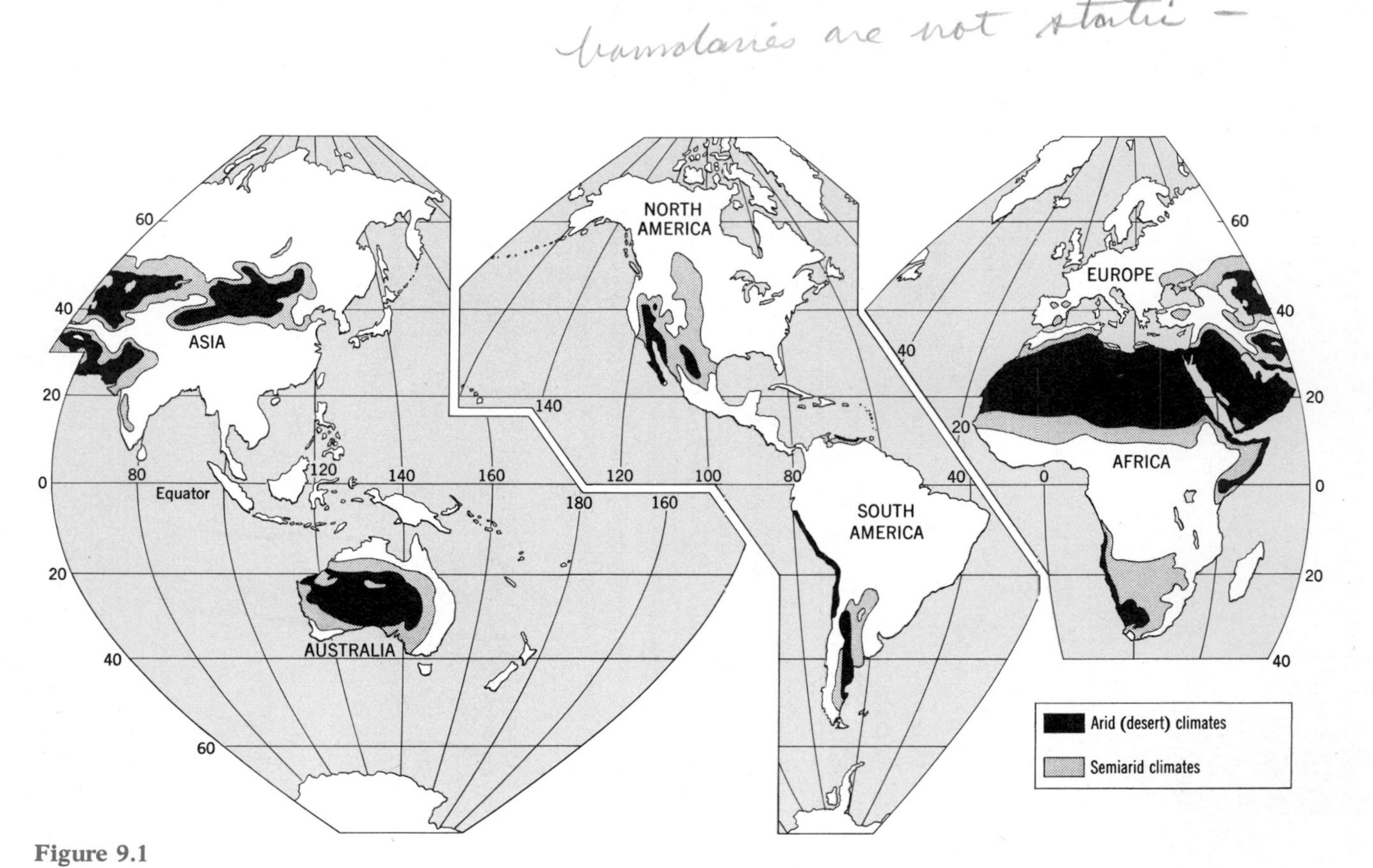


Figure 9.1

**Arid and semiarid climates of the world according to the Koppen–Geiger classification. Very dry parts of the polar regions, a special, cold kind of desert, are not included.**

near the equator than near the poles. The heated air near the equator expands, becomes lighter, and rises by convection, like the rise of boiling water in a teakettle. High up, it spreads outward toward both poles; and on the way it gradually cools, becomes heavier, and sinks. Meanwhile, beneath it still cooler, heavier air, chilled in the polar regions, forms a return flow toward the equator. The returning air replaces the heated rising air and in turn is heated and rises.

This simple circulation, however, is interfered with by Earth's rotation, which sets up a force, named the *Coriolis force* after the 19th-Century French engineer who first analyzed it. The effect is to cause any body that moves freely with respect to the rotating solid Earth, to veer toward the right in the northern hemisphere and toward the left in the southern, *regardless of the direction in which the body may be moving.* Flowing water (such as an ocean current) and flowing air (winds) respond to it, as do smaller bodies such as airplanes, and projectiles of all kinds.

The Coriolis force breaks up the simple general flows of air between equator and poles into sections or belts (Fig. 9.2). Looking first only at the northern hemisphere, we note that at about latitude 30° some of the high-level, north-moving equatorial air is descending toward Earth's solid surface. As it descends it gets warmer; and the warmer it gets the more water it can hold; so this air also becomes drier. This means that near that latitude, right around the world, climates are warm and dry, skies are clear, and rain is scarce.

The descending air spreads out along the surface, toward the north

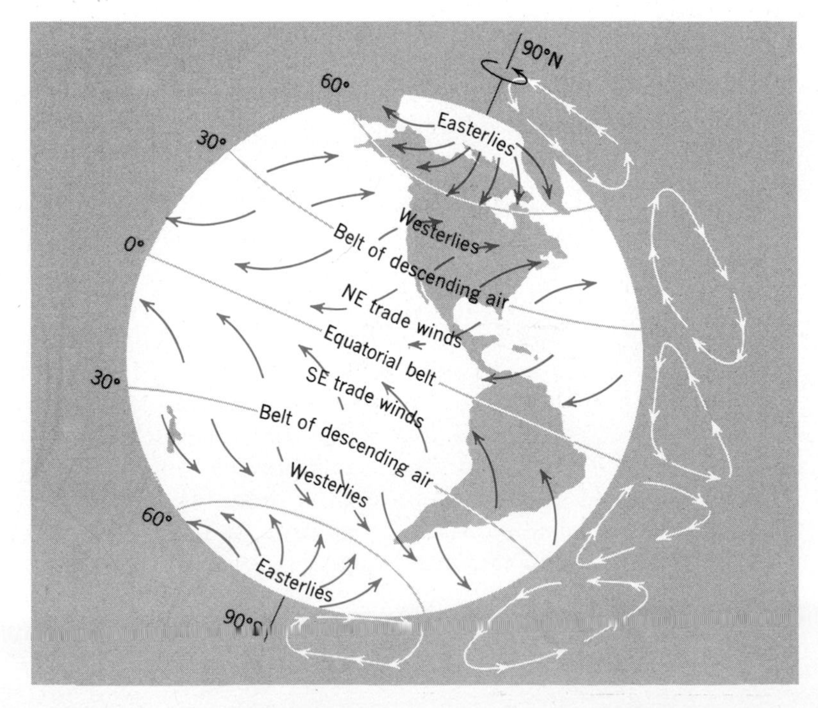


**Figure 9.2**
**Earth's planetary wind belts, shown schematically. At right are vertical cross sections of the six belts. In each cross section the arrows show direction of flow of air.**

and toward the south. In the northern hemisphere the south-flowing part is twisted toward the right (west), so that the south flow becomes a southwest flow. In ordinary language we call such a wind a *northeast* wind, for the direction *from* which it is blowing. As Figure 9.2 shows, these winds form a belt of *northeast trade winds,* extending around the world.

Returning to the northern edge of this belt where air is descending, we now follow the air that is spreading toward the north. This flow too is twisted toward the right, so that northward flow becomes northeastward flow, forming winds that blow from the southwest. These winds are the belt of westerly winds or *westerlies,* again extending around the world.

At higher latitudes, somewhere between 40° and 60°, the westerlies encounter the cold heavy air that flows from the polar region toward the equator. This flow, of course, is also twisted toward the right; so it moves toward the southwest as generally *easterly* winds. Where the warmer and lighter air of the westerlies encounters this cold polar air, some of it tends to rise and return toward the equator. The polar air itself becomes warmer, gradually rises, and returns toward the pole.

To see what happens in the southern hemisphere we need only repeat this description, substituting left turns for right turns throughout. The result is a belt of southeast trade winds, a belt of descending air near 30° south latitude, a belt of westerlies, and a belt of polar easterlies, thus completing a neat pattern for the Earth. In this pattern, what is important for our discussion of deserts is that in both hemispheres most of the arid land (Fig. 9.1) is centered between latitudes 15° and 35°. There, because air is descending and becoming warm, skies are clear and rain is scarce. A perfect place for a desert!

This great system of flow and return flow of air is a cycle, the cycle of circulation of the atmosphere, one of the many systems mentioned in Chapter 4.

### Kinds of deserts

Arid lands fall into three chief classes, the most extensive of which is (1) the one determined by Earth's wind belts. Examples of this class are the Sahara and other deserts in northern Africa. (2) A second sort of desert is found in continental interiors, where heating in summer and dry, cold continental air in winter prevail. Instances are the deserts of central Asia. (3) Yet a third, more local kind of arid area is one that lies in the lee of a mountain range. The mountains act as a barrier to rainfall on the desert beyond. Moist air encounters the mountains, rises over their windward slope, is cooled, drops rain. As it descends over the leeward slope the air becomes warmer and drier, creating a dry climate over the country beyond. The high Sierra Nevada in eastern California is the barrier mainly responsible for the arid climate of the country east of it.

# CLIMATE AND VEGETATION

The arid climate of a desert results from the combination of three factors:

1. *High temperature.* The highest temperature recorded in the United States, 56°C, is at Death Valley, a desert area in southeastern California. The world's record, at a place in the Libyan Desert, is 57.7°C.

2. *Low precipitation.* At Death Valley annual precipitation averages between 20mm and 50mm, and in the Atacama Desert in northern Chile periods of more than 12 consecutive years without rain have been recorded.

3. *Great evaporation.* The higher the temperature, the greater the evaporation, and therefore the more precipitation an area can receive and still be arid. For if most of the precipitated water evaporates, little is left for streams and for vegetation. In parts of the southwestern United States evaporation from lakes and reservoirs amounts to as much as 250mm annually—10 to 20 times more than the annual precipitation.

Besides aridity resulting from these three factors, deserts are generally characterized by frequent strong winds. These commonly result from convection. During daytime hours air over especially hot places is heated and rises, and this allows surface air to move in rapidly and take its place. Desert winds are effective movers of sediment, as we shall see.

The vegetation in deserts is a direct reflection of dry climate. Usually the vegetation is not continuous. Where grass is present, it is likely to be thin and to grow only in clumps. More commonly the plants consist of low bushes growing rather far apart, with bare areas between them. This pattern of vegetation promotes active movement of sediment by the wind and by running water as well.

But regardless of the local effects of wind and of running water, the climates of deserts and of the semiarid regions adjacent to them fluctuate; no one year is exactly like another. Natural year-to-year variations, when accompanied by human interference, can be disastrous. In the country that lies south of the Sahara Desert is a belt of very dry grassland known as the *Sahel,* an Arabic word meaning *border.* On it the annual rainfall is normally only 100 to 300mm, most of it falling during a single short season.

In the early 1970s the Sahel was visited by a drought, the worst in this century. For several years in succession the rains failed, causing the adjacent desert to spread southward—according to one estimate, by as much as 150km. The drought extended from the Atlantic to the Indian Ocean, a distance of 6,000km (Fig. 9.3), affecting a population of at least 20 million people, many of them seminomadic tenders of cattle, camels, and goats. The results of the drought were intensified by the fact that between about 1935

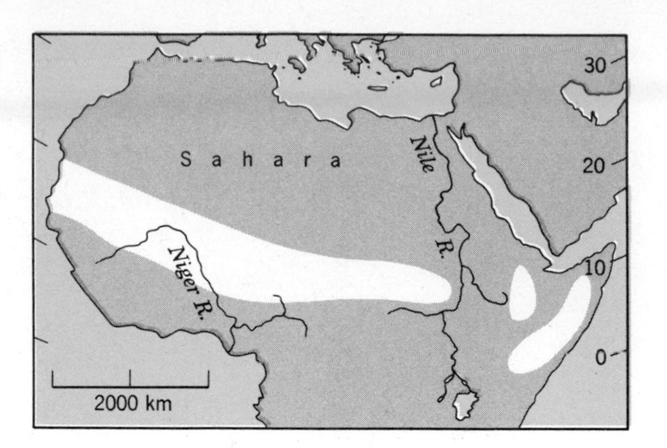


Figure 9.3
**Sketch map of part of Africa, showing the extent of the Sahel (white) and the adjacent Sahara. Compare Figure 9.1. (*After H. Brabyn, UNESCO.*)**

and 1970 the human population had doubled, and along with it that of the livestock. This increase of people and animals had led to severe overgrazing, so that with the coming of drought the grass cover failed almost completely. Some 40 percent of the cattle—a great many millions—died, and millions of people, suffering from thirst and starvation—some to the point of death—migrated southward in search of food and water.

By 1975 the rains had returned, but a huge problem remained, for according to some opinions the grazing lands will be unusable for years to come. Experts recommended that grazing lands be fenced to control the nomads' herds, that deep wells be drilled for watering stock, and that in the less dry southern Sahel where in normal times cropland agriculture is possible, lands should be allowed to lie fallow in alternate years to accumulate moisture in the soil, and should be protected further by creating windbreaks (Fig. 9.24). Irrigation systems are recommended for parts of the southern area.

The story of the Sahel illustrates the important fact that the fertility of most desert soil depends mainly on water. The story also is a clear warning of the danger involved in the intrusion of too many people into a delicately balanced system that involves regolith, hydrosphere, and biosphere and that when not interfered with is in a condition of steady state.

# GEOLOGIC PROCESSES

No major geologic process is restricted entirely to arid regions. Rather, the same processes operate with different intensities in moist and arid regions. As a result, in a desert the forms of the land, the soils, and the sediments show distinctive differences. Let us look at some of the major processes and note the differences.

### Weathering and mass-wasting

In a moist region, regolith covers the ground almost universally, and is comparatively fine textured because it usually contains clay, a product of chemical weathering. It is in motion downslope mainly by creep, and it is covered with almost continuous vegetation. Creep fashions hill profiles into a series of curves.

In a desert the regolith, much of it a product of mechanical weathering, is thinner, less continuous, and coarser in texture. Slope angles developed by downslope creep become adjusted to the average diameter of the particles of regolith; the coarser the particles the steeper the slope required to move them. As the particles created by mechanical weathering tend to be coarse, slopes are generally steeper than in a moist region.

Mechanically weathered chunks of rock tend to break off along joints, leaving steep, rugged cliffs. Hills with cliffy slopes, particularly where layers of bedrock are nearly horizontal, are common in dry regions (Fig. 9.4).

### Streams

One of the characteristics of deserts is that most of the streams which originate in them soon disappear by evaporation and by soaking into the ground, and never reach the sea. Exceptions are long rivers such as the Nile in Egypt and the Colorado in the southwestern United States, which originate in mountain regions with abundant precipitation. Such rivers carry so much water that they keep flowing to the ocean despite great losses where they cross a desert. In a desert the few plant roots are no great impediment to runoff of rain, and the loose dry regolith is eroded easily. Typical violent rainstorms are therefore likely to be accompanied by "flash" floods that move heavy loads of sediment suddenly and swiftly. The loads are deposited as alluvium, forming fans at the bases of mountain slopes (Fig. 9.5) and on the floors of wide valleys and basins. The deposits of some flash floods are spectacular.

Often, streams in flood effectively undercut the sideslopes of their valleys, causing the slopes to cave. Then, as the flood subsides, the load is deposited rapidly, creating a flat floor of alluvium. The result is a steep-sided, flat-bottomed "box canyon" characteristic of many dry regions (Fig. 9.6)

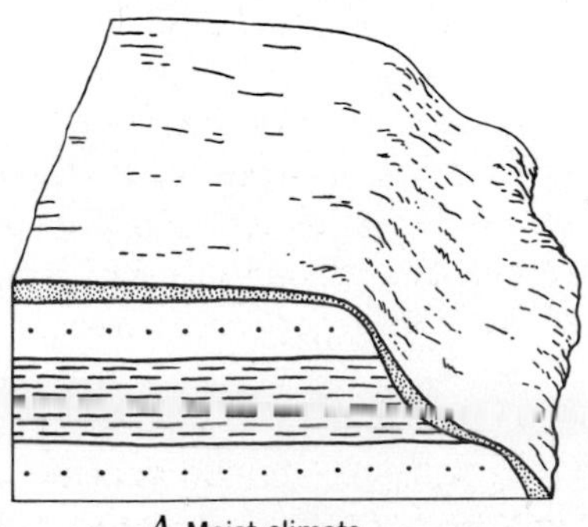

*A* Moist climate

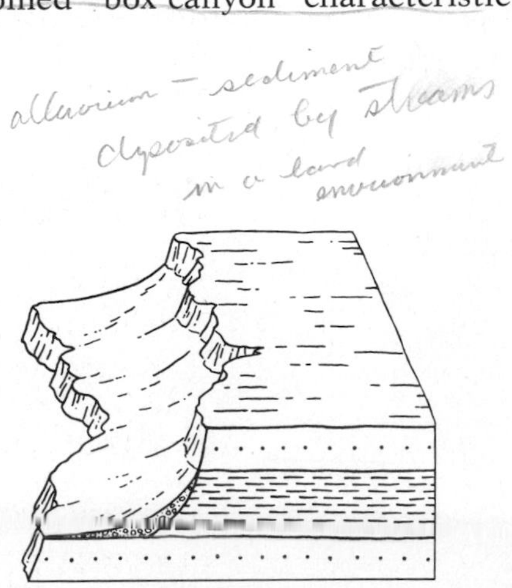

*B* Dry climate

**Figure 9.4**
**Effect of climate on sideslopes of valleys.**
***A.*** **In a moist climate resistant strata are partly masked by a creeping mantle of chemically weathered waste.**
***B.*** **In a dry climate resistant rocks stand out as broad plat forms and steep cliffs, partly concealed by taluses.**

Figure 9.5
**Fan being built out into Death Valley, California, a down-faulted, desert basin with white, salt-incrusted playas. Highway across fan serves as a scale. (*J. S. Shelton.*)**

**Figure 9.6**
**Small "box canyon" cut into silty alluvium. Cornfield Wash, Albuquerque district, New Mexico. (*F. W. Kennon and H. V. Peterson, U.S. Geol. Survey.*)**

**Playa lakes**

In an arid region, only rarely is water abundant enough to flow into a basin and fill it to overflowing. Streams that flow down from a highland rarely last until they reach the center of the nearest basin. But after an exceptionally large rainstorm some of them discharge enough water to convert the basin floor into a shallow lake that may last a few days or a few weeks (Fig. 9.7). In the dry region of the western United States an *ephemeral shallow lake in a desert basin* is called a ***playa lake;*** when *dry* (that is, most of the time) the *lake bed* is a ***playa*** (Fig. 9.8). Many playas are white or grayish because of precipitated salts at their surfaces. But if the lake water can escape downward through the basin floor before evaporation saturates the water with salts, no salts can be precipitated, and the playa sediments consist mainly of clay.

**Ground water**

In a dry region ground water is derived only from the scanty local rainfall plus water that flows in from outside via rivers or via artesian aquifers. Useful supplies are therefore small and the water table lies well below the surface. Under these conditions it is especially important that withdrawal of ground water be kept in balance with rate of recharge.

**Wind**

In dry country, wind is an effective geologic agent. However, contrary to popular belief, deserts are not characterized mainly by sand dunes. Only one-third of Arabia, the sandiest of all dry regions, and only one-ninth of the Sahara are covered with sand. Much of the nonsandy area of deserts is cut by systems of stream valleys or is characterized by fans and alluvial plains. Thus even in deserts more geologic work is done by streams than by wind.

The way in which wind works in deserts and in moister regions is set forth later in this chapter, after we have summarized desert landscapes.

**Figure 9.7**
**Braun's Playa, near Las Vegas, Nevada, nearly 8km in greatest diameter. Mountain crest is 24km distant. Playa lake on April 10 after an unusually large rainfall. The lake is less than 1m deep. (*C. E. Erdmann.*)**

**Figure 9.8**
**Braun's Playa two weeks later, after the water had evaporated. Wind is blowing fine-grained lake sediment (clay and crystals of salts) into dust clouds. Dark spots are desert bushes. (*C. E. Erdmann.*)**

# LAND SCULPTURE

### Pediments

In deserts, the land forms sculptured by erosion differ from those in country with more rainfall. When the terrain unit consists of a mountain range and an adjacent basin, two kinds of situations are common. One (shown in Figures 9.5 and 9.9*A*) consists of a row of fans along the mountain base; the fans merge outward into a general fill of sediment in the basin. The other situation (Fig. 9.9*B*), more remarkable and less easy to explain fully, consists of a sloping surface at the mountain base that closely resembles a merging row of fans. But the surface is not that of fans. Instead of being constructional (built of alluvium), it is erosional and cuts across bedrock. Scattered over it are rock particles, some brought by running water from the adjacent mountains and some derived by weathering from the rock immediately beneath. Downslope, the rock particles gradually merge, forming a continuous cover of alluvium. The bedrock surface, which may be many kilometers long, has passed beneath a basin fill.

The eroded bedrock surface is called a *pediment* because the surfaces adjacent to the two bases of a mountain mass, seen in

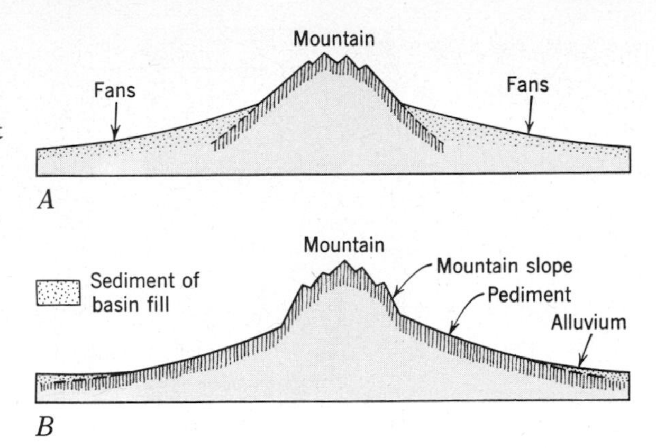


**Figure 9.9**
**Two common relationships at the bases of mountains in desert country.**
***A.* Fans built by streams at the foot of a mountain.**
***B.* Pediment eroded across bedrock at the foot of a mountain.**

profile, together resemble the triangular pediment or gable of a roof (Fig. 9.9*B*). A ***pediment,*** then, is *a sloping surface, cut across bedrock, adjacent to the base of a highland in an arid climate.* The kinds of bedrock on which pediments are cut are those that yield easily to erosion in such a climate. The profile of a pediment (Fig. 9.10), like that of a fan, is concave-up, a form we associate with the work of running water, and the pediment surface is marked by faint, shallow channels. It is likely, therefore, that pediments are the work mainly of running water, flowing as definite streams, as sheet runoff, or as rills, and in any case flowing mainly during rainstorms.

Beyond this general statement we must admit that the exact way in which a pediment develops is not established and is partly a matter of opinion. But we can see that a pediment meets the mountain slope at its head not in a curve but at a distinct angle (Fig. 9.9*B*). This suggests that instead of becoming gentler with time, as they would do in a wet region where chemical weathering and creep of regolith are dominant, mountain slopes in the desert seem to adopt an angle determined by resistance of the bedrock, and to maintain that angle as they gradually retreat under the attack of weathering and mass-wasting. Retreat of the mountain slope lengthens the pediment at its upslope edge. This growth of the pediment at the expense of the mountain should continue until the entire mountain has been consumed. During the whole time of pediment growth, rock particles are transferred downslope intermittently from mountain and pediment to the basin fill beyond.

### Cycle of erosion in deserts

The concept of a cycle of erosion (Chapter 7), in which a land having abundant rainfall is gradually reduced to a peneplain, applies equally well to deserts. Figure 9.11 sketches a desert cycle of erosion in three stages that can be seen today in parts of the western United States.

The stage represented in *A* commonly occurs in northern Nevada, where mountains are being dissected actively. At the mouths of the mountain canyons streams spread out waste in the form of fans

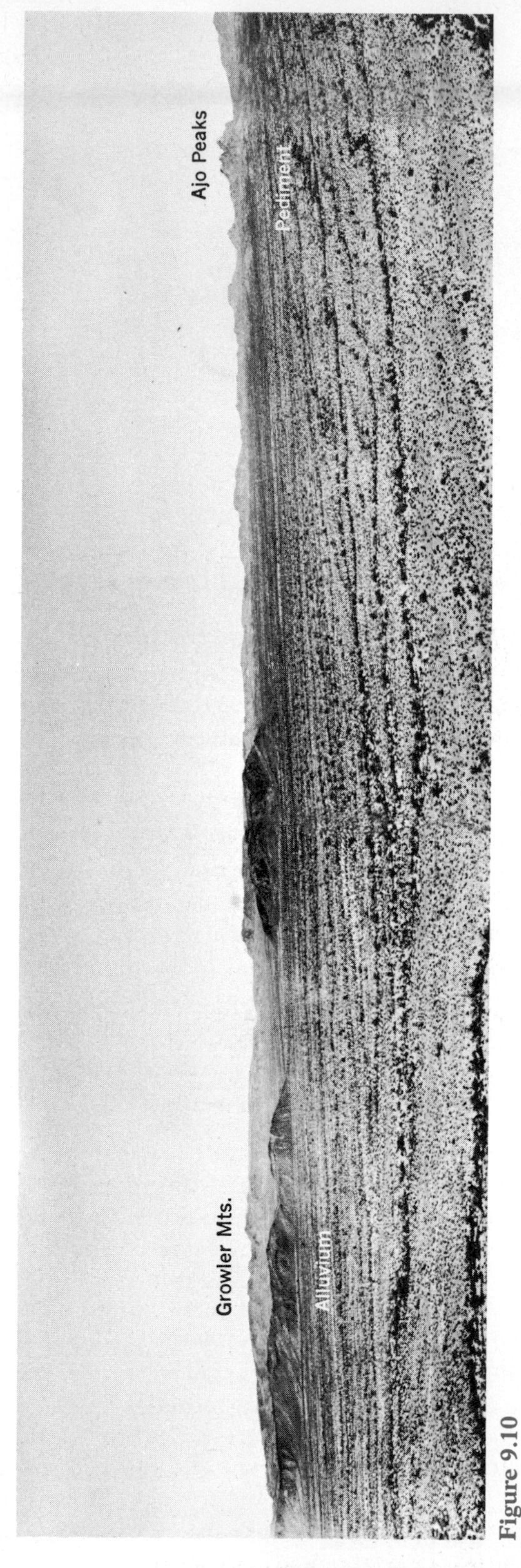


Figure 9.10
**Pediment at south base of Little Ajo Mountains, Arizona. The black butte near center is about 1.5km long. Stripelike rows of bushes are growing along faint, shallow channels. The view represents one-half of what is shown in the section (Fig. 9.11). (*James Gilluly, U.S. Geol. Survey.*)**

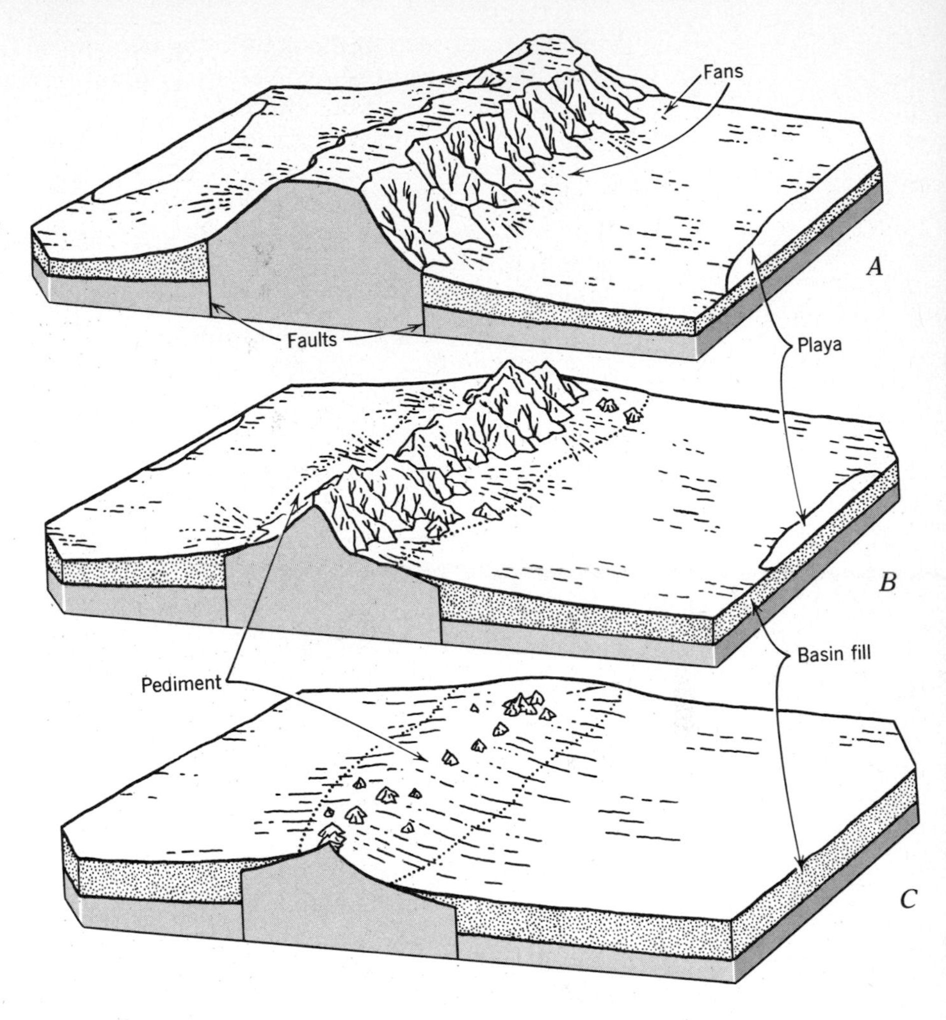


**Figure 9.11**
**Three stages in the sculptural evolution of a mountain range and two basins (originally created by faulting) in an arid climate.**

that grade outward into playas. The wind picks up fine waste, sorts it, heaps sand-size particles into dunes, and lifts some of the finest particles out of the basins altogether. Meanwhile the rocky mountain slopes are being worn back.

In *B*, a stage occurring in southern New Mexico in the country north of El Paso, Texas, the mountain slopes have retreated, exposing a belt of bedrock at each mountain base to weathering and running water and so creating pediments. The steep mountain slopes, covered sparsely with coarse weathered rock waste moving slowly down them, maintain their steepness instead of becoming gentler with time, as would result from soil creep in a moist region.

As the area of the mountains diminishes, the sediment contributed to the streams during rainstorms decreases also. This reduces the loads of the streams, which begin to erode the heads of the former fans, planing them down. In the process the streams cut sideways into the bedrock at and near the mountain front, planing it off and adding to the area of pediment.

In *C*, a stage seen in the country northwest of Tucson, Arizona, the mountains have been reduced by gradual retreat of their steep slopes to a series of big knobs projecting abruptly above the sloping pediment that surrounds them. Outward beyond the pediment is the surface of the basin fill, beneath which the pediment disappears without any break in the smooth, concave slope. As the basin slopes become gradually gentler, water from the mountains reaches the basin centers more rarely, and increasingly winds that sweep across the basins pick up fine sediment and carry it away. The stage shown in *C* is essentially the desert equivalent of a peneplain.

There is reason to believe that in southern Arizona a desert cycle of erosion has been in progress with little interruption for millions of years, and it still has a long way to go.

# WIND ACTION

At this point it is convenient to discuss wind action. However, although the effects of wind action are visible in deserts, we repeat that winds are not the dominant process that shapes desert lands. The foregoing section on land sculpture makes it clear that in deserts as in moist regions running water is the dominant agent.

## TRANSPORT OF SEDIMENT BY WIND

We are now in a position to compare flowing air with flowing water as an agent that erodes and deposits sediment. Wind (flowing air) is normally turbulent. Its velocity, like that of a stream, increases with height above the ground. Like a stream, it can carry coarser particles as bed load and finer ones as suspended load. The two loads are clearly visible in a strong wind blowing across a North African desert. The bed load, a layer rarely as much as 1m thick, consists of sand; it grades upward into clouds of suspended silt and clay particles that may reach great heights.

### Bed load

Experiments with sand blown artificially through glass-sided wind tunnels show that sand grains move in long jumps, as they do in a stream of water (Fig. 9.12*A*; compare Fig. 7.6). The jumps are elastic bounces like those of a ping-pong ball.

A sand grain gets into the air only by bouncing or by being knocked into the air through the impact of another grain. When the wind becomes strong enough, a grain starts to roll along the surface under the pressure of a fast-moving forward eddy. It strikes another grain and knocks it into the air. When the second grain hits the ground, it either splashes up still other grains, making a tiny crater, or bounces into a new jump (Fig. 9.13). In a short time the air close to the ground has become filled with jumping sand

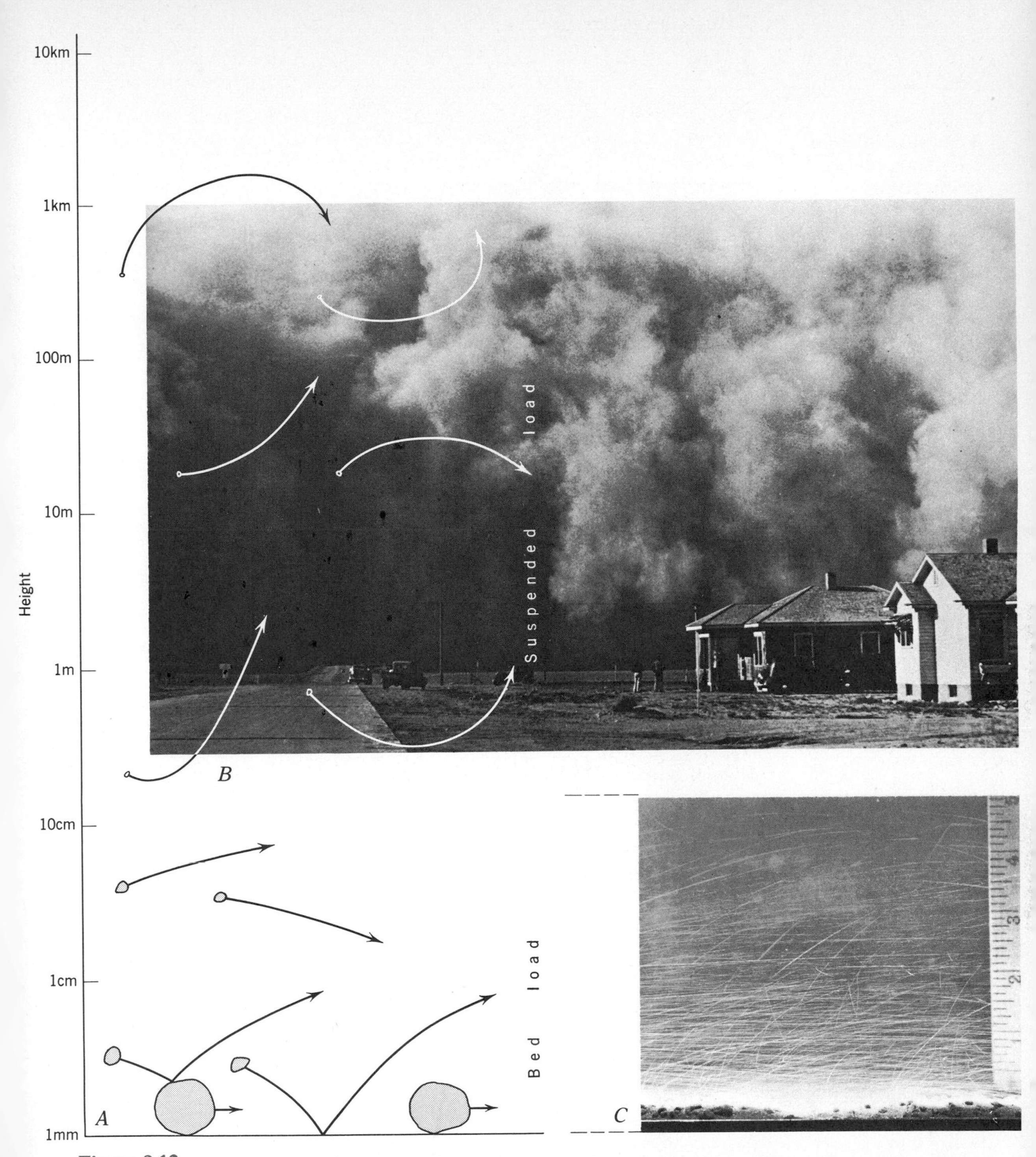


**Figure 9.12**
**Movement of rock particles carried by wind.**

***A.*** **Graph showing vertical distribution, comparative sizes, and movement of particles in bed load and suspended load.**

***B.*** **The "Dust Bowl" during the 1930s. A dust cloud approaches Springfield, Colorado at 4:47 P.M. on May 21, 1937. Total darkness lasted 30 minutes. (*U.S. Soil Conservation Service.*)**

***C.*** **Paths of sand grains being blown through a wind tunnel, photographed in a narrow beam of sunlight. Scale units are inches. Sand grains and small pebbles are visible on tunnel floor. Air current is moving left to right. (*A. W. Zingg, U.S. Dept. of Agriculture.*)**

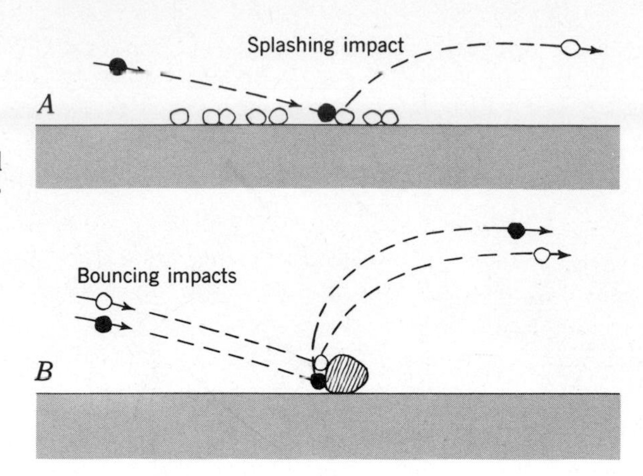


**Figure 9.13**
**A. In splashing impact, jumping sand grain strikes one or more other grains and splashes them into air at slow speeds and to low heights. Splashing happens most commonly when all grains are nearly the same size.**
**B. In bouncing impacts, sand grains strike pebble or other wide surface and bounce up at high speed and to greater heights. Angle of rise depends on inclination of surface of impact.**

grains, which hop and bounce, moving with the wind, as long as wind velocity is great enough to keep them moving.

The jumping sand grains never get far off the ground. They are usually limited to about 10cm, as in the experiment shown in Figure 9.12*C*. In desert country they generally jump no higher than about 45cm, as shown by utility poles, which are sandblasted up to about that height but no higher. Even in the strongest desert winds the height of jump rarely exceeds 1m. This explains why windblown sand rarely moves far except on very smooth surfaces. By being always close to the ground, it is easily stopped by obstacles and heaped into dunes.

### Suspended load

To understand the suspended load of fine particles carried by a wind we must note that obstacles on the ground below, whether sand grains, blades of grass, buildings, or trees, create a very thin layer of "dead," motionless air immediately above the surface (Fig. 9.14). The thickness of the dead layer equals 1/30 the height of the obstacles, regardless of velocity. So over ground that is covered with pebbles 3cm in diameter, the dead-air layer would be 1mm thick. Thin though it is, it plays a controlling part in the movement of all particles finer than medium-size sand grains. The reason why it does not influence the movement of coarser sand grains is that each individual grain constitutes an obstacle that

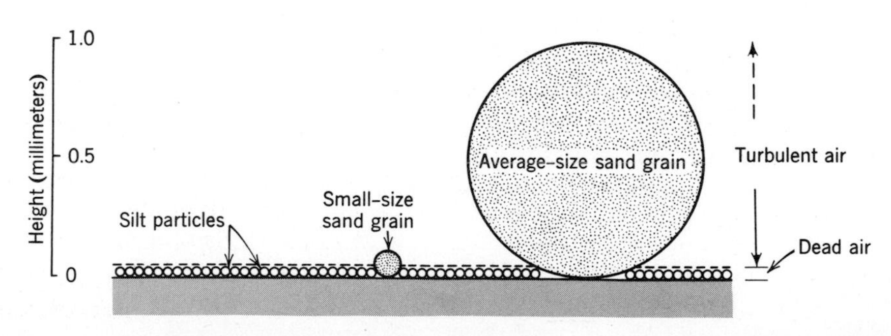


**Figure 9.14**
**Silt particles form a smooth surface that lies within the dead-air layer caused by sand grains and other large obstacles to flowing air. Thickness of dead-air layer is 1/30 the diameter of the average-size sand grain.**

projects far above the dead-air layer. But particles of silt and clay, down within that layer, are crowded so closely together that they do not act as individual obstacles; they present a smooth surface to the wind, which cannot lift them off the ground directly. It spatters them into the air by the impacts of jumping sand grains.

We can see how this happens by looking at a dusty road in dry country on a windy day. The wind blowing across the road generates little or no dust. But a car driving over the road creates a choking cloud, which is blown a short distance before settling once more. The car wheels have broken up the surface of powdery silt that was too smooth to be disturbed by the wind.

Once in the air, fine particles constitute the wind's suspended load. They are continually tossed up by eddies (again as in a stream of water) while gravity tends to pull them toward the ground. Meanwhile they are carried forward, and although in most cases suspended sediment is deposited fairly near its place of origin, strong winds have been known to carry fine particles thousands of kilometers before depositing them. The amount of sediment actually moved by the atmosphere, year in and year out, is probably only a fraction of 1 percent of capacity, for the air is rarely if ever fully loaded.

During the great windstorms in the dry years of the 1930s, however, loads became abnormally heavy. In a particularly great storm on March 20, 1935, when the sky looked much as in Figure 9.12*B*, the cloud of suspended sediment extended 3.6km (12,000 ft) above the ground, and the load in the area of Wichita, Kansas, was estimated at 35,000 metric tons per cubic kilometer in the lowermost layer 1.6km thick. Sampling of sediment on flat roofs of buildings showed that during that day about 280 metric tons of rock particles—around 5 percent of the load suspended in the lowermost layer 1.6km thick—were deposited on each square kilometer. Enough sediment was carried eastward on March 21 to bring temporary twilight over New York and New England, 3,000km east of the principal source area in eastern Colorado. The distance and travel time imply wind velocities of about 80km/hr (50mph).

### Fine-grained volcanic ash

Not all wind-blown sediments originate by being picked up from the ground. Large quantities are shot into the air during explosive eruptions of volcanoes (Fig. 15.7). Although coarse particles fall out quickly, small particles travel long distances. The bulk of the particles that fall out during an eruption tend to form an elongate body of sediment that trails out and thins downwind from the volcano (Fig. 15.3).

## EROSION BY WIND

### Deflation

Flowing air erodes in two ways. The first, *the picking up and removal of loose rock particles by wind,* provides most of the wind's

**Figure 9.15**
**Part of a large area within which a thickness of 1m of deflation can be measured directly.**
**The plant roots in the hummock, just above the dark band (the remains of an ancient soil) mark the position of the surface before deflation. Note recent ripples in the sand surface. The date is 1936, in the "Dust Bowl" period.** ***(R. H. Hufnagle, Philip Gendreau.)***

load and is known as ***deflation*** (from the Latin word meaning "to blow away"). The second, ***abrasion*** of rock by wind-driven rock particles, is analogous to abrasion by running water.

Deflation on a large scale happens only where there is little or no vegetation and where loose rock particles are fine enough to be picked up by wind. The great areas of deflation are the deserts; others are the beaches of seas and large lakes and, of greatest economic significance, bare plowed fields in farming country during times of drought, when no moisture is present to hold the soil particles together.

In most areas the results of deflation are not easily visible, inasmuch as the whole surface is lowered irregularly. In places, however, measurement is possible. In the dry 1930s deflation in parts of the western United States amounted to 1m or more within only a few years—a tremendous rate compared with our standard estimate of rate of general erosion. In Figure 9.15 the turf-like yucca plants held the soil in place, but elsewhere it was deflated.

The most conspicuous evidence of deflation consists of basins excavated by wind. These occur in tens of thousands in the semiarid Great Plains region from Canada to Texas. The lengths of most of them are less than 2km and depths are only a meter or two. In wet years they are clothed with grass and some even contain shallow lakes; an observer seeing them at such times would hardly guess their origin. But in dry years soil moisture evaporates, grass dies away in patches, and wind deflates the bare soil. At the same time, drifting sand accumulates to leeward, especially along fences and other obstructions.

When sediments are particularly prone to deflation, depths of deflation basins can reach 50m, as in southern Wyoming, and even more, as in the Libyan Desert in western Egypt, where the floor of the Qattara basin lies about 125m below sea level. Deflation in any basin is limited finally only by the water table, which moistens the surface, encourages vegetation, and inhibits wind erosion.

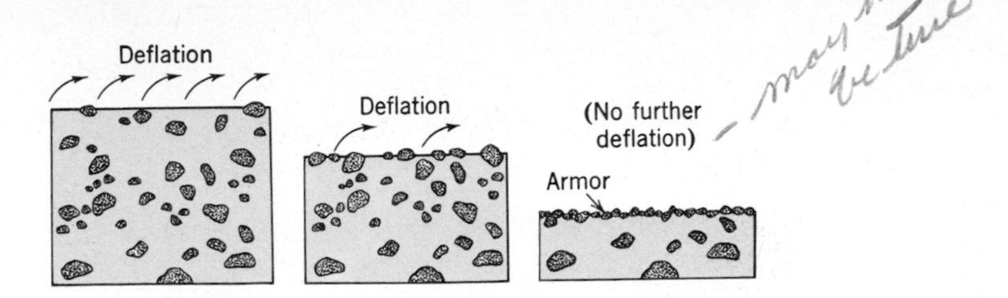


**Figure 9.16**
**Three stages in the development of a deflation armor.**

A natural preventive of deflation is a cover of rock particles too large to be removed by the wind. Deflation of sediment such as alluvium, which generally consists of silt, sand, and pebbles, creates such a cover (Fig. 9.16). The sand and silt are blown away, and in places are carried off by sheet erosion also, but the pebbles remain. When the surface has been lowered just enough to create a continuous cover of pebbles, the ground has acquired a ***deflation armor****, a surface layer of coarse particles concentrated chiefly by deflation.* Such armors are also called *desert pavement,* because long-continued removal of the fine particles makes the pebbles settle into such stable positions that they fit together almost like the blocks in a cobblestone pavement.

### Abrasion: ventifacts

In desert areas, surfaces of bedrock and of loose stones are abraded by wind-driven sand and silt, which can cut and polish them to a high degree. A ***ventifact,*** the name given to *a stone the surface of which has been abraded by wind-blown sediment,* is recognized by polished, greasy-looking surfaces, which may be pitted or fluted, and by facets separated from eachother by sharp edges (Fig. 9.17).

**Figure 9.17**
**Ventifacts.**
**Two basalt cobbles faceted by wind-driven particles of sand and silt. Diameter of compass = 7cm. (*Marion Whitney and R. V. Dietrich.*)**

– may not be true

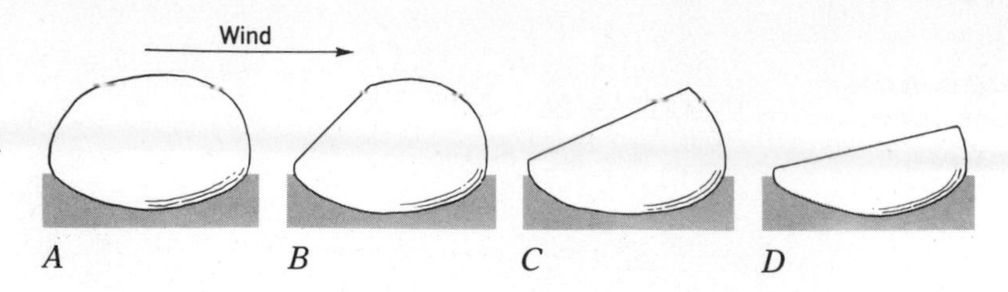


**Figure 9.18**
**Four stages in the cutting of a ventifact. The pebble becomes a ventifact between stage *A* and stage *B*.**

In the laboratory, sand blasting of pieces of plaster of Paris demonstrates that the facets always face the wind (Fig. 9.18). A stone can be worn down flush with the ground by enlargement of a single facet. If the stone is undermined or otherwise rotated or if the wind direction varies from time to time, two, three, or many more facets can be cut on it.

# WIND DEPOSITS

In the wind as in a stream of water, the jumping sand grains of the bed load move rather slowly and are deposited early, whereas the finer particles of the suspended load travel faster and much farther before they drop out. Also, winds commonly deposit sand in heaps or hillocks, whereas finer sediment is generally deposited as smooth blankets. Looking more closely at these two kinds of deposits, we begin with the sand that drops out of the bed load, usually forming dunes.

## Dunes

A ***dune*** is *a mound or ridge of sand deposited by wind.* Generally it is localized by an obstacle small or large (Fig. 9.19), which distorts the flow of air. Velocity within a meter or two of the ground varies with the slightest irregularity of the surface. As it encounters an obstacle, the wind sweeps over and around it, but

**Figure 9.19**
**Wind blowing from left to right deflated the bare dry field (left) and carried away the finer particles in suspension, but dropped sand along the obstacle created by a wire fence with weeds caught in it. The result was a row of dunes along the fence. One way in which flowing air and flowing water resemble each other is that both create ripple marks. Compare Figure 12.12. Raymondville, Texas.** *(Soil Conservation Service, U.S. Dept. Agriculture.)*

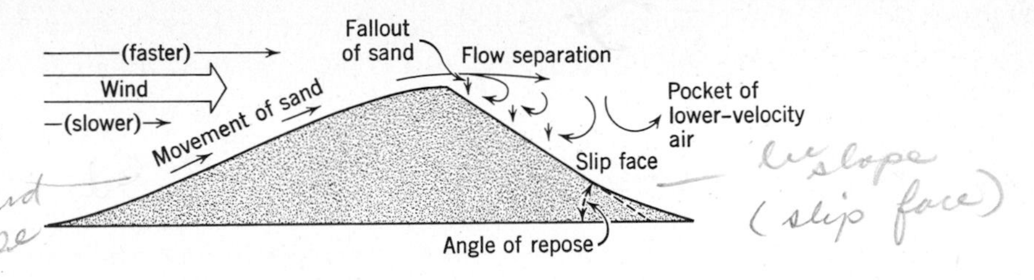


**Figure 9.20**
**Long section through a bare sand dune, showing development of windward and lee slopes.**

leaves a pocket of slower-moving air immediately behind the obstacle and a similar but smaller pocket in front of it. In these pockets of lower velocity, sand grains drop out and form mounds. Once formed, the mounds themselves influence the air flow. As more sand drops out they generally coalesce and form a single dune.

A dune is unsymmetrical. It has a steep, straight lee slope and a gentler windward slope (Fig. 9.20). The wind rolls or pushes sand grains up the windward slope, and at the crest they drop onto the lee slope, building that slope up to the angle of repose, which, for loose sand, is close to 34°. When more sand drops onto it the slope

**Table 9.1**
**Several Kinds of Dunes Based on Form**

| Kind | Definition and Occurrence | Illustration (Arrows indicate wind directions) |
|---|---|---|
| ***Beach dunes*** | Hummocks of various sizes bordering beaches. Inland part is generally covered with vegetation. | Sea<br>Beach |
| ***Barchan dune*** | *A crescent-shaped dune with horns pointing downwind.* Occurs on hard, flat floors in deserts; constant wind, limited sand supply. Height 1m to more than 30m. | |
| ***Transverse dune*** | *A dune forming a wavelike ridge transverse to wind direction.* Occurs in areas with abundant sand and little vegetation. In places grades into barchans. | |
| ***U-shaped dune*** | *A dune of U-shape with the open end of the U facing upwind.* Some form by piling of sand along leeward and lateral margins of a growing blowout in older dunes. | Blowout |
| ***Longitudinal dune*** | *A long, straight, ridge-shaped dune parallel with wind direction.* As much as 100m high and 100km long. Occurs in deserts with scanty sand supply and strong winds varying within one general direction. Slip faces vary as wind shifts direction. | 2 km |

becomes unstable, sand slides ("slips") downward, and the angle is maintained. For this reason *the straight, lee slope of a dune* is known as the ***slip face***, which, as might be expected, meets the windward slope at a sharp angle.

The angle of slope of the windward side varies with wind velocity and grain size, but it is always less than that of the slip face. The asymmetry of a dune with a slip face, then, indicates the direction of the wind that shaped it.

Many dunes grow to heights of 30m to 100m, and some desert dunes reach the great height of 200m. Possibly the height to which any dune can grow is determined by upward increase in wind velocity, which at some level will become great enough to whip sand grains off the top of a dune as fast as they arrive there by creeping up the windward slope.

**Stratification and Migration.** The dropping of sand grains over the crest onto the slip face of a dune produces cross-strata much like the foreset layers in a delta (Fig. 7.12). Erosion of the windward slope continually erodes the layers already deposited, as new ones are added to the slip face (Fig. 9.20).

Transfer of sand from the windward to the lee side of a bare dune causes the whole dune to migrate slowly downwind. Measurements on desert dunes of the barchan type (Table 9.1) indicate rates of migration as great as 10 to 20m/year. The migration of dunes, particularly along coasts just inland from sandy beaches, has been known to bury houses and threaten the existence of towns. In such places, sand encroachment is countered most effectively by planting vegetation that can survive in the very dry sandy soil of the dunes. A good plant cover inhibits dune migration for the same reason that it inhibits deflation: if the wind cannot move sand grains across it, a dune cannot migrate.

A dune covered with grass, however, can be reactivated wherever, by drought or trampling by animals, the grass is killed off in patches and allows deflation to start. This converts the bare patches into irregular, shallow basins known as *blowouts*. A ***blowout*** is merely *a deflation basin excavated in shifting sand*. The view in Figure 9.15 represents a large blowout.

In the table we see that some dunes, of which the barchan is the best example, are built by winds blowing from one direction, whereas others are built by winds that shift direction from one time to another. As long as the dune form exists, we can infer wind direction from the position of the slip face. But even after erosion has destroyed the form, and after what is left of the dune sand has been buried, converted into sandstone, subjected to uplift, and re-exposed to a new cycle of destruction, we can recover the direction (or directions) of the wind that deposited the sand by measuring the direction of dip of the cross-strata, which are always inclined downwind (Fig. 9.21).

**Figure 9.21**
**If wind had uniform direction and velocity, it would build (lower left) simple, parallel cross-strata while dune migrates from AB to A′B′. But variations in direction and velocity are common and result in cross-strata like those at left.**
**Photograph shows cross-strata in wind-blown sand consolidated to form hard sandstone (Jurassic; more than 150 million years old). Parts of lee slopes of a succession of dunes are exposed. The cavities result from dissolution at places where calcium-carbonate cement between the grains of quartz is especially soluble. Monument Valley, Arizona. (*Tad Nichols.*)**

**Figure 9.22**
**Loess is very cohesive and has vertical joints; so it forms cliffs as jointed bedrock does.**
**A typical surface expression of bare loess appears in this cliff, in the east bluff of the Mississippi River valley, Madison County, Illinois. The cliff will stand with little change for many years. (*J. C. Frye.*)**

**Composition.** Virtually all dunes consist of sand-size particles. Since quartz is the most common mineral in sand-size sediments, it is not surprising that quartz predominates in most dunes. But where other minerals are abundant, dunes are built from them. The calcite dunes on Bermuda and the gypsum dunes at White Sands National Monument in New Mexico are examples.

### Deposits of silt

**Loess.** Most of the world's regolith includes small proportions of fine sediment deposited from suspension in the air, but so thoroughly mixed with other materials as not to be distinguishable from them. But over wide areas sediment having this origin is so thick and so pure that it constitutes a distinctive deposit. It is known as ***loess*** (pronounced "less") and is defined as *wind-deposited silt, commonly accompanied by some clay and some fine sand.*

Most loess is not stratified, apparently because many grain sizes were deposited together from suspension, and also perhaps because plant roots, worms, and other organisms turned over and churned up the sediment as it was deposited. Where exposed, loess stands at such a steep angle that it forms cliffs (Fig. 9.22) as though it were firmly cemented rock. This is the result of the fine grain size of loess, in which molecular attraction is strong enough to make the particles very cohesive. Also, because the particles are angular, porosity is extremely high, commonly exceeding 50 percent. Hence loess accepts and holds water and constitutes a basis for productive soils. The corn-growing region that centers in Illinois, Iowa, northern Missouri, and eastern Nebraska is famous for its productivity, partly because most of its soils are developed in loess.

Minerals composing loess are chiefly quartz, feldspar, micas, and calcite. The particles are generally fresh, showing little evidence of chemical weathering other than the slight oxidation of iron-bearing minerals that has occurred since deposition and that gives a yellowish tinge to the deposit as a whole.

**Origin of Loess.** Loess possesses two characteristics that indicate it was deposited by wind: (1) it forms blankets that mantle hills and valleys alike through a wide range of altitudes, and (2) the fossils it contains consist of land plants and animals—mostly air-breathing snails, but mammals too, including elephants.

The world distribution of loess shows that winds carried it from two principal sources: (1) deserts and (2) sediments of streams that originated in the melting of glaciers. Desert loess covers enormous areas that lie to leeward of deserts, an obvious source of mechanically weathered sediments. The loess that covers some 800,000km$^2$ in western China, said to reach a thickness of 200m in some places, was blown there from alluvium on the floors of the great desert basins of central Asia.

Glacial loess is abundant in the middle part of North America (especially Nebraska, South Dakota, Iowa, Missouri, and Illinois) and in the drier parts of Europe. It has two distinctive features: The first is that the shapes and mineral composition of its particles resemble the "rock flour" ground up by glaciers. The second feature is that glacial loess is thickest immediately in the lee of rivers such as the Missouri and Mississippi, which are known to have been swollen with water from melting ice during glacial ages (Chapter 10). Carbon-14 dates, fossil plants, and other kinds of evidence confirm the relation of loess to the presence of great quantities of melting ice. In glacial ages the areas just outside the margins of broad glaciers were cold and very windy. Rivers that were fed by melting ice were filling their valleys with gravel, sand, and silt so rapidly that plants could not gain a foothold, from one year's flood to the next, on valley floors. So the floors remained bare and were easily deflated. Silt particles, splashed into the air by jumping sand grains, were carried to leeward. The silt settled out, forming blankets 8 to 30m thick near the source valleys and thinning downwind to thicknesses of 1 to 2m, spread over thousands of square kilometers (Fig. 9.23).

Why was the silt not picked up again and again and carried even farther by the wind? One reason lies in the stability of a silt surface, which results from its fine grain size (Fig. 9.14). Another reason is that the silt settled out chiefly on grassland and to some extent on woodland. Once on the ground in environments such as these, silt would be in no danger of further deflation.

## ENVIRONMENTAL ASPECTS: SOIL EROSION

Climates fluctuate continually. It therefore happens that regions ordinarily suitable for agriculture experience dry periods during which soil erosion by wind reaches tremendous proportions. As we have noted, during the 1930s an enormous volume of soil was blown away from parched, unprotected plowed fields in the "Dust Bowl" region of the Great Plains (Figures 9.12 and 9.15). Sand-size particles were piled up along fences (Fig. 9.19) and around farm buildings, and finer particles were blown eastward to be deposited over wide areas; a good deal of such material was dropped into the Atlantic Ocean.

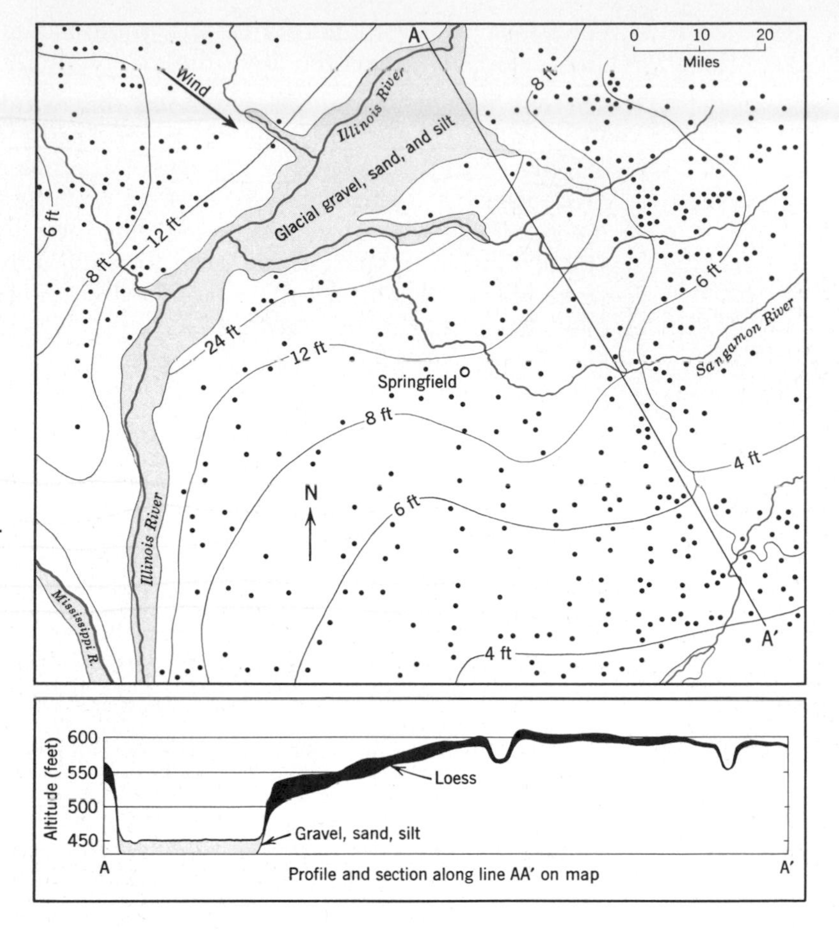


**Figure 9.23** **Loess in central Illinois. Thicknesses were determined by borings at places shown by dots. Lines are isopachs (Appendix D); others appear in Fig. 15.3. Thickness decreases away from river valley in both directions, but loess is thickest on southeast side, the leeward side for prevailing winds. Grain diameters also decrease in both directions away from valley. (*Data from G. D. Smith, 1942.*)**

What determined the location of the "Dust Bowl"? That area closely approximates the largest area in the United States in which *average* wind velocity is great enough both to move sand grains and to keep coarse silt grains in suspension. All that was needed further was a long succession of dry years—the drought in the 1930s. With only very slight changes in Earth's climates, deserts expand and contract and can be created or disappear.

In the "Dust Bowl" area, good practice includes the planting of windbreaks (Fig. 9.24), consisting of bushes and hardy trees set in strips at right angles to the strongest winds, at intervals of 1,500m or so. It also involves planting strips of grass alternating with strips of cultivated grain, for the soil must lie bare during alternate years in order to accumulate moisture sufficient to grow grain. The principle on which the rows of plants act to retard deflation is based on the presence, just about the ground, of the dead-air layer we noted earlier, the thickness of the layer being equal to 1/30 the height of obstacles on the ground. A group of trees and bushes 10m high, although somewhat permeable to wind, should create a dead-air layer nearly 30cm thick, thus effectively preventing deflation beneath it and through some distance to leeward.

**Figure 9.24**
**Windbreaks on dry sandy farmland in northern Texas.** (*U.S. Soil Conservation Service.*)

# SUMMARY

## DESERTS

1. Deserts, constituting about a quarter of the world's land area, are areas of slight rainfall, high temperature, great evaporation, relatively strong winds, sparse vegetation, and interior drainage.

2. No major geologic process is confined to deserts, but in them mechanical weathering, impact of raindrops, flash floods, and winds are very effective. The water table is low.

3. Pediments are a conspicuous feature of many deserts. They are shaped by streams, rills, sheet runoff, and weathering.

4. In the arid cycle of erosion, pediments grow headward at the expense of mountain slopes, which appear to retreat without becoming gentler as they do in a moist region. Wind is a secondary agent. It removes fine rock waste, and in places builds sand dunes.

## WIND ACTION

5. Wind carries a bed load of jumping sand grains close to the ground and a suspended load of fine particles higher up. Sorting of sediment results.

6. Wind erodes by deflation and abrasion, chiefly in dry regions and on beaches. It creates deflation basins, blowouts, deflation armor, and ventifacts.

7. With long-continued wind activity, sand grains become rounded.

DUNES

8. Many dunes are localized by obstacles. Bare dunes have steep slip faces and gentler windward slopes. They migrate downwind, forming cross-strata that dip downwind.

9. Dunes are classified by form into beach dunes, barchan dunes, transverse dunes, U-shaped dunes, and longitudinal dunes.

LOESS

10. Loess is deposited chiefly in the lee of (a) bodies of glacial outwash, and (b) deserts. Once deposited, it is stable and is little affected by further wind action.

# SELECTED REFERENCES

Bagnold, R. A., 1941 (repr. 1954), The physics of blown sand and desert dunes: New York, William Morrow.

Blackwelder, Eliot, 1954, Geomorphic processes in the desert: California Div. Mines Bull. 170, Chap. 5, p. 11–20.

Bryan, Kirk, 1923, Erosion and sedimentation in the Papago Country, Ariz., with a sketch of the geology: U.S. Geol. Survey Bull. 730, p. 19–90.

Cooper, W. S., 1967, Coastal dunes of California: Geol. Soc. America Mem. 104.

Finkel, H. J., 1959, The barchans of southern Peru: Jour. Geology, v. 67, p. 614–647.

Hadley, R. F., 1967, Pediments and pediment-forming processes: Jour. Geol. Education, v. 15, p. 83–89.

Hume, W. F., 1925, Geology of Egypt, v. 1: Cairo, Government Press.

McGinnies, W. G., Goldman, B. J., and Paylore, Patricia, eds., 1968, Deserts of the world: Tucson, Univ. of Arizona Press. (Contains maps of all deserts.)

McKee, E. D., 1966, Structures of dunes at White Sands National Monument, New Mexico: Sedimentology, v. 7, p. 1–69.

Neal, J. T., ed., 1965, Geology, mineralogy, and hydrology of U.S. playas: U.S. Air Force, Office of Aerospace Research, Environmental Res. Paper 96 [Defense Documentation Center, Alexandria, Va.].

Sharp, R. P., 1949, Pleistocene ventifacts east of the Bighorn Mountains, Wyoming: Jour. Geology, v. 57, p. 175–195.

———, 1963, Wind ripples: Jour. Geology, v. 71, p. 617–641.

Thorarinson, Sigurdur, 1954, The tephra-fall from Hekla on March 29th, 1947: Societas Scientiarum Islandica, v. 2. no. 3.

Whitney, M. I., and Dietrich, R. V., 1973, Ventifact sculpture by windblown dust: Geol. Soc. America Bull., v. 84, p. 2561–2582.

MAP

Thorp, James, and others, 1952, Pleistocene eolian deposits of the United States: Boulder, Colo., Geol. Soc. America.

# 10 GLACIERS AND GLACIATION



# GLACIERS

**Glaciers in the water cycle**

If a person living in New York City today were curious to meet a glacier at first hand, he would have to travel some 2,500km west to the Rocky Mountains or an equal distance north to northeastern Canada to find even a few small ones. Yet, between 15,000 and 20,000 years ago glacier ice covered the whole region between northeastern Canada and New York City; the southern margin of the ice followed a continuous line from the Rocky Mountains in Montana to New York and onward into what is now the Atlantic Ocean. Whether or not there were people in North America at that time, they could not have lived at the site of New York, because the locality would have been uninhabitable, buried beneath the glacier.

Similarly, Stone Age people, known to have been living in Europe since long before 20,000 years ago, were driven southward by another ice sheet that spread over northern Europe.

Not only then, but at various other times, some of them millions of years earlier, glaciers existed where none exist today and compelled strong local changes in the biosphere. Part of the cause lies in the drifting of continents, floating, with their plates of lithosphere, from cold latitudes toward warmer ones or vice versa. Another part of the cause lies in changes of climate that have repeatedly affected the whole Earth. What we are dealing with is part of the water cycle, for the ice of glaciers is water in the solid state. Going back to Figure 2.7, we recall that the overwhelmingly great reservoir of Earth's hydrosphere is the ocean, and that today only a tiny 2.1 percent of Earth's stock of water substance exists as a solid in the form of glaciers. If Earth's surface temperature were lowered by even a very small amount, some of the precipitation that now falls as rain would fall as snow, and the amount of ice on the continents would increase. At the time when temperatures were lower and when the site of New York City was buried beneath ice, Earth's stock of ice was probably nearly three times as bulky as it is now.

Also there is reason to believe that at other times in Earth's history, glaciers in one region or another have been fewer and smaller than they are today, and that therefore temperatures must have been higher. Glaciers are well worth our attention because we can draw inferences from them about changes in Earth's climate.

Even though the ice of glaciers is withholding water from the sea, it is not dead. It is very active, receiving nourishment in the form of snowfall and giving up water by melting. So solid water is passing through it all the time, as liquid water passes through a lake. The oldest ice in the great glacier that now covers Greenland is believed to be much more than 25,000 years old; so it has formed part of the glacier for a long time. But as it melts, it is replaced by new ice contributed by the snow that falls on Greenland every year. In other words the glacier is in or near a steady-state condition.

Obviously, then, the fate of this or any glacier depends on how much ice melts and how much is replaced by new snow. That, in turn, depends on *climate:* primarily on temperature and secondarily on rate of snowfall. And climates change continually (albeit slowly), thus affecting the sizes and the number of glaciers.

**Glaciers and Sea Level.** Because most of the world's ice today lies not in the densely peopled middle latitudes of North America and Europe but in high latitudes far to the north and south, the fact that glaciers fluctuate as the climate grows warmer or colder may seem to be of no great concern. Yet it does concern us because of the part glaciers play in the water cycle.

If world climates become colder, snowfall on the continents increases at the expense of rainfall, glaciers increase, and less water flows back from continents into the ocean. As a result, sea level must become lower. This is what happened when glacier ice covered much of North America, and Europe as well. Evaporation of water not replaced by water returned from the continents, lowered sea level by an estimated 100m or so. This caused much of the area of the continental shelves to emerge and become land. The Atlantic coast of the United States then lay some 150km east of the site of New York City. The intervening coastal plain was forested with spruce and pine and its population included mammoths, mastodons, and other mammals now extinct. The Hudson River emptied into the ocean after flowing through a broad valley, 50m deep, that cut right across the continental shelf. The valley is still there today, now drowned by rise of the ocean as the great ice sheets melted away (Fig. 10.1)

At the same time, lowering of sea level joined Britain to France where the English Channel is now. And where the Bering Strait now separates North America from Asia, Alaska was firmly connected with the USSR.

Rise of sea level has re-submerged the shelves, but a good deal of glacier ice still exists on the continents, particularly the Antarctic Continent. If that ice too were to melt, returning its substance to the ocean as water, a layer of water more than 60m thick, worldwide, would be added to the ocean, which of course would submerge coastal lands. This unwelcome possibility is sure to make us respect the potential possessed by remote Antarctic ice for influencing life, not just in Antarctica but throughout the world.

To some extent such submergence has happened in the past. Along certain coasts old beaches, some with fossil shells, and other shore features, lie high and dry at altitudes of a few meters to more than 35m. Their presence implies that when they were made, the sites of coastal cities such as Norfolk, Virginia, were drowned. Such submergences happened at times when glaciers were even less abundant than they are today, so that more water stayed in the ocean. Of course, before we fully accept this inference we must make sure the beaches record real changes in the volume of the oceans, and have not simply been raised by general uplift of parts of Earth's crust.

To get a better understanding of the fluctuations of glaciers and of sea level, let us examine today's glaciers to see how they work.

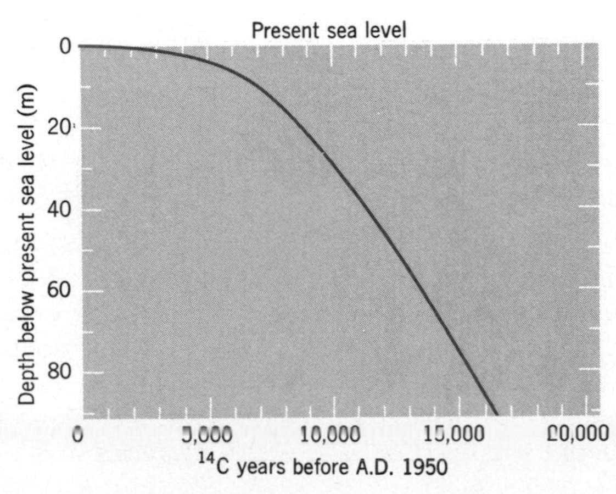


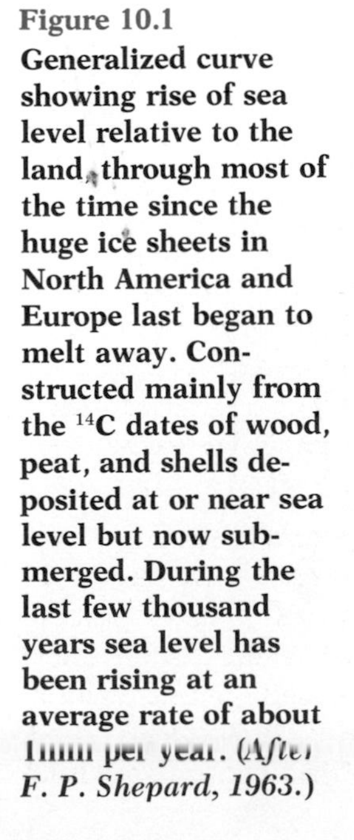
**Figure 10.1**
**Generalized curve showing rise of sea level relative to the land, through most of the time since the huge ice sheets in North America and Europe last began to melt away. Constructed mainly from the $^{14}C$ dates of wood, peat, and shells deposited at or near sea level but now submerged. During the last few thousand years sea level has been rising at an average rate of about 1mm per year.** *(After F. P. Shepard, 1963.)*

### Kinds of glaciers

Defined simply, a ***glacier*** is *a body of ice, consisting mainly of recrystallized snow, flowing on a land surface.* On the basis of their form, we can readily distinguish four kinds of glaciers. Most numerous and possibly most familiar are ***cirque glaciers*** (Fig. 10.11), *very small glaciers that occupy cirques.* In the conterminous United States there are about 950 cirque glaciers, and about 1,500 exist in the Alps.

A ***valley glacier*** (Fig. 10.2) is *a glacier that flows downward through a valley.* Such glaciers vary from no more than a few acres in extent to mighty tonguelike forms many tens of kilometers in length. In the conterminous United States they number about 50; in the Alps their number is larger.

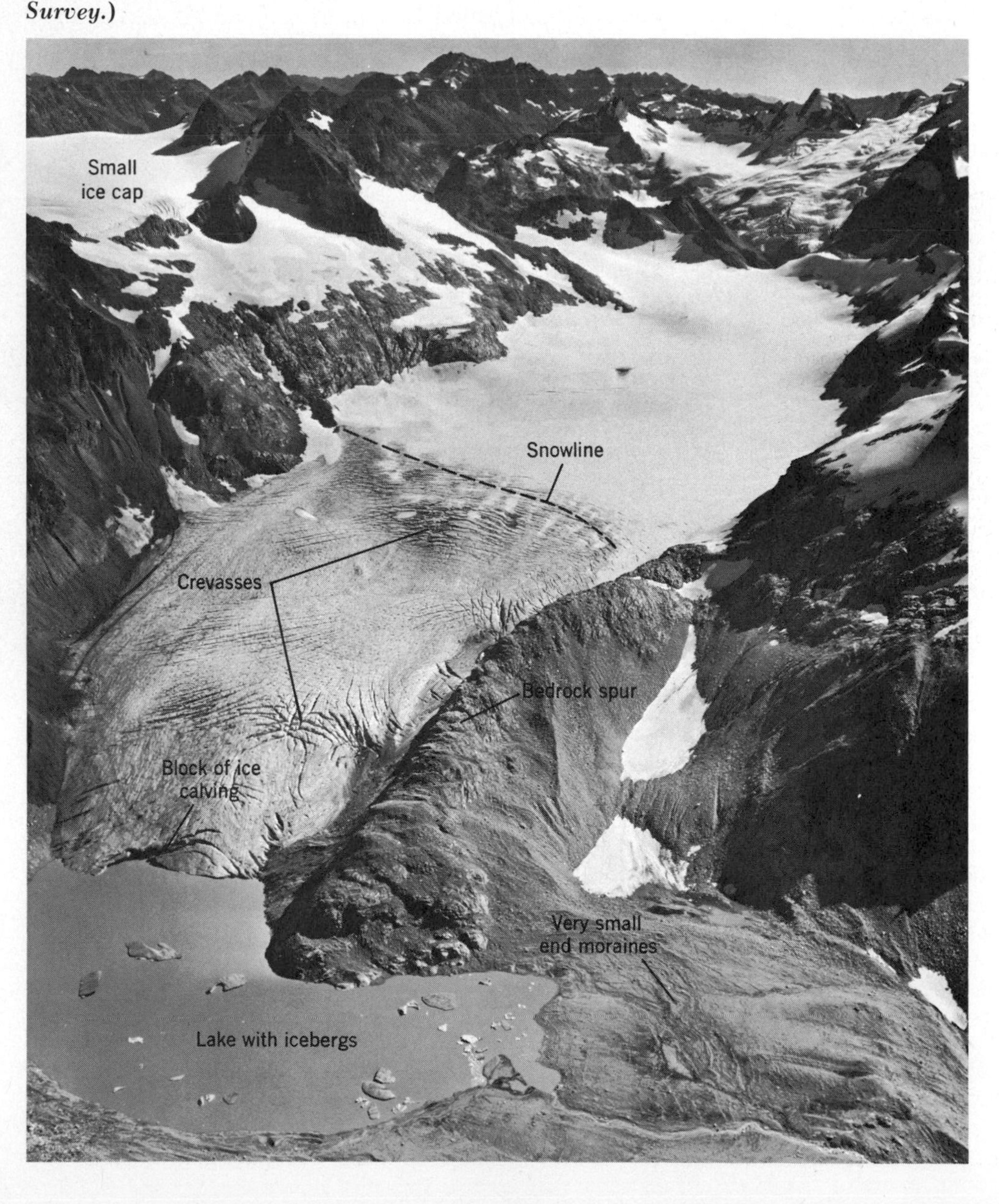


**Figure 10.2**
**South Cascade Glacier, a valley glacier about 3km long, in the North Cascade Range, Washington, September 1960. In view are position of snowline and a small ice cap. Width of glacier near snowline, about 800m. (*A. S. Post, U.S. Geol. Survey.*)**

A ***piedmont glacier*** is *a glacier on a lowland at the base of a mountain, fed by one or more valley glaciers.* It is shaped like a covered frying pan whose narrow inclined handle represents a valley glacier that feeds it.

An ***ice sheet*** is *a broad glacier of irregular shape, generally blanketing*

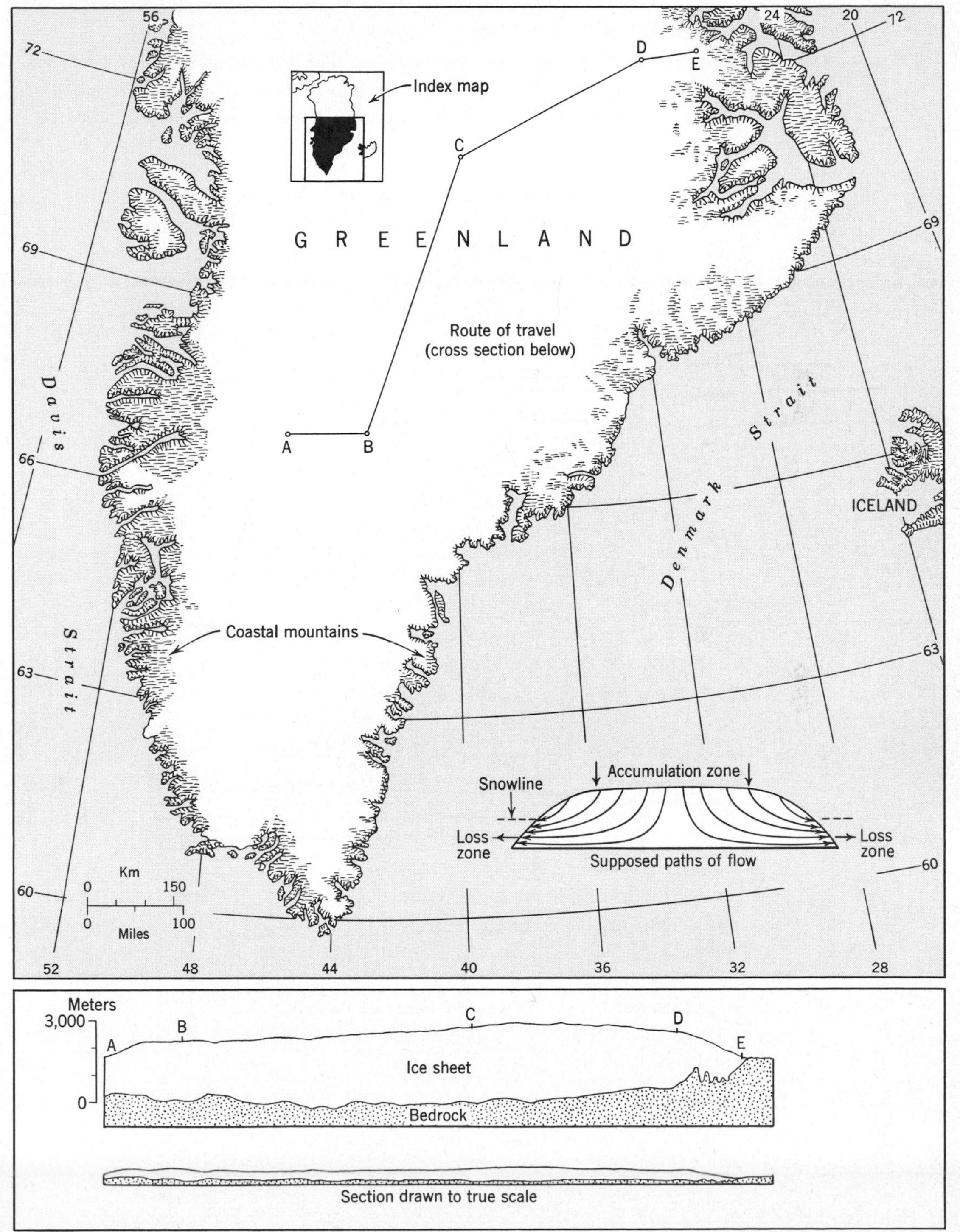


**Figure 10.3**
**Map of south half of Greenland and cross section showing thickness of ice sheet. (*Thickness data from Albert Bauer.*)**

*a large land surface. A small ice sheet* is an ***ice cap;*** some ice caps, such as those in Figure 10.2, are very small.

### Greenland ice sheet

The world now possesses two big ice sheets, the most accessible of which is the Greenland Ice Sheet. Its area is about 1,726,000km$^2$ and its highest part is more than 3km above sea level. Seismic measurement—that is, measurement by a method using artificial-earthquake waves (Chapter 16)—of the approximate thickness of this glacier was made at intervals along certain routes of travel, one of which is shown in Figure 10.3. The profile and cross section along it reveal that in its central area the base of the ice sheet lies below sea level and that the ice is very thick; at one point its thickness exceeds 3km. Near its margins the spreading ice sheet encounters mountains, through which it flows along deep valleys to reach the sea. There the ice terminates, as shown in Figure 10.8. Icebergs floating in the sea are masses broken off from the ends of such glaciers. Calculations show that near its western margin, the ice sheet is spreading seaward at an average rate of 150m/year.

### Antarctic ice sheet

The **Antarctic Ice Sheet** (Fig. 10.4) is far bigger. Its area, including the parts of its margin that float in the ocean, is about 1.4 times the area of the conterminous United States, its highest altitude exceeds 4km, its greatest thickness likewise exceeds 4km in one place, and its volume is estimated at around 24,000,000km$^3$. With this huge volume the Antarctic Ice Sheet alone contains 66 percent of all the fresh water in the world.

Like its Greenland counterpart, the Antarctic Ice Sheet extends below sea level in a number of areas. This fact results, as it does in Greenland, partly from the great weight of the ice, which has caused the rock beneath to subside by as much as one-third the thickness of the ice. But even if the Antarctic Ice Sheet were to melt completely away, permitting its rocky floor to rise again to its old position, much of West Antarctica would still be submerged, so that that part of the continent would be merely a cluster of big islands.

The Antarctic Continent has the highest average altitude, and of course the lowest average temperature, of all continents. Centered on the South Pole, it is truly a polar ice sheet. In fact it is the only one, because the North Pole is positioned near the center of the large, deep Arctic Ocean. Consequently, there can be no glaciers near the North Pole, only very thin floating ice consisting of frozen seawater and snow.

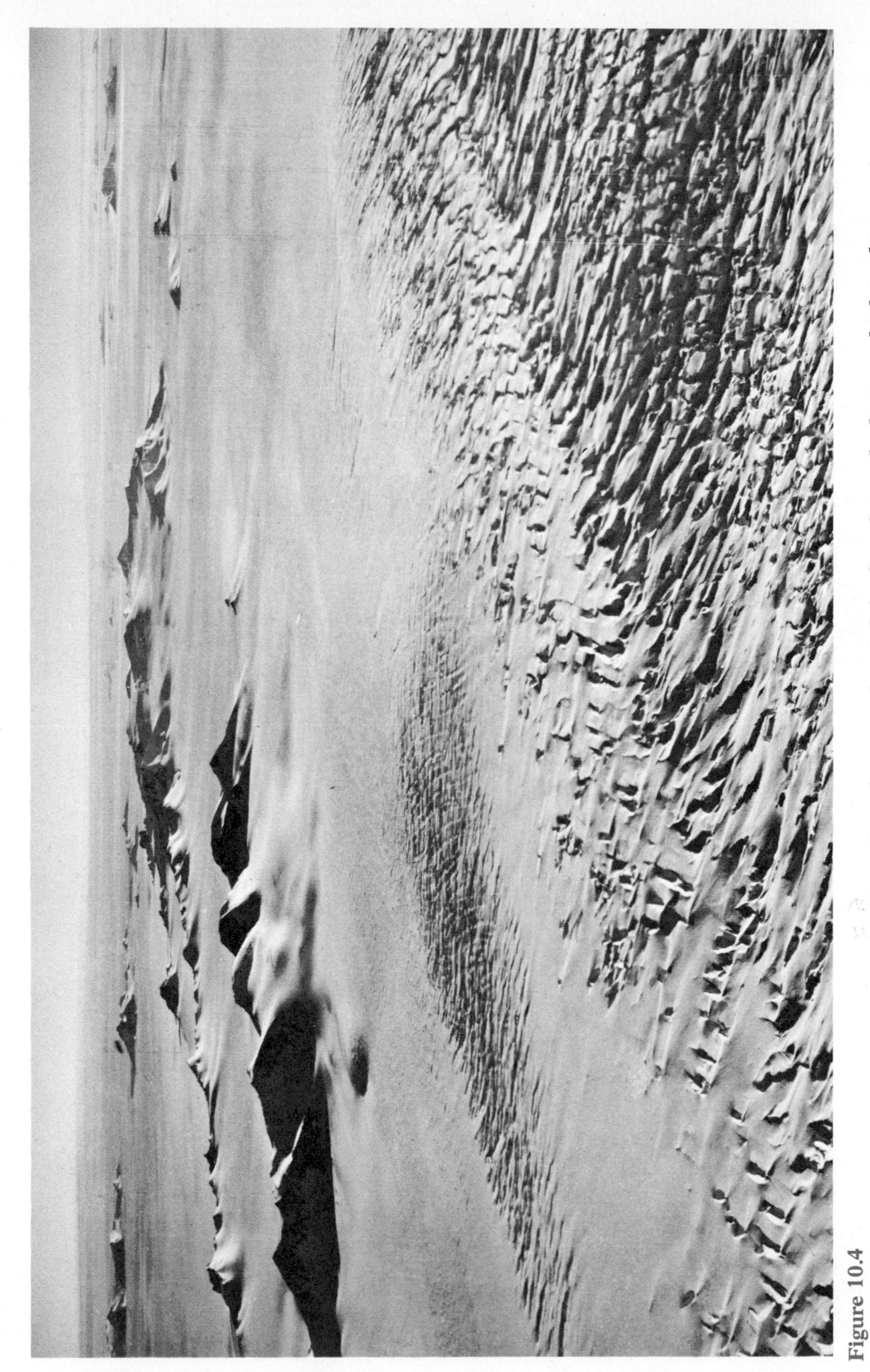

Figure 10.4
**Part of the Antarctic Ice Sheet in Marie Byrd Land, with peaks of Edsel Ford Mountains projecting through its surface. In foreground are crevasses, reflecting a change of gradient of the floor beneath glacier, which in this area is comparatively thin. (*P. A. Siple, U. S. Antarctic Service.*)**

Valley glaciers, piedmont glaciers, and ice sheets constitute a gradational sequence. Indeed the two existing great ice sheets probably began to form in high mountains with the growth of thousands of valley glaciers, which flowed down and spread out in lower areas as piedmont glaciers, and gradually thickened to form bodies of ice higher in places than the mountain peaks themselves.

### Snowline

From what has been said already, evidently two chief requirements for the existence of glaciers are snowfall and low temperature, both of which depend on climate. These requirements are fulfilled in high latitudes and at high altitudes, and occur more commonly in wet coastal areas than in the dry interiors of continents. It is not surprising, then, that the world's two largest glaciers cover lands that are surrounded by water and that the largest glaciers in Alaska are clustered close to the Pacific Ocean. Altogether, glaciers cover an aggregate area equal to about 10 percent of the land area of the world.

Glaciers are related to each other by the ***snowline,*** *the lower limit of perennial snow* (Fig. 10.2). Above the snowline are ***snowfields,*** *wide covers, banks, and patches of snow,* usually in protected places, *that persist throughout the summer season.* The snowline passes across all active glaciers. It rises from near sea level in polar regions to altitudes of as much as 6,000m in tropical mountains, and rises also from coasts toward continental interiors.

### Conversion of snow into glacier ice

When temperatures fall below 0°C some atmospheric water vapor changes into the solid state, forming the hexagonal ice crystals we know as snowflakes. Newly fallen dry snow is feathery, light, and porous. It has much greater internal surface area (Fig. 10.5) than that of ordinary sediment. Air penetrates its large pore spaces. In it, ice evaporates, mostly at the points of crystals, and the water vapor condenses, mostly in the narrow spaces near the centers of the ice crystals. In this way, by evaporation near their edges and condensation near their centers, snowflakes gradually become smaller, rounder, and thicker, and the spaces between them gradually disappear. The whole mass of snow takes on the granular texture we find in old snowdrifts at winter's end. Air is forced out from the spaces between the grains. The snow is being changed from a sediment into a sedimentary rock.

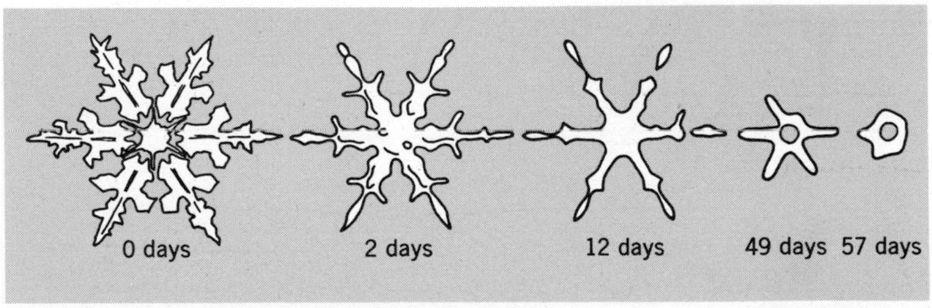

**Figure 10.5**
**Conversion of a snowflake into a granule of ice during nearly two months of evaporation and condensation. (8 times natural size.) (*After Bader and others, 1939.*)**

When the specific gravity of the body of granular snow has reached 0.8, the snow has become so compact that it is impermeable to air and is then said to be *ice.* Such ice is a metamorphic rock, for it consists of interlocking crystalline grains of the mineral ice. It has been metamorphosed by pressure, caused by the weight of the overlying ice itself. Although a rock, such ice has a far lower melting point than any other rock. The specific gravity of glacier ice is about 0.9; so it floats in water. Most of the icebergs in the North Atlantic Ocean consist of ice broken off from glaciers.

### Movement

At some time, depending on the steepness of slope of the surface on which it is lying, compacted snow begins to respond to the pressure caused by the weight of snow above it and starts to move, somewhat as regolith starts to creep down a slope in the process of mass-wasting. The movement of ice, generally spoken of as *flow,* takes place because of the pull of gravity on the ice mass. The motion is of two kinds (Fig. 10.6). One kind consists of *sliding* of the whole mass along the ground, as a blanket of snow slides very slowly down the steeply sloping roof of a building (Fig. 18.8). The other kind is *creep,* much of which occurs through slipping on a tiny scale, along flat surfaces within the ice crystals themselves. The crystals are deformed as the body of snow or glacier ice creeps downslope.

The uppermost or surface part of a glacier, having little weight upon it, is brittle. Where the glacier flows over an abruptly steepened slope, such as the downstream side of a rocky "bump," the uppermost ice is subjected to tension and cracks. The cracks open up, forming ***crevasses****—deep, gaping cracks in the upper surface of a glacier* (Figures 10.2, 10.4, and 10.18). Although sometimes impossible and always dangerous to cross on foot or on a snowmobile, crevasses are rarely as much as 50m deep. At greater depths the glacier flows actively and prevents crevasses from forming. Because it cracks at the surface but flows at depth, we

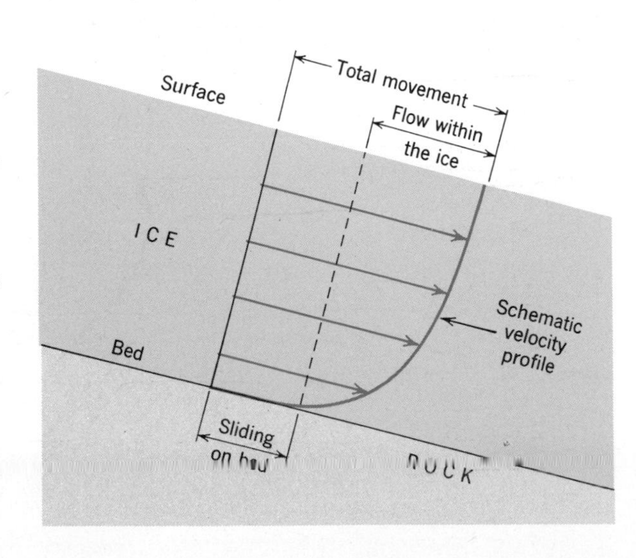


**Figure 10.6**
**The two kinds of motion in a glacier.**
**(1) *Rate of flow* within the ice increases upward while (2) *rate of sliding* on the bed remains the same.**
**The velocity profile, like that in a stream (Fig. 7.5) shows total movement at any depth.**

can liken a glacier to Earth's crust itself, which consists of a surface zone with joints and fractures (the lithosphere) and a deeper zone (the asthenosphere) that can flow.

The surface velocity of a valley glacier is measured by surveying from the sides of the valley, at intervals of time, a line of markers extending from side to side of the glacier. Results show that the central part of the glacier's surface moves faster than the sides, as is true of a stream of water. Velocities normally range from a few millimeters to a few meters per day, about the same slow rates as the rates of percolation of ground water. At such velocities ice takes a long time to travel the length of a glacier. Probably hundreds or even thousands of years have elapsed since the ice now exposed at the downstream end of a long glacier fell as snow upon the upstream part.

Many measurements show that outer parts of the Antarctic Ice Sheet are flowing at rates of around 50m per year (roughly 15cm per day). But within the body of the ice sheet, not readily visible at the surface, are ice streams that follow large valleys in the bedrock beneath the ice sheet. Such ice streams, somewhat analogous to well-defined currents in the ocean, are flowing as much as ten times faster than the ice around them.

### Temperatures

Both the upper surfaces and the bases of glaciers maintain temperatures that are fairly close to 0°C. In summer, solar heat melts at least a little ice at the upper surface, and heat rising continually from Earth's interior melts a little ice at the base. A comparatively thin glacier in a cold, polar climate may be frozen to its bed, at least at times, whereas a glacier in a lower latitude may have a base that is "wet" most of the time. In January 1968, at latitude 80°, a core-drill hole right through the Antarctic Ice Sheet was completed. After cutting through 2,164m of ice, the drill encountered the contact between ice and the rock beneath, and there found water. Although only an extremely thin film, the water was under hydrostatic pressure and so rose in the drill hole, as would happen in an artesian well (Chapter 8).

### Economy

A glacier, like a river, can be thought of in terms of its economy. We see in Figure 10.7 that a glacier consists of an upstream section that collects snow and a downstream section that wastes ice. The two are separated by the snowline. On the upstream section more snow falls than is removed by melting and evaporation. The weight of the extra snow tends to thicken the glacier and promote movement. In the downstream section the reverse is true. Wastage exceeds snowfall, and this tends to thin the glacier. But the thinning is made up for by the ice that flows down from the upstream section. The lines of flow, represented by arrows in the figure, bend upward to make up for the ice lost at the surface.

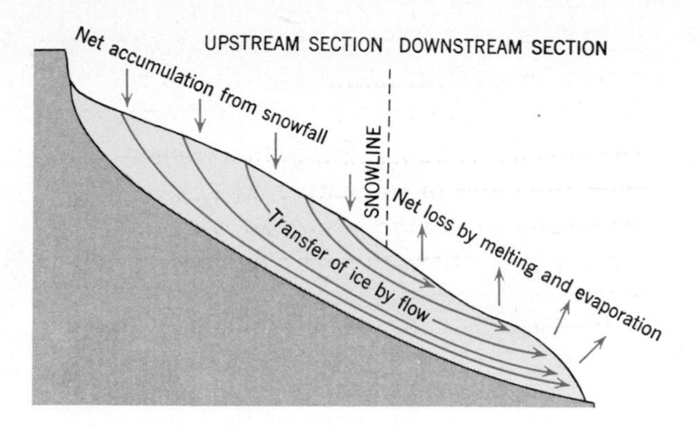


**Figure 10.7**
**Economy of a glacier. Paths of flowing particles of ice are suggested by long, curved arrows somewhat like those of percolating ground water (Fig. 8.3). Compare the economy of an ice sheet (Fig. 10.3, inset section).**

So the two sections of a glacier are like a balanced account at a bank. The spendthrift section downstream, a thoroughgoing waster, draws on the ice account and wastes all its assets, while the indulgent section up above underwrites all the losses. It makes continual payments into the account and keeps it balanced. This is yet another example of a system in a steady state.

Nevertheless the generous depositor does control the rate of spending, for the account cannot be overdrawn. If resupply is abundant because of increased snowfall upstream, the downstream section thickens and spreads and the snowline shifts downward. With decreased snowfall or increased melting, opposite changes take place. Steady state in a glacier is controlled by climate. When the climate changes, the glacier changes too. Glaciers, therefore, are a sort of yardstick for measuring variations of climate. When the climate in any region becomes cooler and more moist, glaciers expand; when it grows warmer and drier, they shrink. A nearly worldwide shrinkage of glaciers during the first half of the present century implies a general, although slight, warming of climates during at least that span of time (Fig. 10.8).

During the shrinkage of a glacier in a warming climate, the ice of course does not reverse the direction of its movement. The glacier continues to flow in the same direction, but the terminus melts back faster than ice can be supplied to it by flow (Fig. 10.17). Shifting of the terminus, either backward or forward, is merely the adjustment between supply and melting, tending to re-establish a steady-state condition after an interruption caused by a change in the climate.

**Surges.** Despite the close relation between climate and the behavior of glaciers, from time to time some glaciers experience violent changes called *surges,* which seem to have little to do with climate. When a surge occurs, a glacier seems to go berserk. It "takes off" and for a few years or decades moves at what, for a glacier, is high speed—in one instance more than 6km per year (nearly 70cm per hour). What makes this happen is not surely known. One likely possibility is that the base of the glacier had been frozen to its bed, and that sudden melting of the base had

**Figure 10.8**
**Combined Guyot (pronounced Gheé-o) and Yahtse Glaciers, seen from the southeast in 1938. The glaciers flow toward the camera and terminate in Icy Bay, Alaska, at West longitude 141°20′.**

**During the last few decades these glaciers, like their neighbors, have been shrinking, principally by calving. Between 1892 and 1938 the terminus receded about 20km. Successive positions of the termini in four later years, determined by instrumental surveys, are shown. Terminal recession during the 25-year period 1938 to 1963 totaled nearly 10km, a retreat more rapid than would be expected if the glacier had ended on land. By 1963 the two glaciers had become separated by the Guyot Hills. The mountains on the skyline are 190 to 210km distant. (*Photo by Bradford Washburn; data from M. M. Miller.*)**

"uncoupled" a sizable part of the glacier, allowing it to slide, rather than flow, rapidly downslope. The sliding part then overrode the part farther downstream, only to freeze up again and put an end to the surge.

# GLACIATION

Landscapes in Canada, northern United States, and northern Europe are different from those farther south. A principal reason for the difference is that they have been glaciated, and so recently that erosion by weathering, mass-wasting, and running water has not had time to alter them much. Like the geologic work of other external processes, ***glaciation****, the alteration of a land surface by massive movement of glacier ice over it,* includes erosion, transport, and deposition.

## GLACIAL EROSION AND SCULPTURE

It has been said that a glacier is at once a plow, a file, and a sled. As a plow it scrapes up regolith and gouges out blocks of bedrock; as a file it rasps away firm rock; as a sled it carries away the load of sediment acquired by plowing and filing, plus additional rock waste fallen onto it from adjacent cliffs.

In Figure 10.9 we see a feature common in northern lands: the result of erosion by a glacier that acted as a plow and a file. The under surface of the ice, studded with rock particles of many sizes, abraded the bedrock over which it moved, and made long scratches (*glacial striations*) and grooves. In places fine particles of sand and silt in the base of the glacier acted like sandpaper and polished rock to a smooth finish.

At the same time the under surface of the ice drags at bedrock, breaking off blocks of it (usually at joints) and quarrying them out, especially on the downstream sides of hillocks. This process, called *plucking,* creates cliffs facing downstream in contrast to the smooth, abraded upstream slopes of the hillocks. This

Figure 10.9
**Glaciated surface of sandstone bedrock, showing both striations and grooves. The sandstone contains pebbles and cobbles of quartz, which resist erosion better than the sandstone matrix does. The "tails" of sandstone behind them stream away from the camera; hence the former glacier must have moved in that direction (south). The straight gouges that parallel the handle of the hammer were made by the teeth of a power shovel. East Haven, Connecticut. (*R. F. Flint.*)**

**Figure 10.10**
**Results of glacial erosion.**
***A.* Knobs of granitic bedrock abraded and striated on one side and plucked on the other, by a former glacier. Unsymmetrical form shows glacier moved toward southwest. Lake Athabaska, Saskatchewan. (*F. J. Alcock, Geol. Survey of Canada.*)**
***B.* Abrasion and plucking in progress.**

unsymmetrical form shows which way the glacier was moving (Fig. 10.10). The great length and straightness of some grooves are made possible by the fact that the flow of ice is not turbulent. In contrast, rock surfaces abraded by the turbulent flow of water look different; they consist of a complex pattern of curves.

### Frost wedging

Although not a process of glaciation, intense frost wedging (Chapter 6) accompanies glaciation because both occur in cold climates. When hills or mountains stand higher than the surface of a glacier, wedged-out blocks of rock roll down onto the glacier, which carries them away. Intense frost wedging is responsible for much of the detail of the sharp, jagged peaks of glaciated mountains (Fig. 10.13*C*).

### Cirques

Among the characteristic features made by glaciation in many mountain areas is the ***cirque*** (pronounced *sirk*). This is *a steep-walled niche, shaped like a half bowl, in a mountainside, excavated mainly by frost wedging and glacial plucking* (Figures 10.11 and 10.12). A cirque begins to form beneath a snowbank or snowfield just above the snowline, and is at least partly the work of frost wedging. On summer days water from melting snow infiltrates openings in the rock beneath the snowbank. At night the temperature drops and the water freezes and expands, prying out rock fragments. The smaller rock particles are carried away downslope by meltwater during thaws. This activity creates a depression in the rock and enlarges it. If the snowbank grows into a glacier, plucking helps to enlarge the cirque still more, but frost wedging continues as water descends the rock wall of the cirque

**Figure 10.11**
**Cirques at the heads of small valleys, altitude about 12,000 ft. Red Table Mountain, near Aspen, Colorado.** **(*J. S. Shelton.*)**

and freezes there. The floors of many cirques are rock basins, some containing small lakes. In summary, small cirques are excavated beneath snowbanks even where no true glacier exists; large cirques are mostly the work of glaciers that continue the excavation process.

### Glaciated valleys

Glaciated valleys differ from ordinary stream valleys in several ways. Their chief characteristics, not all of which are found in every glaciated valley, include (Figures 10.12 and 10.13*C*): (1) a cross profile that is trough-like (**U**-shaped), and (2) a floor that lies below floors of tributaries, which therefore "hang" above the main valley. Both (1) and (2) result from erosion by the sides as well as by the base of the glacier, the thickness of which is far greater than the depth of an ordinary stream. In addition (3) the long profile of the floor is marked by steplike irregularities (Fig. 10.18) and shallow basins. Many of these are related to the spacing of joints in the rock, which determine ease of plucking. Finally (4) the valley head is likely to be a cirque or a group of cirques.

Some valleys are glaciated from head to mouth. Others are glaciated only in their headward parts, and downstream their form has been shaped by streams, sheet erosion, and mass-wasting. This shows that the valleys were cut before the glaciers formed, and leads to the conclusion that glaciers, unlike streams, do not cut their own valleys but occupy and remodel valleys already made.

### Mountain sculpture

If we examine a mountain area that has been glaciated by a large group of valley glaciers (Fig. 10.13), we find cirques, **U**-shaped troughs, hanging tributaries, and (in greater detail) striated and polished bedrock. In addition, the forms of the mountain crests are characteristic and are mainly the result of frost wedging in a cold climate. The forms are combinations of three features, for which

**Figure 10.12 Glaciated valley, Coast Mountains, British Columbia. A former glacier has ground away projecting spurs to create *U*-shaped trough. On the upland, left center, are several cirques containing very small glaciers. (*Canadian Government copyright.*)**

we use names given to them by Alpine mountaineers. An ***arete*** (pronounced "arrette") is *a jagged, knife-edge ridge created where two groups of cirque glaciers have eaten into the ridge from both sides.* A **col** is *a gap or pass in a mountain crest where the headwalls of two cirques intersect eachother.* A **horn** is *a bare, pyramid-shaped peak left standing where glacial action in cirques has eaten into it from three or more sides.* The Matterhorn in the Swiss Alps is a well-known example.

All these features are primarily the work of frost wedging coupled with glacial erosion and transport. All are shown in Figure 10.13.

### Fiords

The deep, bay-like ***fiords*** along mountainous coasts like those of Norway, British Columbia/Alaska, and southern Chile are *glaciated troughs partly submerged by the sea.* The form and depths of many fiords imply glacial erosion of 300m or more. The positions of fiord floors below sea level result in part from rise of sea level since the last glacial age (Fig. 10.1), in part (locally) from subsidence of Earth's crust, and in part from glacial erosion below the sea's surface. Sea level is not a base level for glaciers as it is for streams, because glacier ice 300m thick, with a specific gravity of 0.9, can

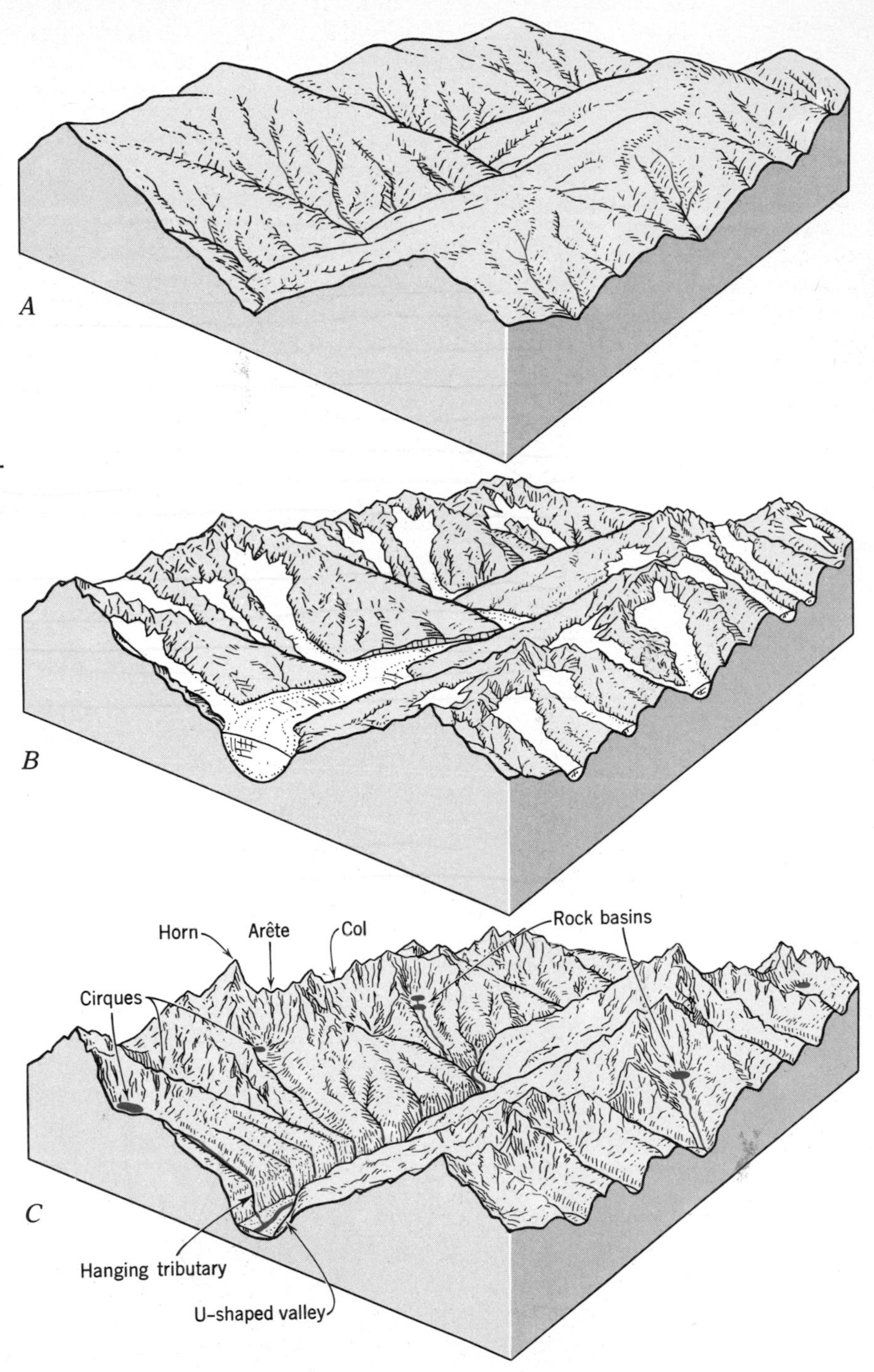


**Figure 10.13 Erosion of mountains by valley glaciers.**
***A*. Mountain region being eroded by streams. Main valley has many curves.**
***B*. Climate grows colder, snowfields form, and small cirques are excavated beneath them. Some snowfields form cirque glaciers, which merge to form a large valley glacier with tributaries. Frost wedging begins to sharpen the mountain summits.**
***C*. Warmer climate melts glaciers and reveals their geologic work. Valleys have been deepened, widened, and straightened, tributaries left hanging above main valley, and empty cirques, some with small lakes, now indent the highest areas. Mountain crests have been frost wedged to form knife-edge ridges with pyramid-shaped peaks.**

continue to erode its bed until it is submerged to −270m, whereupon it floats and bed erosion ends. In contrast to the deep erosion of fiords, some areas that have been glaciated by ice sheets still preserve chemically weathered regolith beneath glaciated surfaces. Because such regolith is ordinarily thin, the thickness of rock removed by glacier ice must have been small, possibly only a meter or two. So the intensity of glacial erosion, like that of stream erosion, varies according to topography, kind of bedrock or regolith, and thickness and velocity of flow.

## GLACIAL TRANSPORT

In the way in which it carries its load of rock particles, a glacier differs from a stream in two main respects: (1) its load can be carried in its sides and even on its top, and (2) it can carry much larger pieces of rock; also it can transport large and small pieces side by side without segregating them into a bed load and a suspended load that must be deposited in order of decreasing weight of the individual particles. Because of these differences, deposits made directly from a glacier are neither sorted nor stratified.

The load in a glacier is concentrated in base and sides (Fig. 10.18), because these are the areas where glacier and bedrock are in contact and where abrasion and plucking are effective. Much of the rock material on the surface of a valley glacier got there by landsliding from clifflike valley sides.

A good deal of the load in the base of a glacier consists of fine sand and silt. Most of the particles are fresh and unweathered, with angular, jagged surfaces. They are the products of crushing and grinding. *Fine sand and silt produced by crushing and grinding in a glacier* are ***rock flour,*** a material that differs from the chemically weathered, more rounded particles found in the sediments of nonglaciated areas.

Much of the transfer of sediment from a glacier to the ground occurs by release of particles as the surrounding ice melts. Therefore most glacial deposition takes place in the downstream part of the glacier, below the snowline, where melting is dominant.

## GLACIAL DEPOSITS

### Drift, till, and stratified drift

*Sediment deposited directly by glaciers or indirectly in glacial streams, lakes, and the sea* together constitute ***glacial drift,*** or simply ***drift.*** The name *drift* dates from the early 19th Century, when it was vaguely conjectured that all such deposits had been "drifted" to their resting places by the Flood of Noah or by some other ancient body of water.

Drift consists of two extreme kinds, *till* and *stratified drift,* which grade into eachother. Drift whose constituent rock particles are not sorted according to size and weight but lie just as they were released from the ice (Fig. 10.14*A*) is known as *till,* a name given it by Scottish farmers long before its origin was understood. ***Till*** is defined simply as *nonsorted drift* and is deposited directly from ice.

*A*

*B*

**Figure 10.14**
**Till and a cobble from till.**

***A*. Till exposed in a road cut near Bangor, Pennsylvania. Its nonsorted character is clearly evident. The fine sediment between the stones consists of rock flour. Surfaces of two large boulders have been faceted and polished. (*Pennsylvania Geol. Survey.*)**

***B*. Limestone cobble, collected from till exposed in northern New York State. The broad surface is a facet. The random striations on it are deep because the calcite of which the rock consists is a soft mineral. (*R. F. Flint.*)**

Probably most till is plastered onto the ground, bit by bit, from the base of the flowing ice near the outer margins of glaciers. We say "probably" because no one has yet devised a means of getting underneath a glacier to observe the process. The surfaces of pebbles and larger fragments in till include facets (Fig. 10.14*B*) joining each other along smoothed or rounded edges; some facets are striated. Facets are made by grinding. As pebbles turn within their matrix of ice, new facets are made. The sand and silt particles in till generally consist of rock flour (Fig. 10.15).

Colluvium, described in Chapter 6, is nonsorted, and some colluvium, on first comparison, looks very much like till. However, the presence of faceted, striated stones, rock particles of distant origin, and, in some places, a grooved or striated pavement underneath, help to identify till.

On the other hand, much drift is stratified, indicating that water from melting ice has moved and sorted rock particles carried in the ice and has deposited them in immediate contact with the ice or beyond the glacier itself. ***Stratified drift***, then, is *drift that is sorted and stratified;* it is deposited not by a glacier but by glacial meltwater.

## Streamline hills

Stratified drift and till are sediments. They occur in various bodies, each having a distinctive topographic form. Some of the bodies are described in the following paragraphs.

In many areas drift is molded by flowing ice near the outer edge of an ice sheet into smooth, nearly parallel ridges and furrows that range up to many kilometers in individual length. These forms resemble the streamlined bodies of airplanes and racing cars; they offer minimum resistance to the ice flowing over them. The best-known variety of streamline form is the ***drumlin***, *a streamline hill consisting of drift, generally till, and elongated parallel with the*

**Figure 10.15**
**Rock flour, forming part of a body of till near Bethany, Connecticut, viewed in a thin section under a microscope. The very angular shapes of particles (mostly quartz and feldspar) are the result of crushing. Particles are not sorted. Note similarity to Figure 10.14 even though the two scales are very different. (*R. W. Powers.*)**

**Figure 10.16**
**A streamline hill. Drumlin near Madison, Wisconsin. See the map, Figure 10.24. (*C. C. Bradley.*)**

*direction of glacier movement* (Figures 10.16 and 10.24). Not all such forms, however, are built up. Some are shaped by glacial molding of pre-existing drift; these too are drumlins. Others, like the much smaller striations and grooves, are shaped by glacial erosion of bedrock, and even though their form is streamline they are not drumlins. Whether made by building up or cutting out, or both, all these forms reflect streamline molding by flowing ice, and therefore their long axes indicate the direction of flow of former glaciers.

### Moraine

*Widespread thin drift with a smooth surface consisting of gently sloping knolls and shallow closed depressions* is ***ground moraine*** (Fig. 10.17). Probably its irregularities result from irregular distribution of rock particles in the base of the glacier.

*A ridgelike accumulation of drift, deposited by a glacier along its margin, is an* ***end moraine*** (Figures 10.17 and 10.18). It can be built by snowplow- or bulldozer action, by dumping off the glacier margin as the ice melts, by repeated plastering of sticky drift from basal ice onto the ground, or by streams of meltwater depositing stratified drift at the glacier margin. End moraines range in height from a few feet to hundreds of feet. In a valley glacier the end moraine is built not only at the terminus but along the sides of the glacier as well, for some distance upstream. The terminal part is a *terminal moraine;* the lateral part is a *lateral moraine;* but both are parts of a single feature (Figures 10.2, 10.18, and 10.24).

### Erratics, boulder trains

Some of the boulders and smaller rock fragments in till are the

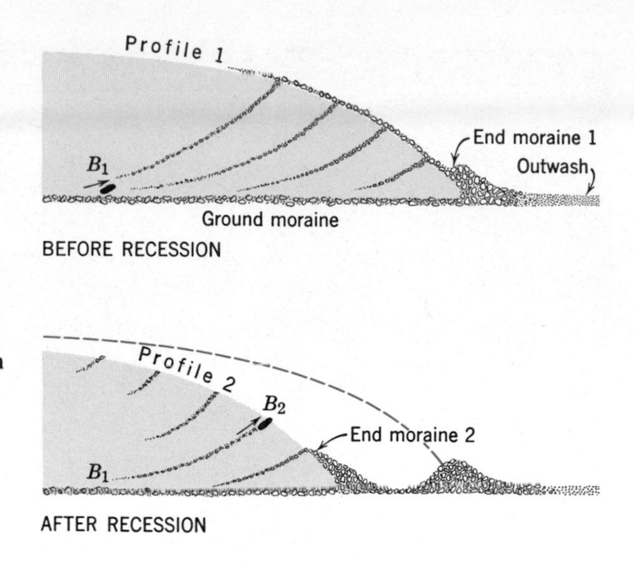


**Figure 10.17**
**With change to a slightly warmer climate, rate of melting increases and the terminus of a glacier recedes from Profile 1 to Profile 2. Continued flow of glacier during recession moves boulder $B$ from $B_1$ to $B_2$. We can expect the boulder to be built into End Moraine 2. Repeated fluctuation can pile up a complicated sequence of end moraines and outwash sediments.**

same kind of rock as the bedrock on which the till was deposited, but many are of other kinds, having been brought from greater distances. *A glacially deposited particle of rock whose composition differs from that of the bedrock beneath it* is an ***erratic.*** The word means simply "foreign," and the presence of foreign boulders was

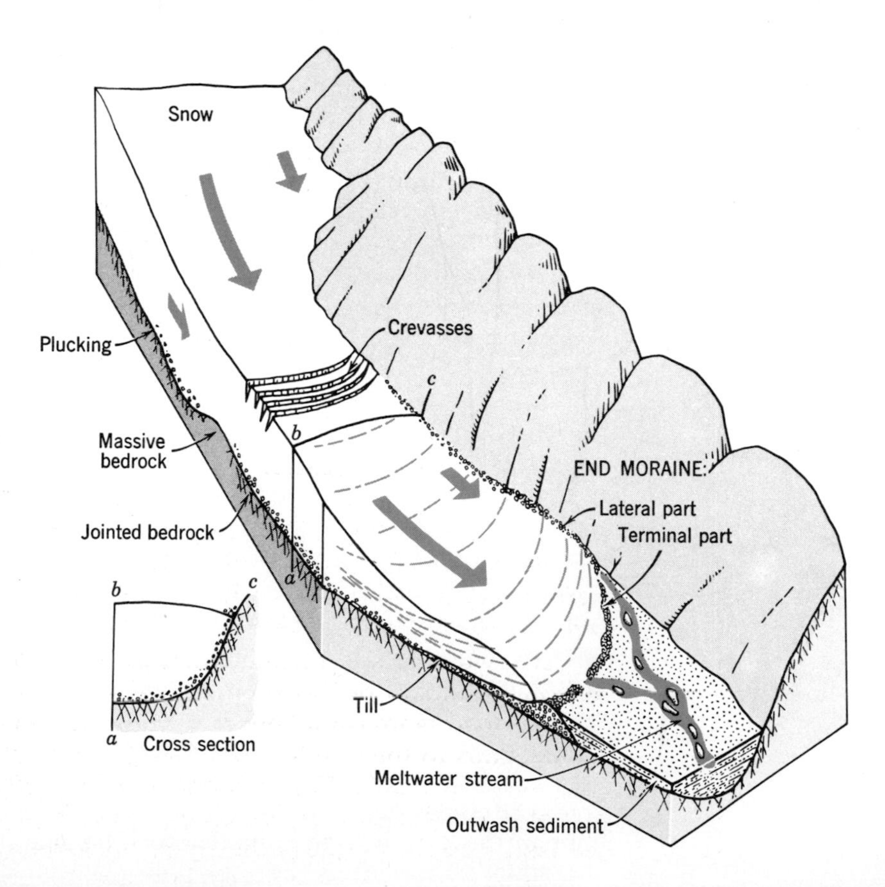


**Figure 10.18**
**Chief features of a valley glacier and its deposits. The glacier has been cut away along its center line; only half of it is shown. Compare Figure 10.2. The crevasses shown result from steepened slope of the bed beneath the glacier. Lengths of arrows are proportional to speed of flow.**

**Figure 10.19**
**Valley train being actively built of outwash sediment by streams of meltwater from glaciers upstream. Tasman Valley, Southern Alps, New Zealand. Tasman Glacier (*upper center*), dark with stony debris, is the principal source of outwash. Hooker Glacier (*upper left*) is contributing a fanlike body of outwash that has pushed the streams from Tasman Glacier to one side of their valley. Other, local fans are visible in left foreground. The valley train has little or no vegetation because a new layer of sediment is deposited on it during each melting season. (*V. C. Browne.*)**

one of the earliest recognized proofs of former glaciation. Some erratics form part of a body of drift; others lie free on the ground. Many erratic boulders exceed 3m in diameter, and some are enormous, with weights estimated in the thousands of tons. Such boulders are far larger than those that can be carried by an ordinary stream of water.

In some areas that have been glaciated by ice sheets, erratics derived from some distinctive kind of bedrock are so numerous and easily identified that they can be readily plotted on a map. Generally the plot shows a fanlike shape, spreading out from the area of outcrop of the parent bedrock and reflecting the spreading of the ice sheet. *A group of erratics spread out fanwise* is a ***boulder train,*** so named in the 19th Century when rock particles of all sizes were called boulders. The boulder train shown in Figure 10.24 consists of quartzite, very conspicuous in an area in which all the bedrock consists of limestone and sandstone.

## Outwash sediment

On the downstream sides of most terminal moraines is *stratified drift deposited by streams of meltwater as they flow away from a glacier.* Sediments of this kind are ***outwash*** ("washed out" beyond the ice). *A body of outwash that forms a broad plain* is an ***outwash plain.*** In contrast, a ***valley train*** (Fig. 10.19) is *a body of outwash that partly fills a valley.* Meltwater generally emerges from the ice as one or more swift streams, turbid with a suspended load of rock flour and with an abundant bed load of pebbles, cobbles, and even boulders. The bed load is invisible, but it is there. The deposition of outwash is analogous to the building of a fan (Fig. 7.9). The stream emerges from an ice-walled valley or tunnel onto a broad smooth surface; part of its load, therefore, becomes excess; so the stream deposits bed load, and the profile of the entire thick deposit is steep like that of a fan. As the coarse particles are dropped first, the average diameter of particles decreases downstream.

Although the bed load of a meltwater stream is invisible, the suspended load has been measured. The rock flour washed out of the big Muir Glacier in coastal Alaska corresponds to a loss of 2cm of bedrock, from the entire area beneath the glacier, every year. Such a rate of erosion is 332 times faster than the 1cm every 166 years estimated to be lost to the United States by weathering, mass-wasting, and erosion by running water.

### Ice-contact stratified drift

When rapid melting and evaporation reduce the thickness of the terminal part of a glacier to 100m or less, movement virtually ceases. Meltwater, flowing over or beside the nearly motionless stagnant ice, deposits stratified drift, which slumps and collapses as the supporting ice slowly melts away. *Stratified drift deposited in contact with supporting ice* is ***ice-contact stratified drift.*** It is recognized by abrupt changes of grain size, distorted, irregular stratification, and extremely uneven surface form (Fig. 10.20). Bodies of ice-contact stratified drift are classified according to their shape: short, steep-sided knolls and hummocks are *kames;* terracelike forms along the side of a valley are *kame terraces;* long, narrow ridges, commonly sinuous, are *eskers* (Fig. 10.21); and basins in drift, created by the melting-out of masses of underlying ice, are *kettles* and in form are complementary to kames.

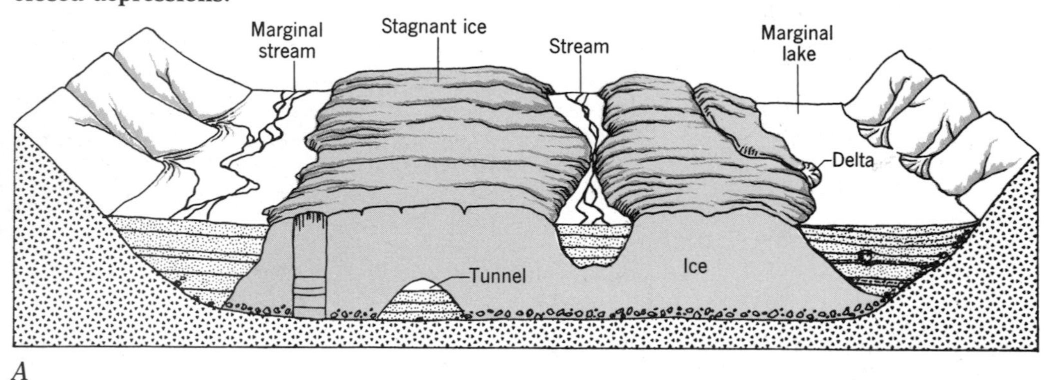


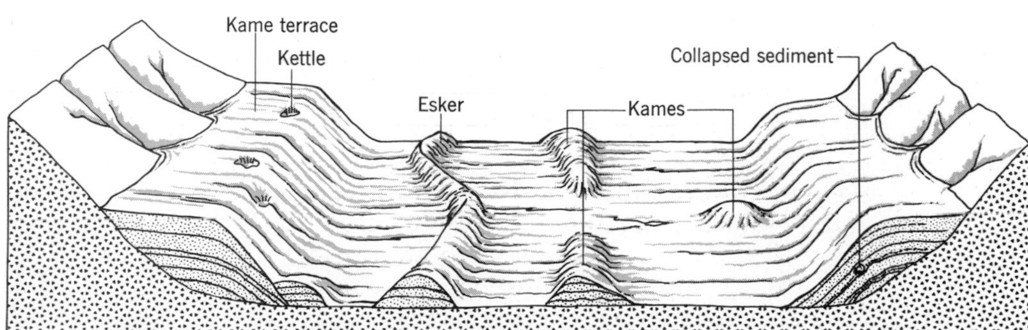


**Figure 10.20**
**Origin of bodies of ice-contact stratified drift.**
***A.*** **Nearly motionless melting ice furnishes temporary retaining walls for bodies of sediment built chiefly by streams of meltwater.**
***B.*** **As ice melts, bodies of sediment slump, creating characteristic knolls, ridges, terraces, and closed depressions.**

Figure 10.21
**This glacial-age esker, overlying ground moraine in Morrison County, Minnesota, consists of gravel and sand deposited in a winding tunnel within an ice sheet, but near its edge. When supporting ice melted away, the deposit was left as a curving ridge 10 to 15m high. (*W. S. Cooper.*)**

# THE GLACIAL AGES

### History of the concept

As early as 1821 European scientists began to recognize features characteristic of glaciation in places far from any existing glaciers. They drew the inference that glacier ice must once have covered wide regions. Consciously or unconsciously, they were applying the principle of uniformity of process and materials. The concept of a glacial age with widespread effects was first set forth in 1837 by Louis Agassiz, a Swiss scientist who achieved fame through his hypothesis. Gradually, through the work of many others, information on the character and extent of former glaciation was added to the growing body of knowledge, so that today we have a basic picture of former glacial times, although many important questions still remain unanswered.

### Extent of glaciers

During the second half of the 19th Century, geologists searched for glacial drift and other characteristic features in order to determine the extent of former glaciers. In most mountain regions, the characteristics are those shown in Fig. 10.13*C*. In most lowland regions, however, they consist generally of rolling ground moraine, end moraines, and related features (Fig. 10.24). By 1900 the general extent of glaciation had been learned. Figure 10.22 shows areas in the Northern Hemisphere that are presently believed to have been covered by glaciers during the glacial ages. Figure 10.23 shows in more detail the extent of glaciation in the northern

**Figure 10.22**
**Areas (white) in the Northern Hemisphere that were glaciated during the glacial ages.**
**Arrows show generalized directions of flow of glacier ice. Coastlines are shown as they were when sea level was 100m lower than it is today. The gray tone with irregular pattern represents floating ice in the Arctic Ocean, extending south into the Atlantic. Recent research suggests that the large area of glacier ice in northern Europe may have extended north across what is now shallow continental shelf to fill the area here left blank.**

United States and southern Canada. Such maps were compiled from observations by hundreds of geologists on the distribution of features characteristic of glaciation. On a world scale, the areas formerly glaciated add up to the impressive total of more than 44 million square kilometers—about 29 percent of the entire land area of the world. Today, for comparison, only about 10 percent of the world's land area is covered with glacier ice, of which 84 percent (by area) is on the Antarctic Continent. If we neglect that continent and consider only the rest of the world, the glacier-covered area on non-Antarctic lands was more than 13 times larger during glacial ages than it is on the same lands today.

### Directions of flow

In most mountain regions, former glaciers simply flowed down existing valleys (Fig. 10.13*B*). In lowland regions, streamline forms and end moraines show that ice sheets spread out in a radial manner (Figures 10.22 and 10.24). Flow directions are determined also by tracing erratics of conspicuous kinds to their places of origin in the bedrock. Native copper found as far south as Missouri has been traced to bedrock on the south shore of Lake Superior. In eastern Finland, copper ore traced backward along the line of ice flow led to the discovery of a valuable copper deposit in the bedrock. Several large diamonds of good quality have been found in drift in Wisconsin, Michigan, Ohio, and Indiana. Their source has not yet been found, but the flow directions of former glaciers suggest it is in central Canada.

### Amount of erosion

The great central parts of the former ice sheets in north-central Canada and northern Scandinavia and Finland removed underlying regolith and bedrock to a depth averaging perhaps 10m to 20m. Much of the load thus acquired by the ice sheets was deposited as drift beneath their broad outer parts. One such belt of drift reaches from Ohio to Montana; another extends from the British Isles to European Russia. In these belts, the average thickness of the drift is believed, from well borings and other measurements, to be as much as 12m. Most of the rest of the load went down the Mississippi River and down various rivers in Europe; some of the fine-grained part was picked up by the wind and blown away to form loess.

Where, as in New England and parts of Quebec, the bedrock is resistant and breaks into large chunks, the glacier ice spread boulders liberally over the surface, creating soil that is very difficult for agriculture. Where, as in the southern Great Lakes region and on the Plains, the bedrock is weak and crumbles easily, the glaciers deposited a thick layer of fine-grained drift, which now supports a rich soil.

### Depression of the crust

We should note at this point the subsidence of Earth's crust beneath the weight of the ice sheets. This effect is described in Chapter 14. Probably the surface of the crust beneath the central part of today's Greenland Ice Sheet would stand as much as 700 to 900m higher if the ice sheet were not there. The Hudson Bay region, which formerly lay beneath the central part of an ice sheet 3km or more in thickness, is today still rising measurably toward its old altitude (Fig. 14.6).

### Repeated glaciation

Most of the glacial drift is fresh and little weathered. From that

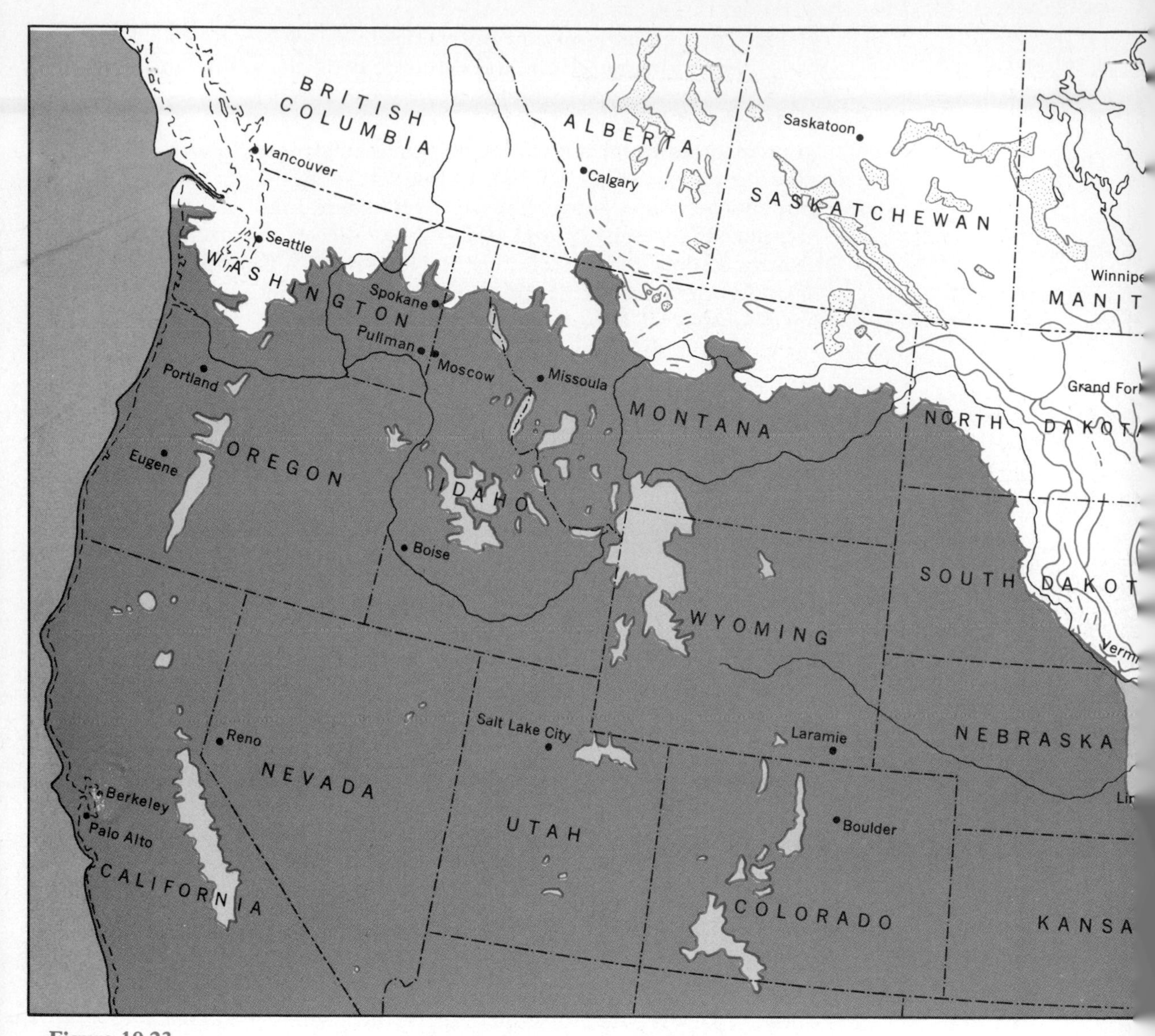


Figure 10.23
**Extent of glaciation in northern United States and southern Canada.**
**Inner, ticked line, during latest glacial age; outer line, during earlier glacial ages. Outwash valley trains in the Mississippi River drainage system are seen in Figure 20.7*B*. (*Compiled from several sources.*)**

fact it was realized very early that the glacial invasion was recent. But in the mid-19th Century geologists began to find exposures showing a blanket of comparatively fresh drift overlying another layer of drift whose upper part is chemically weathered (Fig. 10.25). This led to the inference that there had been two glaciations separated by a period of time long enough to cause weathering to a depth of one or two meters. Before the beginning of the 20th Century, accumulated evidence of this kind had established the fact of not merely two but several great glaciations, each covering approximately the same areas.

Radiometric dating suggests that the latest of these glaciations occurred 20,000 to 15,000 years ago, and the earliest may date from 1 to 2 million years ago. The time that embraced that particular bundle of glaciations is the Pleistocene Epoch (Table

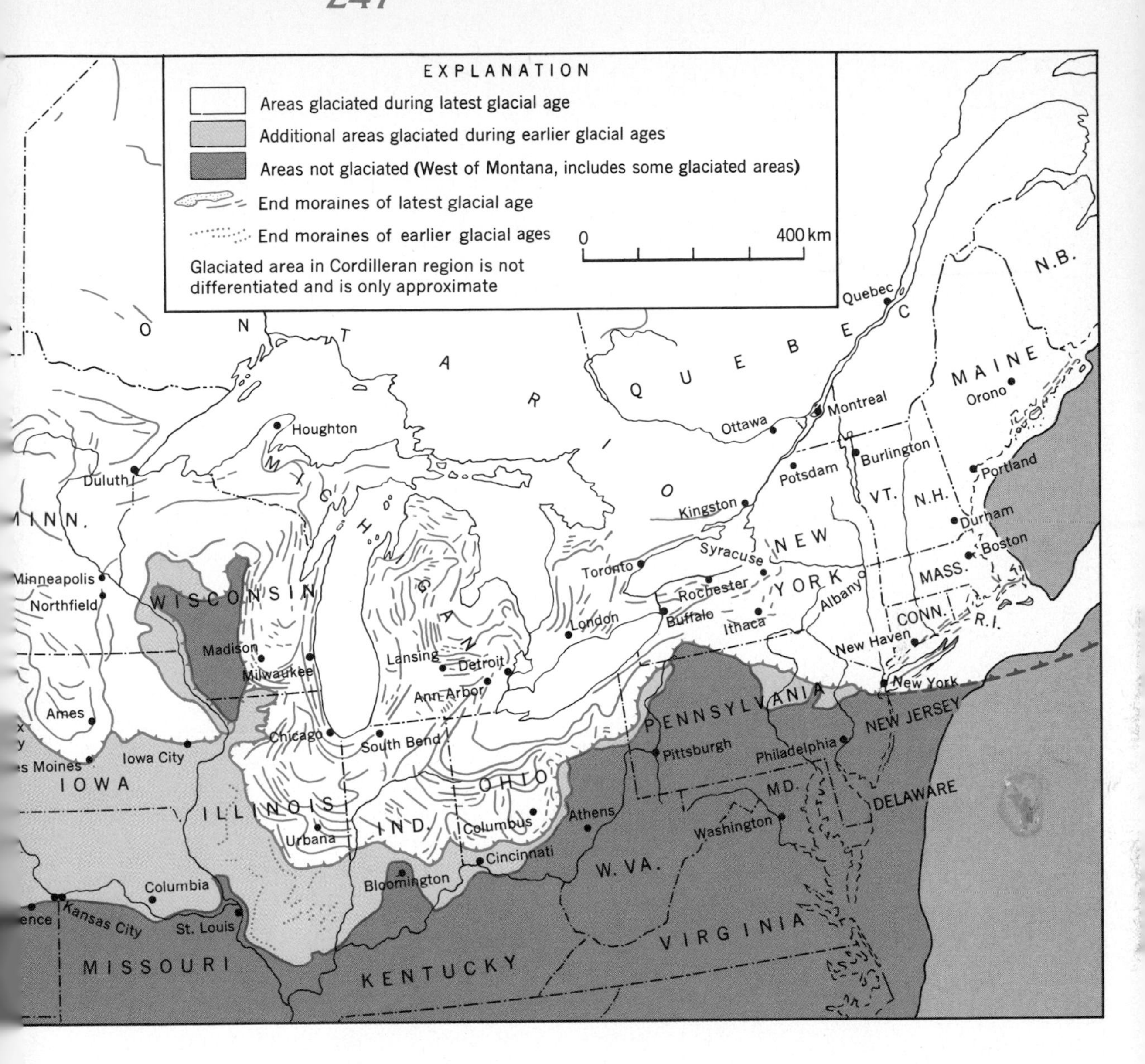


5.1), a period often referred to as the *glacial ages,* notwithstanding the fact that between the glaciations were intervals (*interglacial ages*) in which climates were about as warm as today, and at times even a little warmer than existing climates.

Other, earlier glacial ages, some of them apparently with less extensive glaciers, have appeared again and again throughout much of Earth's history. Much of the evidence of those earlier cold times consists of ***tillite*** (*till converted into solid rock*) in strata of many different ages. In some places the tillite overlies striated surfaces of still older rock. Such evidence tells us that Earth's climate has fluctuated repeatedly, causing glaciers to develop and creating great changes in environments in many parts of the world. This kind of event is indicated at many levels in the geologic column (Table 5.1) and probably has happened, in one region or

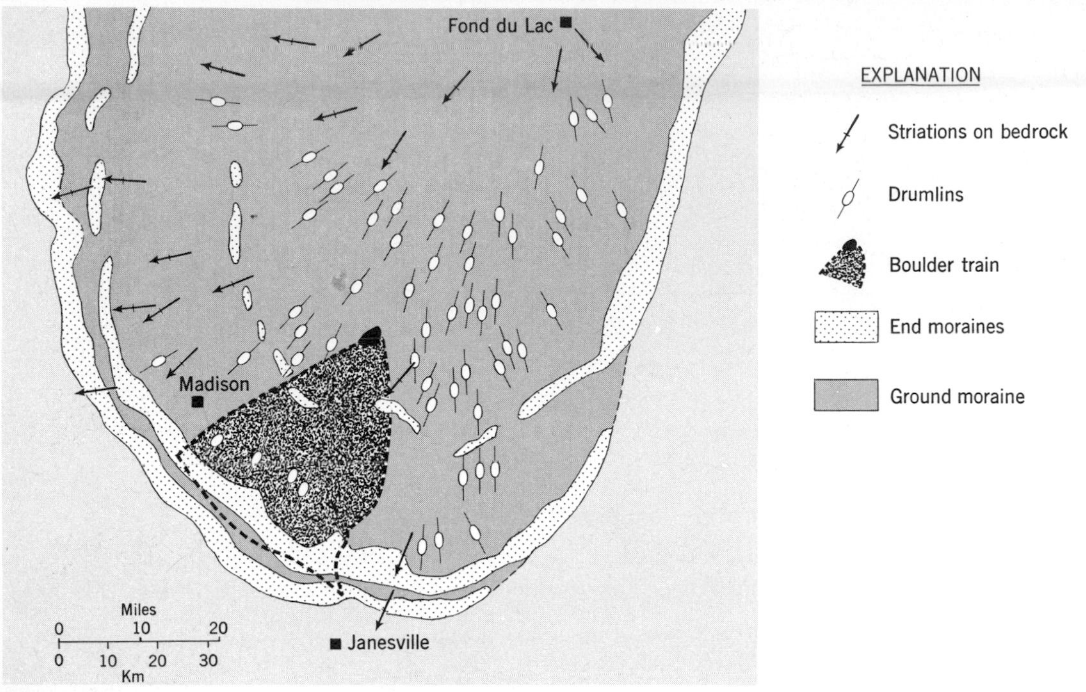


**Figure 10.24**
**Map of lobe-shaped layer of glacial drift in southeastern Wisconsin.**
**Three successive end moraines (both terminal and lateral parts) mark successive positions of the glacier margin. Directions of drumlins, striations on bedrock, and a boulder train all show spreading of ice toward margin of glacier.** ***(Adapted from Flint and others, 1956, in part after W. C. Alden.)***

another, even more frequently. The oldest known glaciation has been radiometrically dated at 2.2 billion years ago—a minimum time equal to nearly half the elapsed life of the planet. In this we see another expression of the Principle of Uniformity. Some of Earth's earlier glaciations—although their beginnings and ends have not yet been specifically dated—seem to have been of long duration. By comparison, the latest glacial age, occupying little more than 15,000 years from start to finish, appears almost ephemeral. Yet the disturbances it caused in the blanking-out of territory, fall and rise of sea level, bending of the crust downward and upward, and severe interference with the biosphere are still reflected in events we can see today. Some of these matters, including details of the latest glaciation, are discussed in Chapter 20.

# SUMMARY

## GLACIERS IN THE WATER CYCLE

1. Glaciers are water in the solid state. They are an integral part of the water cycle.

2. When glaciers form or enlarge, sea level is lowered; when they shrink or disappear, sea level is raised.

## DEFINITION AND FORM

3. Glaciers are accumulations of snow and ice; they flow under their own weight. Their surface parts are brittle; below these parts flow occurs.

**Figure 10.25**
**Evidence of repeated glaciation.**
**A layer of fresh till, weathered only slightly at its surface, overlies a layer of older till that had been weathered deeply before the till above it was deposited.**

4. On a basis of form, glaciers include cirque glaciers, valley glaciers, piedmont glaciers, and ice sheets.

5. Glaciers require low temperature and adequate snowfall. All glaciers are connected with eachother by the snowline, which stands low in polar regions and high in tropical regions.

MOVEMENT

6. The motion of a glacier includes both internal flow and sliding on its bed.

ECONOMY

7. A glacier can be thought of as a system in which thickness, velocity, and other factors adjust to eachother to create a steady-state condition.

GLACIAL EROSION

8. Glaciers erode rock by plucking and abrasion; they transport the waste and deposit it as drift.

9. Valley glaciers convert stream valleys into **U**-shaped troughs with hanging tributary valleys. Cirques form beneath snowbanks, cirque glaciers, and the heads of valley glaciers. Mountain areas that project above glaciers are reshaped by frost wedging into aretes, cols, and horns.

GLACIAL SEDIMENTS

10. The load, carried chiefly in the base and sides of a glacier, includes particles of all sizes, from large boulders to rock flour.

11. Till is deposited by glaciers directly. Stratified drift is deposited by meltwater; it includes outwash deposited beyond the ice, and kames, kame terraces, and eskers deposited upon or against the ice itself.

12. Ground moraine is built beneath the glacier, end moraines (both terminal and lateral) at the glacier margins.

GLACIAL AGES

13. During glacial ages huge ice sheets repeatedly covered northern North America and Europe, eroding bedrock and spreading drift over the outer parts of the glaciated regions. Their weight bent down the crust beneath them.

# SELECTED REFERENCES

Agassiz, Louis, 1967, Studies on glaciers (Neuchâtel, 1840), Translated and edited by A. V. Carozzi: New York, Hafner.

Bentley, C. R., and others, 1964, Physical characteristics of the Antarctic Ice Sheet: American Geog. Soc. Antarctic Map Folio 2.

Charlesworth, J. K., 1957, The Quaternary Era: London, Edward Arnold.

Crowell, J. C., and Frakes, L. A., 1970, Phanerozoic glaciation and the causes of ice ages: American Jour. Sci., v. 268, p. 193–224.

Dyson, J. L., 1962, The world of ice: New York, Alfred A. Knopf.

Flint, R. F., 1971, Glacial and Quaternary geology: New York, John Wiley.

Fristrup, Børge, 1966, The Greenland Ice Cap: Copenhagen, Rhodes.

Sparks, B. W., and West, R. G., 1972, The Ice Age in Britain: London, Methuen.

Wright, H. E., and Frey, D. G., eds., 1965, Quaternary of the United States: Princeton, N.J., Princeton Univ. Press.

GLACIAL-AGE MAPS

Flint, R. F., and others, 1945, Glacial map of North America: Geol. Soc. America Spec. Paper 60. Scale 1:4,555,000.

———, 1959, Glacial map of the United States east of the Rocky Mountains: Geol. Soc. America. Scale 1:750,000.

Wilson, J. T., and others, 1958, Glacial map of Canada: Geol. Assoc. of Canada. Scale 1:3,801,600.

**Scuba-diving geologist monitors a submarine lava flow near Hawaii.**

**Most of the ocean is underlain by basaltic lava with pillow structure. The pillows are rounded masses, like squirts of toothpaste, formed when the cooled crust of the lava splits, as seen in the photo, and hot lava oozes out to form a new pillow. (*Photo taken from "Fire Under the Sea" by Dr. Lee Tepley. Film distributed by Moonlight Productions, Mountain View, California.*)**

# PART THREE
# EXTERNAL PROCESSES IN THE SEA

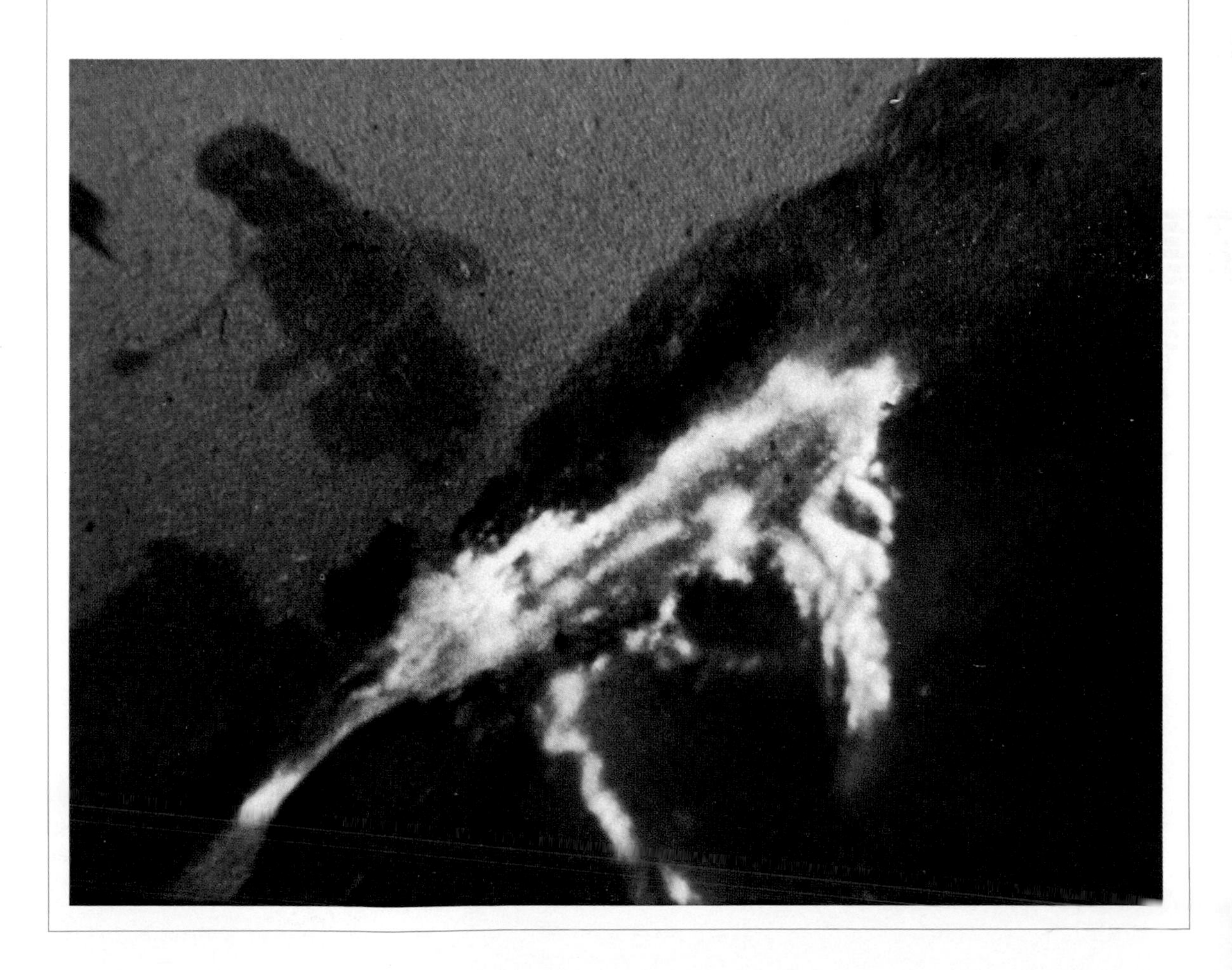

# 11 COASTS AND CONTINENTAL MARGINS



## CONTINENTAL SHELVES

Having examined the various external processes that are continually at work on the continents, we turn now to the margins of the continents. Here are the coasts and continental shelves, where other activities are at work. The continental margins are overlapped by the world ocean, for the deep ocean basins are like great dishes with flattened rims. The dishes are so full that water slops over and floods the rims. Figure 13.6 shows the western rim of the North Atlantic ocean basin, the continental shelf of eastern North America. The same rim of the Atlantic dish is seen in diagram in Figure 11.1.

These flat rims, *the shallow, nearly flat, submerged borders of the continents* are ***continental shelves.*** Take away the seawater and they become very conspicuous (Fig. 11.2). Their flat upper surfaces add about 10 percent to the areas of the continents and their outer slopes, described below, add at least half as much more. The shelves average about 60km in width, but there are local variations, from 1,300km off the Arctic coast of Siberia down to almost zero. Off North America, the Pacific coast has a shelf only a

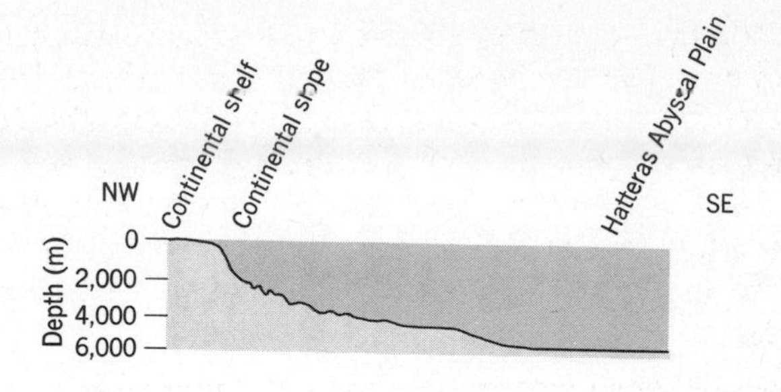


**Figure 11.1**
**Profile across the eastern margin of North America, showing the flat, step-like surface of the continental shelf and the steep continental slope. The slope grades gently out into an abyssal plain. The position of this profile, which runs southeastward from Norfolk, Virginia, can be identified on Figure 13.6.** (*After Heezen, Tharp, and Ewing, 1959.*)

few kilometers wide, and off South America it has hardly any shelf at all.

The seaward edges of the shelves are covered everywhere by at least 100m of water, and in places by as much as 600m. Thus the edge of a shelf is defined, not by depth of water but by a marked steepening of slope.

## Continental slopes

As we see in Figure 11.1, the shelf passes rapidly into the continental slope, a pronounced slope beyond the seaward margin of the continental shelf, leading down into deep water. Continental slopes mark the outer edge of the continental crust where it abuts against the oceanic crust. Many continental slopes, like the one shown in Figure 11.1, grade downward and outward into a region of gentler slopes. This region is a vast, sloping pile of sediment, consisting of waste derived from the adjacent continent and from sedimentary debris that slumps down from the shelf and slope above.

The surface of the continental slope, like that of the shelf, is rather smooth, with a single exception. Cut into the continental slopes are many remarkable valleys, some so deep that they have been called *submarine canyons*. These valleys, as much as 1km deep and with steep sideslopes, are widely spread around the world. Their origins are still a puzzle. The valleys extend down to water depths of 3km or more—far too deep to permit the canyons to have been excavated by ordinary rivers, even at times of lowered sea level. Some, such as those off the mouths of the Hudson, Congo, and Ganges Rivers, line up with valleys that cross the continental shelves, while others seem to have no connection with such valleys. At least some submarine canyons are being actively eroded today, as described in Chapter 12.

## Changing sea level

The surfaces of continental shelves are smooth but not completely so, mainly because during the latest glacial age, when oceans were lowered by the building of glaciers, much of the shelf area was land. Valleys, like the valley off the mouth of the Hudson River (Chapter 10), lead from the coast outward and are extensions of large valleys on the adjacent continents. The submerged valley of the Rhine extends from the Netherlands coast nearly 300km northward across the shelf to an area between Scotland and Norway. Off the New England coast, submarine hills mark the sites of glacial deposits and piles of boulders lie where they were deposited by former glaciers or by floating ice. Former beaches, coastal sand dunes, and similar features mark shorelines built by the rising sea at the end of the glacial age, and later drowned. Shallow depressions scoured by waves and currents reflect marine erosion of still later date.

Most of these features witness the fact that the level of the world ocean, the reservoir for the water cycle, is not fixed; it fluctuates. It is accidental, changing its relation to the continents and altering the shapes and positions of coastlines.

How do the changes come about? Of course small, temporary, local changes are created by rhythmic rise and fall of the tide and by the piling-up of water against a coast by a big storm. But there are greater, longer-term changes. Measuring them involves special problems because we cannot fix the position of sea level relative to the center of Planet Earth. We can fix it only relative to adjacent land. The vertical relationship of sea to land is somewhat like that between a boat and a floating dock to which it is made fast. Both boat and dock can move. By loading either boat or dock we can alter the vertical position of one relative to the other.

The surface of the world ocean may either rise or fall relative to the continents, or some part of a continent may rise or subside relative to sea level. Or both kinds of movement can occur at the same time. The principal cause of long-term rise and fall of sea level is probably the considerable changes in volume of the world ocean caused by the building and later melting of glaciers during glacial ages, described in Chapter 10.

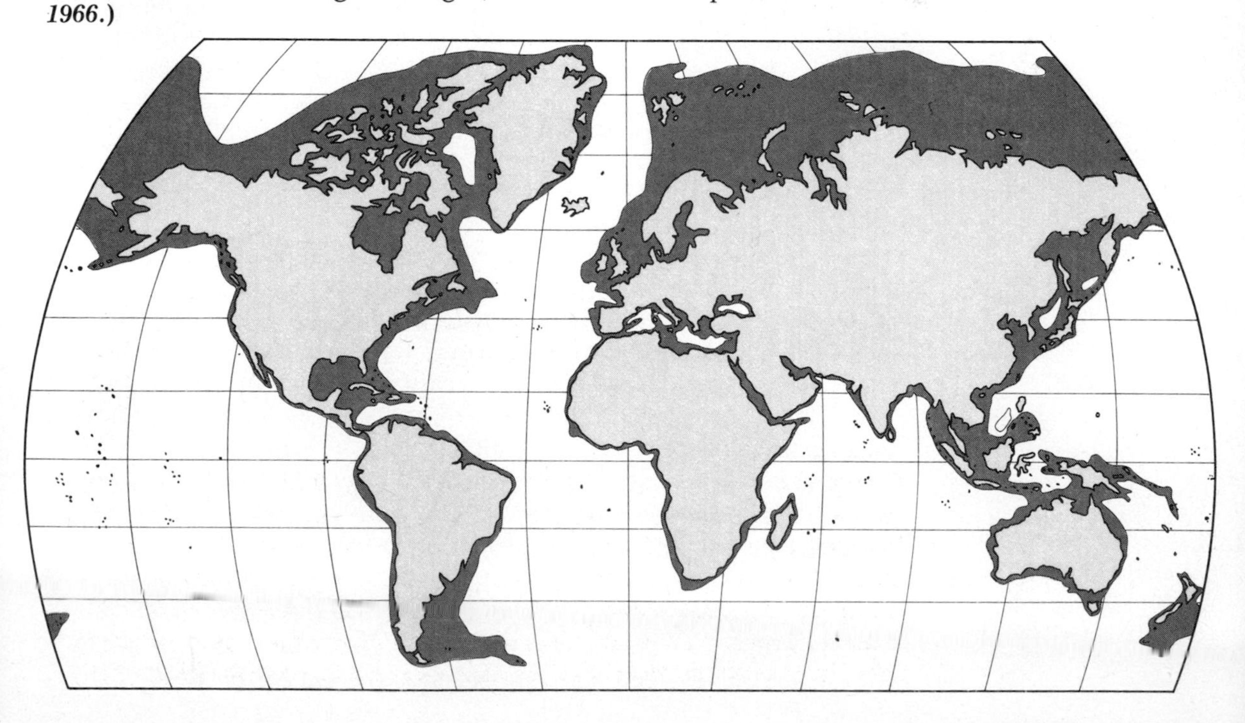

Figure 11.2
**The continental shelves and continental slopes, shown together in dark shading, add some 15 percent to the areas of the continents. The elliptical projection used for the base map (from which the Antarctic continent has been left out) exaggerates the shelves in the Arctic region—but they are still very extensive. (*Data from Menard and Smith, 1966.*)**

Rise and fall of sea level are universal movements, affecting all parts of the world ocean at the same time. But uplift and subsidence of the land, causing emergence or submergence along a coast, are piecemeal movements, generally involving only parts of continents. As we noted in Chapter 10, recent piecemeal uplift of some land areas is occurring today at maximum rates of more than 1cm per year, in just those areas that were overspread by thick ice sheets during the latest glacial age. This uplift has been in progress ever since the ice sheets began to melt away, and has lifted some sea beaches more than 200m above sea level, where of course they were formed. We reason that the weight of the ice sheets, in places 3km or more in thickness, caused the crust beneath them to subside, and that when the extra weight was removed by melting of the ice, the crust slowly "rebounded" toward its former position. The glacial age was so recent that the "rebound" is not yet completed.

Another cause of uplift is the slow slipping of a large plate of lithosphere over another (Chapter 1), the edge of the overriding plate lifting as it moves forward. It is thought that such movement is happening in western North America. Wave-cut benches somewhat like the one in Figure 11.13, along the coast of southern California, have been lifted as much as 250m above the sea. Along parts of the coast several of them stand one above the other, forming a great flight of step-like terraces. The movements that lifted them out of the sea may be related to the jostling of two lithospheric plates, sliding past and scraping one another (Fig. 4.4).

Inland from the Atlantic Coast of the United States from Virginia to Florida are many marine beaches, spits, and barriers, the highest of which reach an altitude of more than 50m. It is thought that as a group, these features owe their present altitude to a combination of two causes. One is upward bending of the crust in coastal North America. The other is lowering of the world ocean through the building of glaciers, following periods when climates were warmer than today's, glaciers were smaller, and sea level was therefore higher.

### Origin of shelves

Why does each continent have a step-like pedestal entirely or partly surrounding it? Looking ahead to Figure 11.10, we might suppose that the shelves are enormous wave-cut benches. Many decades ago, when knowledge was scanty, this origin was thought likely. But today, in the light of many more facts, we realize that bench cutting has played only a minor part in creating the shelves, many of which were built up by deposition of sedimentary layers instead of being carved out of pre-existing rock. Such shelves resemble wave-built terraces (Fig. 11.10) more than they resemble wave-cut benches.

The shelf and slope off North Carolina are a good example of the building-up process in the making of shelves. A section (Fig. 11.3) running southeast from Raleigh, North Carolina to the base of the

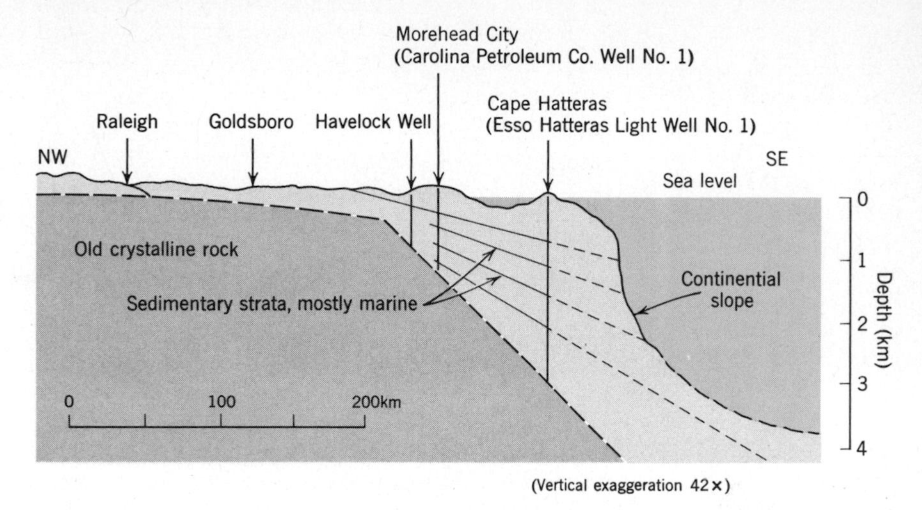


**Figure 11.3 Section through coastal North Carolina and continental shelf, showing underlying strata. (*After W. F. Prouty, 1946; B. C. Heezen and others, 1959.*)**

continental slope passes close to three deep holes drilled for oil. The detailed records of the drill holes show that the shelf consists of a wedge of sedimentary strata, about 3km thick near the slope but thinning landward, overlying a sloping floor of very old igneous and metamorphic rock. The kinds of sediment and fossils in the drill cuttings show that this great wedge consists mainly of sediment that was deposited in shallow seawater. Because the sedimentary layers now lie as much as 3km below sea level, they must have sunk down below the positions at which they were deposited. The layers are in various stages of transformation into rock, some of which is between 70 and 100 million years old. This and similar sections tell us that through long periods the margins of continents have been receiving sediment from rivers and have been gradually subsiding.

The age of the shelf off eastern North America cannot be much more than about 150 million years. Before a shelf could be built, there had to be a North Atlantic Ocean basin into which to build it, and the evidence suggests that the basin began to form at around that time, when the great North American lithospheric plate formed, began to move, and slowly carried North America away from the continents that once joined it on the east. When the break occurred, the severed continental margins—those of the Americas on one side and of Europe and northern Africa on the other—were thinned and bent slightly downward to form a long, narrow basin. Seawater flowed into the basin, creating the infant Atlantic Ocean. In this new ocean, sediment brought by rivers that flowed out of the adjacent continents began to accumulate.

As the drifting movement continued at a rate of a few centimeters per year, North American rivers built deltas into the sea and formed a growing continental shelf. Younger layers extended outward over older ones, and were draped over shelf and continental slope like tablecloths draped over a table, as the margin of the continent slowly subsided.

Looking again at Figure 11.3, we can see that strata that once

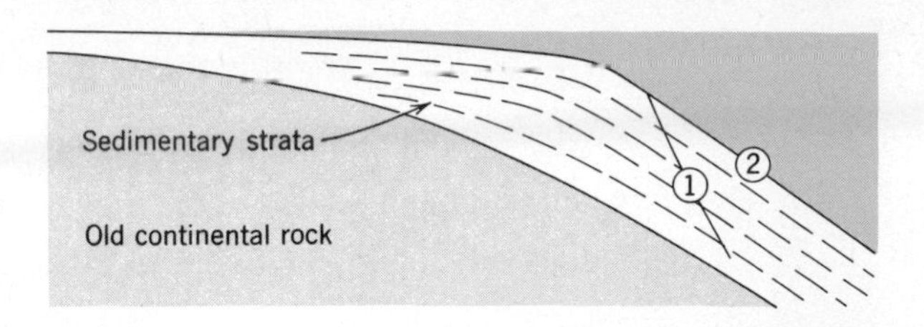


**Figure 11.4**
**Schematic cross section of a shelf like that in Figure 11.3, showing two continental slopes: (1) A slope resulting from erosion. (2) A slope, resulting from original deposition, on which no erosion has occurred.**

formed the continental slope are now cut through by that slope. The most likely explanation is that the slope has been gradually eroded (Fig. 11.4) by repeated slump, turbidity currents, and other processes involved in the creation of the canyons that today gash the continental slope.

Other shelves, like that off southern California, are narrow, irregular in surface form, and apparently consist mainly of bedrock. They seem to have resulted from bending down and faulting of the western border of the North American Continent, accompanied by repeated chiseling by surf to create wave-cut benches (Fig. 11.10).

Shelves play an important part in the rock cycle because they are the resting places of most of the sediment washed out from continental interiors to continental margins. Therefore they are also the chief places where sediment is gradually converted into rock. Besides, the transfer of sediment from centers to margins of continents gradually widens the continents by amounts equal to the widths of the shelves. This process by which continents widen themselves, and the story of what eventually happens to the shelf sediments, are explained in Chapter 18.

# THE SHAPING OF COASTS

The coast of a continent is a great boundary between two realms, land and water. Along this, as along other boundaries, two very different realms must adjust to eachother, and conflict occurs. At a coast, ocean waves that may have traveled unimpeded through thousands of kilometers encounter an obstacle to their further progress. They dash against firm rock, erode it, and pick up the eroded rock particles. Over the long term the results of this conflict are great.

Waves and the currents created by waves are the agents responsible for most of the erosion of coasts.* They are responsible likewise for most of the transport and deposition of the sediment created by wave erosion or washed into the sea by rivers. This sediment is moved outward from the coast and is deposited offshore, mostly in shallow water, in distinct layers. Each layer is a thin fringe around a continent, a fringe continually added to and covered by other layers, to form a thickening continental shelf. The

* The tidal currents mentioned in Chapter 13 likewise erode, but their effect is much smaller.

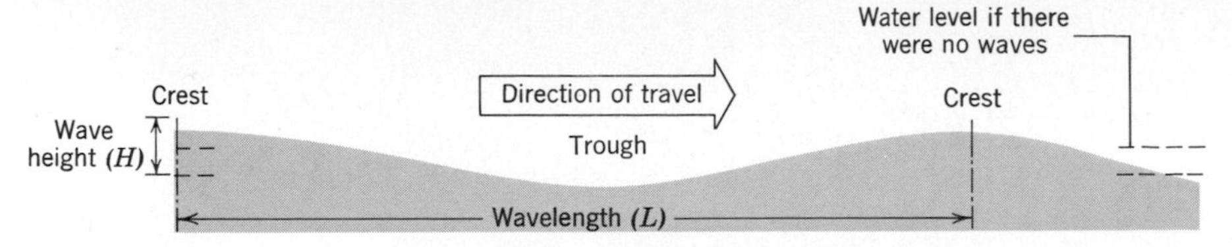


**Figure 11.5**
**Profile of waves in deep water.**

growing shelf adds to the width of its continent, the source from which most of the shelf sediment was derived. The things that happen along a continental coast, therefore, are an essential part of the rock cycle and of the growth of continents. They deserve a much closer look.

# WAVES

### Wave motion

Ocean waves are generated by winds that blow across the surface. Figure 11.5 shows the significant dimensions of a wave traveling in deep water where it is unaffected by the bottom far below. The motion of a wave is very different from the motion of the particles of water within it. As wind sweeps across a field of grain or tall grass, the individual stalks bend forward and return to their positions (Fig. 11.6), creating a wave-like effect. In similar fashion the *form* of a wave in water moves continuously forward, but each water *particle* revolves in a loop, returning, as the wave passes, very nearly to its former position. This loop-like, or *oscillating* motion of the water, first determined theoretically, was later proved by injecting droplets of colored water into waves in a glass tank and by photographing their paths with a movie camera.

Waves receive their energy from wind, and so can receive it only at the surface of the water. Because the wave form is created by loop-like motion of water particles, the diameters of the loops at the water surface exactly equal wave height (Fig. 11.6). But there is progressive loss of energy (expressed in diminished diameters of the loops) downward from the water surface at which energy is received. The rate of decrease of energy downward is so rapid that at a depth equal to only half the deep-water wave length (in other words, at a depth of $\frac{L}{2}$) the diameters of the loops have become so small that motion of the water is negligible.

### Erosion by waves

This means that depth $\frac{L}{2}$ must be the effective lower limit of wave motion, and therefore also the lower limit of erosion of the bottom by waves. In the Pacific Ocean, wave lengths as great as 600m have been measured. For them, $\frac{L}{2}$ equals 300m, a depth almost twice as great as the outer edge of the average continental shelf. Although the wave lengths of most ocean waves are far less than 600m, we can see it is nevertheless possible for very large waves to affect

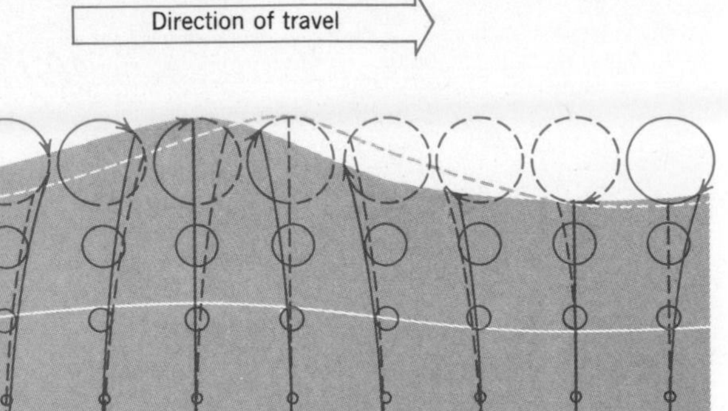


**Figure 11.6**
**Loop-like motion of water particles in a wave of oscillation, in deep water.**
**To follow the successive positions of a water particle at the surface, follow the arrowheads in the largest loops from right to left. This is the same as watching the wave crest travel from left to right. Particles in smaller loops underneath have corresponding positions, marked by continuous, nearly vertical lines. Dashed lines represent wave form and particle positions one-eighth period later. Resemblance to stalks of grain bending in the wind is now apparent.**
**(*After Ph. H. Kuenen, 1950, Marine geology: New York, John Wiley, p. 70.*)**

even the outer parts of continental shelves. What the wave motion does, landward of depth $\frac{L}{2}$, is to lift and drop, endlessly, fine particles of bottom sediment, very slowly moving them seaward along the gently sloping bottom. Such erosion is very slow and not at all spectacular, but by the end of a million years the cumulative result is great.

What happens in the shallow water at the shore is more rapid, sometimes spectacular, and always different in style. When a wave moving toward shore reaches depth $\frac{L}{2}$ it "feels bottom," and when this happens, the form of the wave begins to change. The loop-like paths of water particles gradually become elliptical, and velocities of the particles increase. Interference of the bottom with wave motion distorts the wave by increasing its height and shortening wave length. Often the height is doubled. This means the wave is growing steeper. Because the front of the wave is in shallower water than the rear part, it is steeper than the rear. Eventually the steep front becomes unable to support the wave, the rear part slides forward, and the wave collapses or *breaks* (Fig. 11.7).

When a wave breaks, the motion of its water instantly becomes turbulent, like that of a swift river. Such "broken water" is called ***surf,*** defined as *wave activity between the line of breakers and the shore.*

In turbulent surf each wave finally dashes against rock or rushes up a sloping beach until its energy is expended. Then it flows back. Water piled against the shore returns seaward in an irregular and complex way, partly as a broad sheet along the bottom and partly in localized narrow channels. The returning water is mainly responsible for the currents known to swimmers as "undertow."

Surf possesses most of the original energy of each wave that created it. This energy is quickly consumed in turbulence, in

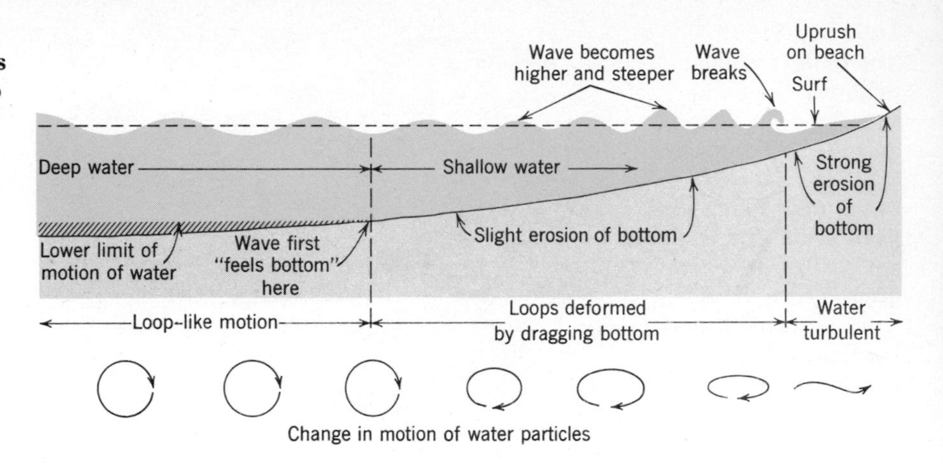


**Figure 11.7**
**Waves change form as they travel from deep water through shallow water to shore. Circles and ellipses are not drawn to scale with the waves shown above. (Compare Figure 11.6.)**

friction on the bottom, and in moving the rock particles that are thrown violently into suspension from the bottom, as in a stream. Ceaselessly the sediment is shifted landward and seaward within the surf zone. Some of the finest particles are carried in suspension into deeper water, where they settle out on the bottom. But they are still lifted, and dropped to the bottom again, wherever and whenever their depth is less than $\frac{L}{2}$. Most of the geologic work of waves, therefore, is accomplished by surf, shoreward of the line of breakers.

How deep below sea level can surf erode rock and move sediment? The answer depends on the depth at which waves break. Most ocean waves break at depths that range between wave height and 1.5 times wave height. Such waves are rarely more than 6m high; so the depth of vigorous erosion by surf should be limited to 6m times 1.5, or 9m, below sea level. This theoretical limit is confirmed by observation of breakwaters and other structures, which are found to be only rarely affected by surf at depths of more than about 7m. The surf zone, the place where high-energy turbulent water cuts into the land, is limited then to the narrow vertical range extending from sea level down to 7m, or a little more, below sea level.

But during great storms surf can strike effective blows well above sea level. The west coast of Scotland is exposed to the full force of Atlantic waves. During a great storm on that coast a solid mass of stone, iron, and concrete weighing 1,200 metric tons was ripped from the end of a breakwater and moved inshore. The damage was repaired with a block weighing more than 2,300 tons, but five years later storm waves broke off and moved that one too. The pressures applied to such erosion were around 27 tons per square meter. Even waves having much smaller force break loose and move blocks of bedrock from sea cliffs, partly by compressing the air in fissures in the rock; the compressed air pushes out blocks of rock.

**Figure 11.8**
**The tools (rounded pebbles), the rock they have smoothed, and the surf itself are all visible on this beach near Pescadero, California. The wave-cut cliff in the background forms one flank of a rock headland. (*Richard Weymouth Brooks.*)**

The vertical distance through which water can be flung against the shore would surprise anyone whose experience of coasts is limited to periods of calm weather. During a winter storm in 1952, again on the west coast of Scotland, the bow half of a small steamship was thrown against a cliff and left there, wedged in a big crevice, 45m above sea level.

Another important kind of erosion in the surf zone is the wearing down of rock by wave-carried rock particles. By continuous rubbing and grinding with these tools, the surf wears down and deepens the bottom and eats into the land, at the same time smoothing, rounding, and making smaller the tools themselves (Fig. 11.8). But as we have seen, this activity is limited to a depth of only a few meters below sea level. The surf therefore is like an erosional knife edge or saw, cutting horizontally into the land.

### Transport of sediment

The rock particles worn from the coast by surf as well as those brought to the coast by rivers are intermittently in motion. They are dragged or rolled along the bottom, lifted in irregular jumps, or carried in suspension, according to their size and to the varying energy of waves and currents. In the surf, as can be seen on almost any beach, sediment is moved to and fro, shoreward and seaward. But because the bottom slopes down in the seaward direction, the net effect is to carry rock particles from the land gradually out to sea.

Seaward of the surf zone, in deeper water, bottom sediment is shifted by unusually large waves during storms and by currents, again with net movement seaward. Each particle is picked up again and again, whenever the energy of waves or currents is great enough to move it; but as the particle gets into ever-deeper water it is picked up more and more rarely. The result of these movements is that the sediment gradually becomes sorted according to diameter, from coarse in the surf zone to finer offshore.

In the gradual building of a continental shelf by sedimentation, the deposited sediments normally grade seaward from sand into mud. This gradation is true not only of the particles eroded from the shore by surf but also of the particles contributed by rivers, whose currents carry suspended loads into the sea. Calculations made along some coasts indicate that the volume of sediment contributed by rivers is much greater than that contributed by surf erosion. The actual proportions, of course, differ, with several varying factors, from place to place.

### Wave refraction

A wave approaching a coast over an undulating bottom cannot "feel bottom" along all parts of its crest simultaneously. As each part "feels bottom," wave length at that part begins to decrease and wave height increases. As a result the wave gradually swings around, part by part, to become parallel with the bottom contours; it is said to be *refracted* (Figures 11.9, 11.13, and 11.19). Thus waves approaching the shore in deep water at an angle of 40° or 50° may, after refraction, reach the shore at an angle of 5° or less. Waves coming in over a submerged ridge off a headland will converge on

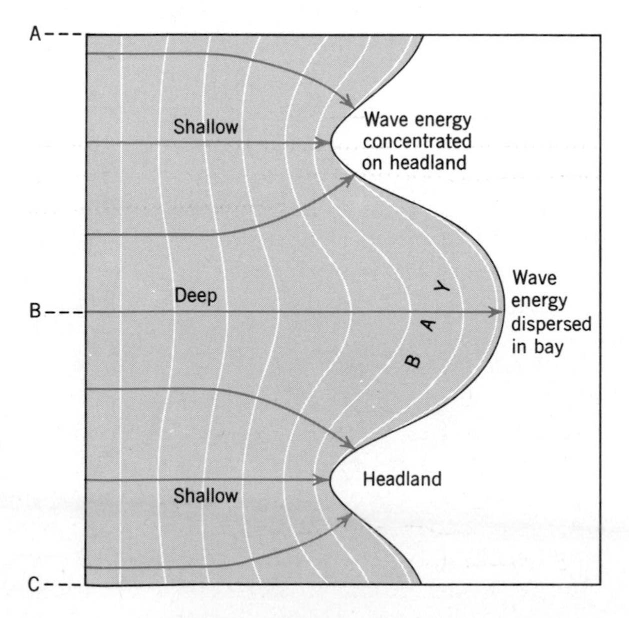


**Figure 11.9**
**Refraction of waves concentrates wave energy on headlands, disperses it at shores of bays. Sketch map showing 8 waves that become more distorted as they approach the shore over a bottom that is deepest opposite the bay.**

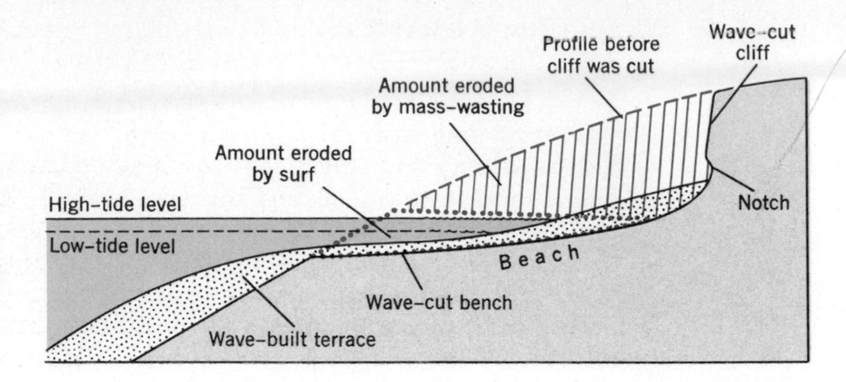


**Figure 11.10**
**Principal features of the shore profile.**
**The comparatively large proportion of erosion by mass-wasting results from notching of the cliff by surf, undermining rock material above the notch.**

the headland. Convergence, plus the increased wave height that accompanies it, concentrates wave energy on the headland. Conversely, refraction of waves approaching a bay will make them diverge, diffusing their energy at the shore. Because of refraction, headlands are eroded more rapidly than are bays, so that in the course of time irregular coasts become smoother and less indented (Fig.11.18).

In summary, ***wave refraction*** is *the process by which the direction of a series of waves, moving in shallow water at an angle to the shoreline, is changed.*

## ON- AND OFFSHORE MOVEMENT AND THE SHORE PROFILE

To understand the changes made by the sea along a coast, we need to look first at what happens at the surface along a line at right angles to the shore, the shore profile, described in this section. Then in the following section we shall look at the forces that act along, parallel to, the shore. We shall then have a three-dimensional picture of coastal activities.

**Figure 11.11**
**Minor erosional features along a rocky coast, seen at low tide.**
**Surf hollows out a sea cave in more erodible part of bedrock. Cave cut through headland becomes a sea arch. Surf tears away parts of bedrock, leaving isolated stack as an "island" on wave-cut bench.**

**Elements of the profile**

Seen in profile, the usual elements of a coast (Fig.11.10) are a wave-cut cliff and wave-cut bench, both the work of erosion, and a beach and wave-built terrace, both the result of deposition.

The ***wave-cut cliff*** is defined as *a coastal cliff cut by surf.* Acting like a horizontal saw, the surf cuts most actively at the base of the cliff. The upper part of the cliff is undermined and crumbles, furnishing rock particles to the surf. A cliff in which undercutting keeps well ahead of crumbling has a *notch* at its base. The notch is a concave part of the cliff profile, overhung by the part above. Other minor erosional features associated with cliffs are *sea caves, sea arches, and stacks* (Fig. 11.11). The cliff as a whole gradually retreats as the surf eats into the land.

The ***wave-cut bench*** is *a bench or platform cut across bedrock by surf.* It slopes gently seaward and grows wider in the landward direction as the cliff retreats. Some benches are bare or partly bare, but most are covered with sediment in gradual transit from shore to deeper water. This is the condition shown in Figure 11.10. At low tide the shoreward parts of some benches are exposed (Fig. 11.12). If the coast has been lifted up by movement of the crust, the bench can be wholly exposed (Fig. 11.13). In some areas the bench is concealed and can be inferred only from maps constructed from soundings.

The *beach* is thought of by most people as the sandy surface above water along a shore. Actually it is much more than this. We define

Figure 11.12
**Shoreward part of wave-cut bench exposed at low tide, and wave-cut cliff 15m high. Beach here is scanty, consisting mostly of boulders, because high-energy waves carry finer rock particles seaward. North of Bonne Bay, Newfoundland. (*R. F. Flint.*)**

**Figure 11.13**
**Emerged and abandoned wave-cut cliff, wave-cut bench, and beach. Portuguese Point, San Pedro, California. Uplift of the crust in this part of Pacific North America has raised these features by 45m. A new set of features are forming along today's shoreline. Note waves being refracted around the headland. (*Spence Air Photos.*)**

a ***beach*** as *the wave-washed sediment along a coast, extending throughout the surf zone.* In this zone, as we have seen, sediment is in very active movement. The sediment of a beach is derived in part from erosion of the adjacent cliff and from cliffs elsewhere along the shore. As we shall see presently, along coasts in general much more of it comes from alluvium contributed by rivers.

The ***wave-built terrace*** is *the body of wave-washed sediment that extends seaward from the breakers.* Its extent in the seaward direction is indefinite. "Terrace" is not a very appropriate name for the feature because it is not everywhere a real terrace or embankment. In places it is hardly more than a sort of carpet. Shoreward, it merges with the beach, as shown in Figure 11.10.

### The steady state along a coast

As we noted at the beginning of this chapter, the line along which water meets land is a scene of conflict that causes erosion and the creation, transport, and deposition of sediment. As a result, the form of the land slowly changes, and the water in motion moves and shapes the sediment derived from the land. The forces that fashion the shore profile—cliff, bench, beach, and terrace—tend to reach and maintain a condition of steady state, a compromise in the water/land conflict.

The compromise is reached in several different ways. On the beach, for instance, the *swash* of a wave running up the beach as a thin sheet of water moves sediment upslope, while gravity pulls it back again. More energy is needed to move pebbles than to move sand grains downslope; so the pebbles brought by the swash remain until the slope becomes steep enough for them to be moved back again. This is why pebble beaches are generally steeper than beaches built of sand.

Again, during storms the increased energy in the surf erodes the exposed part of the beach and makes it narrower. During calm weather, the exposed beach is likely to receive more sediment than it loses and consequently becomes wider. But at all times the beach profile represents an average steady-state condition in the group of forces that shape it.

In the surf zone, on the wave-cut bench, the steady state exists also. Waves tend to move sediment shoreward, in the direction of their travel. But opposed to shoreward movement is the gravitational tendency of sediment to move downslope—seaward. The two tendencies together establish a bottom slope just steep enough to permit slight net movement of sediment down it in the seaward direction. The profile is a product of the steady state among the forces that shape it.

Or consider a delta. The position of the outer limit of a delta, the extent to which it projects seaward from the land, is a compromise between the rate at which the river delivers sediment at its mouth and the ability of surf to erode the sediment and move it elsewhere along the coast. The great size of the Mississippi delta (Fig. 7.13) testifies to the huge volume of sediment carried by its parent river, and to the comparatively small size of the waves of the Gulf of Mexico. Imagine the mouth of a Mississippi-size river on the west coast of Scotland, facing the waves of the Atlantic. The delta would be much smaller.

Whether or not deltas are present, all coasts follow a pattern of evolution that is basically standard. It includes the long, slow retreat of cliffs, the gradual widening of benches, and the transport of sediment seaward, to be deposited on wave-built terraces. These are among the long-term processes by which continental shelves are made.

# MOVEMENT ALONGSHORE; DEPOSITIONAL FEATURES

Up to now we have been describing mainly the erosional effects of surf and the shaping of the shore profile. But the deposits made by surf and by the currents it sets up are equally important, particularly in that they result in large part from movements *along* the shore at right angles to the shore profile. The deposits occur as recognizable forms, which we will now examine.

### Longshore current and beach drift

The most common depositional shore features are beaches. Because it shows only two dimensions, Figure 11.10 may give the impression that the sediment in beaches is derived entirely from the cliff behind it. This is not true. As we indicated earlier, rivers contribute more sediment to most coasts than cliff erosion

**Figure 11.14**
**Movement of sediment along a coast as a result of oblique waves.**
**Longshore current moves fine sand in suspension; zigzag path of a typical coarse sand grain along beach is shown by dotted line.** (*L. Graff from A. Devaney.*)

contributes, but even were there no river-borne sediment, the beach at any point would include more material derived from elsewhere along the shore than particles from the cliff immediately behind it. The same is true of the finer sediment seaward of the beach.

The reason is that despite refraction, most waves reach the shore at an angle, however small. The oblique approach of waves sets up longshore movement of two distinct kinds, both of which move sediment *along* the coast. The first kind consists of a ***longshore current,*** *a current, within the surf zone, that flows parallel to the shore* (Fig. 11.14). Such currents easily move fine sand suspended in the turbulent surf. Meanwhile, on the exposed beach itself, the second kind of movement alongshore is occurring. The swash of each wave is oblique, but the backwash flows straight down the slope of the beach. The result, for sand and pebbles, is *beach drift,* a zigzag movement with net progress along the shore. Both beach drift and a longshore current are shown in Figure 11.14. The greater the angle of waves to shore, the greater the longshore movement. Pebbles tagged and timed have been observed to drift along a beach at a rate of more than 800m per day, transporting in that time a volume of sediment equal to more than 500m$^3$. When the amounts of sand moved by the longshore current are added to those moved by beach drift, the total must be very large, and it has important consequences for beaches used by people, as will be seen later in this chapter.

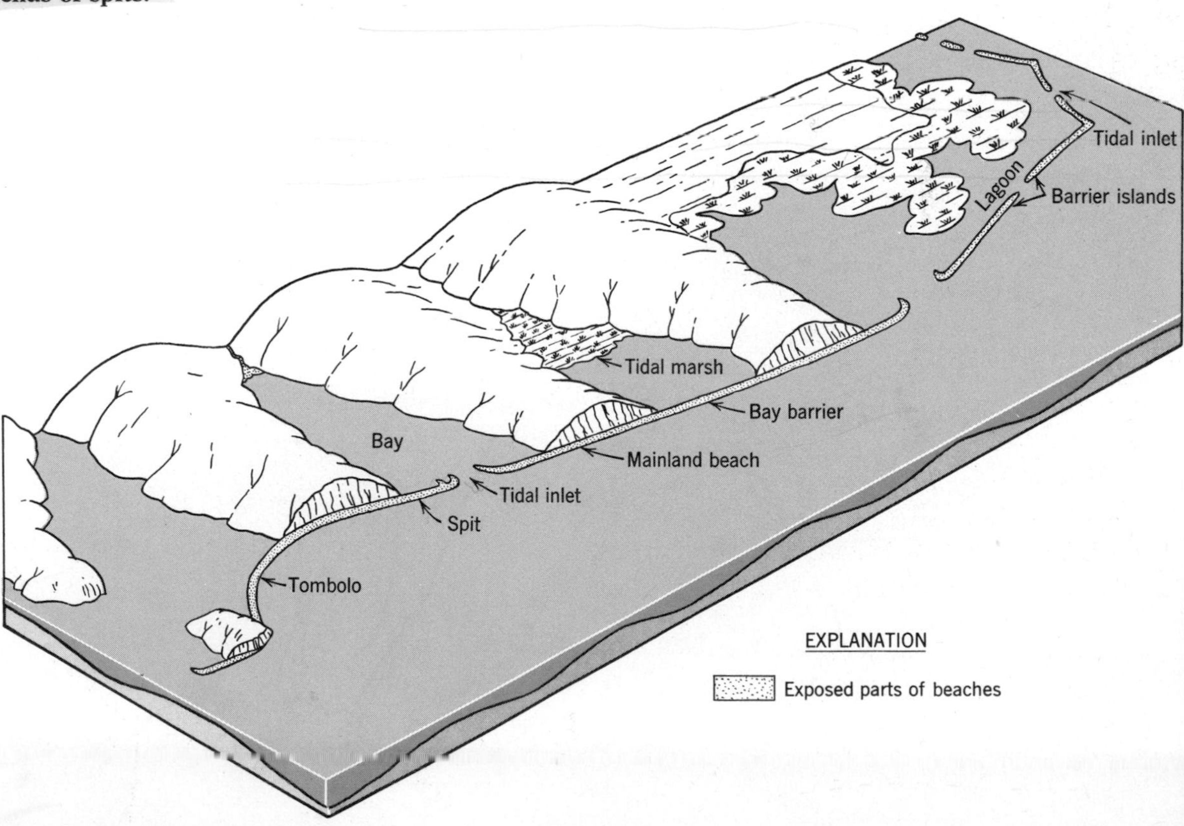


**Figure 11.15**
**This stretch of coast shows several kinds of depositional shore features. Local direction of beach drift is always toward free ends of spits.**

### Spits, bay barriers, and other forms

In addition to beaches, other related forms are conspicuous on many coasts (Fig. 11.15). Common among these is the ***spit,*** *an elongate ridge of sand or gravel that projects from land and ends in open water.* Well-known large examples are Sandy Hook on the southern side of the entrance to New York Harbor, and the northern tip of Cape Cod, Massachusetts. Most spits are merely continuations of beaches, built by beach-drifted sediment dumped at a place where the water deepens, as at the mouth of a bay. When the spit has been built up to sea level, waves act on it just as they would act on a beach. Much of a spit, therefore, is likely to be above sea level, although the tip of it cannot be. The free end curves landward in response to currents created by surf.

A ***bay barrier*** is *a ridge of sand or gravel that completely blocks the mouth of a bay.* It is believed to be formed by the lengthening of a spit, by beach drift, across a bay in which tidal- or river currents are too weak to scour away the spit as fast as it is built.

A ***lagoon*** is *a bay inshore from a line of barrier islands or from a coral reef.* A ***tombolo*** (tóm bōlō) is *a ridge of sand or gravel that connects an island to the mainland or to another island.* It forms in much the same way as a spit does.

### Barrier islands

A ***barrier island*** (Figures 11.15 and 11.16) is *a long island built of sand, lying offshore and parallel to the coast.* Such islands are found along most of the world's lowland coasts. Large-size examples are Coney Island and Jones Beach (New York City's shore-playground areas), the long chain of islands on one of which Atlantic City, New Jersey, stands, the long chain centering at Cape Hatteras, North Carolina, and Padre Island, Texas, 130km long. Although barrier islands originate in various ways, most of them were built during

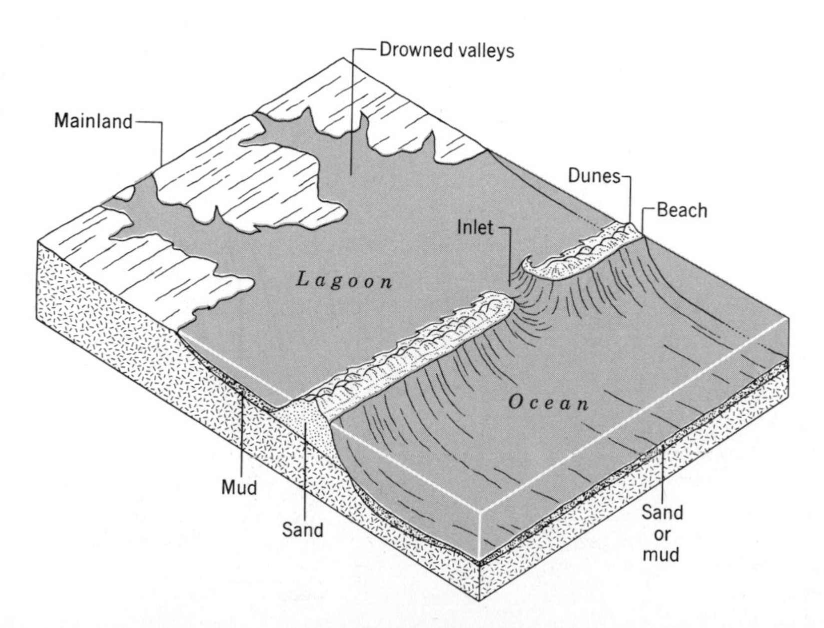


**Figure 11.16**
**Idealized sketch of a typical barrier island. Vertical scale greatly exaggerated.**

the worldwide rise of sea level that began some 15,000 years ago (Fig. 10.1), a fact established by the $^{14}C$ dates of shells in the basal parts of barriers. As sea level slowly rose over a plain of sand or mud that sloped very gently, waves necessarily broke at a considerable distance offshore. There they eroded the bottom and piled up sand to form a long bar. Built up above sea level, the bar became a beach. Thus the barrier island is the result of both on- and offshore movement and beach drift. Wind picks up beach sand and deposits it as a row of dunes (Table 9.1, top). During great storms, surf washes across a low place in the barrier and erodes it, cutting an inlet that may stay open permanently. Finer sediment is washed and blown into the lagoon between barrier and mainland.

### Sediments and their stratification

Now let us examine more closely the sediments themselves. Beaches, barriers, and related features consist of the coarser sizes of whatever range of rock particles is contributed by erosion of cliffs or by rivers. Quartz is the most durable of common minerals in continental rock, and generally occurs as crystals having the diameters of sand grains. It is not surprising, therefore, that the majority of beaches consist chiefly of quartz sand. Bedrock that breaks down into larger pieces, along joints and other surfaces, makes beaches of gravel.

Dragged back and forth by the surf and turned over and over, particles of beach sediment become rounded by abrasion, much as do comparable particles in streams. In fact, we know of no easily recognized difference between the shapes of stream-worn and surf-worn particles.

There are, however, some differences in stratification. Spits, bay barriers, and the exposed parts of beaches, examined where natural erosion or man-made cuts expose them in section, generally are cross-stratified (Fig. 11.17). Their *seaward* parts consist of thin layers, gently inclined at many different angles. By watching calm-weather waves on a beach, we can see that the layers are deposited because part of the uprushing water sinks into the beach, leaving the backwash unable to remove all the sediment brought by the uprush. The varying angles of the layers mostly represent the varying slopes related to sediments of varying diameter. In contrast, the *landward* parts usually consist of foreset layers deposited by high waves that wash entirely over the beach, spit, or barrier and deposit on the far side much of the load carried in the uprush. The result is not unlike the stratification of a sand dune. However, the foreset layers are less variable in direction

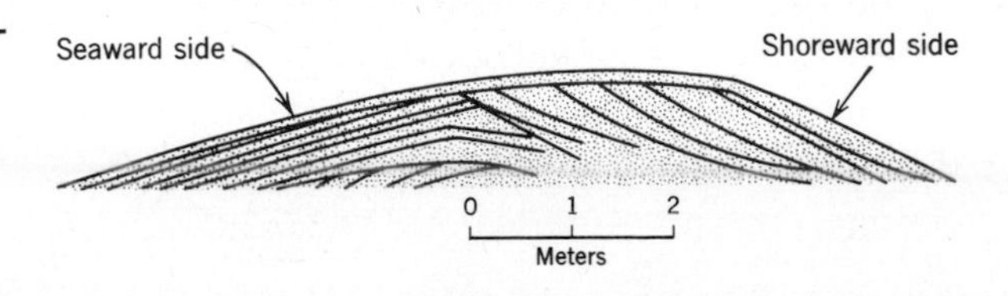


**Figure 11.17**
**Idealized cross section through a small beach, spit, or bay barrier, showing stratification of sand.** ***(Drawn from photographs by W. O. Thompson.)***

than those in a dune because, owing to refraction, the angle the waves make with the beach varies less than the angle between wind direction and the crest of a dune.

## SCULPTURAL EVOLUTION OF COASTS: CYCLE OF COASTAL EROSION

The world's coasts do not fall into easily identifiable classes. Their variety is great because their configurations depend on combinations of at least 4 characteristics:

1 Form of the line along which the sea meets the land—a line that, as we have seen, is accidental, changing from time to time as sea level changes.

2 Steepness of the seaward slope (which determines depths of water offshore).

3. Resistance of the exposed rock materials to erosion by surf and other agents such as mass-wasting.

4 Extent to which the coast has been changed by erosion and deposition along the shore.

Some coasts, like most of the Pacific Coast of North America, consist of mountains or hills with deep valleys between them. Others, like the Atlantic and Gulf Coasts from New York City to Florida and onward into northern Mexico, cut across a broad coastal plain that slopes gently seaward, and are festooned with barrier islands. These coasts represent two extremes, between which are many intermediate kinds of coast. Whatever the nature of their terrain, nearly all coasts have been submerged by the worldwide rise of sea level that has occurred during the last 15,000 years, as glaciers melted (Fig. 10.1). Because of this submergence, most coasts, whether they consist of mountains, hills, or plains, are characterized by drowned valleys, some of them deep.

Comparison of the amounts of erosion that have affected many different segments of coasts shows that a coast, under attack by surf and agents of erosion on land, undergoes a series of gradual and broadly predictable evolutionary changes, eventually reaching a form that is comparatively smooth and straight. Such evolution is analogous to the sculptural evolution of a land mass as it is eroded by streams and mass-wasting. In other words, just as there are cycles of erosion characteristic of the lands under a moist climate (Chapter 7) and under an arid climate (Fig. 9.11), so also there is a cycle of erosion characteristic of the effect of waves and currents along coasts. As with the stream cycle, stages described as youth, maturity, and old age represent continuous transformation of the coast, although they are not sharply marked off from each-other. They serve merely to emphasize the fact of unbroken evolution. A typical sequence is shown in Figure 11.18.

If, during the cycle, the level of the sea relative to the land should change, the shoreline would move seaward or landward and a

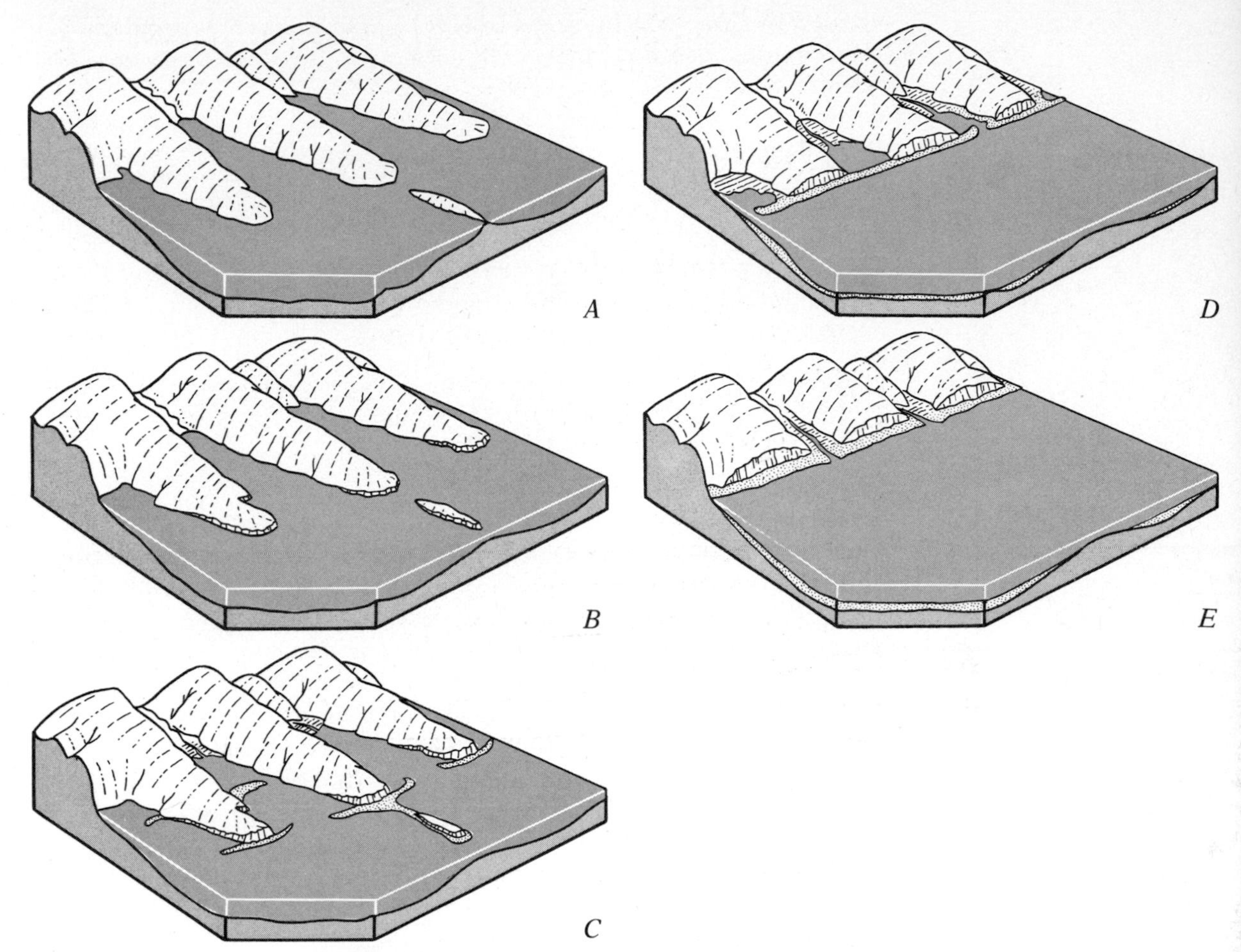


Figure 11.18
**Cycle of coastal erosion: sequence of sculptured forms developed through time.**

*A.* **Coast at start, with headlands, islands, and deep bays.**

*B.* **Headlands cliffed by surf.**

*C.* **Beaches, spits, and a tombolo are built as cliffing continues.**

*D.* **Cliffing reduces headlands to short stumps; spits join to form bay barriers.**

*E.* **Headlands and bays are eliminated; shoreline has become nearly straight.**

whole series of adjustments, involving net erosion in some places and net deposition in others, would occur, bringing the system into a steady-state condition once again.

### Rates of erosion

Rates of erosion are far greater in weak, erodible rock materials than in strong, resistant rock. During the few thousand years since the rising sea level approximated its present position, rates of erosion in three coastal areas in New England have approximated these values, based on measurements made between pairs of years:

| Locality | Rock Type | Average Rate of Erosion cm/year |
|---|---|---|
| Boothbay, Maine | Hard metamorphic rock | Too small for measurement |
| Boston, Mass. | Till | 20 to 30 |
| Edgartown, Mass. | Sand and gravel | 167 |

# PROTECTION OF SHORELINES AGAINST EROSION

The steady state among the forces that operate on coasts is interrupted temporarily by rare, exceptional storms, which erode cliffs and beaches spectacularly. During a single storm in 1944, cliffs on Cape Cod consisting of compact, nonlithified sediment retreated 15m, more than 50 times the normal rate of retreat per year. The bursts of rapid erosion caused by such storms are insignificant in the natural evolution of a coast. But now, more than 75 percent of the population of the United States lives in coastal belts that include only 5 percent of the land area of the nation. The spread of such large numbers of people into coastal areas has brought with it the narrower point of view that such erosion causes intolerable damage to property.

To protect a strip of shore that consists of comparatively erodible material like that on Cape Cod, two things can be done. A cliff can be clad with an armor consisting of tightly packed boulders so large that they can withstand the onslaught of storm waves. Or it can be defended by a strong *seawall* built parallel to the shore on foundations deep enough to prevent undermining by surf during storms. Both structures protect cliffs against ordinary storms, at least, but of them both are expensive.

Because of their great recreational value, beaches in densely populated regions justify greater expense for maintenance than most headlands do. But a beach presents a special sort of problem. Because of beach drift (Fig.11.14), what happens on one part of a beach affects all other parts that lie in the downdrift direction. Thus, for example, a seawall, dock, or other structure built at the updrift end of a beach reduces the amount of sand available for beach drift. The surf becomes underloaded and makes good the loss by eroding sand from along the beach until it becomes loaded again. Small beaches have been completely destroyed by this process in the course of only a few years.

Such erosion can be checked, at least to some extent, by building groins at short intervals along the beach. A ***groin*** *is a low wall, built on a beach, that crosses the shoreline at a right angle* (Fig. 11.19). Many groins contain openings that permit some water to pass

**Figure 11.19 Groins on Westhampton Beach, Long Island, New York. Piling up of sand on far side of each groin indicates drift of sand is toward the camera. Wave refraction by the projecting groins is visible.** (*U.S. Army Corps of Engineers.*)

through them. Groins act as a check on the rate of beach drift and so cause sand to accumulate against their updrift sides. Some erosion, however, occurs beyond their downdrift sides.

Another way of protecting a beach that is being eroded is to bring in sand artificially and pile it on the beach at the updrift end. Surf then erodes the pile and drifts the new sand down the length of the beach. In this method the sand that artificially nourishes a beach must be replenished. Both the feeding of a beach and the construction of groins are expensive, and both require maintenance.

In summary, the need for both remedies is caused by human interference with a natural process whose factors have long been in a steady-state condition.

Beaches in southern California are deteriorating for another reason, likewise the result of human interference. As we said earlier, most of the sand on those beaches is supplied, not by erosion of wave-cut cliffs but by alluvium poured into the sea by streams at times of flood. Because the floods themselves cause damage to man-made structures along the stream courses, dams have been built across the streamways in order to control the floods. But the dams also trap the sand and gravel carried by the streams, thus preventing the sediment from reaching the sea. This in turn has affected the steady state among the factors involved in longshore currents and beach drift. Sand is now in short supply, and the shortage is being made good by erosion of beaches. One scientist has estimated that before 1985 most of the beaches in southern California will have been washed away, leaving only the bare rock of the wave-cut benches that underlie them.

A similar situation has developed along the Black Sea coast of the

USSR. Of the sand and pebbles that form the natural beaches there, 90 percent was supplied by rivers as they entered the sea. During the 1940s and 1950s three things happened: Large resort developments including high-rise hotels were built at the beaches. By construction of breakwaters, two major harbors were extended into the sea. Dams were built across some rivers inland from the coast. All this construction interfered with the steady state that existed among supply of sediment to the coast, processes of beach drift and longshore currents, and deposition of sediment on beaches. By 1960, it was estimated, the combined area of all beaches along the coast had decreased by 50 percent. Then beach-front buildings began to sag or collapse as the surf ate away the beaches. An ironic twist to the chain of events lies in the fact that contractors removed large volumes of sand and gravel from beaches for use as concrete aggregate, to build not only buildings but also the dams that cut off the supply of sediment to the coast.

### Protection of wetlands

In the foregoing paragraphs we have described some of the results of human interference with sandy beaches and rocky cliffs. Apart from beaches and cliffs and as greatly in need of protection are low, plant-covered coastal areas, including marshes. These "wetlands" are critical environments for many kinds of marine organisms, some of which are the starting point of a food chain very important for human nutrition. This fact greatly concerns ecologists and is beginning to concern governments, especially along the east coast of the United States, where low-lands prevail. With the food chain in mind, several states are actively attempting to control the siting of industrial developments along their shores, in places where such development would harm critical environments.

## SUMMARY

1. Most continental shelves consist of thick aprons of sediment washed out from the continents and deposited mostly in shallow seawater.

2. The shelves, with their accompanying continental slopes, add about 15 percent to the combined areas of the continents.

3. The level of the sea relative to the land keeps changing, both by rise and fall of the sea surface and by uplift and subsidence of Earth's crust.

4. In deep water waves have little or no effect on the bottom. At a depth of $\frac{L}{2}$ (usually less than 300m) waves can begin to stir bottom.

5. Most of the geologic work of waves is performed by surf at depths of 9m or less.

6. Wave refraction tends to concentrate wave erosion on headlands and to diminish it along the shores of bays.

7. The shore profile consists of four related elements: wave-cut cliff, wave-cut bench, beach, and wave-built terrace.

8. There is a net gravitational tendency to move sediment seaward, downslope. Sediment becomes finer in the seaward direction.

9. Longshore currents and beach drift transport great quantities of sand along coasts.

10. Depositional shore features include beaches, spits, bay barriers, barrier islands, and tombolos.

11. Most barrier islands are built offshore, where a rising sea advances over a very gently sloping plain.

12. A coast attacked by waves and currents passes through an orderly sequence of forms, a cycle of erosion, ending with a nearly straight form that permits almost uniform distribution of wave energy.

13. Because of movement of the crust and change of sea level, lands of continental size apparently have never been completely reduced to wave-cut benches.

14. A shore cliff can be protected for a time, at least, by a seawall or an armor of boulders. A beach can be protected, at least temporarily, by a series of groins or by importation of sand.

# SELECTED REFERENCES

Bascom, Willard, 1964, Waves and beaches. The dynamics of the ocean surface: Garden City, N.Y., Anchor Books, Doubleday. (Paperback)

Dolan, R. B., and others, 1973, Man's impact on the barrier islands of North Carolina: American Scientist, v. 61, p. 152–162.

Inman, D. L., 1954, Beach and nearshore processes along the southern California coast: Calif. Div. Mines Bull. 170, Chap. 5, p. 29–34.

King, C. A. M., 1959, Beaches and coasts: London, Edward Arnold.

Shepard, F. P., 1973, Submarine geology, 3d ed.: New York, Harper & Row.

———, 1967, The Earth beneath the sea: Baltimore, Johns Hopkins Press.

———, and Wanless, H. R., 1971, Our changing coastlines: New York, McGraw-Hill.

Steers, J. A., 1954, The sea coast: London, Collins.

Zenkovich, V. P., 1967, Processes of coastal development: London, Oliver & Boyd. (Transl. from the Russian by D. G. Fry.)

# 12 SEDIMENTARY STRATA



We have postponed discussing sedimentary strata until after all the processes of erosion and deposition that lead to the making of strata have been set forth. Now we can gather together the many threads we followed in our study of erosion and deposition, with two purposes. The first is to visualize the making of new strata out of the sedimentary waste eroded from continents, in order to see where most of the waste is deposited and why very little reaches the deep sea. The second purpose is to analyze ancient strata, to interpret the environments in which they were formed. It is through such analysis that we are able to define confidently the regions in which strata are deposited. The present chapter, then, consists of two distinct parts: (1) description of regions of deposition and (2) analysis of sedimentary rocks.

## REGIONS OF DEPOSITION

### SEDIMENT IN LAND BASINS AND ON CONTINENTAL SHELVES

Now that we have examined them, we understand how it is that continental shelves are the chief rubbish dumps of the continents, the places where rock waste accumulates as it is continuously swept seaward from continental interiors, mostly by rivers but also by other transporting agents. Although the shelves are indeed built of rubbish, that material is not at all the heterogeneous accumulation of many substances of all sizes and shapes, hastily thrown together, that characterizes rubbish of human origin. Most of it has been partly recycled already, during the very processes of transport and deposition. It is sorted neatly and stratified in an orderly way. It is lying there, in or on the shelves, slowly being compacted and cemented as ground water percolates through it, to become firm sedimentary rock.

Going inland from a shelf into the interior of a continent, we can follow every step in the making of rock waste, its transport, and its deposition and redeposition, as cycles of erosion of land areas run their course. We can see weathered bedrock at the surface, colluvium on its way downslope, gravel and sand accumulated along stream channels, waiting for the next flood to wash them farther downvalley. We can see also broad basins into which fans are being built, at the bases of confining mountains. The basins are sagging, creating and maintaining pockets in which successive layers of alluvium are forming a thick pile. Near the continental margins broad plains of alluvium are being built, the sediment much finer grained in this region of gentle slopes. Beyond, we see rivers discharging into a shallow sea that covers the shelf, the suspended loads forming brownish plumes that spread far out into the sea.

The drill hole through the shelf in North Carolina (Fig. 11.3) shows clearly how such plumes of suspended fine sediment, settling to the bottom, have built a pile of strata as much as 3km thick, through a time 70 to 100 million years long. To build the whole pile, only a small fraction of one millimeter of sediment need have been deposited each year. This includes both sediment discharged from rivers and sediment washed from the coast by surf. A little is blown into the sea by winds.

Despite its fineness, most of this sediment settles onto the shelf and continental slope. If it comes to rest at depths of less than $\frac{L}{2}$ in terms of incoming waves, it will be stirred up and deposited in deeper water, until it lies too deep for even the greatest storm waves. Thereafter it can be moved only by currents, as we shall see in a moment, and by currents it can reach down the continental slopes to the edge of the deep-sea floor. But no more than about 10 percent of land-derived sediment—off some coasts even less —remains in suspension long enough to pass beyond the slope and settle on the deep-sea floor, where water depths average some 4km.

Because the settling-out of sediment upon them is *comparatively* rapid, the shelves are conservers of continental crust. They trap most of the rock waste that reaches them from the adjacent land, keeping it in the family—that is, in the continents to which the shelves belong. So most of the continental crust, lighter in weight than the oceanic crust, is recycled again and again, always within the continental realm.

Of course some continental sediment never travels as far outward as the shelf. It is trapped inland in sagging basins, where it is buried and preserved. But analysis of ancient strata shows that the amount of sediment thus retained in the continent itself is much less than the amount carried to the shelf. And since we have noted already that only around 10 percent of what reaches the shelf and slope ever gets to the deep sea, it is clear that the great bulk of Earth's sedimentary strata are shelf strata whose sediment originated within the continents themselves.

### Effect of fluctuating sea level on shelf sediments

Because most of the sediment on the shelves consists of material derived from the land, we would expect it to become gradually finer in the seaward direction. Much sampling of shelves confirms our expectation. Nevertheless, on some shelves the distribution of particle sizes is patchy, with coarse sediment, which normally belongs near shore, occurring far offshore near the seaward limits of the shelves. The patchy distribution is believed to be partly the work of localized currents that deposit coarse sediment at some distance from shore. But to a greater extent it is the result of change of sea level. When we remember that during each glacial age sea level was lowered, we can see that the shoreline must have migrated seaward across the shelves, exposing new land. Patches of coarse sediment, deposited near shore or even on the land itself, could then have been submerged as the sea level rose again.

## CONTINENTAL SLOPES

### Turbidity currents

The continental slopes (Figures 11.1, 11.3, 13.6) are the sites of the unique submarine canyons and of an unusual process as well. On the deep-sea floor, at the foot of a continental slope, and in places far beyond it, thick bodies of coarse sediment of continental origin lie in places at depths as great as 4 to 5km. Such occurrences were an unsolved mystery until it appeared that their explanation lies in ***turbidity currents,*** *density currents whose excess density results from sediment suspended in them.* The following is a small-scale example of a turbidity current.

In 1935, soon after completion of the Hoover Dam (Fig. 7.14), engineers were surprised to find that from time to time the clear water discharging through the lowest outlet pipes became muddy. Also during the 1930s, researchers at the California Institute of Technology were pouring streams of water, densified by dissolved salts or suspended silt and clay, into a water-filled tank in their laboratory. They found that the dense water, with kinetic energy acquired by flowing down the front of a delta, possessed enough momentum to travel along the bottom on an extremely small slope, past the clear water above (Fig. 12.1). Energy not expended in friction en route even enabled the current to climb part way up the face of a dam at the far end of the tank. The current had distinct boundaries, like a cloud of dust; in fact, someone described it as "a dust storm under water." Indeed, it is essentially the same thing, mechanically, as a dust storm.

It was realized that such a current, climbing part way up Hoover Dam, could explain the muddy water appearing in the discharge pipes. Muddy water entering Lake Mead, far upstream from Hoover Dam, formed turbidity currents that swept along the floor of the lake until they reached the dam, rose up the side, and flowed out through the discharge pipes. Observations in this and other reservoirs confirmed that the suspended loads of rivers were the source of the currents of turbid, dense water, which were then named *turbidity currents.*

As a result of more laboratory experiments, it was learned that turbidity currents could perform a surprising amount of erosion. The densities of natural turbidity currents on gently inclined lake floors are rarely as great as 1.02, and their velocities are less than 30cm/sec. But greater densities and velocities have been produced in the laboratory, and a submarine turbidity current, flowing down a continental slope with an inclination of 40m/km, should develop a velocity greater than that of the swiftest streams on the land. In theory, a current with a density of 1.5, moving at about 1.6km/hr, could move a rock particle 14,000 times the diameter of the largest particle that can be moved by clear water at the same velocity.

**Breaks in Transatlantic Cables.** Turbidity currents, then, have been observed in lakes, can be created in the laboratory, and, according to theory, have enormous potential for the transport of sediment. But do they occur in the sea? In 1952, re-examination of the record of an old earthquake suggested strongly that they do.

On November 18, 1929, a severe earthquake on the continental slope off Nova Scotia broke 13 transatlantic cables in 28 places. At the time it was supposed that the breaks were the direct result of buckling or other movements of the sea floor, the usual cause of submarine earthquakes. Turbidity currents were then unknown,

**Figure 12.1 Turbidity current flowing from left to right, seen through the glass wall of a water-filled laboratory tank. (*U.S. Conservation Service, California Institute of Technology.*)**

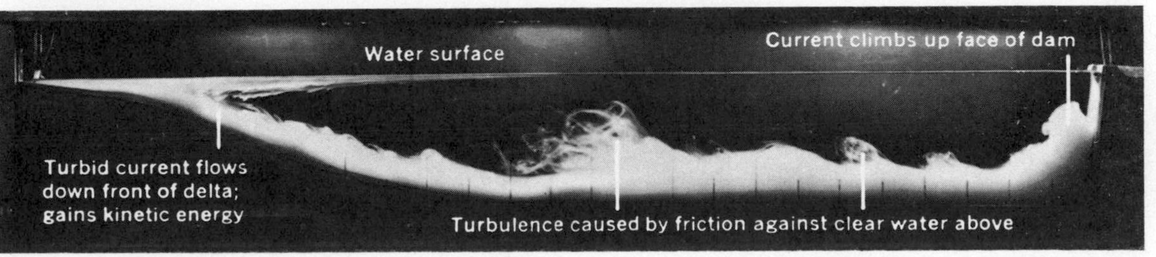

and the event was lost sight of. But many years later the whole story, supported by a thick file of measurements made by the repair ship, was reviewed and studied. There were two odd things about the cable breaks. First, although all the cables on the continental slope and deep-sea floor were broken, not one of the many cables that crossed the continental shelf was damaged. Second, the breaks occurred in sequence, in order of increasing depth, over a period of 13 hours and through a distance of 480km from the earthquake center. Repair ships found that each cable had been broken at two or three points more than 160km apart. The detached segments of cable between the breaks had been carried part way down the continental slope or buried beneath sediment beyond the bottom of the slope.

The whole event was like a huge laboratory experiment with times and distances controlled by measurement. Each break was timed by the machines that automatically record the messages transmitted through the cables, and was accurately located after the quake by the electric-resistance measurements always made to enable repair ships to locate breaks and repair damage.

The only hypothesis yet formulated to explain all these facts is that the quake set off great submarine slides on the continental slope. The slides quickly became turbidity currents, which flowed down the slope, breaking each cable as they came to it. Eight cables on the slope were broken instantaneously at the moment the quake occurred, and five others were broken at times ranging from 59 min. to 13 hr. 17 min. after the quake (Fig. 12.2). The area affected was about 320km wide and at least 650km long.

From the times of the breaks and the distances between them the velocities of the inferred turbidity currents could be calculated. Velocities ranged from 93km/hr at the toe of the continental slope down to 22km/hr at the latest and farthest break. Even at the latter point, the velocity of the current was four times that of the lower Mississippi River.

Theoretical calculations suggest that near the earthquake area, velocities may have approached 135km/hr and that the heights of the turbidity currents may have reached 300m. As the currents lost

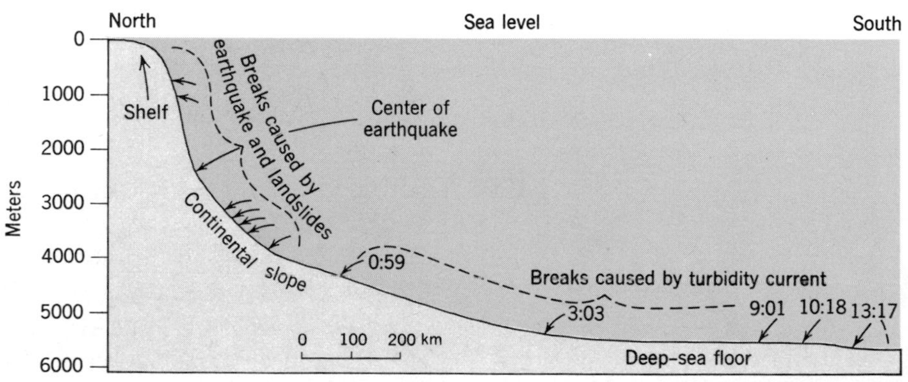


**Figure 12.2**
**Profile of sea floor off Nova Scotia, showing events of the earthquake of November 18, 1929. Short arrows point to locations of breaks in transatlantic cables. Numbers show times of breaks in hours and minutes after earthquake. Vertical scale is exaggerated enormously. (*After B. C. Heezen and M. Ewing.*)**

energy, the sediment eroded from the continental slope was dropped onto the ocean floor, burying some of the long, broken pieces of cable beneath deposits of "sand and small pebbles," the bottom sediment reported by the repair ships. Two core samples collected from the floor, downslope from the cable breaks, show that the floor is blanketed with a layer of silt and muddy sand in the form of a graded layer (Fig. 12.9). In one core the layer is 70cm thick; in the other it is 128cm.

As pointed out later in the present chapter, a graded layer is the result of rapid, continuous loss of energy in the transporting agent. This would occur in a spent turbidity current, but it is not expected in the other sorts of marine currents described in Chapter 13. A graded layer of this kind is termed a ***turbidite***, defined as *sediment deposited by a turbidity current.*

So, although a turbidity current on the continental slope off Nova Scotia on November 18, 1929, is not proved, it is the only hypothesis that seems able to explain all the facts. The record shows that within the past 75 years, similar events have occurred at least twice at each of 40 localities around the world. Some were related to earthquakes; others occurred off the mouths of large rivers, suggesting that the inferred turbidity currents were set off by stream floods or by slump in stream deposits. This evidence is impressive. It leads us to believe that on continental slopes turbidity currents are effective agents, despite the fact that no one has yet seen such a current in actual operation in the sea.

### Deep-sea fans

Several of the submarine canyons that are cut into continental slopes are aligned with the mouths of big rivers such as the Hudson, Mississippi, Amazon, Congo, Ganges, and Indus. The mouths of most such canyons lead into huge fan-shaped features that slope downward and spread outward to the deep-sea floor, merging into the abyssal plains described in Chapter 13. These features are *deep-sea fans.* The surfaces of some are marked by distributary channels, some as much as 200m deep, not unlike those of a fan on the land (Fig. 7.21).

The Amazon deep-sea fan, (Fig. 12.3), one of the largest such features, is about 350,000km$^2$ in area, extends down to a depth of 4,700m, is 1km to 5km thick, and is thought to have been built during at least the last eight million years, mainly during glacial ages, when sea level stood so low that the shore lay along the outer edge of the continental shelf of Brazil, 200km to 250km seaward of the position it occupies today.

The sediments of the fan, sampled by coring, prove to be derived mostly from the land via the Amazon River. They include many graded layers (Fig. 12.8) that consist of mixtures ranging from clay particles up to small pebbles. They are classed as turbidites.

Deep-sea fans, therefore, seem to be the single great exception to the generalization that final deposition of land-derived sediment in

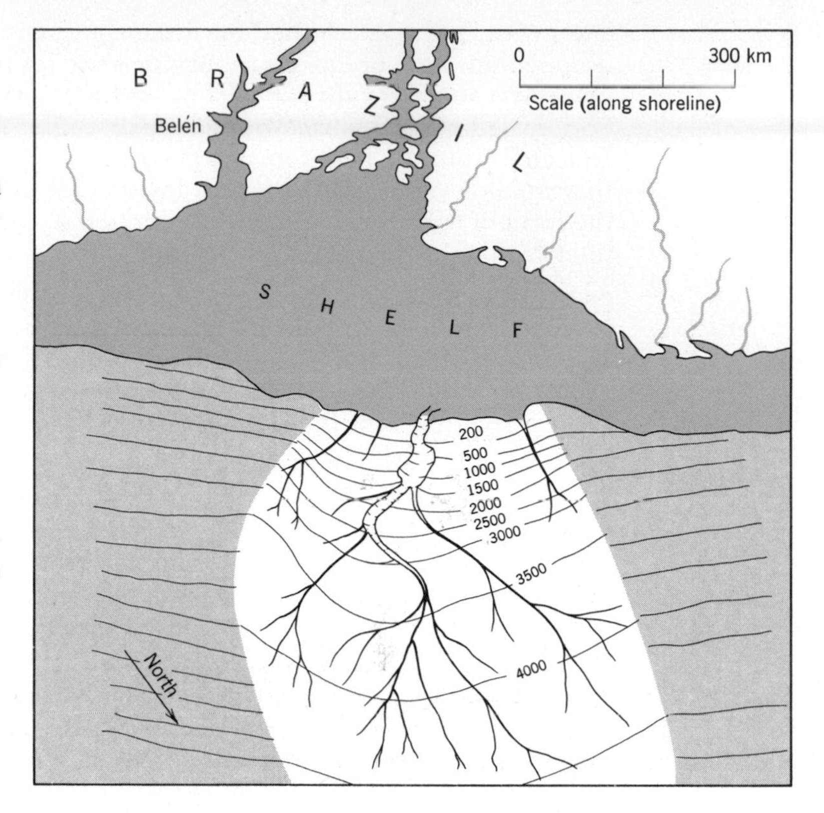


**Figure 12.3**
**A great fan-shaped body of sediment, the Amazon deep-sea fan. Sediment is transported by the Amazon River across the continental shelf, emerges when the glacial-age sea level is low, and is deposited on the continental slope. It is transferred to greater depths by slump and turbidity currents, creating the fan. Only the landward part of the fan is shown. (*Generalized after Damuth and Kumar, 1975.*)**

the ocean is largely confined to the continental shelves. When shelves are emerged at times of lowered sea level and rivers extend across them nearly to the continental slope, the stage is set for the rapid building of deep-sea fans. The sediment deposited on the continental slope becomes unstable in places, and slumps, creating turbidity currents. Very likely, scour by such currents as they race down the slope is the principal means by which submarine canyons are excavated.

# SEDIMENTS ON THE DEEP-SEA FLOOR

### Sampling techniques

The deep-sea floor is mantled with a carpet of loose, fine-grained sediment. Although this carpet covers more than half of Earth's surface, its systematic study dates only from the 1870s, when the British research ship *Challenger,* on a history-making cruise, collected a vast number of samples at carefully chosen positions. So long and thorough was the resulting study that the published reports were not completed until eighteen years after the cruise ended. Most of the samples were obtained by scoops dragged along the bottom, or by small clamshell buckets that take a bite of sediment and then snap shut.

Since the classic cruise of the *Challenger,* other expeditions of

**Table 12.1**
**Origins of Sediment on the Deep-Sea Floor**

1. Terrigenous sediment (Lat. "derived from the lands"). Sea-floor sediment derived from sources on land. Contributed by (a) rivers, (b) erosion of coasts by waves, (c) wind (clay and silt, including volcanic ash), (d) floating ice.
2. Pelagic sediment (Gr. "belonging to the deep sea"). Sediment, on the deep-sea floor, consisting of material of marine organic origin. Shells and skeletons, mostly microscopic, of marine animals and plants.
3. Sediment derived from submarine volcanoes. Volcanic ash.
4. Extraterrestrial sediment (derived from outside the Earth). Meteorite particles, mostly microscopic.

various nationalities have added greatly to the take of samples. Then, in 1947, a tremendous improvement in sampling was made possible through the invention of *coring devices.* One of these is a long metal tube, let down on a cable and then forced into the sea floor by one of several mechanisms. The open lower end of the tube then closes and brings up a core sample in which all except the topmost layers of sediment are undisturbed. Cores as much as 25m long have been obtained. Another coring device is a drill capable of cutting into the sea floor at a water depth of more than 3000m. Within recent years ships such as *Glomar Challenger* (a second and more modern *Challenger*), using this drilling technique, have been able to extract many cores as much as 500m long and some even longer.

**Kinds and distribution of sediments**

Minute analyses of samples brought up by coring devices have made it possible to sort out the various sources from which sea-floor sediment is derived. General sources are listed in Table 12.1: they are the land, the sea, and sources outside the Earth. The table shows sources only, not the sediments themselves. This is because study of great numbers of samples indicates clearly that *all* the sediments are mixtures; no one body of sediment comes entirely from a single source. Therefore we have to classify the sediments according to their chief constituents, the predominant kinds of material they contain. Seven principal kinds are described in Table 12.2, but all are mixtures and grade from one into another. Figure 12.4 is a map on which the distribution of the seven different kinds of sediment is shown. The scale of the map is so small that the sediment areas are greatly generalized. If we compare the areas of various sediments in the western North Atlantic with the detailed topography in Figure 13.6, we can see that a detailed map of sediment distribution would be much more complex. Even so, it could not be as accurate as a map of a

**Table 12.2**
**Classification of Kinds of Sediment on the Deep-Sea Floor**

1. Terrigenous sediment. Mainly on abyssal plains (Chapter 13). Mud, sand, and gravel, varying greatly from place to place.
2. Glacial-marine sediment. Terrigenous sediment, including nonsorted mixtures of particles of all sizes, dropped onto the sea floor from floating ice.
3. Sediment displaced by gravity. Mainly terrigenous sediment, originally deposited on the continental shelf and slope, that has moved to the deep ocean floor under the influence of gravity, by gliding, slump, or flowing.
4. Brown clay (also called pelagic clay) (Fig. 12.5*A*). Confined to the deep-sea floor, mostly in high latitudes or at depths greater than 4000m. Contains, by definition, less than 30 percent calcium carbonate. Chief constituents are clay minerals, quartz, and micas. Since these are the sorts occurring in weathered soils, volcanic ash, and fine windblown material, they are thought to come from such sources. The clay is brown as a result of gradual oxidation during the very slow process of deposition.
5. Calcareous ooze (Fig. 12.5*B*). Contains, by definition, more than 30 percent calcium carbonate, most of it consisting of shells and skeletons. Confined to regions in which surface water is warm and surface organisms exist in myriads; the resulting shells, falling like snow, accumulate on the bottom more rapidly than do the inorganic-clay particles. Because it contains much carbon dioxide, deep-sea water dissolves calcium carbonate. As the shells drift slowly down, they are gradually dissolved, but only at depths of more than 5000m are they completely consumed. Hence calcareous ooze rarely occurs at such depths.
6. Siliceous ooze. Contains a large percentage of skeletons built of opaline silica. Occurs where organisms with calcareous shells are few in the surface water and in areas in which such shells are destroyed by dissolution before they reach the sea floor.
7. Authigenic materials. *Authigenic* means sedimentary deposits *formed in place.* These have not been transported physically. They consist of minerals that crystallized from the seawater itself. The principal authigenic deposits are nodular growths of manganese minerals growing on the sea floor (Fig. 22.9).

comparable land area, because the samples on which it must be based are taken from points very far apart.

In some places the sampling apparatus put down by exploring ships hits not sediment but bare, hard rock. Comparison with topographic records shows that some of the rocky places are cliffs and other steep slopes, from which any accumulating sediment would be expected to slide off. But still others are flat surfaces, and we are not yet sure why they have no cover of sediment. Some such surfaces might be recent lava flows. Others, perhaps, have been scoured bare by currents crossing the floor of the deep ocean.

**Figure 12.4**
**(see opposite page)**
**Map showing distribution of sediments on continental shelves and deep-sea floor.**
***(After F. P. Shepard, Submarine Geology, 2d ed.: 1963. Fig. 198. With permission of Harper & Row, Publishers, New York.)***

Whatever the reasons for the bare places, our knowledge of the sediments themselves is confined mainly to the uppermost 25m beneath the ocean floor, because 25m is the maximum length of most of the cores thus far recovered. Great interest in these cores lies in what the layers reveal about recent glacial ages and therefore about Earth's changes of climate. This research is discussed in Chapter 20.

As for the sedimentary layers that lie beneath the upper 25m of sediment commonly sampled by core tubes, we have two limited sources of information. Waves generated by earthquakes traverse the crust beneath the sea floor, and by recording devices on boats

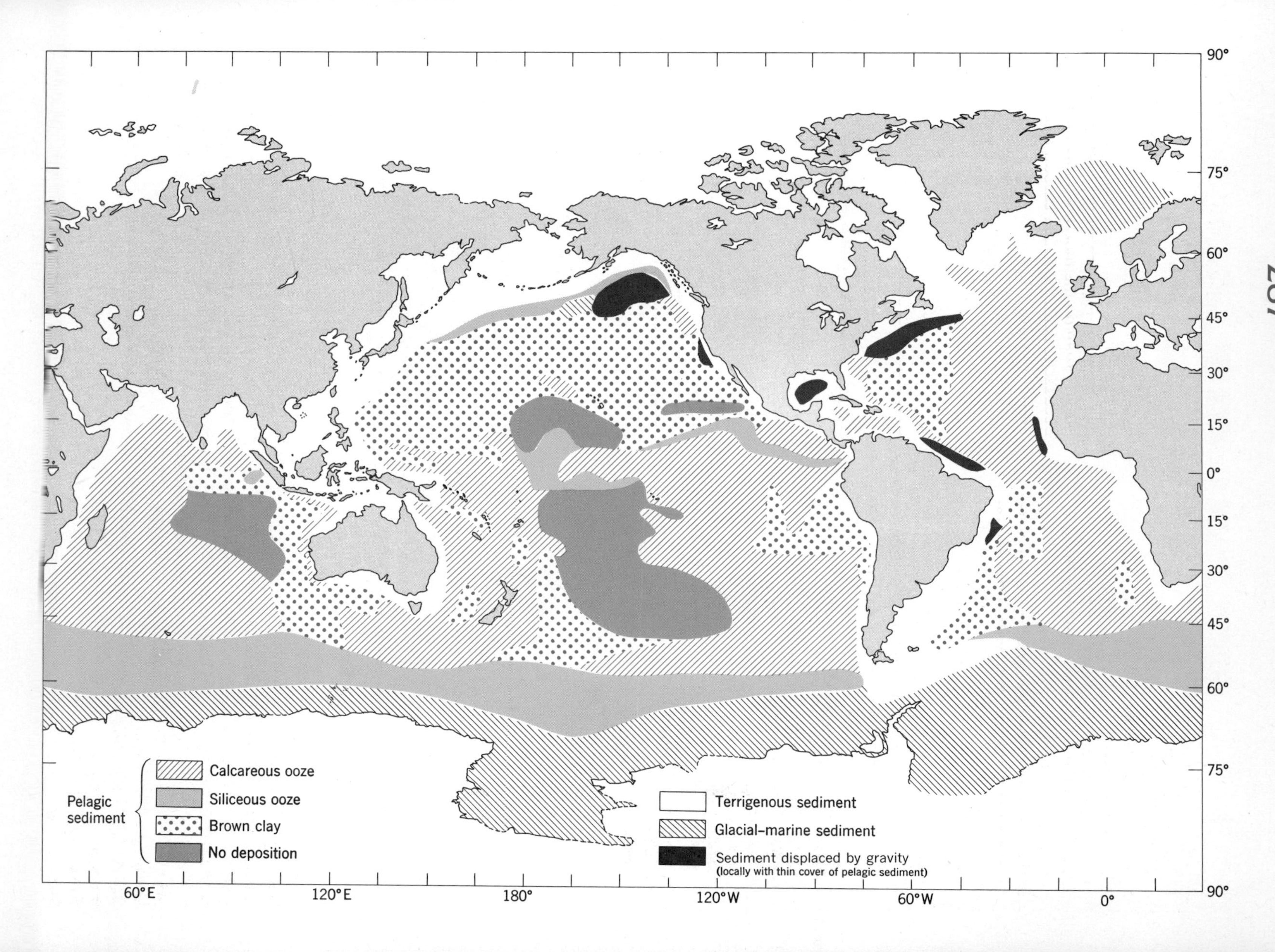
90°
75°
60°
45°
30°
15°
0°
15°
30°
45°
60°
75°
90°
Pelagic sediment
Calcareous ooze
Siliceous ooze
Brown clay
No deposition
Terrigenous sediment
Glacial-marine sediment
Sediment displaced by gravity
(locally with thin cover of pelagic sediment)
60°E
120°E
180°
120°W
60°W
0°

**Figure 12.5**
**Common materials of deep-sea sediment.**
***A.*** **Crystals and particles of various minerals, mostly clay minerals, seen with enlargement about 10,500 times natural size. Brown clay from 1.68m beneath floor of Pacific Ocean, at a depth of more than 4,500m, at a point about 1,450km west of Mexico. (*U.S. Geol. Survey.*)**
***B.*** **Microscopic calcareous shells of tiny aquatic animals called foraminifers. Recovered from the floor of the Caribbean Sea 160km west of Martinique, at a depth of 880m. The animals lived near the surface of the sea; the shells sank to the bottom as they were discarded. (*Yale Peabody Museum.*)**

and on land, reveal that sediments as much as 1000m thick overlie the oceanic crust. Also, at many deep-water sites, samples of sediment have been obtained by drilling from aboard *Glomar Challenger.* Strata as old as the Jurassic Period have been recovered in the drill cores. But sediments older than Jurassic have not been found. Although scientists once thought the oceans contained sediments from Earth's earliest history, drilling into the deep-sea floor proved that theory to be incorrect. The drilling confirmed the theory that plates of lithosphere, together with their mantle of sediment, are continually destroyed when they plunge down through the trenches. The sea floor is therefore being continuously renewed and the older sediments continuously destroyed.

# ANALYSIS OF SEDIMENTARY STRATA

## PRINCIPLES

Sedimentary strata are a key phase in the rock cycle, the phase that links the external processes, which break up rock and deposit sediment, and the internal processes that fracture, bend, squeeze, and melt sedimentary rocks and other rocks as well. This is why the present chapter stands between our discussion of Earth's external activities and that of the internal ones.

Another broad aspect of sedimentary strata is that every feature of sediment that geologists have described occurs also in sedimentary strata. This fact is perhaps the most obvious proof of the principle of uniformity of process.

### Original horizontality

We noted in Chapter 5 that sediment now being deposited in stream valleys and on the floors of lakes and the sea is spread out in layers that are generally almost horizontal. When immersed in water, loose particles are easily moved, and water in motion tends to spread the particles evenly and so to fashion a nearly level surface. In this way, at the end of each stream flood and of each storm offshore, a new sedimentary layer is deposited almost horizontally over the one beneath. Of course some layers of limited extent are deposited in inclined positions (Fig. 7.12), but these represent special conditions and are exceptions to the general rule. It is, then, a principle, recognized as early as the 17th Century, that most layers of sediment are nearly level when formed. We rely on the principle when we analyze sedimentary layers that are no longer horizontal, and conclude that they have been deformed.

### Stratigraphic superposition

We noted in the preceding paragraph that during each new bout of activity a new sedimentary layer is deposited over the one beneath. This fact embodies the *principle of stratigraphic superposition* (Chapter 5), which says that the order in which strata are deposited is from bottom to top. Together with the fossils the strata contain, this principle is fundamental to the building up of the geologic column.

## STRATIFICATION

*A layered arrangement of the particles that constitute sediment or sedimentary rock* is ***stratification.*** An obvious feature of most sedimentary rocks, it is seen also in ancient lava flows. Looking closely at rocks that are stratified distinctly, we can see that the strata differ from one another because of differences in some characteristic of the constituent particles or in the way in which the particles are arranged (Fig. 12.8). Very commonly one stratum consists of particles of different diameter from those in another. In a clastic rock, such changes of diameter result from fluctuations of energy in a stream, in surf, in wind, in a lake current, or in whatever agency is responsible for the deposit. The energy changes, usually small, are not the exception but the rule.

### Sorting

A conspicuous result of the transport of particles by flowing water or flowing air is *sorting* in the deposited sediments. Sorting according to specific gravity is evident in mineral placers (Chapter 22). Particles of unusually heavy minerals such as gold, platinum, and magnetite are deposited quickly on stream beds and on beaches, whereas lighter particles are carried onward. Most of the particles carried in water and wind, however, consist of quartz and other minerals with similar specific gravity. Therefore such particles are commonly sorted, not according to specific gravity but according to diameter. In a stream, gravel is deposited first, whereas sand and silt are carried farther before deposition. Thin, flat particles are carried farther than spherical particles of similar weight. Long-continued handling of particles by turbulent water and air results in gradual destruction of the weaker particles. In this way rocks and minerals that are soft or that have pronounced cleavage are eliminated, leaving as residue the particles that can better survive in the turbulent environment. Very commonly the survivor is quartz, because it is hard and lacks cleavage. In this case sorting is based on durability.

Although sorting is the chief cause of stratification, it is not the sole cause. Successive layers that do not differ from each other in grain size, composition, or degree of compaction can still be separated from each other by surfaces of easy splitting representing minor intervals when no deposition occurred. Again, two adjacent layers, not otherwise distinct, can differ from each-other as to the kind or abundance of cement they contain.

In summary, each stratum nearly always possesses definite characteristics by which it differs from the stratum beneath or above it. With this in mind we can describe two chief kinds of stratification and then examine the particles within a stratum.

### Parallel strata

Layers of sediment fall into two classes according to the geometric relation between successive units. One class consists of ***parallel strata,*** *strata whose individual layers are parallel* (Fig. 12.6). Parallelism indicates that deposition probably occurred in water, and that the activity of waves and (except for the special case of graded layers, described in a following section) currents was

**Figure 12.6**
**Varves in glacial-lake clay and silt. Uppsala, Sweden.**
**In this case each varve is also a graded layer. Thick, pale silt (summer portion) sharply overlies the layer beneath it and grades upward into thinner, darker silty clay (winter portion), thus representing an annual cycle. Pencil (placed vertically at base of exposure) is 15cm long. (*R. F. Flint.*)**

minimal. Indeed, the sediments of lakes and the deep-sea floor occur rather commonly in parallel layers.

A distinctive variety of parallel strata consists of repeated alternations of layers of unlike grain size or mineral composition. Such alternation suggests the influence of some naturally occurring rhythm, such as the rise and fall of the tide or the seasonal change from winter to summer. *A pair of sedimentary layers deposited during the cycle of the year with its seasons* is a ***varve*** (Swedish for "cycle"). Varves occur in many glacial-lake sediments (Fig. 12.6) deposited 12,000 years ago or less, during the melting of the latest great glaciers. Radiocarbon dating of related sediments has confirmed that the pairs of layers are true varves.

Paired layers deposited in deep glacial lakes are generally very distinct, because close to an ice sheet the contrast between summer and winter weather markedly affects the rate of melting of ice. Pairs of laminae very similar to these occur in ancient rocks. Certain claystones in South Africa have been interpreted as varves deposited in glacier-dammed lakes during a glaciation more than 200 million years ago.

Varves of different origin characterize rocks of the Green River Formation, which underlies a vast area in Wyoming, Colorado, and Utah. In each varve one layer consists of calcium carbonate; the other includes dark-colored organic matter. The rhythm is explained as follows: the sediments were deposited in a lake, which warmed in summer, therefore lost carbon dioxide, and precipitated calcium carbonate from solution. During the same warm season floating microscopic organisms reached a peak of abundance. The relatively heavy carbonate sank promptly and formed a summer layer; the lighter organic matter sank much more slowly to form an overlying winter layer. Thus the pair of layers is a varve. It has been estimated from sample counts that between five and eight million varves are present in the Green River Formation; hence we reason that an immense lake occupied the Green River basin for millions of years under remarkably uniform conditions.

### Cross-strata

Very distinct from parallel strata are *strata that are inclined with respect to a thicker stratum within which they occur.* These are ***cross-strata.*** All such strata consist of particles coarser than silt and are the work of turbulent flow of water or air, as in streams, wind, and waves along a shore. As they are driven forward, the particles tend to collect in ridges, mounds, or heaps in the form of ripples and waves, which migrate forward bodily or simply enlarge in the downcurrent direction. Particles continually accumulate on the downcurrent slope of the pile, forming strata with inclinations as great as 30° to 35°;

Cross-strata are seen in deltas (Fig. 7.12), sand dunes (Fig. 9.21), and beaches (Fig. 11.17). We must keep in mind that although the foreset layers in deltas and in dunes are commonly parallel with *one another,* nevertheless they are cross-strata because they are not parallel with the larger strata that enclose them. Cross-strata therefore do not argue against our belief in original horizontality. Indeed the larger strata that enclose the cross-strata support that belief.

Under some conditions, however, no larger enclosing strata are present; instead, all the layers deposited at a locality are inclined. For example, volcanic ash and coarser volcanic particles characteristically accumulate in conical piles surrounding their volcanic source and form layers with steep inclinations (Fig. 12.7). Because *all* such layers are inclined, they are not cross-strata, and so are real exceptions to the idea of original horizontality.

The direction in which cross-strata are inclined is the direction in which the related current of water or air was flowing at the time of deposition. So, by measuring direction of inclination we can learn that one thing about a former environment.

### Arrangement of particles within a stratum

In addition to the relationships of layers to eachother, several kinds of arrangement of the particles within a single layer are possible. Each kind gives information about the conditions under which the sediment was deposited. Chief among these kinds are *uniform layers, graded layers,* and *nonsorted layers.*

**Uniform Layers.** A layer that consists of particles of about the same diameter is called *uniform.* A uniform layer of clastic rock implies deposition of particles of a single size, with little change in the velocity of the transporting agent. This might occur in the bottomset layers of a delta. A uniform layer of nonclastic rock implies uniform precipitation from solution, which produces crystalline particles of a single size. But a layer that is subdivided into thin layers marked off from eachother by differences in grain size suggests a transporting agent the velocity of which fluctuated.

**Figure 12.7**
**These layers of volcanic ash and coarser particles were deposited at a steep angle on the slope of a volcanic cone near Cerro de Camiro, Mexico. Oldest layers are at lower right, youngest at upper left. Each layer represents a single eruption. (*Kenneth Segerstrom, U.S. Geol. Survey.*)**

**Graded Layers.** If we put a quantity of small solid particles of different diameters and about the same specific gravity into a glass jar of water, shake the mixture well, and then let it stand, the particles will settle out and form a deposit on the bottom of the jar. The heaviest and largest particles settle first, followed by successively smaller ones; the finest, if small enough, may stay in suspension for hours or days, keeping the water turbid, before finally settling out. Thus particle size decreases *gradually* from the bottom upward. This arrangement characterizes a ***graded layer,*** defined as *a layer in which the particles grade upward from coarse to finer* (Fig. 12.8). The deposit is, as it were, a one-stroke affair. The water receives energy from the shaking of the jar and becomes turbulent; as a result sediment is lifted above the bottom. When the shaking stops, abrupt loss of energy results in continuous deposition.

Although a graded layer produced by simple shaking in a jar is graded only in the vertical dimension, those produced in nature

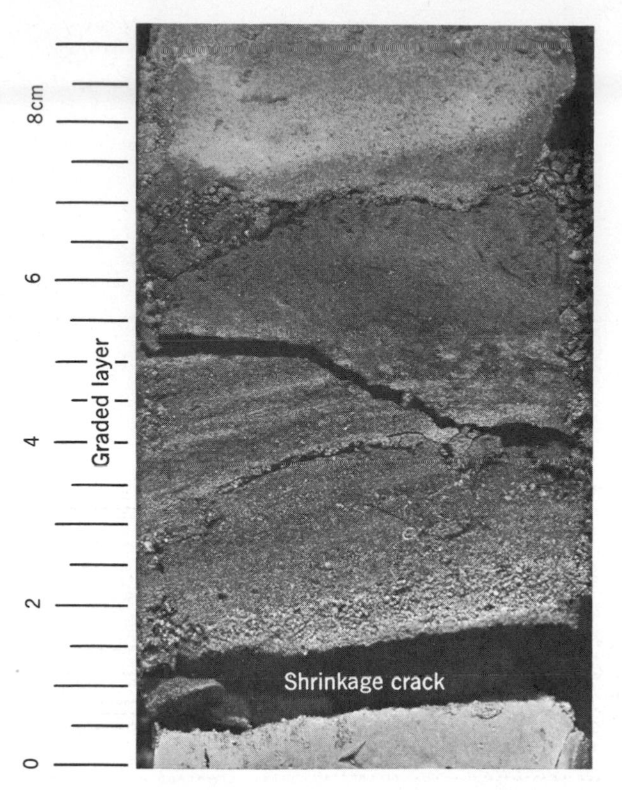


**Figure 12.8**
**Graded layer exposed in a vertical slice through a core taken from sediment beneath the deep-ocean floor at depth 4,074m. The layer, overlain and underlain by nonstratified ooze, extends from about 1 to about 7 on the centimeter scale. (*Lamont Geol. Observatory, courtesy C. D. Hollister.*)**

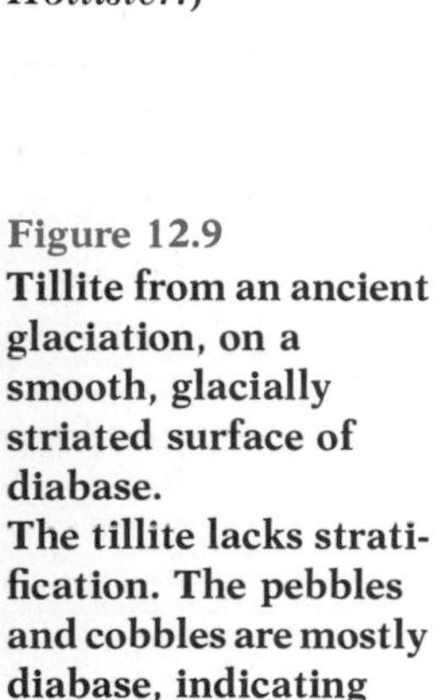

are graded laterally also, because they are made by currents moving from one place to another. Since the heaviest particles settle first, grading occurs laterally in the downcurrent direction.

As we noted earlier in this chapter, graded sediments in the form of turbidites are widespread not only in deep-sea fans but beneath

**Figure 12.9**
**Tillite from an ancient glaciation, on a smooth, glacially striated surface of diabase.
The tillite lacks stratification. The pebbles and cobbles are mostly diabase, indicating glacial erosion of the underlying bedrock. Nooitgedacht, Cape Province, South Africa. (*R. F. Flint.*)**

the great abyssal plains as well. Processes that cause grading, other than turbidity currents, include streams, as they lose energy and deposit bed-load sediment while floods subside, falls of volcanic ash, and dust storms as they die down. Such processes represent rapid, continuous loss of energy, an essential condition for the creation of graded layers.

**Nonsorted Layers.** The particles in some sedimentary rocks are not sorted at all. They consist of mixtures of various sizes arranged chaotically, without any obvious order. Processes that create sediments of this class include avalanche, debris flow, mudflow, solifluction, submarine slump, and transport by glaciers and floating ice. Widely recognized among nonsorted sediments is tillite (Fig. 12.9), of glacial origin.

### Rounding and sorting

As we noted in Chapters 6 and 7, particles broken from bedrock by mechanical weathering and other processes tend to be angular, because breaking commonly occurs along joints, and surfaces of stratification. As they undergo transport by water or air, the same particles tend to become smooth and rounded. Figure 12.10 shows what can happen to pebbles on a beach, and Figure 12.11 indicates how sand grains become shaped during transport. Degree of rounding therefore gives some idea of the distance or time involved in transport by flowing water or flowing air.

**Figure 12.10**
**Rounding of pebbles during transport by surf on a beach.**
**These fragments of basalt were collected at random from a talus and an adjacent beach near Clarence, Nova Scotia. They are arranged here to show what can be expected to happen to a plate-shaped piece (*upper left*) and to a spindle-shaped piece (*upper right*) during progressive abrasion. The end product of each could be the same—a spherical pebble. Although this rounding is the work of beach drift, stream transport produces a similar result. (*J. E. Sanders from Longwell and Flint.*)**

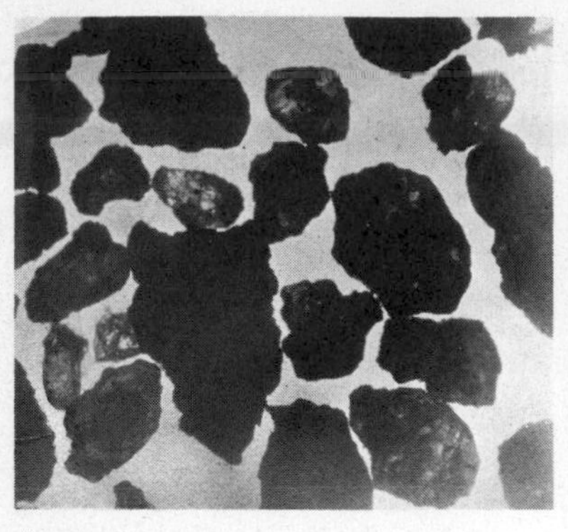
A

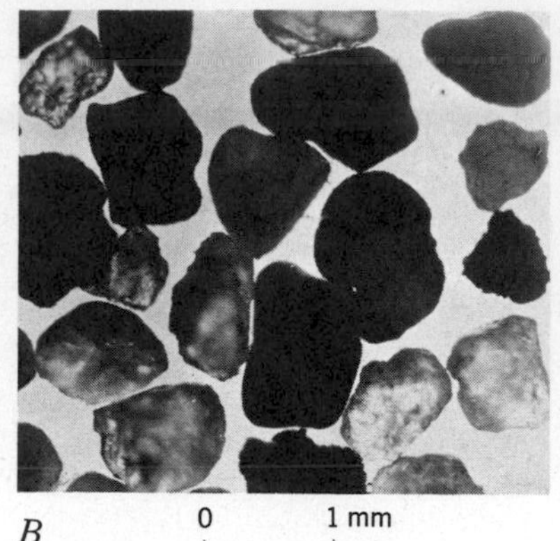

C

**Figure 12.11**
**Rounding and sorting of mineral grains during transport.**
***A.* Mineral grains loosened and separated from igneous and metamorphic rocks by mechanical and chemical weathering before transport. The angular shapes of the individual grains, slightly altered by weathering, are the forms assumed by the minerals as they crystallized from a magma. The aggregate of grains is a nonsorted sand.**
***B.* Sand carried from an area of rock similar to that which yielded the sand in *A*. Some of the less-durable mineral grains have been broken up and lost, leaving a larger proportion of the durable mineral quartz. Battering in transit has partly rounded the grains.**
***C.* Sand transported through a long distance. Grains have become well-rounded and consist almost entirely of durable quartz.**

Glaciers, however, tend to make rock particles irregular in shape by crushing and abrading them. The faceted shapes of ideal glaciated coarse particles, and the scratches on them, are illustrated in Figure 10.14. Flowing air likewise grinds facets on coarse particles, but the facets on ventifacts (Fig. 9.17) are more distinct and meet eachother more sharply than do those made by glaciation. Ventifacts are not rounded because wind does not move them. They lie motionless while fine particles driven by the flowing air abrade them.

We have said that sorting of sediment is responsible for most stratification. Rock particles become increasingly well sorted as they become increasingly rounded. This change with distance makes it possible to draw general inferences about directions and distances of transport of ancient sediments and therefore about geographic relationships at the times when the sediments were deposited.

### Derivation

Through the kinds of minerals or rocks of which they are composed, sediments reflect the kind of parent rock from which they were derived. An uncomplicated example is a boulder train in glacial drift; another is a placer (Chapter 22) of gold, diamond, or other economic substance. Many boulder trains and placer minerals have been traced backward (upstream) to their sources; from a boulder train the direction of flow of a glacier can be inferred.

Also we can draw inferences, broad and general at least, as to the kind of rock from which a widespread body of sediment or sedimentary rock was derived. Going back to the chemical weathering of granite and basalt (Table 6.1), we recall that granite yields quartz and clay minerals, but that basalt can yield no quartz. If, however, either kind of rock were weathered mechanically, it would yield bits of the rock itself, and those bits would include feldspar. If transported and deposited quickly, the feldspar would not be destroyed by weathering. The presence of fresh feldspar in sedimentary strata would suggest one of two things about the origin of the sediment. Either a dry or very cold climate with a minimum of chemical weathering prevailed, or the cutting of valleys by streams occurred on slopes so steep that rate of chemical weathering could not keep pace with rate of erosion by the streams.

### Features on surfaces of layers

In foregoing chapters we discussed the making of ripples by streams, by the wind, and by currents and waves in lakes and the sea. Ripples are preserved in some sandstones and siltstones as *ripple marks* (Fig. 12.12).

Some claystones and siltstones contain layers that are cut by polygonal markings. By comparison with sediments forming today, such as those in roadside puddles following a rain, we infer that these are ***mud cracks,*** *cracks caused by shrinkage of wet mud as its surface becomes dry* (Fig. 12.13). The presence of mud cracks in a rock generally implies at least temporary exposure to air and therefore suggests tidal flats, exposed stream beds, playa lakes, and similar environments. Occurring with some ripple marks and mud cracks, and preserved in a similar manner, are the footprints and trails of animals. Even the impressions of large raindrops made during short, hard showers are preserved in some strata.

Figure 12.12
**Ancient ripple marks made by a current in seawater and now exposed on a big slab of hard sandstone. Capital Reef National Monument, Utah. (*Richard Weymouth Brooks.*)**

**Figure 12.13**
**Mud cracks preserved in sandstone of Mississippian age near Pottsville, Pennsylvania.**
**This is the underside of a slab turned up on its edge. The sand was deposited over a layer of river mud in which the cracks had formed and then hardened. The claystone (the former mud) has been destroyed by recent weathering, but chips of it litter the foreground. Most ancient mud cracks, like these, are really fillings of cracks by sand that was deposited in them. (*Joseph Barrell.*)**

# OTHER FEATURES

### Fossils

The animals and plants (or parts of them) that were buried with sediments protected against oxidation and erosion, and preserved through the long process of conversion to rock, constitute the fossils (Fig. 12.14) that form a very important element in the geologic column. Not only are fossils an important clue to former environments, but they are the chief basis for the correlation of strata and the construction of the geologic column.

**Figure 12.14**
**Fossils in sedimentary rock.**
**Impressions of the shells of marine invertebrate animals in Paleozoic sandstone from eastern New York. (*C. O. Dunbar.*)**

## Color

The colors of sedimentary rocks vary considerably. Some rocks exposed in cliffs are colored only skin-deep by products of chemical weathering. For example, a sandstone that is pale gray on freshly fractured surfaces may have a surface coating of yellowish-brown limonite, developed during weathering by the oxidation of sparse iron-rich minerals included with the grains of quartz sand.

The color of fresh rock is the combination of the colors of the minerals that compose it. Iron sulfides and organic matter, buried with the sediment, are responsible for most of the dark colors in sedimentary rocks. Microscopic examination of red and brown rocks shows that their colors result mainly from the presence of ferric oxides, as powdery coatings on grains of quartz and other minerals, or as very fine particles mixed with clay.

These oxides get into strata in at least two different ways. In most red and brown rocks probably the colored oxides are secondary, created in the strata by alteration of ferromagnesian minerals such as hornblende and biotite, during weathering and during conversion of sediment into rock. Such chemical alteration can take place in any warm climate, dry or moist.

In some strata the colored oxides are believed to be primary, having been deposited as components of the sediments themselves. Such oxides may have been derived from the erosion of red-clay soils like those forming today in the warm climate of the southeastern United States. Washed down from uplands into rivers, the oxides are deposited in places on swampy floodplains, where decaying plant matter creates strong reducing agents that reduce the red ferric oxide to ferrous oxide, which is not red. Rivers carry the rest of the red sediment into the sea, where organic matter on the sea floor likewise reduces it. Probably, therefore, the red color can be preserved only if the sediments escape reduction after deposition. This they could do if deposited in basins where, for one reason or another, organic matter does not accumulate in amounts sufficient to reduce the ferric iron.

Without detailed research, therefore, all we can tell about the climate where and when an ancient red-colored stratum was deposited is that the climate was warm, either there or in the region from which the sediment was derived.

## Concretions

Enclosed in some sedimentary strata are bodies called *concretions.* They range in diameter from less than a centimeter to two meters or more, and in shape from spherical through a variety of odd shapes, many with remarkable symmetry (Fig. 12.15), to elongate bodies that parallel the stratification of the rock. Concretions are composed of many different substances, including calcite, silica, hematite, limonite, siderite (iron carbonate), and pyrite. Small concretions are dredged up from the sea floor (Fig. 22.9), showing they are forming there today as sediments are deposited. This origin, contemporaneous with the enclosing sediments, is indicated

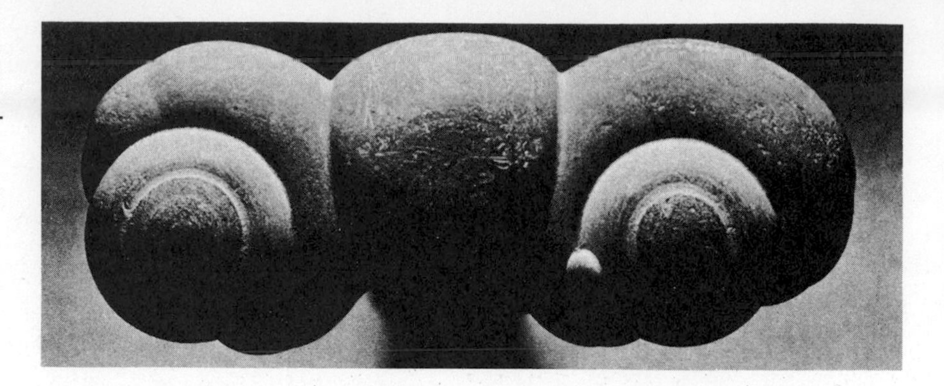

**Figure 12.15**
**Concretion 7cm long, consisting of calcium carbonate. It was embedded in claystone. (*Andreas Feininger.*)**

also by the shapes of some concretions and by their relation to the stratification of the surrounding sedimentary rock. Others formed after the deposition of the sediments, as, for example, those that retain the stratification of the surrounding rock (Fig. 12.16).

The substances of which concretions are made show that these objects are the result of localized chemical precipitation of dissolved substances from seawater, lake water, or ground water. Once precipitation starts around a fossil or other body that differs from the enclosing rock, the concretion thus formed continues to grow. Indeed, in some rocks, perfectly preserved fossils are found at the centers of concretions.

Because we do not yet understand concretions fully, the best definition we can suggest is that a ***concretion*** is *a localized body having distinct boundaries, enclosed in sedimentary rock, and consisting of a substance precipitated from solution, commonly around a nucleus.* Some geologists use the word *nodule* as a synonym for concretion; others restrict it to concretions of small size.

# STRATIGRAPHIC INTERPRETATION

## Environments of deposition

The study of strata is *stratigraphy.* One of its broader aspects concerns our interpretation of the environments in which sediments are deposited.

One of the objectives of research in geology is to reconstruct the *paleogeography* (the ancient geographic relations) of a region at the time when a particular sequence of strata was deposited as

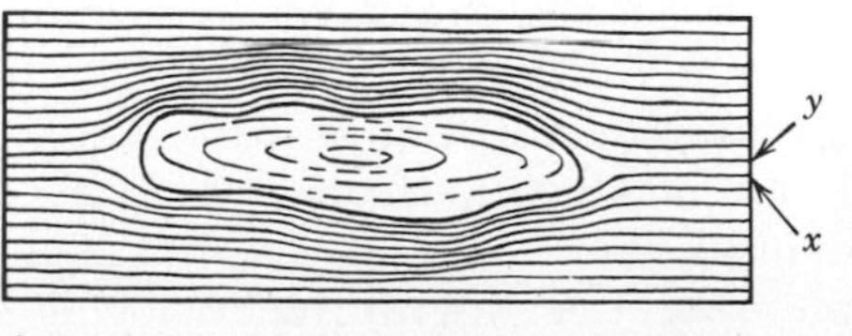


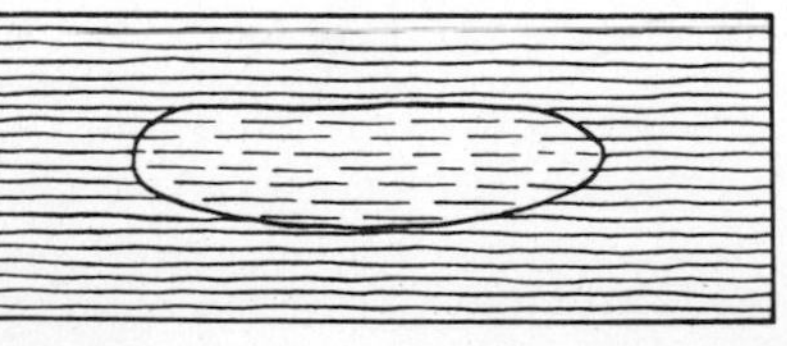


**Figure 12.16**
**Vertical cross sections of two concretions, each 15 to 20cm long, in claystone, showing different times of origin relative to deposition of the enclosing strata.**

***A.*** **Concretion formed after layer *x* was deposited but before layer *y* was laid down. Hence the concretion is contemporaneous with deposition of the body of sediment.**

***B.*** **Concretion transected by layers of the enclosing rock. Hence it was formed after all the layers shown had been deposited.**

sediment. The reconstruction would show the distribution of land and sea, streams with their directions of flow, mountains and lowlands, deserts, glaciers, and possibly indications of the climate then prevailing. Such information is commonly assembled in the form of a *paleogeographic map*, representing a series of inferences drawn from the physical characteristics of the strata and from the fossils they contain. The inferences mainly concern the environment in which each kind of sediment was deposited and are based on analogy with the environments of today's sediments, by use of the principle of uniformity of process.

Looking at modern sediments, we conclude that sediments deposited along streams, in lakes, and in the sea are deposited primarily in basins, where the chances of preservation are better than on high places and on steep slopes. To distinguish among the various environments in which a group of strata may have been deposited, we must examine exposures, following the west–east linc marked by numbers on the map, Figure 12.17.

1. In the area numbered (1) we see conglomerate consisting of little-rounded and little-sorted cobbles and pebbles, with very irregular cross-strata dipping east. In places the conglomerate is interbedded with thin layers of turbidite. These characteristics suggest a fan or fans at the foot of the steep slope of a highland.

2. Farther east we find better-sorted sandstone with cross-strata associated with elongate bodies of siltstone and broad thin layers of claystone. Fossils might include freshwater mollusks and leaves and branches of trees. This assemblage suggests a floodplain on a lowland, with meanders, natural levees, and overbank-flood sediments.

3. Still farther east is well-sorted, well-rounded sandstone, with long, parallel, gently dipping foreset beds overlain by stream-channel cross-strata. Fossils include plant debris and a few brackish-water mollusks. Evidently we have passed from a former piedmont environment, across a coastal plain drained by sizeable rivers, to a low-lying seacoast.

4. Assuming that good exposures continue eastward, we can see sandy beach sediments overlain in places by well-sorted sandstone with cross-strata dipping steeply west, suggesting beach dunes of wind-blown sand.

5. Still farther east we find the texture becoming finer, with sandstone grading into siltstone and claystone that contain fossil mollusks of offshore-marine kinds.

From this information, clearly indicating several environments that differ from one another but that are logically related, we can sketch the crude paleogeographic map shown in Figure 12.17. In a similar manner we can identify former lakes, estuaries, lagoons, and reefs, as well as sandy deserts and country overrun by glaciers, each feature representing a distinctive environment of deposition.

These environments no longer exist in the district we are studying. But they can be recaptured by virtue of the fact that the various sediments formerly deposited in them were preserved and

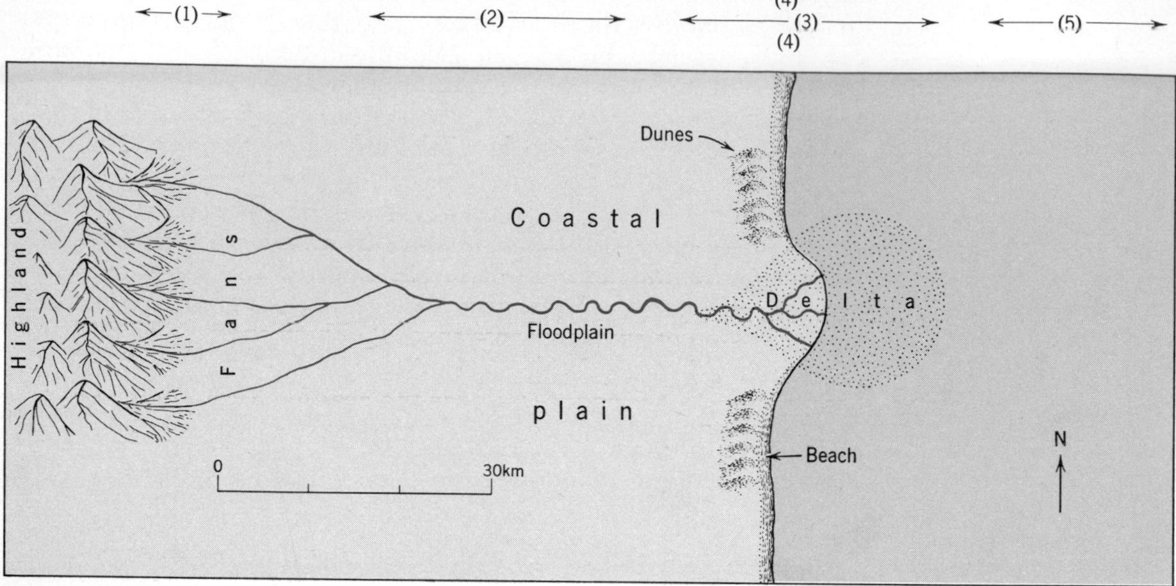


**Figure 12.17**
**Paleogeographic sketch map showing several former environments, reconstructed from characteristics of strata. Numbers indicate positions along a west–east line, described in text.**

converted into rock, as they subsided and were covered by other sediments. In short, they became a part of the geologic record.

### Sedimentary facies

Implicit in our discussion of environments is the fact that most strata change character from one area to another. In the foregoing discussion we followed a unit or group of strata, all parts of which were deposited in the same interval of time, from the base of a highland across a coastal plain and into a shallow sea that deepened offshore. The changing environments are represented by changing grain size, grain shape, stratification, depositional structures, and fossils in the unit. *A distinctive group of characteristics, within a rock unit, that differ as a group from those elsewhere in the same unit,* is a sedimentary ***facies*** (''aspect''). Two facies merge laterally into eachother either gradually or abruptly, depending on the relations between the two former environments of deposition.

If a sedimentary unit were exposed in section from end to end of its extent, it could be identified as a unit despite changes in its facies. But if, as is usual, only widely separated parts are exposed and if each part represents a different facies, its contained fossils would be needed for correlation. A difficulty arises here because the assemblages of fossils in two facies may not be exactly the same, even though the organisms they represent lived at the same time. This happens because the environments in which the organisms lived were different. In the same sea, deep-water shellfish are unlike shallow-water kinds; and on land, animals living in deserts are unlike those living at the same time in moist, forested regions. These variations of fossils with varying facies do not, however, make correlation impossible; they only make it more difficult.

In Figure 12.18 are two strata, each grading from a pebbly-sand facies (*left*) into a sandy-silt facies (*right*). Each stratum represents a beach that graded seaward into finer sediment offshore. We can infer that sea level must have been lowered from position 1 to position 2. By boring a hole, *B*, we can see the beach facies of the younger stratum overlying the offshore facies of the older one.

**Sequences of strata**

Now that we have followed a single stratum laterally through two or more facies, the next step is to visit an area and examine a vertical sequence of several strata. Our objective is to reconstruct the succession of environments through which the area has passed—in other words, to synthesize a piece of geologic history. This is done in Appendix D, under the heading *Field Study of a Sequence of Strata.*

**Rock units: formations**

It is easy to identify the rock in Figure 12.13 as sandstone, but a thorough study must distinguish it from other sandstones. One respect in which any given stratum differs from all others is its position in the vertical sequence of strata. Hence we give it a designation by which its position is fixed and by which it can be catalogued and referred to. The basic rock unit to which such a designation is applied is the *formation* (Appendix D). A formation must constitute a mappable unit; that is, a unit (1) thick and extensive enough to be shown to scale on a geologic map, and (2) distinguishable from the strata immediately above and below, not just at one exposure but generally wherever the unit is exposed. Within these requirements a formation can be thin or thick, to suit the geologist's convenience. Its thickness is likely to depend on the degree of detail of the field study and correspondingly on the scale of the map to be made. In North America each formation is given a name, typically the name of a locality near which it is exposed (Lexington Limestone, Fox Hills Sandstone, Green River Formation). Not only sedimentary rocks but igneous and metamorphic rocks as well are identified as formations (examples: San Juan Tuff, Conway Schist). Figure D.9 is a series of geologic sketch maps showing three formations.

Formations can be subdivided and also grouped into larger rock units. But because we are concerned here only with the way in which strata are identified, the formation alone serves our purpose. For more about this see Appendix D.

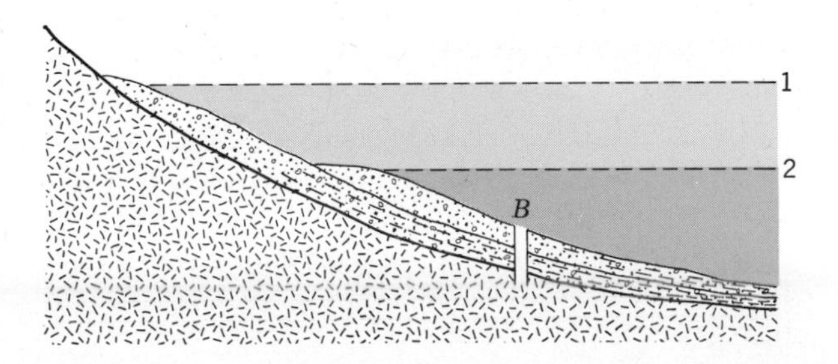


**Figure 12.18**
**Reconstruction of two former sea levels from the facies in each of two strata encountered in a borehole *B*, as explained in text.**

### Matching rock units by physical characteristics

Once the layered rock units have been identified in vertical sequence, the extent of each (that is, the area it underlies) must be determined as closely as possible. With few exceptions, layers of sediment are deposited in basins located on land, beneath lakes, or in the sea. A layer of sediment may cover all or only a part of a basin. If deposited in the sea or in a lake, the stratum is likely to extend over the whole basin floor. However, during the time, perhaps very long, since conversion from sediment into sedimentary rock, the stratum may have been eroded so much that only parts of it remain. An example is the Pittsburgh Coal Seam, an easily recognized formation, whose original area of perhaps 50,000km$^2$ has been reduced by long-continued erosion to approximately 15,000km$^2$. One of the responsibilities of a geologist is to study the remnants so thoroughly that he can determine as nearly as possible the original extent of each formation.

This is done by matching the remnants, preserved in the hills left by erosion, on a basis of physical characteristics such as grain size, grain shape, mineral content, kind of stratification, and color. Matching on this basis is likely to be reliable through short distances, but generally becomes less reliable through longer ones because the physical characteristics tend to change in lateral directions.

### Correlation by means of fossils

The usefulness of matching by physical characteristics is virtually limited to the area of the basin in which the strata were originally deposited. In order to determine equivalence in the ages of strata in different basins, and even in different continents, geologists compare fossils, the chief basis of stratigraphic correlation (Fig. 5.5)

### Correlation by means of radiometric dates

As we learned in Chapter 5, the ages of fossil-bearing strata are fixed by the radiometric dates of igneous bodies closely related to them. Such dates are useful also in correlation. Where, as is not uncommon, sedimentary layers contain no fossils, we can fix their ages approximately through the radiometric ages of related igneous rocks. In some cases this enables us to correlate dated strata that lack fossils with dated strata that contain them, thus gradually enlarging the known extent of strata whose positions in the geologic column have been fixed by fossils.

We have now examined what strata themselves indicate about geologic history. In Chapter 14 we shall see what strata can reveal about geologic structure.

## SUMMARY

1. Most sedimentary strata are built of continental waste, deposited in basins on continental shelves and slopes. They are the result of complex processes of transport and sorting.

2. The sediment is recycled again and again, nearly always within the continental realm.

3. By depositing turbidites, turbidity currents have built huge deep-sea fans beyond the mouths of several submarine canyons.

4. Chief classes of sediment on the deep-sea floors are brown clay, calcareous ooze, siliceous ooze, and sediment of continental origin.

5. Stratigraphy is the study of stratified rock. From it most of our knowledge of Earth's history has been learned.

6. Various arrangements of the particles in strata are seen in parallel strata and cross-strata, uniform layers, graded layers, and nonsorted layers.

7. Particles of sediment become rounded and sorted in transport by water and air but not in transport by glaciers and by mass-wasting.

8. Ripple marks, mud cracks, and fossils in sedimentary strata give evidence of environments of deposition.

9. An extensive unit of strata may possess several facies, each determined by a different environment of deposition.

10. The basic physical rock units are formations, each identified by a locality name.

11. By means of fossils, and in some instances by radiometric dates, strata can be correlated through long distances.

12. Correlation takes account of the two principles: that strata were horizontal when deposited and that they were formed in sequence from bottom to top.

# SELECTED REFERENCES

Damuth, J. E., and Kumar, Naresh, 1975, Amazon Cone: morphology, sediments, age, and growth pattern: Geol. Soc. America Bull., v. 86, p. 863–878.

Dunbar, C. O., and Rodgers, John, 1957, Principles of stratigraphy: New York, John Wiley.

Hatch, F. H., and Rastall, R. H., 1965, The petrology of sedimentary rocks, 4th ed., revised by J. T. Greensmith: London, Thomas Murby.

Krinsley, D. H., and Smalley, I. J., 1972, Sand: American Scientist, v. 60, p. 286–291.

Krumbein, W. C., and Sloss, L. L., 1963, Stratigraphy and sedimentation, 2d ed.: San Francisco, W. H. Freeman.

Lahee, F. H., 1961, Field geology, 6th ed.: New York, McGraw-Hill.

Shrock, R. R., 1948, Sequence in layered rocks: New York, McGraw-Hill.

Walker, T. R., 1974, Formation of red beds in moist tropical climates: a hypothesis: Geol. Soc. America Bull., v. 85, p. 633–638.

# 13 THE OCEAN AND OCEAN BASINS



## THE WORLD OCEAN

If, by some mysterious means, it were in our power to remove briefly all the water from the oceans and then to view the dry Earth from a space ship, we would see continents standing abruptly above the deep-sea floor. We would observe that a continent ends where its steep continental slope meets the deep ocean floor. If we swooped down in our space ship and examined the rock in such a place, we would find that the foot of the continental slope is the place where continental crust meets oceanic crust. Exploring further, we would observe that the world ocean actually occupies several great basins, each floored with oceanic crust and rimmed with continental crust. As we noticed in Chapter 11, the water in the great ocean basins actually just overfills them; so the world ocean we see today spills over onto the continental shelves.

Within the ocean, beyond the continental slope, lies the strange, unseen world of the deep-ocean floor. With newly perfected devices for sounding the sea bottom and for sampling its sediment, teams of oceanographers and seagoing geologists have explored the ocean floor and have now put submarine geology on a basis of knowledge almost as solid as that already established on the land. Scuba-diving geologists have visited, photographed, and mapped areas of sea floor at depths as great as 70m, and observers in special submarine chambers and deep-diving submarines have visited the deepest known places in the ocean.

Because of this intensive research, involving many nations, the oceans are gradually giving up their mysteries. The romanticist that is in each one of us cannot help regretting that beliefs built up through more than three thousand years of human legend have vanished: the singing mermaids, the strange and threatening gods of the sea, the fabled cities and castles believed to have sunk into watery deeps, the monsters of seafarers' tales. These and other poetic mysteries have faded away under the hard glare of knowledge. But in return that knowledge can help us protect the fragile values of a priceless environment that covers more than two-thirds of Earth's surface and is responsible for a large part of Earth's biological heritage.

### Dimensions

Seawater covers 71 percent of Earth's surface. The 29 percent that is land is not evenly distributed. In the Northern Hemisphere, often called the land hemisphere, 40 percent of the surface is land, while in the southern hemisphere only 20 percent is land. As we shall see, the uneven distribution of land plays an important part in determining the paths along which water circulates in the ocean.

The greatest ocean depth yet measured is slightly more than 11km, near the island of Guam in the western Pacific. If Mount Everest were dropped into the sea at that point, 1.6km of water would cover its peak. The average depth of the sea, however, is about 3.8km, compared to an average height of the land of only 0.75km.

Knowing the area of the sea and its average depth, we can compute the present volume of seawater to be about 1,350 million $km^3$. We say *present* volume because, as we read in Chapter 10, the volume fluctuates with the growth and melting of glaciers. But if we consider the entire hydrosphere (Chapter 2), the volume of total water substance appears to be constant.

### Composition

About 3.5 percent of average seawater, by weight, consists of dissolved salts (Table 13.1)—enough to make the water undrinkable. It is enough also, if precipitated, to form a layer of solid salts, about 56m thick, over the entire sea floor.

The measure of the sea's saltiness is termed *salinity*. We commonly express salinity in parts per thousand, rather than in percent (parts per hundred). Average seawater therefore has a salinity of 35 parts per thousand. The principal elements that contribute to the salinity of the sea are sodium and chlorine; when seawater is

**Table 13.1**
**The Major Constituents of Seawater (Constituents Are Listed as the Principal Ions in Solution as an Indication of the Kinds of Compounds That Form When Seawater Evaporates)**

| Ion | Percentage of All Dissolved Matter |
|---|---|
| Chloride ($Cl^{-1}$) | 55.07 |
| Sodium ($Na^{+1}$) | 30.62 |
| Sulfate ($SO_4^{-2}$) | 7.72 |
| Magnesium ($Mg^{+2}$) | 3.68 |
| Calcium ($Ca^{+2}$) | 1.17 |
| Potassium ($K^{+1}$) | 1.10 |
| Bicarbonate ($HCO_3^{-1}$) | 0.40 |
| Bromine ($Br^{-1}$) | 0.19 |
| Strontium ($Sr^{+2}$) | 0.02 |
| Total | 99.97 |

evaporated, more than 75 percent of the dissolved matter is precipitated as common salt (NaCl). But seawater contains most of the other elements as well—many of them in such low concentrations that they can be detected only by super-sensitive analytical devices. As we can see in Table 13.1, more than 99.9 percent of the salinity is caused by only nine ions.

Where do the dissolved salts come from? Each year rivers and streams carry 2.5 billion tons of new salts to the sea. The salts are soluble materials leached from rock during chemical weathering, together with a small amount of soluble material carried up from the mantle and spewed out in volcanic gases. Through millions of years, the amount of dissolved salts added by rivers has far exceeded the salts now dissolved in the sea. Why, then, isn't the sea more salty than it is? The reason, as we saw in Chapter 5, is that the composition of the ocean is in a steady state; material is being removed at the same rate at which it is added. Some of the elements, such as silicon, calcium, and phosphorus, are withdrawn from seawater by aquatic plants and animals to build their shells or skeletons. Other elements, such as potassium and sodium, are absorbed and removed by clay particles and other minerals as they slowly settle to the sea floor. Still others, such as copper and lead, are precipitated as sulfide minerals in claystones and mudstones rich in organic matter. The net result of all these processes of extraction, taken together with what is being added, is that the composition of seawater remains constant.

### Age and origin

The world ocean is very old. Earth's oldest rocks include water-laid sedimentary strata similar to those we see being deposited today. Therefore, we are sure that as far back in history as we can see, which is 3.8 billion years, Earth has been provided with large bodies of water. Indeed, water has probably been present at the surface almost as long as Earth has existed as a solid body.

Where did the water come from? We can be sure the ocean was created between 4.6 billion years ago (when Earth formed) and 3.8 billion years ago, when the oldest known rock was made, but we cannot be sure *how* it formed. Most probably it was baked and sweated out of minerals in crust and mantle, arriving at the surface as steam during primeval volcanic eruptions. The process must have been gradual; so the world ocean enlarged through time. The enlargement seems now to have stopped. Water is, of course, slowly removed from the hydrosphere by being buried along with sediment; so it eventually becomes incorporated into igneous and metamorphic rock. On the other hand, water is still added to the hydrosphere by volcanoes and by the slow compaction of sedimentary strata. As the subtraction and addition of water appear to balance out, we can say that the hydrosphere has now reached a steady state.

Where did the first dissolved salts in the ocean come from, and has the ocean always been salty? The best evidence of the sea's saltiness in the past is the presence, in marine strata, of salts precipitated by evaporating, isolated bodies of seawater. Strata containing marine evaporites are common in young sedimentary basins, but are not known from rocks older than about a billion years. This happens mainly because the soluble evaporite minerals dissolve in ground water and pass rapidly through the rock cycle. Probably evaporites existed long ago but have since been destroyed. We can be sure, therefore, that the sea has been salty for a billion years, and we suspect but cannot yet prove that it has always been salty.

# MOVEMENTS OF SEAWATER

The restless sea is always in motion, at generally slow rates to be sure, though in places at velocities comparable with swift rivers. Although there are several immediate causes of motion, the chief motive power is solar energy, with some help from the energy of Earth's rotation and from the gravitative pull of Sun and Moon. We can group the several kinds of currents and waves in this way:

A. Geologically important along coasts:

1. Waves
2. Longshore currents
3. Tidal currents

B. Geologically important in the deep oceans:

4. Surface ocean currents
5. Density currents

### Waves and longshore currents

Waves and longshore currents, described in Chapter 11, are caused by wind blowing across seawater. They are the chief agents by which the sea erodes a coast and deposits sediment along the shore.

### Tidal currents

Tidal currents are caused by the twice-daily tidal bulges that pass around the Earth (Chapter 2). In the open sea, tides are small and do not cause currents, but in bays, straits, estuaries, and other narrow places, tides can cause rapid currents. As a tidal bulge in the open ocean approaches a restricted inlet, it becomes confined and the water level rises rapidly. Tidal flows reach 25km per hour in places, and tidal heights of more than 16m are known (Fig. 13.1). Such fast-moving currents, though restricted in extent, readily move sediment around.

### Surface ocean currents

***Surface ocean currents*** are *broad, slow drifts of surface water.* They are set in motion by the prevailing surface winds. Air that flows across a water surface causes waves, but it also drags the water slowly forward, creating a current of water as broad as the current of air but rarely more than 50 to 100m deep. The marked effect of winds on the ocean is evident when we compare a map of surface ocean currents (Fig. 13.2) with the positions of the belts of prevailing winds (Fig. 9.2). In low latitudes surface seawater moves westward with the trade winds. The generally westerly direction of the North and South Equatorial Currents (Fig. 13.2) is reinforced by Earth's rotation. Both north and south of the equator, the westerly moving currents eventually become movements of rotation, caused partly by deflection where a current encounters a coast, and partly by the Coriolis force (Chapter 9).

A

B

**Figure 13.1**
**The Bay of Fundy, between Nova Scotia and New Brunswick, experiences an extreme tidal range and rapid tidal currents.**
***A.* Hall's Harbor, Nova Scotia, at high tide.**
***B.* The same place at low tide. (*Russ Kinne, Photo Researchers.*)**

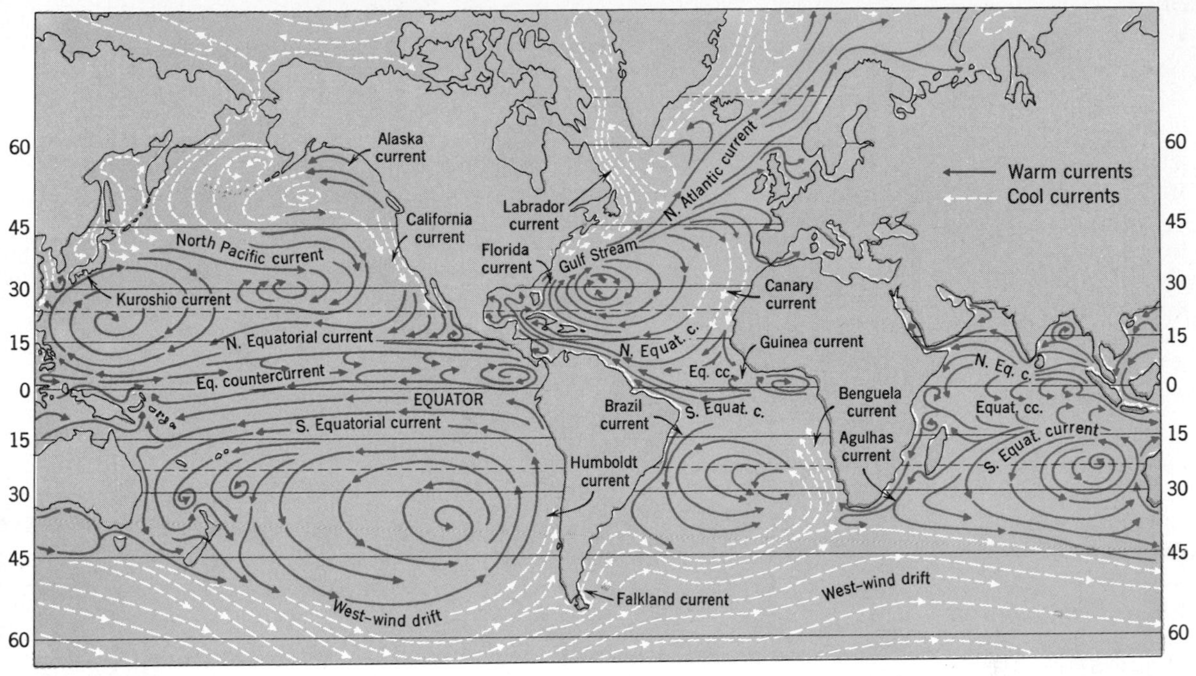


**Figure 13.2**
**Surface ocean currents form a distinctive pattern, curving to the right in the Northern Hemisphere and to the left in the Southern.** (*After a map by U.S. Navy Hydrographic Office.*)

In the North Atlantic, then, the North Equatorial Current flows west. Deflected water flows north and is represented by the Florida Current the Gulf Stream, and the North Atlantic Current. The currents transfer warm equatorial waters to higher latitudes and therefore have considerable effect on climates. Part of the North Atlantic Current eventually moves into the Arctic Ocean, taking heat with it, and part is deflected south along the European–African coast as the Canary Current. This water has by then lost so much heat that it has become cooler than the surrounding tropical water. Approaching the equator, it has completed its circulation and once more begins to be dragged westward by the trade winds.

We can follow a similar pattern in the northern Pacific Ocean, where we have this sequence: North Equatorial Current, Kuroshio Current, North Pacific Current, and the cool California Current. In the Southern Hemisphere we find similar great circular movements of surface seawater, but the direction of rotation is counterclockwise, while north of the equator it is clockwise. There is another major difference between circulations north and south of the equator. The far-southern oceans are not impeded by continents; this makes possible a major globe-circling movement of water, the West-wind Drift. This moves water from one ocean to another, mixing their waters, a process that is completed every 1,800 years.

Although rates of movement are generally slow, in narrowly confined areas surface ocean currents become rapid. In the narrow strait between Florida and Cuba the rate approaches 5km per hour.

### Density currents and circulation in the deep ocean

Throughout the ocean, beneath the great currents set up by winds in the shallow surface zone, deeper—and much slower—circulation

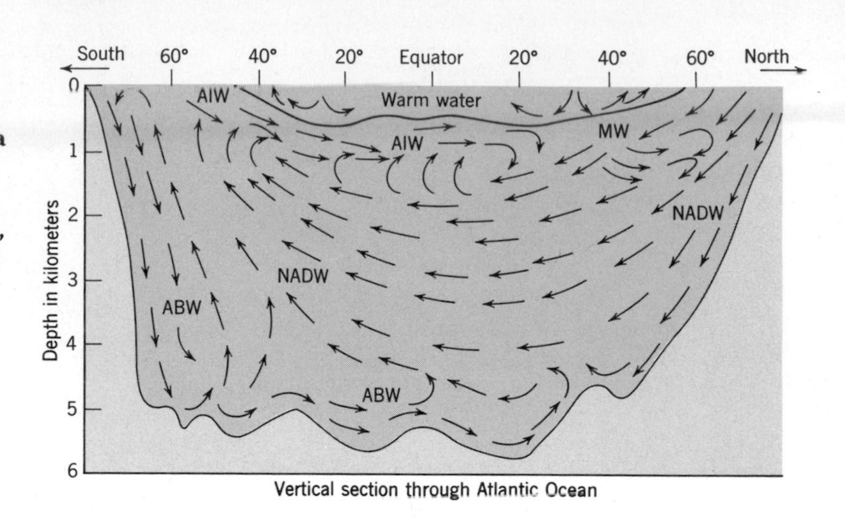


**Figure 13.3**
**Circulation of deep water in the Atlantic Ocean, shown along a north–south plot.**
**ABW is Antarctic Bottom Water, dense, cold water that sinks in the Weddell Sea and slowly flows north to form the cold bottom water throughout most of the Atlantic Ocean.**
**NADW is North Atlantic Deep Water, which sinks in the North Atlantic off Greenland, and eventually flows far south, where it meets Antarctic Intermediate Water (AIW), cold, near-surface water flowing north.**
**MW is dense, saline water flowing into the Atlantic from the Mediterranean Sea.**
**(*K. K. Turekian, 1968.*)**

of water is also occurring. It is caused by differences in density. A ***density current*** *is a localized current, within a body of water, caused by dense water sinking through less-dense water.* When seawater gets colder or becomes saline, its density increases. Dense water tends to sink, displacing less-dense water below.

In polar regions surface water becomes chilled by the cold atmosphere. Also it acquires increased salinity by freezing to form sea ice. As the ice (nearly pure $H_2O$) forms, the salts remain in solution in the residual seawater, which becomes more saline. As polar water grows denser by cooling and formation of sea ice, it sinks, then slides slowly along the sea floor toward the tropics. Deep, cold polar water may even cross the equator into the opposite hemisphere.

In the Atlantic Ocean four clearly defined sources of cold bottom waters have been identified (Fig. 13.3). (1) Water in the Gulf Stream and North Atlantic Current is highly saline as a result of evaporation in low latitudes. Being saltier than most seawater, when it reaches the Arctic region and becomes chilled, it sinks to the bottom. This cold North Atlantic Deep Water (NADW in Fig. 13.3) again flows south, almost into the Antarctic region, before it is obscured by mixing. (2) In the Antarctic region, water is chilled during the winter, and at the same time becomes saltier through the formation of sea ice. The cold, dense Antarctic Bottom Water (ABW) sinks to the ocean floor and flows northward, crossing the equator and reaching intermediate northern latitudes before being displaced by the cold North Atlantic water. (3) Antarctic Intermediate Water (AIW) forms by chilling of the highly saline waters of the Brazil Current. (4) The Mediterranean Sea is an enclosed basin with rapid evaporation. Mediterranean surface water becomes very saline, sinks and eventually flows, as a deep current (MW), through the Strait of Gibraltar and down into the Atlantic. To counterbalance the flow of deep water out of the Mediterranean, a fast-moving surface current flows in through the

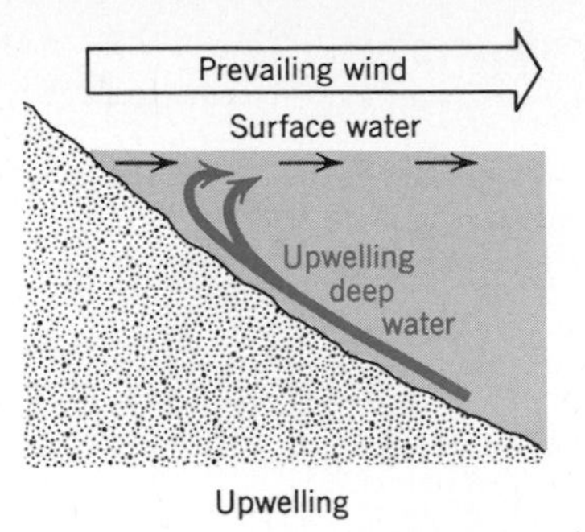


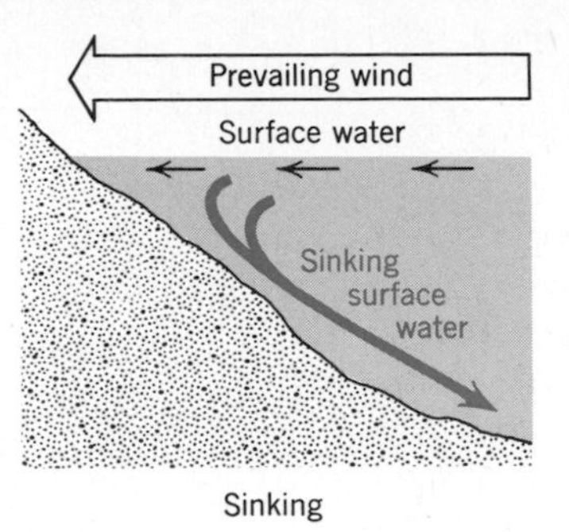


**Figure 13.4**
**Cross section of a coast, showing the effect of wind on vertical movement of seawater.**
**The prevailing wind blows surface water away from the coast, causing deep water to well up. The opposite effect—sinking—happens when the prevailing wind blows toward shore.**

Strait of Gibraltar. So, because of differences in density between two great water bodies, the Strait of Gibraltar carries two currents, one above the other, flowing in opposite directions.

Deep circulation in the Pacific and Indian Oceans differs from Atlantic circulation in that all the deep water comes from the Antarctic region. In the northern Pacific, there is no large source of deep, cold water because a shallow barrier at the Bering Strait prevents deep Arctic water from breaking through. It is possible for deep water originating in Antarctica to flow as far north as California and Japan.

### Upwelling and sinking

Prevailing winds, blowing offshore or onshore, cause vertical movements of seawater (Fig. 13.4). Upwelling is caused by offshore winds. It is the principal way in which cold, deep water is brought to the surface. Sinking is the opposite effect and is caused by onshore winds. Deep water is colder than surface water. When it wells up and reaches the surface it cools the air and creates fog. This is the source of the great summer fog banks along the Pacific coast of North and South America. Deep water also tends to be rich in nutrients such as phosphorus and nitrogen. When it reaches the sunny surface where organisms can thrive, microscopic plant life blooms in abundance and fish populations that feed on the plants expand too. Some of the world's greatest fishing fields, such as those off the coast of Peru, are in areas of upwelling.

## TOPOGRAPHY OF OCEAN BASINS

As we sail over the sea or fly above the broad oceanic expanses, we receive few hints of the complex sea-floor topography beneath. Yet the topography is just as varied, irregular, and fascinating as the familiar land topography we see around us. Present on the sea floor are long mountain chains, valleys and canyons, featureless plains, great escarpments, and steep-sided volcanoes (see back endpaper). If we compare a topographic profile across the Atlantic Ocean basin with one across North America (Fig. 13.5), we see that submarine terrain can be just as rugged as that on land and that relief can be even greater (see back end paper).

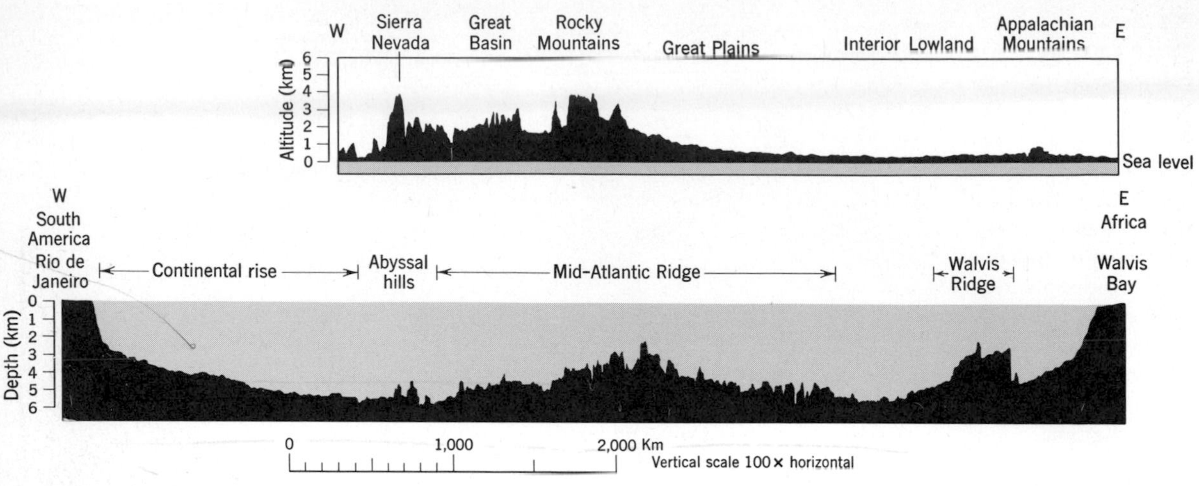


**Figure 13.5**
**Accurate profiles show that the topography of the Atlantic Ocean basin is fully as rugged as the surface of North America. The horizontal scale is, of course, compressed and the vertical scale exaggerated to emphasize the topography. (*After Shepard, 1963.*)**

## Oceanic ridge

The most striking feature of the submarine profile in Figure 13.5 is the Mid-Atlantic Ridge. This great mountain chain would be one of the most eye-catching features we would see if we could view a dry Earth from a space ship. The ridge dominates the artistic view in Figure 13.6 and is one segment of the globe-encircling ***oceanic ridge,*** *a continuous rocky ridge on the ocean floor, many hundred to a few thousand kilometers wide, with a relief of more than 600m.* The oceanic ridge, also known by the equivalent names *mid-ocean ridge* and *oceanic rise,* is a chain of mountains 60,000 kilometers long, that twists and branches in a complex pattern through the ocean basins, and that marks the spreading edges of the moving plates of lithosphere shown in Figure 4.4.

Figure 13.6 shows many prominent features. Perpendicular to the ridge is a parallel series of long, straight valleys, gashes or fractures in the ocean floor. These are the transform faults shown in Figure 4.4 and discussed in Chapter 14. Looking at the ridge itself, we see it consists of a system of ridges paralleling a narrow,

**Figure 13.6**
**(See opposite page)**
**This artist's view shows the features we would see in the Atlantic Ocean basin if all the water were removed. Slopes appear steeper than they actually are because the vertical scale has been exaggerated for emphasis.**

**Along the center of the basin is the Mid-Atlantic Ridge, a great chain of volcanic mountains, in places broken and offset by huge fractures. The fractures are zones of intense breaking of the crust; they form steep-walled valleys. Away from the ridge lie seamounts, a few of which reach up above sea level and form islands. Farther away are abyssal plains, smooth-floored parts of the ocean floors that are bounded by the continental slopes and nearly flat continental shelves. The small square in the center of the ridge marks the site of project FAMOUS, an international study effort in 1973 and 1974 in which teams of divers studied the ridge and its central rift from deep-diving submarines. (*From a painting by Heinrich Berann; courtesy Aluminum Company of America.*)**

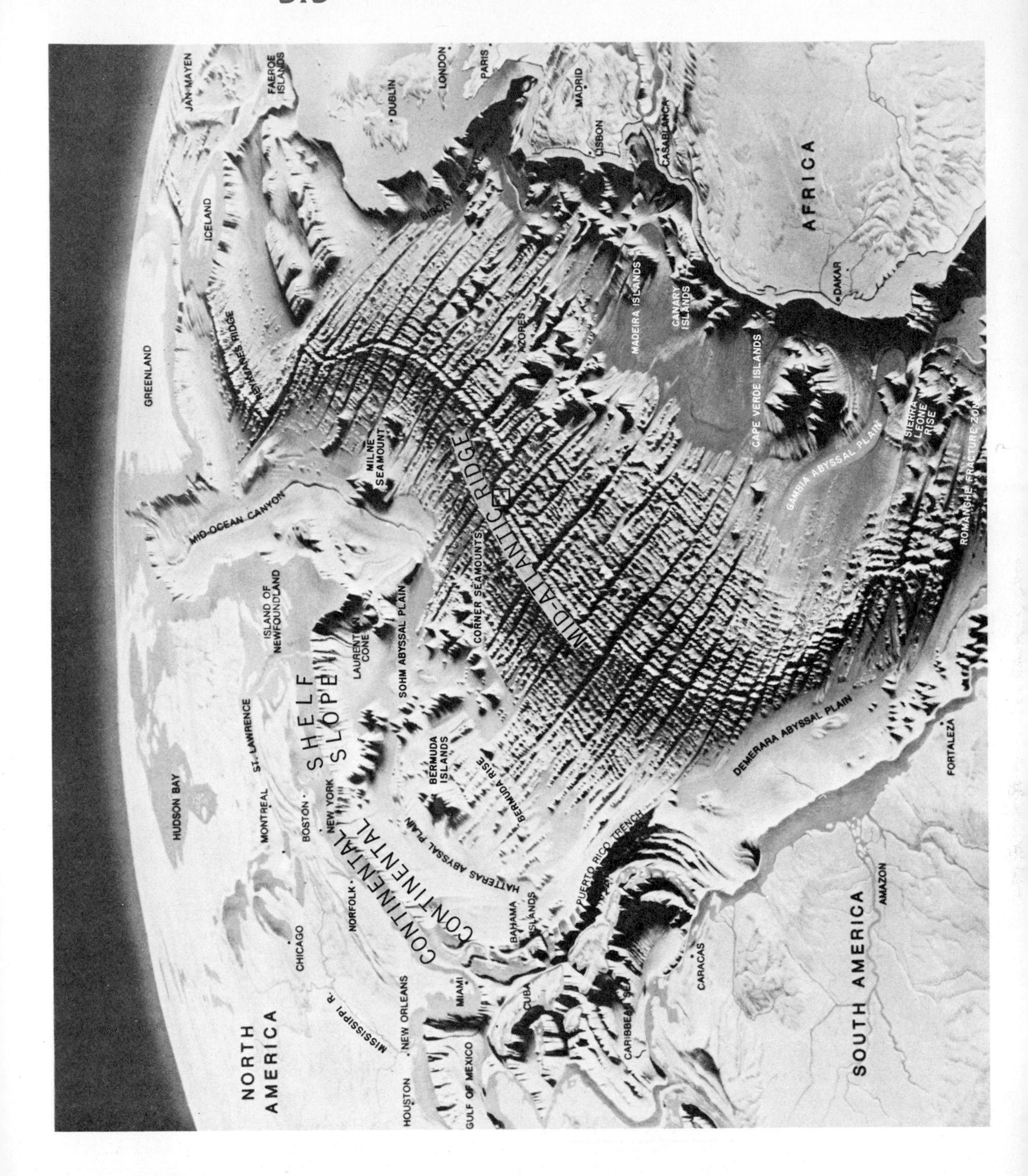
JAN MAYEN
FAEROE ISLANDS
LONDON
PARIS
DUBLIN
MADRID
LISBON
CASABLANCA
AFRICA
ICELAND
BISCAY ABYSSAL
GREENLAND
REYKJANES RIDGE
AZORES
MADEIRA ISLANDS
CANARY ISLANDS
DAKAR
CAPE VERDE ISLANDS
MILNE SEAMOUNT
MID-ATLANTIC RIDGE
GAMBIA ABYSSAL PLAIN
SIERRA LEONE RISE
ROMANCHE FRACTURE ZONE
MID-OCEAN CANYON
ISLAND OF NEWFOUNDLAND
CORNER SEAMOUNTS
LAURENTIAN CONE
SOHM ABYSSAL PLAIN
HUDSON BAY
ST. LAWRENCE
SHELF
SLOPE
MONTREAL
BOSTON
NEW YORK
BERMUDA ISLANDS
BERMUDA RISE
CONTINENTAL
CONTINENTAL
HATTERAS ABYSSAL PLAIN
PUERTO RICO TRENCH
DEMERARA ABYSSAL PLAIN
FORTALEZA
CHICAGO
NORFOLK
BAHAMA ISLANDS
NORTH AMERICA
MISSISSIPPI R.
NEW ORLEANS
MIAMI
CUBA
CARIBBEAN SEA
CARACAS
AMAZON
SOUTH AMERICA
HOUSTON
GULF OF MEXICO

**Figure 13.7**
**Rift valley at Thingvellir, Iceland.**
**The center of the Mid-Atlantic Ridge bisects Iceland. The two halves of Iceland are moving apart from each other, creating a region of tension marked by long parallel fractures. The two outermost fractures, here visible in the lower-left and upper-right corners of the photograph, define the limits of the rift valley. (*Iceland Photo and Press Service.*)**

steep-walled valley or rift that runs down the center of the ridge. During 1973 and 1974 this rift was the object of intense international study. A portion of the rift in the small area outlined in Figure 13.6 was examined by French and American scientists from deep-diving submarines. They found that the rift is bounded

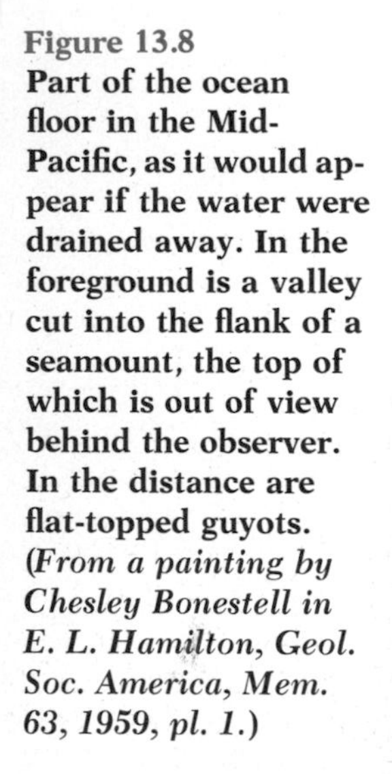

**Figure 13.8**
**Part of the ocean floor in the Mid-Pacific, as it would appear if the water were drained away. In the foreground is a valley cut into the flank of a seamount, the top of which is out of view behind the observer. In the distance are flat-topped guyots. (*From a painting by Chesley Bonestell in E. L. Hamilton, Geol. Soc. America, Mem. 63, 1959, pl. 1.*)**

by steep walls and broken by long, parallel fractures in the sea floor. The central rift is also the site of intense submarine volcanism as magma wells up spasmodically from the mantle to fill the constantly forming rift. The rift is continually renewed because it is the gap between two plates of lithosphere that move continually apart from each other.

At several places around the world the oceanic ridge reaches sea level and forms oceanic islands. Most famous of these is Iceland, which is split by the center of the Mid-Atlantic Ridge; there we can see part of a rift valley on land (Fig. 13.7).

### Seamounts and oceanic islands

Looking again at Figure 13.6, we see that the ocean floor is dotted with many steep-sided mountains, some alone, some in groups. Where these mountains reach the surface of the sea, as in the Azores, they form volcanic islands. Where a submarine mountain does not reach the surface, but is *an isolated volcanic hill standing more than 1,000m above the sea floor,* we call it a ***seamount.***

*A seamount with a conspicuously flat top well below sea level* is a ***guyot*** (Fig. 13.8) thought to be a seamount with its top eroded by

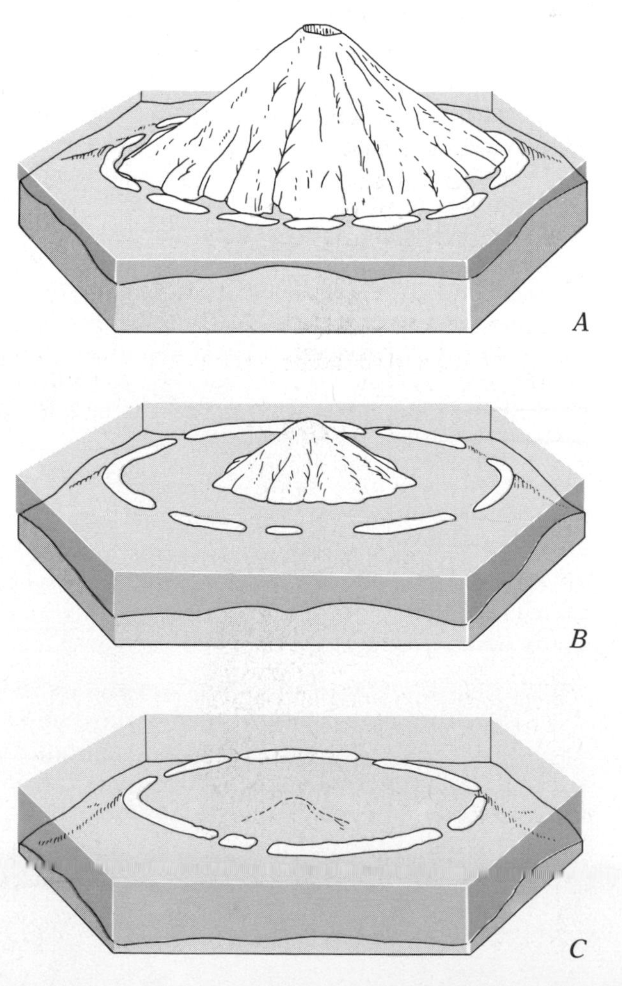


Figure 13.9
**Chief kinds of coral reefs.**
***A*. Fringing reef, attached to the shore of a land mass.**
***B*. Barrier reef, formed offshore from a land mass.**
***C*. Atoll. A reef that forms a nearly closed figure within which there is no land mass.**

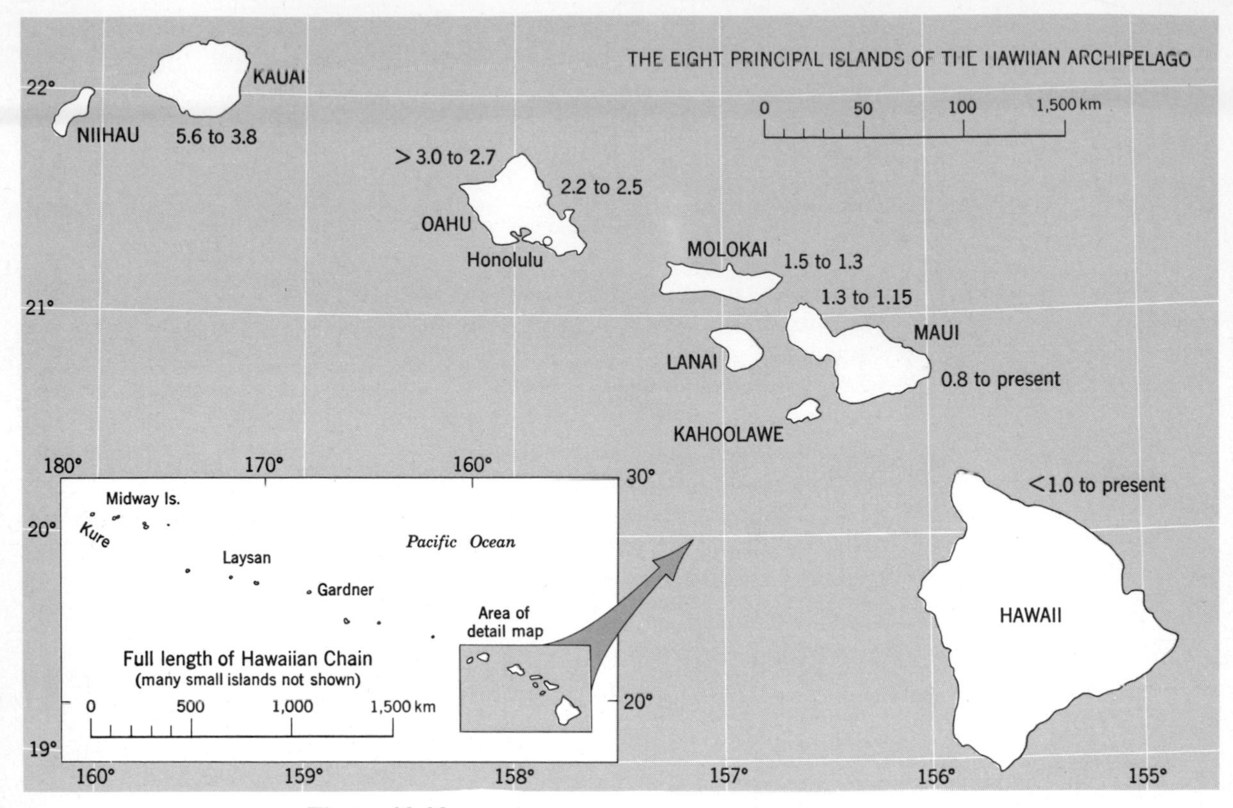


**Figure 13.10**
**In the Hawaiian Archipelago, volcanism has ceased on all islands except the two most southeasterly ones, Maui and Hawaii. Activity on Maui is very infrequent. On Hawaii, three volcanoes are still alive. The tops of the most northwesterly islands, such as Midway and Laysan, have been eroded to positions below sea level and are now crowned by coral atolls. Beyond Midway are many seamounts that have now sunk below the sea. Numbers beside larger islands are K/Ar dates (in millions of years) of basalts that form the cones. Evidently volcanism has moved steadily from northwest to southeast, each island apparently taking little more than one million years to grow from the sea floor to its ultimate height as the Pacific Plate moved over what appears to be a hot place in the mantle below. (*Data from Ian McDougall, 1963.*)**

surf; in other words, the flat tops are wave-cut benches. Because most of the tops are now covered by at least 1km of water, they must have sunk below sea/level after they were eroded. Despite their great depth, the tops of some guyots yield shallow-water shells when dredges are dragged across them.

Support for the idea that guyots are sunken islands comes from three different types of *coral reef, a ridge of limestone built by colonial marine organisms.* Coral reefs abound in tropical seas; they are built by vast colonies of tiny colonial organisms that secrete calcium carbonate. Because of temperature, light, and oxygen requirements of the coral organisms, reefs are built only at or close to sea level. The different kinds of reef are depicted in Figure 13.9.

It is not difficult to see that if a volcanic island slowly subsided, its fringing reef would have to grow upward so that the organisms could continue to live at sea level. Eventually the island would disappear, leaving an atoll. This is believed to be the origin of many atolls in the Southwest Pacific.

More than 2,000 seamounts and guyots have been discovered, mostly in the Pacific, and some oceanographers believe the total will eventually reach as high as 20,000. Where oceanic islands and seamounts have been dated by radiometric means, they often turn out to be younger than the surrounding ocean floor. This means they could not have formed at oceanic ridges and then have been rafted by moving lithosphere to their present sites. Instead, it must mean that volcanism occurs at spots on the sea floor away from the spreading edges of the plates. The discovery that some lines of oceanic islands and seamounts get progressively older (Fig. 13.10) suggests that local hot spots in the mantle continually generate magma. As a plate of lithosphere moves over the hot spot, a succession of oceanic volcanoes forms.

### Sea-floor trench

The greatest depth in the ocean occurs in a ***sea-floor trench,*** *a long narrow, very deep basin in the sea floor.* As much as 200km wide and 25,000km in length, such trenches mark the places where moving plates of lithosphere plunge back into the mantle. Most trenches lie in the Pacific Ocean (Fig. 13.11). Some, such as the Aleutian Trench and the Tonga-Kermadec Trench, are far from continental

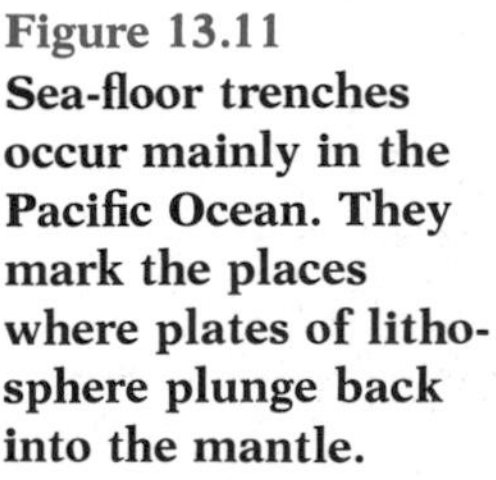

**Figure 13.11**
**Sea-floor trenches occur mainly in the Pacific Ocean. They mark the places where plates of lithosphere plunge back into the mantle.**

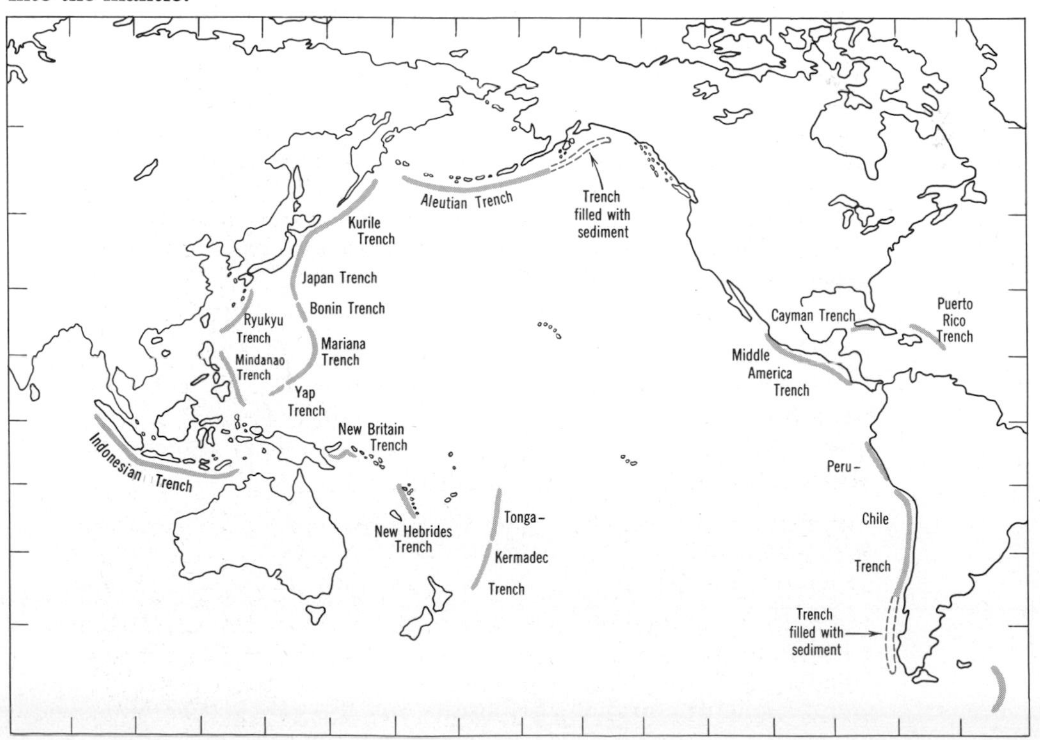

margins. Others, such as the Peru-Chile Trenches, are immediately adjacent to continental margins and in places are filled with sediment from the nearby land mass.

### Abyssal plains

The final major topographic feature we must mention is the ***abyssal plain,*** *a large flat area of deep-sea floor having slopes less than about 1m per km.* Abyssal plains are far removed from oceanic ridges. They tend to be close to land, which is a source of sediment. They occur because the original sea-floor topography has been completely buried beneath a mantle of sediment, much of which has been transported by turbidity currents.

## ROCK OF OCEAN BASINS: OCEANIC CRUST

Beneath the thin veneer of sediment that mantles the ocean floor lies the oceanic crust. Drilling and dredging at sea indicates the crust is everywhere composed of igneous rock.

### Composition

Where oceanic rock can be sampled, as in Iceland and in dredged samples, it is usually found to be basalt. It commonly displays a unique pattern called *pillow structure* or *pillow lava* (Fig. 13.12). Pillow structure forms when basaltic magma is extruded under water, and the surface of the lava is quickly chilled. The brittle, chilled surface cracks, making an opening for the still-molten magma inside to ooze out like a strip of toothpaste. The newly oozed strip chills in turn, its surface cracks, and the process continues. The end result is a pile of lava pillows, each with a quickly chilled, glassy skin that resembles a jumbled pile of sand bags.

Because oceanic crust is many kilometers in thickness, we cannot see whether it is all made of pillow basalt, nor can we sample its structure entirely by drilling and dredging. But nature has provided a few samples of old oceanic crust that we can examine on land. They exist because two masses of continental crust, each riding on its own plate of moving lithosphere, have collided with each other. During the collision, fragments of oceanic crust were broken off and caught up in the crumpled edges of continental crust.

The best preserved *fragment of ancient oceanic crust,* called an ***ophiolite complex,*** is exposed on the island of Cyprus. A generalized diagram of an ophiolite complex is shown in Figure 13.13. At the top is a thin veneer of sediment deposited after the igneous activity had ceased. Beneath the sediment are layers of basaltic pillow lavas; beneath the lavas are many sills consisting of gabbro. Cutting through the basalts and the gabbros are thousands of vertical dikes of diabase. Gabbro, diabase, and basalt have the

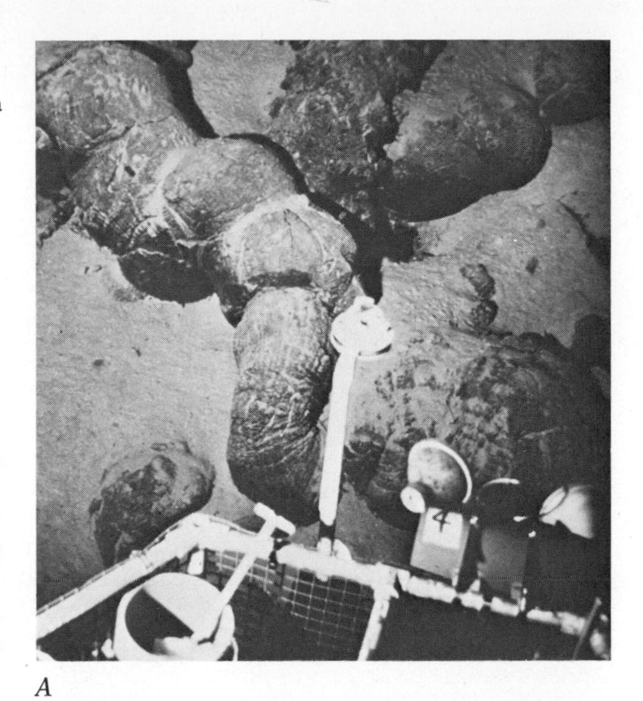

A

B

**Figure 13.12**
**Pillow lava is a characteristic form of lava extruded beneath the sea.**
*A*. **Tubular shaped pillow of basalt photographed in 1974 in the central rift of the Mid-Atlantic Ridge, at a water depth of 2,900m, by scientists of project FAMOUS. Equipment in foreground is used for collecting samples.** (*J. R. Heirtzler, Woods Hole Oceanographic Instn.*)
*B*. **Spectacular pile of basalt pillow lavas exposed in the Sultanate of Oman, where an ophiolite complex (Fig. 13.13) has been thrust up from an ancient sea floor.** (*E. H. Bailey, U.S. Geological Survey.*)

same composition and presumably formed from identical magma.

Beneath the gabbros is rock of quite different composition, rock such as peridotite, that is characteristic of the upper mantle. The whole array of rocks therefore includes not only a sample of oceanic crust, but also a small sample of the upper mantle as well.

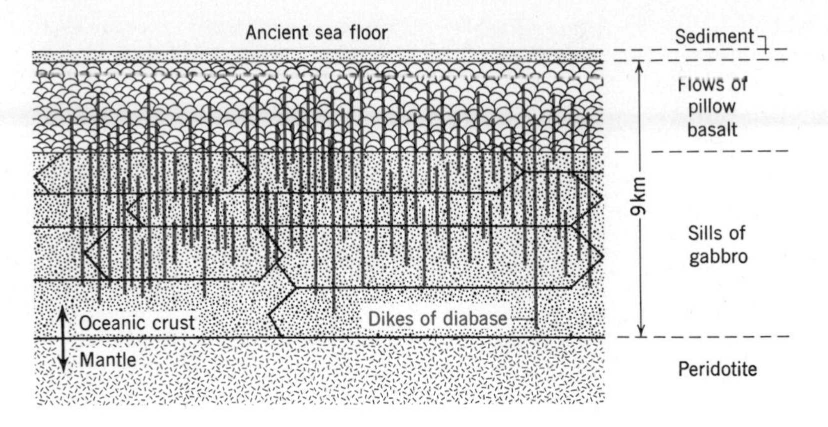


**Figure 13.13 Idealized section through an ophiolite complex.**
**Once part of the sea floor, ophiolite complexes are believed to have formed at an oceanic ridge, moved away from the ridge and eventually been preserved for examination when part of the sea floor was caught in a collision between continents moving on their rafts of lithosphere. Ophiolite complexes consist of flows of pillow basalts overlying a thick pile of gabbro sills. Both are intruded by a vast number of nearly vertical, diabase dikes (blue). The boundary between oceanic crust and the mantle below is the place where gabbro comes into contact with peridotite.** ***(Adapted from data by Moores and Vine, 1971.)***

## Origin

How does oceanic crust form? Beneath the oceanic ridge and deep in the mantle, hot solid material is believed to be slowly rising upward. Whether or not this is what happens, and how and why it does happen, is one of Earth's still-unsolved mysteries. And equally unresolved is the question of how rising masses beneath the oceanic ridge are coupled with movement of lithosphere away from the ridge. Regardless of how it happens, near the top of the mantle a small portion of the rising mass melts to form the basaltic magma from which the oceanic crust forms. Some of the unmelted residue is visible in the peridotite of the upper mantle.

When the newly formed magma rises and reaches the sea floor, it forms layers of pillow basalt. The magma that does not reach the sea floor is intruded as sills and solidifies as gabbro. Finally, as the newly formed crust moves outward away from the ridge, the movement causes deep, vertical fractures. Into these the dikes of diabase are intruded.

Measurements show that flow of heat out of the Earth is greater along the oceanic ridge than elsewhere. Indeed the entire central region of the ridge is comprised of hot rock; perhaps this is what makes the oceanic ridge stand up as a mountain range: the great heat simply expands the rock. Then, as the lithosphere moves sideways, the hot rock slowly cools, contracts, and subsides back to the general level of the deep-sea floor.

## Age of oceanic crust: earth's magnetism

The means of dating the oceanic crust came indirectly out of measurements of the magnetism of rock. The results are surprising because they indicate that all of the oceanic crust is very young.

The magnetism of rock is derived from magnetism of the Earth as a whole. The exact cause of Earth's magnetism is not known, but is believed to be caused by motions in the liquid iron of the outer core. Those motions, in turn, are caused by Earth's rotation; they make the core act like a huge dynamo. The dynamo sets up a magnetic field that has two poles—one pointing north, one south—just as if there were a huge bar magnet buried inside the

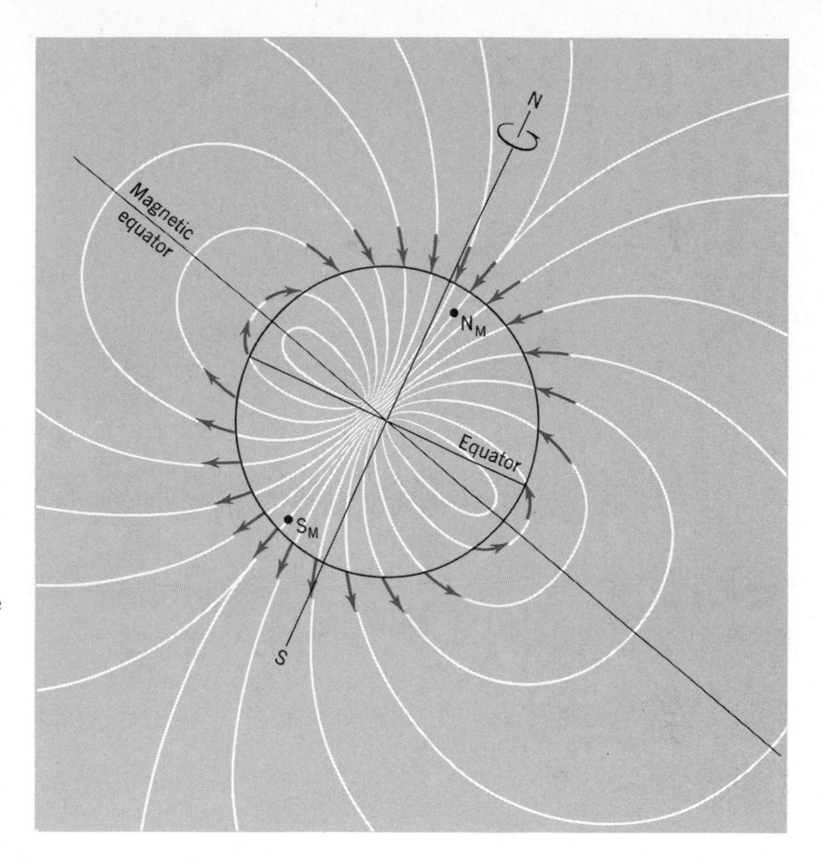


**Figure 13.14** ***The magnetic line of force surrounding the Earth*** **(white) define the** ***magnetic field.*** **A free-swinging magnetic needle would point along the nearest line of force, with the north-seeking pole in the direction of the arrows.** ***The angle with the horizontal assumed by a magnetic needle*** **is the** ***magnetic inclination. The clockwise angle from true north assumed by a magnetic needle*** **is the** ***magnetic declination.*** **The axis of the magnetic field does not coincide exactly with the axis of Earth's rotation. Where the axis of the magnetic field intersects Earth's surface, a magnetic needle stands vertical and we define the points as the north and south magnetic poles. The north magnetic pole ($N_M$) lies in the arctic islands of Canada; the south magnetic pole ($S_M$) lies in Antarctica, south of Tasmania.**

Earth. The north and south magnetic poles do not coincide exactly with the north and south poles of Earth's axis of rotation (Fig. 13.14). There is no reason why fluid motions should create a magnetic axis coinciding exactly with the Earth's axis of rotation, but the two axes should be close; so the observed lack of perfect coincidence is believed to be evidence in support of the dynamo theory. Nor is there any reason why the fluid motions should be constant, and therefore why the magnetic field we measure at the surface should be constant. One of the many variations that occur in the magnetic field is that it dies down at irregular intervals then starts up again but with its poles reversed.

Pole reversal has not happened during historic times, but evidence of many reversals of the magnetic field is locked up in rocks. A few minerals such as magnetite, ilmenite, and pyrrhotite are natural magnets. If all the tiny mineral magnets in a rock are fixed in space randomly, their effects cancel each other out, so that the rock as a whole is not magnetic. However, if all or most of the tiny magnets are parallel, their effects reinforce each other; this makes the rock magnetic.

The manner in which minerals develop a magnetic orientation depends on the kind of rock in which the minerals occur. Each magnetic mineral has a ***Curie point,*** *a temperature above which all magnetism is destroyed.* As an igneous rock cools below the Curie

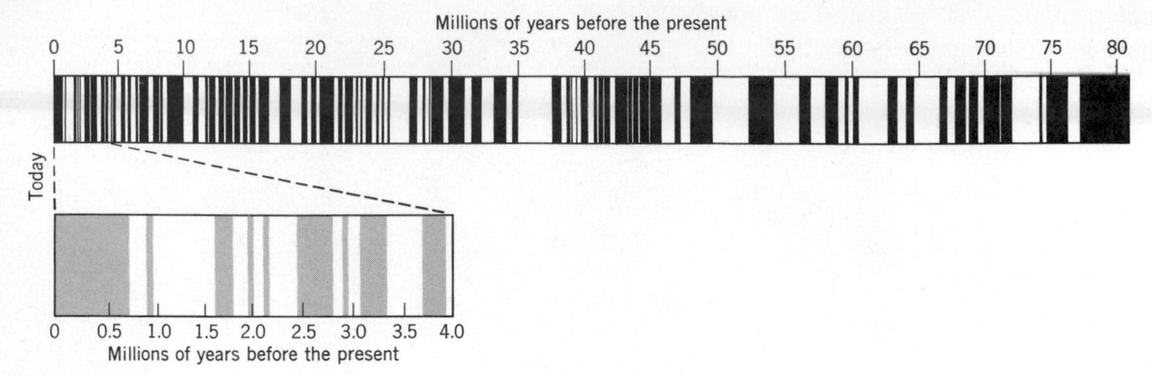


**Figure 13.15**
**During the past 80 million years, Earth's magnetic field has experienced 78 periods when the polarity was exactly the reverse (white) of the present polarity (black). Frequency of reversal has been highly irregular. Although the average is approximately one reversal every million years, there have been 9 periods of reversed polarity in the last 4 million years. (*After Cox, 1969, and Berggren, Opdyke and Watkins, 1975.*)**

point, any mineral magnets that are present acquire the magnetic orientation of Earth's field. The igneous rock thereby becomes weakly magnetic. When a sediment forms, an entirely different process occurs. Grains of sediment that are magnetic do the same as a compass needle: they tend to orient themselves parallel to Earth's field as they settle through water. The oriented-mineral magnets eventually convert the sedimentary rock as a whole layer into a weak magnet.

Although this magnetic effect in rocks is very weak, it can be measured accurately. Also, rock magnetism is stable; it normally does not tend to change with time. If, therefore, we collect a sample of igneous or sedimentary rock and determine the direction of its magnetism, we have an indication of the positions of Earth's magnetic poles at the time the rock formed. By careful radiometric and fossil dating of rocks whose magnetic orientation has been measured, we get an accurate record of the times when the magnetic field had the present, or normal orientation, and when it was reversed. In strata deposited during the last 80 million years, evidence of 78 magnetic reversals has been found. The irregularity of the 9 reversals that have occurred within the last 4 million years, (Fig. 13.15) is typical of the pattern they make.

When lava is extruded at the oceanic ridge, or when gabbro and diabase are intruded, they become magnetized and acquire the magnetic polarity that existed at the time they cooled through the

**Figure 13.16**
**Schematic diagram of oceanic crust. Lava extruded along an oceanic ridge forms new oceanic crust. As lava cools, it becomes magnetized with the polarity of Earth's field. Successive strips of oceanic crust have alternate normal polarity (black) and reversed polarity (white).**

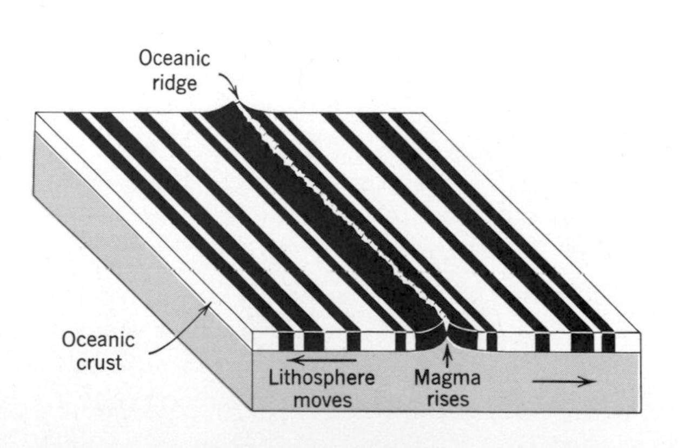

Right & left halves mirror images of each other.

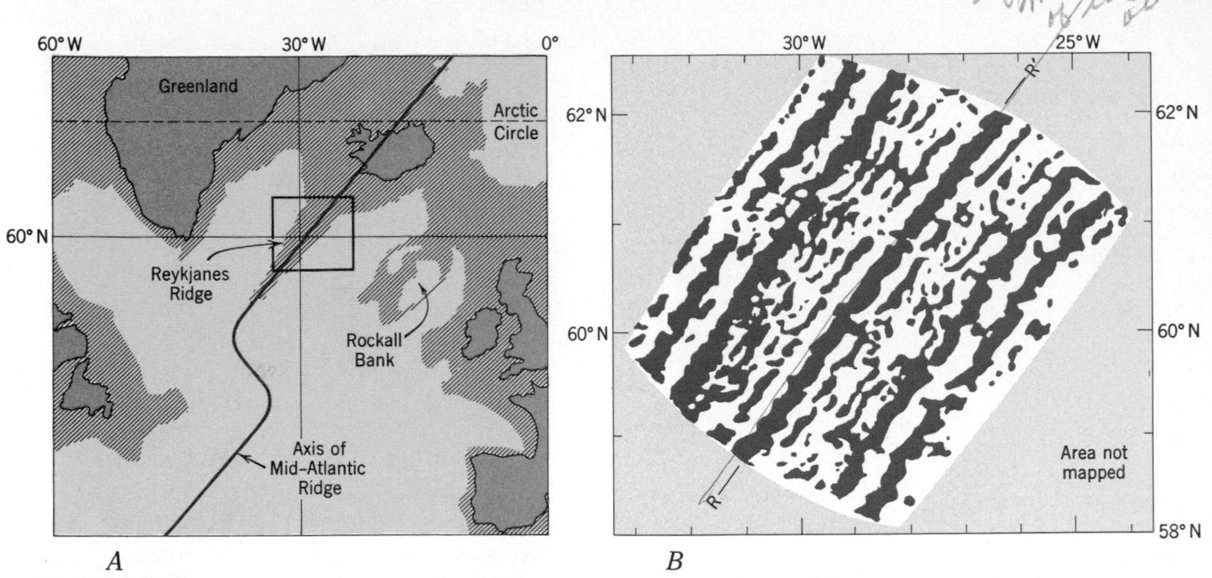


**Figure 13.17**

**A. Index map showing location of Reykjanes Ridge, a portion of the Mid-Atlantic Ridge southwest of Iceland.**

**B. Map of the magnetic striping of rock on the sea floor. RR′ is the center line of Reykjanes Ridge. Strips of rock with normal polarization (black) alternate with reversely polarized rock (white). (*After Heirtzler, Le Pichon and Baron, 1966.*)**

Curie point. Because new lava, new gabbro, and new diabase are forming and are continually moving away from the oceanic ridge, the oceanic crust contains a continuous record of Earth's magnetic polarity. The crust is, in effect, a very slowly moving magnetic tape recorder in which successive strips of oceanic crust are magnetized with normal and reversed polarity (Fig. 13.16). Sea-floor magnetism can easily be measured with instruments carried in ships or airplanes. An example of the results is given in Figure 13.17. It is a simple matter to match the sort of pattern observed in Figure 13.17*B* with the record of magnetic polarity, such as that

**Figure 13.18**

**Age of the ocean floor in the central, North Atlantic, deduced from magnetic striping, increases regularly away from the axis of the Mid-Atlantic Ridge. Numbers give ages in millions of years before the present. The Kane Fracture Zone, observed near the center of the oceanic ridge, seems to continue across the Atlantic and to cause a consistent offsetting of the age contours. (*After Pitman and Talwani, 1972.*)**

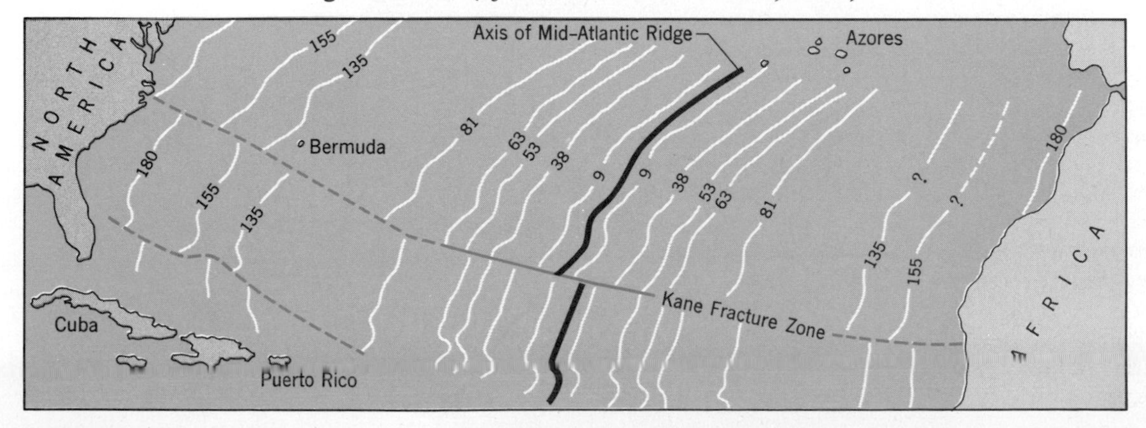

shown in Figure 13.15. The distinctive magnetic striping allows the age of any place on the sea floor to be determined.

The entire crust beneath the world's oceans is young. Within it, no rock older than Jurassic has been found. The pattern of the increasing age of oceanic crust away from the oceanic ridge is very distinctive (Fig. 13.18). It leads directly to the question: how is old sea floor destroyed? We must defer the answer to Chapter 17, because it involves deformation of solid rock and melting of rock to form new magma. These are vital parts of Earth's internal processes and are discussed in the next three chapters.

# SUMMARY

1. Oceans cover 71 percent of Earth's surface. Beneath them lies a topography as rugged and diverse as the topography of the continents.

2. The volume of Earth's seawater is nearly constant, but fluctuates slightly as the amount of glacier ice on land changes. The total amount of water in the hydrosphere is in a steady state.

3. Salts have been in the sea for as long as seas have existed on Earth. They are added continually to the sea by rivers and streams, which derive them from chemical weathering. Salts are extracted by a variety of processes; so the composition of the sea is in a steady state.

4. Surface seawater circulates as currents in a number of huge, circular cells that rotate clockwise in the Northern Hemisphere, counterclockwise in the Southern. Surface ocean currents are driven by winds and move warm equatorial water toward polar regions.

5. Ocean circulation takes the form of density currents set up by salty water with increased density caused by chilling or evaporation. The water sinks in polar regions and moves slowly toward, or even across, the equator.

6. Topographic features of the deep-sea floor include submarine valleys and canyons, sea-floor trenches, abyssal plains, seamounts, and guyots.

7. Probably most seamounts are volcanic cones. Guyots and some coral reefs indicate subsidence of the crust beneath the sea floor.

8. Oceanic ridges occupy all the ocean basins as a world-circling, sinuous chain of sea-floor mountains. Oceanic ridges are the lines along which magma from the mantle adds new rock to the edges of growing plates of lithosphere. Trenches are the places where older parts of the lithosphere plunge back into the mantle.

9. Ocean basins are underlain everywhere by oceanic crust that is basaltic in composition.

10. Earth's magnetic field magnetizes rocks and provides a record of changes in the polarity of the magnetic field. At times in the past, the field has been alternately reversed and normal (as it is today).

11. Magnetism in rocks of the oceanic crust produces a distinctive pattern of normal and reversed polarity. The pattern can be used to determine the age of all parts of the ocean floor.

12. The ocean floor is young. No part of the present-day ocean floor has been shown to be older than Jurassic.

# SELECTED REFERENCES

Bullard, Edward, and Cann, J. R., 1971, A discussion on the petrology of igneous and metamorphic rocks from the ocean floor: Royal Society London, Phil. Trans., ser. A, vol. 268, p. 381–745. (A series of 27 papers presented at a conference held in 1969.)

Gross, M. G., 1972, Oceanography. A view of the earth: Englewood Cliffs, N. J., Prentice-Hall.

Heezen, B. C., and Hollister, C. D., 1971, The face of the deep: New York, Oxford University Press.

Heirtzler, J. R., and Bryan, W. B., 1975, The floor of the Mid-Atlantic Rift: Scientific American, vol. 233, no. 2., p. 79–91.

Hill, M. N., ed., The sea. Ideas and observations on progress in the study of the seas: New York, Interscience Publishers; v. 1, 1962, Physical oceanography, Chap. 5, Section III; v. 2, 1963. The composition of sea water, comparative and descriptive oceanographs, Chaps. 2, 4, 10, 11, 12, 17, 18, and 23; v. 3, 1963. The Earth beneath the sea. History, Chaps. 4, 5, 12, 14, 17, 19, 20, 25, 26, 27, 28, 30, 31, 33, and 34.

King, A. M. C., 1975, Introduction to marine geology and geomorphology: London, Edward Arnold.

Menard, H. W., 1964, Marine geology of the Pacific: New York, McGraw-Hill, Chaps. 1, 2, 7, 8, 9, and 10.

Moore, J. G., 1975, Mechanism of formation of pillow lava: American Scientist, vol. 63, p. 269–277.

Shepard, F. P., 1973, Submarine geology, 3rd ed.: New York, Harper and Row.

Shepard, F. P., and Dill, R. F., 1966, Submarine canyons and other sea valleys: Chicago, Rand-McNally.

Sverdrup, H. V., Johnson, M. W., and Fleming, R. H., 1942, The oceans, their physics, chemistry, and general biology: Englewood Cliffs, N. J., Prentice-Hall.

Turekian, K. K., 1976, Oceans 2nd ed.: Englewood Cliffs, N. J., Prentice-Hall.

**A lake of basaltic lava on the floor of a caldera at the summit of Kilavea Volcano, Hawaii. Out of the lake flow streams of lava, each a few meters wide.** ***(Richard S. Fiske, U.S. Geol. Survey.)***

# PART FOUR
# INTERNAL PROCESSES

# 14 DEFORMATION OF ROCK



## HOW IS ROCK DEFORMED?

The preceding chapters have dealt largely with external activities that we can see and study as they happen. The next three chapters discuss internal activities that we cannot see in progress, such as bending of rock, movement of magma, earthquakes, and making of mountains. To decipher what happens inside the Earth we must use our best detective-like skills.

Internal and external activities just offset each other, and so are in a steady state. This must be true because the rock cycle brings new rock to the surface as fast as old rock is removed by erosion. The balance of activities is indirect evidence that materials inside the Earth must be capable of movement, for without internal movement to counteract erosion, how could continents remain above sea level? Yet Earth is a vast, solid mass. How can movement occur within it?

Solids cannot flow as liquids do. But they can be deformed, so that their shapes change permanently. Change of shape involves movement of matter, and we can therefore say that movement within solids indicates deformation. The shapes of solids can be changed in three different ways: a solid can be (1) fractured and shattered; this has happened to the rock in Figure 2.6, (2) folded and bent; this has happened to the rock in Figure 2.5, and (3) as we can see in Figure 2.4, under some conditions, a solid can apparently flow like soft butter.

### What controls deformation?

Fracturing of rock is a process easy to understand. In ordinary experience rock is strong but brittle. Building stone, for example, keeps its shape under heavy loads but breaks like glass under sharp blows of a hammer, or with very forcible bending.

To understand folding and flowing we must have in mind some elementary properties of solids. Any solid, including rock, can be deformed as if it were a mass of stiff dough—under certain conditions. The essential conditions are (1) confining pressure, (2) temperature, and (3) time. The higher the confining pressure and the higher the temperature, the weaker and less brittle a solid becomes. A rod of iron or of glass is difficult to bend at room temperature. If we try too hard, both will break. But both can be readily bent if we heat them to redness over a flame. The effect of confining pressure (steady squeezing from all directions) is less familiar in common experience. Confining pressure tends to keep a solid mass in the form of a single body and hinders the formation of fractures. At high confining pressures, therefore, it is easier for rock to bend and flow than to break. Weakening of rock by the high temperatures associated with deep burial, and at the same time loss of brittleness caused by high confining pressures, are therefore part of the reason why solid rock can be bent and folded.

The effect of time on deformation of rock is vitally important, but as with confining pressure, it is not obvious from common experience. A force applied to a solid is transmitted by all the constituent atoms of the solid. If the force exceeds the strength of the bonds between atoms, either the atoms must move to relieve the force, or the bonds must break. But atoms in solids cannot move rapidly. So if the force is sharp and quickly applied, the solid breaks. Nevertheless, if the force is applied slowly and is maintained for a long period, the atoms have time to move, and the solid can slowly readjust and change shape by folding and flowing. Forces within the Earth tend to be applied slowly—there has been abundant time for even the strongest and most massive rock to be folded and deformed.

Deformation occurs and movement results. These two events are intimately connected. Let us examine first some of the evidence of movement and deformation in the crust today, and then evidence of deformation in the past.

## MOVEMENT IN PROGRESS

Even though we cannot observe directly events that happen deep inside the Earth, we can sometimes detect movement of the land surface that is caused by deep-seated events. Apparently many movements happen so slowly that they cannot be measured over a few tens or even a few hundreds of years. Other movements can be detected, however, and for convenience we divide them into two groups: *abrupt movements,* in which blocks of the crust suddenly move a few centimeters or a few meters in a matter of minutes or hours; and *gradual movements,* in which slow, steady motions occur without any abrupt jarring.

### Abrupt movement

We call *a fracture along which the opposite sides have been displaced relative to each other* a ***fault.*** Movement along faults tends to be abrupt rather than a slow, steady sliding. Forces build up slowly until friction between the two sides of the fault is overcome. Then abrupt slippage occurs. If the forces persist, the whole cycle of slow buildup, culminating in an abrupt movement, repeats itself. Although the extent of movement on a large fault may eventually total many kilometers, it is the sum of numerous small, sudden slips. Each sudden movement may cause an earthquake and, if the movement occurs near Earth's surface, may disrupt and displace surface features, and in doing so, leave clear evidence of the amount of movement.

During the San Francisco Earthquake of 1906, abrupt horizontal movement occurred along a large fault, the San Andreas Fault. Roads and fences that crossed the fault were offset by as much as 7m. In 1940 another earthquake occurred, again with horizontal movement, along the same fault, this time in the Imperial Valley, nearly 800km southeast of San Francisco. The displacement, as much as 5.5m, was registered accurately by offset rows of fruit

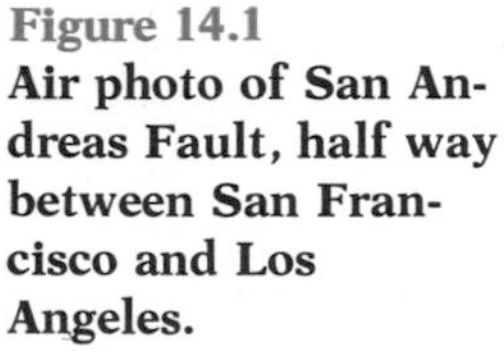

Figure 14.1
**Air photo of San Andreas Fault, half way between San Francisco and Los Angeles.**
**Fault is marked by prominent, straight valley running from bottom to top of photo. Ground to left of fault is moving north (toward top of view) relative to ground on right. Streams flowing from left to right, are being continually offset by slow, horizontal movement of fault. Prominent stream, upper left, meets fault valley, flows toward camera and eventually leaves fault valley in an offset stream channel more than 1km distant. The prominent black line along left side of photo is not a geologic feature; it is tumbleweed caught along a fence. (*R. E. Wallace, U.S. Geological Survey.*)**

**Figure 14.2**
**On December 25, 1965, a fault abruptly cut one of the main roads near Kilauea Volcano, Hawaii. The difference in height of the two sides of the fault is only 1m, but the damage was increased greatly by slump along the fault interface. Faulting occurred when large volumes of magma, previously held in deep-lying magma chambers, were rapidly erupted. (*R. S. Fiske, U.S. Geol. Survey.*)**

trees as well as by broken fences. Horizontal movement, then, seems to be a habit of the San Andreas Fault (Fig. 14.1). As we shall see in Chapter 17, this is because that fault marks a boundary along which two plates of the lithosphere are sliding past each other. In this case the surface movement is a record of deformation caused by forces that operate hundreds of kilometers down within the Earth.

Abrupt vertical movements are more obvious than horizontal movements. During the eruption of Kilauea Volcano, Hawaii, in December 1965, underground movement of magma caused abrupt deformation of the surface. Fault cliffs as much as 1m high were formed; one of them cut a main road (Fig. 14.2). Most faults, however, display a combination of horizontal and vertical motion (Fig. 14.3).

The largest known abrupt displacement occurred in 1899 at Yakutat Bay, Alaska, during an earthquake. A stretch of the Alaskan shore, with beach, barnacle-covered rocks, and other telltale features was suddenly lifted as much as 15m above sea level. This visible displacement may be less than the total amount, because the fault is hidden offshore and the block of crust on the other side of it, entirely beneath the sea, possibly moved downward, thus adding to the total displacement.

### Gradual movement

Movement along faults is not always abrupt, nor is it always accompanied by earthquakes. Measurements along the San Andreas Fault reveal places where slow, steady slipping occurs, sometimes reaching a rate as high as 5cm a year. Similarly, subterranean movement of magma does not always cause sudden faulting such as that shown in Figure 14.2. Scientists who have studied Kilauea Volcano observe that the top of the mountain slowly rises as magma flows upward prior to eruption. The magma is apparently driven upward under such pressure that it causes the whole mountain to be inflated, as compressed air inflates a balloon (Fig. 14.4). Following eruption, the mountain becomes deflated and the surface slowly falls again, but never to exactly the position from which it started.

A classic example of slow changes in the level of the land is seen in the ruins of an ancient Roman marketplace known as the Temple of Serapis, west of Naples. Three columns left standing have been bored into by a peculiar marine clam to a height about 6m above the floor (Fig. 14.5); the shells of the clams still line some of the borings. Along the shore near the ruin is sediment that contains abundant shells of ordinary clams like those now living in the adjoining bay; these deposits are exposed in bluffs as much as 7m above present sea level. At first thought, one might try to explain these observations by a worldwide rise and later sinking of sea level while the land remained stationary. But such fluctuation would have left a record on all the world's coasts. As the evidence cited above is found only within a limited area near the old ruin, we conclude that at some time after the Romans built the Temple, this part of the Italian coast slowly sank, and then, within more recent time, was re-elevated. According to local records, the re-elevation started not long after A.D. 1500.

The examples just cited involve very small portions of the crust.

Figure 14.3
**Scarp formed by a fault that crosses 9th Avenue, Anchorage, Alaska. Appearing during the earthquake of March 27, 1964, the fault shows vertical displacement of approximately 2m and horizontal displacement of about 1m. The position of the roadway shows that the upper block has moved from left to right relative to the lower block. The small fault in the foreground likewise shows horizontal movement of about 0.5m. Fault displacement is more recent than the snow.** *(Alaska Pictorial Service.)*

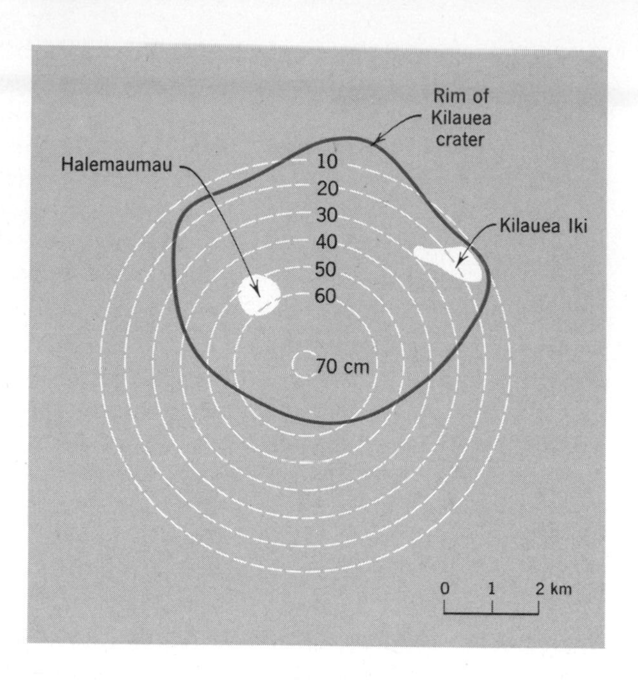


**Figure 14.4**
**Inflation of Kilauea Volcano, Hawaii, by magma rising to erupt. Inflation was expressed by a rising land surface during a 22-month period, from January 1966 to October 1967. Contours, showing the total rise in centimeters, are not quite centered over the crater at the summit of Kilauea, nor do they coincide exactly with two of the recently active vents, Halemaumau and Kilauea Iki. From the amount of inflation, and the shapes and positions of the contours, scientists calculate that the magma accumulated in a chamber 2 to 3km below Kilauea Crater, and then moved quickly upward at the time of eruption.** ***(After R. S. Fiske and W. T. Kinoshita, 1969, Science, v. 165, p. 341.)***

**Figure 14.5**
**Columns in a Roman ruin at Pozzuoli, Italy, as they appeared in 1828. Borings made by marine clams indicate former submergence.** ***(Drawing from Lyell, 1875, Principles of Geology, 12th ed.)***

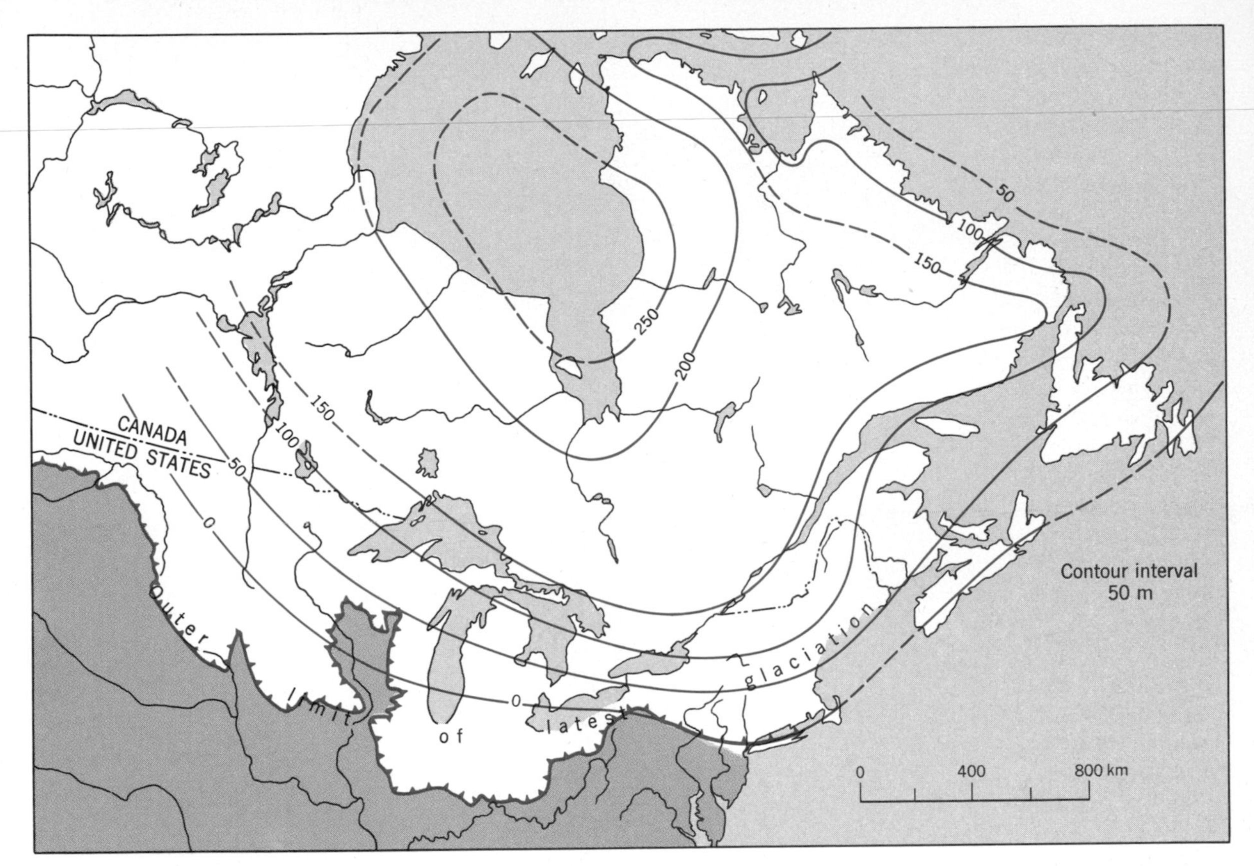


**Figure 14.6**
**An ice sheet centered on Hudson Bay during the latest glacial age caused downward buckling of the crust in northeastern North America. As the ice melted, the land surface rose again. Beaches that formed at sea level before melting are now many meters above it, and so provide a measure of uplift. Contours show amounts of uplift. Dates of beaches, determined radiometrically with $^{14}C$, prove uplift is still in progress, at rates as fast as 1m per century. (*After P. B. King, 1965.*)**

Movement of much larger areas also can be well-documented. One example is an effect of the latest of the glacial ages (Chapter 10). Any large weight placed on Earth's surface deforms the crust, causing it to sag downward. The great ice masses that covered large parts of continents had just this effect; in some areas they pushed down the crust by at least several hundred meters. When, later, the ice melted, the pressure was relieved; the crust began to rise back to its original position, and although at least 7,000 years have passed since the ice masses disappeared, northeastern North America is still rising. Near Hudson Bay, where the ice was thickest, beaches that were at sea level a few thousand years ago are now hundreds of meters above sea level (Fig. 14.6) and the surface is still going up.

It may be that across the entire land surface of the United States, no spot is completely stationary. Measurements by U.S. Government surveyors over the past 100 years reveal great areas where the surface is slowly rising, and others where it is slowly sinking (Fig. 14.7). Not all the causes of these vast, slow movements are well understood, but the movements do prove that the solid Earth is not as solid as it seems at first sight, and that great internal forces are continually deforming its crust.

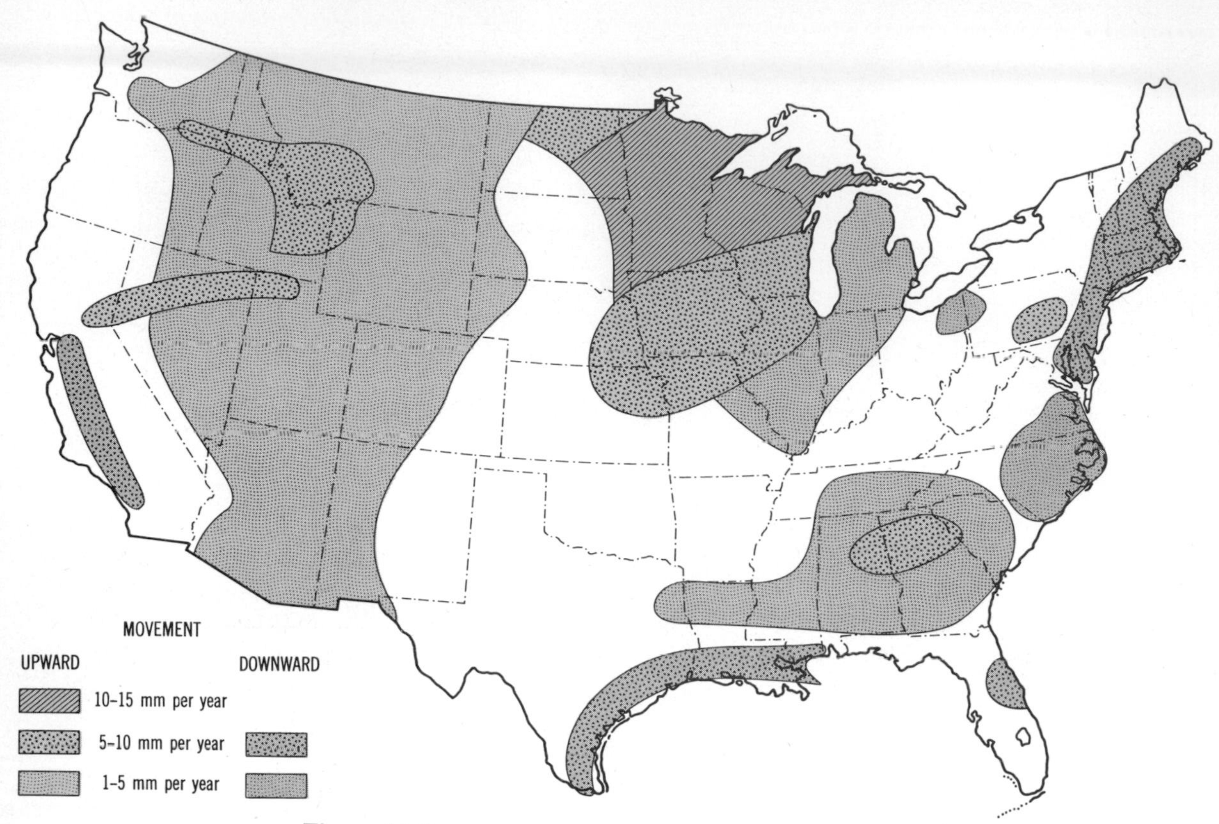


Figure 14.7

**Accurate measurements over a 100-year period show that in large areas of the United States, the surface is slowly moving up or down. Subsidence along the coasts of California and the Gulf of Mexico is believed to have been caused in part by withdrawal of gas, oil, and water, which allow subsurface reservoirs to collapse. Uplift near the Great Lakes is a rebound effect following melting of the last ice sheet (Fig. 14.6).**

**The causes of movements in other areas are not known with certainty. Those areas in which no movement is shown are not necessarily stationary. They are simply areas in which measurements are very few. (*After S. P. Hand, National Oceanic and Atmospheric Administration, 1972.*)**

# EVIDENCE OF FORMER MOVEMENT

With so much evidence of movement and deformation of Earth's crust, we might reasonably expect to find a great deal of evidence of former movements. Indeed we do. Studies of land and sea-bottom topography provide abundant evidence of vertical movements; and in some areas the distributions of various kinds of rock provide clear evidence that horizontal movements through distances as great as several hundred kilometers have occurred. Let us examine a few examples of such evidence, beginning with topographic features.

## Topographic features

In many parts of the world, well-developed wave-cut benches stand

**Figure 14.8 Wave-cut benches far above sea level on San Clemente Island, off the coast of Southern California. In this air photo taken in 1960, ten benches are visible. The height of the cliff behind each bench varies, indicating that some upward movements were greater than others. (*J. S. Shelton.*)**

one above another like stairsteps (Fig. 14.8). Some of the lowest steps, still decorated with barnacles, terminate inland against typical wave-cut cliffs. Along the southern California coast and nearby islands, the highest recognizable terraces are now more than 450m above the sea. Because these terraces are found along only a part of that coastal area, we reason that a segment of the Coast Ranges has risen in a succession of pulses, separated by pauses long enough to allow surf to create cliffs and beaches. Also, the altitude of a terrace varies appreciably from place to place, telling us that the amount of uplift was irregular. Apparently the uplift of coastal California is caused by pressures developed between two moving plates of lithosphere. The suture between the Pacific and American Plates passes very close to the region of uplift (Fig. 17.5).

Some topographic evidence indicates large-scale subsidence. Accurate surveys of the sea floor north of the Aleutian Islands reveal a submarine topography of high ridges and hills separated by valleys that unite in what appears to be a well-developed drainage system. The best explanation seems to be the submergence of a broad land area that was shaped by mass-wasting and stream erosion. Was this drowning of a former landscape caused by rise of sea level or by sinking of the land? We know the level of the sea was raised by the return of water that had been locked up in ice sheets during the glacial ages. The total rise from this cause is estimated to have been about 100m, but the drowned hills and valleys near the Aleutian Islands lie at depths greater than 400m. We infer, therefore, that a large part of the submergence was caused by sinking of land.

Additional evidence of vertical movements can be found in volcanic islands on the ocean basins. We read in Chapter 13 that the wave-cut tops of many guyots are now covered by seawater more than 1km deep, indicating large subsidence. Indeed, the gross topography of the entire ocean basin indicates that the sea floor must be subsiding as it moves away from the elevated oceanic ridge toward the abyssal plains.

**Bedrock features**

When we examine the distribution of various kinds of rock, we find so much evidence of former movement that one striking example will suffice. A remarkable fault, the Great Glen Fault, crosses Scotland from southwest to northeast. The trace of the fault is a line of easy erosion, and a valley containing a string of lakes now marks its path. One lake is Loch Ness, home of the famous Loch Ness Monster.

When, during the Paleozoic Era, the Great Glen Fault was active, it severed a large granite mass into two fragments. The fragments now lie on opposite sides of the fault, approximately 100km apart, and thus give striking evidence of large-scale horizontal movement (Fig. 14.9).

Not all evidence of movement and deformation observed in bedrock is as striking and obvious as the Great Glen Fault. But once we learn to recognize it, evidence of deformation is seen to be very widespread—so much so that a special branch of geology, *structural geology,* has the study of rock deformation as its primary purpose. In order to understand Earth's internal activities, we must also be able to recognize and evaluate evidence of rock deformation. Let us first examine deformation by fracture.

# DEFORMATION BY FRACTURE

We have learned that it is the brittle properties of rock that lead to fracture. Rock in the crust, especially rock close to the surface, tends to be brittle; so it is cut by innumerable fractures. The fractures aid erosion, serve as channels for the circulation of ground water, provide entryways along which magma is intruded, and in many places serve as the openings in which veins of valuable minerals are deposited.

Most fractures are too small for slipping to have occurred along

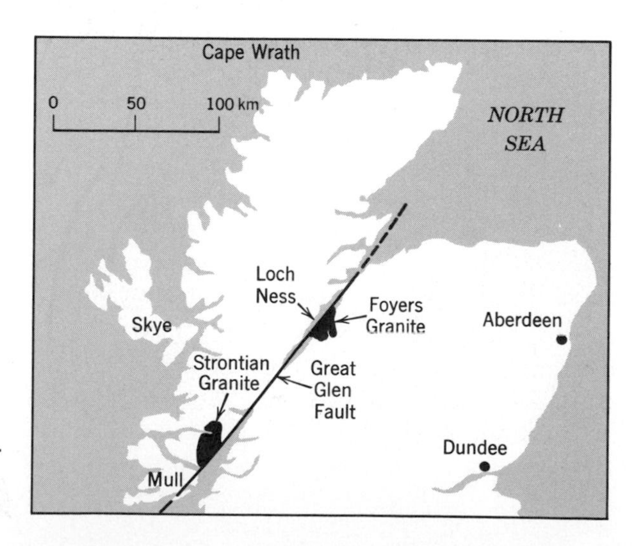


**Figure 14.9**
**The Great Glen Fault in Scotland cuts a granite mass into two separate segments, the Strontian Granite and the Foyers Granite. Horizontal movement along the fault has separated the masses by approximately 100km. (*After W. Q. Kennedy, 1946.*)**

them. Along many fractures, however, movement has occurred, converting the fractures into faults.

Generally there is no way of telling how much movement has occurred along a fault nor which side of the fault has moved. Even if a crystal or a pebble in the rock has been cut through by the fracture and the halves carried apart a measurable distance, we cannot know whether one block stood still while the other moved past it, or whether both sides shared in the movement. Precise surveys of points on the ground before and after movement on an active fault can, of course, indicate which side of a fault is moving. Such a check is made continuously along the San Andreas Fault. But most faults with which we have to deal are old features whose former expression at Earth's surface was destroyed by erosion long ago. In classifying fault movements, then, we can only speak of *apparent* and *relative* displacements.

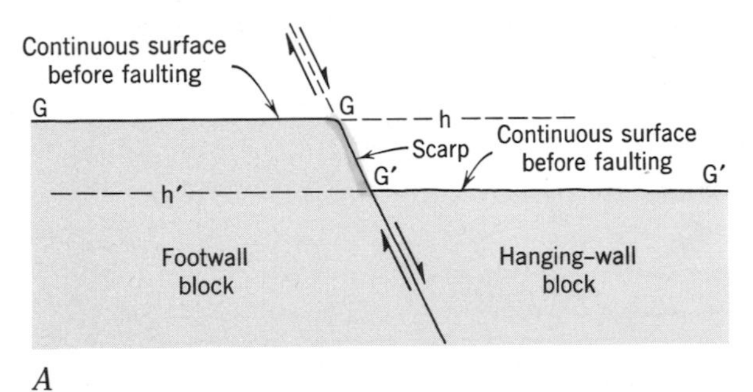

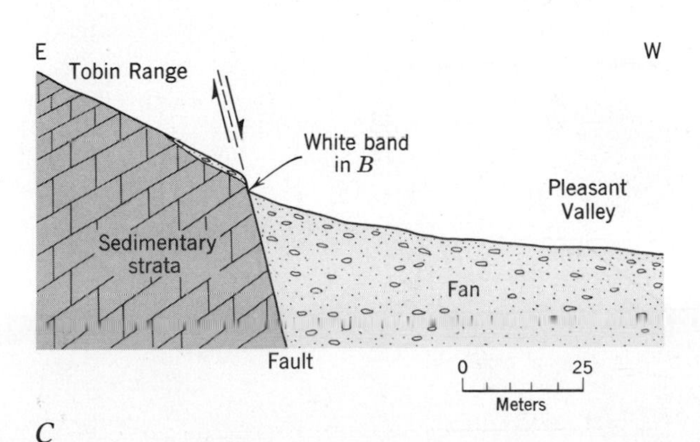


**Figure 14.10**
**Fault that displaces surface of ground.**
***A*. Scarp made by displacement of a formerly level surface G-G′ may mean that the hanging-wall block moved down from position *h* to *h′*; that the foot wall block moved up from position *h′*; or that both blocks moved to some degree, to create the net displacement shown.**
***B*. Scarp at west base of Tobin Range, Nevada, formed in 1915 by abrupt movement along a fault that generated an earthquake. The whitish band of newly exposed rock marks displacement at the top of the fan, several miles from the camera (*Eliot Blackwelder*).**
***C*. Vertical section across lower part of Tobin Range to Pleasant Valley, approximately in the direction of the photographer's line of sight in *B*. Half arrows indicate relative movement of crust blocks.**

Most faulting occurs along fractures that are inclined. Because many veins of metallic ore lie along faults, we have inherited some terms used by the miners who dug out the veins. From the miner's viewpoint, one *wall* of an inclined vein overhangs him, while the other is beneath his feet. Because veins of ore commonly occupy openings created by faults, we adopt the old miner's terms. The ***hanging wall*** is *the surface of the block or rock above an inclined fault; the surface of the block of rock below an inclined fault* is the ***footwall*** (Fig. 14.10). These terms, of course, do not apply to vertical faults.

**Figure 14.11**

**Principal kinds of faults, the directions of forces that cause them, and some of the topographic changes they cause.**

| Block diagram | Name of fault | Definition |
| --- | --- | --- |
| 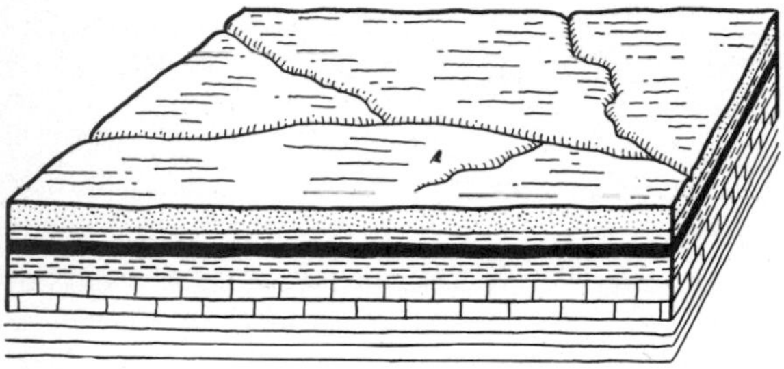  |  | Reference block before faulting. Drainage is from left to right. |
| 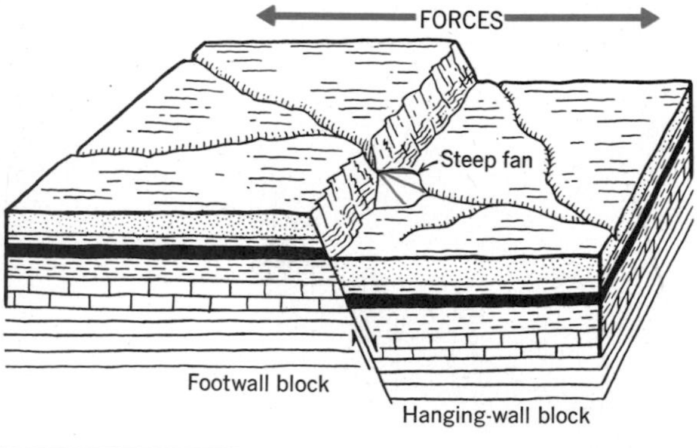 FORCES<br>Steep fan<br>Footwall block<br>Hanging-wall block | ***Normal fault*** | *A fault, generally steeply inclined, along which the hanging-wall block has moved relatively downward.* |
| 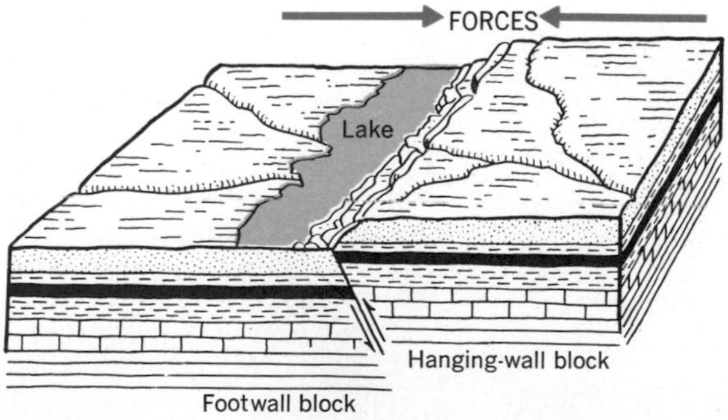 FORCES<br>Lake<br>Hanging-wall block<br>Footwall block | ***Reverse fault*** | *A fault, generally steeply inclined, along which the hanging-wall block has moved relatively upward.*<br><br>A normal or reverse fault on which the only component of movement lies in a vertical plane normal to the strike of the fault surface is a *dip-slip fault.* |

Faults are grouped into classes according to (1) the inclination of the surface along which fracture has occurred and (2) the direction of relative movement of the rock on its two sides. The common classes of faults, together with the changes in local topography they sometimes create, are listed in Figure 14.11. Our standard planes of reference in classifying faults are the vertical and the horizontal. Along some faults, movement is confined to these two reference planes, although, as we saw in Figure 14.3, along other faults both vertical and horizontal movements occur. Not all movement on faults is in straight lines. Sometimes fault blocks are rotated, and, as seen in Figure 14.11, we classify such faults as

| Block diagram | Name of fault | Definition |
|---|---|---|
| FORCES | **Strike-slip fault** | *A fault on which displacement has been horizontal.* Movement of a strike-slip fault is described by looking directly across the fault and by noting which way the block on the opposite side has moved. The example shown is a *left-lateral fault* because the opposite block has moved to the left. If the opposite block has moved to the right it is a *right-lateral fault.* Notice that horizontal strata show no vertical displacement. |
|  | **Oblique-slip fault** | *A fault on which movement includes both horizontal and vertical components.* See also Fig. 14-3. Forces are a combination of forces causing strike-slip and normal faulting. |
|  | **Hinge fault** | *A fault on which displacement dies out (perceptibly) along strike and ends at a definite point.* Figure 14-3 shows a small example located in the foreground of the photograph, between the viewer and the man walking away from the camera. Forces are the same as those causing normal faulting. |

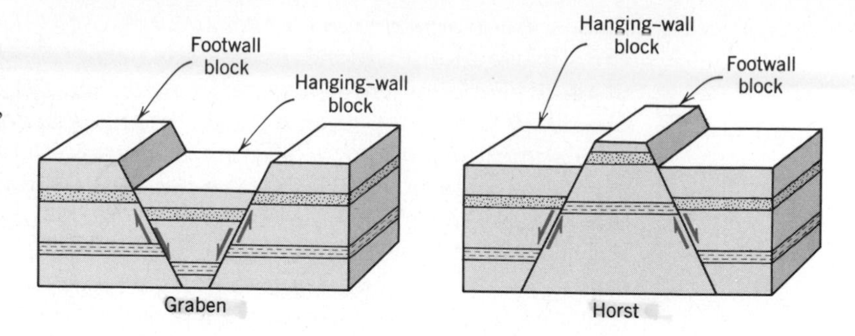


**Figure 14.12**
***Graben, a trench-like structure bounded by parallel normal faults,* formed when the hanging-wall block that forms the trench floor moves downward relative to the footwall blocks. *Horst, an elevated elongate block bounded by parallel normal faults,* formed when the elevated footwall block moves upward relative to the hanging-wall blocks.**

*hinge faults.* Many normal and reverse faults become hinge faults as they approach the point where they die out.

### Normal faults

Normal faults are caused by tensional forces that tend to pull the crust apart, and also by forces tending to expand the crust by pushing it upward from below. In the crust are many zones that have been deformed repeatedly by normal faulting. Commonly two or more similarly trending normal faults enclose an upthrust or downdropped segment of the crust. As shown in Figure 14.12, a downdropped block is a *graben,* an upthrust block a *horst.* The central, steep-walled valley that runs down the center of the oceanic ridge (Fig. 13.6) and cuts through Iceland (Fig. 13.7) is a graben. Perhaps the world's most famous graben is the African Rift Valley (Fig. 14.13), which runs north-south through more than 6,000km. Within parts of it are volcanoes, where magma has formed channelways that lead upward along the fault surfaces. The Rift Valley is important for many reasons, one of which is that it seems to mark the place where the spreading edge of a new plate of lithosphere is passing beneath a continent, and is pulling the continent into two giant fragments by tensional forces.

Normal faults are innumerable. Horsts and grabens also are very common, although none is as spectacular as the Rift Valley. The valley in which the Rhine River flows through western Europe follows a series of grabens, and Lake Baikal in central Asia, Earth's deepest lake, is located in a very deep graben. A spectacular example of normal faulting is found in the Basin and Range Province in Utah and Nevada. There, movement on a series of parallel and subparallel, north–south, normal faults has formed horsts and grabens that are now mountain ranges and sedimentary basins. The province starts with the Wasatch Range in the east and continues westward to the eastern edge of the Sierra Nevada.

### Reverse faults

Reverse faults arise from compressive forces. Movement on reverse faults tends to push older rocks over younger ones, thereby shortening and thickening the crust.

A special class of reverse faults, called ***thrust faults*** and generally known as ***thrusts,*** are *low-angle reverse faults with dips generally less*

**Figure 14.13**
**African Rift Valley in Ethiopia.**
**The eastern wall of the African Rift Valley, a giant graben several thousand km in length. The valley floor was originally at the same height as the plateau above, but has been lowered by movement on a normal fault. The fault surface has been modified by erosion; so the valley walls are no longer straight.** ***(George Gerster. Rapho-Photo Researchers.)***

*than 45°*. Such faults, common in great mountain chains, (Figs. 18.7 and 18.10), are noteworthy because along some of them the hanging-wall block has moved many kilometers over the footwall block. In most cases the hanging-wall block, thousands of meters thick, consists of rocks much older than those adjacent to the thrust on the foot-wall block (Fig. 14.14). The strata above some thrusts lie nearly parallel to those beneath, and so may appear, deceptively, to represent an unbroken sequence.

### Strike-slip faults

We have already mentioned two strike-slip faults, the San Andreas Fault and the Great Glen Fault (Fig. 14.9). The San Andreas is a right-lateral strike-slip fault (Fig. 17.5). Apparently movement has been occurring along it from at least as long ago as Cretaceous time to the present day—in other words, through at least 65 million years. The total movement is not known, but some evidence suggests it amounts to more than 600km. The Great Glen Fault is a left-lateral strike-slip fault; it is no longer active.

### Transform faults

Strike-slip faults appear to be most commonly caused by movement between adjacent plates of lithosphere. A special kind of strike-slip fault is widespread in the oceanic crust. For example, many of the great fracture zones that cut the oceanic ridge system are strike-slip faults. In Figure 13.6, each of the numerous fracture

*A*

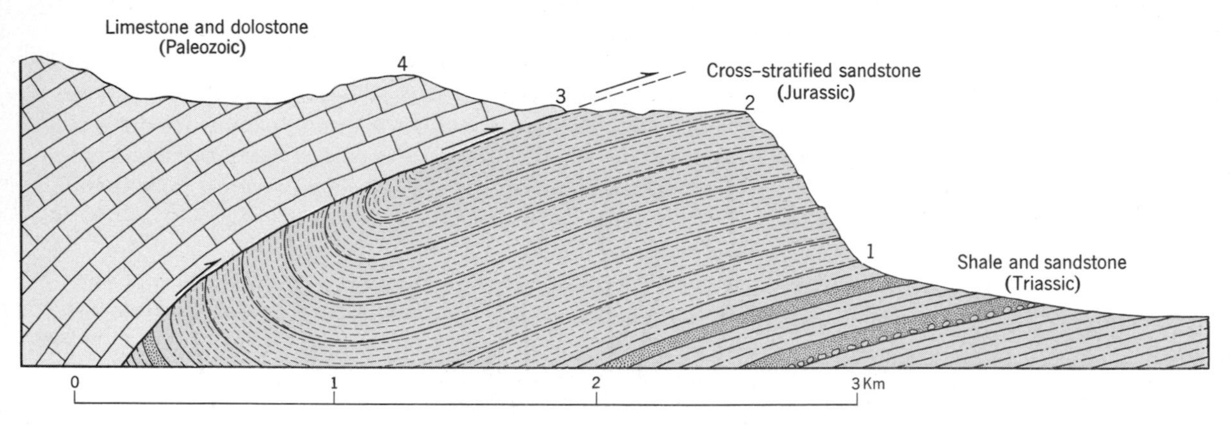


*B*

zones that offset the Mid-Atlantic Ridge is a strike-slip fault. Indeed, so common are they that it has been suggested that the major structural forms marking sites of deformation of Earth's crust are sea-floor trenches, oceanic ridges, and strike-slip faults. These features link together to form continuous networks encircling the Earth. When one feature terminates, another commences; their junction point is called a *transform*. J. T. Wilson,

**Figure 14.14**
**(see opposite page) Keystone Thrust, west of Las Vegas, Nevada.**
***A*. Air view northward shows the fault clearly defined by a color contrast in the strata adjacent to it. Light-colored Jurassic sandstone, forming a cliff nearly 600m high (right), lies below the fault; dark-colored Paleozoic limestones and dolostones (left) lie above it. Numbered points correspond to similarly numbered points on the section below. (*J. S. Shelton.*)**
***B*. Section drawn along white line in photograph and extending somewhat farther east and west. Canyons crossing the fault reveal that it steepens downward toward the west and crosses overturned layers of the sandstone. Farther east the fault becomes essentially parallel to the sedimentary layers, both below and above. (*After C. R. Longwell.*)**

a Canadian scientist who first recognized the network relation, proposed that *a special class of strike-slip faults that links major structural features* occurs at many junctions; he called them ***transform faults.*** Close study of the strike-slip faults that offset the oceanic ridges proved Wilson's suggestion correct. As seen in Figure 14.15, movement along transform faults is a consequence of the continuous addition of new crust material along oceanic ridges, the lateral movement of older crust away from the ridge and its consumption beneath sea floor trenches.

## Evidence of movement along faults

Often we find fractures in rock but cannot tell at first glance whether or not movement has occurred along them. For example, in uniform, even-grained rock such as granite, or in a pile of thin-bedded strata, no one of which is unique or distinctive, we would not see displacement of any obvious features. However, examination of the fault surface, or of rock immediately adjacent to it, commonly reveals signs of local deformation, indicating that movement has occurred. Under special circumstances, even the direction of movement can be deciphered.

Adjacent to some faults, bending of strata or other internal features can be seen. Large and small structures created by bending adjacent to faults are known collectively as *fault drag* (Fig. 14.16).

**Figure 14.15**
**Transform faults, a special class of strike-slip faults forming a globe-encircling network with oceanic ridges and sea-floor trenches.**
***A*. An oceanic ridge is offset by a transform fault. Crust on both sides of the two ridge segments is moving laterally away from the ridge. Between segments of the ridge, along *AA′*, movement on the two sides of the fault is in opposite directions. Beyond the ridge, however, along segments *AB* and *A′B′*, movement on both sides of the fault is in the same direction. A transform fault does not, therefore, cause the ridge segments to move continuously apart.**
***B*. A transform fault joins an oceanic ridge with a sea-floor trench. New crust formed at a spreading edge (the oceanic ridge) moves laterally away from the ridge and plunges back into the mantle at a place marked by the sea-floor trench. The triangles on the trench point in the direction of movement of the downward-plunging crust.**
***C*. Transform fault joining two sea-floor trenches.**

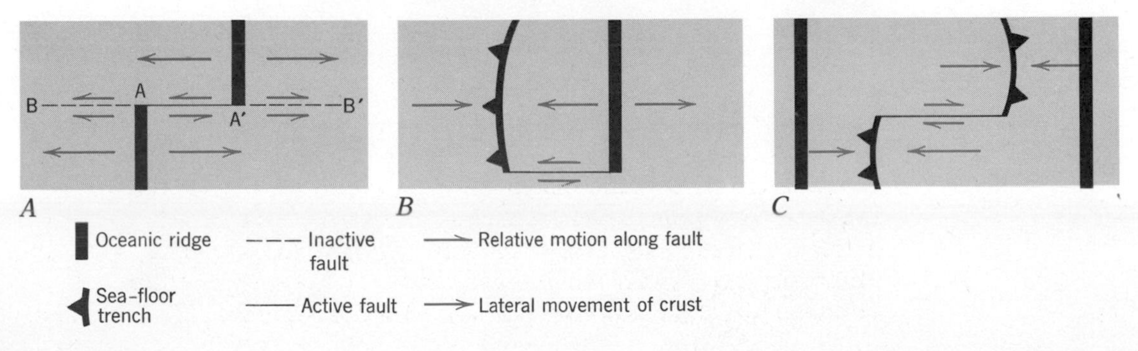

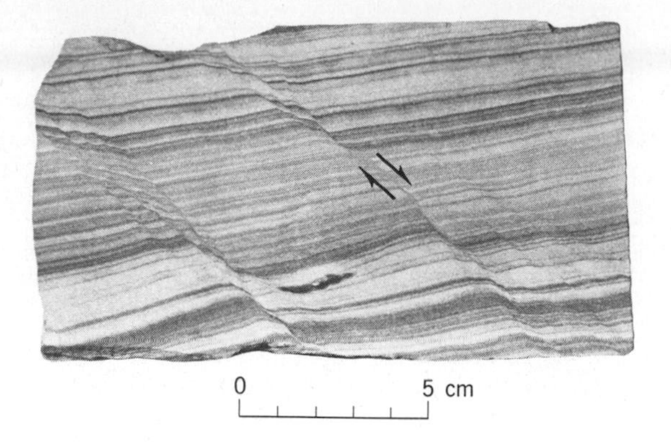


**Figure 14.16**
**Fault drag.**
**Near small faults that cut sandstone, thin layers have been bent during movement. Direction of bending indicates that direction of last movement was as shown by arrows. The drag in this case indicates these are normal faults. (*B. M. Shaub.*)**

Movement of one mass of rock past another can cause the fault surfaces to be smoothed, striated, and grooved. *Striated or highly polished surfaces on hard rocks, abraded by movement along a fault* are ***slickensides.*** Parallel grooves and striations on such surfaces record the direction of latest movement (Fig. 14.17).

Not all fault surfaces are slickensides. In many instances, fault movement crushes rock adjacent to a fault into a mass of irregular pieces, forming *fault breccia.* More intense grinding breaks the fragments into such tiny pieces they may not be individually visible even under a microscope.

## DEFORMATION BY BENDING

Bending may consist of broad, gentle warping that extends over hundreds of kilometers, or close, tight flexing of microscopic size.

**Figure 14.17**
**Slickensides in diabase, Mount Tom Range, Westfield, Massachusetts. The direction of the striations on this hanging-wall block and the presence of rough steps breaking off the striation indicates that the footwall block moved in the direction of the arrow. (*B. M. Shaub.*)**

**Figure 14.18**
**A monocline in southern Utah that interrupts the generally flat-lying sedimentary strata of the wide Colorado Plateau. On both sides of the monocline the exposed layers are nearly horizontal. View looking south. (*J. S. Shelton.*)**

Regardless of volume of rock involved or degree of warping, we refer to the bending of rocks as *folding*. Before we discuss folds and folding, we must become familiar with the few terms used to describe them.

The simplest fold is a ***monocline,*** *a one-limbed flexure, on both sides of which the strata either are horizontal or dip uniformly at low angles* (Fig. 14.18). An easy way to visualize a monocline is to lay a book on a table, and drape a handkerchief over one side of the book and out onto the table. So draped, the handkerchief forms a monocline.

Most folds are more complicated than monoclines. *An upfold in the form of an arch* is an ***anticline*** (Fig. 14.19). *A downfold with a troughlike form* is a ***syncline.*** As we see in Figure 14.19, *the sides of a fold* are the **limbs,** and *the median line between the limbs, along the crest of an anticline or the trough of a syncline,* is the ***axis*** of the fold. *A fold with an inclined axis* is said to be a **plunging fold,** and *the angle between a fold axis and the horizontal* is the **plunge** of a fold. *An imaginary plane that divides a fold as symmetrically as possible, and that passes through the axis,* is the ***axial plane.***

Many folds, like those in Figure 14.19, are nearly symmetrical. Others, however, are not symmetrical, and strong deformation may create very complex shapes. The common forms of folds are shown in Figure 14.20. If we have only fragmentary exposures of bedrock, it is apparent we might have trouble in deciding whether a given fold is overturned or not. We must know whether a given layer is right-side up or upside down in order to decide which limb of a fold it is in. This is not always possible, but in some cases sedimentary structures, such as mud cracks and graded layers, do record this. In other examples only careful, thorough mapping of all bedrock exposures can give us the answer.

Folds are commonly so large that when we examine a single exposure we are not even aware we are seeing folded rock. Nevertheless, when all the exposures of a particular rock body are

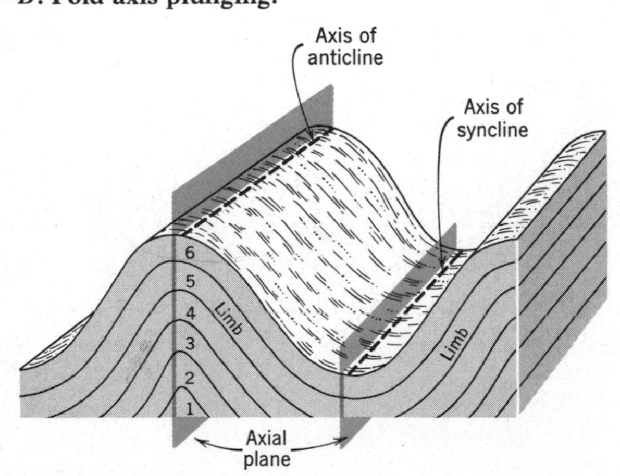


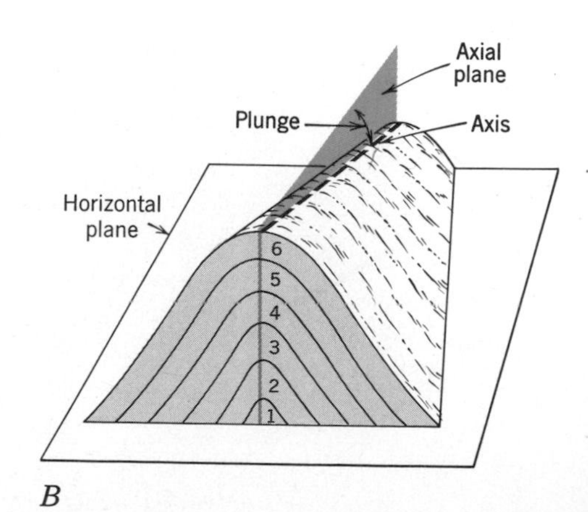


**Figure 14.19**
**Features of simple folds. Upper surface of the youngest layer (6) slopes *toward* the axis of the syncline but *away from* the axis of the anticline.**
***A*. Fold axis horizontal.**
***B*. Fold axis plunging.**

Figure 14.20
**Five kinds of folds.**

| Name | Description |
|---|---|
| Symmetrical | Both limbs dip equally away from the axial plane. 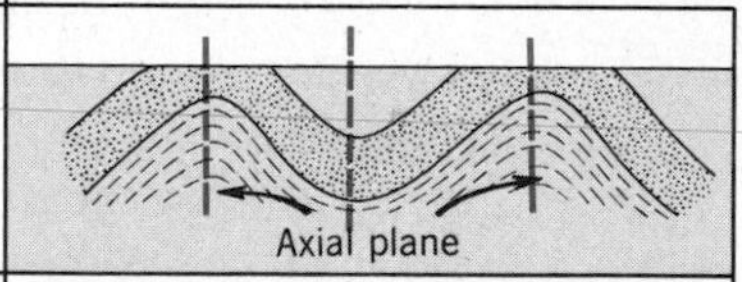 |
| Asymmetrical | One limb of the fold dips more steeply than the other. 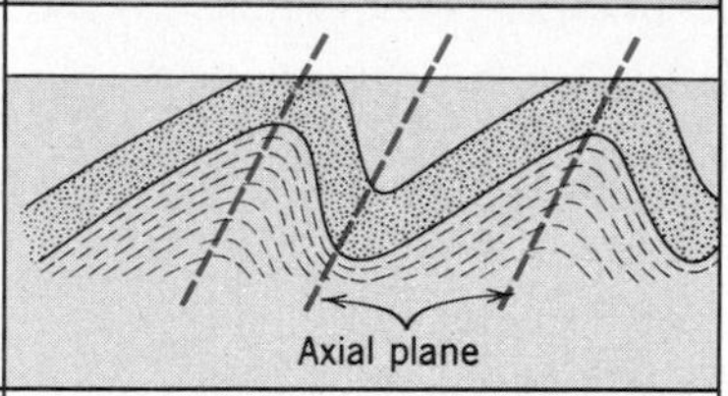 |
| Overturned | Strata in one limb have been tilted beyond the vertical. Both limbs dip in the same direction, though not necessarily at the same angle. 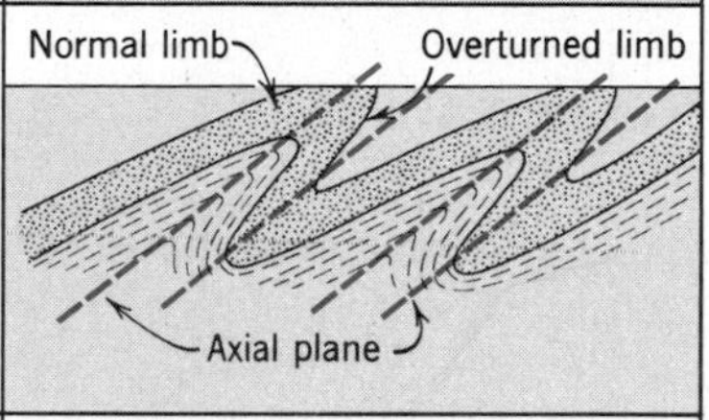 |
| Recumbent | Axial planes are horizontal. Strata on the lower limb of anticline and upper limb of syncline are upside down. 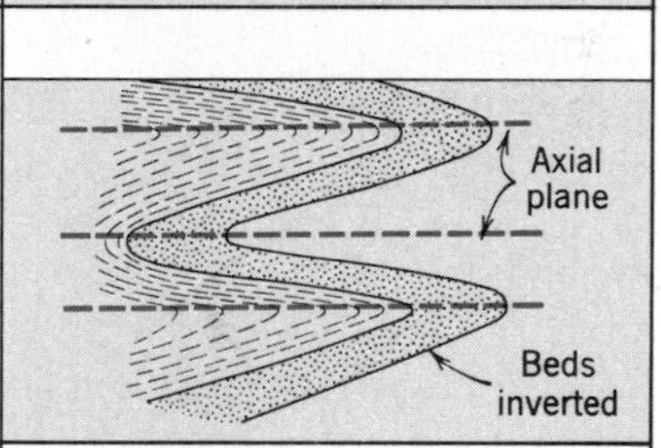 |
| Isoclinal | Both limbs are essentially parallel, regardless whether the fold is upright, overturned, or recumbent. 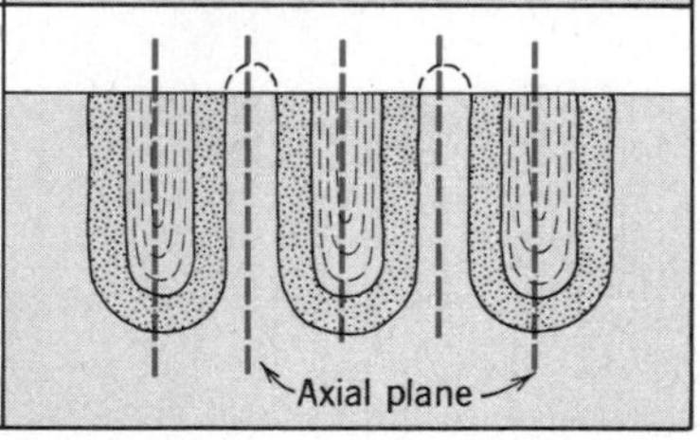 |



plotted, and a geologic map is prepared (Appendix D), folds can be recognized from the distribution of the various rock types. Differences of erodibility in adjacent strata can also lead to distinctive topographic forms by which we can recognize the presence of folds (Fig. 14.21).

Gentle warping of the crust, upward or downward, which is really gentle folding on a very large scale, is common. It is likely to be most evident in arid regions where, because of the absence of concealing regolith and the presence of deep dissection by streams, wide areas of rock layers are exposed. Most commonly, however, vegetation and regolith prevent direct observation, and the presence of warping may not even be apparent from examination of several exposures. Inclinations of strata may be only 1° or 2°, and such small slopes are not conspicuous. Yet if a stratum has a persistent dip of 2°, the altitude of the layer changes by about

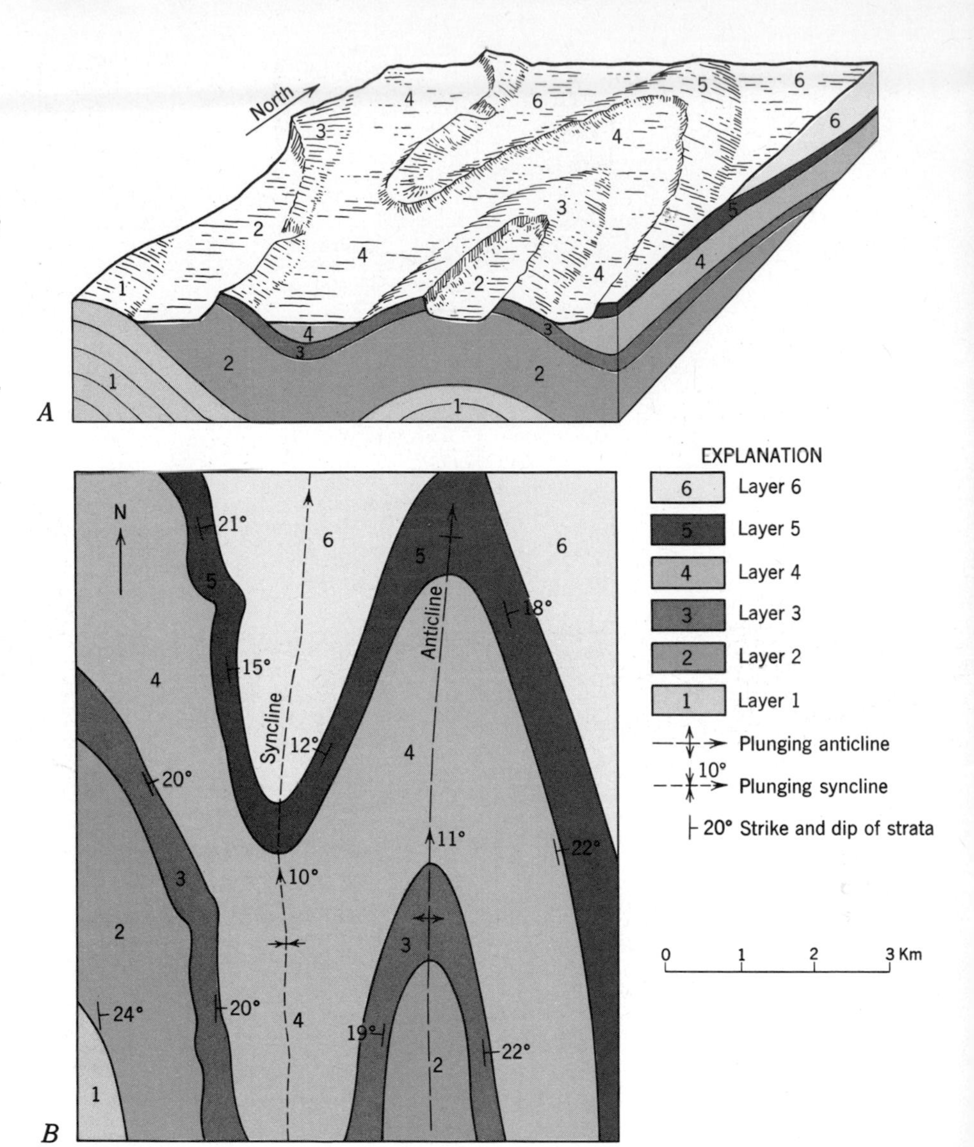


**Figure 14.21**
**Distinctive topographic forms and distinctive patterns in the distribution of various kinds of rock reveal the presence of plunging folds.**
***A*. Block diagram showing topographic effects.**
***B*. Geologic map. (Appendix D gives further details on the preparation and reading of geologic maps.)**

300m in a horizontal distance of 8km. Careful mapping therefore reveals even the most gentle warping. Upwarping forms a dome, and downwarping creates a basin. Domes are particularly important because some of them are the sites of large accumulations of oil and gas.

### Relations between faults and folds

Folds and faults can be related to each other in various ways. As we have just seen (Fig. 14.16), strata near active faults may be folded by the effects of frictional drag along the planes of movement. A steeply inclined fault may pass, either upward or laterally, into a fold and thus die out. The hinge fault shown in Figure 14.22 is an example of how a high-angle fault can become a monocline. Typically, the projected continuation of the fault surface coincides with the axial plane of the related fold.

*C*

***C*. Air view of a doubly plunging anticline. Sheep Mountain, a high ridge 24km long, in the Bighorn Basin, Wyoming, is formed by resistant older strata, which lie along the axis of the anticline. Younger strata, mostly alternating layers of sandstone and shale, have been eroded from the crest of the fold but make jagged, low ridges along the steep limb. These low ridges curve around each end of the mountain, the curve pointing in the direction of plunge. (*J. S. Shelton*.)**

Many thrusts in sedimentary rock form after the strata have been compressed and folded into overturned folds. Continued compression finally causes one of the folds to break, generally near the axial plane, and the former folds become a thrust fault. Apparently the Keystone Thrust (Fig. 14.14) was formed in this fashion.

Not only have strata been deformed as the result of active forces of compression, but some thrusts have themselves been folded as if they were strata (Fig. 18.9). A few thrusts have even been overturned. Deciphering the sequence of events in such a case can be a challenging and difficult task.

## UNCONFORMITY

Movement along faults brings rocks of different kinds into contact with each other. It can also bring into contact two groups of layered rocks that diverge widely as to dip. When we see such features exposed, therefore, we might think first of faulting as the cause. But close examination commonly proves our initial assumption wrong. An example can be found in the disconformable interface in Figure 12.20. Another example can be seen in the contact between the two formations shown in Figure 14.23. The contact is highly irregular in detail and lacks evidence of

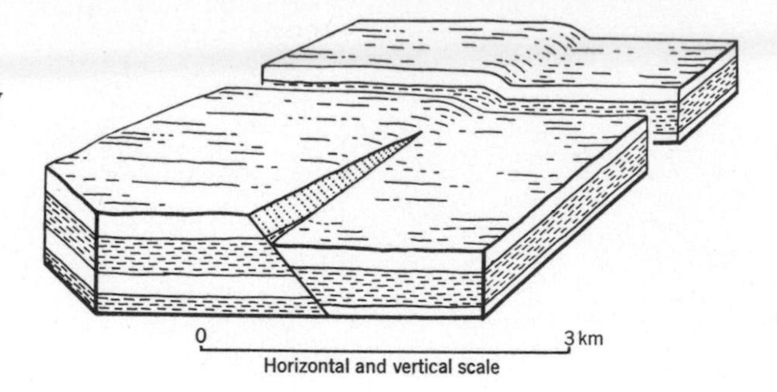


**Figure 14.22**
**Hinge fault (front block) passes laterally into monocline (rear block).**

movement along it. Furthermore, in the lower strata of the upper group, we see pebbles and cobbles consisting of distinctive kinds of rock derived from the steeply dipping layers beneath. Marine fossils, abundant in the tilted layers, show that those layers were formed on the sea floor during the Cambrian Period. Bones of land mammals, together with the nature of the sediments around them, tell us that the flat-lying layers were deposited by streams in the Pliocene Epoch. Clearly, then, we are looking not at a fault but at a feature that was created by both deformation and the deposition of sediment. At some time after the thick marine layers were formed, a strong disturbance tilted the layers and raised them above sea level. After a large part of the uplifted mass had been destroyed by erosion, a blanket of nonmarine sediment accumulated.

The Pliocene alluvium in Figure 14.23 does not *conform* with the lower Cambrian strata beneath it. The relationship between them is one of ***unconformity,*** meaning there is *a lack of continuity of sedimentary deposition between two units of rock in contact, corresponding to a gap in the geologic record.* The example in Figure 14.23 is of ***angular unconformity,*** which is *unconformity marked by angular divergence between older and younger rocks.*

Unconformity between a stratum and older rocks beneath it is commonly a record of movement of the crust. Such a record gives a few important details; it tells us that the older rocks were brought into the realm of erosion (this suggests some kind of uplift) and that at some later time conditions became favorable for deposition of new sediments. The record does not necessarily tell us why conditions became favorable for sedimentation. Possibly it

**Figure 14.23**
**Surface of unconformity between nearly horizontal, gravelly alluvium containing fossil bones of Pliocene age, and tilted and eroded carbonate rocks containing marine fossils of Cambrian age. Close examination shows this is not a fault contact. It was formed by deposition of Pliocene sediments over an old erosion surface. Meadow Valley Wash, Lincoln County, Nevada. (*C. R. Longwell.*)**

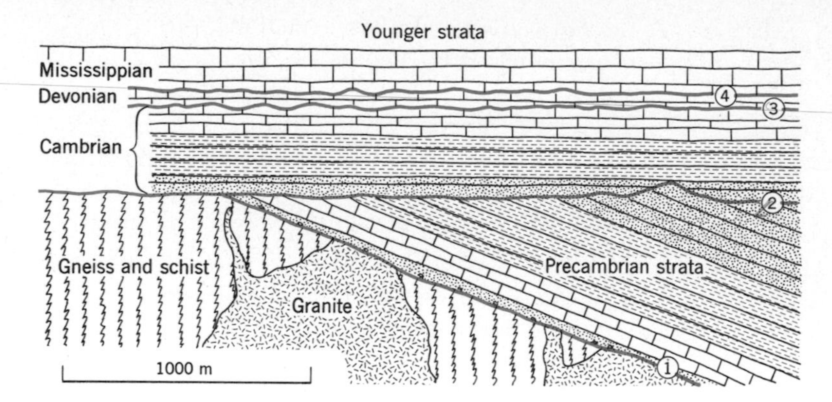


**Figure 14.24**
**Meaning of unconformity.**
**The sedimentary record in the Grand Canyon is broken by large gaps. Angular unconformity indicated at (1) and (2) represent large-scale movements of the crust followed by long intervals of erosion. Surfaces of erosion indicated at (3) and (4), nearly parallel to the sedimentary layers, record broad upwarping followed by erosion, and renewed deposition of sediment.**

was because the land sank, or because movement, lava flows, or landslides obstructed drainage and started deposition by streams or in a lake. If marine beds lie above and below the contact, the information is more definite: (1) older beds were lifted up and eroded, (2) submergence brought back the sea, and (3) another uplift made possible the erosion which has exposed both sets of beds to our view.

### Kinds of unconformity

Possible combinations of crustal movement, erosion, and sedimentation are almost countless, and there are many variations among the examples of unconformity we find in rocks. But we recognize only a few differences of major importance. Angular unconformity has been presented first because the divergence of beds above and below the surface of unconformity makes it easy to recognize. More common but less conspicuous are examples of unconformity between two beds that are essentially parallel. In the walls of the Grand Canyon is a layer of limestone that contains marine fossils of Devonian age. Through nearly the full length of the canyon, about 320km, this layer overlies another marine limestone that contains fossils of Cambrian age (Fig. 14.24). On close inspection, we find the contact between the two limestones is slightly irregular, with shallow, valley-like depressions cutting into the Cambrian limestone. Hence erosion of the Cambrian layer occurred before the Devonian layer was deposited. There was ample time for erosion, because the combined lengths of the Ordovician and Silurian Periods, which lie between the end of the Cambrian and the beginning of the Devonian, total more than 100 million years. Were sediments deposited in that area during those periods and then completely eroded before the Devonian sea came in? Or was the area above sea level and not receiving deposits during two long periods? In either case the surface of unconformity between Cambrian and Devonian strata in the Grand Canyon represents a gap in history. This sort of unconformity, involving a gap in the record but no obvious deformation, is called ***disconformity*** to indicate *unconformity that is not marked by angular divergence between two groups of strata in contact.*

Every unconformity testifies to a "lost interval" in the geologic record at the locality, an interval of time in which no deposition of sediments occurred and in which erosion destroyed part of an earlier record. Critical evidence on the basis of which we recognize any unconformity, then, is of two kinds: absence of sedimentary record for a part of geologic time, long or short, and presence of a buried erosion surface. Recognition of this surface is made easier by difference in structure above and below it, but that difference is not essential.

In many places we find sedimentary strata overlaying metamorphic and igneous rocks. The contact of sandstone on intrusive igneous rock is recognized as sedimentary if the sandstone fills small valleys and contains pebbles and mineral grains clearly derived from the underlying igneous rock. Rock of any kind, then, may be involved in an unconformity.

### Unconformity and earth history

On the continents the deposition of sediment has been interrupted again and again by uplift and erosion. Indeed such erosional destruction makes the record of former events more complete. In the Appalachians, the Rocky Mountains, and the Alps we see great thicknesses of sedimentary beds folded, faulted, and cut by deep valleys, full of information about geologic history. These three mountain chains were formed at different times, and so the records they reveal partly supplement each other. But every thick sedimentary section includes surfaces of unconformity, each representing a gap in the record, and each recording some important event, such as folding and thrusting in a mountain belt or upwarping of a wide continental area.

Within certain limits we can date the events that create unconformity. If the youngest layers in a sequence of folded strata are Permian and the oldest nonfolded beds above are Triassic, we know that the folding occurred near the close of the Paleozoic Era. If the gap in the geologic record is a long one, however, the date may be very imprecise. In central Connecticut the youngest strata in the hanging wall of a great normal fault are of late Triassic age, and the strata deposited across the eroded fault are Pleistocene. Without further information, we cannot fix the date of faulting within the 210-million-year span of post-Triassic, pre-Pleistocene time.

A study of surfaces of unconformity brings out the close relationship between crustal movements, erosion, and sedimentation. All of Earth's land surface is a potential surface of unconformity. Some of today's surface will be destroyed by erosion, but some will be covered by future sediment and preserved as a record of the present. Vigorous erosion is now going on where recently there has been uplift. Erosion by streams is laying bare the records in old rock, and in doing this, it is slowly destroying some of the records. Meanwhile the eroded material is carried away and deposited somewhere else, perhaps on a land surface recently sunk below sea level. And so building up in one place compensates tearing down in another. The many surfaces of unconformity within the geologic column are records of former surfaces of the Earth and testify that the interactions between internal and external processes have been going on throughout Earth's long history.

## SUMMARY

1. Rocks are deformed by fracturing, folding, and flowing.

2. The strength of a rock and the manner in which it is deformed depend on temperature, pressure, and the length of time during which the forces are applied.

3. Evidence of present-day deformation in the crust is found in abrupt movements along faults as well as in areas of the surface that are slowly sinking or rising.

4. Movement on a fault is cumulative, in successive small increments. Some total displacements are hundreds of meters vertically and hundreds of kilometers horizontally.

5. Durable evidence of former crustal movements is registered in deformation of bedrock. The bending and breaking of layered rocks give the clearest record of these movements.

6. Geologic structures, in bedrock, that result from deformation include domes, basins, folds, and faults.

7. Thrust faults, involving great crustal blocks that have moved across gently dipping surfaces, commonly are related to folding in mountain belts.

8. In many areas new rock materials have been deposited as sediment after parts of deformed older bedrock were removed by erosion. The resulting surfaces of unconformity are clear records of crustal movements.

## SELECTED REFERENCES

Anderson, D. L., 1971, The San Andreas Fault: Scientific American, v. 225, no. 11, p. 53–68.

Anderson, E. M., 1951, The dynamics of faulting and dyke formation: Edinburgh, Oliver and Boyd.

Billings, M. P., 1972, Structural geology, 3rd ed.: Englewood Cliffs, N.J., Prentice-Hall.

Compton, R. R., 1962, Manual of field geology: New York, John Wiley.

Dennis, J. G., 1972, Structural geology: New York, Ronald Press.

Hills, E. S., 1972, Elements of structural geology: 2nd ed.: New York, John Wiley.

Lahee, F. H., 1961, Field geology, 6th ed.: New York, McGraw-Hill.

Sitter, L. U. de, 1964, Structural geology, 2nd ed.: New York, McGraw-Hill.

# 15 IGNEOUS ACTIVITY AND METAMORPHISM



## PROPERTIES OF MAGMA

Igneous activity and metamorphism seem like strangely different topics to be discussed in the same chapter. Yet often they are closely related. Like giant invading armies that change forever the complexion of an invaded country, great bodies of fluid magma intrude the crust, metamorphosing and changing the rocks with which they come into contact. Conversely, under conditions of intense metamorphism, rock may melt, forming new magma that rises upward, intruding metamorphic rock above. Granite batholiths, for example, always intrude metamorphic rocks. These activities, plus all other activities of magma on and in the crust, are embraced within the term *igneous activity*.

**What is magma?**

In an earlier chapter we defined magma as molten silicate material beneath Earth's surface, including crystals derived from it and gases dissolved in it. As this complex mixture of liquid and crystals cools and crystallizes, it reaches a point at which so many crystals have formed that the mixture can no longer flow. The magma begins to behave as a solid igneous rock. When flow ceases, the amount of liquid that remains varies from one body of magma to another, but probably is never less than 10 to 15 percent. Magma is characterized by (1) a range of compositions in which silica ($SiO_2$) is always predominant, (2) high temperatures, and (3) ability to flow.

### Composition

We saw in Figure 3.16 that six minerals or mineral groups—olivine, pyroxene, amphibole, mica, feldspar, and quartz—make up the bulk of all common igneous rocks. The elements contained in these minerals must therefore be the principal elements in magmas. They are Si, Al, Ca, Na, K, Fe, Mg, H, and O. For convenience, we usually express compositional variations in terms of oxide ions, such as $SiO_2$, $Al_2O_3$, CaO, and $H_2O$. The most abundant ion, and the important one for controlling the properties of magma, is $SiO_2$.

The best samples of magma on Earth's surface are newly erupted lavas. Analyzing them, we find compositions within the range of 45 to 75 percent $SiO_2$ by weight. A few lavas reach compositions as low as 30 percent $SiO_2$ and others as high as 80 percent. But these are varieties that form when magma assimilates fragments of sedimentary and metamorphic rock causing its compositions to change, or when magma differentiates on cooling, a process discussed later in this chapter.

Chemical analyses of all kinds of igneous rock indicate that three specific composition ranges predominate over all others. This suggests that three types of magma must be more common than any others. The first contains about 50 percent $SiO_2$ and forms basalt, diabase, and gabbro. The second contains about 60 percent $SiO_2$ and forms andesite and diorite, while the third contains about 70 percent $SiO_2$ and forms rhyolite and granite, (Fig. 15.1). Of all the igneous rock in the crust (oceanic and continental combined) approximately 80 percent forms from basaltic magma, 10 percent from andesitic magma, and 10 percent from rhyolitic magma. By comparison, other magma compositions are insignificant.

The gases dissolved in a magma are important in determining its properties. Despite their importance, however, the amounts and compositions of dissolved gases are difficult to determine. As a magma rises toward Earth's surface, and the confining pressure decreases, dissolved gases bubble out of solution much as carbon dioxide bubbles out of an open bottle of soda. The principal gas is apparently water vapor, which, together with carbon dioxide, accounts for more than 90 percent of all gases emitted from volcanoes. Others are nitrogen, chlorine, sulfur, and argon, plus many species present in amounts less than 1 percent. Although vast quantities of gases are emitted by volcanoes, we cannot always be sure how much has actually been released by the magma, and how much is merely gas that has leaked into the magma from adjacent rock and has been boiled off. For example,

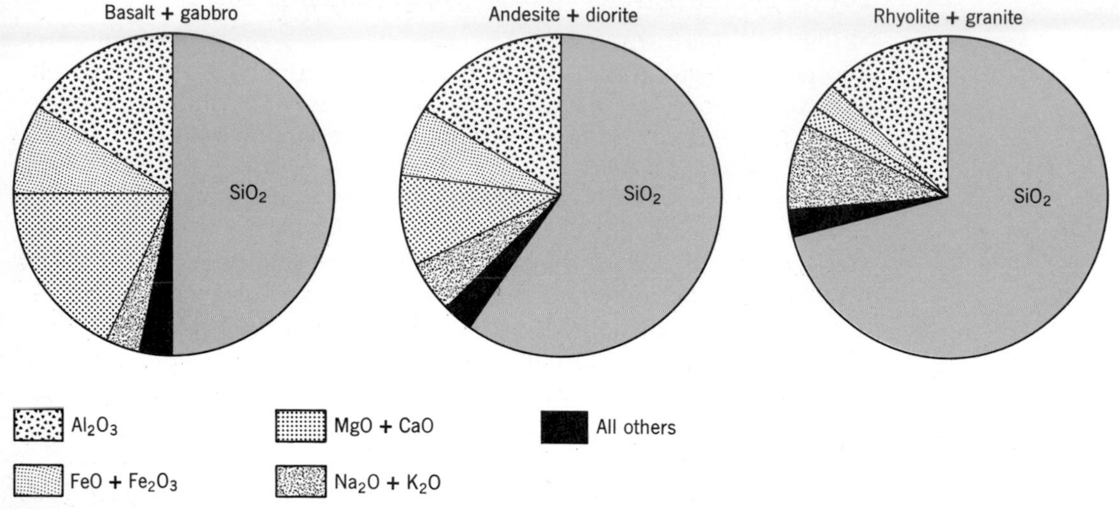


**Figure 15.1**
**Average composition, in weight percent, of the most common types of igneous rock. The three distinct groups are believed to indicate the three fundamental kinds of magma on Earth.** *(After Ronov and Yaroshevsky, 1969, and McBirney, 1969.)*

at the height of its activity in May 1945, the volcano Parícutin, in Mexico, released an estimated 116,000 metric tons of material a day, of which 16,000 tons (14 percent) were water vapor, and 100,000 were lava and ash. But the maximum amount of water vapor it is possible to dissolve in laboratory melts having the same composition as Parícutin magma is only 10 percent. We must conclude, therefore, that the gases emitted by Parícutin contained not only the gases released from the magma, but also an unknown quantity of ground water heated by the rising magma.

Despite uncertainties, estimates of gases dissolved in magma can be obtained from analyses of glassy volcanic rock. Such rock is erupted and chilled so quickly that dissolved gases cannot escape completely. Typical volcanic glasses have water contents of 0.2 to about 3 percent, and most scientists believe that the dissolved gas contents of magmas are probably not much greater.

### Temperature

The temperature of lava as it comes out of a volcano is likewise difficult to measure. Volcanoes are dangerous places and scientists who study them are not anxious to be roasted alive. Measurements must be made from great distances. Using a pyrometer (an optical measuring device), scientists have obtained a number of measurements. They range from 1,040° to 1,200°C. Once extruded, however, a flowing lava cools, and temperatures as low as 800°C have been measured in some lava flows that had nearly ceased to move.

### Mobility

Although magma may seem to be very fluid, many of its properties are more akin to solids than to liquids; lava is more like asphalt on a road than like water. With rare exceptions, lava flows slowly and with difficulty. The *internal property of a substance that offers resistance to flow* is called ***viscosity.*** The more viscous a lava, the less fluid it is. Viscosity of a lava depends on composition (especially the $SiO_2$ and dissolved-gas contents) and on temperature.

The effect of temperature is simple to understand. As with asphalt or thick oil, high temperature leads to greater fluidity. The higher the temperature, therefore, the lower the viscosity and the more readily a magma flows. A very hot magma erupted from a volcano may flow readily, but it soon begins to cool, becoming more viscous and eventually slowing to a complete halt. Figure 3.11*A* is an example of a hot, fluid lava with low viscosity; Fig. 3.11*B* shows very viscous lava of similar composition but presumably lower temperature and possibly containing less dissolved gas.

The effect of silica content on viscosity is not so obvious as the effect of temperature. The same $(SiO_4)^{-4}$ ions that occur in silicate minerals (Chapter 3) occur also in magmas. Just as they do in minerals, the $(SiO_4)^{-4}$ ions in magmas have a tetrahedral shape and link together by sharing oxygens. But unlike the tetrahedra in minerals, those in magma form irregular groupings. As the number of tetrahedra in the groups becomes larger, the magma behaves more and more like a solid and becomes more and more resistant to flow. The number of tetrahedra in the groups depends on the silica content of the magma. The higher the silica content, the larger the groups. Magma composition, therefore, exerts a direct and strong control on viscosity (Fig. 15.2). Magma with the composition of basalt sometimes flows rapidly. Basaltic magma moving down a steep slope on Mauna Loa, in Hawaii, during an eruption in 1850, was clocked at an average speed of 16km per hour. Such fluidity, however, is very rare. Flow rates are more commonly measured in meters per hour or even meters per day.

Magmas that have rhyolitic composition, containing 70 percent or

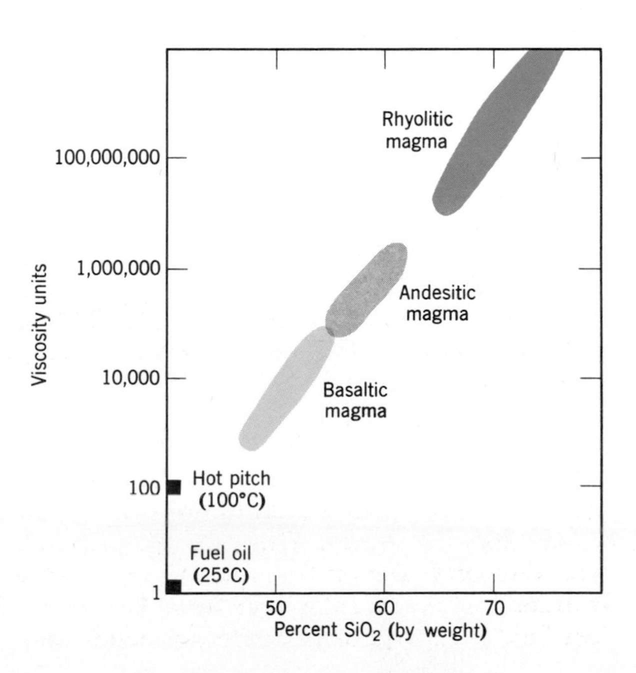


**Figure 15.2**
**Viscosity of magma is strongly controlled by silica content. The more viscous a magma, the less tendency it has to flow. Viscosities of fuel oil at 25°C and pitch at 100°C are plotted for comparison.**

isopach - lines on a map that connect points of equal thickness of a rock unit.

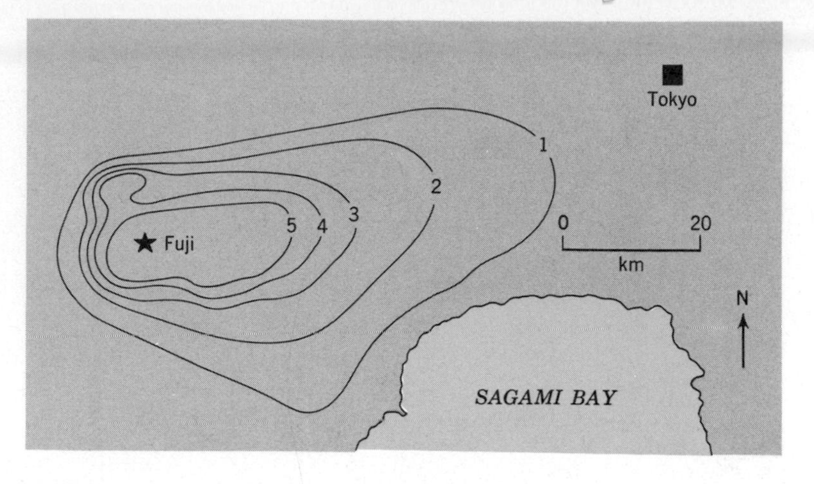


**Figure 15.3**
**Map showing thickness of volcanic ash deposited by Fuji volcano during a single phase of eruption. Lines are isopachs scaled in meters. Most ash is deposited within the first hour of an eruption. (*Data from Machida, 1967.*)**

more $SiO_2$, are so viscous and flow so slowly that their movement can hardly be detected. Their high viscosity even makes it difficult for gas bubbles to escape. As viscous magma cools, the pressure of the trapped gases may become so great that the whole mass of sticky magma simply explodes into billions of tiny, hot, glassy fragments. The mass of fragments and hot gases then erupt as volcanic ash, widely covering the surrounding country (Fig. 15.3).

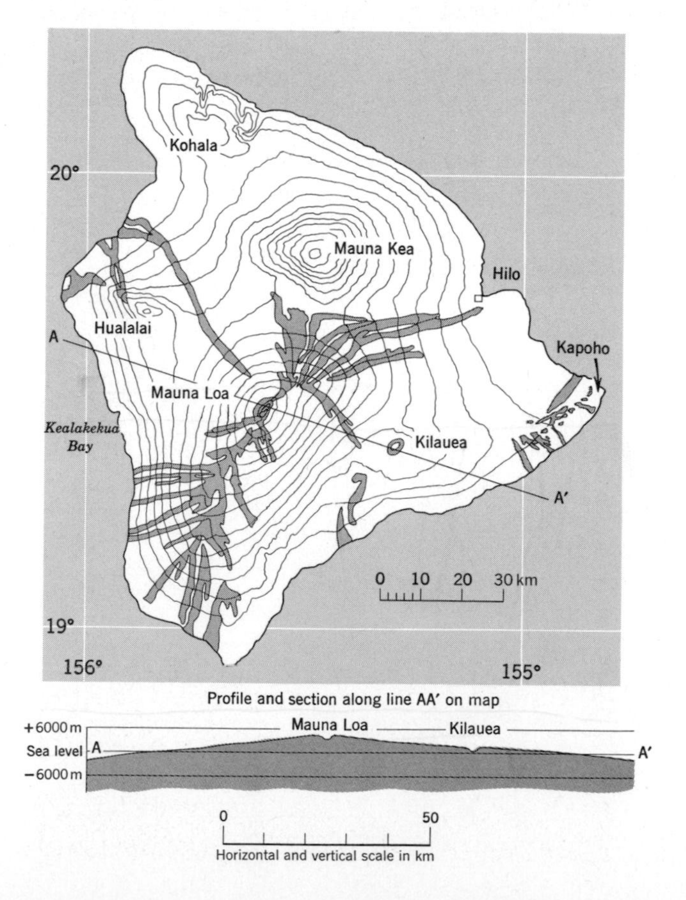


**Figure 15.4**
**The island of Hawaii is composed five overlapping shield volcanoes, of which the largest are Mauna Loa and Mauna Kea. The topographic map shows the major volcanic centers and the longest lava flows erupted since 1750. The contour interval is 300m. A profile and section along the line A-A′ (lower diagram) illustrates the gentle slope characteristic of shield volcanoes. (*After H. T. Stearns and G. A. MacDonald.*)**

# VOLCANOES

Before we consider the origin of magma, we must discuss volcanoes a little further, because they are the only places on Earth where we can study magma directly. We noted in Chapter 3 that a volcano is a vent from which molten igneous matter, solid rock debris, and gases are erupted. A ***fumarole*** is *a vent that emits only gases*. A vent is commonly surrounded by either a cone-shaped or sheet-like pile of volcanic material. Most people refer to both the vent and the volcanic pile as a volcano, and we will do so too, although this is not strictly correct in terms of our definition. The shapes of many volcanic piles are so similar, and the processes that build them so nearly alike, that special names have been given to the common kinds of volcanoes.

### Kinds of volcanoes

The volcano easiest to visualize is one built up by successive flows of lava. Such volcanoes are characteristically built by very fluid lavas, capable of flowing great distances down gentle slopes and of forming thin sheets of nearly uniform thickness. Eventually the pile built up in this fashion develops a shape resembling a shield, convex-side up. A ***shield volcano***, then, is *a volcano that emits fluid lava and builds up a broad, dome-shaped edifice, convex upward, with a surface slope of only a few degrees* (Fig. 15.4). Shield volcanoes are characteristically formed by the eruption of basaltic lava; the proportions of ash and other fragmental debris is small. Because basalt is the igneous rock of the ocean basins, shield volcanoes are characteristically oceanic. Hawaii, Iceland, Samoa, the Galapagos, and many other oceanic islands are the upper parts of large shield volcanoes.

Andesitic and rhyolitic lavas are so highly viscous that they congeal even on steep slopes. As we have discussed, gas that bubbles out of viscous magma commonly does so with great violence, ejecting vast quantities of pumice fragments, volcanic ash, and other fragmental debris (Fig. 15.5). In many instances

**Figure 15.5**
**Explosive eruption of volcanic ash and rock fragments from Parícutin Volcano, Mexico. In 1943, when this photo was taken, Parícutin was only a few months old. It was born February 20, 1943, when it suddenly burst into action before the startled eyes of a farmer ploughing his cornfield. (*W. F. Foshag.*)**

**Figure 15.6**
**Shishaldin Volcano, a giant composite volcano of andesitic composition, towers 2,857m above sea level on Unimak Island in the Aleutians. (*U.S. Navy.*)**

andesitic and rhyolitic volcanoes erupt larger volumes of fragmental debris than of lavas. A ***composite volcano*** (also called a ***stratovolcano***) is, then, *a volcano that emits both fragmental material and viscous lava and builds up a steep conical mound.* The surface slope of a composite volcano may reach 30° or more near the summit, decreasing to about 6° near the base. This is due to the extreme stickiness of the lavas they erupt. The beautiful steep-sided cones of composite volcanoes are among Earth's most picturesque sights (Fig. 15.6). The snow-capped peak of Mt. Fuji in Japan has inspired poets and writers for centuries. Mount Rainier in Washington and Mt. Hood in Oregon are majestic examples in North America. Andesites and rhyolites are most commonly found on continents; so composite volcanoes are more common on continents than in the ocean basins.

Associated with both shield and composite volcanoes are numerous features that give volcanic terrains a special and unique character. Fractures may split the cone so that lava, ash, or both emerge along its flanks. Small cones then develop, peppering the slope of the main mountain like so many small pimples. Gases emerge from small vents, altering and discoloring nearby rocks, and hot springs may form, bubbling off evil-smelling, sulfurous gases. Near the summits of most volcanoes is a ***volcanic crater***, *a funnel-shaped depression from which gases, fragments of rock, and lava are ejected.* The summit of Shishaldin Volcano is crowned by a small crater that is just visible in Figure 15.6; a little puff of steam is rising from one edge of it.

Many volcanoes are marked, near their summits, by a striking and much larger depression than a crater. This is a ***caldera***, *a roughly circular, steep-walled basin several kilometers or more in diameter.* Calderas originate through collapse because magma has been withdrawn from below. Following eruption, the chamber from which the lava and volcanic ash were emitted becomes empty or partly empty. The now-unsupported roof of the chamber simply collapses under its own weight, like a snow-laden roof on a shaky barn, dropping downward within a ring of steep normal faults. Subsequent volcanic eruptions commonly occur along these faults, thus creating roughly circular rings of small cones. Crater Lake, Oregon, occupies a circular caldera 8km in diameter (Fig. 15.7), formed after a great eruption about 6,600 years ago. That eruption pulverized and blew off the top of a composite volcano, posthumously named Mount Mazama. What remained of the summit of Mount Mazama then collapsed into the partly empty magma chamber.

The foregoing volcanic features originate when volcanic materials issue from nearly circular vents. Some lava and ash reach the surface via elongate fractures which, when the walls are spread apart, become fissures. *Extrusion of volcanic materials along an extensive fracture* is a ***fissure eruption***. Such eruptions are characteristically associated with fluid basaltic magma, and the lavas that emerge from the fissures tend to spread widely and to build up flat plains. The fissure eruption of Laki, Iceland, in 1783 took place along a fracture 32km long. Lava flowed 64km outward from one side of the fracture and nearly 48km outward from the other side; altogether it covered an area of 558km$^2$. The volume of lava extruded has been estimated at 12km$^3$, making this the largest single eruption observed during recorded history. There is evidence that many larger eruptions have occurred in the past, however. The Roza flow, a great sheet of basalt in eastern Washington, can be traced over 22,000km$^2$ and shown to have a volume of 650km$^3$.

Large as the Crater Lake caldera may seem, it is tiny by comparison with calderas observed in some regions of widespread volcanism. There, complexes of volcanoes and fissure eruptions may cover more than 100,000 square kilometers with lava and volcanic ash. Yellowstone National Park is a striking example. Following several earlier periods of volcanic activity, a catastrophic eruption occurred in Yellowstone about 600,000 years ago. Approximately 900km$^3$ of rhyolite lava and hot ash were rapidly erupted from a shallow magma chamber, whose roof, no longer supported from below and suddenly loaded with newly erupted lava above, collapsed to form a caldera 70km long and 45km wide (Fig. 15.8). Most of the hot springs and other features for which Yellowstone Park is now so famous lie within the caldera. Scientists have shown that a few kilometers beneath its floor a huge mass of rhyolitic magma still occupies the remains of the magma chamber. The scientists are divided on whether future eruptions are in store.

**Frequency of eruption**

Some volcanoes give off steam and gases almost continuously. Others lie dormant, seemingly devoid of all volcanic activity through periods of hundreds or even thousands of years; then suddenly they erupt with violent activity. When we consider the frequency of volcanism, two questions arise. How long does activity last at any volcanic center, and how frequent are eruptions? Neither question can be answered with certainty. The best way to attempt an answer is to describe the behavior of volcanoes in three well-known volcanic areas.

First, the Hawaiian volcanoes. Referring again to Figure 13.10, we see that the Hawaiian Archipelago is a line, nearly 2,500km long,

*A*

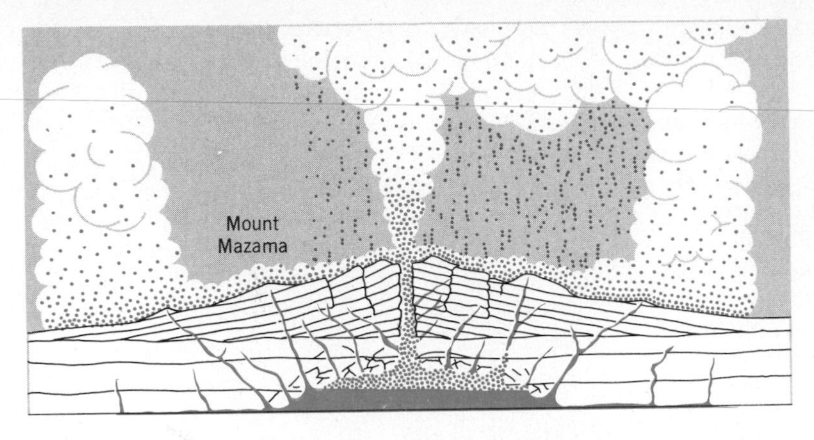


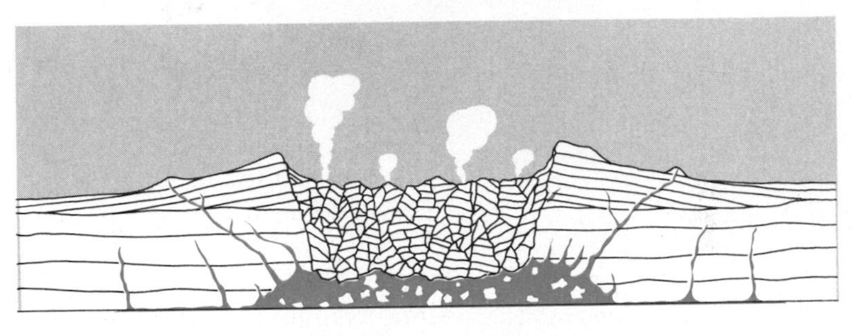

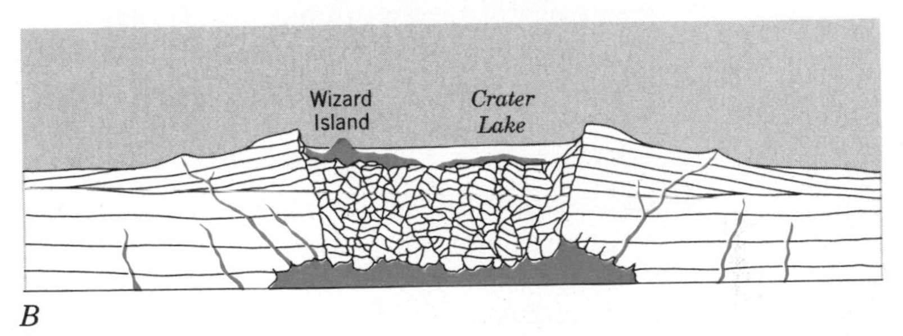


*B*

**Figure 15.7**
**Crater Lake and ancient Mt. Mazama.**
***A.*** **(see opposite page) Crater Lake, Oregon, occupies a caldera 8km in diameter, which crowns the summit of a once-lofty composite volcano, posthumously named Mount Mazama. Wizard Island (in the foreground), a small cone of volcanic cinders, capped by a circular crater, formed after the collapse. (*Washington Air National Guard, H. Miller Cowling.*)**
***B.*** **Following a violent eruption 6,600 years ago, the caldera formed when what remained of the summit of ancient Mount Mazama collapsed into the partly evacuated magma chamber. (*After Howel Williams, 1942.*)**

of small volcanic islands. Each island consists of one or more shield volcanoes, and each volcano is built of innumerable flows of basalt. The volcanoes become progressively younger toward the southeast and apparently are caused by the Pacific Plate moving northwesterly over a hot source of magma in the mantle below. K/Ar dates indicate that the exposed parts of most islands were constructed in less than 500,000 years, and suggest that each volcano grew upward from the sea floor to its maximum height in little more than a million years.

The island of Hawaii, at the southeast end of the chain, has three volcanoes active in recent years (Fig. 15.1). Kilauea, most active of the three, has been emitting steam and other gases almost

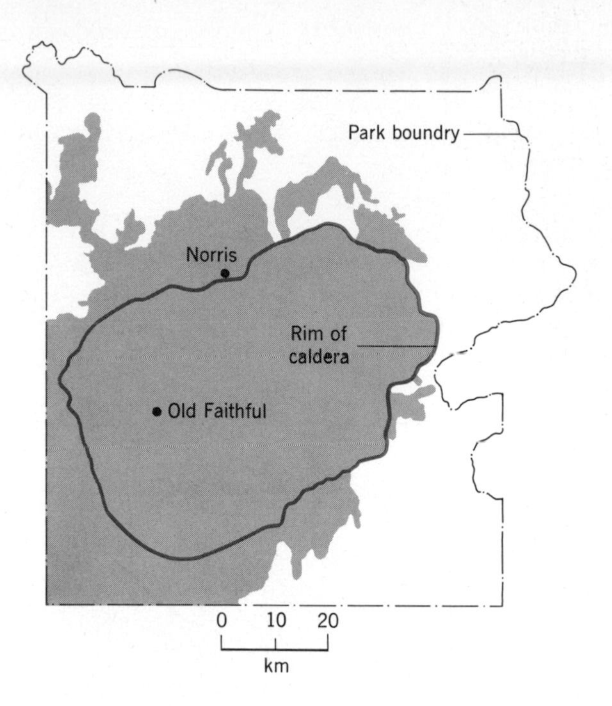


**Figure 15.8**
**Yellowstone Park caldera.**
**Much of Yellowstone National Park is occupied by a huge caldera that formed 600,000 years ago. Lavas and ash flows (gray) that preceded formation of the caldera extend beyond Park boundaries to south and west.** *(Data from Eaton and others, 1975.)*

continuously since systematic observations began in 1912. Frequency of lava eruptions also is high. Eruptions have occurrred, on an average, every two years, although periods of almost ten years have sometimes passed without eruption. Off-setting these quiet periods are times of great activity. From 1968 to 1973 Kilauea was in a state of nearly continuous eruption.

Another example of a basaltic volcano in nearly continuous eruption is Stromboli, in the Tyrrhenian Sea between Sicily and Italy. Since the beginning of recorded history, Stromboli has been hurling clots of red-hot lava upward at 10- to 15-minute intervals. These fall back into the crater and mingle with a pool of molten lava. Sometimes the pool overflows, and lava pours gently down the mountain, but at all times the luminous masses thrown up by the crater can be seen from great distances by men at sea, thus earning for Stromboli the friendly title, "Lighthouse of the Mediterranean."

An entirely different kind of eruptive style was observed at Mt. Katmai on the Alaskan Peninsula. Until 1912 Katmai, a rhyolitic volcano, was not even recognized as being active. In that year it exploded with great violence, blasting the top off its steep-sided cone and leaving a caldera 3.5km long and 2.5km wide. At the same time, from a vent on its lower flank, Katmai erupted a great sheet of pumice, glassy volcanic ash and pulverized rock fragments. When they had settled, the fragments formed a red-hot blanket of tuff 33km long, up to 5km wide, and as much as 30m thick. It took many years for this hot blanket to cool, during which time rain falling on it caused it to fume and steam from innumerable cracks, leading to the now-famous name, "Valley of Ten Thousand Smokes." Katmai has again lain dormant for 60 years. When and if it will erupt again are unanswered questions. Why it erupted in 1912 is also an unanswered question.

### Volcanoes and people

Some aspects of volcanism are helpful to mankind. For example, weathering converts volcanic ash, with great rapidity, to exceptionally fertile soils. In some parts of the world crops can be grown beginning as soon as one year after eruption. In other places, such as Italy, New Zealand, Iceland, and California, volcanic steam is tapped by deep drill holes and is used to drive electric generators. Volcanic power of this sort is *geothermal power.* Despite these helpful aspects, volcanoes are most commonly thought of as danger spots. Underground movement of magma can trigger destructive earthquakes. Flowing lava, such as that shown in Figure 3.11, destroys everything in its path, riding over fields and villages alike. Yet flowing lava moves so slowly that one can easily get out of its way; in spite of the frequency of lava flows on Hawaii, loss of life is exceedingly rare.

The violent eruptions associated with andesitic and rhyolitic volcanism present a different situation; loss of life is more common. Great incandescent clouds of volcanic ash and rock debris erupt explosively, and buoyed by superheated steam and other volcanic gases, form hot, deadly avalanches that roll down steep mountain slopes at high speed. This is the kind of eruption that occurred at Katmai in 1912, but fortunately no people were nearby when the eruption took place. But a devastating eruption of this kind occurred on Mont Pelée, Martinique, in 1902. After 50 years of quiescence, Mont Pelée burst into life, emitting steam and erupting ash for several months. On May 8, a hot cloud of ash roared down the mountainside at 60m per second, searing everything in its path. Eight kilometers away from the mountain lay St. Pierre, capital of Martinique, a community of 30,000 people. The fiery cloud engulfed St. Pierre, killing all but two persons, one a prisoner in a dungeon. All buildings were destroyed (Fig. 15.9), ships in the harbor capsized, and the sea was converted to a boiling mass in which fish and other creatures were scalded to death.

Another tremendous and destructive volcanic explosion, the largest of modern times, occurred in 1815 when Tamboro, an Indonesian volcano, decapitated itself. After a long period of apparent dormancy, Tamboro sprang to life, and with a cataclysmic eruption blew 150km$^3$ of volcanic ash and pulverized rock high into the air. Debris was thickly scattered over a roughly circular area 550km in diameter, the mountain lost 1,300m in height and after the top of the mountain collapsed into the space left by eruption of magma, a caldera nearly 11km in diameter remained as mute evidence of the event. The vast number of people killed in heavily populated Indonesia has never been estimated closely. But people all around

**Figure 15.9**
**Devastated remains of St. Pierre, Martinique.**
**On May 8, 1902, Mont Pelee, 8km away, erupted a fast-moving sheet of hot volcanic ash and gases. The sheet removed all vegetation, reduced the town to utter ruin, and killed 30,000 people.** (*Underwood and Underwood.*)

the world felt the effects of Tamboro. Dust from the eruption was blasted high into the outer atmosphere; some of it remained suspended and circled the Earth for years. Suspended volcanic dust of this sort prevents so much sunlight from reaching the surface that it lowers temperatures world-wide through periods several years long.

The geologic record is filled with evidence of eruptions of the Pelée, Katmai, and Tamboro types. During the last few million years, for example, such eruptions have occurred repeatedly in California, Oregon, Washington, Nevada, New Mexico, and Alaska. What would be the magnitude of the catastrophe if Lassen Peak in California, or one of the supposedly dormant volcanoes in the Cascade Range in Oregon or Washington, should suddenly erupt with the violence of Tamboro?

During the last thousand years about 520 volcanoes have been observed in eruption. They are widely distributed around Earth's surface; somewhere on Earth a volcano is always erupting. Many others that have not erupted during the past thousand years will almost certainly erupt during the next thousand. The timing and frequency of eruptions are erratic, making predictions uncertain. Yet a promising new development may help. Scientists have noted that prior to many eruptions, slight warming of the ground occurs in the immediate vicinity of the volcano. With films sensitive to radiation in the infra-red region, cameras mounted on orbiting satellites can post a continuous watch for such developing hot spots, warning of areas of likely activity so that closer appraisals can be made and endangered populations warned.

### Thermal springs and geysers

When volcanism finally ceases, igneous rock in the old magma chamber remains very hot for possibly a million years or more. Descending ground water that comes into contact with the hot igneous rock becomes heated and tends to rise again toward the surface, along a fault or other avenue, where it forms a *thermal spring.*

Although most thermal springs are associated with volcanism, some are not. Ground water sometimes descends so deeply into the Earth that it is warmed by the general internal heat. Altogether there are more than 1,000 thermal springs in the United States, most of them in volcanic regions in the western states. Even larger numbers exist in other parts of the world.

Water temperatures in thermal springs range all the way up to the boiling point. Because dissolution is more rapid in warm water than in cold, thermal springs are likely to be unusually rich in mineral matter dissolved from rocks with which they have been in contact. In some springs the mineral content has medicinal properties.

*A hot spring equipped with a system of plumbing and heating that causes intermittent eruptions of water and steam* is a ***geyser.*** The name comes from an Icelandic word meaning *to gush,* for Iceland is the home of many geysers. Most of the world's geysers that are not in Iceland are in New Zealand or in Yellowstone National Park. In all these regions there is evidence of volcanic activity late in geologic time, and the heat for geysers probably consists of masses of hot rock down below the surface.

The feature that marks a geyser is that it erupts not continuously but intermittently. No two geysers behave in exactly the same way, and we cannot observe and study the system of underground passages that supplies any one of them. However, they are probably all alike in that they are fed by ordinary ground water derived from rainfall. The water occupies a natural tube, probably crooked, that extends downward from the surface. It is heated by contact with hot rock until its temperature is nearly at the boiling point. In a straight tube, convection would occur as it does in a teakettle, and would equalize water temperature throughout the tube. But in a crooked tube convection cannot occur effectively, and so from top to bottom of the tube the water is at its boiling point.

Here we come to a basic principle, which probably explains the on-again off-again character of a geyser: pressure increases with depth and the boiling point rises with increasing pressure. So at the bottom of our tube, pressure is greatest and boiling temperature is highest. When the condition is reached where all the water in the tube is at its boiling point, a very slight decrease of pressure or increase of temperature will make the bottom water boil. The resulting steam pushes the overlying water up through the tube and a little is forced out at the top. The loss of water reduces the pressure below, so that through most or all of the tube the water is suddenly converted into steam and a violent eruption occurs at the surface.

Old Faithful in Yellowstone Park, the most famous American geyser, erupts for a few minutes about once an hour, throwing a jet of steam high in the air. During the intervals between eruptions the emptied tube is refilled with water, which is then heated to the critical point at which the next eruption is triggered off.

## Origin of magma

Making a magma is like cooking in the kitchen, where both ingredients and a source of heat are necessary. First consider the source of heat. In the kitchen it is the stove. In Earth it is the great store of heat in the mantle and lower regions of the crust. But temperatures have to be very high indeed before rock melts—so high, in fact, that we usually think of rock as being fireproof and indestructible. How, then, does temperature increase with depth and at what point does rock start to melt? To answer these questions we must first consider the ***geothermal gradient,*** *the rate of increase of temperature downward in the Earth.* Gradients beneath continental crust differ from those beneath oceanic crust, because the distribution of radioactive elements, the principal sources of heat, differs in the two regions. Nevertheless, as we see in Figure 15.10, temperatures in both cases rise above 1,000°C at rather shallow depths. We already know that some lavas are fluid at 1,000°C. Why then isn't Earth's mantle entirely liquid? The answer is that pressures are too great. As pressure rises, the temperature at which a compound melts also rises. For example the mineral albite, which melts at 1,104°C at Earth's surface where pressure is 1kg per cm$^2$, must be raised to 1,440°C at a depth of 100km below the surface before it will melt, because the pressure is 35,000 times greater. Therefore, whether a given rock melts and forms magma at any specified depth in the Earth depends on the geothermal gradient and the effect of pressure on its melting temperature.

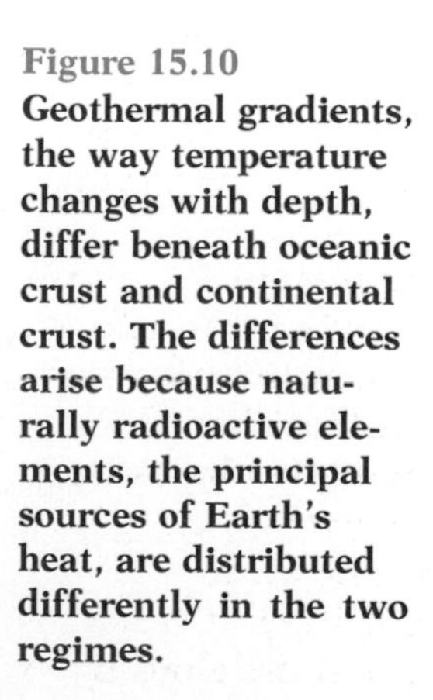

**Figure 15.10**
**Geothermal gradients, the way temperature changes with depth, differ beneath oceanic crust and continental crust. The differences arise because naturally radioactive elements, the principal sources of Earth's heat, are distributed differently in the two regimes.**

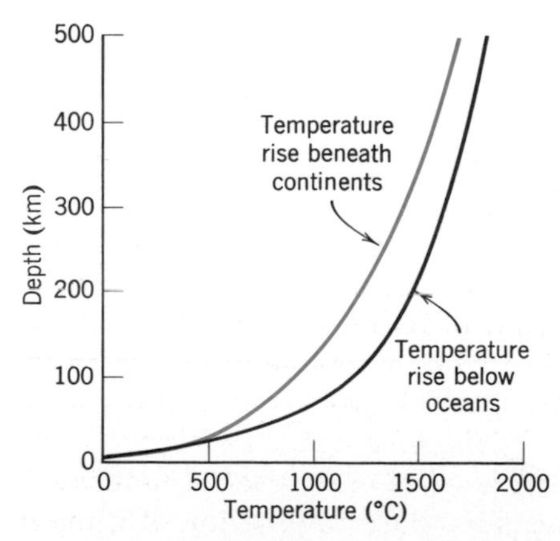

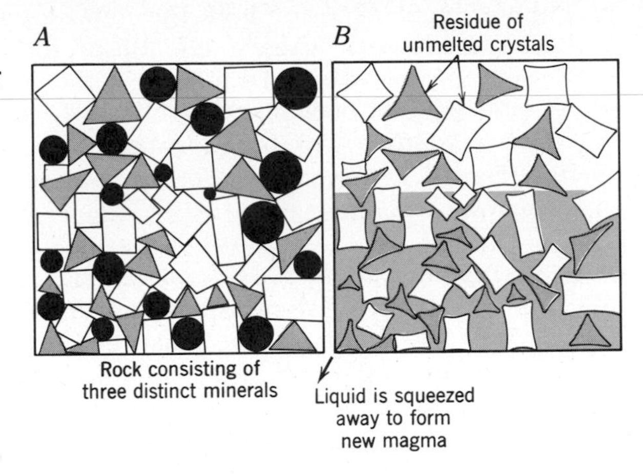


**Figure 15.11**
**Creation of magma by partial melting of rock.**
***A*. Rock consisting of three distinct minerals.**
***B*. As rock is heated, one mineral melts and dissolves small portions of the others. The composition of the newly formed liquid, therefore, differs from the composition of the now-melted mineral and of the remaining minerals. When the liquid is squeezed out, it forms magma of one composition and leaves a residue of unmelted crystals having a different composition.**

The effect of pressure on melting is easy to understand, provided the rock is dry. When water or water vapor is present, however, a complication enters and another effect occurs, an effect similar to that of salt on an icy road. Salt causes ice to melt by forming a salty solution that can freeze only at temperatures below the freezing point of pure water. We say that salt *depresses* the freezing point of water. Similarly, the presence of water dissolved in magma depresses the freezing point of magma. Or, to say it another way, wet rock will melt at lower temperatures than dry rock of the same composition. Furthermore, as the pressure rises, the effect of water also rises. Increasing pressure therefore decreases still further the temperature at which a wet rock starts to melt. This is exactly opposite to the effect of pressure on the melting of a dry rock.

### How rock melts

Let us now return to our analogy between magma formation and cooking. We must still discuss the ingredients. For cooking, we need familiar things such as flour, eggs, and sugar; the recipe for magmas, however, calls for rocks and minerals.

If a whole rock melts, the resulting magma will have the same composition as its parent. But rock is a complex mixture of minerals, and it does not melt at one specific temperature as a pure compound would do. Once a rock reaches a temperatue at which melting starts, and the temperature continues to rise, first one mineral melts, and then another. At any instant the aggregate of unmelted crystals has a composition different from the already-formed liquid. Suppose now that the liquid from a partially melted aggregate is squeezed out of the remaining pile of unmelted crystals. The liquid is a magma, with a composition that differs from that of its parent rock, just as the composition of water squeezed from a sponge differs from the overall composition of the wet sponge. *The process of forming magmas with differing compositions by the incomplete melting of rocks* is known as ***magmatic differentiation by partial melting*** (Fig. 15.11). It is not difficult to see how the composition of a magma that develops by partial melting depends on both the composition of the parent rock and the percentage of it that melts. Basalt forms, apparently, by partial melting of rock in the mantle; some people estimate that as much as 40 percent of the parent rock in the mantle may melt at one time or another. Granite, too, appears to form by partial melting, but by melting of rock of the continental crust rather than of the mantle.

### Basaltic magma

Basalt is the characteristic igneous rock of the oceanic crust. Immediately below the oceanic crust lies the mantle; availablc evidence points to the mantle as the source of basaltic magma. Basalt contains water-free minerals such as pyroxene, olivine, and feldspar. This fact, plus observations that basaltic magma in Hawaii and elsewhere contains little water, suggests that basalt is essentially a dry or water-poor magma. All evidence suggests that the water content of basaltic magma rarely exceeds 0.1 percent. Basaltic magma must therefore originate by some sort of dry-melting process. From such considerations as Earth's total mass we know that the mantle does not have the composition of basalt. We surmise, therefore, that basalt must originate by some sort of dry partial melting within the mantle.

Much scientific debate has centered on the exact composition of the mantle and what fraction of that composition melts to form basalt. One possibility is that the composition of the upper part of the mantle resembles the peridotite found in ophiolite complexes (Fig. 13.13). At a depth of about 350km the temperature on the geothermal gradient just reaches a point at which the effect of pressure on the dry melting of periodotite is overcome (Fig. 15.12) and a small amount of liquid with basaltic composition could develop. The melting process may not be completely dry, because trace amounts of water vapor may indeed occur in the mantle. The presence of even tiny amounts of water would allow melting to start at even lower temperatures and shallower depths. We can therefore take 350km as a maximum depth at which basaltic

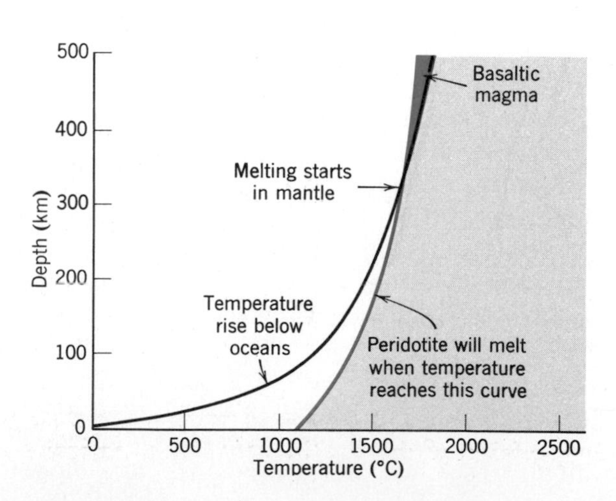


**Figure 15.12**
**Depth/temperature curve for the beginning of melting of peridotite (blue). The composition of peridotite is believed to be similar to that of the upper mantle. Where the melting curve crosses a curve (black) outlining the geothermal gradient below oceanic crust, a small amount of liquid would develop by partial melting of the peridotite. The magma so developed would have the composition of basalt.**

magma is formed. Some evidence suggests that small amounts of basaltic magma can form even at depths as little as 100km. Although opinions differ as to the exact composition of the mantle, and how much must melt to form basalt, it is generally agreed that the formation of basalt commences with partial melting of dry or nearly dry rock in the mantle.

Once a small amount of liquid is formed in the mantle, how does it ever reach the surface? Most liquids are less dense than the solids from which they form. A small amount of liquid developed by partial melting, therefore, will slowly float and force its way up toward the surface. Furthermore, as the liquid rises, the pressure on it continually decreases. As can be seen in Figure 15.12, therefore, a basaltic magma remains liquid as it approaches the surface, even though it may be cooled a little during its upward passage. Most basaltic magma actually reaches the surface and forms lava. Basalts are common rocks, but gabbros, their deep intrusive equivalents, are much less common.

We stated earlier in this Chapter that 80 percent of all magma is basaltic. This high percentage must reflect the huge volume of the mantle that is the source of the magma. Andesitic and rhyolitic magmas, as we shall see, apparently are formed by the melting of oceanic and continental crust respectively. The volume of the crust is tiny compared with that of the mantle.

Most of the basaltic magma that reaches Earth's surface does so along the oceanic ridge. With rare exceptions, as in Iceland, we cannot see the eruptions themselves. Some localized spots in the mantle, not near the ridge, also seem, for reasons still unknown, to be hot spots that are continuing sources of basalt magma. We have discussed the possibility that the Hawaiian Archipelago, and lines of seamounts and guyots as well, form when plates of lithosphere move over a fixed hot spot in the mantle and lava escapes continually upward.

When moving lithosphere places a mass of continental crust over a hot spot in the mantle, spectacular eruptions of basalt can also occur on land. Eruption usually occurs through fissures, and the resulting *fissure basalt that forms widespread, nearly horizontal layers on continents* is called ***plateau basalt.*** Two spectacular examples are seen in Figure 15.13. Even if basaltic magma does not break through the continental crust to form plateau basalt, it can cause extensive volcanism. The volcanic field in Yellowstone National Park, discussed earlier in this chapter, is an example. Scientists have shown that Yellowstone lies above a hot, partly molten mass of basaltic magma in the mantle. It is surmised that the magma has heated the crust, causing part of it to melt, forming a rhyolitic magma. Most of the magma erupted in the Yellowstone region is rhyolitic; only a small fraction is basalt.

### Andesitic magma

The chemical composition of andesite is close to the average composition of the continental crust. Andesite and its equivalent

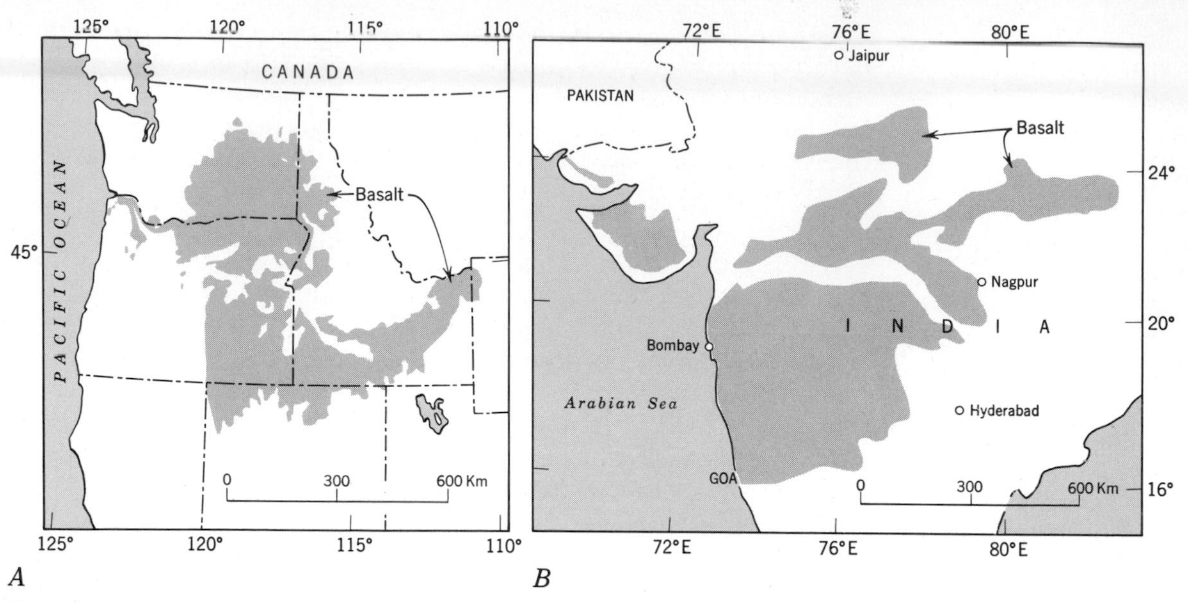


**Figure 15.13**
**Maps of two great basalt plateaus. Areas of basalt (gray) are only remnants of the original flows, reduced by erosion and covered by younger sediments.**

***A.* Columbia Plateau, northwestern United States.** (*After A. C. Waters, 1955, and Geologic Map of the United States, U.S. Geol. Survey.*)

***B.* Deccan Plateau, India.** (*After H. D. Sankalia, 1964.*)

intrusive rock, diorite, are commonly found on the continents. From these two facts we might suppose that andesitic magma forms simply by the complete melting of a portion of continental crust. Some andesitic magma may indeed be generated in this way, but this cannot be the origin of all such magma. Because andesitic magma is extruded from some oceanic volcanoes that are far from continental masses, it must somehow be developed from the mantle or from oceanic crust. Laboratory experiments provide a possible answer. In them, partial melting of wet basalt yields, under some conditions, a magma of andesitic composition. An interesting hypothesis suggests how this might happen. When a moving plate of lithosphere plunges back into the mantle, it carries the wet oceanic crust with it. The plate heats up and eventually the wet rock melts. A small degree of wet partial melting would produce a liquid having the composition of andesitic magma.

The validity of this suggestion remains to be proved, but one piece of evidence indicates it may be correct. In Figure 15.14, the locations of many active volcanoes in and around the Pacific basin are plotted, together with a well-defined line that separates regions where andesite occurs from regions where it does not occur. The line is called the *Andesite Line.* Inside this line and inside the main ocean basin, andesite is unknown. All active volcanoes inside the line erupt basaltic magma, and all the volcanic rock associated with dormant volcanoes is basalt. Outside the line, andesite is common. The Andesite Line coincides closely with the distribution of deep-sea trenches (Fig. 13.11), which, as we learned in Chapter 13, mark the very places where plates of lithosphere plunge back into the mantle. If, therefore, andesites form by partial melting of wet basalt, the process occurs at the very places on Earth where wet basalt is probably present in the mantle. Close examination of the geographic relation between deep-sea trenches and the

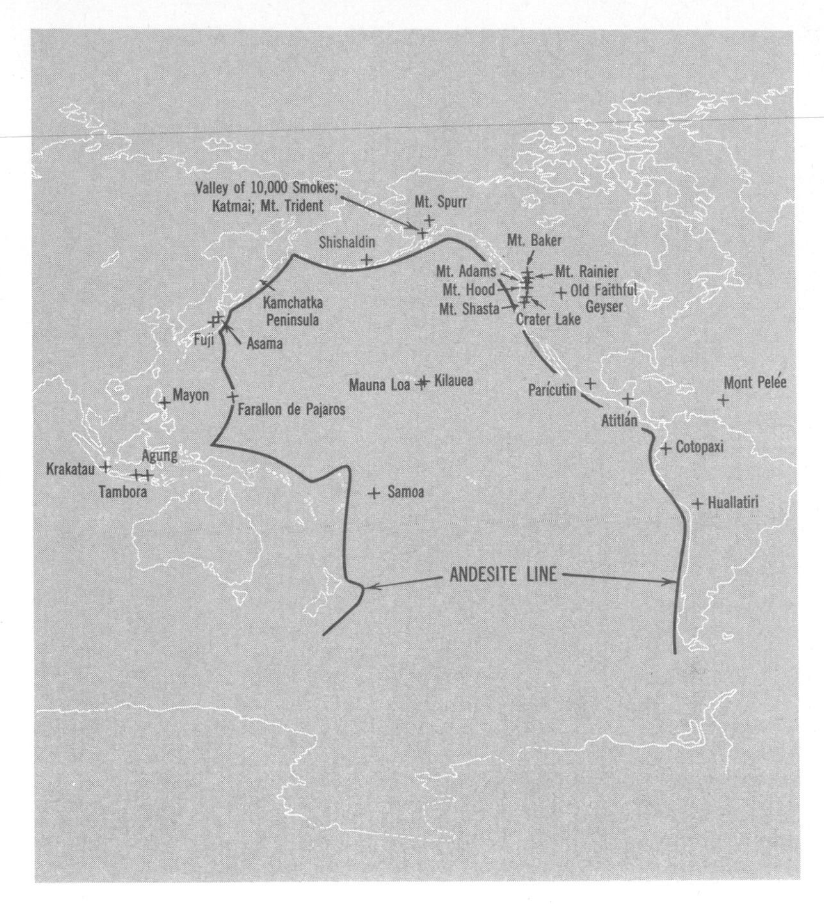


**Figure 15.14**
**The Andesite Line, surrounding the Pacific Ocean basin, separates areas within the basin where andesitic rocks are not found, from areas where andesites are common. Volcanoes, such as Mauna Loa, that are inside the line, erupt basaltic magma but not andesitic magma. Those outside the line may erupt basaltic magma too, but they also erupt andesitic magma.**

arc-shaped belts of volcanoes that lie parallel to the trenches provides more support. The relationship is illustrated in Figure 15.15 using the Japanese islands as an example.

### Rhyolitic magma

Two important observations suggest a unique origin for rhyolitic magma. First, modern volcanoes that extrude rhyolitic magma are confined to regions of continental crust. Similarly, the distribution of ancient rhyolites and their equivalent intrusive rocks (granites and granodiorites) is confined to continents. Second, rhyolitic volcanoes give off a great deal of water vapor, and granitic rocks contain significant amounts of water-bearing minerals such as mica and amphibole. These two points of evidence strongly suggest that the source of rhyolitic magma lies within the continental crust, and that its origin involves some sort of wet melting. Laboratory experiments bear this suggestion out. When, in the laboratory, water-bearing rocks of the continental crust start to melt, the composition of the first liquid that forms is that of granite. As seen in Figure 15.16, the wet-melting curve for granitic magma intersects the geothermal gradient at a depth of 35km to 40km, a depth near the base of the continental crust. Ordinary water-bearing crustal rocks, buried to these depths, will start to melt and the liquid they form is granitic. Similarly, when a heat

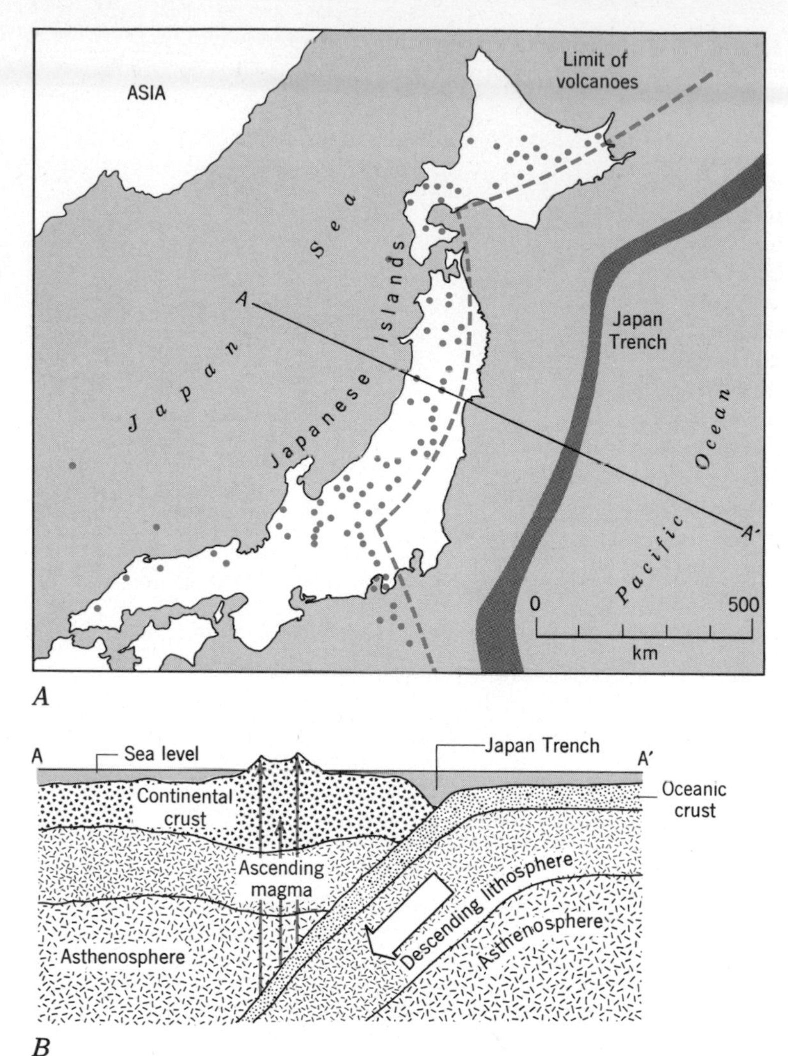


**Figure 15.15**
**Relations among ocean trenches, island arcs, partial melting of oceanic crust, and andesitic volcanism.**
***A*. Arc shaped Japanese islands parallel the Japan Trench. Volcanoes, many of them andesitic and active during the last one million years, are also confined behind arcuate boundaries. (*Adapted from Matsuda and Uyeda, 1971.*)**
***B*. Cross section along line A-A′, demonstrating generation of andesitic magma by partial melting of oceanic crust. At a depth between 100 and 150km, oceanic crust carried downward by descending lithosphere commences partial melting. Andesitic magma rises, creating volcanoes on Japanese islands.**

source such as a rising body of basaltic magma locally raises the temperature of the crust, partial melting will lead to formation of rhyolitic magma. This is what has happened beneath Yellowstone Park.

Although granite and granodiorite are very common, rhyolite is not a common rock. Why should extrusive rock be rare but intrusive rock of the same composition be common, exactly opposite to the relation between gabbro and basalt? Scientists have long been puzzled by this observation. Figure 15.16 illustrates a possible explanation. Once a granitic magma has formed, it starts to rise. But as it rises, the pressure on it decreases and, as we learned earlier, the effectiveness of water in reducing the melting temperature is diminished by reduced pressure. A rising magma formed by wet partial melting must therefore increase in temperature or it will solidify, forming an intrusive igneous rock.

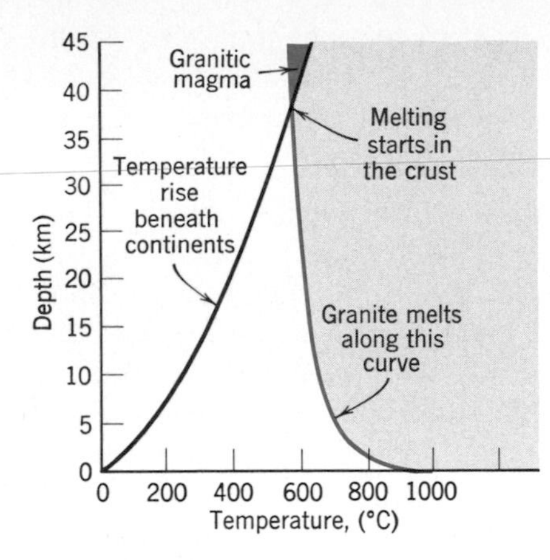


**Figure 15.16**
**Depth/temperature curve for the beginning of melting of granite in the presence of water.**
**At a depth of 35km to 40km below the surface of a continent, the temperature on the geothermal gradient is just sufficient to cause granitic magma to start forming from average crustal rocks. Compare Figure 15.11.**

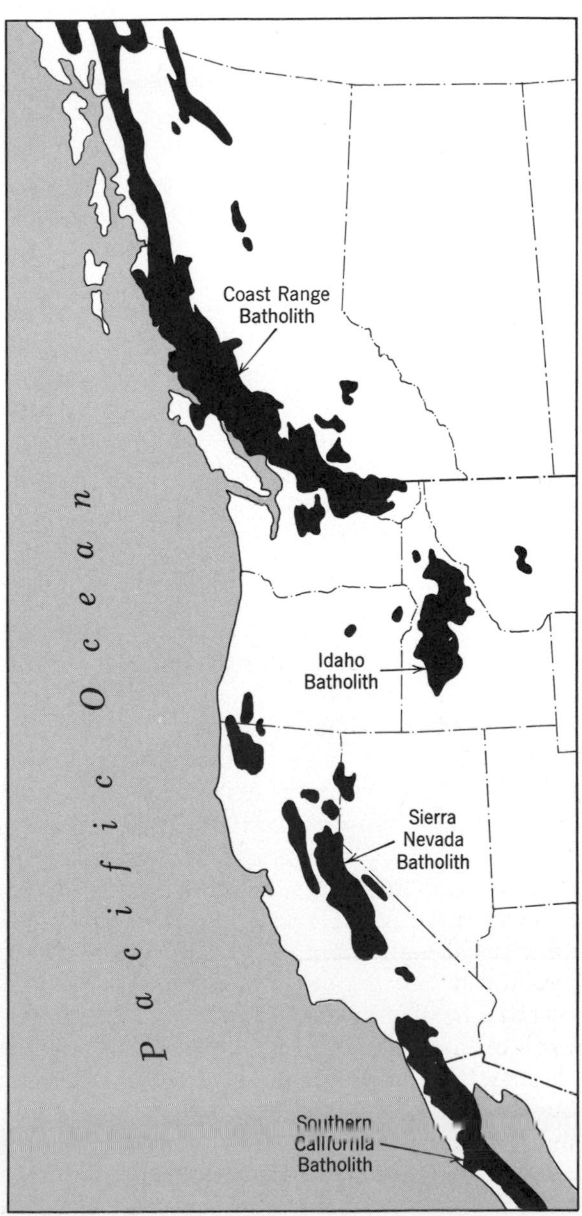


**Figure 15.17**
**The Idaho, Sierra Nevada, and Southern California Batholiths, largest in the United States, are dwarfed by the Coast Range Batholith in British Columbia and Southern Alaska. Each of these gigantic batholiths is believed to have formed from magma generated by partial melting of continental crust, and each intrudes highly metamorphosed rocks.**

There is no way for a rising magma to get hotter; so the generation of magma by partial melting of wet rocks leads naturally to the making of large volumes of intrusive igneous rock and only small volumes of volcanic rock.

One further observation is explained by the origin proposed for granitic magma. When rock becomes buried deep in the crust, it is metamorphosed. When melting starts near the base of the crust, newly formed magma rises and invades the metamorphic rock above. We should therefore expect to find that bodies of granitic rock are closely associated with metamorphic rock. This indeed is the case. Although granitic batholiths may be thousands of kilometers in individual length (Fig. 15.17), they are invariably intruded into highly metamorphosed rock.

# COOLING AND CRYSTALLIZATION OF MAGMA

Although only three common types of magma exist, there are literally hundreds of different kinds of igneous rock. Most are rare, but the fact that they exist emphasizes an important point: a single magma can crystallize into many different kinds of igneous rock. This is true because magma is a complicated liquid. It does not crystallize, like water freezing to form ice, into a single compound. Freezing magma forms several different minerals, and, again unlike water, the minerals crystallize at different temperatures. The process is just the opposite of partial melting. As the temperature slowly falls and a magma freezes, first one mineral crystallizes, then another. Therefore, a freezing magma soon consists of a mixture of already-crystallized minerals and still unfrozen liquid. The combination is like a partly frozen bottle of cider. When cider cools, crystals of ice form from the water it contains. All the other ingredients—sugar, alcohol and flavorings—become concentrated in the remaining liquid. Similarly, in cooling magma, the first minerals that crystallize have different compositions from the remaining liquid. Because different minerals begin to crystallize at different temperatures, the composition of the remaining liquid changes continually as the temperature changes. If, at any time during crystallization the remaining liquid becomes separated from the crystals, the liquid can continue to cool as a magma with a brand-new composition, and the crystals left behind form an igneous rock with an entirely different composition.

### Bowen's reaction series

The great importance of separation between crystals and liquid in magmas was first recognized by the American scientist N. L. Bowen. In 1922, Bowen published the results of experiments that showed that a wide range of igneous rocks could be developed from a single magma, depending on whether or not the early-formed crystals remained in, or were separated from, the

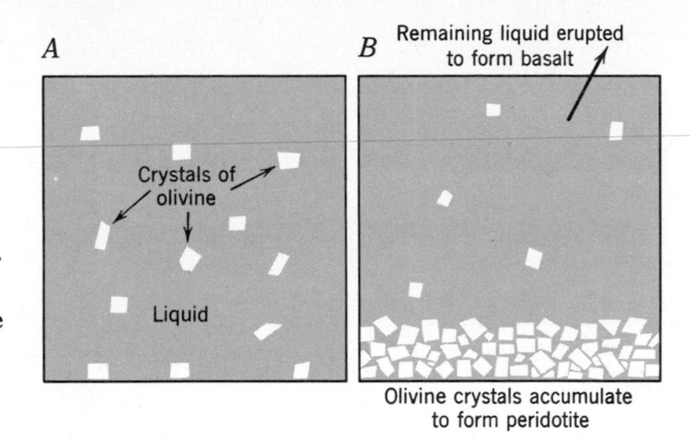


**Figure 15.18**
**Cooling of magma to form different kinds of igneous rock.**
**A. Magma in an underground chamber cools and begins to crystallize. In this example, the first crystals that form are olivine. Heavier than the remaining liquid, the olivine crystals sink.**
**B. Olivine crystals accumulate to form the igneous rock peridotite. The remaining liquid, if erupted on the Earth's surface, crystallizes to form basalt, different in composition from peridotite.**

remaining liquid. We call *the compositional changes that occur in magmas by the separation of early-formed minerals from residual liquids,* ***magmatic differentiation by crystallization.*** Differentiation occurs most commonly when crystals simply sink to the bottom of a pool of magma, leaving a crystal-free liquid above. In the process, two kinds of igneous rock are formed. Crystals that sink and accumulate form one kind; the remaining liquid eventually crystallizes and forms the other (Fig. 15.18).

Bowen also discovered another important property of cooling magma. His experiments proved that once a crystal forms in a cooling magma, it does not remain immune from further changes. As a crystal plus a liquid mixture cools, early-formed crystals react with the residual liquid, continually changing their composition. Therefore, the kind of igneous rock that is formed by sinking crystals, and the kind of residual magma from which the crystals have separated depends very much on when, during crystallization, the separation occurs.

Reactions between crystals and liquids in cooling magmas are very complicated, but Bowen grouped them into two series. In the first, as crystallization proceeds, some minerals such as plagioclase feldspar continually change their compositions but not their crystal structures. These reactions he called a *continuous reaction series.* In the second series, early-formed minerals later react with the cooling liquid to form entirely new minerals. For example, olivine reacts with the cooling residual liquid to form pyroxene. Bowen called such reactions, that lead to the transformation of early-formed crystals into entirely new minerals, a *discontinuous reaction series.*

Figure 15.19 is an example selected by Bowen to demonstrate a continuous and a discontinuous reaction series operating side by side in the same magma. In a magma with the composition of basalt, the earliest crystals to form are olivine and anorthite (the calcium-rich plagioclase feldspar). As cooling continues, the anorthite grains continuously change their composition, becoming richer in sodium. Thus, the feldspar always retains the feldspar crystal structure, and the composition changes continually by solid solution (Chapter 3). Olivine, however, is part of a discontinuous

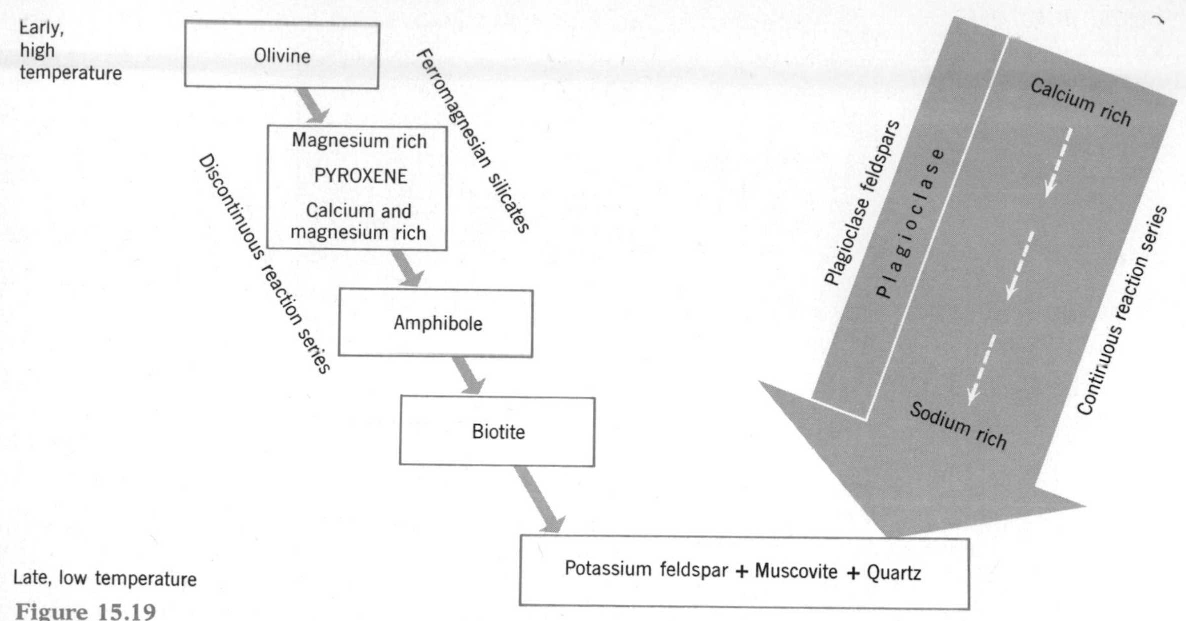


**Figure 15.19**
**Bowen's reaction series.**
**The earliest minerals that crystallize from a cooling magma of basaltic composition are olivine and plagioclase. As crystallization proceeds, olivine (upper left) reacts with the remaining liquid to form a new mineral, pyroxene. Pyroxene reacts to form amphibole, and amphibole reacts to form biotite. The early plagioclase (right) also reacts with the remaining liquid, but, instead of forming a new mineral, continuously changes its composition.**

reaction series. Olivine forms, and, provided it remains in contact with the cooling residual liquid, it eventually reacts to form pyroxene. The pyroxene in turn reacts to form amphibole.

Bowen's reaction series has many consequences. The most important for us is that the residual liquid in a cooling magma contains more silicon, sodium, potassium, and water than the earlier formed crystals. Bowen demonstrated that continuing separation of crystals and liquid in a cooling basaltic magma could even lead to formation of a small residue of granitic magma. Most granites formed during the last 3 billion years were the result of wet partial melting of continental crust, as we have already seen. But the example demonstrates the way magmatic differentiation by crystallization can form igneous rocks that have a wide range of composition, and it also demonstrates how the first granites on Earth probably formed. This happened as long ago as 3.8 billion years, when the continental crust was just beginning to form.

# METAMORPHISM

We learned in Chapter 3 that metamorphic rock is formed within the crust by the transformation of pre-existing rock. Metamorphism is apparently a widespread, common process among Earth's internal activities, for metamorphic rock constitutes roughly 15 percent of the crust. Metamorphism is like cooking, but with the added wrinkle that pressure as well as temperature can cause transformation. The changes are as dramatic as those that occur when a mixture of flour, salt, yeast, and water is baked to form a loaf of bread. They involve both mineral compositions and textures, and may be so profound that it is sometimes difficult to guess what a rock was before metamorphism.

We have already seen that intense metamorphism can lead to melting and the generation of magma. Long before that stage is reached, however, less dramatic changes can be seen. The two processes that cause the changes are *mechanical deformation* and *chemical recrystallization*.

Mechanical deformation includes grinding, crushing, and the development of new textures such as rock cleavage and foliation. Chemical recrystallization includes all the changes in mineral composition, in growth of new minerals, and in loss of $H_2O$ and $CO_2$ that occur as rock is heated. Deformation and chemical recrystallization generally proceed simultaneously, as in slate, when mica and other new minerals grow as parallel grains and so produce cleavage. Purely mechanical efffects do sometimes occur, however, without any changes in mineral chemistry. But they are rare and usually very localized. For example, adjacent to faults, coarse-grained massive rock such as granite may be broken, and individual mineral grains may be shattered and pulverized (Fig. 15.20*A*). This sort of deformation occurs in brittle rocks. When confining pressures are high, so that brittle properties are suppressed, flattening and elongation without associated chemical recrystallization can occur also (Fig. 15.20*B*). But chemical recrystallization is so much more common than simple mechanical deformation that our discussion must center on chemical effects. Thus, we shall now examine the two principal kinds of metamorphism: contact metamorphism and regional metamorphism.

### Contact metamorphism

Contact metamorphism occurs adjacent to hot bodies of igneous rock intruded into cooler rock of the crust. It happens in response

Figure 15.20
**Purely mechanical deformation leads to distinctive textures.**
***A*. Thin section of granite from the vicinity of a fault near Branford, Connecticut. Mineral grains have been shattered, broken, and pulverized. (*B. M. Shaub.*)**
***B*. Deformation under high confining pressure has caused the pebbles in this conglomerate to become flattened and elongate; pocket-knife indicates scale. (*J. W. Ambrose, Geol. Survey of Canada.*)**

*A*

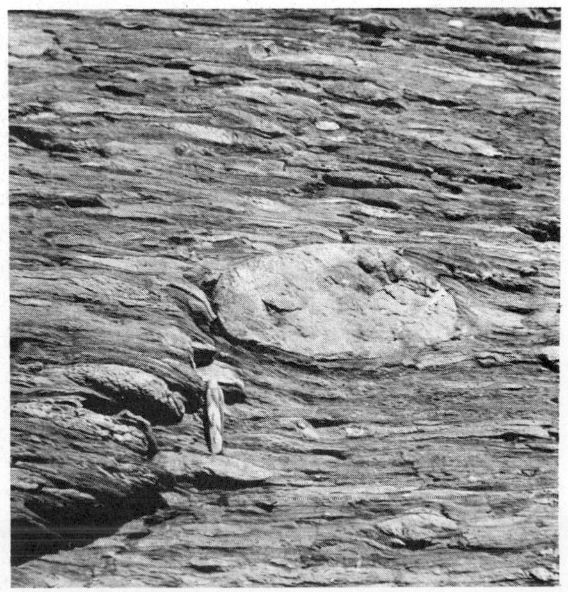
*B*

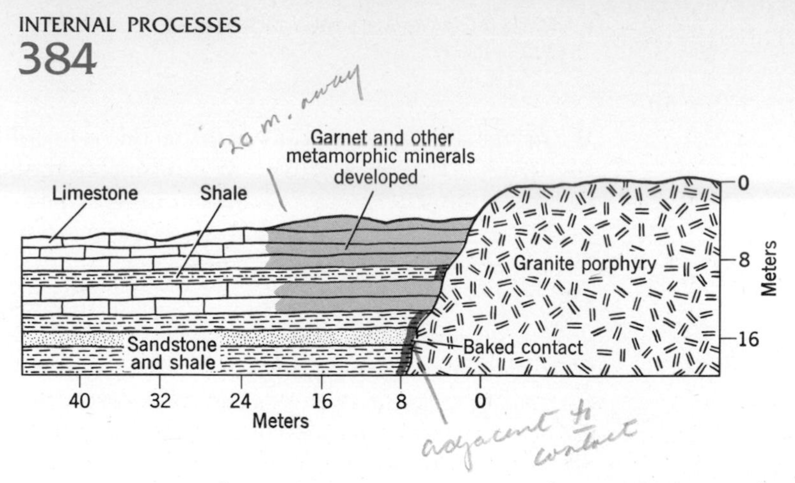


**Figure 15.21 Metamorphism around an intrusive mass of granite porphyry near Breckenridge, Colorado. Sandstones and shales have been baked to fine-grained hornfels immediately adjacent to the contact. However, in limestone, a much more reactive rock, new metamorphic minerals such as garnet have been developed as far as 20m away from the contact.**

to a pronounced increase in temperature but without mechanical deformation. The temperature of the intrusive may be as high as 1,000°C, that of the intruded rocks only 200 to 300°C. Rock adjacent to the intrusive becomes heated and metamorphosed, developing a well-defined shell, or *metamorphic aureole* of altered rock (Fig. 15.21). The width of the aureole depends on the size of the intrusive body. With a small intrusive, such as a dike or sill a few meters thick, the width of the metamorphic aureole may only be a few centimeters. But when the intrusive is large, perhaps a kilometer or more in diameter, the aureole may reach a hundred meters or more in width. The size of an aureole also depends on how susceptible the rocks are to change.

When heated, all kinds of rock react in some fashion, but some are more reactive than others. Those consisting of a single mineral such as sandstone made of pure quartz or limestone made of pure calcite, are simply recrystallized to quartzite and marble respectively. Most commonly, however, two or more minerals are present and can react to form new minerals. For example, grains of quartz in limestone might react with calcite to form wollastonite, $CaSiO_3$, releasing $CO_2$ in the process.

The cook who puts a cake in either a cool oven or a very hot oven knows that the final result will probably not be edible. For correct results the ingredients have to be cooked at the right temperature. The final result of rock cooking depends on temperature too. Close examination of a contact-metamorphic aureole usually reveals that several different and roughly concentric zones of mineral reactions can be identified. Each zone is characteristic of a certain temperature and pressure. Immediately adjacent to the intrusive we find water-free minerals such as garnet, pyroxene, wollastonite, and andalusite. Beyond them we find water-bearing minerals such as epidote and amphibole, and beyond them in turn micas and chlorites. The exact assemblage of minerals we find in each zone depends, of course, on the chemical composition of the intruded rock as well as the temperatures and pressures reached during metamorphism. Because temperature and pressure are the important variables in metamorphism, we refer to *contrasting assemblages of minerals that reach equilibrium during metamorphism within a specific range of physical conditions* as belonging to the same ***metamorphic facies.*** To continue our analogy with cooking, think of a large roast of beef. When it is carved, one sees that the center is rare, the outside well done, and in between there is a region of medium-rare meat. The differences occur because temperature was not uniform throughout. The center or "rare-meat" facies, is a low-temperature zone; the outside, or "well-done" facies is a high-temperature zone.

Clearly, then, each of the zones observed within a contact-metamorphic aureole belongs to a specific metamorphic facies. Furthermore, whenever a rock of specific composition occurs within a given metamorphic facies, the same mineral assemblage appears. We can therefore reason in reverse and study (or examine) mineral assemblages, determine the distribution of metamorphic facies, and then reconstruct the temperatures and pressures at the time of metamorphism. This is done most conveniently by selecting key minerals such as andalusite, sillimanite, garnet, or biotite, and noting where they first appear in the aureole. *A line on a map, connecting points of first occurrence of a given mineral in metamorphic rocks,* is an ***isograd.***

### Regional metamorphism

Contact metamorphism is always local and therefore does not affect large portions of the crust. Nevertheless, study of contact metamorphism is helpful because it allows us to decipher the general processes involved in other kinds of metamorphism. The most common metamorphic rocks occur through areas of thousands of square kilometers and are therefore called *regional metamorphic rocks.* Unlike contact metamorphism, regional metamorphism involves a considerable amount of mechanical deformation in addition to chemical recrystallization. As a result, regional metamorphic rocks tend to be strongly layered and distinctly foliated. Slate, phyllite, schist, and gneiss are the most common varieties, and as we learned in Chapter 3, each has a characteristic texture.

Regional metamorphism occurs when large volumes of Earth's crust are buried during mountain building, and are subjected to the high temperatures and pressures that characterize deep parts of the crust. How such burial occurs is discussed in Chapter 18.

What happens when a pile of strata, formerly at or near the surface, becomes deeply buried? Strata that were originally horizontal become faulted, folded, and buckled. In short, mechanical deformation commences. As depth of burial increases, the strata are subjected to increasing pressure and temperature. New minerals start to grow. But rocks are poor conductors of heat; so the heating-up process is very slow. The temperature reached by a buried pile of strata depends on both depth and duration of burial. If burial is very slow, heating of the pile keeps pace with the temperature of adjacent parts of the crust; that is, a normal continental geothermal gradient is maintained. But if burial is very fast, the pile has insufficient time to heat up; so conditions of high pressure but rather low temperatures prevail. The minerals that

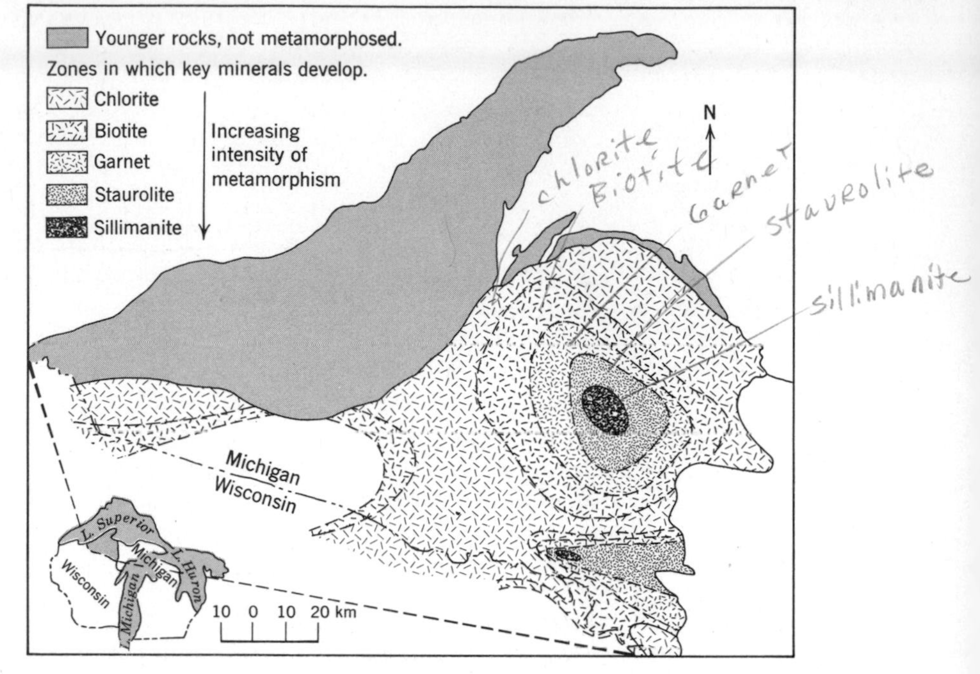


**Figure 15.22** **Zonal sequences of minerals and textures developed in regionally metamorphosed rocks reflect gradients of temperature and pressure during metamorphism. Metamorphic zones in an area of Michigan are reflected by the first appearance of key metamorphic minerals. Metamorphism occurred about 1.5 billion years ago. Rocks that were deposited later than the metamorphism are unchanged. Two clearly defined centers of intense metamorphism are identified by the presence of sillimanite. (*After H. L. James, 1955.*)**

grow are controlled both by temperature and pressure; so the mineral assemblages we observe depend on rates of burial.

In a buried pile of strata both temperature and pressure vary. The larger the pile, the greater the variations. Regional-metamorphic rocks, like their contact-metamorphic equivalents, develop zonal sequences of minerals and textures in response to the variations of temperature and pressure (Fig. 15.22). Unlike metamorphic aureoles, however, zones of regional metamorphism tend to be broad and undulating, showing both horizontal and vertical changes. Where aureoles resemble a series of concentric cylinders surrounding an intrusive, zones of regional metamorphism are analogous to a domed pile of blankets, each blanket representing a specific metamorphic facies. Both the concept of facies and the concept of isograds apply equally well to contact- and regional metamorphism. But the sequence of index minerals that develops differs from case to case, reflecting the many different geothermal gradients that are possible.

### Role of volatiles in metamorphism

So far we have discussed metamorphism as if the only important variables were temperature and depth of burial. But other variables also play a part. These are the amounts of $H_2O$ and $CO_2$ that are present, and the ease with which these substances can move around in rocks.

At depths greater than about 10km, the only significant amounts of $H_2O$ and $CO_2$ in rock are combined in mineral structures. But the kinds of reactions that take place during metamorphism commonly involve the release of $H_2O$ and $CO_2$. If they cannot escape after

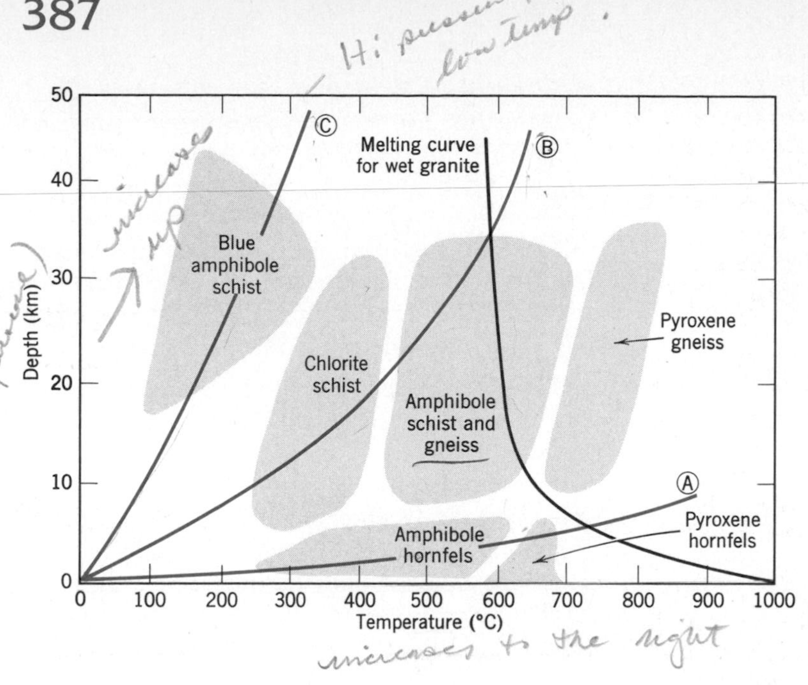


**Figure 15.23**
**The kind of rock that develops from metamorphism of shale depends on temperature and depth of burial. Curve A is a typical thermal gradient around an intrusive igneous rock that is causing contact metamorphism. Hornfels, a dense, nonfoliated, even-grained rock, is created. Curve B is a normal continental geothermal gradient and it indicates the rock types developed during slow burial of a pile of strata. Curve C is the geothermal gradient developed in a sequence of strata that are buried so rapidly it cannot maintain thermal equilibrium.**

release, the reaction cannot proceed. In a sense, this is analogous to saying you cannot dry wet clothes if you lock them up in a tight closet. How a regional-metamorphic reaction occurs, therefore, depends on whether the $H_2O$ and $CO_2$ can escape.

### Physical conditions of metamorphism

Despite the influence that $H_2O$ and $CO_2$ must exert, common metamorphic rock types and common mineral assemblages are closely related to temperature and depth of burial. One attempt to systematize common metamorphic rocks in this fashion is shown in Figure 15.23. In this figure also is plotted the melting curve for wet granite that occurs in Figure 15.16. Clearly, if sufficient water is present in the rock down near the base of a pile that is being

**Figure 15.24**
**Example of migmatite, a part-way stage in the progression from metamorphic to igneous rock. Complexly folded veins of granite are enclosed by intensely metamorphosed rock. The granite represents the fraction of the rock that melted at the climax of metamorphism. Skeena River, British Columbia. (*D. M. Shaub.*)**

metamorphosed, magma can start to form while amphibole schists and gneisses are forming also; the result is *a composite rock containing both igneous and metamorphic portions,* called a ***migmatite.*** In many places it is possible to see evidence of the phenomenon of partial melting and formation of migmatite (Fig. 15.24). Still further melting causes large volumes of magma to develop and to rise through the pile of metamorphic rock above. As discussed earlier, this event leads to the close connection that exists between metamorphic rocks and granitic intrusives.

We have discussed the sort of changes that occur during metamorphism, and how certain igneous activity is connected with metamorphism. But an important part of the story concerning both metamorphic and igneous rocks remains to be discussed. Igneous and metamorphic rocks are not randomly distributed like raisins in a pudding. They occur together in elongate belts, and are closely associated with mountainous regions and zones of intense deformation of Earth's crust. In subsequent chapters we shall see how this association comes about.

# SUMMARY

1. The principal controls on the compositions and physical properties of magmas are their content of $SiO_2$ and $H_2O$.

2. High temperature and low content of $SiO_2$ result in fluid magma, such as basaltic magma. Lower temperature and high content of $SiO_2$ result in viscous magma, such as andesitic and rhyolitic magma.

3. The sizes and shapes of volcanic edifices depend on the kind of material extruded, viscosity of the lava, and explosiveness of the eruptions.

4. Volcanoes that dispense sluggish lava, generally rich in silica, tend to be explosive; those that extrude fluid lava, generally low in silica, erupt less violently.

5. Widespread sheets of basaltic rock have resulted from fissure eruptions of fluid lava.

6. The composition of lavas extruded from volcanoes within ocean basins is not the same as that of lavas extruded from volcanoes lying at the margins of ocean basins.

7. Basaltic magma, characteristic of oceanic crust, is formed by the partial melting, under dry or nearly dry conditions, of rock in the mantle.

8. Andesitic magma is believed to be formed by partial melting of water-saturated basaltic rock that is being carried downward to the mantle by descending plates of lithosphere.

9. Rhyolitic magma is formed by partial melting of crustal rock in the presence of water.

10. Processes that separate remaining liquid from already-formed crystals in a cooling magma lead to the formation of a wide diversity of igneous rocks.

11. Heat given off by intrusive igneous rock creates contact-metamorphic aureoles.

12. Mechanical deformation and chemical recrystallization are the two processes that affect rock during metamorphism.

13. Rocks having the same chemical composition, and subjected to identical metamorphic environments, always react to form the same mineral assemblages.

# SELECTED REFERENCES

## IGENOUS ACTIVITY

Cotton, C. A., 1952, Volcanoes as landscape forms, rev. ed.: New York, John Wiley.

Eaton, G. P., and others, 1975, Magma beneath Yellowstone National Park: Science, v. 188, p. 787–796.

Ernst, W. G., 1969, Earth Materials: Englewood Cliffs, N.J., Prentice-Hall. Chapter 5.

Hamilton, Warren, and Myers, W. B., 1967, The nature of batholiths: U.S. Geol. Survey, Prof. Paper 554-C, p. C1–30.

Keefer, W. R., 1971, The geologic story of Yellowstone National Park: U.S. Geol. Survey Bull. 1347.

MacDonald, G. A., 1972, Volcanoes: Englewood Cliffs, N.J., Prentice-Hall.

Smith, R. L., 1960, Ash flows: Geol. Soc. America Bull., v. 71, p. 795–841.

Tabor, R. W., and Crowder, D. F., 1969, On batholiths and volcanoes—intrusion and eruption of late Cenozoic magmas in the Glacier Peak area, North Cascades, Washington: U.S. Geol. Survey Prof. Paper 604.

U.S. Geol. Survey, 1971, Atlas of volcanic phenomena: Folio of 20 sheets.

Wyllie, P. J., 1971, The Dynamic Earth: Textbook in Geosciences: New York, John Wiley.

## METAMORPHISM

Ernst, W. G., 1969, Earth materials: Englewood Cliffs, N.J., Prentice-Hall. Chapter 7.

Fyfe, W. S., Turner, F. J., and Verhoogen, J. 1958, Metamorphic reactions and metamorphic facies: Geol. Soc. America Mem. 73.

Turner, F. J., 1968, Metamorphic petrology: New York, McGraw-Hill.

Winkler, H. G. F., 1974, Petrogenesis of metamorphic rocks: 3d ed., New York, Springer-Verlag.

# 16 EARTHQUAKES AND EARTH STRUCTURE



In marked contrast to the accessibility of Earth's surface, the interior regions are remote and hidden from view. We have yet to drill to the base of the crust, let alone into the mantle, and we have no means by which we can examine directly Earth's internal structure. To discover what lies between us at the surface and Earth's unseen center, we must use indirect means. How this is done is one of the exciting success stories of science.

The volume of the Earth was calculated long ago from measurement of its circumference. Earth's mass has long been known from such measurements as the gravitational pull between Earth and Moon. Knowing volume and mass, we can calculate that Earth's mean density is 5.5gms per $cm^3$. Yet rock in the crust has densities in the range of 2.6 to 3.1gms per $cm^3$. Comparing the two densities, we are therefore sure that part of the interior must consist of material more dense than 5.5gms per $cm^3$. We could not easily proceed beyond this inference without some way of "seeing" inside the Earth. But there is a way. The vibrations or waves caused by earthquakes have the ability to travel completely through the Earth. By making careful measurements we can use those waves to obtain a sort of giant X-ray picture of what is inside.

# EARTHQUAKES

## Origin

Earthquakes, and the vibrations they cause, happen when Earth is suddenly jolted, as if struck by a giant hammer. Make an experiment yourself. Have a friend hit one end of a wooden plank with a hammer while you press your hand on the other end. You will feel vibrations, set up in the plank by the energy of the hammer blow. The harder the hammer blow, the stronger the vibrations. Fortunately giant hammers don't hit the Earth, but a bomb blast or a violent volcanic explosion will serve just as well. So too will sudden slipping of rock masses along a fault, causing two hard, rocky surfaces to slide suddenly past each other.

Sudden movement along faults is the cause of most earthquakes. But it cannot be that simple sliding occurs every time pressure is applied to a fault; some earthquakes are millions of times stronger than others. The energy that in one case will be released by thousands of tiny slips and tiny earthquakes will, in another case, be stored and released in a single giant earthquake. The answer seems to be that energy can be stored in bodies of rock that are being deformed elastically, just as in a steel spring that is compressed. The word *elastic* indicates that when the force is removed and the energy released, the deformed body returns to its original shape.

Evidence supporting the idea of energy being stored in elastically deformed rocks came first from studies of the San Andreas Fault. During long-term field observations in central California, beginning in 1874, scientists from the U.S. Coast and Geodetic Survey determined the precise positions of many points both adjacent to and distant from the fault. As time passed, movement of the points revealed that the crust was slowly being bent. For some reason the fault was locked and did not slip. On April 18, 1906, the two sides of the fault shifted abruptly. The stored energy was released as the bent crust snapped back to a former position,

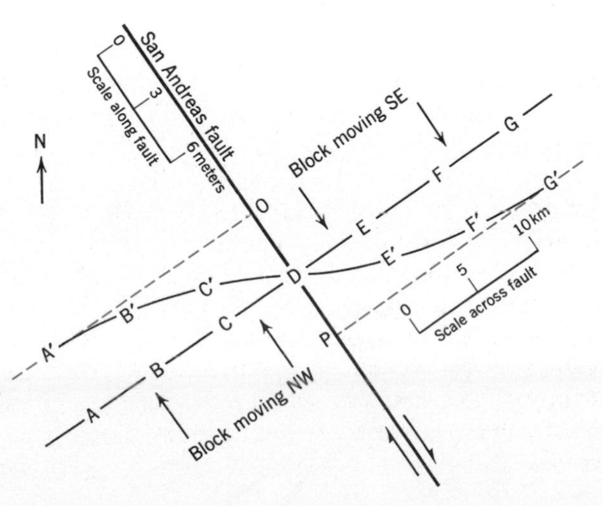


Figure 16.1
**Sketch based on detailed surveys near the San Andreas Fault, California, before and after the abrupt movement that caused the earthquake of 1906. The 7 survey points, A to G, were originally lined up. Slowly, movement of the two fault blocks bent the crust and displaced the points to new positions, A′ to G′. Then suddenly the two sides of the fault moved and the surveyed points lay along the blue lines A′O and PG′. The sudden offset along the fault, distance OP, was 7m.**

thereby creating a violent earthquake. Repetition of the survey then revealed that the bending had disappeared (Fig. 16.1).

Most earthquakes occur in the brittle rock of the lithosphere. The word *brittle* refers to a tendency for a solid to fracture when the deforming force exceeds the limits of elasticity. At great depth, temperatures and pressures are too high for brittle fracture to happen. Under such conditions bodies do not fracture, but instead undergo permanent changes of shape, even after the deforming forces have been removed. Deformation of this kind is *plastic* deformation.

### Seismic (earthquake) waves

*The point of first release of the energy that causes an earthquake* is called the ***earthquake focus.*** The focus generally lies far below the surface; so for convenience we define *the point, on Earth's surface that lies vertically above the focus of an earthquake* as the **epicenter** (Fig. 16.9). A convenient way to describe the location of an earthquake focus is to state its epicenter and its depth.

How is the energy of an earthquake transmitted from the focus to other parts of Earth? As with any vibrating body, waves (vibrations) spread outward from the focus. The waves are commonly called *seismic waves* after the Greek word for earthquake. They spread out in all directions from the focus, just as sound waves spread in all directions when a gun is fired. Seismic waves are elastic disturbances; so the rocks through which they pass return to their original shapes after the waves go by. We cannot therefore tell, by examining a rock, whether seismic waves have passed through it at some time in the past. Seismic waves must be measured and recorded while the rock is still vibrating. For this reason, many continuously operated recording stations are scattered around the world.

Seismic waves are of two kinds. *Compressional waves* deform solids by change of volume in the same way that sound waves do, and consist of alternating pulses of compression and rarefaction acting in the direction of travel (Fig. 16.2). Compressional waves travel

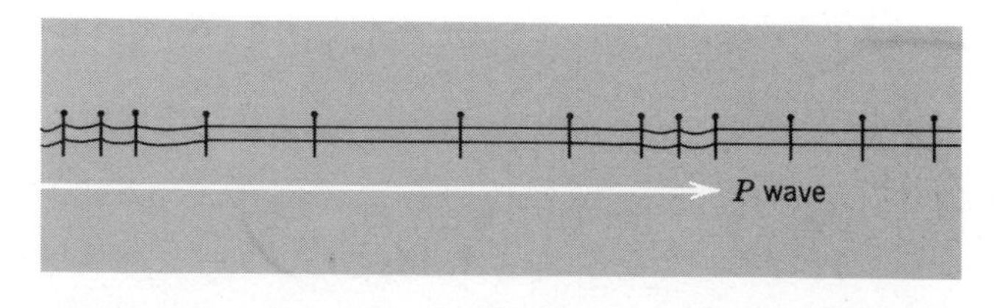


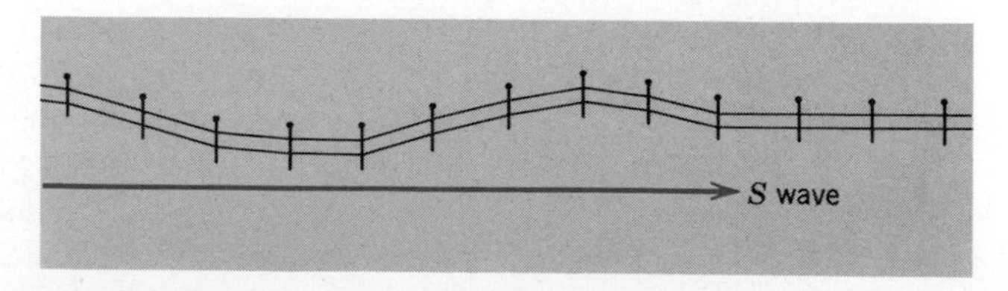


**Figure 16.2** **The difference between seismic waves of the *P* and *S* types is demonstrated by the disturbances they cause in a fence. *P* waves give rise to alternate compressions and rarefactions, causing the fence to be alternately stretched and squeezed. *S* waves give rise to lateral distortions, causing wave-shaped sidewise motions in the fence. Arrows show direction of wave travel.**

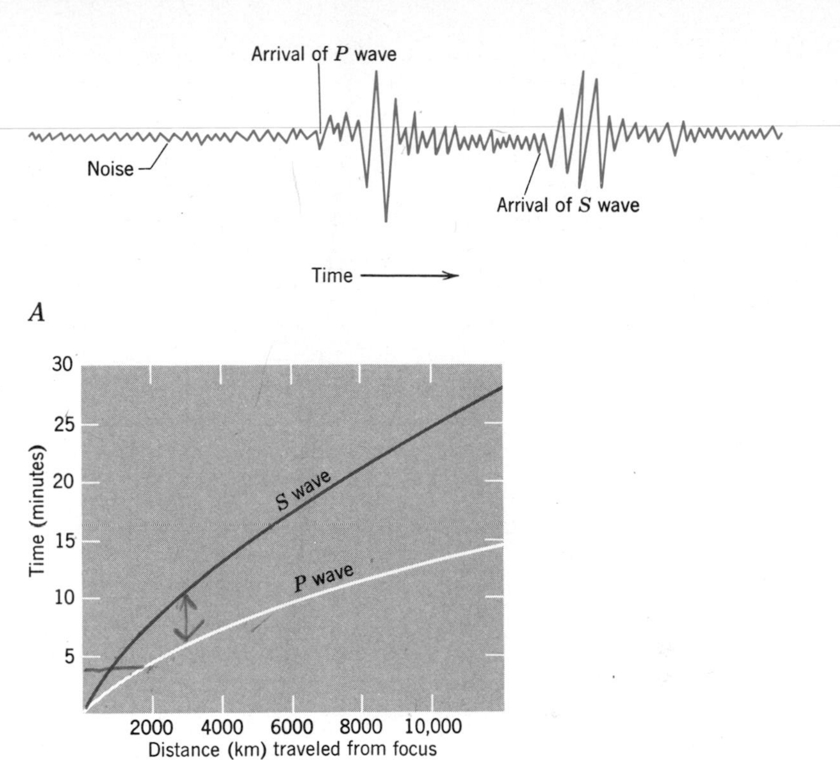


**Figure 16.3 Differing travel times of *P* and *S* waves. *A*. Typical record made by a *seismograph,* the instrument used to record seismic waves. The *P* and *S* waves leave the earthquake focus at the same instant and travel outward in all directions. The fast moving *P* waves reach the seismograph first, and, some time later, the slower-moving *S* waves finally arrive. The delay in arrival times is proportional to the distance traveled by the waves. *B*. Average travel-time curves for *P* and *S* waves in the Earth (*after Bullen, 1954*). When the arrival times of *P* and *S* waves are recorded by seismographs at three different locations, the exact location of the earthquake focus can be calculated.**

more rapidly than other seismic waves, and are therefore called *P* (for *Primary*) waves. *Shear waves* deform solids by change of shape but not change of volume. They are like electromagnetic waves, consisting of an alternating series of sidewise movements, each particle in the deformed solid being displaced perpendicular to the direction of wave travel. Being slower than *P* waves, shear waves are called *S* (for *Secondary*) waves (Fig. 16.3).

*P* and *S* waves are not the only elastic vibrations generated by earthquakes. Strange as it may seem, a large earthquake can make Earth shake and oscillate freely like a giant bell or a huge tub of jelly. As seen in Figure 16.4, the free oscillations that occur when this happens are of two kinds. *Spheroidal* oscillations involve mainly a change in Earth's total volume, while *torsional* oscillations involve only a change in Earth's shape. The properties of free oscillations are therefore similar to those of *P* and *S* waves, and to an observer at the surface the oscillations indeed appear very similar to ordinary *P* and *S* waves. For this reason, the oscillations of the Earth are usually called *surface waves;* those caused by spheroidal oscillations are called Rayleigh waves, and those by torsional oscillations, Love waves, after the English scientists who first recognized them.

We stated earlier that the velocity of a wave passing through a solid depends on the physical properties of the solid. The velocities of *P* and *S* waves change in different ways as physical properties change. This is the very property we need to explore Earth's interior, because the different interior layers affect the *P* and *S*

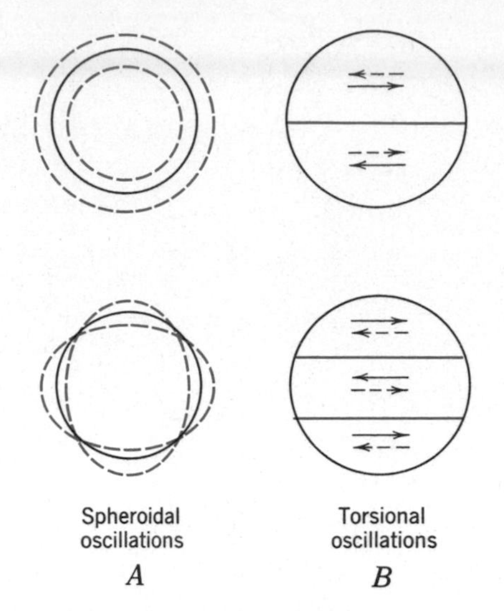


**Figure 16.4**
**Simple free oscillations of a vibrating sphere, such as Earth, are of two kinds: spheroidal oscillations, in which both volume and shape may vary, and torsional oscillations, in which volume is unchanged.**
***A*. The simplest spheroidal oscillations occur when Earth simply expands and contracts radially, and when the sphere becomes deformed to a football shape.**
***B*. The simplest torsional oscillations occur when the northern hemisphere is twisted in the opposite direction from the southern hemisphere. The next-simplest oscillation divides Earth into three twisting slices.**

waves in different ways. Before discussing Earth's interior, however, we will consider briefly the dangers that earthquakes pose to people.

## Earthquakes and people

The dangers of earthquakes are profound and the havoc they can cause is often catastrophic. Their effects are of four principal kinds.

1. *Ground motion* that results from the passage of seismic waves through surface rock layers and regolith can damage and sometimes completely destroy buildings (Fig. 16.5). Proper design of buildings can do much to prevent such damage, but in a very strong earthquake no structure is safe.

2. A secondary effect, but one that is sometimes a greater hazard than moving ground, is *fire*. Ground movement displaces stoves, breaks gas lines, and loosens electrical wires, starting fires. In the disastrous earthquakes that struck San Francisco in 1906, and Tokyo and Yokohama in 1923, probably more than 90 percent of the damage to buildings was caused by fire.

3. In regions of hills and steep slopes, earthquake vibrations may cause regolith to slip and start rapid mass-wasting movements (Chapter 6). This is particularly true in Alaska and parts of southern California (Fig. 16.5*B*). Houses, roads, and other structures are destroyed by rapidly moving regolith.

4. Finally, there are *seismic sea waves* (sometimes called by their Japanese name, *tsunami*) that occur following violent movement of the sea floor. Seismic sea waves, often incorrectly called tidal waves, have been particularly destructive in the Pacific Ocean. About 4½ hours after a severe submarine earthquake near Unimak Island, Alaska, in 1946, such a wave struck Hawaii. With a crest 18m higher than normal high tide, this destructive wave demolished nearly five hundred houses, damaged a thousand more, and killed 159 people.

A

B

**Figure 16.5 Effects of earthquakes.**
***A*. Destruction of school in Long Beach, California, by ground motion during earthquake of 1933. Fortunately the school was not in session. (*Wide World Photos*.)**
***B*. Near Anchorage, Alaska, a landslide triggered by the earthquake of March 27, 1964, destroyed these houses. (*Steve McCutcheon, Alaska Pictorial Service*.)**

Every year Earth experiences many hundreds of thousands of earthquakes. Fortunately only one or two are sufficiently strong, or are close enough to major population centers, to cause serious loss of life. The most disastrous earthquake on record occurred in 1556, in the Shensi Province in China, where an estimated 830,000 people died. These people lived in cave dwellings excavated in loess (Chapter 9), which collapsed as a result of the quake. The next most disastrous earthquake hit Calcutta, India in 1737. An estimated 300,000 people perished. In 1920 another quake, in the Kansu Province of China, killed 180,000 people, and the Japanese earthquake of 1923 killed 143,000 in Tokyo and Yokohama. Since 1900 there have been 32 earthquakes, worldwide, in each of which 500 or more people lost their lives.

No locality at Earth's surface is free from earthquakes, but in some regions the quakes that do occur are weak and consequently not dangerous to people or dwellings. For example, scientists believe that in southern Florida, southern Texas, and parts of Alabama and Mississippi, the probability of damaging earthquakes is almost zero. All other parts of the United States have experienced damaging quakes in the past, and more can be expected to occur in the future (Fig. 16.6).

How frequent are earthquakes? Astonishing as it may seem, about a million are recorded by seismographs every year. Fortunately most are too small to be felt or to cause damage. To compare the strengths of earthquakes, both large and small, we need a comparative measuring scale. The scale most widely used, named after its inventor, is the *Richter magnitude scale.* The strength or magnitude of an earthquake is gauged by the strength of the *P* and *S* wave signals recorded by seismographs. Because the wave signals vary in strength by factors of a hundred million or more, it is impractical to use a scale divided in equal increments. Instead, the Richter scale divides the wave signals into steps called magnitudes, starting with magnitude 1 and increasing upward. Each unit increase in magnitude corresponds to a tenfold increase in the strength of the earthquake. Thus, a Magnitude 2 earthquake is ten times stronger than a Magnitude 1, and a Magnitude 3 a

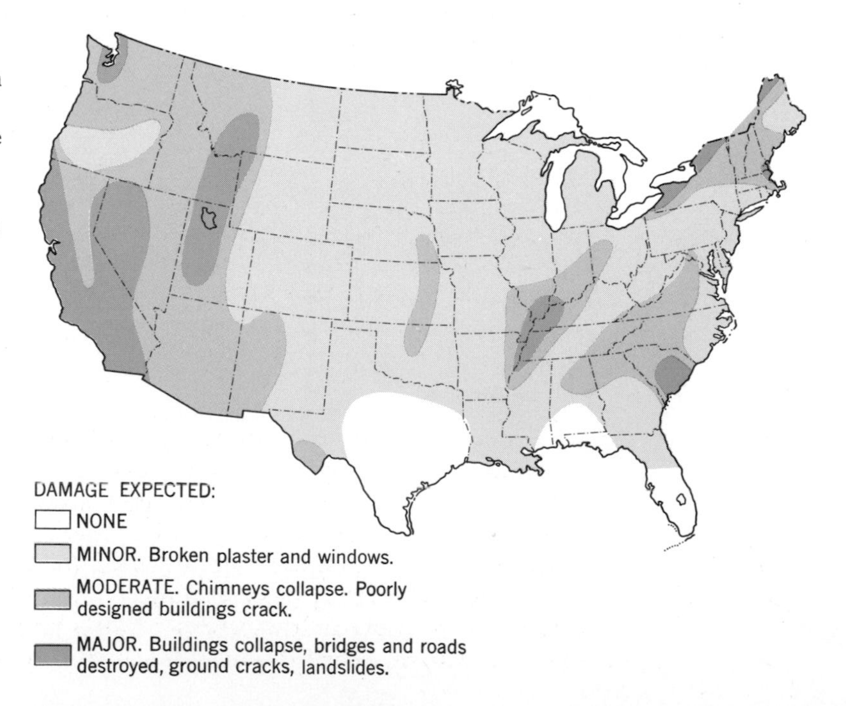


Figure 16.6
**Seismic-risk map of the United States. Zones refer to maximum earthquake intensity and therefore maximum destruction that can occur. The map does not indicate frequency of earthquakes. For example, frequency in southern California is high, but in eastern Massachusetts it is low. Nevertheless, when earthquakes occur in eastern Massachusetts, they can be as severe as the more frequent quakes in southern California. (*After Algermissen, 1969, Fourth World Conf. on Earthquake Eng., Proc., v. 1, p. 14–27.*)**

**Table 16.1**
**Earthquake Magnitudes and Frequencies for the Entire Earth, and Damaging Effects.**
*(After B. Gutenberg.)*

| Magnitude | Number per Year | Effects in Populated Areas |
|---|---|---|
| Less than 3.4 | 800,000 | Recorded only by seismographs |
| 3.5 to 4.2 | 30,000 | Felt by some people |
| 4.3 to 4.8 | 4,800 | Felt by many people |
| 4.9 to 5.4 | 1,400 | Felt by everyone |
| 5.5 to 6.1 | 500 | Slight building damage |
| 6.2 to 6.9 | 100 | Much building damage |
| 7.0 to 7.3 | 15 | Serious damage. Bridges twisted, walls fractured. |
| 7.4 to 7.9 | 4 | Great damage. Buildings collapse. |
| More than 8.0 | One every 5 to 10 years | Total damage |

hundred times stronger. The relation between the magnitude of an earthquake and the damage caused in the area of the epicenter is given in Table 16.1. The strongest earthquakes ever recorded had magnitudes of 8.5. Fortunately an earthquake of Magnitude 8 or higher occurs somewhere on Earth no more commonly than once every 5 to 10 years.

# EARTH'S STRUCTURE

We have already learned that *P* and *S* waves travel through rock at differing velocities and that they respond to changing properties of rock in different ways. The arrival times of *P* and *S* waves at seismographs stationed around the world give us records of waves that have traveled through the Earth along different paths. From such records it is possible to calculate where the boundaries between layers having different compositions are located within the Earth.

### Layers of differing composition

If Earth's composition were uniform, velocities of *P* and *S* waves would increase smoothly with depth because increasing pressure increases the elastic properties of rock. With a uniform Earth, it would be easy to predict how long it would take seismic waves to pass through. But observed travel times differ greatly from such predictions. The only way the discrepancies can be accounted for is to suppose that velocities change because composition changes with depth.

To find out where the supposed composition changes happen, we turn to another property of waves. Seismic waves behave like light waves and sound waves. Wherever they encounter a boundary between two different substances, the waves are reflected or refracted. For example, light waves are reflected by a mirror and are refracted when they pass through a glass of water. Both *P* and *S* waves are strongly influenced by a pronounced boundary at a depth of 2,900km. When *P* waves reach that boundary, they are

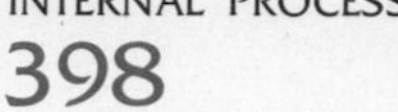


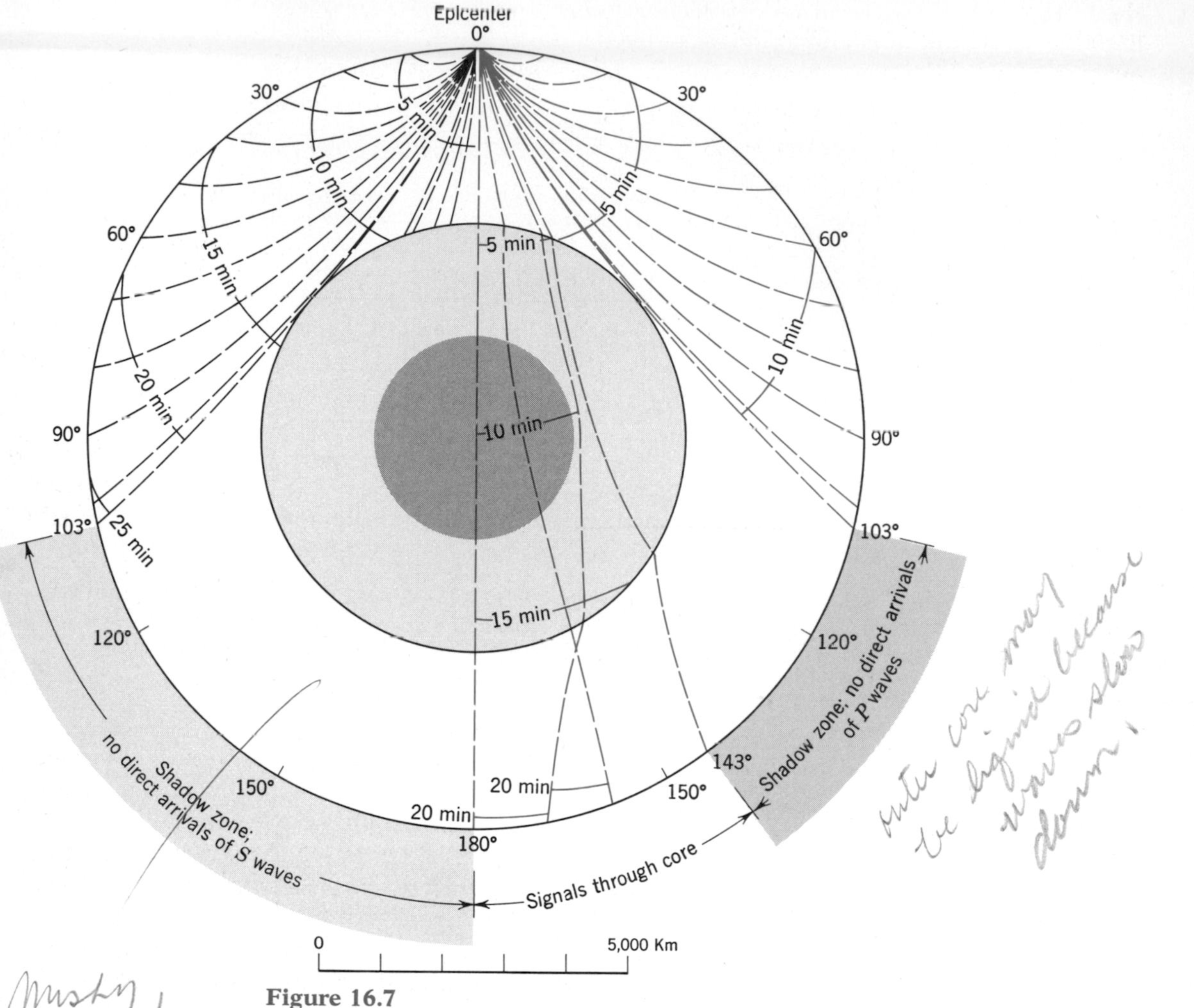


**Figure 16.7**
**Section through Earth showing mantle (white) outer core (light gray) and inner core (dark gray).**
**Paths of *P* waves (blue) from an earthquake focus with epicenter at 0° (top) shown in right half only. Paths of *S* waves (black) shown in left half. Distances reached by waves at 5-minute intervals are indicated. Reflection and refraction of *P* waves at the mantle/core boundary create a *P* wave shadow zone from 103° to 143°. Because *S* waves cannot pass through a liquid, an *S* wave shadow exists between 103° and 180°. (*After B. Gutenberg, Internal Constitution of the Earth, 1950. By permission, Dover Publications, Inc., New York.*)**

reflected and refracted so strongly that the boundary actually casts a *P* wave shadow over part of the Earth (Fig. 16.7). Because the boundary is so pronounced, we infer that it is the place where the comparatively light silicate material of the mantle meets the dense metallic iron of the core. The same boundary casts an even more pronounced *S* wave shadow, but the reason is not reflection or refraction. Shear waves cannot traverse liquids. The huge *S* wave shadow therefore lets us infer that the outer core is liquid.

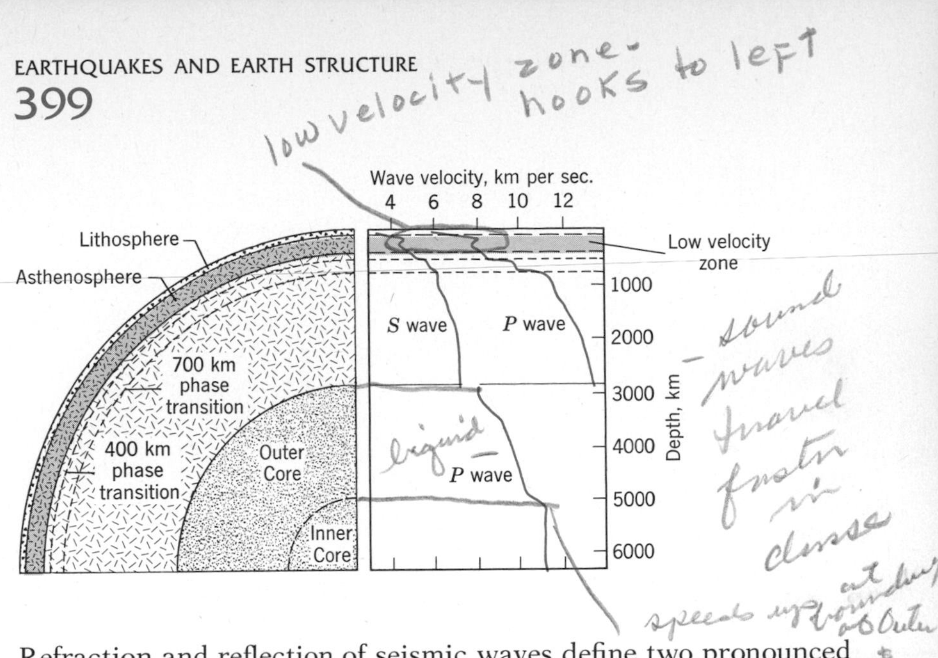


**Figure 16.8 Variation of seismic-wave velocity within the Earth. Abrupt changes occur at the boundaries between crust and mantle and between mantle and core.**

Refraction and reflection of seismic waves define two pronounced boundaries separating three fundamental zones within the Earth. The zones correspond to crust, mantle, and core. The same zones also show up prominently when the *P* and *S* wave velocities are plotted against depth in the Earth. Figure 16.8 demonstrates clearly how seismic-wave velocities differ with composition.

**The crust.** Early in the 20th Century the reality of Earth's crust was demonstrated by a scientist named Mohorovičić, (Mō-hō-rō-vĭtch´-ĭck) who lived in what today is Yugoslavia. He noticed that in measurements of seismic waves arriving from an earthquake whose focus lay within 40km of the surface, seismographs within 800km of the epicenter recorded *two* distinct sets of *P* and *S* waves. He concluded that one pair of waves must have traveled from the focus to the station by a direct path, whereas the other pair represented waves that had arrived slightly later because they had been refracted. Evidently these later waves had penetrated a deeper zone of higher velocity, had traveled within that zone, and had then been refracted upward to the surface (Fig. 16.9). From his conclusions he hypothesized that a distinct compositional boundary, strongly influencing seismic waves, separates the crust from an underlying zone. Scientists now refer to this boundary as the ***Mohorovičić discontinuity*** and recognize it as *the base of the crust.* Since the name is a tongue twister, the feature is commonly called the ***M-discontinuity,*** and in conversation is shortened still further to ***moho.***

By seismic methods we can determine the thickness of the crust.

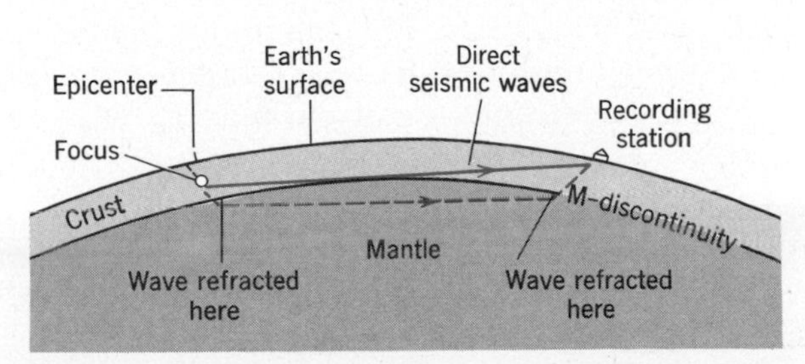


**Figure 16.9 Travel paths of direct and refracted seismic waves from shallow-focus earthquake to nearby recording station.**

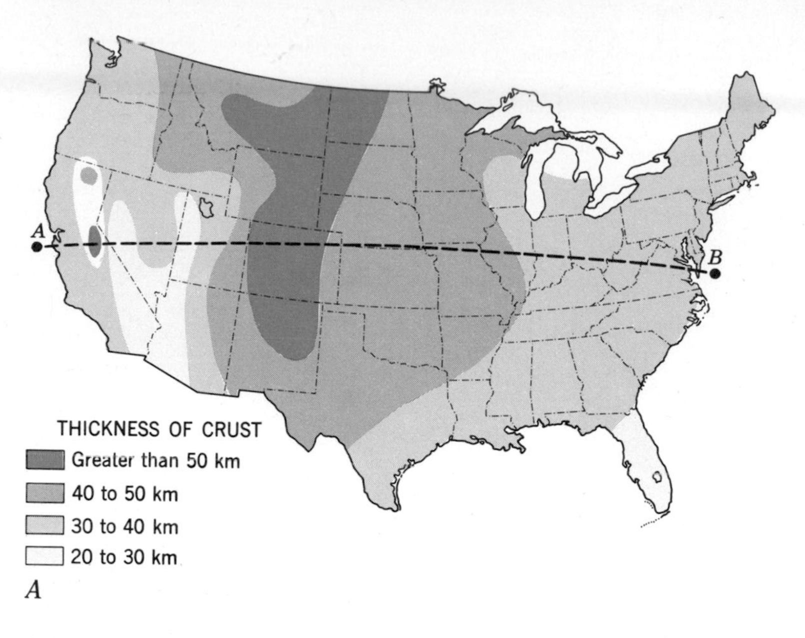


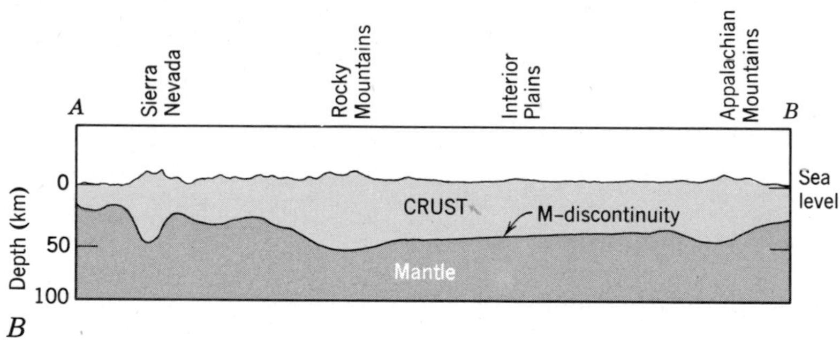


**Figure 16.10 Crust beneath the United States.** ***A*. Thickness of crust beneath United States, determined from measurements of seismic waves.** ***B*. Section through the crust along the line *AB* (above). The crust tends to thicken beneath major mountain masses such as the Sierra Nevada, the Rocky Mountains, and the Appalachians.** **(*After Pakiser and Zietz, 1965.*)**

Also from wave velocities and the elastic properties they indicate, we can estimate the probable composition of the crust. Beneath ocean basins the crust is thin, in most localities averaging only 10km. Elastic properties of the oceanic crust are those characteristic of basalt and gabbro. But in the continental crust both thickness and composition are very different. From 20km to nearly 60km thick, the continental crust tends to be thickest beneath major mountain masses (Fig. 16.10), a fact discussed further on in this chapter. Velocities in the continental crust are distinctly different from those in the oceanic crust. They indicate elastic properties like those of rock such as granite and andesite, although at some places just above the M-discontinuity, velocities close to those of oceanic crust are observed. These conclusions agree well with what is known about the composition of the crust from other lines of evidence. The agreement gives us confidence in drawing conclusions about the mantle, where these other lines of evidence are scarce.

**The mantle.** The mantle is still an enigma. It is huge, it controls much of what happens in the crust, yet we cannot see it. In Figure 16.8 it is apparent that $P$ wave velocities in the crust vary between

6 and 7km per second. Beneath the M-discontinuity, velocities are greater than 8km per second. Careful laboratory tests show that rocks common in the crust, such as granite, gabbro, and basalt, have *P* wave velocities of 6 to 7km per second. But rocks such as peridotite, that are rich in the dense minerals olivine, pyroxene, and garnet, have velocities greater than 8km per second. These minerals, we therefore suppose, must be among the principal materials of the mantle. This inference is consistent with what little is known about the composition of the upper part of the mantle—for example, with the ophiolite complex shown in Figure 13.13, and with rare samples of mantle rocks from ***kimberlite pipes*** (*narrow, pipe-like masses of igneous rock, sometimes containing diamonds, that intrude the crust but originate deep in the mantle*).

From the *P* wave curve in Figure 16.8, we see that there are two places in the mantle where velocity increases sharply. One is about 400km deep, the other 700km. But in those zones, *S* wave velocity is not so strongly affected. The sudden increase in *P* wave velocity is apparently not the result of abrupt changes in composition; the cause must be something else. A probable explanation comes from laboratory experiments. When olivine is squeezed at a pressure equal to that which exists at a depth of 400km, the atoms rearrange themselves into a more dense polymorph (Chapter 3). This process of *atomic repacking caused by changes in pressure or temperature* is called a ***phase transition.*** It is likely that the increase in seismic wave velocities at 400km is caused by a phase transition and not by compositional changes. If so, at that depth composition remains unchanged but density increases by about 10 percent.

The velocity increase at 700km is more difficult to explain but is probably the result of another phase transition, with another density increase of about 10 percent. Some scientists suggest that the increase at 700km results from a polymorphic change in the pyroxenes that are present in the mantle. Others have suggested alternative changes, such as a breakdown of silica tetrahedra to create a more dense structure in which each silicon atom is surrounded by six oxygen atoms. The various possibilities are still being tested by experiments in laboratories around the world.

Below 700km compression causes seismic-wave velocities to increase slowly and regularly until the abrupt boundary with the core is reached at 2,900km.

**The core.** Seismic waves indicate the way in which density increases with depth. Aided both by increasing pressure and by phase transitions, density increases slowly from about 3.3gms per cm$^3$ at the top of the mantle to about 5.5gms per cm$^3$ at the base of the mantle. We already know that the mean density of the whole Earth is 5.5gms per cm$^3$. To balance the less-dense crust and mantle, therefore, the core must be composed of material with a density of at least 10 to 11gms per cm$^3$. The only common substance that comes close to fitting this requirement is iron. By itself, pure iron is a little too dense. But small amounts of silicon, sulfur, nickel, and other elements dissolve in molten iron and could together give the required density.

*P* wave reflections indicate the presence of a solid inner core enclosed within the molten outer core. These two parts—outer and inner—appear to be identical in composition. The reason for the change from a liquid to a solid lies in the effect of pressure on the melting temperatures of iron. As the center of the Earth is approached, pressure rises to values millions of times greater than atmospheric pressure. Temperature rises too, but not steeply enough to offset the effect of pressure. From the base of the mantle (at a depth of 2,900km) to a depth of 5,350km, temperature and pressure are so balanced that iron is molten. But from 5,350km to the center of the Earth, rising pressure overcomes rising temperature, so that the iron is frozen, creating the solid core.

### The low-velocity zone (=asthenosphere)

Recent studies have confirmed an interesting idea proposed many years ago, although at first not widely accepted. Seismic-wave velocities increase from the base of the crust to the core/mantle boundary. The *P* wave velocity, for example, increases from about 8km per sec. to nearly 14km per sec. The increase in seismic-wave velocities is not smooth, and at depths ranging from 100km to 350km, there is a zone in which the velocities actually decrease slightly. The zone is not sharply defined, but it can be seen as a small trough, or blip, in the *P* and *S* wave velocity curves in Figure 16.8. There is no evidence suggesting the zone of reduced wave velocities is a zone where density decreases. To account for the velocity changes, therefore, we infer that the zone between 100 and 350km is, by some means, less rigid and more plastic than the regions above and below.

A possible explanation of this unusual behavior is that between these depths the geothermal gradient reaches temperatures at which partial melting of mantle rock could occur (Fig. 16.11). If the explanation is correct, a small amount of liquid would develop and form very thin films around the mineral grains, thus serving as a lubricant. Of course the amount of melting must be very

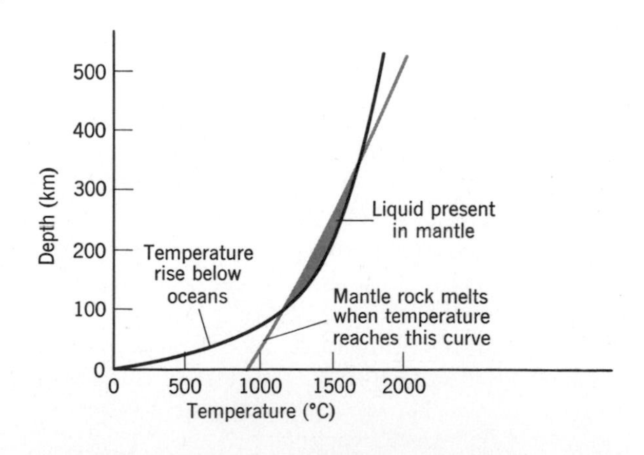


Figure 16.11
**A possible explanation of the low-velocity zone (=asthenosphere).**
**When the melting curve (blue) for mantle rock crosses the curve (black) defining the geothermal gradient, a small amount of liquid would develop by partial melting. A similar state of partial melting, determined experimentally for peridotite, is shown in Figure 15.12. The amount of liquid created could be tiny—only a small percentage of the total volume—but it would be enough to account for the plastic properties of the low-velocity zone.**

small, because the low-velocity zone does transmit *S* waves, and we already know that *S* waves cannot pass through liquids. Thus, rock in the low-velocity zone remains solid. The liquid, like a thin film of oil, merely serves to lubricate the grains, and at the same time reduce wave velocities by reducing the elastic properties.

Following the suggestion that plates of lithosphere slide over a somewhat plastic zone in the mantle, the importance of the low-velocity zone immediately became apparent. The top of that zone coincides with the base of the lithosphere. Furthermore, as we read in Chapter 15, basaltic magma apparently originates by partial melting of rock in the mantle at depths of 100km to 350km. Thus, the low-velocity zone is identical with the asthenosphere.

We have discussed the asthenosphere as if it were a region of uniform thickness, lying everywhere at a constant depth of 100km. But it is already clear that the asthenosphere is not constant either in depth or in thickness. Beneath the oceanic crust the top of the asthenosphere rises, in places to depths as shallow as 60km, being closest to the surface near an oceanic ridge but progressively deeper away from the ridge. Beneath the continents the top of the asthenosphere is closer to 100km, but it sinks as deep as 200km beneath mountain chains or beneath sea-floor trenches. The schematic section through the upper portions of the Earth, shown in Figure 16.12, illustrates our present understanding of the variability of the asthenosphere.

## WORLD DISTRIBUTION OF EARTHQUAKES

Now that we have seen the way in which seismic waves are used to build a picture of Earth's interior, we can turn to the pattern of occurrence of earthquakes. The pattern tells us a great deal about the shape and activities of plates of lithosphere.

Although no part of Earth's surface is exempt from earthquakes, several ***seismic belts,*** or *large tracts,* are *subject to frequent earthquake shocks* (Fig. 16.13). Of these the most obvious is the

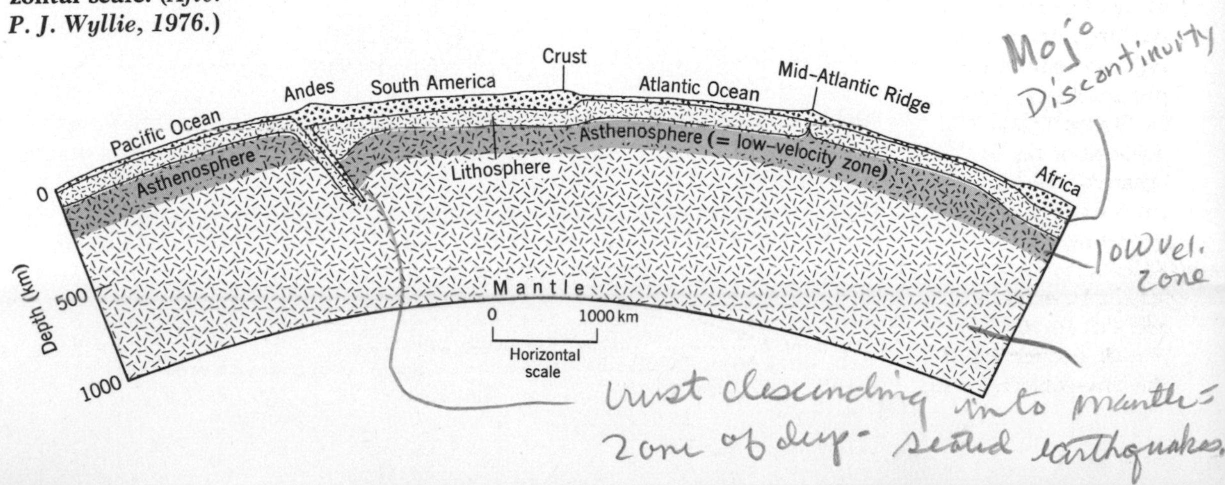


**Figure 16.12**
**Asthenosphere varies in thickness and depth.**
**A section through crust (black) and upper mantle shows that the low-velocity zone corresponding to the asthenosphere (blue) is deeper beneath continents than beneath ocean basins and dips sharply down beneath the Andes. The section appears distorted because the vertical scale is twice the horizontal scale. (*After P. J. Wyllie, 1976.*)**

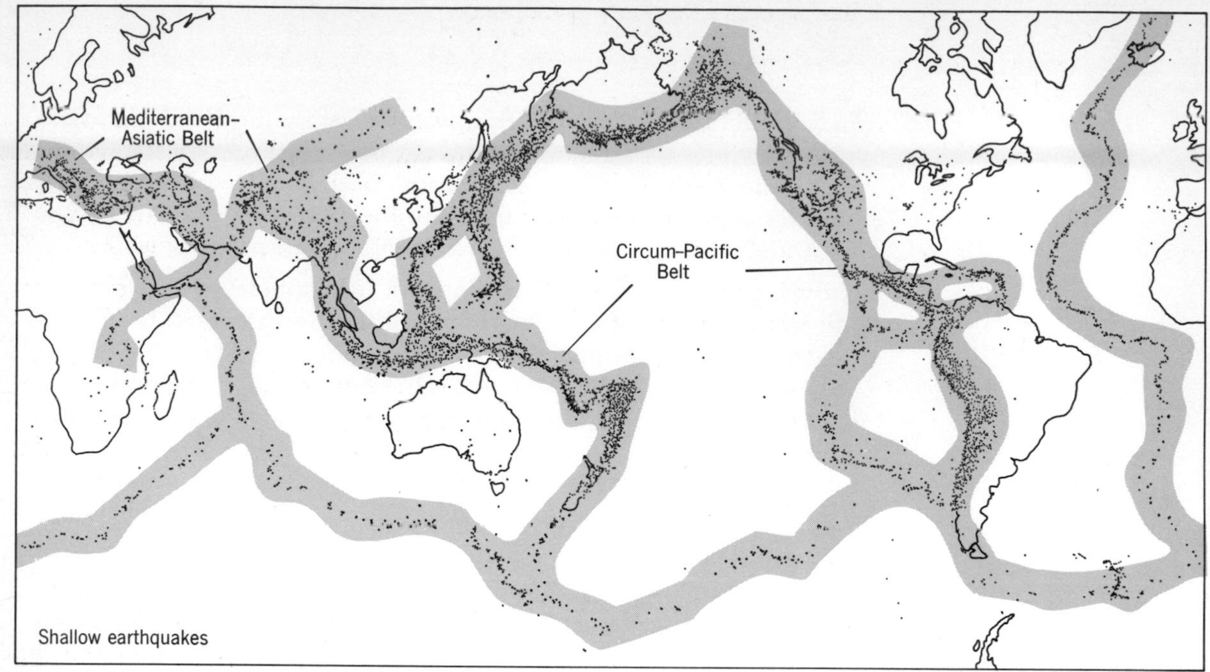


*A*

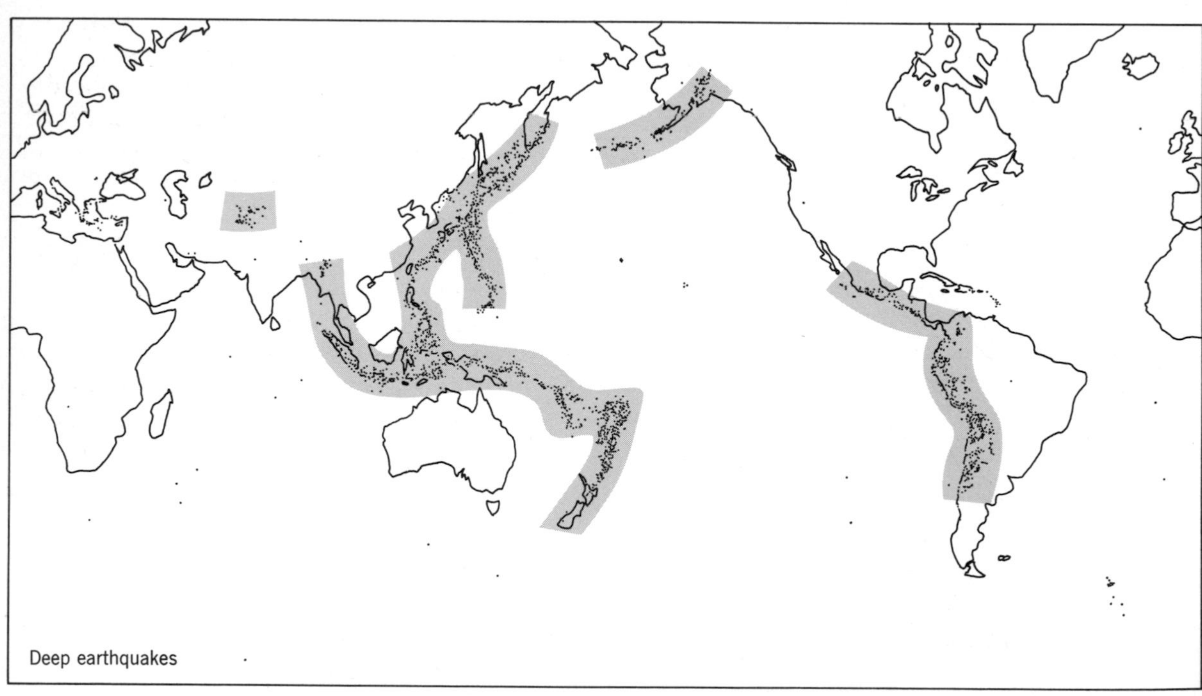


*B*

**Figure 16.13**

**Seismic belts and epicenters of earthquakes recorded by the U.S. Coast and Geodetic Survey between 1961 and 1967. Each dot represents a single earthquake.**

**A. Earthquakes of all depths are plotted. Most are shallow, with foci within 100km of the surface. The epicenters fall into well-defined seismic belts (dark blue shading) that coincide closely with the margins of plates of lithosphere.**

**B. Epicenters of earthquakes having foci deeper than 100km. They form belts that coincide closely with the sea-floor trenches. (*After Barazangi and Dorman, Bull. Seismol. Soc. America, v. 59, p. 369, 1969.*)**

*Circum-Pacific belt,* in which about 80 percent of all earthquakes originate. It follows the mountain chains in the western Americas from Cape Horn to Alaska, and crosses to Asia, where it extends southward down the coast and finally loops far southward to New Zealand. Next in prominence, giving rise to 15 percent of all earthquakes, is the *Mediterranean-Asiatic belt,* extending from Gibraltar to southeast Asia. Lesser belts follow the mid-ocean ridges.

Seismic belts are places where a lot of energy from internal activities is released. We might therefore expect other products of internal activities to appear in these belts. Indeed some of them do, most prominently the margins of plates of lithosphere. Compare Figure 16.13*A* with Figure 4.4, to see that the earthquake belts outline the plate boundaries.

The depths of earthquake foci around the edges of the plates also have a story to tell. Most foci are no deeper than 100km, because, as we have already seen, earthquakes occur in brittle rocks and the brittle lithosphere is only 100km thick. Some earthquakes do, however, originate at greater depths. The epicenters of deep earthquakes, with foci deeper than 100km, are plotted in Figure 16.13*B*. It is noteworthy that the deep earthquakes are not associated with oceanic ridges. When Figure 16.13*B* is compared with Figure 13.11, however, it is immediately apparent that deep earthquakes are closely related to sea-floor trenches. Those trenches, as we have already learned, mark the places where lithosphere plunges down into the mantle.

Detailed study of deep-earthquake foci beneath a sea-floor trench (Fig. 16.14) shows that the foci follow a well-defined zone, sometimes called a *Benioff zone.* This important observation suggests strongly that deep earthquakes originate within the relatively cold, downward-moving plate of lithosphere. Because some earthquake foci can be as deep as 700km, we conclude that

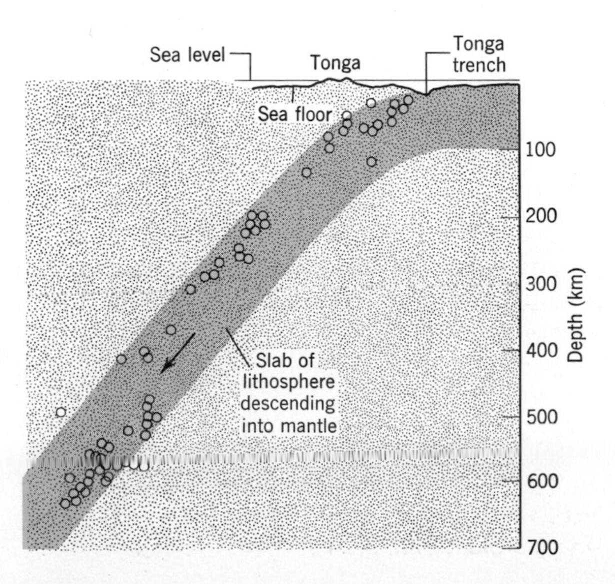


Figure 16.14
**Earthquake foci beneath the Tonga Trench, Pacific Ocean, during several months in 1965. Each circle represents a single earthquake. The earthquakes are believed to be generated by downward movement of a comparatively cold slab of lithosphere (blue band) that is plunging slowly back into the mantle. (*After Isacks, Oliver, and Sykes, Jour. Geophys. Res., v. 73, p. 5855, 1968.*)**

rapidly descending lithosphere can retain at least some brittle properties to that depth.

Locations of earthquakes tell us a great deal about the structure of the moving lithosphere; seismic waves generated by the earthquakes serve as a sort of giant X ray that let us infer Earth's internal structure. But we still have to pose the question: "Why is Earth round?"

## GRAVITY AND EARTH'S SHAPE

The answer to the above question is that Earth's spherical shape results from (1) gravity, the universal force of attraction, and (2) Earth's overall plastic, deformable nature.

The force of gravitational attraction was first explained by Isaac Newton in 1666, when he was only 24 years old. Newton expressed it in the following way.

$$F \text{ is proportional to } \frac{M_1 M_2}{d^2}$$

where $F$ is the force of gravitational attraction, $M_1$ and $M_2$ are the masses of two attracting bodies, and $d$ is the distance between $M_1$ and $M_2$.

Clearly, the larger $M_1$ and $M_2$ are, and the smaller $d$ is, the greater $F$ (the force of attraction) will be. Nevertheless, unless the attracting bodies are very large, gravity does not seem to be a very strong force. We cannot sense the attraction between small objects, like knives and forks, or even somewhat larger objects like automobiles and houses. If bodies are very large, however, as Earth is, gravitational attraction becomes large too. Even the smallest objects fall to Earth's surface because the gravitational pull is so great.

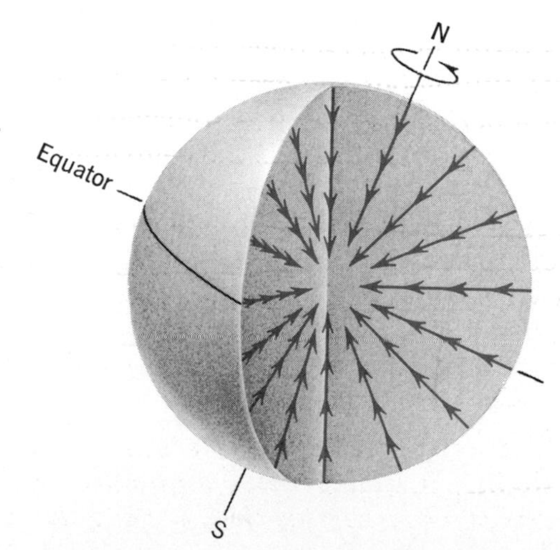


**Figure 16.15 Cut-away view of Earth, demonstrating how the force of gravity pulls constantly at every object, toward the center of mass. The direction of pull is radial and the result tends to give Earth a spherical form.**

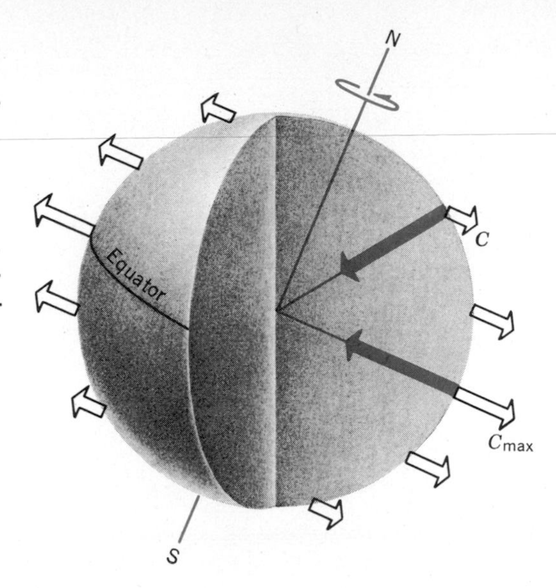


**Figure 16.16**
**The force of gravity, shown in blue arrows, acts along a radius. The centrifugal force caused by Earth's rotation (white arrows) acts in a direction perpendicular to the axis of rotation. The centrifugal force is maximum at the Equator and zero at the poles. As a result of the opposing forces, Earth bulges out at the Equator but is flattened at the poles.**

Every part of planet Earth is affected by ***Earth's gravity,*** *the inward-acting force with which Earth tends to pull all objects toward its center* (Fig. 16.15). This force acts equally in all directions and attempts to make the planet a perfect sphere. It would succeed were it not for Earth's rotation.

Rotations give rise to a force that opposes gravity. This is centrifugal force. As children, we have whirled a weight on a string around our heads and have felt the strong pull exerted by the weight. Perhaps we also noticed that a weight on a long string exerted a stronger pull than one on a short string. The pull is centrifugal force. The faster we rotated the string, and the farther the weight was from the axis of rotation (meaning the longer the string), the stronger was the pull. Earth rotates; so every particle on Earth is rotating, and every particle is subject to a centrifugal force. The force acts directly outward from the axis of rotation. The distance from Earth's surface to the axis of rotation is similar to the length of the string. It varies because Earth is spherical, being maximum at the Equator and zero at the poles. Earth's actual shape is a result of interactions between the two opposing forces, centrifugal pulling-out and gravitational pulling-in (Fig. 16.16), and is not a perfect sphere. It is somewhat flattened. Although the pull of gravity greatly exceeds the centrifugal pull, Earth's radius is 21km less at the poles than at the Equator. This departure from a perfect sphere has an interesting effect on weight; a man who weighs 200 pounds at the Equator, weighs 201 pounds at the North Pole. But we should not get a wrong picture of Earth's departure from a perfect sphere. If we could shrink our planet to the size of a basketball and keep its exact shape, it would seem to be a perfect and highly polished sphere.

The pull of gravity differs between Equator and poles because Earth is slightly flattened. Using a ***gravimeter*** (or ***gravity meter***), *a sensitive device for measuring the force of gravity at any locality,* we

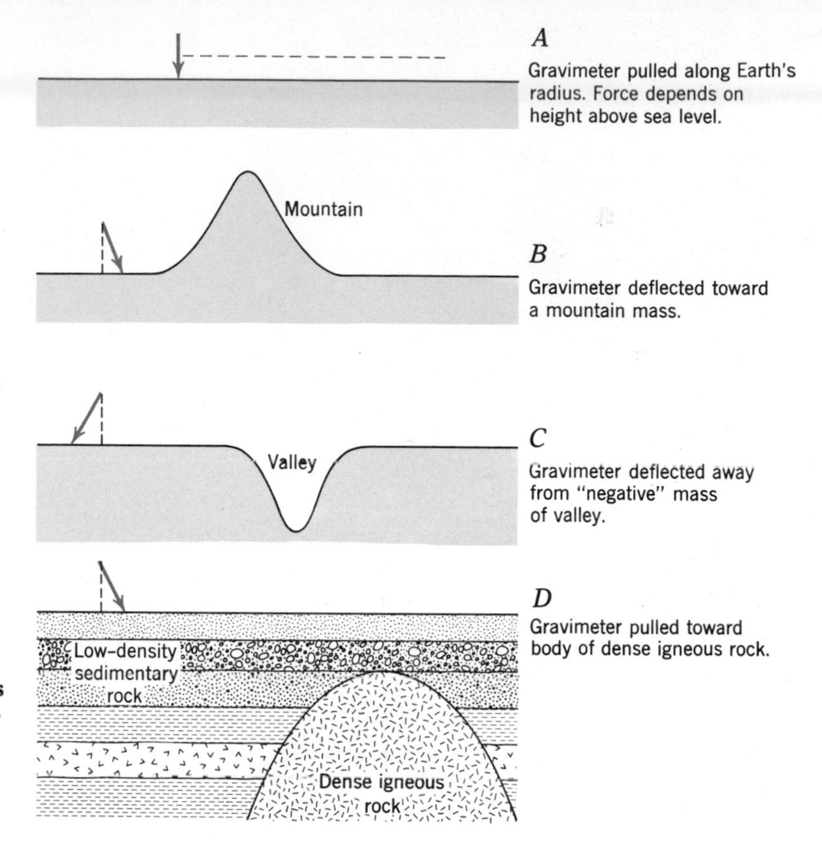


Figure 16.17
**Local variations in the force of Earth's gravity are measured with a gravimeter (a heavy weight suspended on a sensitive spring).**
***A*. On a perfectly smooth surface, the gravimeter is attracted directly toward Earth's center along a radial line. The force exerted on the gravimeter decreases with increasing height above sea level.**
***B*, *C*. When the surface is not smooth, the gravimeter is affected by local masses such as mountains, or "negative" masses such as valleys. In both cases, the direction of strongest pull is deflected slightly from a radial line.**
***D*. A buried mass of rock more dense than its surroundings attracts the gravimeter, so that the direction of strongest pull is not radial.**

can easily check for any other variations on Earth's surface. Local variations turn out to be common, and although the variations are small compared to the overall pull of gravity (sometimes varying by less than one part in a million), they can be measured accurately and can tell us a lot about Earth's internal properties and structures.

Variations in the force of gravity result from three principal effects:

1. Altitude above sea level or depth below it, because altitude controls our distance from Earth's center.
2. Topographic irregularities. A large mountain mass tends to attract the gravimeter, while a large valley is a "negative" mass and has the opposite effect (Fig. 16.17).
3. Most important, bodies of rock with different densities occur beneath the surface, and the gravimeter is attracted by them more or less depending on whether they are more or less dense than surrounding rocks.

Measurements of gravity variations have led to a very important conclusion about the lithosphere. This is that lithosphere can bob up and down like a floating iceberg. It acts in this way because it is "floating" on an easily deformed substratum. We have already

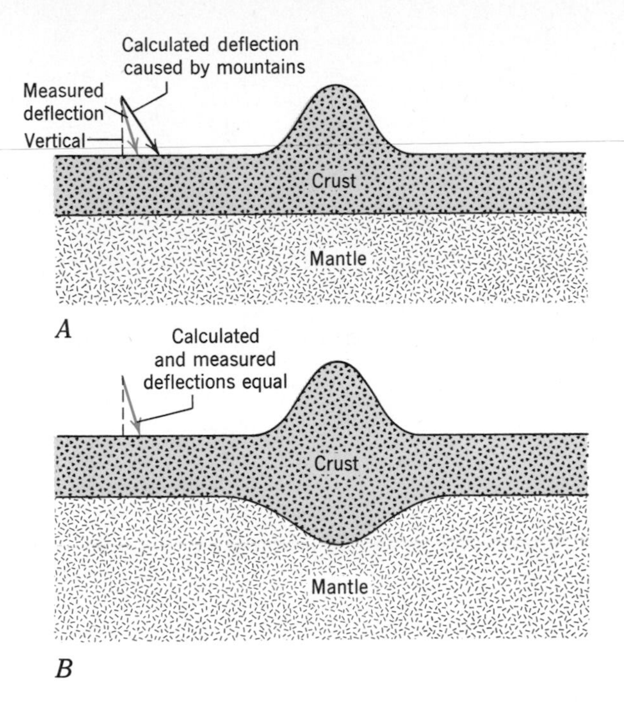


**Figure 16.18**
**Deflection of plumb-bob line by mountain mass.**
***A*. If a mountain is a mass of low-density crust overlying a dense, rigid, smooth-surfaced mantle, the calculated deflection is greater than that observed.**
***B*. But if a "root" of low-density crustal rock projects down into the mantle beneath the mountain mass, calculated deflection equals the observed deflection.**

seen evidence of this in Chapter 14. The weight of a large continental glacier can depress the crust, but when the mass is removed by melting, the land surface slowly rises again (Fig. 14.6). We refer to *the ideal property of flotational balance among segments of the lithosphere* as ***isostasy.*** The property is important, and demands more discussion.

## Isostasy

During the last century, surveyors working in India used a vertical plumb-bob line as a means of leveling their surveying instruments. But when their survey approached the Himalaya Mountains, the mountain mass pulled their plumb bobs slightly off vertical, so that unexpected errors crept into their results. Archdeacon Pratt, a British cleric whose hobby was mathematics, analyzed the measurements and found that the increase in gravitational attraction was actually less than would be expected if the mountains, containing low-density rocks, simply sat on top of a dense but rigid mantle. He quickly recognized that the discrepancy could be explained, provided the extra mass of the mountain was offset by the presence of low-density rock projecting into the mantle below the mountains (Fig. 16.18).

Pratt made another interesting suggestion. Instead of being uniform in density, he suggested, the outer portion of the Earth actually consists of blocks having different densities. He suggested further that the bottom of each block floats, at the same level, on a fluid substratum, and that the height of each block is controlled by its density. Blocks of low density would stand high and form mountains, while blocks of high density would be low, forming

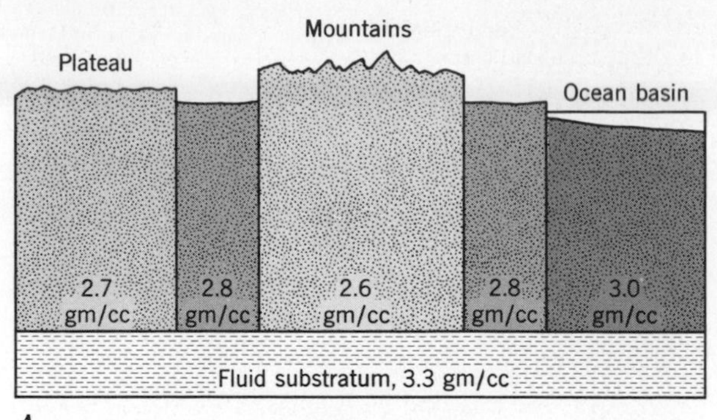


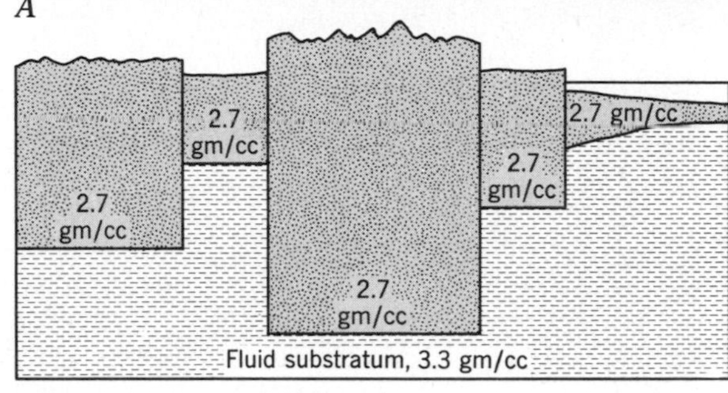


**Figure 16.19**
**The two theories of isostasy:**
***A*. Pratt suggested that the crust consists of blocks of differing density, all floating in a dense, fluid substratum, so that the height reached by each block must depend on its density.**
***B*. Airy suggested that the crust has the same density throughout, but varies in thickness. Hence, beneath high mountains we should find deep roots. (*After Bowie, 1927, and Longwell, 1925.*)**

plateaus and ocean basins (Fig. 16.19). The situation suggested by Pratt is similar to a balloon and a block of wood floating side by side in a tub of water. The balloon floats high in the water because it is light. The block of wood floats low because it is dense and heavy. The actual depth of the fluid substratum, on which Earth's blocks float, was not specified by Pratt, but he presumed it to be many kilometers down and called it the *depth of compensation.* Pratt's theory was soon challenged by another English mathematician, Sir George Airy, who pointed out that the gravity observations would also be explained if the crust, floating as proposed by Pratt, consisted of blocks of the same density but of different thicknesses, like a group of icebergs floating in the sea (Fig. 16.19). The biggest icebergs float deepest in the sea, but their tops are also highest above the water. According to Airy, the extra mass of a high mountain mass should be offset by deep "roots" of the same material.

We now know that both Pratt and Airy were partly correct. The Pratt theory applies to the entire lithosphere but not to the crust. His suggestion of different densities was correct because densities of different parts of the lithosphere do differ. Oceanic crust is about 10 percent more dense than continental crust, and ocean basins are indeed much lower than continents. Furthermore, it is easy to calculate that at a depth of approximately 100km, pressure caused by the weight of the overlying lithosphere, regardless of whether the lithosphere is capped by oceanic or continental crust, is everywhere the same. Thus, Pratt's depth of compensation is 100km. A depth of 100km puts us at the top of the asthenosphere. The correspondence between the depth of compensation and the asthenosphere is an important piece of evidence in support of the theory that plates of lithosphere float in the asthenosphere. Thus the Pratt model applies to features as large as continents and ocean basins. But it is not strictly correct for the crust. There, Airy won the day. The continents do not consist of separate blocks of rock, each with a different density. The continental crust behaves as if its density were nearly uniform. Yet beneath high mountains, where the crust projects into the mantle, there are certainly deep roots. Beneath the Rocky Mountains and the Sierra Nevada are roots, plainly visible in Figure 16.10. In this sense Airy was correct. To explain the details of isostasy, therefore, we must borrow from both models, Pratt's as well as Airy's.

The important points to be drawn from this discussion of isostasy are that the continental crust acts as if it were floating at the top of the lithosphere, while the lithosphere in turn is floating on the asthenosphere. Floating is not exactly the correct word, because we are discussing materials that are solid. But the system acts as if it were floating or *nearly* floating. Sometimes we observe that a mountain has too little root for its mass; sometimes, as in the sea-floor trenches, we observe that light-weight crust has been dragged down to form a root without a mountain mass above it. These and many other situations lead to local gravity measurements that are anomalous. But the anomalies are not very great, and overall the Earth is either in, or moving toward, an isostatic balance. An important consequence of isostatic balance is that if the weight of a mountain mass is removed, the depressed crust will slowly "bob" back up. Thus, as erosion removes material from a mountain, thereby reducing the mountain's mass, isostatic adjustment will cause the entire mountain to rise slowly. As we shall see in the next two chapters, such isostatic rebounding is a vital feature in the development and history of Earth's crust.

# SUMMARY

## EARTHQUAKES

1. Abrupt movement on faults is responsible for most earthquakes, many of which cause destructive damage to dwellings and other man-made structures.

2. Ninety-five percent of all earthquakes originate in the Circum-Pacific belt and the Mediterranean-Asiatic belt. The remaining 5 percent are widely distributed.

3. Energy released at an earthquake focus radiates outward as *P* (compressional) waves, and as *S* (shear) waves. Earthquake energy also makes Earth vibrate like a giant bell.

4. From the study of seismic waves, scientists infer the internal structure of Earth by locating boundaries or discontinuities in its composition. Pronounced compositional boundaries occur between crust and mantle and between mantle and outer core.

5. Within the mantle there are two zones, at depths of 400 and 700km, where sudden density changes produce seismic-wave discontinuities. The changes are produced by phase transitions in the minerals of the mantle.

6. The lithosphere is approximately 100km thick and rigid; it overlies a plastic zone within which seismic waves have low velocities. The low-velocity zone coincides with the asthenosphere.

7. The base of the crust is a pronounced seismic discontinuity called the M-discontinuity. Thickness of the crust varies from 20km to 60km in continental regions, but is only about 10km thick beneath the oceans.

8. The two forces that give Earth its spherical shape are gravity, acting inward, and centrifugal force caused by rotation, that acts outward. Earth is not a perfect sphere. It bulges at the equator and is flattened at the poles.

9. Both crust and lithosphere are in approximate isostatic balance; in other words, they act as though they were huge icebergs floating in fluid substrata.

# SELECTED REFERENCES

Anderson, D. L., 1962, The plastic layer of the Earth's mantle: Sci. American, v. 207, no. 1, p. 52–59.

Bolt, B. A., 1973, The fine structure of the Earth's interior: Sci. American, v. 228, no. 3, p. 24–33.

Clark, S. P., 1971, Structure of the Earth: Englewood Cliffs, N.J., Prentice-Hall.

Davidson, Charles, 1931, The Japanese Earthquake of 1923: London, Thomas Murby.

Hodgson, J. H., 1964, Earthquakes and Earth structure: Englewood Cliffs, N. J., Prentice-Hall.

Nichols, D. R. and Buchanan-Banks, J. M., 1974, Seismic hazards and land-use planning: U.S. Geol. Survey Circ. 690.

Press, Frank, 1975, Earthquake prediction: Sci. American, v. 232, no. 5, p. 14–23.

Wyllie, P. J., 1974, The Earth's mantle: Sci. American, v. 232, no. 3, p. 50–63.

**Folded strata in the Valley-and-Ridge Province, central Appalachians, Pennsylvania, and northern Maryland.**

**Image recorded in red and infra-red wavelengths by ERTS satellite from approximately 900km above the Earth. Prominent ridges were created by erosion of folded Tuscarora Sandstone of Silurian age. The sandstone resists erosion; easily eroded shale and limestone underlie the valleys. Terrain west of the Valley-and-Ridge Province is underlain by flat-lying Pennsylvanian strata; it is the site of rich bituminous coal fields. (*NASA*.)**

# PART FIVE THE PLANETS

# 17 DYNAMICS OF THE LITHOSPHERE



A revolution always makes people re-examine old ideas; the scientific revolution loosed by the concept of plate tectonics is no exception. Begun in the 1960s, this revolution shows no signs of slowing down. Every corner of geology has had to be looked at through new eyes, every conclusion rethought, and every question asked again. Throughout this book we have tried to integrate the abundant fruits of the revolution; they appear in almost every chapter, but we must remember that plate tectonics is only a theory. Nevertheless, it is a theory so well-constructed that daily it comes closer to proof. Also it is so widely important that we are going to devote this chapter entirely to plate tectonics.

## SEARCH FOR A SOLUTION

One cannot but wonder why continents have their peculiar shapes, why ocean basins are where they are, why mountain ranges, earthquake belts, volcanoes, and many other major features occur where they do. Such wonderings have prompted many scientists to think that there might be a single, underlying cause for the whole array of Earth's major features. Scientists speculated for more than a hundred years about possible causes, before the idea of plate tectonics was suggested.

During the 19th Century, people favored the idea that planet Earth was originally hot and that it has been gradually cooling and contracting. They pointed to mountain ranges and seismic belts as the places where most of the contraction has occurred. Contraction did explain some features, but it did not help with questions about the shapes and the distribution of continents. Also, as we now know, the theory of contraction has a fundamental flaw: Earth is *not* cooling down.

When it was realized that Earth's internal heat was caused by radioactive decay, some scientists suggested that our planet Earth might actually be heating up. A much smaller Earth, they suggested, was once covered largely by continental crust. Heating would cause the Earth to expand and the continental crust would then crack, and break into fragments. As expansion continued, the cracks would grow into ocean basins; through the cracks basaltic magma would well up from the mantle to build new oceanic crust. Although the theory of an expanding Earth does not account for mountain ranges, it does offer a plausible explanation of the approximately parallel coastlines of adjacent continents, such as Africa and South America. Nevertheless, the expansion theory also has other, fundamental flaws. There is no evidence (such as an increase in the rate of production of magma) to suggest that Earth is heating up. Nor can the theory explain evidence that continents have not only moved laterally but also have rotated relative to each other. To get around the flaws in both the expansion theory and the contraction theory, the effects of other forces on the crust were examined. Such forces include Earth's rotation and the Moon's gravitational pull. The centrifugal force due to rotation and the gravitational pull were both suggested as the driving forces that deformed continents, just as they are forces that influence ocean currents. Calculations showed, however, that both forces were too weak, so the ideas were abandoned. By the middle of the 20th Century all the reasonable suggestions concerning the shapes and positions of continents seemed to have been exhausted. The time was ripe for a totally new approach.

Curiously, the ideas that eventually led to the plate-tectonics concept were proposed early in the present century. Yet, because of many oversights and prejudices, a long time passed before the concept was enunciated and accepted. The key suggestion was made early in the 20th Century, soon after the contraction theory had collapsed at the end of the 19th Century. Alfred Wegener, a German scientist, suggested in 1912 that continents drift slowly across the surface of the Earth, sometimes breaking into pieces and sometimes colliding with each other. His *theory of continental drift* was originally suggested to explain the striking parallelism of the edges of the continental shelves on the two sides of the Atlantic Ocean, but other bits of favorable evidence were soon found. These bits of evidence supported the idea that the world's land masses had once been joined together in a single great supercontinent, which was dubbed *Pangaea* (pronounced pan-jée-ah), meaning "all lands" (Fig. 17.1). According to the theory, Pangaea was somehow disrupted and its fragments, the continents of today, slowly drifted to their present positions. Proponents of the theory likened the process to the breaking-up of a sheet of ice that floats in a pond. The broken pieces, they argued, should all fit back together again, like pieces of a giant jigsaw puzzle. Figure 17.1 *A* shows that a jigsaw reconstruction works very well.

Some of the more impressive types of evidence presented by Wegener and his colleagues are: fossils of identical land-dwelling animals are found in South America and in southern Africa, but nowhere else; fossils of identical trees are found in South America, India, and Australia; fragments of what is apparently a single mountain system are observed on both sides of the North Atlantic Ocean (Fig. 4.6). But the most impressive clue consists of unmistakable evidence that, about 300 million years ago, a continental ice sheet covered parts of South America, southern Africa, India, and southern Australia (Fig. 17.1*B*). The ice sheet resembled the one that covers Antarctica today; evidence of its existence is so well-preserved that thousands of glacial striations (Chapter 10) reveal the directions in which the ice flowed. But if, 300 million years ago, continents were in the positions they occupy today, the ice sheet would have had to cover all the southern oceans, and in places would even have had to cross the equator! A glacier of such huge size could only mean that the world climate was exceedingly cold. But if the climate had been cold, why has no evidence of glaciation at that time been found in the northern hemisphere? The dilemma is explained neatly by continental drift. Three hundred million years ago the regions covered by ice lay in high, cold latitudes. Indeed those regions were adjacent to the South Pole (Fig. 17.1*A*). At that time, therefore, Earth's climates need not have been greatly different from those of today.

Despite the impressive evidence that favors drifting of continents, many scientists remained unconvinced, because Wegener and his supporters could not explain how and why the continents moved. The fluid-like properties of the asthenosphere had not been discovered, and Wegener suggested that continental crust somehow slid over oceanic crust. Opponents quickly argued that because of the great friction such sliding would generate, rigid continental crust simply could not slide over rigid oceanic crust without both crusts disintegrating. The process is like trying to slide two sheets of coarse sandpaper past each other.

Wegener died in 1930. Although debate continued, its pace slowed down. True, the geological evidence of glaciers and fossils suggested that continents might have moved; but geophysicists, studying Earth's physical properties, could not suggest how the movement could happen. The situation became a stalemate, and more and more scientists discarded the theory of continental drift.

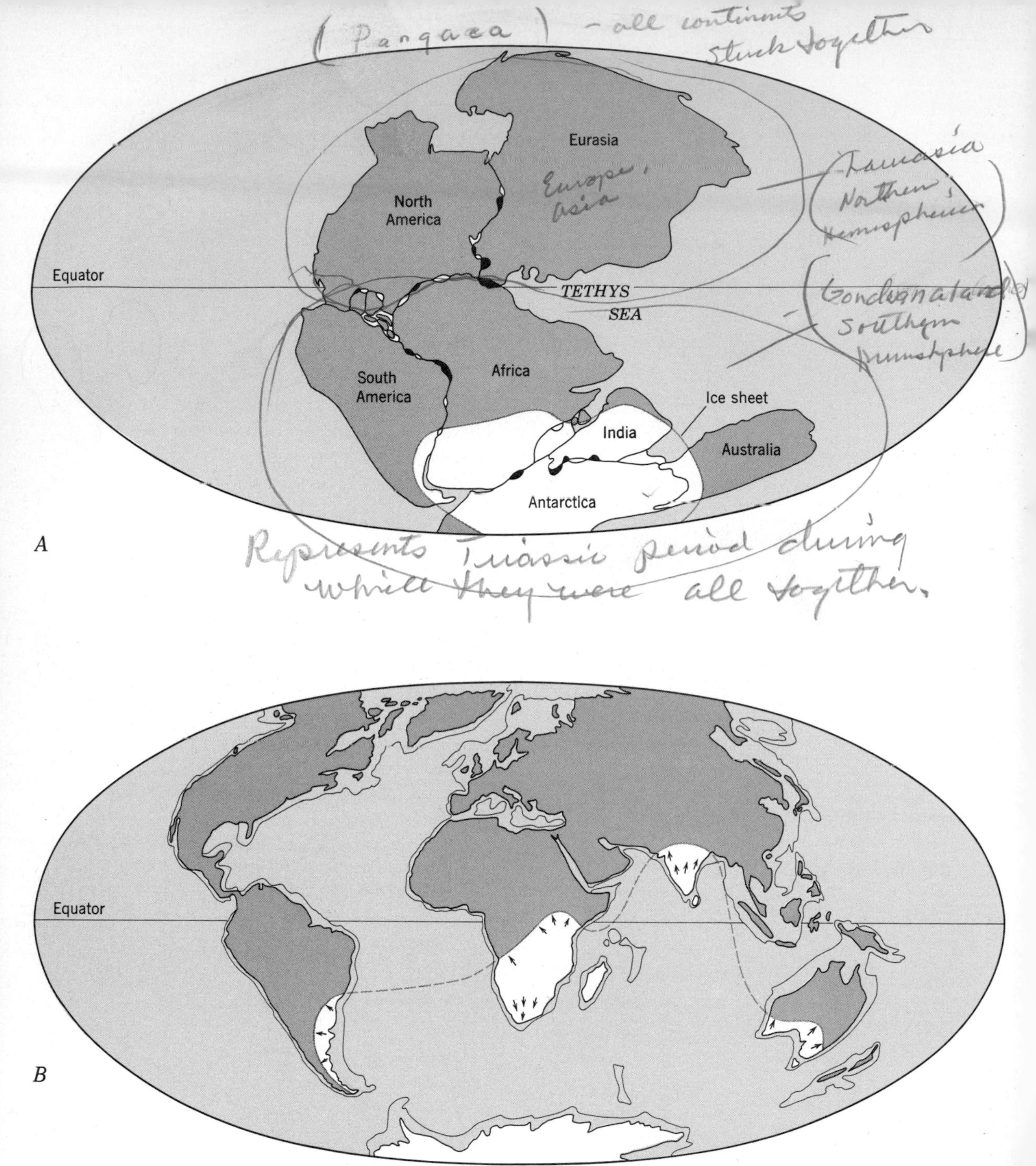


Figure 17.1

**The continents attained their present shapes when Pangaea broke apart 200 million years ago.**

***A.*** **Shape of Pangaea, determined by fitting together pieces of continental crust along a contour line 2,000m below sea level. This line coincides with the foot of the continental slope. It is the line along which continental crust meets oceanic crust. In a few places some overlap (black) occurs; in others, small gaps (white) are found. These are places where existing maps are poor or where later events have modified the shapes of continental margins. Shaded area is the region affected by continental glaciation 300 million years ago.**

***B.*** **Present continents and the 2,000m contour below sea level. Shading outlines areas where evidence of the old ice sheets exists. Arrows show directions of movement of the former ice (compare Fig. 10.22). The dashed line joining the glaciated regions indicates how large the ice sheet would have to be if the continents were in their present positions at the time of glaciation. (*After Dietz and Holden, 1971, and Du Toit, 1937.*)**

In the 1950s the debate reached a turning point. As we have already seen, some rocks, when they form, become weakly magnetized (Chapter 16), and the direction of magnetization preserves a "fossil" record of the direction of Earth's magnetic field and the position of the magnetic pole at the time. *Magnetism in ancient rock* is called ***paleomagnetism.*** Very detailed paleomagnetic measurements prove that a magnetic field has existed on Earth as long as rock has been forming. They also reveal an extraordinarily important piece of information: the positions of the magnetic poles seem to have wandered all over the Earth. The group of geophysicists who first measured the *paths of apparent polar wandering* in the 1950s were puzzled by the apparent wandering. Earth's magnetic field, they knew, was caused by motions in the liquid outer core, and the motions in turn were caused by Earth's rotation.

Although the magnetic poles might wobble a little, they must always remain close to the poles of rotation. When it was discovered that the path of apparent polar wandering measured in North America differed from that in Europe (Fig. 17.2), the geophysicists were even more puzzled. Somewhat reluctantly, they concluded that because the magnetic poles could not have moved, it must be that the continents and the magnetized rocks had moved. In this way the theory of continental drift was revived, but a mechanism to explain how the movement occurred was still lacking.

Then help came from an unexpected quarter. All the early debate about continental drift had centered on evidence drawn from the continental crust. Before 1950, the 70 percent of Earth's surface that is covered by oceanic crust was still largely unknown and unexplored. During the 1950s the mysteries of the ocean floor were revealed. In 1962, the revelation led Professor H. H. Hess of Princeton University to suggest that the sea floor spreads and

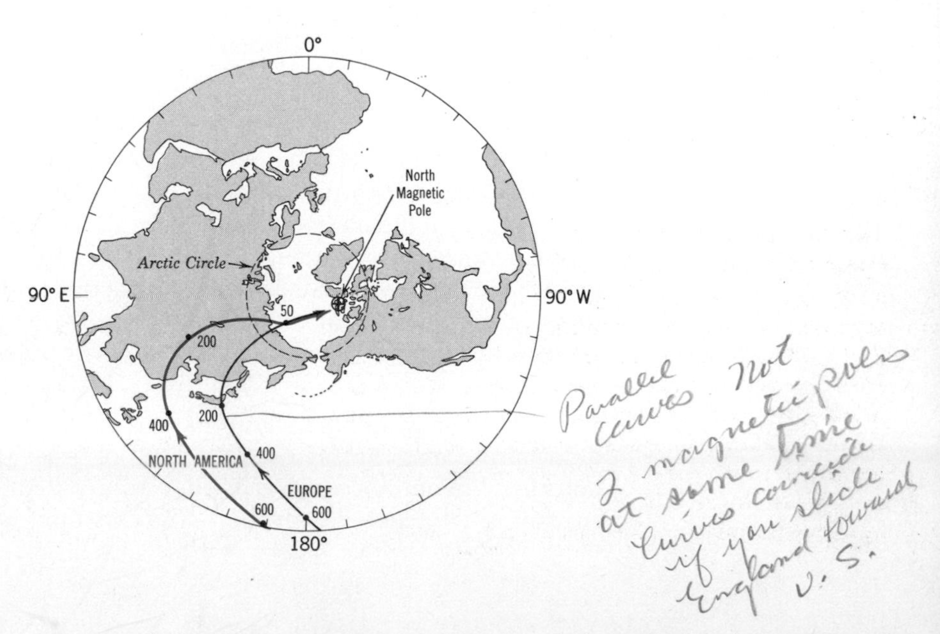


**Figure 17.2**
**Curves tracing the apparent path followed by the north magnetic pole through the past 600 million years. Numbers are millions of years before the present. The curve determined from paleomagnetic measurements in North America (blue curve) differs from that determined by measurements made in Europe (black curve). Wide-ranging movement of the pole is unlikely; therefore, we believe that it is not the pole, but the continents that have moved. Divergence of the North American and European curves indicates that at times the two continents have moved independently. (*After Northrop and Meyerhoff, 1963.*)**

moves sideways, away from the oceanic ridges. This suggestion, which became the theory of sea-floor spreading, was soon proved correct by the discovery and explanation of magnetic striping in the sea floor (Figures 13.16, 13.17, and 13.18). Other discoveries followed quickly; for example, seismic studies proved the existence of the asthenosphere. As a result, by 1967 the theory of plate tectonics had been formulated. No single person claims credit for all aspects of the theory. A great many geologists, geophysicists, paleontologists and others contributed to the events that led up to and culminated in the most comprehensive and all-embracing theory ever proposed for the Earth. The same people are now combined in an effort to prove the theory.

# THEORY OF PLATE TECTONICS

### The plate mosaic

The theory of plate tectonics maintains that the surface of the Earth is covered by six large and several small plates of lithosphere, each about 100km thick, sliding over the fluid-like asthenosphere (Fig. 4.4). The plates are believed to be rigid, moving as single coherent units; that is, the plates do not crumple and fold like wet paper, but act more like rigid sheets of plywood floating on water. The plates may flex slightly, causing gentle up- or down-warping of the crust, but the only places where intense deformation occurs is at edges along which plates impinge on each other. Such plate margins are *active zones;* plate interiors are *stable regions.*

### Classification of plate boundaries

Plates have three kinds of margins: *divergent,* margins along which two plates move apart from each other; *convergent,* margins along which two plates move toward each other; and *fault,* margins along which two plates simply slide past each other. Each margin creates distinctive topography in its vicinity, and is associated with a distinctive kind of earthquake activity and volcanism. These features are summarized in Table 17.1 and are briefly discussed below.

**Divergent margin.** Also called a ***spreading edge,*** a ***divergent margin*** is *a line along which two adjacent plates move apart from each other, and along which new lithosphere is made.* Divergent margins are places where crust is being stretched, and the stretching forces are tensional forces. The kinds of faults associated with tensional forces are invariably normal faults (Fig. 14.11). Therefore, normal faults and grabens (Fig. 14.12) are very common along divergent margins.

The topography along a plate margin depends not only on the deforming forces but also on the kind of crust that is being deformed. Where oceanic crust lies adjacent to oceanic crust, the spreading edge is beneath the sea and is marked by an oceanic

**Table 17.1**
**Kinds of Plate Margins and Characteristic Features**

| Crust on Each Plate | Feature | Kind of Margin | | |
|---|---|---|---|---|
| | | Divergent | Convergent | Transform Fault |
| Oceanic–Oceanic | Topography | Oceanic ridge with central rift valley | Sea-floor trench | Ridges and valleys created by oceanic crust |
| | Earthquake | All foci less than 100km deep | Foci from 0 to 700 km deep | Foci as deep as 100km |
| | Volcanism | Basaltic pillow lavas | Andesitic volcanoes in an arc of islands parallel to trench | No volcanism |
| | Example | Mid-Atlantic Ridge | Tonga-Kermadec Trench; Aleutian Trench | Kane Fracture (Fig. 13.18) |
| Oceanic–Continental | Topography | — | Sea-floor trench | — |
| | Earthquake | — | Foci from 0 to 700km deep | — |
| | Volcanism | — | Andesitic volcanoes in mountain range parallel to trench | — |
| | Example | (No examples) | West coast of South America | (No examples) |
| Continental–Continental | Topography | Rift valley | Young mountain range with folded crust | Fault zone that offsets surface features |
| | Earthquake | All foci less than 100km deep | Foci as deep as 300km over a broad region | Foci as deep as 100km throughout a broad region |
| | Volcanism | Basaltic and rhyolitic volcanoes | No volcanism. Intense metamorphism and intrusion of granitic plutons | No volcanism |
| | Example | African Rift Valley | Himalaya, Alps | San Andreas Fault |

ridge. Running down the center of the ridge is a narrow, steep-walled graben; both the ridge and the graben can be seen in Figure 13.6. When the spreading edge is bounded on both sides by continental crust, the feature is again marked by a graben, but it is a broad rift valley, such as the African Rift Valley (Fig. 14.13).

Earthquakes along spreading edges tend to have low magnitudes and foci shallower than 100km. This is so because the plastic asthenosphere comes close to the surface beneath a spreading edge. The kind of volcanic activity along a spreading edge is almost always basaltic. Along oceanic ridges the basalt usually consists of pillow lava (Fig. 13.12).

**Convergent margin.** Also called an ***edge of consumption*** or a ***subduction zone***, a **convergent margin** is *a line along which two plates move toward each other and where old lithosphere is subducted, that is, forced downward into the mantle,* where it is eventually remixed. Because convergent margins are margins of compression, the kinds of faults present are reverse and thrust faults (Fig. 14.11), those associated with compressional forces.

A convergent margin bounded on both sides by oceanic crust is marked by a sea-floor trench. In order to keep Earth's surface area constant and to balance the production of new lithosphere at divergent margins, one plate at a convergent margin dips downward beneath the other at an angle of 30 to 45 to the horizontal (Figures, 4.5 and 16.14). As the plate descends into the mantle, it is heated up and eventually reaches a temperature at which partial melting commences. This process (Chapter 15) forms andesitic magma. Rising to the surface, it forms a chain of volcanic islands (Fig. 17.3*A*). The chain of volcanoes is parallel to the sea-floor trench and separated from it by a distance of 200 to 300km, the distance depending on the angle of dip of the descending plate. A convergent margin bounded on one side by oceanic crust and on the other by continental crust is similarly marked by a line of volcanoes. Partial melting again forms andesitic magma, but instead of a chain of volcanic islands, a chain of andesitic volcanoes forms on the continental crust (Fig. 17.3*B*). The chain of volcanoes that runs down the center of the Andes, from Ecuador to Chile, is a modern example.

When the movement of plates brings two masses of continental crust into contact at a convergent margin, a gigantic collision occurs (Fig. 17.3*C*). The continental crust cannot be carried downward with the descending lithosphere; it is too light. Instead, the descending tongue of lithosphere breaks along the line where oceanic crust joins continental crust. The broken fragment capped by dense oceanic crust continues its slow downward plunge, while

**Figure 17.3**
**(see opposite page)**
**Three kinds of convergent plate boundaries.**

***A*. Boundary where lithosphere on both sides is capped by oceanic crust. One plate is subducted into the asthenosphere. Partial melting creates magma and gives rise to an arc of volcanic islands parallel to the sea-floor trench.**

***B*. Boundary where lithosphere is capped on one side by continental crust, on the other by oceanic crust. Oceanic plate is subducted beneath continental plate. Partial melting creates magma and gives rise to a chain of volcanic mountains.**

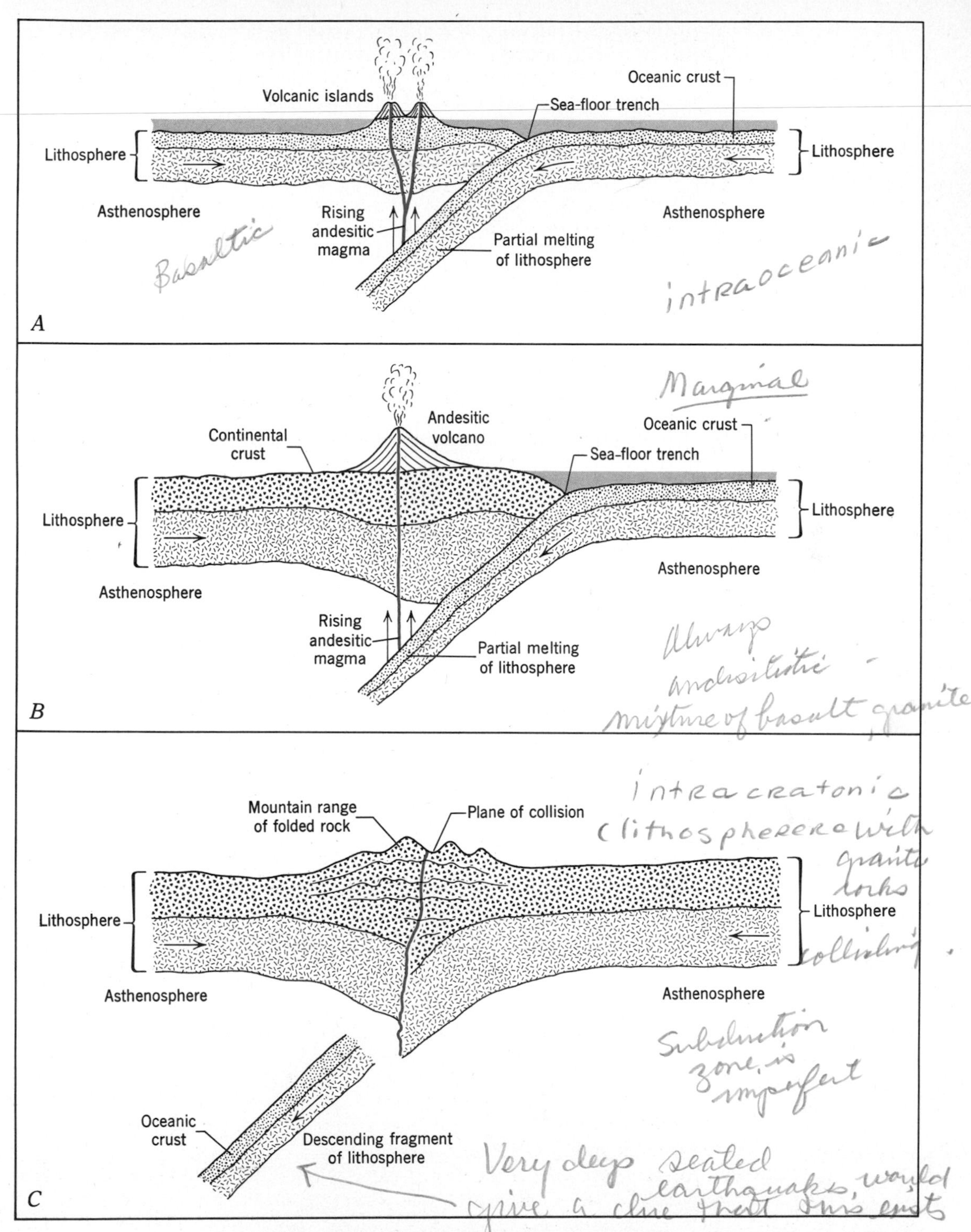


*C*. **Boundary where both plates are capped by continental crust. Because light weight, continental crust cannot be subducted, a tongue of lithosphere capped by oceanic crust breaks off and continues to sink into the mantle. The two continental masses collide and produce a mountainous belt of folded rock.**

at the surface the two colliding masses of less dense continental crust are buckled and deformed, creating a great, mountainous belt of folded and metamorphosed rock. In places the base of the crumpled rock mass reaches temperatures high enough to cause partial melting and the formation of rhyolitic magma. The magma rises, intrudes the crumpled continental crust above, and forms huge granitic batholiths (Fig. 15.17).

Earthquakes associated with convergent plate margins are commonly very strong, and because descending lithosphere retains some brittle properties to depths as great as 700km, the deepest earthquake foci on Earth occur there (Figures 16.13 and 16.14).

**Fault margin.** The faults at the margins of plates are transform faults. They are huge, vertical, strike-slip faults cutting down through the entire lithosphere. They form when either a new divergent or a convergent margin fractures the lithosphere (Fig. 17.4). Neither compressional nor tensional forces are associated with the faults. They are simply margins along which two plates slide past each other. The sliding margins smash and abrade each other like two giant strips of sandpaper; so the faults are marked by zones of intensely shattered rocks. Where the faults cut oceanic crust, they make elongate zones of narrow ridges and valleys on the sea floor (Fig. 13.6). When transform faults cut continental crust, they do not make distinctive topographic features; instead they are marked by parallel or nearly parallel faults, in a zone as much as 100km wide.

No volcanic activity is associated with transform faults, but the sliding movement causes a great many earthquakes, some of them of high magnitude.

Probably the best-known transform fault is the San Andreas Fault in California. The many earthquakes that disturb California are caused by movement along it. Some of those earthquakes, including the one that leveled San Francisco in 1906, have been particularly destructive, and it is possible that future quakes will be just as devastating. As long as the plates continue to move,

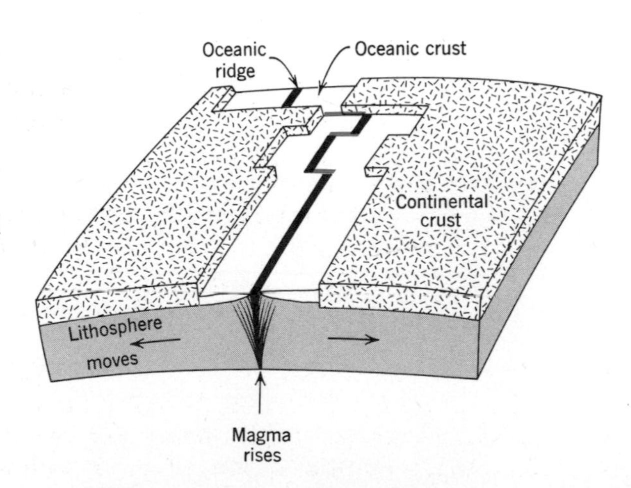


**Figure 17.4**
**Transform faults (blue) form when an oceanic ridge first forms. Shapes of continental margins reflect faults now found along oceanic ridge.**

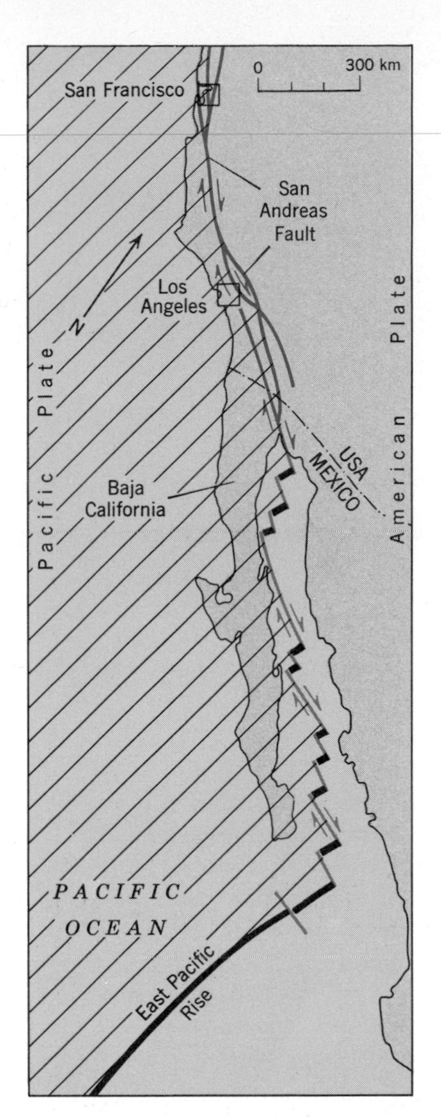


**Figure 17.5**
**The San Andreas Fault is one of several transform faults that offset an oceanic ridge (East Pacific Rise). The faults and ridge segments separate the Pacific Plate (left) from the American Plate (right). The two plates are sliding past each other, causing frequent earthquakes in California. (*After Elders and others, 1972.*)**

activity must occur along the San Andreas Fault, and residents of California can expect more earthquakes.

The San Andreas Fault is the largest of several faults that offset the segment of oceanic ridge called the East Pacific Rise. Figure 17.5 shows how the transform faults and segments of the East Pacific Rise separate the American Plate from the Pacific Plate. Figure 17.5 also shows that the San Andreas is only one of several faults that break the continental crust. The others are subsidiary to the San Andreas, however, and are part of a fault zone that is roughly parallel to the main fault. Movement along the San Andreas Fault arises from movement between the American and Pacific Plates. The peninsula of Baja California and the portion of the state of California that lies west of the San Andreas Fault lie on the Pacific Plate. That plate is moving northwest, relative to the American Plate, at a rate of several centimeters per year. In about 10 million years Los Angeles will have moved far enough north so as to be

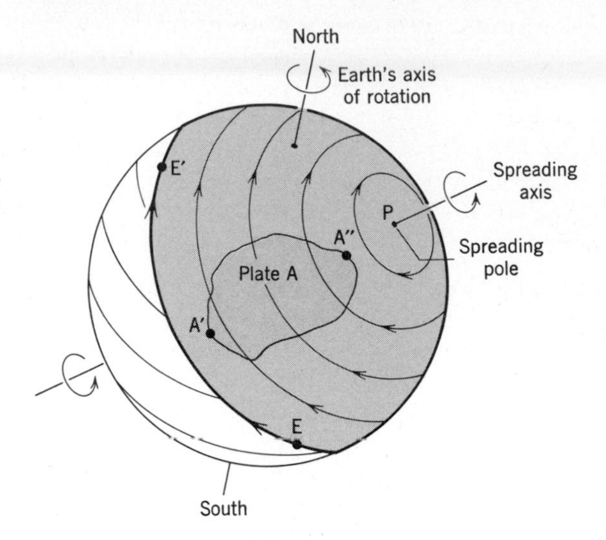


**Figure 17.6**
**Rotation of a hemispherical cap that fits over a sphere. Movement of Plate A, a part of the cap, is controlled by the rate of rotation around the spreading axis. The movement of each plate of lithosphere on Earth's surface can be described as a rotation about a spreading axis.**
**(*Adapted from P. J. Wyllie, 1976*.)**

opposite San Francisco. In about 60 million years, the segment of continental crust on which Los Angeles lies will have become separated completely from the main mass of continental crust that comprises North America.

### Movement of plates

The speeds at which plates move have been mentioned several times. As we discuss later, speeds can only be calculated, but they can be calculated quite accurately. We might think, intuitively, that all points on a plate move with the same velocity. This is incorrect. Our intuitions would be correct if plates of lithosphere were flat like sheets of plywood, and if they moved over a flat asthenosphere like plywood floating on water; then, all points on the plate *would* move with the same velocity. But plates are pieces of a shell on a spherical Earth, so they are not flat, but curved. In the geometry of a sphere, any movement on the surface can be described as a rotation about an axis of the sphere. A consequence of rotation, and therefore of a curved plate moving over the surface of a sphere, is that different parts of the plate move with different velocities.

To picture how points on a plate move with different velocities, imagine a plate so large that it forms a hemispherical cap covering half the Earth (Fig. 17.6). The cap moves independently of Earth's rotation and rotates instead about an axis of its own, colloquially called a *spreading axis.* In the figure, point P, where the spreading axis reaches the surface, is a *spreading pole.* Point P has no velocity of movement because it is the fixed point around which the hemispherical cap rotates. However, point E, at the edge of the cap, has a high velocity because it must move completely around the Earth, along path EE′, during a single revolution of the cap. Any point on the cap between points P and E has an intermediate velocity that is slower if the point is close to P, faster if it is close to E.

No plate is large enough at present to cover half the Earth, nor does any plate rotate around a spreading pole in the center of the plate. But the principle is the same for a small plate as it is for a hemispherical cap. Consider Plate A, which is only a small portion of the hemispherical cap. The motion of Plate A is from east to west around the spreading axis. Point A'', close to the spreading pole, must move more slowly than point A', more distant from the pole.

The motion of each of Earth's plates can be described in terms of rotation around a spreading axis, and the velocity of each point on the plate depends on its distance from the spreading pole. One consequence of differing velocities of motion is this: The width of new oceanic crust that borders a divergent margin between two plates increases with distance from a spreading pole (Fig. 17.7). A further consequence is that each segment of an oceanic ridge lies on a line of longitude that passes through the spreading pole, and each transform fault that offsets the oceanic ridge lies on a line of latitude around the spreading pole. The relation between transform faults and spreading poles can be used to determine the position of the spreading pole of each plate. Also the positions of old transform faults can be used to determine the positions of former spreading poles and therefore to determine if a plate has, at some time, changed its direction of motion.

Velocities on a large plate such as the American Plate (Fig. 4.4) vary from as little as 1cm per year to as much as 4cm per year. On the Pacific Plate, velocities as high as 6cm per year are determined. Although the velocities cannot be measured directly, they are calculated from the magnetic stripes in the oceanic crust (Fig. 13.18). If we know the radiometric age of a particular magnetic stripe, and also the distance the stripe has moved from the spreading edge, it is a simple matter to calculate the average velocity of the moving lithosphere. When velocities are calculated for successive magnetic stripes, it is found that plates have been moving with constant velocities for at least 4 million years. The

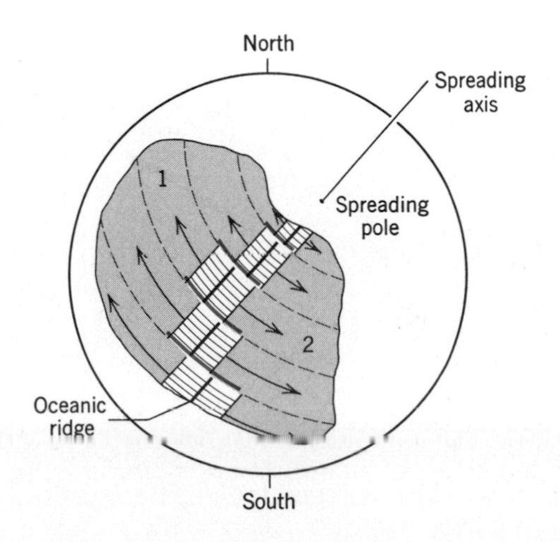


**Figure 17.7**
**Relation between spreading axis, oceanic ridge, and transform faults in two adjacent plates.**
**Plates 1 and 2 have a common spreading edge offset by transform faults (blue). Each segment of the oceanic ridge lies on a line of longitude that passes through the spreading pole. Each transform fault lies on a line of latitude with respect to the spreading pole. The width of new oceanic crust (striped shading) increases away from the spreading pole.**

ages of magnetic stripes older than 4 million years are known only approximately; so we cannot be sure whether the velocities have been constant through longer times. But even the approximate ages determined for stripes suggest that constant velocities have been operating through much longer times. This must mean, therefore, that whatever force it is that causes the plates to move is apparently a steady driving force, and not an irregular or stop-and-go affair. This brings us back to the question of what causes the motion.

# CAUSES OF PLATE TECTONICS

Just as Alfred Wegener could not explain what made continents drift, we are still unable to say *why* plates of lithosphere move. Until we can explain the driving force, plate tectonics must remain a theory. But meanwhile we can speculate about the causes of the motion.

The lithosphere and asthenosphere are inevitably bound together. If the asthenosphere moves, it will make the lithosphere move just as movement of sticky molasses will move a piece of wood floating on its surface. So too will movement of the lithosphere cause movement in the asthenosphere below. Such is our state of uncertainty that we cannot yet say whether the asthenosphere makes the lithosphere move or whether it is the other way around. Let us examine both possibilities.

### Movement of the asthenosphere

The asthenosphere is solid rock, but it is so hot and so weak that it can be easily deformed and made to flow like a very sticky, viscous liquid. Like a liquid, too, the asthenosphere must be subject to convection currents if a local source of heat causes a mass of rock to become heated to a higher temperature than surrounding rock. The heated mass expands, becomes less dense and rises very slowly, at rates as low as 1cm per year. To compensate for the rising mass, cooler, more dense material must flow downward. Thus a convection cell starts to operate in the asthenosphere.

Two kinds of convection cells within the asthenosphere have been suggested. The first kind is a cell in which all movement is confined to the asthenosphere. There would be several convection cells and their size and shape are indicated by the size and shape of the plates. The masses of hot rock are supposed to rise vertically beneath oceanic ridges, then turn sideways, flowing horizontally beneath the lithosphere and dragging the plates as they move. The farther the flowing asthenosphere moves away from the ridge, the cooler it gets. Eventually the asthenosphere becomes so cool and so dense that it sinks, dragging the lithosphere with it. The place where sinking occurs is beneath a sea-floor trench (Fig. 17.8*A*). The big problem with the suggestion that convection cells are confined

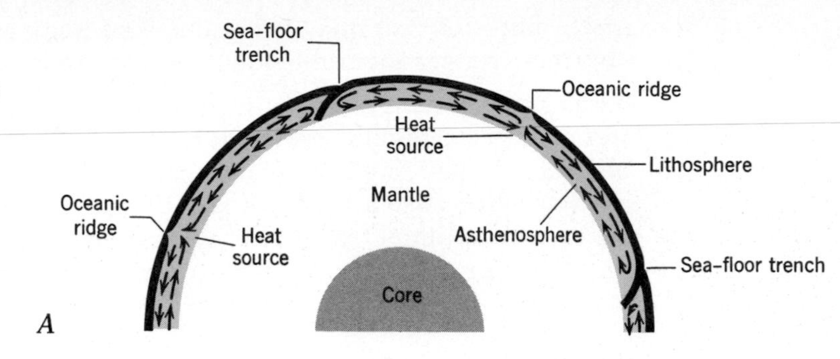


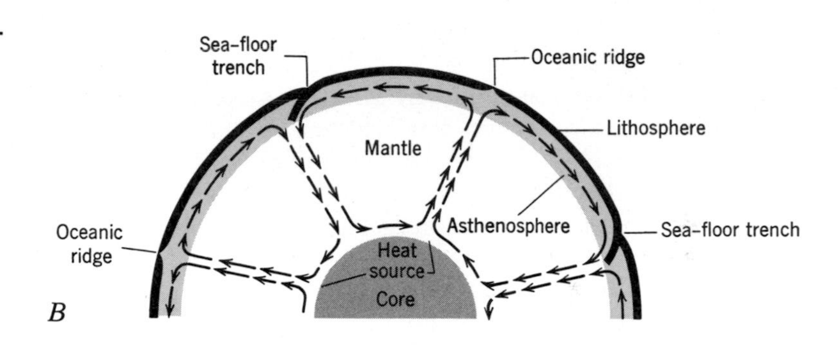


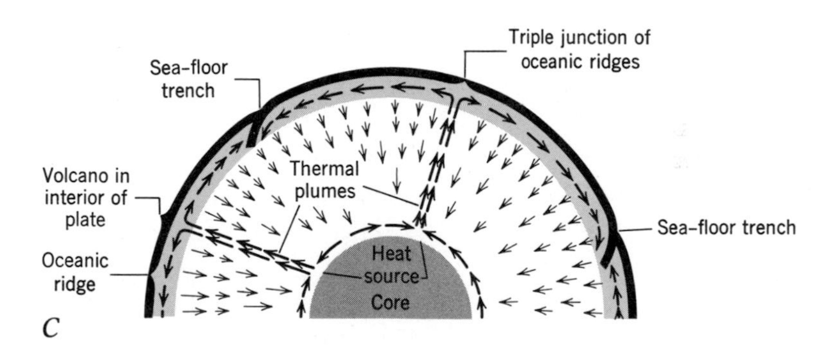


**Figure 17.8** **Three suggested mechanisms by which convection and flow in the asthenosphere might move plates of lithosphere.**
***A*. Convection is confined to the asthenosphere.**
***B*. Convection involves the entire mantle.**
***C*. Thermal plumes rise from the mantle-core boundary and cause local hot spots at Earth's surface.**

to the asthenosphere is that we do not know of sources of heat, such as enrichments in radioactive elements, that could cause localized heating in the asthenosphere. A second suggestion, one that avoids the problem of heat sources in the asthenosphere, is that convection cells involve the entire mantle, and that heat is somehow supplied from hot regions in the outer core (Fig. 17.8*B*). Just how hot regions develop, and how localized transfer of heat takes place between core and mantle, are not known. The flowing motions and sideways dragging of lithosphere are the same in the two convection models. The main differences are the sizes of the convection cells in the two cases, and in the latter case, the notion that convection can involve not only the weak asthenosphere, but also the stronger mantle below.

Some scientists suggest that if local hot regions do occur on the core-mantle boundary, they will be small, roughly circular spots. Instead of producing a large convection cell, they maintain, a small hot spot will cause a long cylinder of hot rock, a few hundred kilometers in diameter, to rise. They refer to the vertically rising cylinder as a *thermal plume* and suggest that places of long-continued volcanism, like the Hawaiian Islands, and the many spots where three oceanic ridges meet in a *triple junction,* such as beneath the Azores Islands in the Atlantic and beneath the Galapagos Islands in the Pacific, might possibly be above the tops of plumes. All upward motion could be accounted for by no more than 20 thermal plumes. When a plume reaches the base of the lithosphere, it creates a local hot spot, then spreads out and flows horizontally in all directions. Return flow to balance the concentrated upward flow in the plumes does not involve well-defined down-flowing plumes, but is accomplished by slow downward movement of the entire mantle. Movement of plates of lithosphere is more difficult to visualize by thermal plumes, but outward spread of hot rock, away from a hot spot, must cause motion. Most of the motion arises from the plumes that rise beneath triple junctions in the oceanic ridge. The three plates that meet at the triple junction will each be carried away from the junction by the flowing asthenosphere.

All of the preceding discussion about convection is speculation. There is not yet any direct evidence that convection of any sort occurs beneath the lithosphere. For this reason, some scientists suggest that the motion of the lithosphere is due to entirely different processes.

### Movement of the lithosphere

If a force causes lithosphere to move, the plastic asthenosphere must also move. Thus the surface may actually move the material beneath it. This might happen in one of three ways. First, rising magma at a spreading edge creates new lithosphere and pushes the lithosphere sideways (Fig. 17.9*A*). Once the process is started, it tends to keep itself going. The problem is to get it started.

A second way by which lithosphere can be made to move is by dragging rather than pushing. Proponents of the dragging idea point out that a descending tongue of cold lithosphere must be more dense than the hot mantle surrounding it. Because rock is a poor conductor of heat, they urge, the temperature at the center of a descending slab of lithosphere can be as much as 1,000°C cooler than the mantle at depths of 400km to 500km. The dense slab of lithosphere must then sink under its own weight, and, acting like a weight, exerts a pull on the entire plate. This is somewhat like a heavy weight that hangs over the side of a bed and is tied to the edge of a sheet. The weight falls and pulls the sheet across the bed. To compensate for the descending lithosphere, there is a slow flow

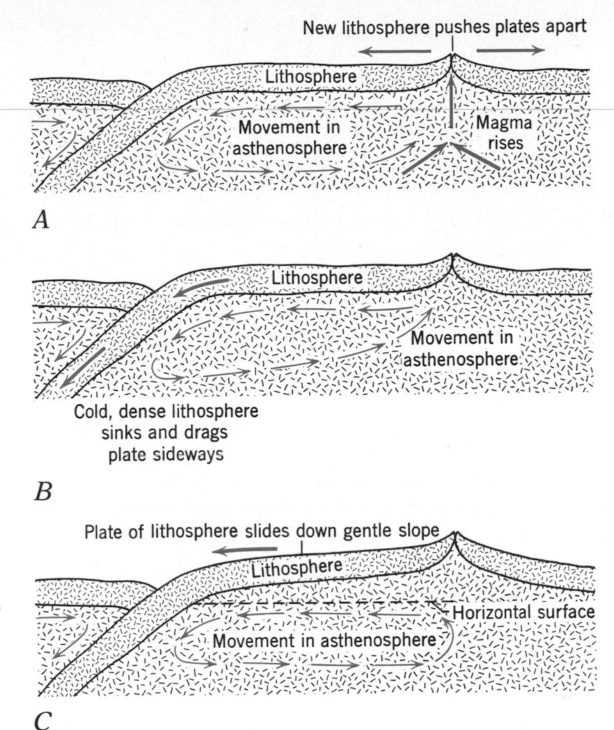


**Figure 17.9** **Three suggested mechanisms by which lithosphere might move over the asthenosphere.**
***A*. Magma, rising at a spreading edge, exerts enough pressure to push plates of lithosphere apart.**
***B*. A tongue of cold, dense lithosphere sinks into the mantle and pulls the rest of the plate behind it.**
***C*. A plate of lithosphere slides down a gently inclined surface of asthenosphere.**

in the asthenosphere back to the spreading edge (Fig. 17.9*B*). Both the pushing and the pulling mechanisms have a serious problem. Plates of lithosphere may be brittle, but they are too weak to transmit the pushing and pulling forces without major deformation occurring. We do not see the deformation.

A third, more likely mechanism for movement of a plate of lithosphere is for it to slide downhill away from the spreading edge. The lithosphere grows cooler and thicker away from a spreading edge. As a consequence, the bottom surface of the lithosphere will slope away from the spreading edge. If the slope is as little as 1 part in 3,000, the weight of the lithosphere could cause it to slide down the slope at a rate of several centimeters per year (Fig. 17.9*C*).

At present there is no way to test and choose between the three asthenosphere and three lithosphere mechanisms. Possibly several or even all six operate, so that the entire process is more complicated than we now imagine. Only future research will resolve the question. Nevertheless, without any waiting we can answer many questions about the role played by plate tectonics in the history of continents. So we turn at once to the evolution of continental crust.

# SUMMARY

1. Abundant evidence proves that continents have not remained fixed on Earth's surface, but have moved repeatedly from place to place.

2. The lithosphere is broken into six large and several smaller plates, each about 100km thick, and each slowly moving over the top of the solid, but fluid-like, asthenosphere beneath it.

3. Each plate is bounded by three different kinds of margins. Divergent margins are those where new lithosphere forms. Plates move away from them. Convergent margins are lines along which plates compress each other and where lithosphere capped by oceanic crust is subducted back into the mantle. Fault margins are lines where two plates slide past each other.

4. Movement of a plate can be described in terms of rotation across the surface of a sphere. Each plate rotates around a spreading axis. The spreading axis does not coincide with Earth's axis of rotation.

5. Because plate movement is a rotation, the velocity of movement varies from place to place on the plate.

6. Each segment of oceanic ridge that marks a divergent margin of a plate lies on a line of longitude passing through the spreading pole. Each transform fault margin of a plate lies on a line of latitude of the spreading pole.

7. The forces that drive a moving plate are not known, but might result from either convection in the mantle or movement of the lithosphere.

## SELECTED REFERENCES

Anderson, D. L., 1971, The San Andreas Fault: Sci. American, v. 225, No. 5, p. 52–68.

Dewey, J. F., 1972, Plate tectonics: Sci. American, v. 226, No. 5, p. 56–68.

Elders, W. A. and others, 1972, Crustal spreading in southern California: Science, v. 178, p. 15–24.

Hallam, Anthony, 1972, Continental drift and the fossil record: Sci. American, v. 227, No. 5, p. 56–66.

Hallam, Anthony, 1973, Revolution in the earth sciences: Oxford, England, Clarendon Press.

McKenzie, D. P., 1972, Plate tectonics and sea-floor spreading: American Scientist, v. 60, p. 425–435.

Matthews, S. W., 1973, This changing earth: National Geographic, v. 143, p. 1–37.

Takeuchi, H., Uyeda, S. and Kanamori, H., 1970, Debate about the Earth: (revised ed.), San Francisco, Freeman, Cooper.

Tarling, Don, and Tarling, Maureen, 1971, Continental drift: Garden City, N.Y., Doubleday.

Vine, F. J., 1966, Spreading of the ocean floor: new evidence: Science, v. 154, p. 1405–1415.

Wilson, J. T., (compiler), 1972, Continents adrift: Readings from Sci. American, San Francisco, W. H. Freeman and Co.

Wyllie, P. J., 1976, The way the Earth works: New York, John Wiley.

# 18 CONTINENTS AND CONTINENTAL CRUST



We have now examined the theory of plate tectonics. Most of the evidence supporting the theory comes from a study of oceanic crust, and no portion of the oceanic crust has been found to be older than about 200 million years. Older oceanic crust has been subducted into the mantle. The continental crust, however, contains rock as old as 3.8 billion years; so it is continental crust to which we must turn for evidence of Earth's history between 0.2 and 3.8 billion years ago. We can tackle the evidence, even though it is complex, fragmentary, and only partly understood.

Let us work back from the supercontinent Pangaea (Fig. 17.1), which began to break into pieces when the existing plates of lithosphere started to move. Why movement and fragmentation happened at that moment, 200 million years ago, remains an unanswered question. What preceded Pangaea? Were there possibly earlier episodes when plates of lithosphere rafted continents around, alternately fragmenting them and then welding them together again? Answers to questions such as these are still very hesitant, and in order to attempt them, we must look more closely at the structure of the continents themselves.

# STRUCTURE OF CONTINENTS

Every continent contains two main kinds of building units. The first kind is a core or nucleus of ancient rock. Commonly called a ***continental shield,*** each nucleus is *a mass of Precambrian rock around which a continent has accumulated.* Large continents such as Africa and South America have two continental shields, suggesting that they grew around two nuclei (Fig. 18.1). Surrounding each continental shield are the second kind of building unit—mountains of intensely folded and faulted strata as well as mountainous masses of volcanic rock. The mountains themselves differ in age, history, origin, and size. Some, like the volcanic cones Kilimanjaro in East Africa and Mount Ararat in Asia Minor, are single isolated masses. Others form a linear sequence which we call a ***mountain range,*** meaning *an elongate series of mountains belonging to a single geologic unit.* Excellent examples are in the Sierra Nevada in eastern California and the Front Range in Colorado. *A group of ranges similar in general form, structure, and alignment, and presumably owing their origin to the same general causes,* constitute a ***mountain system.*** The Rocky Mountain System is a great assemblage of ranges, all formed within

Figure 18.1
**Continental shields and mountain chains formed since the beginning of the Paleozoic Era. Each shield is itself a patchwork of remnants of much older mountain chains. (*Shield areas after Dunbar, 1966.*)**

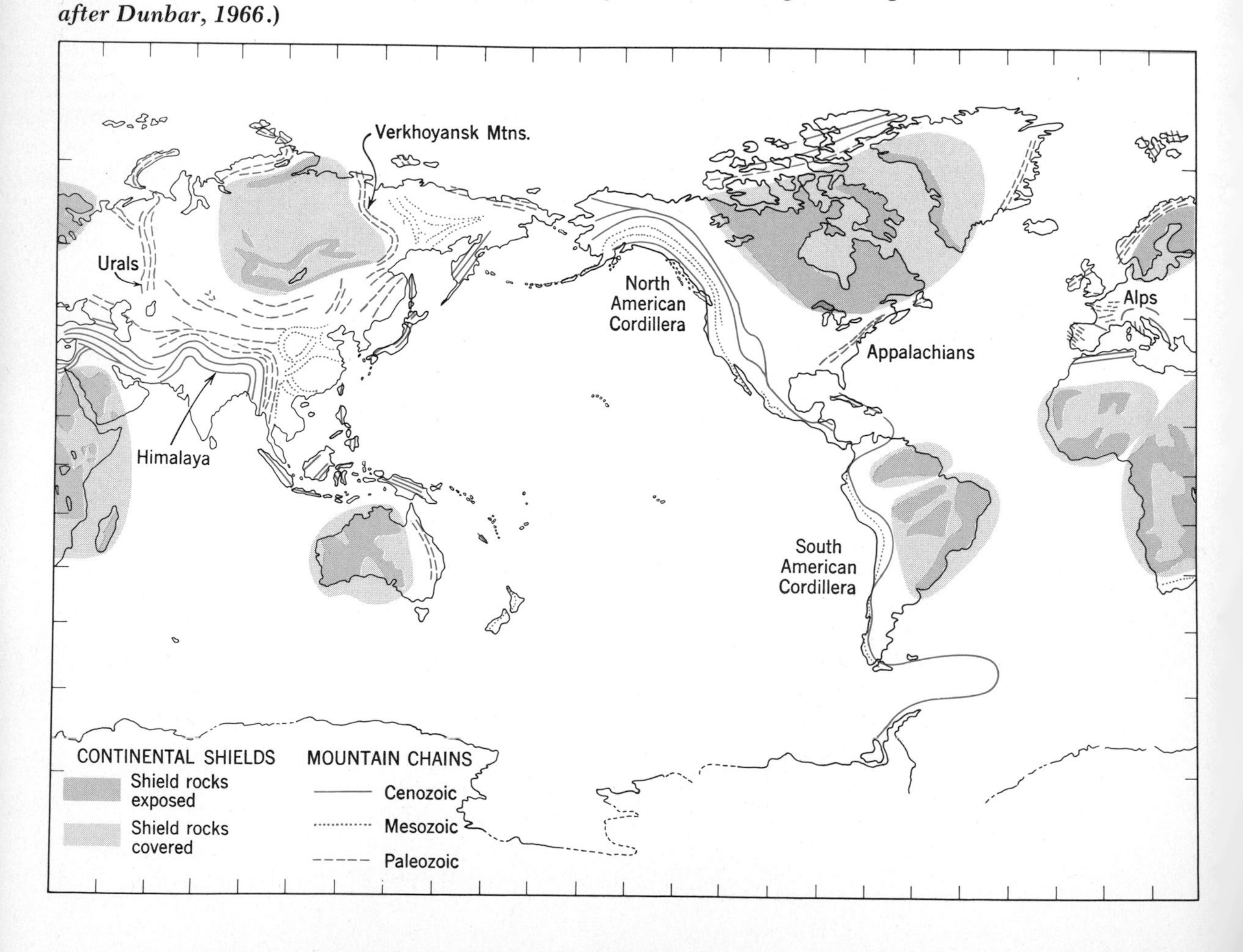

a few million years of each other, that extend from near the Mexican border northward through the United States and western Canada. The term ***mountain chain*** is used somewhat more loosely to designate *an elongate unit consisting of numerous ranges or systems, regardless of similarity in form or equivalence in age.* An example is the gigantic mountain chain that runs along the western edge of the Americas, from the tip of South America to northwestern Alaska, and that includes all the systems and ranges in between. This broad belt of ranges is also called the *American Cordillera.*

All mountain chains formed during the last 600 million years (that is, since the beginning of the Paleozoic Era) are depicted in Figure 18.1. There is abundant evidence that much older mountain chains once existed on Earth because even in the continental shields we find belts of intensely deformed and folded rocks, now deeply eroded. They remain as mute evidence that great mountain ranges once towered above. Indeed, it is probable that most fragments of continental crust were once part of mountain ranges, and that continents have somehow grown to their present sizes by the welding of younger and younger mountain chains onto the growing continental shields.

### Continental shields

Within the continental shields are Earth's most ancient rocks. These were originally lava, similar in composition to modern basalt. Although these ancient rocks have been changed by metamorphism (into schists containing greenish minerals such as chlorite and amphibole; hence the schists are often loosely called *greenstones*), it is still possible to find well-preserved pillow structures within them (Fig. 13.12). This indicates that Earth's oldest rocks were extruded as submarine lavas. Interlayered with the ancient lavas are thin strata very rich in silica ($SiO_2$), now metamorphosed to chert, that were precipitated chemically in the ancient ocean. We conclude, therefore, that at the time these oldest rocks formed, the ocean probably covered the entire Earth. No fragment of crust projected above the water; so there was no erosion and no detrital sediment. When greenstones are dated by radiometric means, they yield ages of 3.4 billion years or older. The most ancient date obtained for any rock so far is 3.8 billion years; it was measured on a granitic rock that intruded greenstone. We must conclude, then, that even earlier than 3.8 billion years ago, the ocean had formed and flows of pillow basalt covered the ocean floor. Perhaps fragments of even more ancient rock remain still undiscovered in the continental shields. For the present, however, 3.8 billion years is a time barrier we cannot penetrate.

The earliest crust must have been similar to but much thinner than the present oceanic crust. We think it was thin because 3.8 billion years ago many more radioactive atoms were present, creating at least twice as much heat in the mantle as is produced today. Geothermal gradients must therefore have been steeper, crust and lithosphere must have been thinner, the asthenosphere must have been nearer the surface, and the source of magma must therefore have been shallower that it now is. What, then, made the crust thicken and project above sea level to become a source for detrital sediment? We are not certain, but a possible answer is indicated by piles of andesite that are found here and there among the ancient basalts. The andesites are similar to those in modern volcanic islands formed above subducted oceanic crust. This suggests that the origins of ancient and modern andesites may be the same, and that subduction and partial melting of basalt in ancient oceanic crust was the source of the andesite. Eventually, some of the andesitic volcanoes rose above sea level, and so were exposed to erosion (Fig. 18.2). By 3.8 billion years ago, this had probably happened more than once.

Another part of the story of ancient crust involves bodies of granite that intrude ancient lavas. Earth's oldest granite, as we just read, is 3.8 billion years old. This most ancient granite, and others like it, seems to have formed by magmatic differentiation in the mantle (Chapter 15). Although, by contrast, younger granites seem to have formed by partial melting of continental crust, the ancient granites could not have formed by partial melting because there was no continental crust to melt. The granitic magmas that formed in the mantle rose slowly upward, thickening and doming up the crust. These magmas likewise metamorphosed the ancient lavas they intruded, and in places caused migmatites to form (Chapter 15). By 3.4 billion years ago, a few patches of crust had become thick enough and rigid enough to be able to support basins in which sediment could accumulate (Fig. 18.2*C*). Some shields today contain old basins still almost completely filled with sedimentary

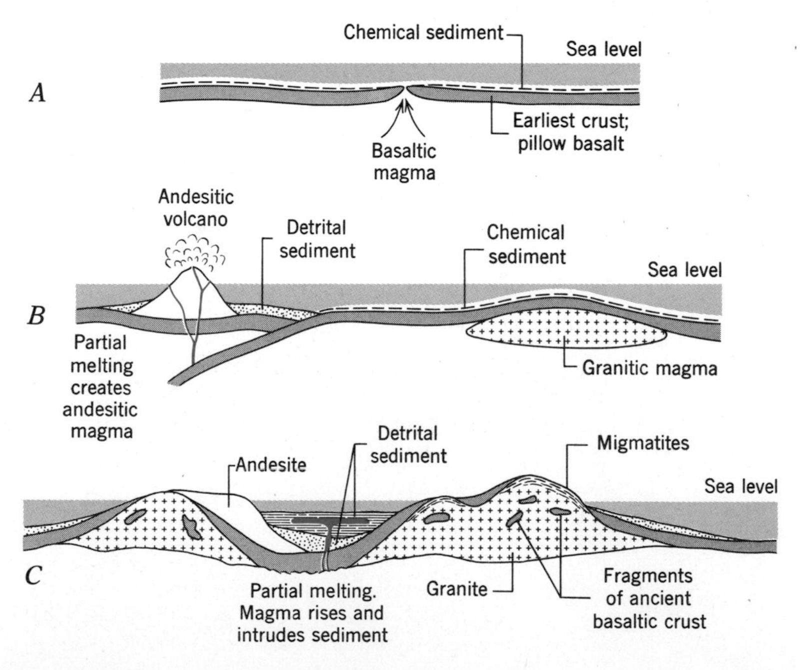


**Figure 18.2**
**Schematic sections showing how the earliest continental crust may have developed.**
***A*. Before 3.8 billion years ago, the earliest crust was thin; it consisted of pillow basalt and siliceous chemical sediments.**
***B*. At a later time subduction of basaltic crust and partial melting created andesitic volcanoes. After the volcanoes rose above sea level, they were eroded, creating detrital sediment. The earliest granites formed in the mantle by magmatic differentiation.**
***C*. Granite intruded the crust, thickening, doming, and deforming it. Migmatites formed along the margins of the granites. Depressions in the crust formed basins in which sediment accumulated. (*After Anhaeusser, 1973*.)**

rock. The ancient patches of basalt, granite, andesite, and primitive sediments are the cores of today's continental shields. Oceanic crust probably once lay between the patches of continental crust, but on this point we can only speculate, because no fragments of it remain. The ancient continental cores, now highly deformed, apparently stopped growing about 3 billion years ago. But the stop was only a pause because growth soon started again.

Continental shields today are areas of low relief. Long-continued erosion has worn them down close to sea level, and the ancient mountains have ceased to rise. The shields seem to be both stable (neither rising nor sinking) and in isostatic balance. But continental shields are only small fragments of continents. Continuing from about 2.5 billion years ago to the present, the continents have increased considerably in size. When we look at the structure of North America, we get a clue to the way in which this might have happened. In central and northern Canada there is a shield within which all rocks give radiometric dates that are 2.5 billion years old or older (Fig. 18.3). Surrounding the shield like a twisted collar is a belt of complexly deformed strata, a belt created during the period 1.0 to 2.5 billion years ago. The collar is not visible everywhere because it is sometimes concealed beneath a veneer of younger, undeformed strata. Nevertheless, all the samples obtained by drilling through the covering strata, and all samples from places where the old deformed collar pokes through, give radiometric dates falling between 2.5 and 1.0 billion years.

The twisted collar consists of many deeply eroded mountain systems. Surrounding it is an even younger belt, somewhat like a showy necklace, of young mountain systems that have not yet been worn down. These are our familiar mountain systems such as the Appalachians, the Rockies, and the Sierra Nevada. All these have

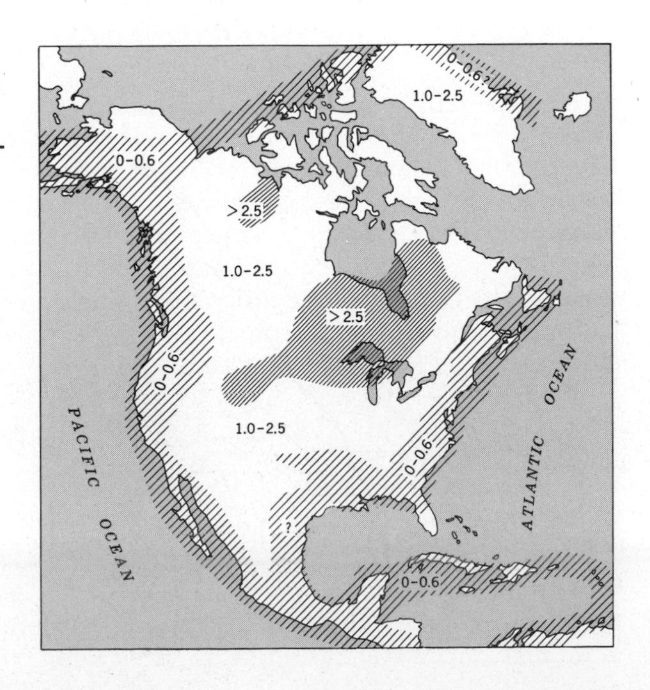


**Figure 18.3**
**Ages of belts of continental crust in North America, based on isotopic dates of metamorphic and igneous rock formed during ancient episodes of mountain building. Numbers are ages in billions of years. (*After Engel, 1968.*)**

been formed during the last 0.6 billion years. The structure of North America suggests, therefore, than an ancient shield forms a nucleus of continental crust, and that the continent has been slowly growing as successive mountain systems were formed along its margins.

There are two schools of thought concerning the growth of continents and the amount of continental crust. One suggests that the total volume of continental crust has remained constant, through the last 2.5 billion years, and is being continually recycled. The second school of thought, to which the authors of this text are attracted, maintains that the continental crust is slowly increasing by addition of magma derived from partial melting of subducted oceanic crust. As an example, the Andes in South America are thought to be now growing larger by addition of magma from the mantle. To counter this argument, the 'constant volume of crust" school maintains that a small amount of continental crust must somehow be subducted into the mantle. The question of growth or constancy is still open and is a fascinating area for research.

How well does the North American pattern fit that of other continents? The answer is quite well, because all shields are surrounded by belts of younger rock. But some of the patterns are confused; some shields seem to have been broken into two fragments and some old mountain systems seem to be incomplete as though they had been sliced off. Evidence of this sort lets us infer that there may have been several periods in the past when plates of lithosphere moved their loads of continental crust around on the surface, and that fragmentation of continental shields and truncation of mountain systems occurred when continents were split apart. Before discussing these inferences further, we must examine more closely the structure and origin of mountain chains.

# CLASSIFICATION OF MOUNTAINS

Although we all know that mountains are rocky masses standing above the surrounding terrain, it is not easy to classify mountains because they display a great variety of rocks and structures; no two are identical. If we concentrate on the details, we are in danger of seeing only the foliage and missing the forest. The most helpful way to organize our thinking about mountains and to see through the foliage is to identify some single, most characteristic feature, and use it for classification. On this basis we can identify three principal kinds of mountains:

1. Fold mountains
2. Volcanic mountains
3. Fault-block mountains

Each kind merits separate discussion.

### Fold mountains

Fold mountains are spectacular, complex structures. Occurring in great arc-shaped systems a few hundred kilometers wide, they commonly reach several thousand kilometers in length. The word *fold* clearly indicates their most characteristic feature. Strata have been compressed, folded, and crumpled, commonly in an exceedingly complex manner. Although folding is the key feature, other kinds of mountain-building processes participate in the making of fold-mountain systems too: faulting, metamorphism, and igneous activity are always present. Examples are widespread: the Appalachians, the Alps, the Urals, the Himalaya, and the Carpathians are all fold-mountain systems. Indeed the Alps, Carpathians, and the Himalaya belong to a gigantic fold-mountain chain formed during the Mesozoic and Cenozoic Eras (Fig. 18.1).

**Geosynclines.** All fold-mountain systems share another feature related to their folded strata. They develop from exceptionally thick piles of sedimentary strata, commonly 15,000m or more in thickness. Early in the study of mountain ranges, it was realized by the American scientist James Hall that so huge a thickness called for an unusual sort of basin in which the sediments could accumulate. Another famous 19th-Century American scientist, J. D. Dana, discussed this fact in his analysis of the history of the Appalachians. He pointed out that both subsidence of the crust to form an unusually deep basin and filling of the basin with sediment must have preceded the final deformation and uplift that created the mountains. Dana coined the name *geosyncline* for the basin in which the Appalachian sediments accumulated. Studies of other fold mountains show that deep, sediment-filled basins preceded all of them. The term ***geosyncline*** therefore has wide application and describes *a great trough that has received thick deposits of sediment during its slow subsidence through long geologic periods.*

The strata of fold-mountain systems are predominantly marine; we can draw this conclusion from the presence in them of marine fossils. In systems such as the Alps, the marine strata are mostly of deep-water origin. In others, such as the Appalachians, the sediments apparently accumulated in shallow water. Regardless of water depth, the kinds and thicknesses of sediments found in geosynclines lead us to two important conclusions. First, geosynclines are oceanic features. Second, the great thicknesses of sediment, which commonly exceed the greatest depths of the ocean, indicate that the basin *must* have been sinking while it was being filled with sediment.

Some geosynclines occur in pairs. An excellent example is the Appalachians. Two elongate geosynclines, roughly parallel, once occupied the region from Newfoundland southwest to Alabama. One geosyncline lay directly adjacent to the continent and became filled with shallow-water sediment, which we now see as the limestone, sandstone, and other strata that are common in most of the Appalachians. The other geosyncline lay farther east and farther offshore. It became filled with deeper-water sediment of the sort we see today at the foot of the continental slope off the east coast of North America. The deep-water geosyncline also contained some volcanic rock. We do not want to give the impression that the two geosynclines were completely separate. They were not. They merge, one into the other, and overlap along their margins. Nowhere is there a break in sedimentary strata. How, then, do geosynclines form?

**Origin of geosynclines.** Two modern, side-by-side geosynclines occur along the present-day Atlantic coast of North America. One, the shallow-water geosyncline, is a wedge of sediment piled on the continental shelf. It is underlain by continental crust. The other geosyncline is a great pile of deep-water sediment, adjacent to and abutting the continental slope. It is partly underlain by oceanic crust (Fig. 18.4). The locations of geosynclines, therefore, are stable continental margins where continental crust joins oceanic crust. Because the joins are in the centers of plates, we believe the sediment in all geosynclines accumulates in stable plate interiors rather than in active marginal zones. But then why is the continental margin depressed below sea level?

To answer the question, picture what happens when a spreading edge splits continental crust and begins to form a new ocean basin (Fig. 18.5). Rising magma bends the crust upward, heating the entire lithosphere, expanding it, and distending it so that an elevated mass of continental crust and a central rift valley form. Thermal expansion can result in increased height of as much as 2.5km. Erosion wears away the expanded crust, perhaps bringing the land surface down to its original height. But at the same time the edges of continental crust are moving away from the spreading edge. As they move, both crust and the lithosphere below it become cooler. The cooling crust and lithosphere contract and the margins of the continents sink below sea level, forming a natural resting place for sediment. New geosynclines have been formed.

Geosynclines are an essential phase in the formation of fold mountains. Before we approach the next step in the process of forming fold mountains, let us describe briefly the geology of some typical fold-mountain systems and compare them with that of other kinds of mountain systems. Then we shall be in a position to predict how new fold mountains may one day rise along the present Atlantic-coast margin of North America.

**The Appalachians.** The Appalachians are a fold-mountain system 2,500km in length bordering the east coast of North America, and

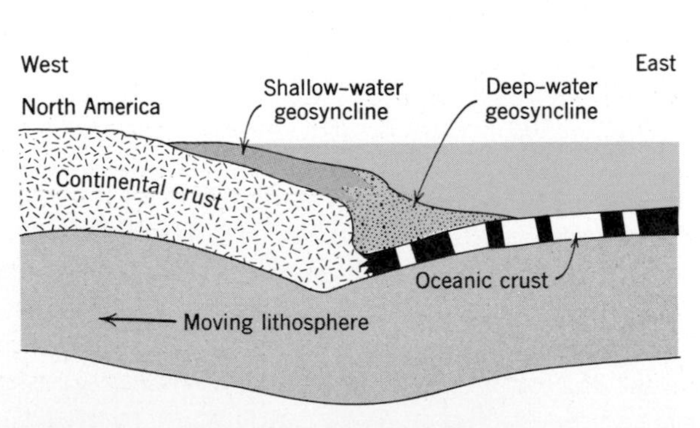


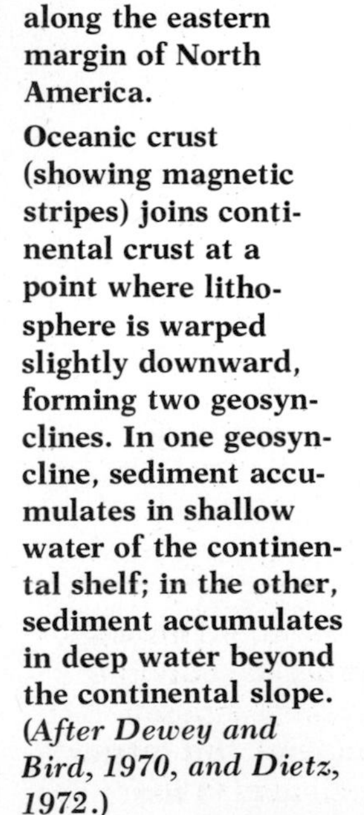

**Figure 18.4**
**Schematic section through lithosphere along the eastern margin of North America.**
**Oceanic crust (showing magnetic stripes) joins continental crust at a point where lithosphere is warped slightly downward, forming two geosynclines. In one geosyncline, sediment accumulates in shallow water of the continental shelf; in the other, sediment accumulates in deep water beyond the continental slope.** *(After Dewey and Bird, 1970, and Dietz, 1972.)*

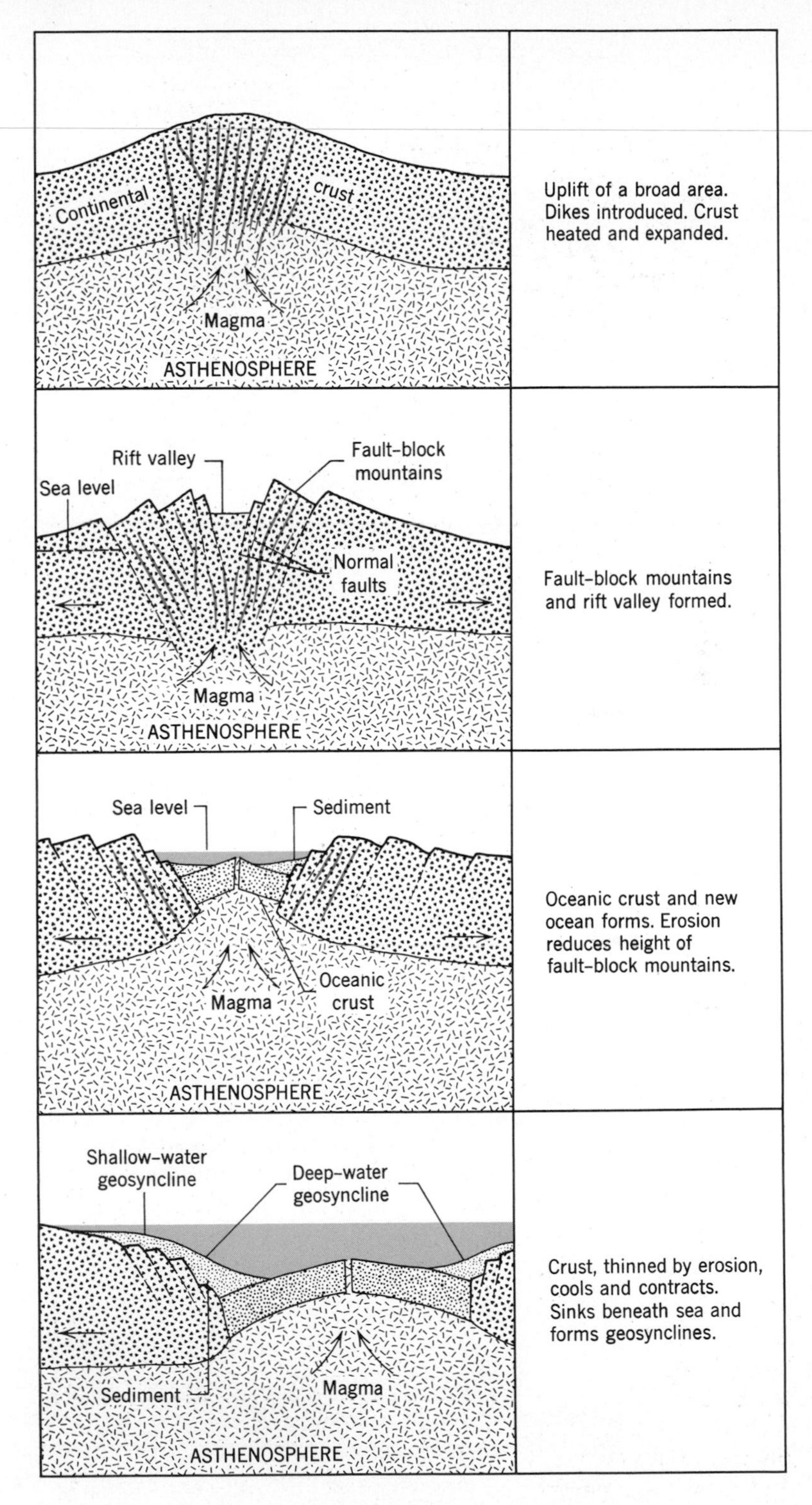


**Figure 18.5**
**Formation of geosynclines at continental margins.**

consisting of folded and faulted strata of Paleozoic age. Although sediment apparently accumulated in two geosynclines, only the mountains formed from the western one remain well exposed. Portions of the eastern basin remain in New England, but farther south the equivalent region is now deeply eroded, and so lies

*A*

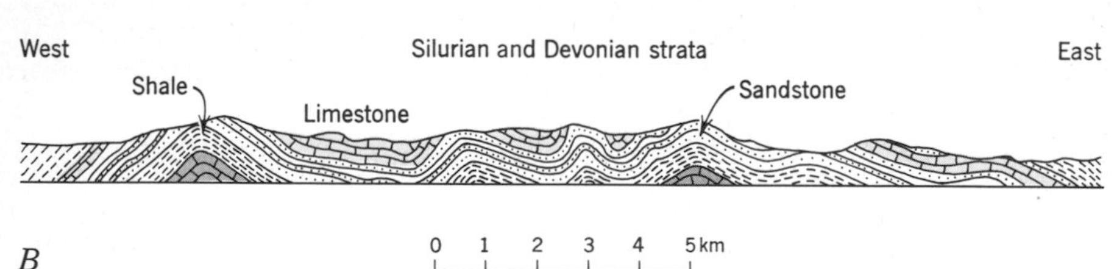


*B*

**Figure 18.6**

**Valley-and-Ridge Province in the central Appalachians.**

**A. View in eastern Pennsylvania, looking southeast. Resistant strata underlie wooded ridges; more erodible strata underlie cultivated lowlands. Distance between Tuscarora Mountain and Blue Mountain is 37km.** (*J. S. Shelton. From Geology Illustrated by John S. Shelton, W. H. Freeman & Co., © 1966.*)

**B. Section through the ranges having the kind of gentle folding found in the Valley-and-Ridge Province. Compare Figure 18.7.** (*After U.S. Geol. Survey Folio 59.*)

offshore beneath sediment of the continental shelf, hidden from view. The shallow-water sediment in the western geosyncline contains mud cracks, ripple marks, fossils of shallow-water organisms, and in places even freshwater materials such as coal. The sediment was deposited on a basement of old metamorphic and igneous rock, and becomes markedly thicker away from the former western shore; that is, it thickens from west to east.

Most, but not all, of the sediment in the western geosyncline has now been deformed. Today, if we approach the Appalachians from western New York and western Pennsylvania, we see, first, the former sediment occurring as essentially flat-lying, undisturbed strata. Continuing eastward, we notice the same strata thicken and become gently folded and thrust-faulted. In eastern Pennsylvania, in the region known as the Valley-and-Ridge Province, the strata have been bent into broad anticlines and synclines. The province gets its name because valleys have been developed by erosion in the most erodible strata such as limestones, dolostones, and claystones. The valleys alternate with prominent ridges formed by very resistant strata, chiefly sandstone (Fig. 18.6). At the northern and southern ends of the Appalachian system, strata on the western margin of the mountains have been deformed more by thrust faulting than by folding. The direction of thrusting is always the same: hanging-wall blocks have been thrust westward, away from the center of the range. The difference in style of deformation between the different regions along the western margin of the Appalachians is seen when we compare Figure 18.7 with Figure 18.6*B*. Farther eastward, toward the core of the Appalachian system, we find deep-water sediments that were deposited in the eastern geosyncline. These strata are increasingly metamorphosed and deformation becomes increasingly intense the farther east we go. Folds become isoclinal and then overturned, and faulting is prevalent. In places, fragments of the old basement are seen thrust up over younger sedimentary strata. Finally, we reach a region where intense metamorphism has occurred and where granite batholiths have been emplaced.

Metamorphism and deformation are most intense where the pile of sedimentary strata is thickest. Because type of metamorphism can be equated with depth of burial (Chapter 15), we conclude that the most intense metamorphism, maximum deformation, maximum depth of burial, and maximum uplift all occurred in the same belt. This association indicates that the core of a fold-mountain system coincides with what was once the deepest part of a geosyncline. When the force that forms the geosyncline and causes deformation of the strata is removed, the deeply buried pile of strata begins to rise, and is apparently kept rising until isostatic equilibrium is reached. Upward movement of the mountain core suggests how folding and thrusting may have occurred in the marginal areas. As the core rose, the still-undeformed strata on the flanks were tilted. Under the influence of gravity the strata slid westward. In the center of the

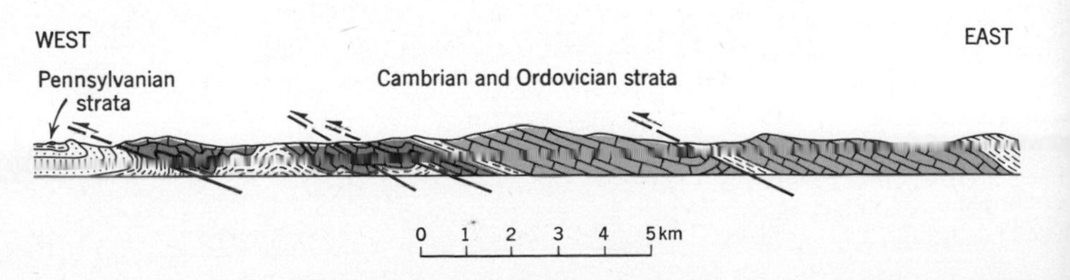


**Figure 18.7**
**Thrust faults and folds in the southern Appalachians. Many of the thrusts originated from the stretching and breaking of limbs of folds. Compare Figure 18.6, where the style of gentle folding found in the central Appalachians is depicted. (*After U.S. Geol. Survey Folio 61.*)**

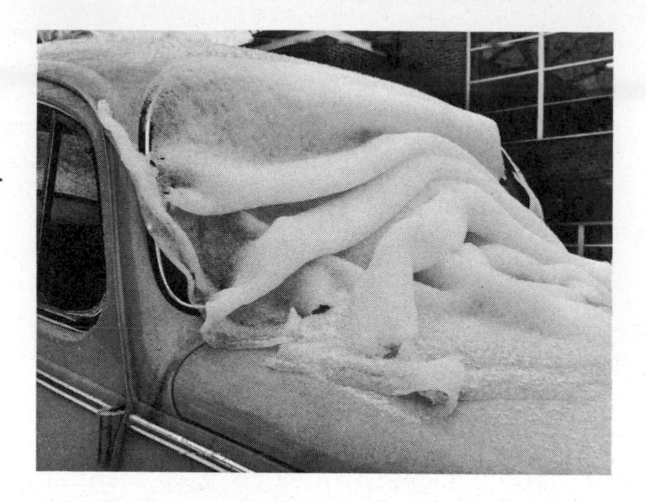

Figure 18.8
**Folds in snow that slid down an inclined windshield. Strata that slide because of the pull of gravity display similar folding. (*Yonkers Herald-Statesman, courtesy Westchester-Rockland Newspaper Group.*)**

system, corresponding to the Valley and Ridge Province, the sliding strata were somewhat plastic and became folded. A more familiar example of folding by gravity sliding is shown in Figure 18.8. At both ends of the mountain range, strata were apparently more brittle; instead of folding they broke into giant thrust sheets, which slid westward like a series of playing cards in a deck, each one riding up over the next sheet to the west.

**The Alps.** We naturally ask how well the Appalachian picture can be applied to other fold mountains. The answer is that similar features are found in all of them. The Alps and associated mountain ranges in southern Europe were formed later than the Appalachians, during the Mesozoic and Cenozoic Eras. Nevertheless, the two systems have many features in common. For instance, the Jura Mountains, lying along the northwestern edge of the Alps, have the same folded form and the same origin as the Valley-and-Ridge Province. Also, the Jura formed from shallow-water sediments. In the high Alps, which correspond to the now deeply eroded and topographically unimpressive Appalachians we observe in eastern Virginia and Maryland, thrusting appears to have developed on a much grander scale than in the Appalachians (Fig. 18.9). The high Alps are composed of deeper-water marine sedimentary strata.

**Canadian Rocky Mountains.** The Canadian Rocky Mountains, a magnificent mountain system much less eroded than the

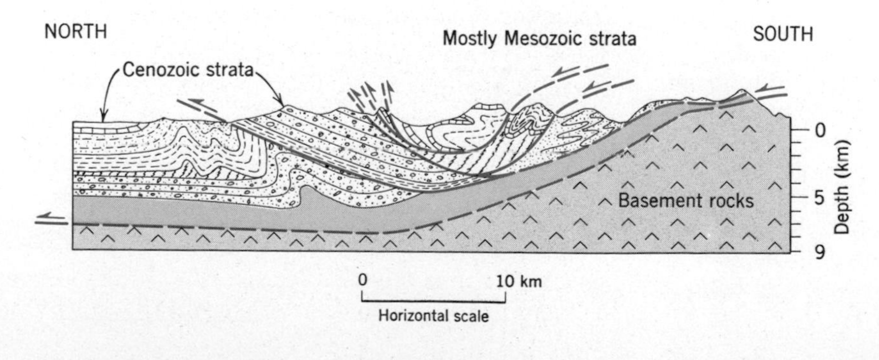


Figure 18.9
**Section through the Alps in central Switzerland. Strata have moved northward along great thrust faults (blue lines) which later were themselves folded. Major thrust separates overlying strata from basement rocks. (*After Albert Heim, 1922: Geologie der Schweiz, v. 2, pl. 27.*)**

Appalachians, can also be compared. A section through the Canadian Rockies at about the latitude of Calgary, Alberta (Fig. 18.10) reveals all the features we have described for the Appalachians. A central or core zone has been intensely metamorphosed. In it, parts of the older basement rocks have been thrust upward, and in the marginal region folding and extensive thrust faulting are evident. The thrust sheets have moved eastward, away from the core zone. It is apparent that each sedimentary unit becomes thinner as it is followed from west to east, indicating that the core zone coincides with the thickest part of the old geosyncline.

### Volcanic mountains

Many fold mountains include a great variety of volcanic rocks. The Andes, for example, contain enormous thicknesses of volcanic rock; yet they owe their origin chiefly to the folding and faulting associated with uplift. But when we say *volcanic mountains,* we mean mountains that have been built entirely by volcanic extrusion.

Most volcanic mountains are built on the sea floor. In some of them, as in the Hawaiian chain, the higher peaks protrude above sea level. In others, as in the oceanic ridges, the entire chain is submerged.

A special class of volcanic mountain chain, ***island arc,*** is mentioned in foregoing chapters. It is *a great arcuate belt of andesitic and basaltic volcanic islands* formed over a subduction zone in oceanic crust (Fig. 17.3*A*). Some island arcs are 2,000km or more in length. The Aleutian Islands are a conspicuous island arc; another arc runs from Kamchatka through the Kurile Islands and down through Japan; yet another consists of the islands of Sumatra, Java, Soemba, and Timor.

Volcanic mountain ranges on land are not common. The only one in the United States is the Cascade Range in Washington and Oregon. This is a range of huge young, andesitic volcanoes, running from Lassen Peak, California, at the south end to Mount Baker, Washington, more than 900km farther north (Fig. 18.11). These snow-covered giants, all active during the past few million years, and some still active today, were erupted onto a platform of older, folded, and deeply eroded rocks. The trend of the line of volcanoes cuts directly across many other geologic features, and for this reason scientists believe that building of the mountains is controlled by partial melting of subducted oceanic crust, as shown in Figure 17.3*B*. The piece of crust that is being subducted and is forming the Cascade Range is part of the tiny Gorda Plate (Fig. 4.4).

### Fault-block mountains

In many parts of the world isolated mountain ranges stand abruptly above surrounding plains. Study reveals that these ranges are separated from intervening lower land by normal faults of

LIZARD RANGE
MOUNT BROADWOOD
WIGWAM RIVER

*A*

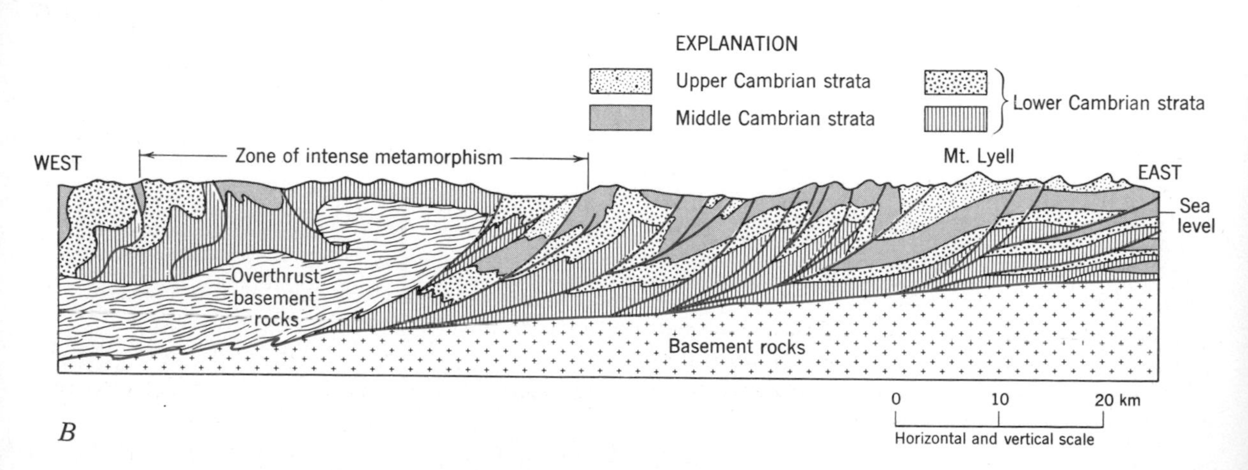
EXPLANATION
Upper Cambrian strata
Middle Cambrian strata
Lower Cambrian strata
WEST
Zone of intense metamorphism
Mt. Lyell
EAST
Sea level
Overthrust basement rocks
Basement rocks
0
10
20 km
Horizontal and vertical scale

*B*

**Figure 18.10 (see opposite page)** ***A*. Thrust faulting in Canadian Rocky Mountains; view looking northwest. Two thrusts (blue lines) can be seen. Rock underlying Mount Broadwood has moved from left, like a giant sled, riding up over steeply dipping strata on right. Lizard Range, in background, is joined by rock that was thrust many kilometers from the west. (*Courtesy Chevron Standard, Ltd. After Haderson and Dahlstrom, Bull. American Assoc. Petroleum Geologists, v. 43, p. 641–653, 1959.*)**

***B*. Section through the Canadian Rocky Mountains at about the latitude of Calgary, Alberta. The zone of intense metamorphism coincides with the region of maximum uplift and maximum deformation. Farther east, where strata become progressively thinner, the pile has been greatly thickened by movement along thrust faults. Movement on the thrust faults is such that each fault block has moved toward the east, riding over the block beside it. (*After Price and Mountjoy, Geol. Assoc. Canada Spec. Paper 6, 1970.*)**

great displacement, and seem to be giant pieces of crust punched upward from below. These are fault-block mountains. Rock within the mountains commonly contains evidence that former fold mountains once occupied the same sites, but that erosion had worn them down before the fault blocks formed.

**Basin-and-Range Province.** One of the most extensive of these fault-block mountain systems lies in Nevada, western Utah, and parts of Oregon, Idaho, Arizona, and California. Known as the *Basin-and-Range Province,* it contains a spectacular development of uplifted fault blocks. The province is underlain by sedimentary strata deposited during the Paleozoic Era. Following a period of fold-mountain formation, the region was deeply eroded during Mesozoic and Cenozoic times, when it supplied sediment to geosynclines that later developed into the Coast Ranges of California and into the Rocky Mountains. Starting about 25 million years ago, and accompanied by extensive volcanic activity, the region broke up into a series of blocks 30km to 40km in width and as much as 150km long, bounded by steeply inclined normal faults. Tilting of the blocks is pronounced (Fig. 18.12).

# ORIGIN OF MOUNTAINS

The formation of all mountains results, in one way or another, from movement of plates of lithosphere. The forces necessary to fold strata and force fault blocks upward can only be explained by movement of lithosphere. The sites of formation of most mountains lie along the active margins of plates, but a few mountain chains do form in the interior regions of plates. To simplify discussion, therefore, we will consider separately the three regions in plates where mountains form—interior regions, divergent margins, and convergent margins—and classify the origin of mountains accordingly. This classification is summarized in Table 18.1.

### Mountains in plate interiors

Perhaps the most puzzling group of mountains are fault-block mountains. They are only found in continental crust and they form in the interior regions of plates, far removed from the varied activities at plate margins.

One way in which fault-block mountains may form is suggested by the Basin-and-Range Province, which we can take as a model. We know that there, an initial period, during which the crust was stretched by upwarping, coincided with the onset of volcanic activity. Possibly such upwarping could be caused by heating from below, as continental crust moves over a hot spot in the mantle. Later the volcanism wanes and the distended crust cools and collapses, along a series of steeply dipping, arc-shaped faults (Fig. 18.13). Collapse and tilting of the blocks results in the relief we see today.

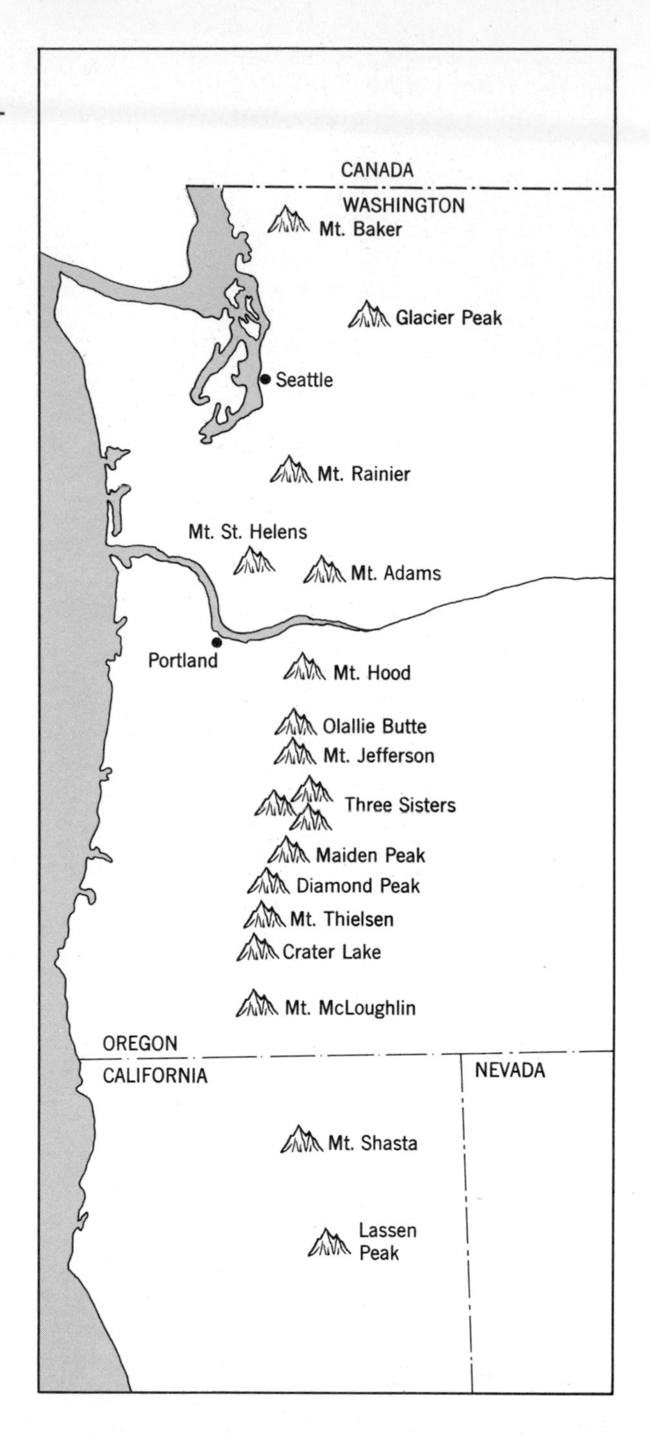


**Figure 18.11**
**Volcanoes of the Cascade Range. Each volcano has been active during the last 2 million years. (*After Tabor and Crowder, U.S. Geol. Survey Prof. Paper 604, 1969.*)**

Many fault-block mountains, however, have no associated volcanic activity; so they cannot be explained by the Basin-and-Range model. One of these is the Adirondack Mountains in northern New York. Originally a Precambrian fold-mountain range, the Adirondacks were eroded down to a surface of low relief and had been stable for hundreds of millions of years. Then, in the Cenozoic Era, the region again became active and the modern Adirondacks

**Figure 18.12**
**Frenchman Mountain, Nevada, is a tilted fault block. Marine strata of Paleozoic age, more than 1,500m thick, have been elevated on steeply dipping normal faults that trend north–south. Blue line marks trace of a small east–west fault that has offset strata.** (*William Belknap, Jr.*)

**Table 18.1**
**Genetic Classification of Mountains**

| Where Formed | | Kind of Mountain and/or Characteristic Feature | Example |
|---|---|---|---|
| Interior of plate | Continental crust | Fault block. Normal faults. | Basin-and-Range Province |
| | Oceanic crust | Chain of basaltic volcanic islands. Submarine volcanism. | Hawaiian Archipelago |
| Divergent plate margin | Continental crust | Crustal highland with a central graben. Volcanoes along edge of graben. | African Rift Valley |
| | Oceanic crust | Oceanic ridge. Submarine volcanism. | Mid-Atlantic Ridge |
| Convergent plate margin | Oceanic crust subducted beneath continental crust. Few or no geosynclinal sediments. | Andesitic volcanoes on continental crust. | Cascade Range |
| | Oceanic crust subducted beneath continental crust. Thick wedge of geosynclinal sediments. | Fold-mountain system along margin of continent. | Andes |
| | Oceanic crust subducted beneath oceanic crust. | Andesitic volcanic islands. | Aleutians |
| | Continental crust collides with continental crust. | Fold-mountain system in interior of continent. | Himalaya |

**Figure 18.13 Possible model for formation of structures in Basin-and-Range Province. Warping of the crust and extrusion of large quantities of lava and volcanic ash were followed by collapse along a series of steeply inclined normal faults.** (*After Mackin,* in *Roberts, UMR Journal, no. 1, p. 101–119, 1968.*)

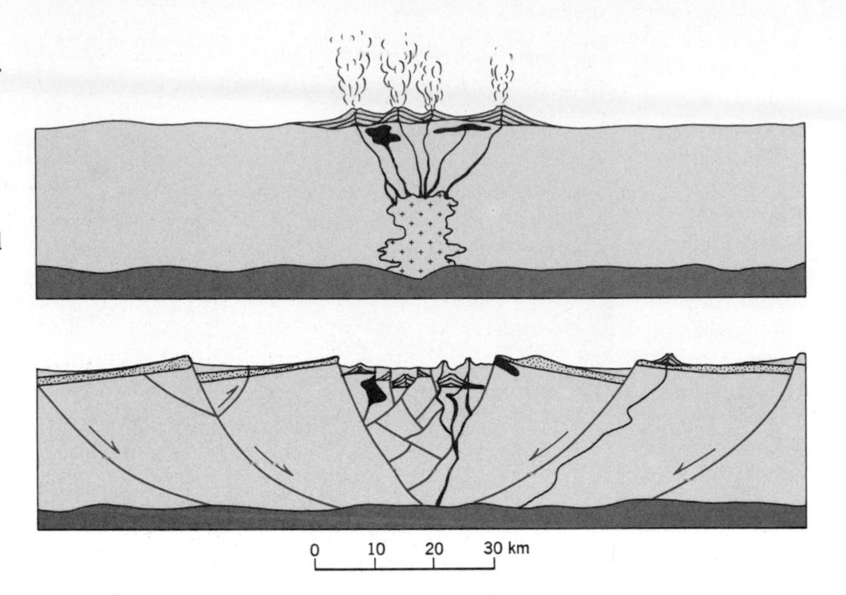


were lifted up as a series of great fault blocks. *How* and *why* are a puzzle. One suggestion is that eroded remnants of the old fold-mountain range were not isostatically balanced. This could happen if the faults were locked, preventing the blocks from rising. If this were the case, a slight warping or twisting of the continental crust, due to movement of the lithosphere, could serve as a trigger and unlock the faults. The fault blocks could then rise and restore isostatic balance.

By contrast with fault-block mountains on continental crust, the kind of mountains that rise above the oceanic crust are giant basaltic volcanoes. The volcanoes are the largest mountains on Earth. Some stand higher above the sea floor than the tallest mountain on continental crust, Mt. Everest, stands above the surrounding continent. Mt. Everest stands 8,850m above sea level and Mauna Loa, tallest of the volcanoes in the Hawaiian Archipelago, stands only 4,300m above sea level. But Mauna Loa has another 7,900m of elevation that we cannot see between the sea floor and sea level. Chains of volcanic mountains form, as we saw in Figure 13.10, when lithosphere capped by oceanic crust moves over a local source of magma in the mantle. The volcanic mountains grow in about 1 million years; then the moving lithosphere separates them from the magma source and they become dormant.

### Mountains at divergent plate margins

The reason that mountains form along divergent margins is the same for oceanic crust as it is for continental crust. Divergent margins are elevated regions because (1) magma rising from below tends to push the plate edges upward, and (2) the hot magma heats the plate edges and causes them to expand. The actual line along which divergence occurs is marked by a rift valley; in continental crust the rift valley is broad and gentle like the African Rift Valley

(Fig. 14.13), while in oceanic crust the rift valley is narrow and apparently steeper walled than its continental counterpart (Fig. 13.7).

We have already discussed the formation of elevated regions when continental crust occurs at divergent margins. The elevation of the crust is the first step in the process that leads, eventually, to geosynclines (Fig. 18.5).

Along divergent margins in oceanic crust we find oceanic ridges. They are the most spectacular mountain chains on Earth and they tower above the deep ocean floors. As the plates move away from the divergent margins, the rock beneath the ridge cools and contracts, and so develops the relief we see between the deep ocean floor and the top of the ridge.

### Mountains at convergent plate margins

A great many activities occur at convergent plate margins, but two can be separated out as being especially important in the formation of mountains. First, subduction and partial melting of oceanic crust produces andesitic magma and therefore andesitic volcanoes. Second, the thick wedges of sediment in geosynclines become folded and elevated. We will discuss the formation of fold mountains first and, in order to do so, we will return to the modern-day geosynclines of the Atlantic Ocean.

The Atlantic Ocean is now slowly growing larger, and the depth of sediment in both the shallow- and deep-water geosynclines lying along its flanks is increasing. The geosynclines that flank the Atlantic coast of North America are 180 to 200 million years old, having formed when the North Atlantic Ocean first appeared. The geosynclines in the South Atlantic appear to be somewhat younger, because this part of the Atlantic formed later than the North Atlantic did. But the accumulation of sediment apparently does not go on forever; eventually all the geosynclines in the Atlantic Ocean will become fold mountains. Eventually, for some reason we still do not understand, the lithosphere beneath the deep-water geosyncline breaks at the point where oceanic crust meets continental crust. The old oceanic crust then begins a plunge down into the mantle as a new subduction zone is formed. The downward-sliding plate of lithosphere butts into the deep-water strata, crumples them, and rams them against the edge of the continent (Fig. 18.14). Strata near the bottom of the geosyncline will be pushed still deeper by the moving plate. Becoming heated and metamorphosed, the strata of sedimentary rocks will be transformed to schists and gneisses. The kind of metamorphism that occurs is Case B in Figure 15.23. The deepest strata may even begin to melt. This melting forms granitic magma that rises and intrudes the strata above it. In this way the deep-water geosyncline becomes the core of a newly made chain of fold mountains.

Strata in the shallow-water geosyncline are compressed and pushed aside by the mass of deformed and metamorphosed rock in the new mountain core. The shallow-water strata become folded

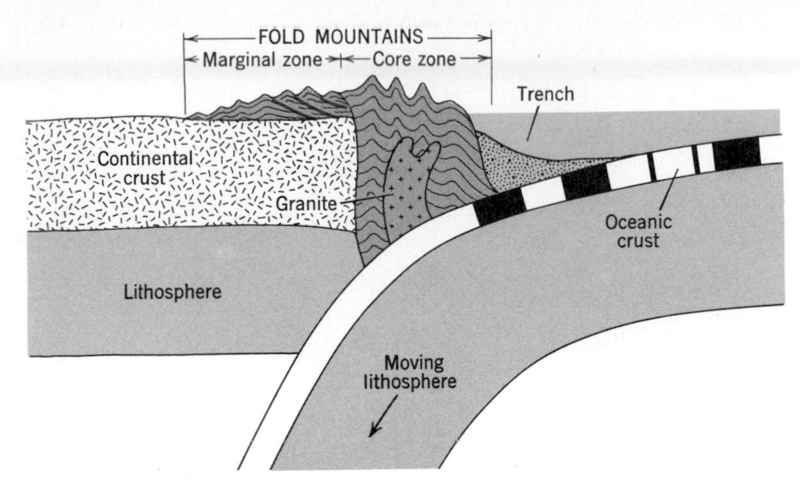


**Figure 18.14 Schematic section showing how fold mountains are formed when a plate of lithosphere plunges downwards beneath the edge of a continent. Strata originally accumulated in a deep-water geosyncline are crumpled, metamorphosed, and intruded by granite, which forms the core of mountains. Strata of shallow-water geosyncline are pushed and slid sideways to form marginal zone of mountains. On seaward side of the mountains, a sea-floor trench forms. (*After Dewey and Bird, 1970.*)**

and deformed, by thrust faulting, to create the marginal ranges of a fold-mountain system.

In many parts of the world, plates of lithosphere are now plunging downward beneath the edges of plates of continental crust, and fold-mountain systems are still growing actively. One such zone is along the west coast of South America (Fig. 18.14): Imagine that we are looking southward at that figure. If we draw a section through the western edge of South America, the downward-plunging plate of lithosphere in the figure is equivalent to either the Nazca Plate or the Antarctic Plate (Fig. 4.4), the fold-mountain system is equivalent to the Andes, and the sea-floor trench is equivalent to the Peru–Chile Trench. At some time in the past, the west coast of South America probably resembled the present Atlantic coast of North America. We can infer that at some unknown time in the future the Atlantic coast of North America will come to resemble the Pacific coast of South America, and a great new mountain system will form.

Once the process of subduction has begun and a sea-floor trench is created, sediment accumulates in the trench. Then, the oceanic crust beneath, continuing to move, smashes the accumulating sediment into *a chaotic mixture of rocks* called a ***mélange.*** The mélange, dragged downward by the moving plate, is metamorphosed. The cold sedimentary rocks are dragged down so rapidly that they remain cooler than adjacent rock at the same depth. The kind of metamorphism that is common in many mélange zones, therefore, is that shown in Case C of Figure 15.23—a high-pressure, low-temperature metamorphism distinguished by blue-colored schists. The color comes from a bluish amphibole called glaucophane. We have already seen that the kind of metamorphism that occurs in the old, deep-water geosyncline sediments, by contrast, is shown in Case B in Figure 15.23. It is a normal metamorphism of the kind that happens when the temperature of buried sediment increases along a normal geothermal gradient. The relations between the two kinds of metamorphism are depicted in Figure 18.15.

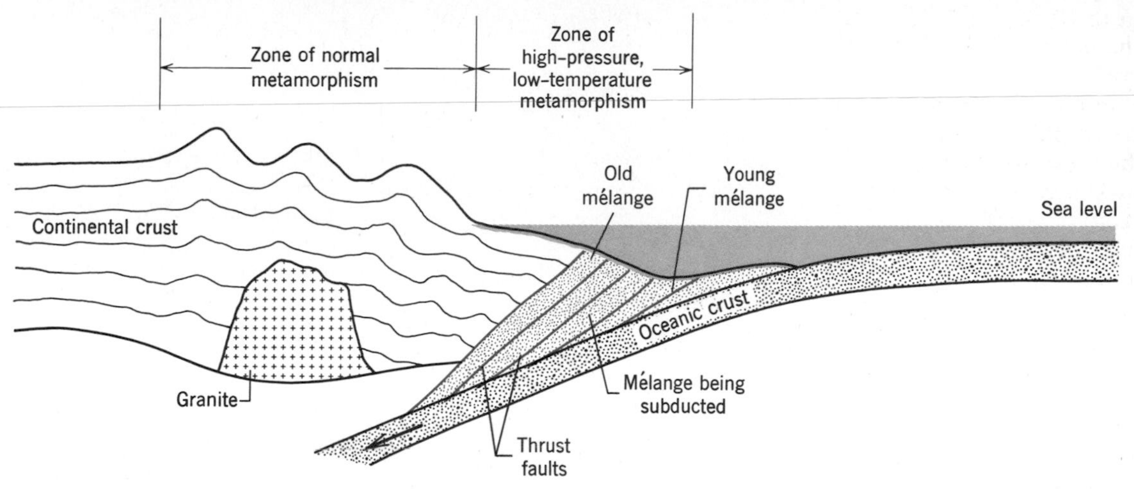


**Figure 18.15**
**Relation between kinds of metamorphism and subduction zone.**
**Mélange is formed when young sediment in a trench is smashed by moving lithosphere and dragged downward in slices bounded by thrust faults. As successive slices are dragged down, older mélange, closer to continent, is pushed back up. The process is like lifting a deck of cards by adding new cards at the base of the deck.**

Why don't we find fold-mountain systems above every subduction zone? The answer is that a fold-mountain system can only form when a geosyncline is full of sediment. When oceanic crust is subducted beneath oceanic crust, no geosyncline is present and the result is a chain of andesitic volcanic islands. The islands form from magma produced by partial melting of the subducted oceanic crust. When oceanic crust is subducted beneath continental crust, andesitic volcanoes are formed on the continental crust. Commonly, as in the Andes, the volcanoes simply cap a large fold-mountain system. However, if a fold-mountain system were missing because a geosyncline had not been first filled with sediment, the volcanoes would be the only prominent features. This seems to be the origin of the Cascade Range.

Subduction suggests a way by which the volume of continental crust might have increased through the ages. When the plate of lithosphere slides down into the mantle, the basalt in the oceanic crust undergoes partial melting, and yields andesitic magma. The magma is then erupted onto the continental crust, where it builds volcanoes that cap the fold-mountain range. We see the process occurring today in the Andes. Erosion will eventually destroy the volcanoes, whose debris will be deposited in the nearby sea-floor trench. The sediment will be incorporated in the mélange, metamorphosed, and added to the continental crust. The process is presumably happening today beneath the Peru–Chile Trench. The process of making fold mountains and andesite seems to imply, therefore, that material from the mantle is being slowly added to the continental crust.

According to the inferences just drawn, the building of fold mountains and the addition of new material to the continental crust occurs along the edges of continents. But not all systems of fold mountains occur in such positions. So let us examine fold-mountain systems in the interiors of continents, to see whether the theory of plate tectonics can account for them as well.

Two prominent fold-mountain systems, formed within the last 200

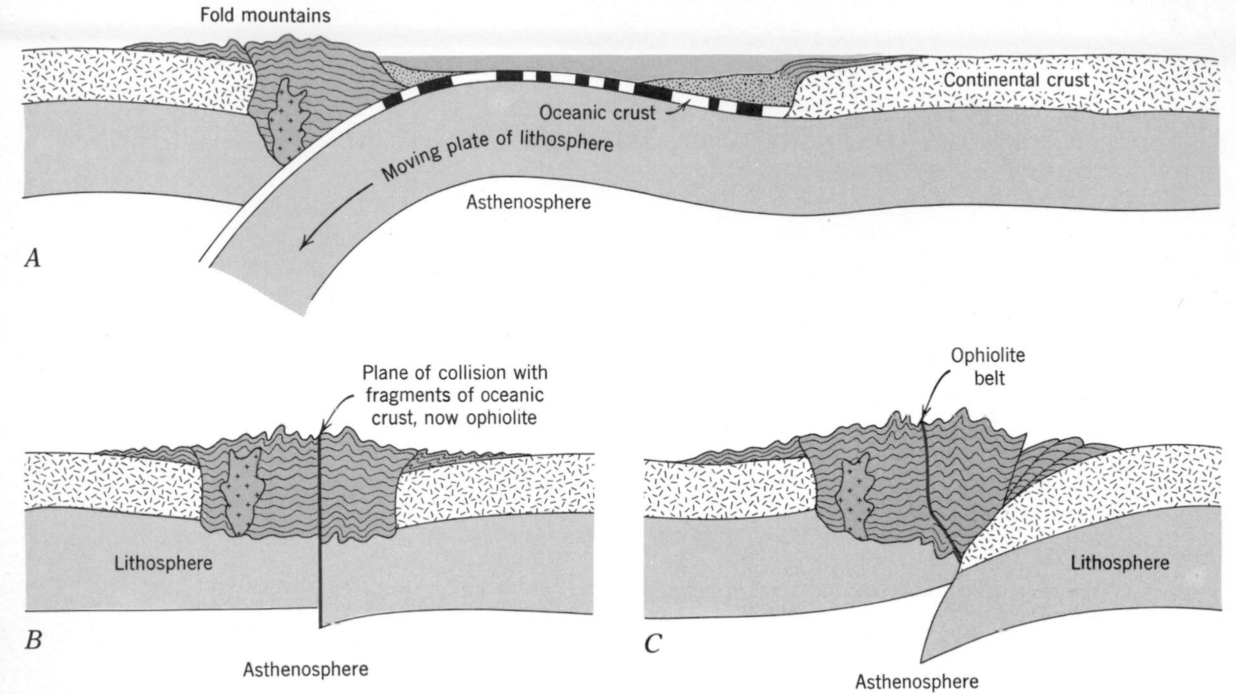


Figure 18.16
**System of fold mountains formed by collision between continents.**
**A. Continental margin, of the type now found on the west coast of South America, forms a chain of fold mountains. Moving plate of lithosphere carries a second continent on a collision course.**
**B. Collision between continents crumples geosynclines and increases size of mountain system.**
**C. The downward-moving plate of lithosphere becomes detached, but the edge of the remaining segment of plate is partly thrust under the edge of the stationary plate, causing further elevation of the fold mountains. (*After Dewey and Bird, 1970.*)**

years, occur in the interiors of continents: the Alps and the mighty Himalaya. Both were formed by collisions between masses of continental crust in the manner depicted in Figure 18.16. The collision planes, or suture zones, can be identified because they have now become belts of ophiolites. The Alps were formed by a collision between the two plates on which Africa and Europe respectively ride. The Himalaya were formed when a small plate, carrying India, rammed into Asia. In that process, sediment laid down in the ancient Tethys Sea (Fig. 17.1) was crumpled, folded, and faulted. The Indian collision started about 38 million years ago (Fig. 18.17) and seems to be continuing today because the Himalaya is still rising and the region north of it, in Tibet and China, is still subject to intense earthquakes.

These are "recent" collisions. But mountain systems more than 200 million years old occupy the interiors of some continents and are important clues to plate movements older than those involved in the "recent" collisions. In the interior of Asia are two fold-mountain systems, the Ural and Verkhoyansk Mountains (Fig. 18.1). We believe these systems, which involve Paleozoic strata, formed by continental collisions that happened several hundred million years ago. Perhaps this earlier period of movement was the one that assembled the former supercontinent of Pangaea. Evidence that supports that idea lies on the east coast of North America.

Apparently the Appalachians were once joined to mountains of similar age in western Europe. But if North America was joined to western Europe, how did a geosyncline form where the Appalachians now stand? Was there a previous North Atlantic Ocean, one that antedates the present ocean 200 million years old?

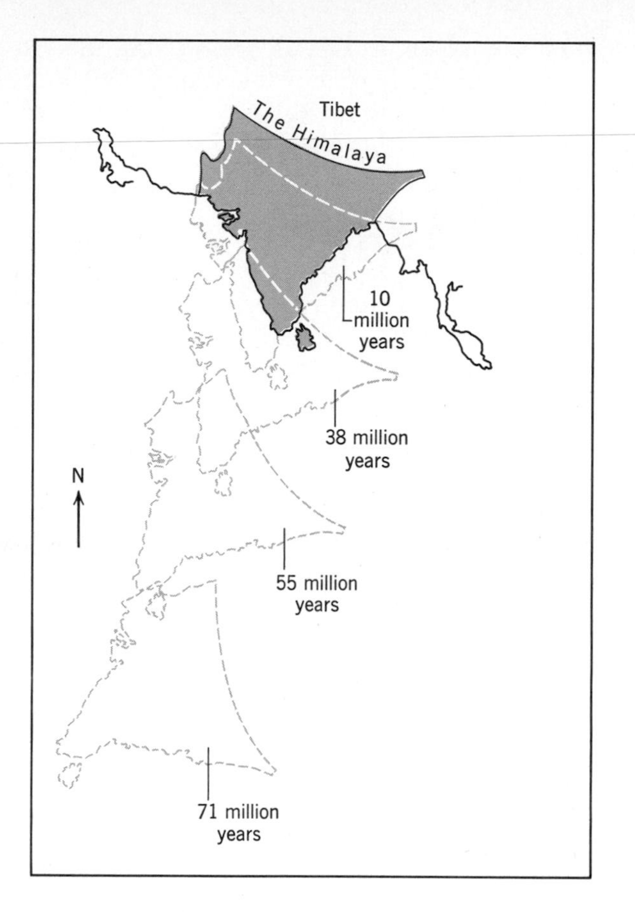


**Figure 18.17**
**Position of India relative to Tibet during last 71 million years. Starting about 38 million years ago, India moved north and collided with the main mass of Asia. Movement continues, resulting in intense crumpling that has created The Himalaya and has thrust it upward. The movement has also caused general deformation and thickening of the crust beneath Central Asia.** (*After P. Molnar and P. Tapponnier, 1975, Science, v. 189, p. 419–426.*)

Evidence strongly suggests there was. The sequence of events that formed the Appalachians could have been as follows. Many hundreds of millions of years ago geosynclines, formed in an ancient North Atlantic Ocean. The strata in them were crumpled and elevated when the continental margin of ancient North America looked like the Andean margin of South America of today. The plate of lithosphere on which Europe rode then began to move westward. Eventually, about 500 million years ago, Europe and North America collided. The collision created a great system of fold mountains in the interior of the supercontinent Pangaea. The system included the Appalachians and many of the Paleozoic mountains of western Europe. For a long time—at least 300 million years—the continents remained welded together and the Appalachians were part of a mountain system in the continental interior. When movement began again, one of the new spreading edges formed along the old suture line that lay near the center of the system of fold mountains.

Evidence from mountain systems strongly suggests, therefore, that at least one episode of plate movement preceded the present one, the one that began 200 million years ago. We cannot be sure, but it is tempting to guess, that there may have been several periods during Earth's long history when plates of lithosphere slid around on top of the mantle. In the process, continents may have been repeatedly split apart, rammed together, and moved past each other, increasing in size as new mountain systems were created. After each episode of movement and collision, there may have been a period of quiet that ended when new convection cells formed and when plates of lithosphere started to move again.

## THE FUTURE

When will the present episode of movement stop? We cannot be sure of that either. It might continue for a hundred million years or more. In the meantime, continents will continue to ride piggyback on their plates and the map of Earth's surface will continue to change. One prediction of what the map could look like in about 50 million years is seen in Figure 18.18.

Where might the next fold mountains appear? The answer to that question also is uncertain. But when we examine the modern distribution of sediment, it is apparent that present-day geosynclines are to be found along many continental margins (Fig. 18.19). If our inferences are correct, many new fold-mountain ranges will eventually arise from these geosynclines and the shapes of continents will change again.

## A WORD OF CAUTION

The theory of plate tectonics seems to provide answers to so many questions that we are tempted to accept it as the long-sought theory that explains the lithosphere. But we must be careful. Other theories, too, have seemed overwhelming in their promise, yet in the long run have proved incorrect. The theory of plate tectonics is still only a theory. Likewise it is still so new that we have had

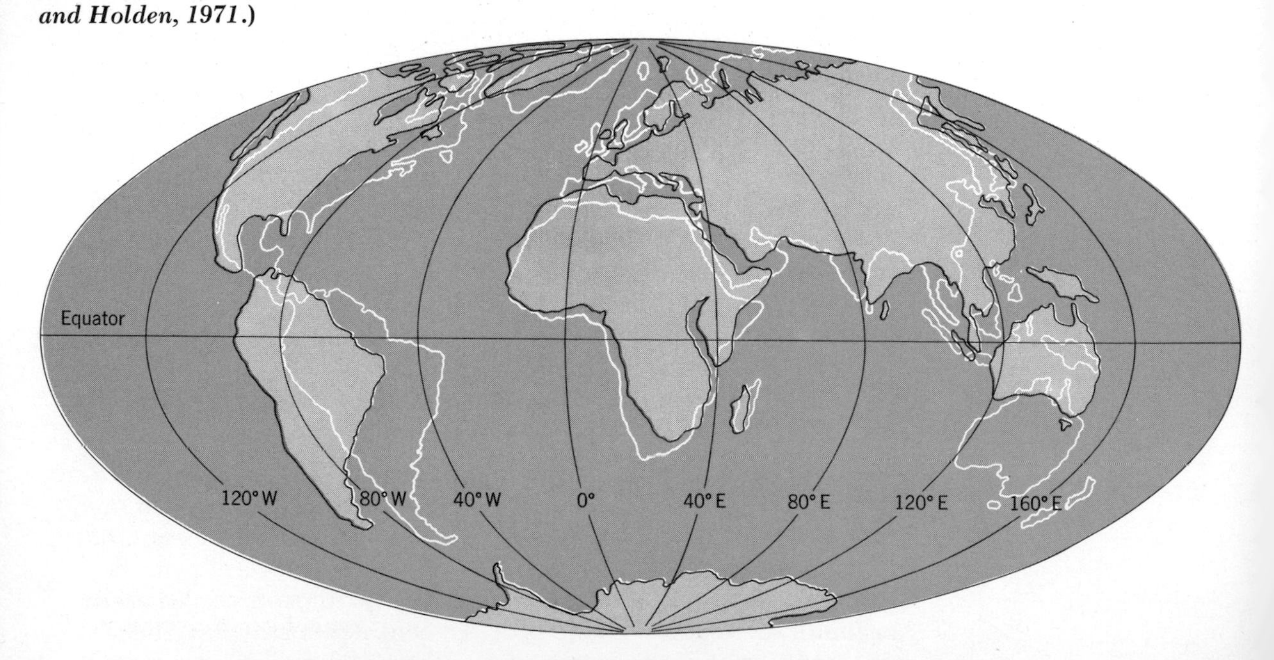


**Figure 18.18**
**If the present motions of plates should continue for 50 million years into the future, the map of the world would show some notable changes. Today's map outlines (white) will be strikingly changed in Africa, where a large fragment in the northeast corner will have almost separated from the rest of the continent. Australia and New Guinea will have moved north, many islands in the East Indies will have vanished, and open seaways will have replaced both the Panama and Suez Canals. (*After Dietz and Holden, 1971.*)**

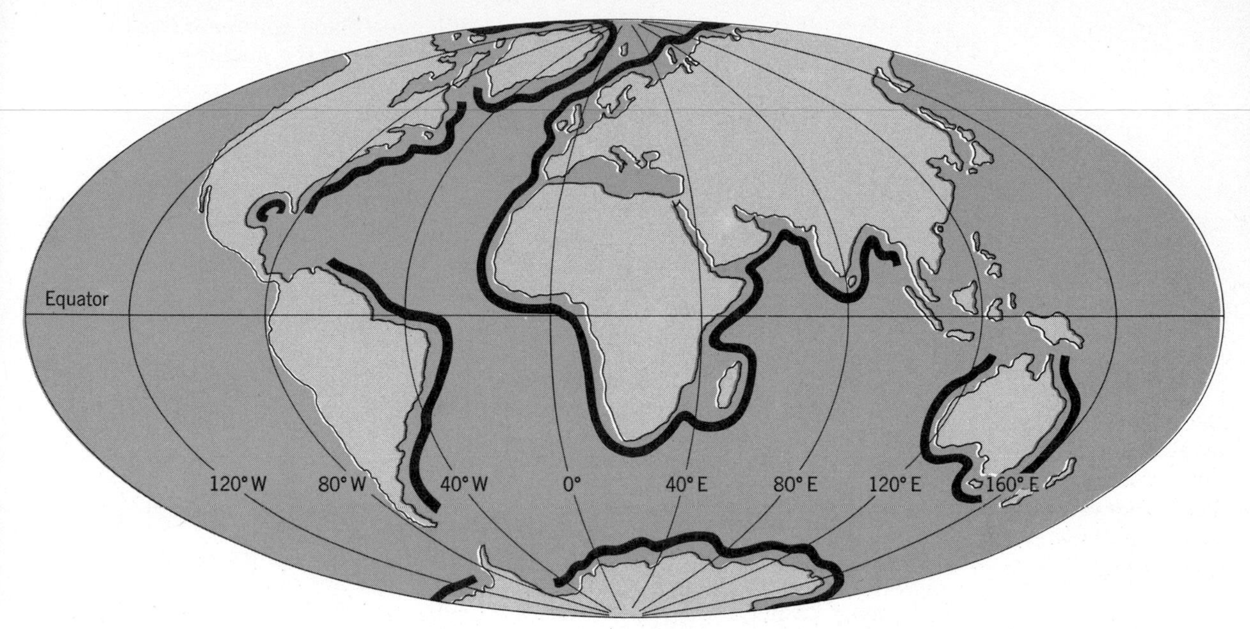


**Figure 18.19**
**Continental margins on the interiors of plates where present-day geosynclines (black) are accumulating sediment.**

insufficient time to examine and test it fully. Perhaps there are things the theory cannot explain, so that it may have to be changed. A few scientists have already found evidence that they believe strains the theory. Only with time and much more work can we decide whether the amazing sequence of events discussed in this chapter and the previous one actually happened as we have tried to describe them.

The theory does not tell us much about the first billion years of Earth's history. Perhaps future studies will show that movement of plates occurred only during the more recent part of Earth's life and that what happened in earlier times was entirely different. We have to face the fact that some of the vital evidence of Earth's earlier history has been obscured by the results of long-continued erosion. But the earliest histories of all the smaller planets are thought to be similar to that of planet Earth. We should therefore examine Earth's sister planets, where erosion may not have obscured the evidence. Perhaps Moon, Mars, Mercury, or Venus hold the key to Earth's earliest days. In the next chapter we discuss the many exciting clues that are coming out of our exploration of space.

# SUMMARY

1. The most ancient rocks so far discovered on Earth are 3.8 billion years old.

2. Ancient continental crust consists of pillow basalt extruded as lavas on the sea floor. Andesites and granites overlie and intrude the basalt and are likewise very ancient.

3. The first detrital sediments were deposited when andesitic volcanoes grew upward, projected above sea level, and were subjected to erosion.

4. Continents have grown larger through the ages. Each continent has a nucleus or shield at least 2.5 billion years old. Around the nucleus are rings of progressively younger fold mountains.

5. There are three basically different kinds of mountains: fold mountains, volcanic mountains, and fault-block mountains.

6. Fault-block mountains are most commonly found in continental crust and form in the interior regions of plates.

7. Volcanic mountains form at plate margins. Oceanic ridges are volcanic mountains formed at divergent margins. At convergent margins, volcanic mountains are either island arcs or chains of andesitic volcanoes on continental crust. The andesitic magma forms by partial melting of subducted oceanic crust.

8. Geosynclines form along stable margins of continents. When a margin becomes unstable and subduction commences, the sediments in the geosynclines are folded, metamorphosed, partially melted, and thrust up to form systems of fold mountains.

9. The present period of moving plates of lithosphere commenced about 200 million years ago, disrupting an old supercontinent we call Pangaea.

10. It is still uncertain whether there were periods of plate movement prior to the present one, but the distribution of ancient mountain systems suggests that prior movements have occurred.

# SELECTED REFERENCES

Anhaeusser, C. R., 1973, The evolution of the early Precambrian crust of South Africa: Royal Soc. London, Phil. Trans. v. A273, p. 359–388.

Bailey, E. B., 1935, Tectonic essays, mainly Alpine Oxford, England, Clarendon Press.

Clark, T. H., and Stearn, C. W., 1968, Geological evolution of North America: New York, Ronald Press.

Engel, A. E. J., 1963, Geologic evolution of North America: Science, v. 140, p. 143–152.

Gilluly, James, 1967, Chronology of tectonic movements in western United States: American Jour. Sci., v. 265, p. 306–331.

King, P. B., 1959, The evolution of North America: Princeton, N.J., Princeton Univ. Press.

# 19 PLANET EARTH AND ITS NEIGHBORS



The first person to turn a telescope to the Moon and nearby planets was Galileo. The year was 1609 and his homemade device was crude by comparison with a modern child's toy telescope. Galileo was astonished to see mountains and what to him looked like seas on the Moon, to see several moons circling around Jupiter, to see that Venus, like the Moon, had phases—full, half, quarter, and to see that huge dark spots appear every now and then on the surface of the Sun. These discoveries electrified Galileo's contemporaries, just as visits to the Moon by astronauts electrified the present generation.

Interest in the visible stars dates back to prehistory. Our ancestors noticed that a few stars seem to wander in completely different paths from the annual progression of most of their fellows across the sky. The Greeks mapped the paths of these strange objects and called them *planetai,* or wanderers. The Romans named the objects after their gods: Saturn, Jupiter, Mars, Venus, and Mercury. We now know that the curious wandering paths of the planets result from orbital motions around the Sun, that the other objects in the heavens are other suns, so far distant that they seem to occupy fixed places in the sky, and that other planets—Uranus, Neptune, and Pluto were unknown to our ancestors because they are visible only through telescopes. Finally, we know that other planets besides Earth and Jupiter have moons of their own. A few vital statistics of the planets are given in Figure 19.1.

Commencing in 1957, when the first artificial satellite was placed in orbit around the Earth, a brand-new scientific specialty has arisen. Called ***planetology,*** and devoted to *a comparative study of Earth with Moon and with the other planets,* the new specialty has taught us much about Earth's earliest days, that time, before 3.8 billion years ago, of which all record seems to have been erased by the operation of the rock cycle. In the present chapter we can give only a brief account of the successes of planetology. But they have been enormous and have motivated scientists to seek a unifying theory of origin of the suns and their planets. If the scientists are successful, it may one day be possible to say which of the billions of other suns visible in the heavens have planets like Earth, and

**Figure 19.1 Properties of the planets.**
**Arranged in order of their relative positions, outward from the Sun, the planets are shown in their correct relative sizes. The Sun, 1.6 million kilometers in diameter, is 13 times larger than Jupiter, the largest planet.**

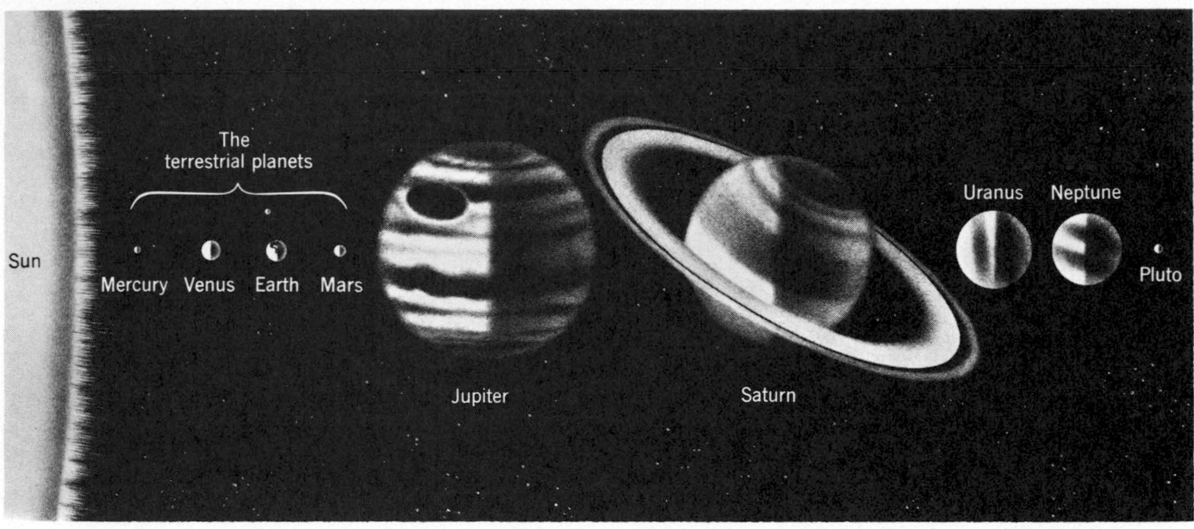


| | Mercury | Venus | Earth | Mars | Jupiter | Saturn | Uranus | Neptune | Pluto |
|---|---|---|---|---|---|---|---|---|---|
| Diameter (km) | 4880 | 12,104 | 12,756 | 6787 | 142,800 | 120,000 | 51,800 | 49,500 | 6000? |
| Mass (Earth = 1) | 0.06 | 0.81 | 1 | 0.11 | 317.9 | 95.2 | 14.6 | 17.2 | ? |
| Density (water = 1) | 5.4 | 5.2 | 5.5 | 3.9 | 1.3 | 0.7 | 1.2 | 1.7 | ? |
| Number of moons | 0 | 0 | 1 | 2 | 13 | 10 | 5 | 2 | 0 |
| Length of day (in Earth hours) | 1416 | 5832 | 24 | 24.6 | 9.8 | 10.2 | 11 | 16 | 153 |
| Period of one revolution around Sun (in Earth years) | 0.24 | 0.62 | 1.00 | 1.88 | 11.99 | 29.5 | 84.0 | 165 | 248 |
| Average distance from sun (millions of kilometers) | 58 | 108 | 150 | 228 | 778 | 1427 | 2870 | 4497 | 5900 |

perhaps even to say which, if any, of those planets may support biospheres of their own.

## THE BIRTH EVENT

The ***Solar System*** is *the system comprised of the Sun and the group of planetary bodies that revolve around it,* held by gravitational attraction. The birth throes of our Sun and its planets were the same as those of any other sun. Birth began with space that seemed to be empty, yet was not entirely empty. Atoms of various elements were present everywhere, even though thinly spread; they formed a tenuous, turbulent, swirling gas. When the gas thickened by a slow gathering together of all the thinly spread atoms, the Sun was formed. The kinetic energy of those turbulent gas swirls eventually gave rise to the rotations of Sun and planets. The gathering force of the gas was gravity, and as the atoms slowly moved closer together, the gas became hotter and denser. As one part of the gathering process, Earth and the other planets formed. Exactly how this was accomplished is still being studied, but the principal features of the process can be identified.

More than 99 percent of all atoms in space are atoms of hydrogen and helium, the two lightest kinds of atoms. Near the center of the gathering cloud of gas, the atoms became so tightly pressed and so hot that atoms of hydrogen and helium began to fuse together to form heavier elements. The gas cloud rotated because it inherited the turbulent energy of the original gas cloud. Fusion of light elements to form heavier ones causes heat energy to be released; the hydrogen and helium undergo *nuclear burning* (Appendix A). When, in the gas cloud that formed the solar system, nuclear burning commenced, the Sun was born. The time was about 6 billion years ago. But nuclear burning was confined to the center of the gas cloud. A vast, rotating envelope of less-compressed gas still surrounded the Sun.

We noted in Chapter 16 that rotation gives rise to centrifugal force. While gravity tended to pull the gaseous envelope inward toward the Sun, centrifugal force tended to pull it outward. As a result of the two opposing forces, the gas cloud slowly became *a flattened, rotating disk of gas* surrounding the hot Sun. Such a disk is a ***planetary nebula*** (Fig. 19.2).

At some stage the cool outer portions of the planetary nebula became compacted enough to allow solid objects to condense, in the same way that ice condenses from water vapor to form snow. The solid condensates eventually became the planets. Planets nearest the Sun, where temperature was highest, contain only compounds that are capable of condensing at high temperatures. Those compounds consist of elements such as iron, silicon, magnesium, and aluminum; we call them *refractory elements.* Planets distant from the Sun, where temperatures were lower, contain not only refractory elements but also *volatile elements* such as hydrogen, helium, and sulfur, that readily form gases even at low temperature.

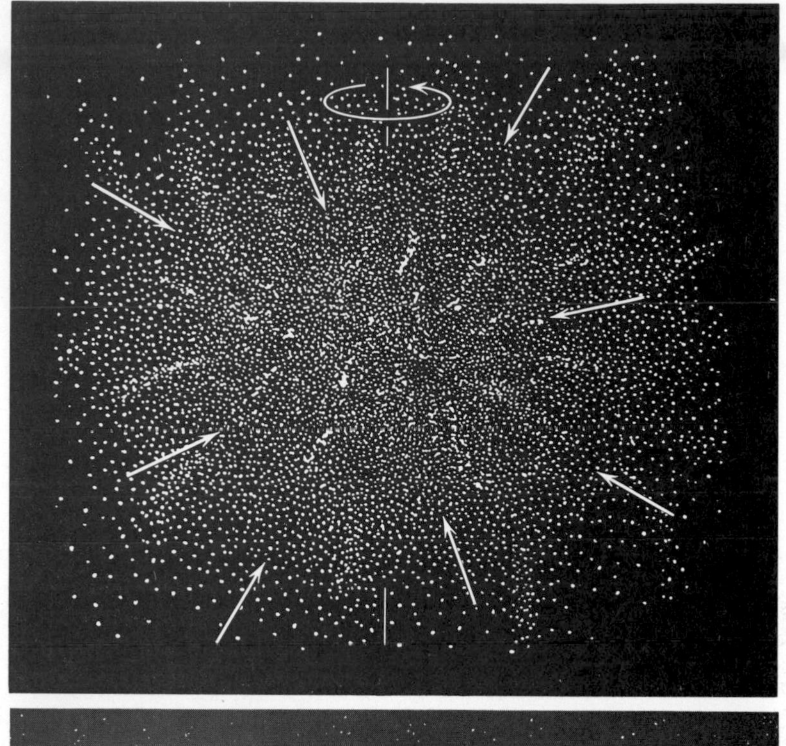

**Figure 19.2 The gathering of atoms in space created a rotating cloud of dense gas that eventually became the Sun, and a surrounding disk-shaped planetary nebula. The planets formed by condensation in the planetary nebula. (*After Turekian, 1972.*)**

The size of a planet is related to its distance from the Sun. A large-diameter ring of gas contains more atoms than a small-diameter ring. Planets close to the Sun formed from comparatively small amounts of gas, and are therefore smaller than more distant planets that condensed from large rings of gas. The small planets Mercury, Venus, Earth, and Mars are dense and rocky, and all four are similar in composition. We call them the *terrestrial planets* because they are similar to *terra* (Latin for Earth). Planets distant from the Sun are much larger than the terrestrial planets; the masses of Saturn and Jupiter are more than a hundred times greater than the mass of Earth. Because the large planets contain large amounts of the volatile elements, especially hydrogen and helium, their densities are low.

The large planets, then, are very different from the terrestrial planets. We still know little about them, because only two space vehicles have visited Jupiter and none has yet reached Saturn or the more distant planets. Perhaps when we do get close to them, we will learn much more about the birth of the Solar System and the childhood of the planets. But until then we must seek our evidence from closer neighbors.

### Condensation of a planetary nebula

Condensation is analogous to the cooling of magma. First one solid compound condensed, then another. A common example of condensation of a solid is the formation of snowflakes from water vapor. In the case of snow, only one kind of solid (ice) condenses from the gas. In a planetary nebula there are many different gases present, so many different solids condense. The earliest compounds that condensed from the planetary nebula gathered together to form a primitive core, and later compounds, as they condensed out of the nebula, were added to it in successive layers. A widely held theory states that the earliest materials were metallic iron and nickel. They correspond to Earth's core. As the planetary nebula cooled, other materials started to condense, mantling the core with compounds that contain, in addition to iron, elements such as silicon, magnesium, and oxygen. These compounds correspond to Earth's mantle. As condensation proceeded and as the planetary nebula cooled still further, increasing amounts of more volatile elements such as sodium, potassium, aluminum, sulfur, and carbon condensed. These are elements we now find concentrated in Earth's crust. The structure of a planet and the distribution of elements within it thus reflect the way in which the planet formed.

The condensation of Earth and the other planets was completed by 4.6 billion years ago. But there still remained around Earth an atmosphere of hydrogen and helium. These gases are no longer with us, but the primordial gases that surrounded Jupiter and Saturn are still present, at least in part. How the terrestrial planets lost their primordial atmospheres is still in doubt. One widely held theory involves the Sun. Astronomers have observed that early in a sun's life cycle the sun undergoes a short period of violent turbulence during which intense winds of charged atomic particles are blown out into space. According to the theory, the wind from our Sun, about 4.6 billion years ago, swept the remaining gases of the planetary nebula away into outer space, leaving the terrestrial planets atmosphere-free. Because of their great masses and huge gravitational attractions, the giant planets retained some of their hydrogen and helium.

The initial layering of compounds within a terrestrial planet is modified by later events. Compounds that condense from a planetary nebula are solids; so the initial planetary body is made up of solids, layer on layer. The planet will certainly be hot, but not hot enough to melt and form magma. Yet radioactive elements trapped within the planet will immediately start to cause internal heating. As time passes, rising temperatures will reach a point at which magma can form (Fig. 19.3). Opinions differ as to whether

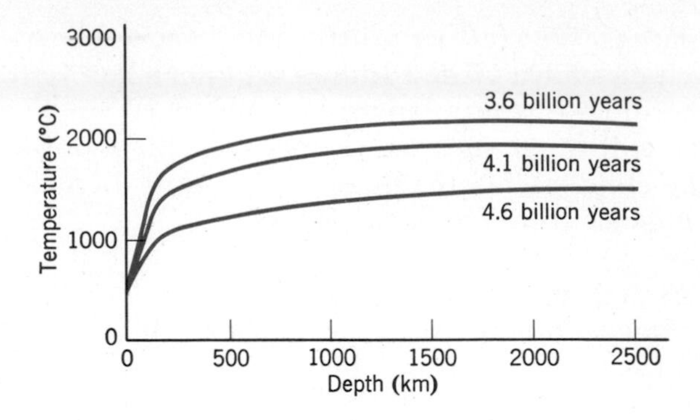


**Figure 19.3 Earth's changing geothermal gradient. The initial geothermal gradient, 4.6 billion years ago when the Earth condensed to form a solid body, was slowly raised by heat released during radioactive decay. Eventually iron in the core melted and magmas formed in the mantle.** (*After Hanks and Anderson.*)

the entire planet would melt. Probably it wouldn't, but partial melting must certainly occur. Any metallic iron in the mantle would melt and sink toward the core. And the fraction of the mantle with the lowest melting point would form magma and rise upward. In this fashion a planet could undergo profound differentiation that would enhance the initial layering. Metallic iron, being heavy, would become concentrated in the core, whereas magma, being light, would move upward, carrying with it most of the volatile elements in the mantle. This would make it possible for a crust to form. Possibly, too, convection in the mantle would begin and plates of lithosphere would start to move. That is what happened to Earth. As internal heating continued and as more and more magma was extruded, volcanoes gave off gases. It was these later gases, mainly water vapor, carbon dioxide, methane, and possibly ammonia, that eventually gave rise to the present atmosphere of the terrestrial planets. From the same source came the water we now find in Earth's hydrosphere.

It is not surprising, perhaps, that many questions concerning Earth's earliest history cannot yet be answered. One question we ask about condensation in a planetary nebula cannot be answered from evidence collected on Earth. This concerns the exact way in which the condensates form. Did Earth start with a single, tiny nucleus and grow larger by addition of layer on layer of atoms, like so many layers of paint? Or did condensation occur like millions of small snowflakes, and did the planetary snowflakes then accumulate, by gravitational attraction, into a planetary snowball? Although Earth does not answer this question, the Moon does. So we begin our discussion of Earth's neighbors with the Moon.

## EARTH'S MOON

Moon is unique in the Solar System. Because its diameter is 3,476km, only a little smaller than that of Mercury, Moon is often described as a small terrestrial planet, and the Earth–Moon pair is described as a double planet. The moons of Jupiter and Saturn, although as big as or even bigger than our Moon, are tiny by comparison with the sizes of their giant neighbors. And the moons of the other terrestrial planets are only small bodies, a few kilometers in diameter.

The exact place of Moon's birth in the planetary nebula is not yet known. It may have formed where it now is—close to Earth and attracted to it by gravity. Or it may have condensed as a separate planet, moving in its own orbit around the Sun. Then, when the separate paths of Earth and Moon brought them momentarily close together, Earth's gravitational pull may have plucked the Moon from its path, causing it to move into a new path and to orbit the Earth endlessly. Although we still cannot choose between these two possibilities, this fact does not influence an important conclusion we *can* draw: because Moon is a small, dense, rocky planet, it must have condensed in the inner regions of the planetary nebula, just as the other terrestrial planets did. If our ideas about the formation of planets from a planetary nebula are even partly correct, Moon's structure and composition should be similar to those of the other terrestrial planets. Information about the Moon, therefore, should help us in our understanding of Earth.

### Structure

Each time astronauts visited the Moon, they made measurements that yield clues about its structure. When they departed, they left behind instruments that continued the measurements and that still transmit the results to Earth. The most informative measurements are of four kinds: measurements of seismic waves, of magnetism, of the heat that flows out of the Moon, and of gravity.

**Seismic evidence.** Compared to earthquakes, moonquakes are weak; also they are infrequent. Moonquakes large enough to be detected by instruments carried to the Moon by astronauts number fewer than four hundred a year; on Earth the same instruments would record nearly a million quakes a year. Despite the infrequency of moonquakes and the weakness of the seismic waves they generate, the quakes tell us a good deal about lunar structure. Some of the most important points are: (1) on the side of the Moon that faces the Earth—at least on those parts that have been visited by astronauts—there is a crust about 65km thick; (2) the crust is layered (Fig. 19.4); (3) covering the surface is a layer of regolith that ranges from a few meters to a few tens of meters thick; (4) below the regolith is a layer, about 2km thick, of shattered and broken rock. (How the rock was broken is explained later in this chapter); (5) below the broken rock zone is about 23km of basalt, then a further 40km of a feldspar-rich rock; (6) at a depth of 65km the velocities of seismic waves increase rapidly, indicating that the lunar crust overlies a mantle.

The exact composition of the lunar mantle is not known. Wave velocities suggest it is possibly similar to that of Earth's mantle, and that the lunar mantle is considerably colder and more rigid than Earth's mantle. Whether or not the lunar mantle is layered is not yet known either. Nor is it yet known if Moon has a core, because moonquakes are too weak to pass through the Moon and resolve the question.

The comparative scarcity of moonquakes, and their weakness,

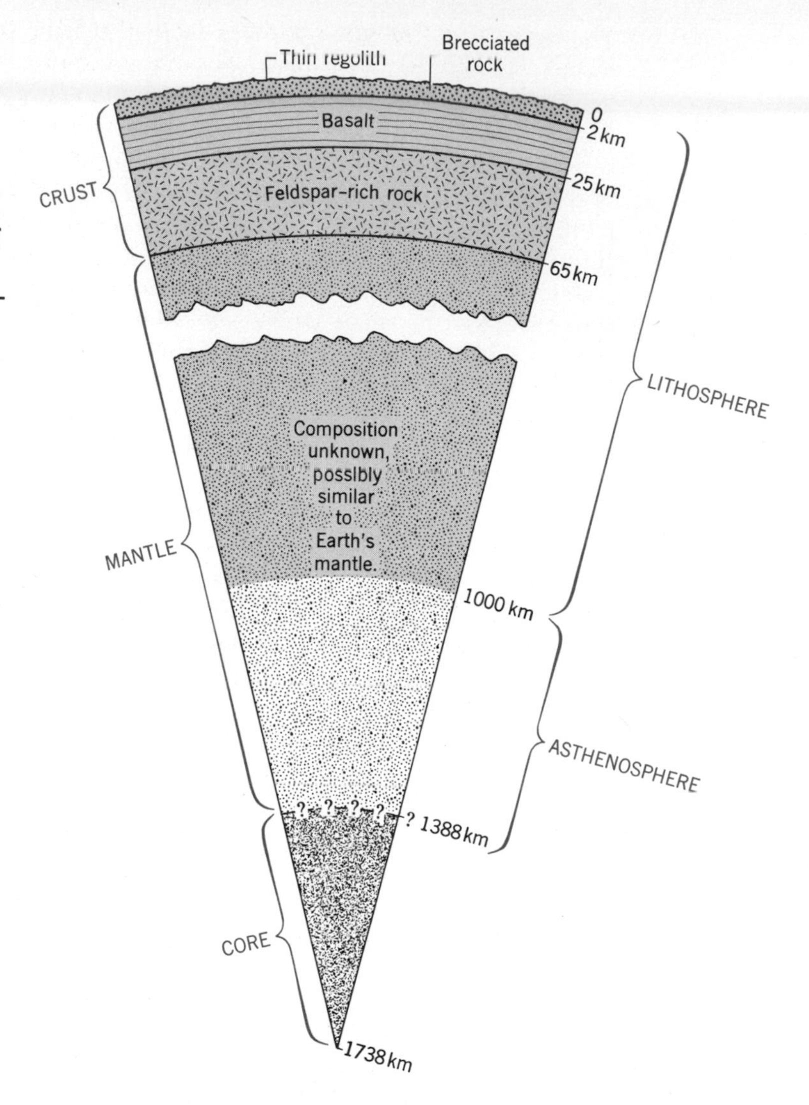


**Figure 19.4**
**Section through the Moon, showing its probable layered structure. The structure of the crust is known with certainty only in the vicinity of the astronauts' landing sites, and the existence of the small core is still uncertain. Depth of boundary between lithosphere and asthenosphere is known only approximately.**

immediately suggest that processes such as volcanism and plate tectonics, the causes of most earthquakes, are not happening on the Moon. Some moonquakes are caused by meteorites hitting the Moon. But most quakes occur in groups, and at the times when its elliptical orbit brings the Moon closest to Earth, the very moment when gravitational forces between Earth and Moon are strongest. This suggests that most moonquakes result from the gravitational pull that Earth exerts on the Moon. The pull causes slight movement along cracks, each one causing a tiny moonquake. Foci of moonquakes have been recorded as deep as 1,000km; this observation too is informative. It means that unlike Earth, Moon is rigid enough at 1,000km for elastic effects to happen. This in turn means that the lunar lithosphere must be at least 1,000km thick and that the asthenosphere lies very deep within the Moon's body. The thick, rigid lithosphere means there can be no present-day tectonic activity on the Moon.

**Magnetism.** Although the Moon lacks a magnetic field like that of Earth, it is weakly magnetic. The magnetism comes from a surprising source: bodies of igneous rock, all of which retain paleomagnetism. Therefore, although there is no lunar magnetic field today, a field must have existed at the time when the igneous rocks solidified from magma. Unless a magnetic field can be generated in some way that we know nothing about, we can infer that Moon has a core, that the core once created a magnetic field, and that it is no longer capable of doing so. We can infer also that the core must be small. Moon's density is 3.35gms per $cm^3$, while that of Earth is 5.52gm per $cm^3$. If we assume that Moon's core, like Earth's, is mainly iron, and is therefore very dense, we can state that its radius can be no larger than about 700km. If the core were larger, Moon's overall density would have to be greater than it is.

**Heat flow.** The amount of heat flowing outward from the Moon was measured during the Apollo 15 and 17 missions, and was found to be surprisingly high—about half the rate on Earth. A high rate of heat flow and a thick lunar lithosphere seem, at first, to be contradictory. But they need not be so; they lead us to conclude that the lunar crust contains a high concentration of heat-producing radioactive elements. From this we can infer that some kind of magmatic differentiation (Chapter 15) must have been involved in the formation of the lunar crust, as a result of which the radioactive elements became concentrated in a thin, near-surface layer.

Indirect measurement also yields information about interior temperatures. The Sun continually emits electrically charged particles, which stream away into space. Those streams are similar to electrical currents, and as they reach the Moon they cause secondary electrical currents to flow within it. The currents are weak—so weak that they cannot be felt—but they are detected by sensitive measuring devices. The strength of an electrical current depends on the composition and temperature of the substance in which it flows; for rock material the principal control is temperature. The strength of Moon's electrical currents indicates that at a depth of 1,000km the temperature of the lunar mantle can be no greater than 1,000°C to 1,100°C. This means, therefore, that the temperature at the base of the lunar lithosphere can only be 1,100°C. This in turn suggests that temperatures in the underlying lunar asthenosphere are probably too low for partial melting to occur. The lunar asthenosphere, therefore, seems to have resulted from causes different from those that created Earth's asthenosphere.

**Gravity.** We can divide Moon's surface into two general categories: (1) *highlands,* mountainous areas that appear to us as light-colored patches; and (2) *maria* (the "seas" seen by Galileo), smooth lowland areas that appear to us as dark-colored regions (Fig. 19.5).

In the highlands, mountains soar tens of thousands of meters above

**Figure 19.5**
**Face of Moon as seen from Earth.**
**Light-colored areas, pitted with craters caused by impacts of meteorites, are lunar highlands. Dark-colored, lowland areas are *maria* (plural of *mare*), formed when basaltic lava flowed out to fill the craters made by exceptionally large impacts. Copernicus, Kepler, and Tycho are three prominent, young meteorite craters that formed after the mare basins were filled. The impacts that made them splashed bright-colored rays of rocky debris over ancient basalt. Sites of the six Apollo lunar landing missions are indicated. (*Mosaic LEM-1, 3rd edition, 1966, made from telescopic photographs by U.S. Air Force.*)**

the maria, and in places stand even higher than Earth's mountains. Because lunar mountains are big, one way to obtain clues about Moon's interior regions is to learn whether isostasy operates as it does on Earth, and whether the great lunar mountains have roots. The test is best made by measuring variations in Moon's gravitational pull (Chapter 16). Such measurements were made continually while spacecraft orbited the Moon; they indicate that the mountains must have roots because the highlands are isostatically balanced. We can infer, then, that when the lunar mountains formed, at least one of Moon's outer layers must have been sufficiently fluid-like and plastic to enable the highlands to float.

But Moon's gravity also has an unexpected surprise and an additional clue. Some of the maria are not isostatically balanced; they have too much mass near the surface, and they produce *a region with an anomalously high gravity pull* called a ***mascon*** (abbreviated from *mass concentration*). Apparently mascons resulted from the piling-up, on the surface, of flows of dense basalt. The excess mass of the surface leads us to draw an interesting inference. When the maria formed, Moon could no longer have been capable of attaining isostatic balance. As appears later, the highlands formed before the maria did. Thus the Moon had once, but later lost, the capacity for isostatic adjustment.

Astronauts not only measured Moon's structures; they returned to Earth with a treasure-trove of samples of rock and regolith. From these, too, we have obtained many clues about Moon's composition and theory.

### Rock and regolith

Astronauts brought back three kinds of material: (1) a variety of igneous rocks, (2) breccias (Appendix C), and (3) regolith, popularly called Moon dust (Fig. 19.6).

**Igneous rock.** The most interesting samples are the igneous rocks; in terms of age and composition, there are three different kinds. The first and oldest consists of feldspar-rich rocks such as anorthosite, a variety of igneous rock formed by extreme magmatic differentiation, and consisting largely of calcium-rich plagioclase. Radiometric dates of these oldest rocks, which come from the highlands, indicate they could have been formed as long ago as 4.5 billion years, only 100 million years after the Moon condensed from the planetary nebula. The second kind of igneous rock is basalt that contains high concentrations of potassium and phosphorus; this too is 4 or more billion years old. The potassium-rich basalts likewise come from highland areas, and they are lava flows that seem to represent the last igneous activity in the highlands. The may possibly be the rocks that cause the high heat flow, because they are richer in radioactive elements than any of the other rocks found on the Moon. The third kind of igneous rock is also basalt, but it is rich, not in potassium but in iron and titanium. It has been found only in the maria, and is dated

*A*

*B*

*C*

**Figure 19.6**
**Three kinds of lunar samples brought back by astronauts.**

**A. Basalt, containing numerous vesicles formed when gases escaped during cooling and crystallization of lava. Collected during the Apollo 12 mission, this sample is the kind of basalt found in the maria. (*Photo: W. Sacco.*)**

**B. Breccia, a rock composed of fragments of igneous rock and glassy fragments similar to those found in the lunar regolith. Collected on the Apollo 15 mission. (*NASA.*)**

**C. Regolith, a mixture of many rock and mineral types, together with glassy fragments made by bombardment of the lunar surface by meteorites. Only coarser grains from the regolith are shown in the picture. (*Photo: J. A. Wood, Smithsonian Astrophysical Observatory.*)**

radiometrically at 3.3 to 3.7 billion years. We infer it is the material that underlies each mare to a depth of about 25km. Mare basalt, then, formed several hundred million years after the highlands had formed. Even though magmas do not seem to occur on the Moon today, the mare basalt proves that magmas similar to those on Earth existed on the Moon a long time ago.

**Regolith and breccias.** What clues can be extracted from samples of the other two kinds of Moon rock, regolith and breccia? The lunar regolith is a mixture of gray pulverized rock fragments and small particles of dust, many of which are glassy. Its composition is essentially that of the lunar igneous rocks. Regolith covers all parts of the lunar surface like a gray shroud (Fig. 19.7), as if giant hammers had crushed the surface rock. Indeed it *is* apparently the product of hammering—by ***meteorites,*** *small solid bodies formed during condensation of the planetary nebula* that crash onto Moon's surface. On Earth, most meteorites burn up in the heat of the friction generated as they come speeding into the atmosphere. But on the atmosphere-free Moon even small meteorites reach the solid surface, pulverizing bedrock and continually creating more regolith. Presumably the layer of shattered rock, 2km deep, beneath the regolith was created in the same way. The breccia brought back by astronauts consists of compacted aggregates of rock fragments, mineral chips, and regolith. They, too, seem to have been formed by the impact of meteorites. Apparently some meteorites compress the regolith and broken rock so greatly that the particles are welded together again to form new breccia. The process is one by which new rock is created from regolith.

The samples of regolith and breccia, then, provide clues about

**Figure 19.7**
**Photograph of Hadley Rille, a deep graben-like valley, near the landing site of Apollo 15.**
**Origin of the graben is not known. The valley walls are cloaked with dusty regolith several meters deep. The numerous small pits on the slope, left rear, were caused by meteorite impacts. Diameter of the largest boulder is 15m. (NASA.)**

erosion on the lunar surface—erosion caused by the impacts of meteorites. Erosion of the kinds we are familiar with of Earth does not exist, because the Moon lacks both an atmosphere and a hydrosphere.

**History of the Moon**

From the clues mentioned above we can construct a history of the Moon. The story begins about 4.6 billion years ago when the Moon condensed. Although one popular theory proposes that the Moon formed a giant fragment split apart from the Earth, Moon's layered structure makes such an origin very unlikely. Moon's origin, like Earth's, was the result of condensation from the planetary nebula. But condensation could not have been an atom-by-atom and layer-by-layer affair. It must have involved a gathering together of millions of small solid particles. As particles condensed from the planetary nebula, gravity continually gathered them together. We know this to be the case because the final stages of the gathering process left scars that can still be seen on Moon's surface. These are the countless craters, large and small, left by falling meteorites.

*The process by which solid particles gather together to form a planet is* ***accretion.*** By 4.6 billion years ago, the Moon had accreted to about its present size, in the process apparently developing a core and a mantle. But as the Moon accreted, there occurred a phenomenon we have not mentioned before. As the Moon grew larger, the strength of its gravitational attraction increased, so that the speeds at which accreting meteorites reached the lunar surface grew ever greater. Eventually speeds became so great that each impact generated a large amount of heat. Near the end of the accretion process, so much heat was generated that apparently an outer layer, 150km to 200km thick, of the Moon's body was melted. The Moon thus had a solid interior but a molten outer shell. The period of rapid accretion soon ended, and the outer layer of magma began to cool and crystallize. The crystals that formed earliest in the magma seem to have been plagioclase feldspar, and because they were lighter than the parent magma, they floated in it. Thus magmatic differentiation was operating, and soon a thick crust, rich in feldspar crystals, floated in the remaining liquid. The dense residue became the upper part of the lunar mantle. Meteorites fall on the lunar surface today; so we can infer that meteorites must also have been falling while the crust was forming. Larger impacts must have broken the crust, letting liquid from below ooze out to form the ancient potassium-rich lava flows. Those ancient products of magmatic differentiation are the materials the astronauts found in the lunar highlands.

The layer of magma, 200km thick, would have cooled and crystallized within about 400 million years. By about 4 billion years ago, therefore, the crust had formed, and the major activity on the Moon had become the incessant rain of meteorites that pitted and pocked the surface. Some meteorites, larger than others, made exceptionally large impact scars. The scars, huge circular

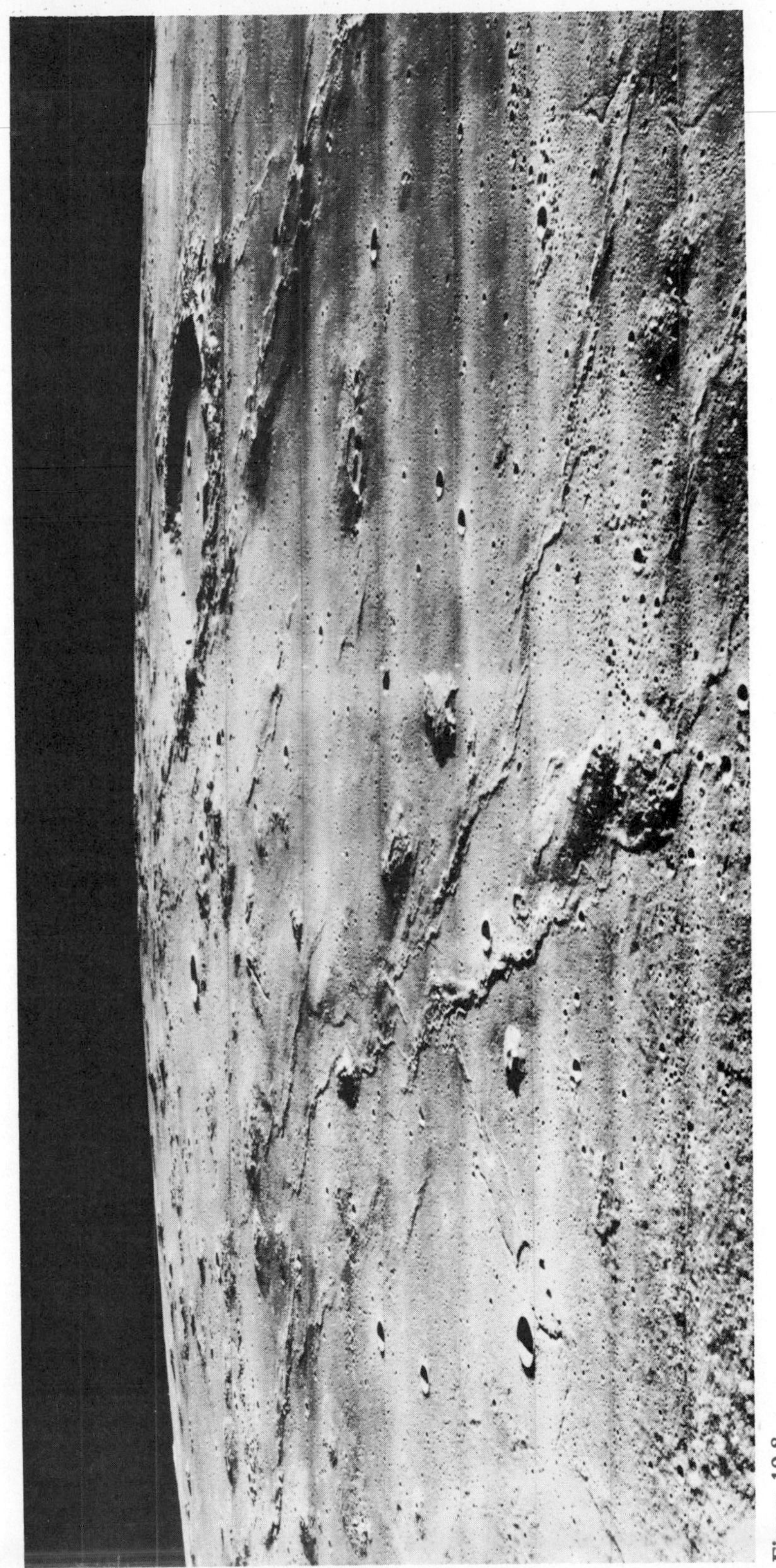

**Figure 19.8**
**Photograph of a mare.**

**The photograph, taken in 1966, shows a portion of the surface of Oceanus Procellarum, with the crater Marius (Fig. 19.5) near the horizon. The dome-shaped objects are believed to be volcanoes; the irregular ridges, looking like long sand dunes, are ancient lava flows. The craters were caused by meteorite impacts. The area of the photograph is a region as large as the states of Connecticut, Massachusetts, and Rhode Island combined. Each of the two prominent volcanic domes in the central foreground is about 20km in diameter. (*NASA*.)**

*A*

*B*

**Figure 19.9**
**How the "front" side of the Moon probably looked at two times in the past. The figures, which should be compared with the Moon's aspect today (Fig. 19.5), were prepared by starting with a photograph of the Moon in its present appearance, and painting in a surface reconstructed from geologic maps.**
***A.* As Moon probably looked about 3.9 billion years ago, before the mare basalt flows filled the largest craters. The crater that later became Mare Imbrium looks like a giant bulls-eye at the upper left. Virtually the entire Moon was then covered with anorthosite. The few dark areas represent old potassium-rich lavas.**
***B.* As Moon appeared about 3.0 billion years ago. The basaltic lava flows that filled the mare basins between 3.7 and 3.0 billion years have ceased. Except for the young impact craters, the Moon looked much as it does today. (*Wilhelms and Davis, 1971.*)**

cavities, eventually became the mare basins, and the tilted crust around their margins became Moon's highest mountains. If we study closely the photograph of the "front" side of the Moon (Fig. 19.5) we see that all the mare basins are roughly circular and have raised rims.

While Moon's surface was being bombarded by meteorites, its interior regions were slowly heating up. If we judge from meteorites (because they too are samples of the condensed planetary nebula), the amount of radioactivity is small, but nevertheless it was sufficient to cause partial melting in the upper mantle, beginning about 3.7 billion years ago. The magma so formed worked its way up to the surface along fractures caused by impacts of the largest meteorites. When it reached the surface, the magma filled the impact basins and formed the basalt flows now seen in the maria (Fig. 19.8). About 3 billion years ago, the extrusion of mare basalt flows ceased and the Moon probably looked as it is depicted in Figure 19.9. We infer that from 3 billion years ago to the present no further magmatic activity has occurred. Except for the continuing rain of meteorites, Moon has remained a dead planet.

### Lessons for the Earth

Did Earth ever look like the Moon? Probably it did. The Moon supports the idea that the terrestrial planets formed by accretion of solid bodies. Earth is larger than the Moon; so its gravitational pull is stronger. Presumably, therefore, an even thicker layer of magma covered the Earth at the end of the accretion process. Perhaps the earliest crust began to form through cooling of that magma. But all traces of Earth's primitive crust have been lost. The reason is not hard to find. Because radioactive heating of Earth has continued to create magma, probably the earliest crust has been remelted and reabsorbed.

In one respect we are sure Earth and Moon have shared a common experience: both have been scarred by meteorite impacts. Despite the speed with which erosion removes features from Earth's surface, many ancient meteroite scars remain today (Fig. 19.10).

Now let us see whether the other terrestrial planets tell a story similar to the story of the Moon.

## MARS

Mars, with a diameter of 6,787km, possesses a mass only one-tenth of Earth's mass. Yet despite its small size, Mars is Earthlike in many ways. It rotates once every 24.6 hours; so the length of the Martian day is nearly the same as the length of an Earth day. Also, Mars has an atmosphere, although it is only one one-hundredth as dense as Earth's and consists largely of carbon dioxide. Mars has polar "ice" caps (Fig. 19.11), consisting mostly of frozen carbon dioxide ("dry ice"). Like Earth, Mars has seasons, and the diameters of the caps alternately grow and shrink with the coming of winter and summer.

**Figure 19.10**
**Scars left on Earth's surface by the impacts of meteorites.**
***A.*** **Wolf Creek Crater in the northern part of Western Australia. The exact age of the crater is not known, but we infer it is not great because the crater has not been eroded deeply. Scale can be judged from the two single-lane roads that approach it (upper right-hand corner). (*Pickands-Mather Co.*)**
***B.*** **Clearwater Lake, west of Hudson Bay in northern Quebec, marks the site where a meteorite struck the surface of the Earth during Precambrian time. The crater must have looked originally like the Wolf Creek Crater (above), but erosion has removed the rim, leaving only the shallow circular feature we see here. (*Royal Canadian Air Force.*)**

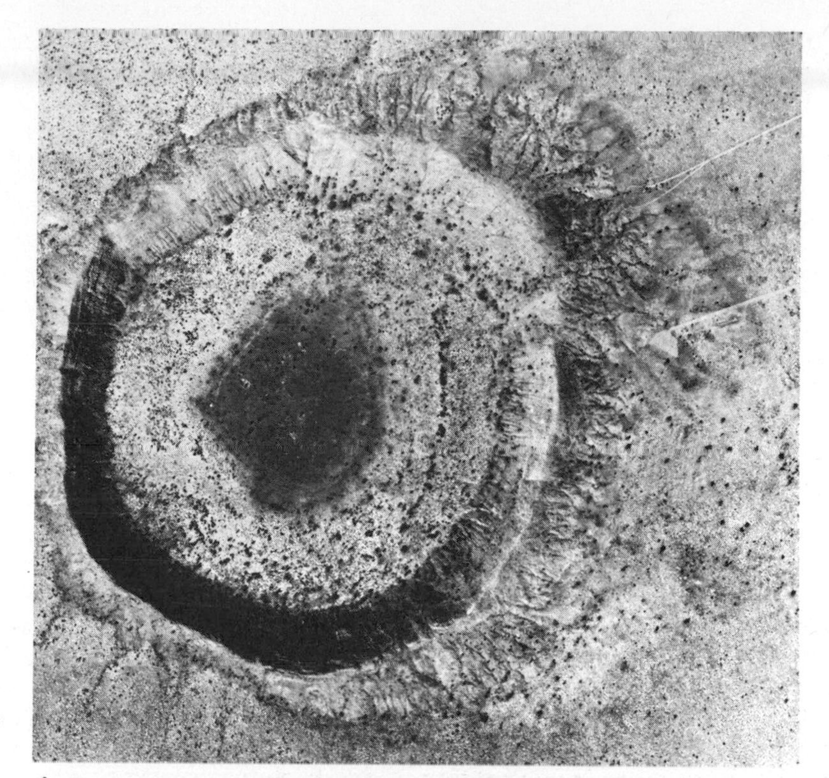

*A*

*B*

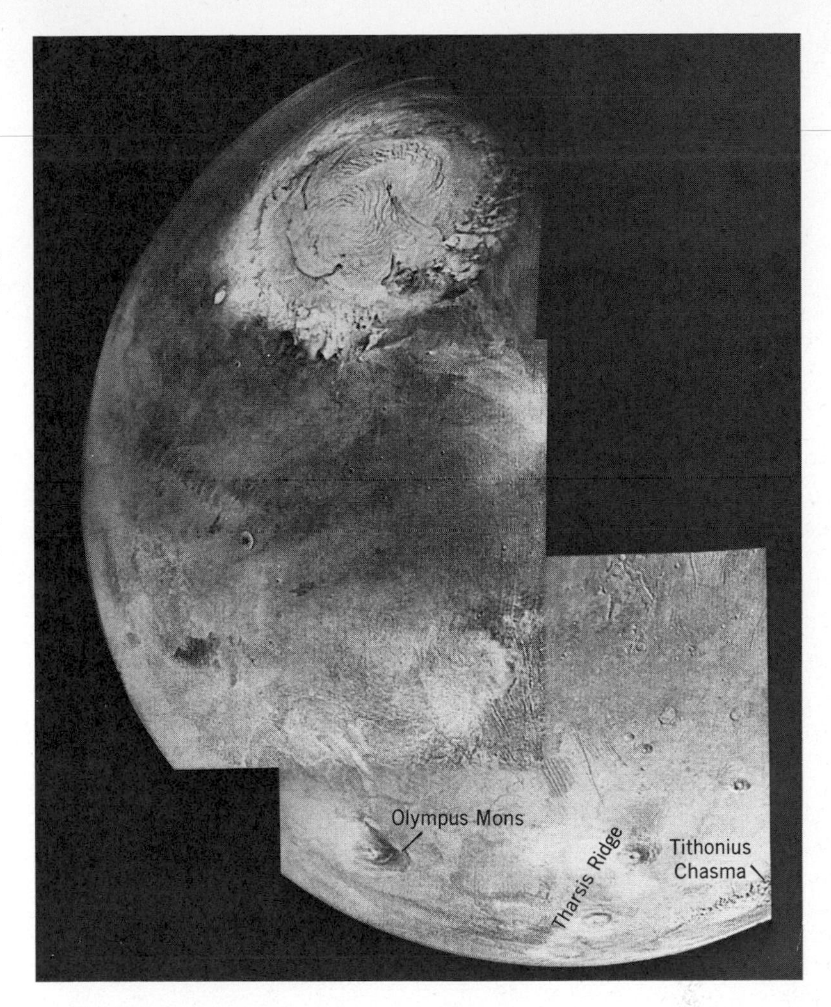


Figure 19.11
**Mars seen from a distance of 13,700km. This figure is a mosaic of photographs taken by Mariner 9, in the late spring of the northern hemisphere; the north polar cap is plainly visible as a circular, white feature, greatly shrunken from its winter maximum. The origin of the spiral-shaped structures in the cap is unknown. Near the base of the view are several shield volcanoes, including the giant Olympus Mons. The great canyon-like structure, lower left, is Tithonius Chasma (see text and Fig. 19.12). (*NASA.*)**

We can infer little about the internal structure of Mars because we have few observations on which to base inferences. Like the Moon, Mars lacks a magnetic field. Its density is 3.96gm per $cm^3$, which means that if an iron core is present, it must be small. Mars must be layered, because it wobbles as it rotates, and the way it wobbles is characteristic of a layered body, not a homogeneous one. But all details of the layering remain unknown.

### Surface features

At the time this book was written, six attempts to send spacecraft to Mars had been successful, and one of the six, after landing on the surface, had sent information back to Earth. The vehicles were aptly named Mariners and Vikings, for they sailed through distant space to reach their destinations. A one-way trip lasts at least five months. The spacecraft (particularly Mariner 9, which started orbiting Mars on November 13, 1971, and Viking 1, which landed on the Martian surface on July 20, 1976) radioed back measurements and photographs that reveal a number of interesting things about Mars and its history.

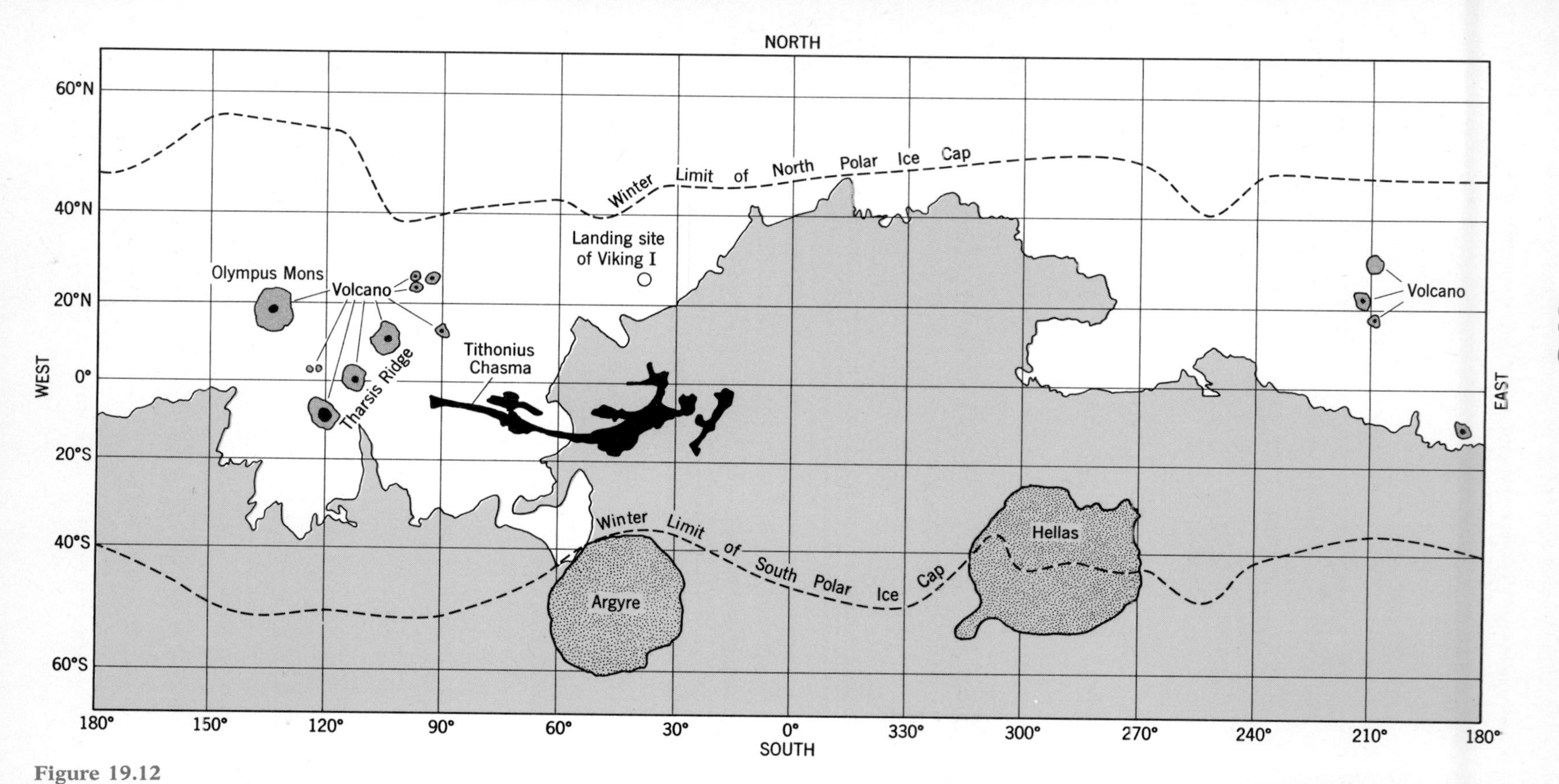


Figure 19.12

**The principal features of Mars shown on a Mercator projection (App. E). Some features on this map are visible also in Figure 19.11. The southern hemisphere is occupied by much-crater terrain (blue) believed to be ancient crust. The northern hemisphere is smoother, and the crust there is believed to be younger. It was on the edge of the smooth, northern hemisphere, that Viking I landed on July 20, 1976. Several large shield volcanoes, each crowned by a caldera, occur in the younger crust, while two very large impact structures, Argyre and Hellas, occur in the ancient crust. The feature labeled Tithonius Chasma is the largest canyon in a region of vast canyons and gorges (Fig. 19.11). The winter limit of the northern polar cap is much farther south than the edge of the cap seen in later spring in Figure 19.11.** (*Adapted from a map, based on Mariner 9 photographs, prepared by U.S. Geol. Survey.*)

A map of Mars, with the locations of the most prominent features, is shown in Figure 19.12. Most of the southern hemisphere looks very like the Moon because it is pitted and pocked by innumerable meteorite craters. Two craters, Argyre and Hellas, are as large as maria on the Moon, and presumably were created by very great impacts. Because of the large numbers of impact craters, we infer that most of the southern hemisphere is covered with rock like that of the lunar highland—primitive crust, older than four billion years.

The northern hemisphere presents a very different picture. Here we see vast areas relatively free of craters and so we think the surface there is younger than that of the southern hemisphere. The clue to the origin of the northern hemisphere comes from photographs that reveal the presence of at least 20 huge shield volcanoes. The giant among them is Olympus Mons (Fig. 19.13), whose basal diameter is 540km, approximately the distance from Boston to Washington; it stands an estimated 22km above the surrounding plain. Olympus Mons is at least twice as wide and twice as high as the largest shield volcano on Earth. Mauna Loa, in Hawaii, one of Earth's largest volcanoes, is only 225km across and stands only 9km above the sea floor. We cannot be sure whether Olympus Mons and the other volcanoes are still active, but the almost complete absence of impact craters on their slopes suggests that they could be.

Adjacent to Olympus Mons, cutting both young and ancient terrain, and running generally parallel to the Martian equator, is a region of

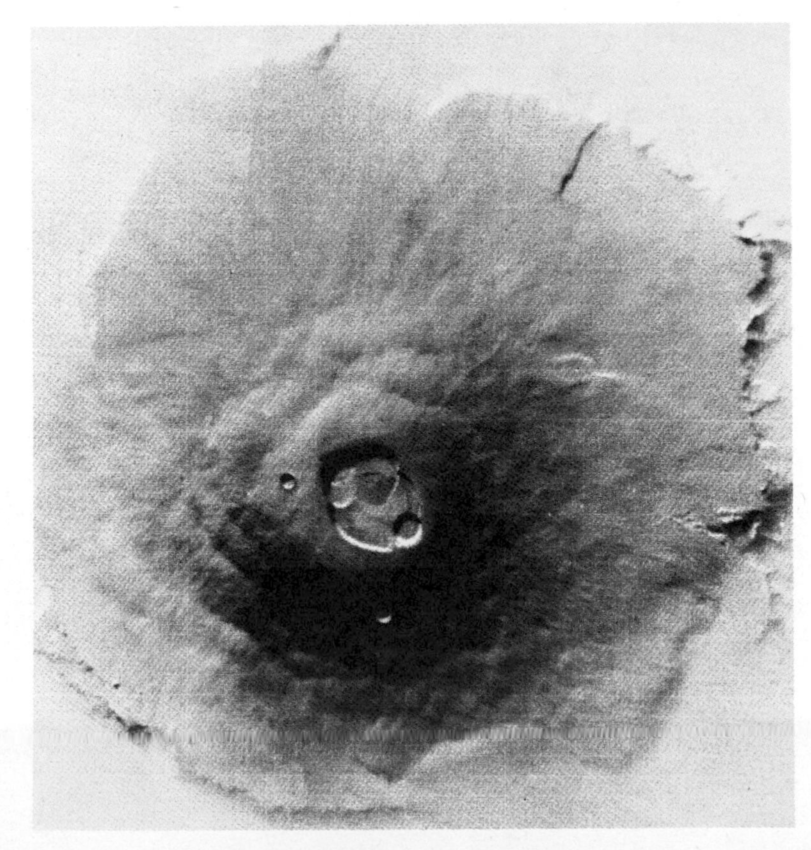

Figure 19.13
**Olympus Mons, giant shield volcano on Mars, is the largest volcano known on any planet. Its diameter is about 540km, it is capped by a caldera 65km across, and its base is marked by steep cliffs believed to have been cut by intense wind erosion during fierce Martian sandstorms. Within the main caldera, smaller calderas and a circular crater can be seen. The two small circular structures on the flanks of the volcano, to the left of and below the caldera, are believed to be volcanic craters. (*Courtesy of NASA.*)**

extraordinary canyons. Here Mariner 9 photographed Tithonius Chasma (Fig. 19.14), the longest canyon discovered on any planet. At least 4,000km long, 100km to 250km wide, and 7km deep, Tithonius Chasma dwarfs the Grand Canyon in Arizona. If the same feature were present on Earth it would stretch from San Francisco to New York. The origin of the great canyons is not known, but they are thought to be a series of giant grabens (Chapter 14). Possibly they formed when great crustal warping occurred, or when the crust subsided into openings left empty when magma was extruded to build the huge volcanoes.

### Erosion

Remarkable as the giant volcanoes and canyons are, even more remarkable features are seen in some of the Mariner 9 and Viking 1 photographs. Viking 1 landed in the northern hemisphere of Mars on a relatively flat plain called Chryse (Fig. 19.12). The plain is covered by angular rocks that look similar to the impact ejecta that covers the Moon (Fig. 19.15). Presumably, therefore, meteorite impacts have been important forces of erosion on Mars too. But photographs taken high above the Martian surface reveal features that seem to have been caused by erosion of unfamiliar kinds.

When Mariner 9 neared Mars, a sandstorm of spectacular proportions was raging. Dust and sand particles were observed as high as 55km above the surface, winds reached velocities of 300km per hour, and the storm raged unabated for several months. Under conditions such as these, sand particles could abrade rock effectively; it seems likely that on Mars, wind-driven sand is now the main agent of erosion. Peering into one of the impact craters near Hellas, Mariner 9 photographed a series of spectacular sand dunes (Fig. 19.16) that suggest what probably happens when the great storms abate. Probably all craters and all low places on Mars contain piles of sand. Every time a storm blows up, the sand is swirled into the atmosphere. When the storms die down, gentle winds slowly sweep the sand back into all the sheltered hollows on the surface.

But still more spectacular than the Martian sand is evidence that

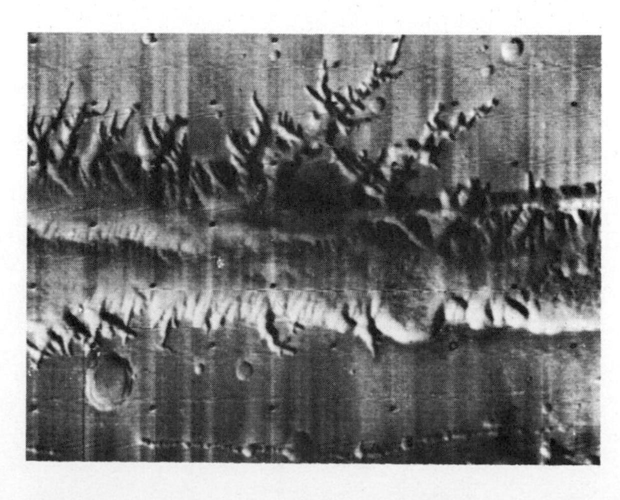

**Figure 19.14**
**Part of the giant canyon Tithonius Chasma. The segment shown here is 480km long and 7km deep. The canyon is believed to be a huge graben. The branching structures that run into the canyon look like water-cut stream channels but their origin remains a mystery. Possibly the channels were cut by fast-moving winds that are believed to roar up and down the canyon. (*Courtesy of NASA.*)**

**Figure 19.15**
**Surface of Mars surrounding the landing site of Viking I in the region called Chryse. The angular, rocky fragments, from a few centimeters to a meter or more in diameter, look as if they were formed when meteorites hit the Martian crust, spreading the rocky, impact debris widely over the surface. (*Courtesy of NASA.*)**

**Figure 19.16**
**Dark-colored area of sand dunes on Mars, near Hellas. The distance from crest to crest of adjacent dunes is about 1.5km. Numerous dark areas have been observed on Mars, many of them inside large craters, suggesting that dunes may be common features of the Martian landscape. (*Courtesy of NASA.*)**

water or some other liquid has influenced the surface of Mars (Fig. 19.17). There are valleys that look very like those cut by intermittent desert streams on Earth. They meander, they branch, and they have braided patterns and other features characteristic of valleys made by running water. Some of the features (Fig. 19.17C) look as if they were caused by gigantic floods. Yet Mars now lacks rainfall, streams, lakes, and seas. The sparse water that does occur on Mars is in the atmosphere, or exists near the poles condensed as ice. The Martian surface is too cold for water to exist as a liquid. Some have suggested that ice might be present beneath the surface dust as permafrost, and that at times the Martian climate warms up, causing the ice to melt and create torrential floods. Others believe that Mars simply does not have sufficient ice in polar caps or as permafrost to melt and to cut great stream channels. The origin of the Martian stream channels, then, remains a mystery.

A

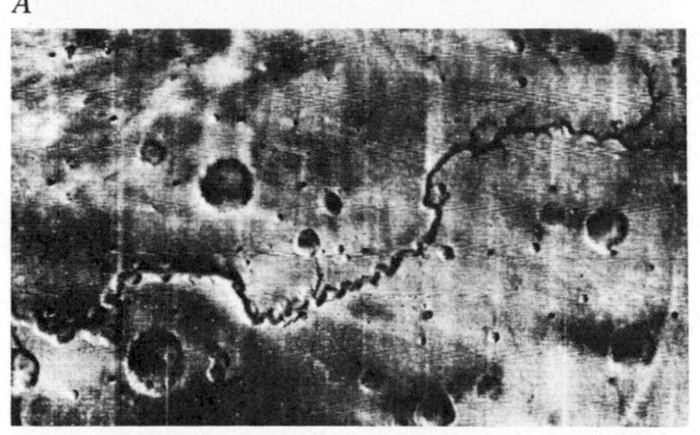

B

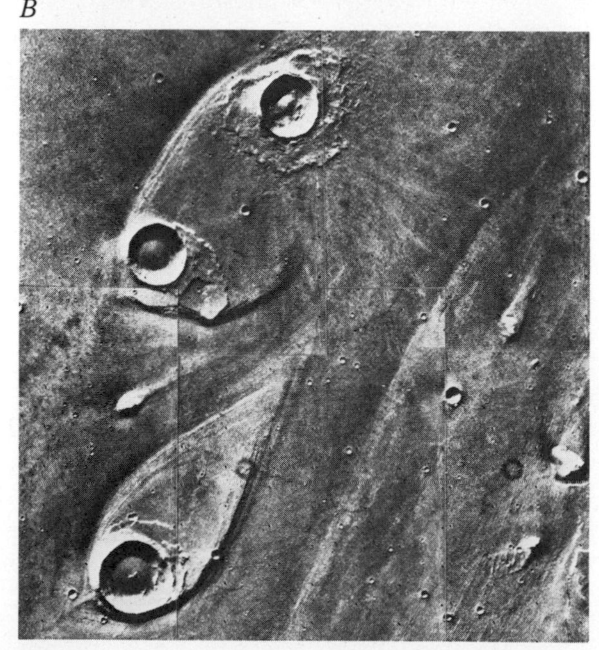

C

Figure 19.17
**Part of the surface of Mars, showing features that have been interpreted as ancient stream channels.**
***A.*** **Channels that branch in a dendritic pattern. The small impact craters are about 2km in diameter.**
***B.*** **Valley approximately 6km wide, showing meanders and branching tributaries.**
***C.*** **Mosaic of five photographs of the Chryse region showing teardrop-shaped plateaus behind resistant craters. Plateaus may have formed when flood waters moved from lower left to upper right. Largest crater is about 10km in diameter. (*Courtesy of NASA.*)**

### History

The early history of Mars must have been much like that of Earth and Moon; the terrain of the southern hemisphere, formed in the early days, is probably equivalent to the lunar highlands. Then, as in both Earth and Moon, radioactive heating inside Mars started to create magma by partial melting. Because Olympus Mons and its mates are shield volcanoes, we can infer that the magma they extruded was of low viscosity and therefore is probably basalt. This means that magmatic differentiation has occurred, but we cannot yet say when the volcanism started. To be able to do that we must have radiometric-age measurements. But the fresh-looking surfaces of the big cones suggest that the volcanoes might be active still and that Mars, like Earth, is still continuing to make magma.

Measurements made from Mariner 9 show that Mars has several high plateaus. Gravity measurements, also made from Mariner 9, suggest further that Martian lithosphere is also in isostatic balance. There are deviations, however, and this suggests that, like Earth, some force may be continually disturbing the balance. Presumably, Mars has an asthenosphere as well as a lithosphere. The gravity observations have suggested to some that convection in the Martian mantle might be causing plates of lithosphere to move around, and that the elevated plateaus are blocks of low-density continental crust of the kind we have on Earth. It has also been suggested that because Mars is smaller than Earth, its heat leaks away faster, and the frequency of its tectonic events may be very much lower than Earth's frequency. Some scientists who believe this is the case suggest that all motion in the Martian mantle is confined to a few thermal plumes, and that movement of the lithosphere—plate tectonics—has yet to happen. Proponents of the idea of plumes and a low frequency of tectonic events point to the huge size of Martian volcanoes as an indication that the volcanoes must have been situated above a magma source for a very long time. Volcanoes on Earth do not grow to such large sizes because moving lithosphere separates them from their source of magma. There can be no doubt that heat leaks away faster on Mars than on Earth, however, so it has been suggested that we should look on Mars as a sample of what Earth may have looked like at a much earlier period in its history—perhaps 2.5 to 3 billion years ago, when tectonic activity was still in an early phase of its evolution. Perhaps too, if evidence of life is ever found on Mars, it will be like the primitive form of life that existed on Earth 2.5 billion years ago.

## VENUS

Venus (diameter 12,042km; density 5.25 per $cm^3$) is the planet most like Earth in size and mass. But the similarities are fewer than the differences. Venus is enveloped by a cloudy atmosphere of carbon dioxide, a hundred times more dense than Earth's atmosphere, that hides the solid surface from view.

Russian scientists have sent to Venus several spacecraft, of which three landed successfully. The spacecraft reported that the surface of Venus is astonishingly hot, with a temperature of about 500°C. At that temperature metals such as lead, zinc, and tin are in a molten state. The explanation of the high temperature is evident. First, Venus is closer to the Sun than Earth is. But more important is the fact that the carbon dioxide in Venus's atmosphere acts like the glass of a greenhouse: It lets the Sun's rays through to heat the surface, but serves as a barrier that prevents heat from leaving. One Russian vehicle sent back a small-scale photograph of the surface of Venus that showed a mass of broken rock fragments, each about 20cm across, covering the surface. Unfortunately, the photograph doesn't tell us anything about the composition of the rocks.

Visiting spacecraft showed that Venus lacks a magnetic field. Because the density of Venus is very close to Earth's density, Venus must have an iron core, and presumably, like Earth's, at least part of it is molten. The reason for the lack of a magnetic field is probably to be found in the slow rate of rotation of Venus about its axis—once every 243 days. The rotation is apparently too slow to cause fluid motion in the core.

Sensitive radar measurements of Venus's shape show that the planet is nearly smooth, with relief apparently much less than that measured on Earth or Mars. Nevertheless, relief as great as 2.5km exists, and features resembling eroded mountain ranges and giant impact craters have been detected. Perhaps Venus once had great mountains, the product of tectonic activity, so that what we now detect are merely the eroded remains of an active past. Even the composition of its atmosphere tells something about Venus. Because gases react quickly with hot solids, the composition of the Venusian atmosphere must be controlled by chemical reactions with hot surface rock. Measurement of the composition of the atmosphere, then, should indicate the probable composition of the surface. By such reasoning we infer that the compositions of Venusian surface rocks seem to be closer to those of andesite and granite rather than to those of basalt. We can further infer, then, that Venus has undergone magmatic differentiation.

We can speculate that the history of Venus may be very like Earth's history. Apparently magmas were created, perhaps convection occurred in the mantle, and possibly there was even a period of moving lithospheric plates. Some have suggested that Venus is even more advanced in its history than Earth is, and that further study of Venus will help us peer into Earth's future. Perhaps; but such suggestions are at best speculations, and we must await future spacecraft and more reliable observations before we can draw any definite conclusions.

## MERCURY

One final terrestrial planet, Mercury, has been visited and photographed by an unmanned space vehicle. Mercury, with a diameter of 4,880km, is barely larger than the Moon, rotates very slowly about its axis, and has a density of 5.4gms per $cm^3$. This high density is puzzling. We know from seismic measurements on Earth that the density of the mantle increases with depth due to compression from the weight of rock above. The same must be true for all planets. Because Mercury is so small, the increased density of its interior layers must be very much less than the equivalent effect on Earth. To account for the high overall density of Mercury, therefore, it is necessary to suppose that the core of Mercury is 3,600km in diameter and that it contains almost 80 percent of the planet's mass. Mercury's core alone must be about the size of the Moon.

Photographs taken by Mariner 10 in 1974 show that Mercury has many similarities to the Moon. It is heavily pockmarked by ancient impact craters, it lacks an atmosphere, and it shows no evidence of moving plates of lithosphere. The largest impact basins are filled with what are apparently basaltic lava flows; so Mercury certainly did have a period of magmatic activity. But the lava plains are not crumpled and deformed; so the magmatic activity was not followed by tectonic activity.

The extraordinary feature about Mercury is the existence of a magnetic field only about 1/100 as strong as Earth's. The field is dipolar and the magnetic axis coincides with the axis of rotation. This indicates two things: the magnetism probably results from motions in a fluid core, and probably the motions are caused by the rotation of Mercury about its axis. These indications pose many dilemmas. If Mercury has a huge core of molten iron, why is it not tectonically active? How can a slow rate of rotation—once every 59 days—give rise to motion in a molten core? And if slow rotations can cause magnetism, why doesn't Venus have a field? These questions remain to be answered. The most important fact we have learned about Mercury is that there is another terrestrial planet besides Earth with a magnetic field. We can look forward to much research in the attempt to solve the puzzle.

# CONCLUSION

Earth, Moon, and other terrestrial planets seem to have similar histories. Each formed by condensation in the planetary nebula and each bears marks from the rain of meteorites accreting to its surface. Each terrestrial planet seems to have passed through a period of radioactive heating that produced kinds of magma familiar to us on Earth. Some planets possess mountains and volcanoes, and both Mars and Venus may be tectonically active. Yet the terrestrial planets differ vitally among themselves. For example, Earth's atmosphere and hydrosphere are unique; and the relative importance of different kinds of erosion differ from planet to planet. The terrestrial planets, then, are similar in some respects but different in others. By examining the differences we can gain a clearer idea of how Earth's internal processes occur.

Our research on the similarities and differences between Earth and its neighbors has led us to a conclusion of peculiar importance: Earth's sister planets are not hospitable dwelling places for us. Nor is it likely that they contain great mineral riches such as deposits of gold, copper, lead, and silver. As we shall see in Chapter 22, the very existence of mineral deposits of various kinds involves the presence of a hydrosphere and an atmosphere. The absence of either or both would make the formation of such deposits unlikely. Earth remains the planet on which we must rely for our food, and for all the other resources we need to keep our civilizations going. For this reason, in the final part of this book it is appropriate to discuss the distribution and abundance of Earth's mineral resources, Earth's climate, and our uses and misuses of planet Earth.

## SUMMARY

1. The Sun formed when atoms in space became sufficiently compacted for nuclear burning to begin.

2. The planets formed by condensation from a disk-shaped envelope of gas (the planetary nebula) that rotated around the Sun. All planets condensed at the same time, about 4.6 billion years ago.

3. Planets, close to the Sun, such as Earth and Venus, are small, dense, rocky bodies. Planets farther away, such as Saturn and Jupiter, are large, low-density bodies.

4. Earth, Moon, and the other terrestrial planets have layered structures. The layered structures probably formed when the planets condensed from the planetary nebula.

5. Each of the terrestrial planets went through a period of internal radioactive heating that led to generation of magma.

6. Earth, and possibly Mars and Venus, are still producing magma from radioactive heating; and Mars and Venus may be tectonically active, as the Earth is.

7. Moon probably has a small core surrounded by a thick mantle, and is capped by a crust 65km thick.

8. On the Moon, magma was formed early, but is no longer generated. Moon is a dead planet.

9. The highlands of the Moon are remnants of ancient crust built by magmatic differentiation more than 4 billion years ago.

10. The maria (lunar lowlands) are vast basins created by the impacts of giant meteorites and later filled in by lava flows.

11. Mars seems to be geologically active. Olympus Mons, a shield volcano on Mars, is the largest volcano yet found in the Solar System.

12. The principal eroding agent on Mars is wind-driven dust, but water or some other flowing liquid may have cut stream channels at some time in the past.

13. Venus has about the same size and density as Earth. It has a dense atmosphere of carbon dioxide and a surface temperature of about 500°C.

14. None of the other terrestrial planets have climates hospitable to human life.

# SELECTED REFERENCES

## SOLAR SYSTEM

Bergamini, David, 1969, The Universe: New York, Time Incorporated.

Cameron, A. G. W., 1975, The origin and evolution of the Solar System: Sci. American, v. 233, no. 3, p. 32–41.

Carr, M. H., ed., 1970, A strategy for geologic exploration of the planets: U.S. Geol. Survey, Circular 640.

Jastrow, Robert, and Thompson, M. H., 1972, Astronomy: fundamentals and frontiers: New York, John Wiley.

Kuiper, C. P., and Middlehurst, B. M., 1961, The Solar System. Vol. 3, Planets and satellites: Chicago, Univ. of Chicago Press.

Sagan, Carl, 1975, The Solar System: Sci. American, v. 233, no. 3, p. 22–31.

## MOON

Kosofsky, L. J., and El-Baz, Farouk, 1970, The Moon as viewed by lunar orbiter: NASA Publ. SP-200.

Levinson, A. A., and Taylor, S. R., 1971, Moon rocks and minerals: New York, Pergamon Press.

Mutch, T. A., 1970, Geology of the Moon: Princeton, N.J., Princeton Univ. Press.

Taylor, S. R., 1975, Lunar science: a post-Apollo view: New York, Pergamon Press.

Wilhelms, D. E., and McCauley, J. F., 1971, Geologic map of the near side of the Moon: U.S. Geol. Survey Map I-703.

Wood, J. A., 1975, The Moon: Sci. American, v. 233, no. 3, p. 92–105.

## MARS

McCauley, J. F., and others, 1972, Preliminary Mariner 9 report on the geology of Mars: Icarus, v. 17, p. 289–327.

Murray, B. C., 1973, Mars from Mariner 9: Sci. American, v. 228, no. 1, p. 49–69.

NASA, 1974, Mars as viewed by Mariner 9: NASA SP-329, U.S. Govt. Printing Office.

Pollack, J. B., 1975, Mars: Sci. American, v. 233, no. 3, p. 106–117.

Sharp, R. P., and Malin, M. C., 1975, Channels on Mars: Geol. Soc. America Bull., v. 86, p. 593–609.

## VENUS

Lewis, J. S., 1971, The atmosphere, clouds and surface of Venus: American Scientist, v. 59, p. 557–566.

Young, A., and Young, L., 1975, Venus: Sci. American, v. 233, no. 3, p. 70–81.

## MERCURY

Murray, B. C., 1975, Mercury: Sci. American, v. 233, no. 3, p. 58–69.

**The climate became colder and wetter, and then changed back to warmer and dry. The record is preserved, in the shorelines of pluvial Lake Bonneville, along the mountainside back of Salt Lake City.** ***(John S. Shelton.)***

# PART SIX
# MAN AND EARTH

# 20 CLIMATE



Abundant fossil bones and teeth of hippopotamus—of the same species that lives in East Africa today—have been found in southeastern England. Herds of hippos lived there about 100,000 years ago. Much earlier—100 million years ago—crocodile-like reptiles were numerous in shallow seas throughout northern Europe. Clearly the climates of those times were warm enough to support such animals. The fossil hippos and reptiles, like many other fossil animals and plants in many strata, are out of adjustment with today's environments and are therefore *relict* from former conditions.

The chief difference between present and former conditions is probably ***climate***, *the average weather conditions of a place or area over a period of years.* Climate is determined by several factors, of which temperature, precipitation, and winds are those that interest us most. As Figure 2.8 shows, the biosphere adjusts closely to climate. Thus, when a significant change in climate occurred in England at some time less than 100,000 years ago, that country became too cold for hippos. Some moved south, others died and thereafter were relict, occurring only as fossils.

Besides fossils, other sorts of out-of-place geologic features give evidence of change of climate. A list would include (1) glacial features in warm territory, (2) desert sand covered with vegetation, (3) beaches and other lake sediments in deserts, (4) stones, on the sea floor in a warm mid-ocean, dropped from floating ice, (5) dried-up systems of streams with their tributaries, and (6) relict soils. Such features are relict today in many parts of the world. In addition, there is a way, which we will mention presently, of using geochemical evidence drawn from oxygen isotopes to measure former temperatures at a given place.

All these kinds of evidence are geologic; the problem of identifying former climates is therefore a geologic problem. Because climate determines habitability and food production, the problem is vital. If we hope to predict change of climate in the near future, we must try to understand climatic changes in the past.

## FACTORS THAT DETERMINE CLIMATE

Rereading the description near the beginning of Chapter 9 and looking again at Figure 9.2, we recall that solar heat warms the Earth, both continents and oceans, and drives winds that then distribute the heat. The winds drive ocean currents (Fig. 13.2) that also distribute heat. Seawater in the tropics is evaporated and carried by winds toward higher latitudes; the vapor condenses, forms clouds, and is precipitated as rain and snow. We recall also that climates are arranged in belts roughly parallel with latitude lines, but the belts are interfered with and distorted by the patterns of oceans, lands, and high mountains. Hence temperature and precipitation vary greatly from one place or region to another. When a change, however slight, occurs in this vast and complex system, the climate in the area affected must change also.

As we shall see, the exact causes of climatic change are still not understood. But we do know that continents, rafted on plates of lithosphere, move from one climatic zone to another. But the continental movements are slow, and the geologic record tells us that climatic changes often happen much too rapidly to be explained by the movement of land masses. In this chapter we will concentrate on the evidence that climates change.

Our discussion emphasizes glaciation, because it was the evidence of former glaciation that first led to the realization that climates have not always been as they are today. Also such evidence is still our clearest indication that very different climates have existed at times in the past.

There are also records that the climates of deserts have fluctuated, but much of the evidence pertains to change in amounts of rainfall and is hard to separate from that of the temperature changes that probably occurred at the same times.

# EVIDENCE THAT CLIMATES ARE CHANGING TODAY

The fact that last winter was colder than the winter before, or that last summer was rainier than the present one does not prove that the climate is changing. Variation of climate must be based on changes in averages through a series of years no two of which are ever identical. Even a five-year trend is hardly an adequate basis for inferring a change, but a trend that persists through 10 years begins to be significant. A consistent 50-year trend, persisting despite the presence of small, short-term fluctuations, definitely indicates a change. The freely sketched gray band through the middle parts of the zigzags in the curve, Figure 20.1, smooths out the small fluctuations and shows a trend of rising temperature from before 1880 through about 1920, after which it accelerated to a peak between 1940 and 1950. Thereafter, temperature began to fall. The curves in the figure, which show the short-term fluctuations, are based on records of instrumental readings of temperature at networks of stations that extend around the world. This evidence tells us that during the six decades before the 1940s, climates changed; since 1880 Earth's surface has become slightly warmer. The amplitude of the change is less than 0.5°C, hardly a tenth of the change that occurred from the beginning to the peak of a glacial age; yet it was a general warming that extended around the world. The change is reflected in the facts that during those six decades, glaciers in most parts of the world shrank, some of them conspicuously (Fig. 10.8), and that floating ice in the Arctic Ocean decreased about 10 percent in area and about one-third in thickness. Many small shallow lakes in western United States shrank or dried up altogether.

The biosphere, as we might expect, was affected also. From 1880 to 1940 summer growing seasons increased in length and crop yields improved. The territory normally occupied by various plants and animals enlarged toward the poles. The common codfish, for example, almost unknown in Greenland before the 20th Century,

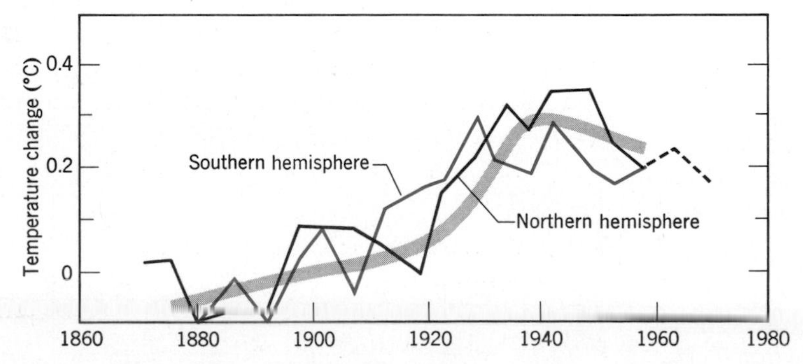


**Figure 20.1**
**Fluctuation of annual temperature (five-year means) since 1870. Data from stations within two belts of latitude: 0° to 60°N and 0° to 60°S. Dotted line: extension of N curve to 1970, based on data from a belt 0° to 80°N.** (*Courtesy J. M. Mitchell.*)
**The wide gray band, added by the authors, suggests a general trend (northern hemisphere only).**

began to migrate northward. Year by year it was observed at places farther north; its rate of migration was clocked at 9° of latitude in 27 years.

The chilling of climates that began in the 1940s is continuing. There is abundant evidence that on the northern fringe of the zone in which agriculture is easily possible, the growing season is shortening, and some kinds of birds and fish are now deserting the northern parts of their ranges.

The information that climates have been changing in recent decades is a new realization. Even as recently as a few decades ago, it was thought that the climate of any region does not change (apart from the catastrophic changes represented by a glacial age, for which there was no acceptable explanation). But now, in the knowledge that climate is changing in the present day and has done so throughout the past, we have another illustration of the Principle of Uniformity. Let us now increase our sampling period and try to get an idea of changes of climate during the last thousand years.

# CLIMATIC CHANGES DURING THE LAST THOUSAND YEARS

### Compilation of historic records

In trying to reconstruct climatic fluctuation during the last millenium, we can use instrumental records only for the latest part of that time. To go further we must rely on careful compilations of historic records of various kinds. Such records, although they lack the precision of instrumental data, confirm each other and together give a good general picture of what has happened to the climate, at least in the middle latitudes of the northern hemisphere where educated people tended to record in writing the things they saw.

The best reconstruction for this period is shown in Figure 20.2. From it we see that in central England, mean annual temperature has fluctuated through an extreme range of 1.7°C, more than four times the range of more recent fluctuation in the northern hemisphere visible in Figure 20.1. The curve indicates two major changes: a period of warmth between A.D. 1000 and 1400 and a cold period from 1500 into the 18th century. During the early warm time, agricultural conditions were good and harvests were abundant. Grapes were grown and wine was made as far north as

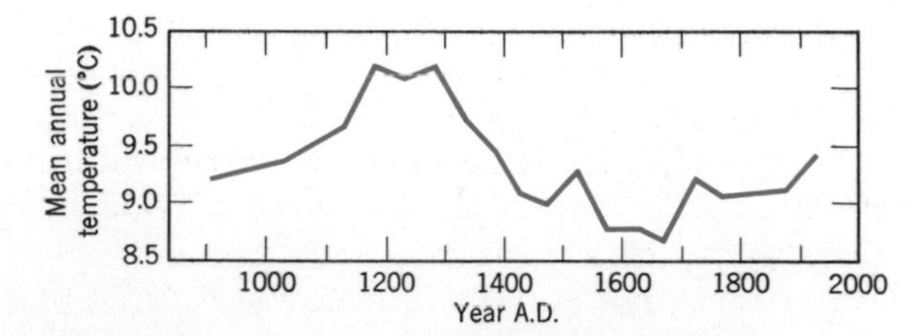


Figure 20.2
**Fluctuation of mean annual temperature in central England, by 50-year averages from A.D. 900 to about 1930. Data derived indirectly from historical records; after 1680 from instrumental measurement, in part. (*After H. H. Lamb*.)**

southern England, an enterprise impossible there during the following period. This time coincided with the Viking period, when people from Scandinavia sailed in small open boats all over the North Atlantic. Such travel was possible for them partly because high temperatures melted the floating sea ice and thus cleared the sea far north up the east coast of Greenland.

It was during this warm time of easier navigation that Norsemen discovered North America. The voyages to "Vinland" shortly after the year 1000, made by Leif Ericson and his companions, are dated—one of them, at least—by a $^{14}C$ measurement on charcoal from a cooking fire, on a shore in northern Newfoundland. At a much later period the ice and the weather might have been too much even for Norsemen, thus leaving the discovery to be made via a warmer route farther south.

The cold period that began about 1500 has been called the "little ice-age." It was characterized by cold winters and wet summers. In parts of northern Europe harvests failed. All around the world, glaciers expanded and in the Alps and Iceland they even covered up farms and villages that had been inhabited since medieval time. (Since 1960 about 30 percent of the glaciers in the Swiss Alps, after nearly a century of shrinkage, have again begun to expand, faithfully following the temperature curve in Figure 20.1. If the trend continues, some Alpine farms will again be overwhelmed by ice.)

1500- little ice age

At various times during the "little ice-age" there were series of winters when rivers and canals in Holland and England froze, and people skated on them in numbers. In one winter the western Baltic Sea froze so solidly that passage from Germany to Sweden was open to traffic by wheeled vehicles. In America the winter hardships faced by the early Pilgrims have become folklore.

### The Greenland colonies

The period of warmth and the following cold period are illustrated dramatically by the history of the Norse colonies in Greenland. In the summer of 985, Erik the Red, father of Leif Ericson, sailed up a fiord in southwest Greenland with several hundred emigrants from Iceland, bringing with them their livestock and household goods. They established two colonies that later numbered 4,000 or more people. They raised sheep and cattle, exported butter and cheese, and even tried unsuccessfully to grow grain. Trade routes were well established; the sailing directions for them survive. In the earlier years, floating ice is not mentioned, but later, ice begins to appear. Traffic becomes increasingly difficult; sailing directions are altered. The growing season becomes shorter and cooler, with serious effects on farming. By 1350 the more northerly colony could not hold out and was abandoned. By about 1500 the southerly settlement had been deserted too. The stone houses and churches remain, and modern scientific excavation of more than 100 burials (Fig. 20.3), in ground now frozen perennially, has helped reconstruct the story. Early burials were deep, and in

**Figure 20.3 Skeletons of Greenland colonists who died probably early in the 14th Century. Excavated from a churchyard near Godthaab, lat. 64°. (*Jørgen Meldgaard.*)**

coffins made of imported wood; as the colony became poorer, later burials were shallow, and in shrouds only. The medieval cloaks that served as shrouds were pierced by roots of plants that in the later cold years could not have grown there because burials and ground were solidly frozen. The later-buried skeletons are of young people, suggesting a very short life expectancy; all are less than 160cm (5 ft. 3 in.) in height and are misshapen, suggesting deficient diet; teeth are worn down extraordinarily, suggesting coarse vegetable food. The skeletons are not at all like those of the healthy Norse ancestors who died during the 12th Century. Overtaken by the "little ice-age," the later people had to endure appalling living conditions that were marginal at best. Apparently the late survivors died of starvation or were so weakened that they could not withstand illness of other kinds. Now, with the passing of the cold period, conditions have improved. Parts of southwestern Greenland have again become possible for farming.

**Physical changes**

Summarizing, we can say that in the northern hemisphere, high temperature 1,000 to 800 years ago gave way 500 to 200 years ago to temperatures nearly 2° lower. Various lines of research show that the change to warmer climates was accompanied by other recognizable changes in the hemispheric circulation system, changes of at least four kinds:

1. The general atmospheric circulation was intensified (winds became stronger).

2. In the belt of westerly winds, established tracks followed by storms shifted toward the equator by 5° to 10° of latitude, bringing a wetter climate to western Europe.

3. The Gulf Stream (Fig. 13.2) shifted southward toward the equator.

4. The area of floating ice in the Arctic Ocean (Fig. 10.22) enlarged southward into the Atlantic.

5. Glaciers expanded generally.

These changes are consistent with each other, and are likewise what we could expect to occur in the early part of a glacial age.

# CHANGES DURING THE LAST 10,000 YEARS

We have discussed climatic change measured in tens of years, and then changes measured in hundreds of years. We come now to changes more difficult to reconstruct, those measured in thousands of years—changes that have taken place during and since the latest full-scale glacial age. A time about 10,000 years ago is a convenient date to assume as the end of the latest glacial age. Of course the date is arbitrary, for the transition from glacial to postglacial was gradual, and so its date differed in different latitudes. But whatever date is chosen, we want to examine now what Earth's climates have been doing during the last 10,000 years, more or less.

**Evidence from fossil plants and pollen**
In the late 19th Century, before the climatic fluctuations we have been discussing were recognized, it was vaguely supposed that the cold glacial-age climate must have warmed continuously until it gave place to the climate of today. The supposition was wrong, and the error was first revealed by fossil plants in Scandinavia. Plant leaves and stems, embedded in postglacial sediments of ponds and bogs, were discovered and identified first, then, later, fossil pollen grains began to be analyzed. The pollen grains are a more useful fossil than leaves, because they are microscopically small, occur in uncounted millions, and are each enclosed in a waxy coating that resists chemical weathering. Much pollen is transported by winds, from which it falls out over wide areas (Fig. 20.4). The pollen grains that fall into lakes, ponds, and bogs are trapped there, protected by the surrounding water from weathering by oxidation. Mixed with plant remains of other kinds, they form strata that slowly accumulate year by year. Pollen has two great advantages for stratigraphic study: It identifies the plants from which it came and, being organic, it can be dated by $^{14}C$ measurements (Chapter

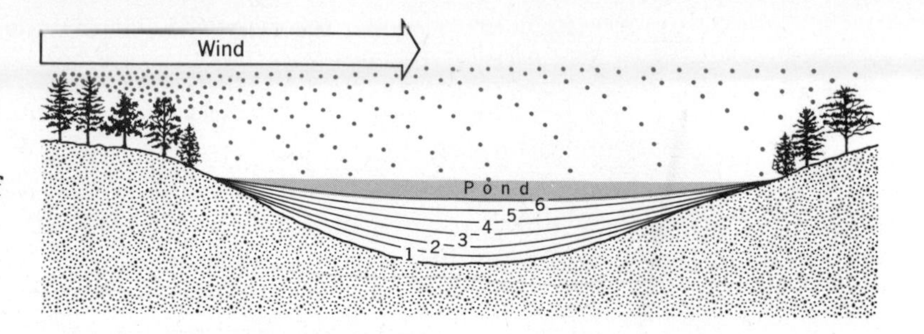


Figure 20.4
**Fallout of windborne pollen grains derived mostly from nearby trees, into a pond, forming a sequence of thin strata. (Schematic.)**

5). Thus a core sample taken by a tubular device punched into the pond sediments yields a huge number of very tiny fossils. These are identified, grouped by species of plant, counted, and treated statistically. By these means the pollen grains reveal the particular assemblage of plants that flourished in the neighborhood when the layers in which they occur were deposited. The assemblage is compared with the nearest similar assemblages now living (in many cases much farther north). Thus former temperature and rainfall at the fossil site can be estimated. Finally the fossil assemblage can be fixed in time by $^{14}C$ dating.

Botanists who study the sediments from lakes and bogs in Scandinavia have found that at most sites, the bottom of the sequence consists of till or stratified drift left by the melting ice sheet. Above the drift are the leaf- and pollen-bearing sediments, in many places extending right up to the surface, where they form the deposits of the present day. From a series of samples taken at close intervals from bottom to top, a continuous history of climate since the glacial age has been read. The climatic changes revealed by the plants are confirmed by fossil beetles, many of which are also sensitive indicators of temperature. The results when first deciphered, were unexpected. Instead of a slow, progressive postglacial warming, there were several fluctuations of climate

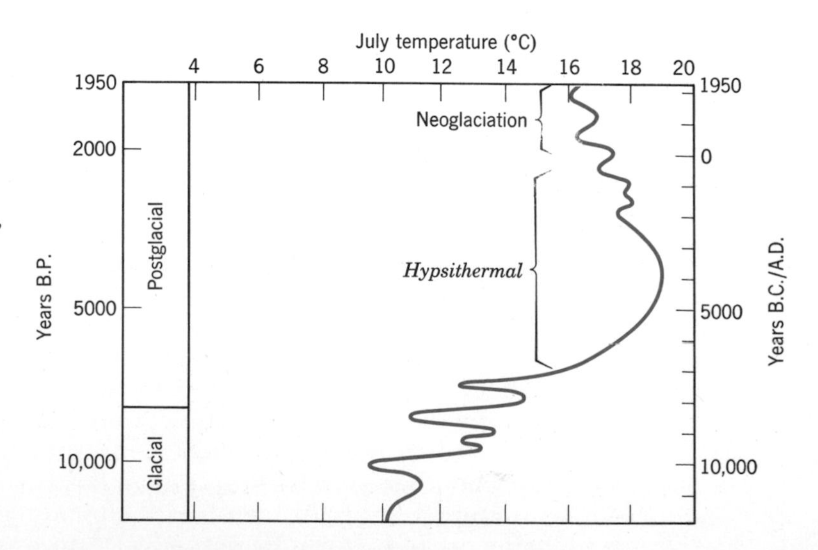


Figure 20.5
**Changes of temperature in Denmark through the last 12,000 years, as inferred from glacial deposits, fossil pollen, and recurrence horizons. (*Modified from J. Iversen, 1973.*)**

**Figure 20.6**
**Recurrence horizons (ticks at left) exposed in a Danish peat bog. Heather peat (dark) alternates with sphagnum-moss peat (light). Divisions on the surveying rod are 20cm long. Radiocarbon dates of peat show that horizons I, II, IV, and V are present in this bog, but I is not in view, having been dug up during cutting of peat. (*Bent Aaby, Geol. Survey of Denmark.*)**

including a time when mean annual temperatures in Scandinavia were as much as 1.5°C higher than they are even today, and when the climate was drier as well. That time goes by the name *Hypsithermal* (Greek "High heat"). It reached its peak at a time near 5,000 years ago (Fig. 20.5).

Later, smaller fluctuations are indicated clearly in artificial exposures cut down into peat bogs. The exposures reveal horizontal layers of heather peat, each layer abruptly overlain by sphagnum-moss peat (Fig. 20.6). The heather represents warmer, drier climate, the sphagnum cooler, wetter climate. The contact

**Table 20.1**
**Recurrence Horizons Widely Recognized in Northern Europe, and their approximate dates, determined from archeologic data.[a]**

| Horizon | Approx. date |
|---|---|
| I | A.D. 1200 |
| II | A.D. 400–600 |
| III | 600 B.C. |
| IV | 1200 B.C. |
| V | 2300 B.C. |

[a] Many other recurrence horizons, recognized only locally, have been dated by $^{14}C$.

between the two types is called a *recurrence horizon*, because it represents the recurrence of cool climate following a warmer one. At least five such horizons have been identified, notably in Sweden and Denmark but also in other northern-European countries as well (Table 20.1). Radiocarbon dates of samples from the bases of the various sphagnum layers are about the same in most bogs. Thus the recurrence horizons can be correlated like sedimentary-rock strata, and record climatic changes that affected a wide region, as indicated in Figure 20.5. Confirmation of these changes comes from other parts of the world. Six thousand kilometers west of the Scandinavian sites, in northern Canada west of Hudson Bay, $^{14}C$-dated pollen layers collected from beneath lakes show that the boundary between forest and treeless tundra shifted far north during the warm Hypsithermal time and moved southward during the subsequent colder period.

### Fluctuation of glaciers—neoglaciation

Not only did glaciers expand during the "little ice-age," they increased also around a time about 2,500 to 3,000 years ago, as indicated by end moraines, some of them enclosing the debris of trees knocked over by advancing glacier termini, recovered, and dated by $^{14}C$. That event (if it was a single cold event) represents the start of what has been called the Neoglaciation ("new glaciation"), a term that embraces the "little ice-age" as well. In the Neoglaciation, many small glaciers in mountains of western North America were reborn after having melted away during the Hypsithermal warmth. Such glaciers, instead of being shrunken relicts of a major glacial age, are comparatively recent, reborn at the time of the most conspicuous recurrence horizon listed in Figure 20.6. The fact that end moraines built during the Neoglaciation have been identified in several parts of the world strongly suggests that the climatic change that caused the glacial readvance was worldwide.

## THE LATEST GLACIAL AGE

### Advance of the ice sheet in Ohio

Following climatic changes back into the time before 10,000 years ago, we immediately run into the latest glacial age. The maps, Figures 10.23 and 10.22, show the extent of glaciers in the northern hemisphere between 20,000 and 15,000 years ago. As the climate cooled about 25,000 years ago, the extensive ice sheet that had been covering a large part of Canada during many thousands of years previously, began to expand. Its southern margin flowed over the Great Lakes region, enveloping the site of Cleveland, Ohio, as it expanded southward.

The rate of advance of its margin is known, because the region south of Lake Erie was marked at that time by scattered groves of spruce trees in an otherwise treeless, tundra-like landscape. The advancing glacier knocked the trees down, buried them in till, and overrode the wreckage. The trees or rather logs, hundreds of them, are found exposed today in the sides of stream valleys. They are bent and twisted, indicating that they were alive when the glacier destroyed them. Some even retain their bark, and some of the logs lie pointing alike, parallel with the streamline flow of the ice sheet. When sawed sections of logs are examined, it is found that their outer (latest) growth rings are progressively thinner than the inner rings, indicating that all the trees were struggling with an increasingly inhospitable climate as the ice sheet approached.

The $^{14}C$ date of the outermost wood in a log is the approximate date when the ice arrived and the tree was killed. The dates are consistent with the places of occurrence of the dated logs: they become younger from north to south. Logs at or near the extreme southern limit of the former glacier date about 18,000 years, not only in Ohio but farther west along the margin of the former ice. Dividing the distance between the localities of two successive samples by the difference between their dates yields an average rate of advance of the ice margin across that distance. Comparison of many such pairs of dates gives results that vary between 25 m per year and a little more than 100 m per year. The results are consistent with measured rates of advance of some existing glaciers, and are surprisingly rapid—much more rapid than rates of many other geologic processes.

### Vegetation zones

Fossil spruce logs in Ohio are only one of the many kinds of plant material from which vegetation zones during the latest glacial age are reconstructed. For reasons already given, the most useful samples consist of fossil pollen; from these, assemblages of plants are set up and stratigraphic boundaries are drawn between them. The assemblages occur in belts or zones closely related to temperature and rainfall. The zones can be compared with the pattern of zones that exists today.

Looking first at the vegetation pattern in North America, as it was just before European man changed it radically (Fig. 20.7*A*), we see several belts that succeed each other with increasing distance from the North Polar region. These belts reflect, principally, increase of temperature from pole toward equator. Superposed on this is a change from moist forest on the east to dry grassland on the west, where the Rocky Mountains create a barrier over which westerly winds must rise, leaving east of it a region of little rain (Fig. 2.8). Zones comparable to those in North America occur in Eurasia also.

The glacial-age zones (Fig. 20.7*B*) form a different pattern. Ice covers the visible part of Canada and northern United States. Just south of the ice is a narrow strip of tundra with clumps of spruce; an extension of that strip projects southward along the high, cold tops of the Appalachians. South of the tundra is a wide belt of forest in which spruce at the north gives way southward to other trees, mostly broad-leaved kinds. To the west, where there are grassy plains today, is woodland consisting of widely spaced pines

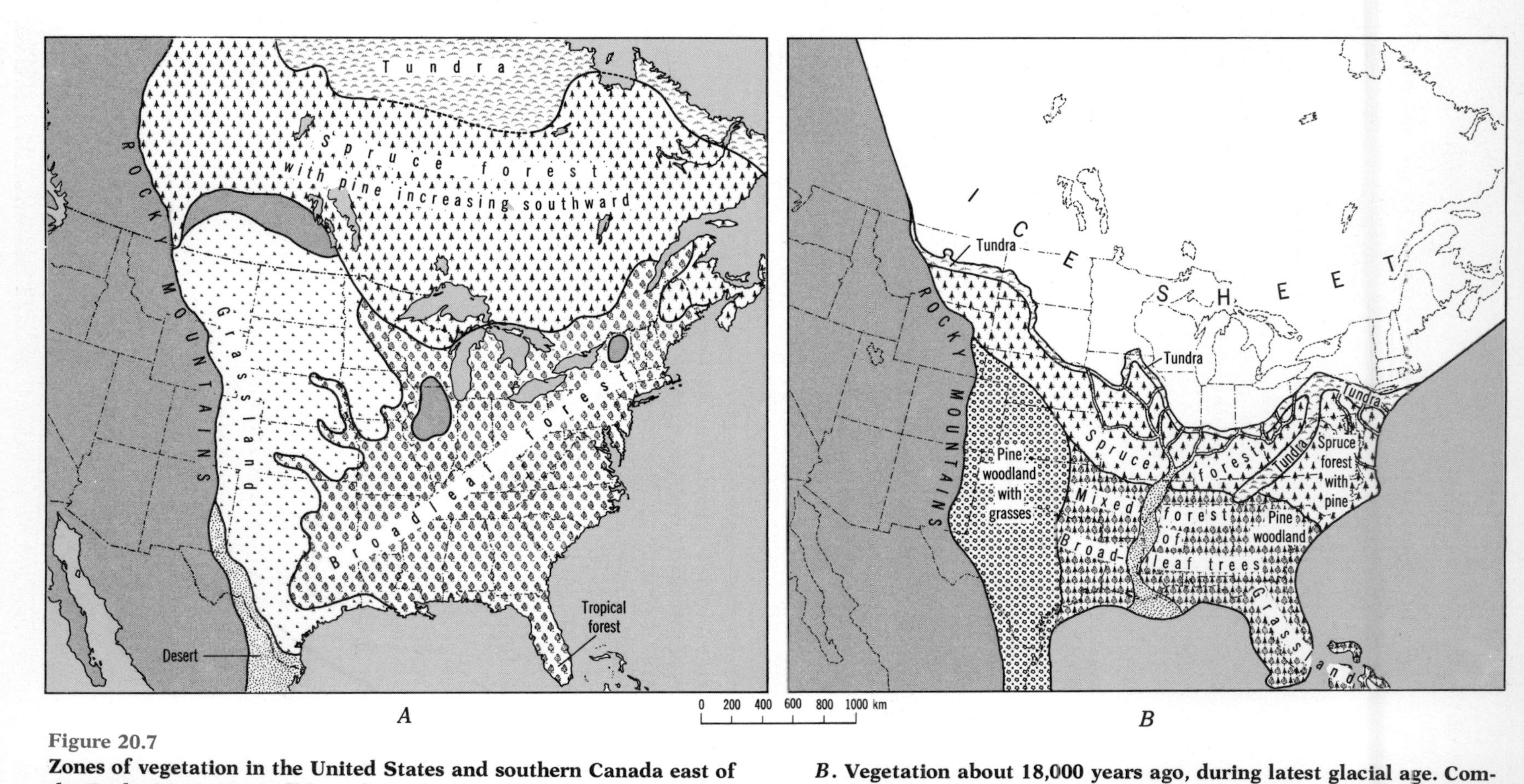


Figure 20.7

**Zones of vegetation in the United States and southern Canada east of the Rocky Mountains. All boundaries are gradational.**

***A*. Vegetation at time of arrival of human population from Europe.**

***B*. Vegetation about 18,000 years ago, during latest glacial age. Compiled mainly from pollen data. Form and position of coastline are different from those in *A* because sea level was then about 100m lower than it is now.**

with grass between them. Thus, in the glacial age, tundra and spruce forest grew farther south than they do now, consistent with a colder climate; and today's dry grassland was then woodland, suggesting that in that zone at least, the climate was moister. It was once supposed that as the glacier spread, the zones seen on today's map crept southward, each maintaining its own character. But the fossil pollen shows that the change was more complicated. The various kinds of plants did shift toward the south, but in different numbers and at different rates, forming new mixtures. This is why the zone labels for the two maps (Fig. 20.7) are not identical.

In Europe, likewise covered with ice in the north, a comparable change occurred, but with one big exception. In North America, migrating plants driven south by the ice sheet could inhabit relatively warm lowlands that extended down to the Gulf of Mexico. But in Europe the Alps, with their own ice sheet 800km long, constituted a high, cold barrier north of the Mediterranean Sea. Many plants were "frozen out," squeezed between the big ice sheet at the north and the Alpine ice at the south. Thus western Europe, which before the glacial ages had an abundant variety of trees, now has only 30 species, and in the north has only six species of broad-leaf trees. North America, with no Alps standing between the Great Lakes and the Gulf of Mexico, has 130 species.

The presence of mountain ranges breaks up the simple circum-polar pattern of vegetation zones, because each highland

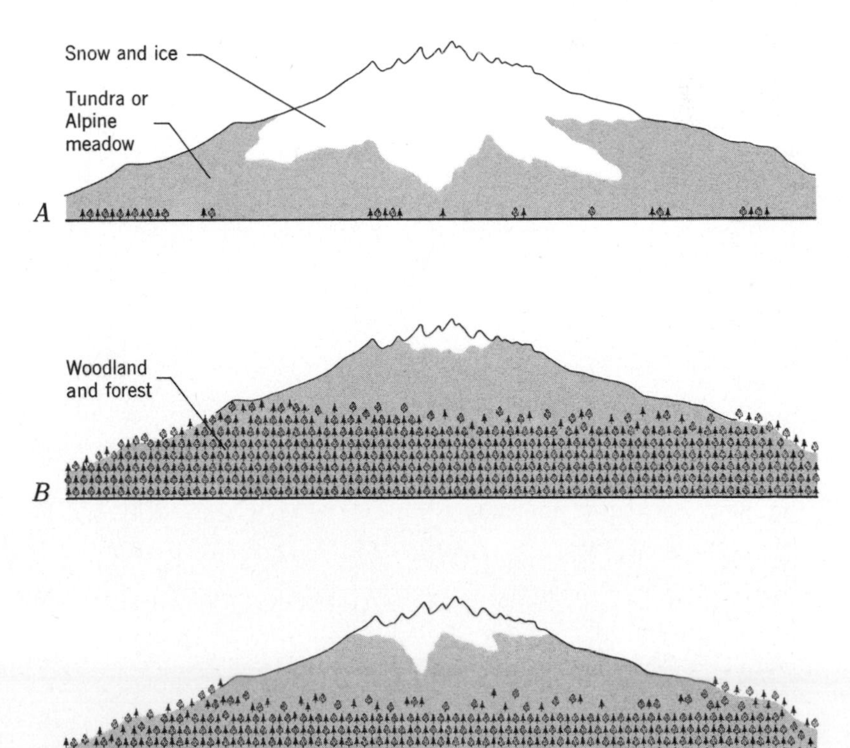


**Figure 20.8**
**Idealized sketch showing shift of vegetation zones, on a high mountain in western North America, through the last 20,000 years. (Not to scale.)**
***A*. 20,000 years ago.**
***B*. About 5,000 years ago.**
***C*. Now.**

has its own belts of vegetation arranged according to altitude and temperature (Fig. 2.8). This is why mountainous western North America cannot be shown on our small-scale maps; its pattern is far too intricate. Figure 20.8 gives an idea of what happened on individual highlands.

### Pluvial lakes and deserts

In the western Americas and in many other regions now, dry, glacial-age climates resulted in the creation or enlargement of lakes (Fig. 20.9).

The basin of Great Salt Lake, Utah, was occupied formerly by the gigantic Lake Bonneville, more than 300m deep and with a volume comparable to that of Lake Michigan. Beaches, deltas of tributary streams, and bottom sediments are all there to tell the story. Such lakes are ***pluvial lakes.*** None exists today because, by definition, they are *lakes that existed under a former climate, when rainfall in the surrounding region was greater than today's.* Although the name implies a pluvial (rainy) climate, we realize that reduced evaporation caused by reduced temperature was a large factor in the changed water economy, but the name *pluvial* still sticks to the ancient lakes. These lakes were, then, the result of (1) lower temperature, and (2) increased rainfall in the regions where they existed.

Direct geologic evidence and $^{14}$C dates show that some pluvial lakes were contemporaneous with the most recent of the glacial ages. This does not mean, though, that such lakes were fed by water derived from the melting of glaciers. Some of them were, but

Figure 20.9
**Existing and former lakes in the dry region of western United States. (*O. E. Meinzer and other sources.*)**

*A* Lakes existing today.

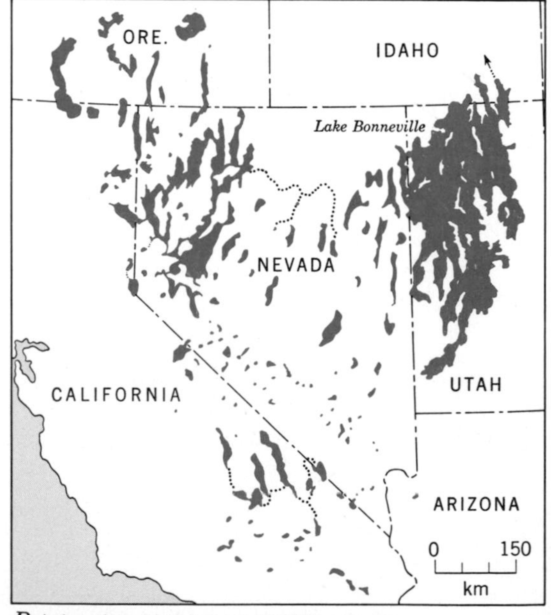


*B* Lakes that existed about 15,000 years ago.

only by coincidence. Most of the lakes occupied basins to which glacial waters had no access.

Pluvial lakes were abundant also in other dry regions of the world. The desert area in the Andes of Bolivia, Chile, and Argentina contained many of them. Numerous pluvial lakes were present also in the Saharan region of North Africa and in western and central Asia. The largest in the world was an ancestor of today's Caspian Sea; another that is historically famous was an ancestor of the present Dead Sea. Although some of these former lakes probably owed their existence to glacial-age climates, others, especially some in Africa, may have originated in other ways. One of these, ancient Lake Chad, occupied the western part of the Sahel region (Fig. 9.3), and around 5,000 years ago was almost as large as the state of California—a striking contrast with the drought described in Chapter 9.

North of the Sahel is the vast region of deserts and scattered mountains together known as the Sahara, comprising an area of more than 9,000,000km$^2$ and extending from the Atlantic to the Red Sea (Fig. 9.3). Today most of the region receives less than 50mm of rainfall yearly. In the latest glacial age the Sahara received more rain, although it was still semiarid. Streams cut channels and deposited alluvium where today no streams flow, and springs made deposits of calcium carbonate where today no ground water emerges. Trees and other plants spread from the Mediterranean coast as far as 1,500km south across what is now desert, and a variety of mammals lived in the region. Many places where water existed were inhabited by Stone-Age people.

The glacial age was followed by climatic drying, which increased gradually but with fluctuations. Mediterranean plants were replaced by desert vegetation. Most of the mammals disappeared, and the human inhabitants were reduced to small, sparsely distributed groups.

### Spreading of permafrost

A third effect of the ice-age climates was the spreading of ***permafrost**, ground that is frozen perennially.* Actually it is not the ground but rather the ground water that has frozen. It forms a firm cement in the pores of regolith and bedrock, and makes serious trouble where a building is built on it in such a way as to cause the permafrost beneath it to thaw.

Today permafrost extends through a wide belt of far-northern country that adds up to 20 percent of the land area of the northern hemisphere. Over that belt average air temperatures are at or below freezing during most of the year, and loss of ground heat to the atmosphere is the cause of freezing of the ground water. During the short summer melting season only a thin surface zone, usually no more than about 1m thick, thaws out. The thawed layer flows down slopes by solifluction, carrying with it the arctic vegetation that grows at its surface.

Drilling has shown that in places permafrost is more than 300m thick, representing a long period of freezing. During glacial ages the southern limit of northern permafrost crept southward to much lower latitudes, reaching into central Europe. Today, under climates that have been growing generally warmer, permafrost is slowly shrinking, as the big ice sheets and the pluvial lakes have done already. But if climates should once more grow colder, permafrost would spread again.

### Temperatures

At the height of the latest glacial age, average annual temperatures were lower than today's by around 5° to 8°C in mid-latitude coastal regions and 10° to 15°C in continental interiors. We cannot expect the lowering to have been uniform everywhere because temperature is influenced by many factors.

Such estimates are arrived at in several ways, of which the following three are the most important.

1. By identifying the microscopic plant or animal fossils in a stratum known to be of glacial age, and applying to them the range of temperature within which the same kinds of organisms (or their nearest living relatives) exist today. Pollen from lakes and bogs and foraminifers (minute, surface-water sea creatures) found in cores raised from the deep-sea floor are commonly dealt with in this way.

2. By a method based on precise measurement of the ratio between two isotopes of oxygen ($^{18}O$ and $^{16}O$) in the calcium-carbonate shells of formaminifers and other microscopic organisms from deep-sea cores. Because this ratio depends partly on the temperature of the surface seawater when the shells were formed, it is possible (with some error) to estimate the surface-seawater temperature (Fig. 20.10) and thereby to estimate the temperature of the air above the sea.

3. A crude method that nevertheless gives fairly consistent results is estimation of the altitude of the glacial-age snowline in a mountain district, from the altitudes of glaciated and nonglaciated summits, respectively, or from the altitudes of the floors of cirques now empty. A typical calculation:

| | |
|---|---|
| Today's snowline (by measurement), altitude | 2,000m |
| Glacial-age snowline (by estimate) | 1,000m |
| Difference | 1,000m |

The general formula for rate of decrease of temperature with increasing altitude is 0.6°C/100m. Applied to the difference above, the formula gives 6°C for the temperature difference between then and now.

**Interglacial ages.** The downward swings of temperature in the curve, Figure 20.10, extend to various depths, but the upward swings reach more nearly uniform heights, thought to average no more than 2° higher than today's temperatures. The upward swings are believed to represent *interglacial ages,* times when climates resembled those of today and when glaciers in mid-latitudes shrank greatly or disappeared. There is a good deal of evidence,

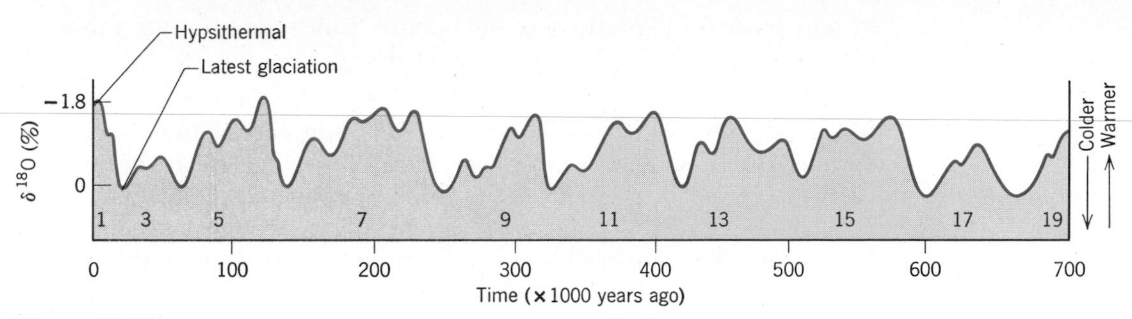


**Figure 20.10**
**Curve indicating fluctuation of temperature of surface water in the Caribbean Sea through the last 700,000 years. Although scaled in terms of the $^{18}O/^{16}O$ ratio mentioned in the text, the ordinate implies fluctuation of temperature through a few degrees, from peaks (identified by odd numbers; high temperatures) to troughs (not numbered; low temperatures). The latest glaciation and the Hypsithermal are clearly apparent. At least eight earlier cold times, perhaps glaciations, can be seen. Between 5 and 7 there is considerable uncertainty in the curve. (*Emiliani and Shackleton*.)**

from fossils and from relationships like those shown in Figure 10.25, that interglacial ages occurred repeatedly. Only those that occurred within the last 700,000 years are reflected in Figure 20.10.

# EARLIER GLACIAL AGES

If a reduction of Earth's average annual temperature by only a few degrees could bring on the latest glacial age, it seems likely that similar fluctuations of temperature have occurred repeatedly throughout the long time represented by the geologic column. And indeed there is evidence on the continents, and in sediment cores from the deep-sea floor, of many glacial ages during the last 15 to 20 million years. The most recent of these are suggested in the curve, Figure 20.10.

A common characteristic of many of the cores taken from tropical regions is an alternation of layers of sediment of two kinds. One is calcareous ooze, which contains the shells of minute organisms that today live only in warm surface water. The other is brown clay, which contains fossil organisms that today live in higher, cooler latitudes, and lacks the warm-water kinds. The alternation of "warm" and "cool" sediments has led to the inference that the layers are a record of nonglacial and glacial ages respectively. Again, cores taken right across the floor of the North Atlantic, from Canada to Britain, consist of alternating layers of (1) calcareous ooze, and (2) silt, sand, and small pebbles believed to have been dumped from hosts of melting icebergs. Similar alternations of ice-rafted sediment and nonglacial sediment have been found in many parts of the southern hemisphere, implying glacial and nonglacial ages. So many other cores from widely separated parts of the sea floor contain alternating layers of various kinds that fluctuating climate is thought to have been responsible for them.

Also it has been found possible to determine the dates of at least the higher, younger layers of deep-sea sediment by measurement of $^{14}C$ and other radioactive isotopes contained in them. In this way the beginning of a calendar has been established. The record has been extended backward in time by use of the magnetic-field reversals described in Chapter 13. Magnetic minerals align themselves parallel to Earth's magnetic field as they settle through the water. Because, in the oceans, sediment is deposited continuously, careful measurements in a core reveal the exact layer at which each magnetic reversal occurs. Since we know the times of reversal (Fig. 13.16), we can construct a calendar for the layers in the core.

Measurements of oxygen isotopes in deep-sea cores from the southern hemisphere are now beginning to reveal the climatic history of Antarctica. They show that the Southern Ocean was growing irregularly colder, with fluctuation, through Eocene and Oligocene time, some 50 to 25 million years ago. During the process many glaciers formed on the higher parts of the Antarctic Continent. Then, in Miocene time, some 12 to 10 million years ago, temperatures fell still lower and glaciers increased, forming a large ice sheet like that of today. The presence of so large a body of ice reduced average temperatures at Earth's surface. From that time onward, large glaciers formed in Alaska and other regions in high northern latitudes. Although the data are still very sketchy, it appears that large ice sheets did not form in middle latitudes until between two and one million years ago.

If this sequence of events is correct even only approximately, glacial ages have affected Earth's lands gradually, the Antarctic region first, then high northern latitudes, and most recently middle latitudes—but always interrupted by fluctuations.

Glacial ages, recorded by tillite in the strata, are known in the older parts of the geologic column. Tillites are not abundant, partly perhaps because fewer of the old strata are exposed to view, and also because a large portion of the older rocks consist of metamorphic and igneous kinds, in which any sediment earlier deposited by former glaciers would have been destroyed. Nevertheless glacial ages are represented at various positions in the Precambrian, the earliest known one dated at 2.2 billion years ago. Others have been identified at several positions in the Paleozoic, especially in the Pennsylvanian and Permian Periods, where glaciers were extraordinarily widespread. Evidence like that in Figures 12.9 and 20.11 implies that glaciation has occurred again and again even in the far-distant past. In the Mesozoic, evidence of glacial ages is scarce or absent. These points in the geologic column can be visualized on Table 5.1. Possibly there are many more, as yet undiscovered.

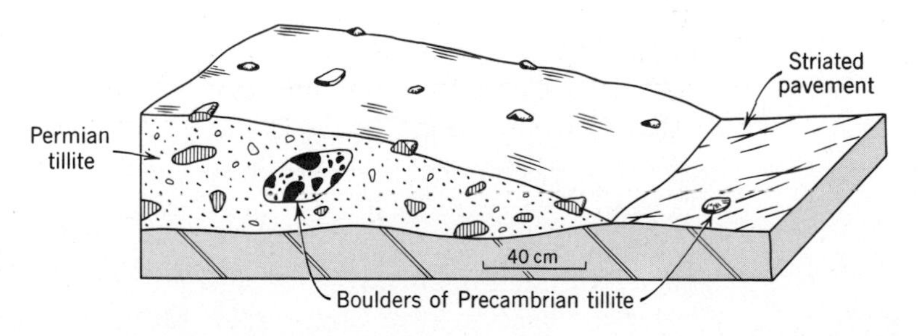


Figure 20.11
**Sketch of exposure near Adelaide, South Australia, showing a body of Permian tillite about 250 million years old, containing a boulder broken from Precambrian tillite more than 600 million years old. This unusual occurrence demonstrates that Australia has experienced at least two very old glacial ages.**

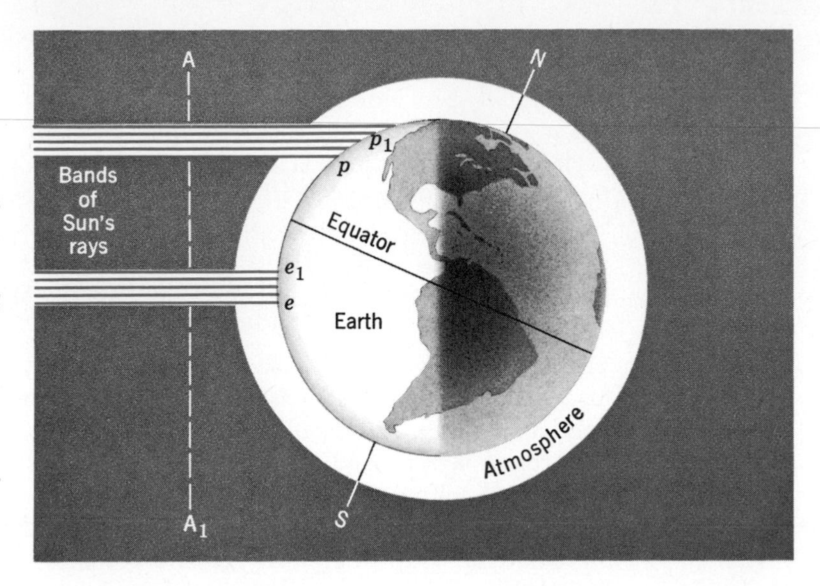


**Figure 20.12 Solar radiation reaching Earth's surface varies with latitude. Radiation crossing plane $AA_1$ amounts to $2.88 \times 10^3$ calories/cm$^2$ every 24 hours. A square centimeter of rays crossing $AA_1$ strikes a square centimeter of Earth's surface when the rays are perpendicular to the surface ($ee_1$), but because of Earth's curvature are spread over a larger area in high latitudes ($pp_1$).**

# WHAT MAKES CLIMATES CHANGE?

What is the cause of the climatic chilling and heating that recur apparently at irregular intervals, and work great changes in the activities of Earth's external processes? The search for the answer to this question is a huge undertaking. Tens of thousands of pages of reasoning and discussion have been published, but we still have reached only a very small part of the answer.

### Changes in the positions and shapes of continents

The part on which we have a fairly good, although still embryonic, theory concerns physical/geographical changes that affect the continental crust. The changes include: (1) lateral movement of continents as they ride along on their plates of lithosphere, (2) uplift of continental crust where one plate overrides another, (3) creation of a high mountain chain where two plates collide, squeezing and mashing vulnerable shelf strata in the zone of collision, and (4) opening or closing of seaways between moving lands, thereby affecting the directions followed by ocean currents such as the Gulf Stream and the west-wind drift.

The effect of such movements on climate is illustrated by the fact that low temperatures are found, and glaciers tend to form and persist, in two kinds of situation: high latitudes (Fig. 20.12) and high altitudes (Fig. 2.8), especially where the existing winds can bring abundant moisture evaporated from a nearby ocean. Earth's largest existing glacier is centered on one of the poles, where receipt of solar heat is minimal, and is centered likewise on the continent that has the greatest average altitude. The only existing glaciers that lie on or close to the equator are at extremely high altitudes.

From Figure 17.1 we learned that toward the close of the Paleozoic Era an enormous complex of ice sheets (Fig. 17.1), and other glaciers, evidence of which persists widely today, was centered in the southern part of the former supercontinent of Pangaea, in very high latitude. Later, in Triassic time, Pangaea broke up and its pieces drifted outward, with Antarctica later centering on the South Pole while the other pieces drifted into lower, warmer latitudes. Large fragments of the former glaciated terrain ended up in four and possibly five different continents, spread today through at least 70 degrees of latitude.

This history gives us a reasonable explanation of the creation and later melting of the glacier ice, which was "passed under the broiler"—carried into regions too warm for it. In contrast, Antarctica was passed into the refrigerator—shifted into a polar position in which, in Cenozoic time, it became the site of a huge ice sheet. Again, in Cenozoic time, when the Alps and the Himalaya were squeezed up to great height by collisions of the Africa and India Plates with Eurasia, glaciers formed on the new mountains where, before the collisions, altitudes would have been too low for snow and ice. And in some such situations new deserts are created on the leeward sides of new highlands (Chapter 9).

### Other changes

These examples suggest that the dynamics of lithospheric plates give us an explanation of *some* climatic changes. But such changes are not worldwide; they are confined to certain parts of certain continents and they are slow, developing through millions of years as plates slide sideways and as highlands are created and then later destroyed by erosion.

In a different category are changes of climate that affect two (or all) continents at the same time and that happen quickly, on time scales of tens of thousands, thousands, and even mere hundreds of years. These demand other explanations. Of the many ideas suggested, two that are widely discussed are outlined here.

**Astronomic Theory.** One of them, sometimes called the Astronomic Theory, is based on the fact that as Planet Earth travels around the Sun, its orbit undergoes regular changes and its axis wobbles in a regular fashion. These movements have periods measured in tens of thousands of years, but of different lengths. Their combined effect on Earth's climates is that the share of solar heat received at any given point on Earth's surface varies irregularly by a small amount, although the total heat remains unchanged. The local variation can be expressed by an irregular curve. Although the variations of temperature are real, it is uncertain whether they are large enough to account, alone, for the changes read from the geologic record.

**Theory of a fluctuating Sun.** An obvious theory of climatic change is based on the idea that the Sun's output of heat fluctuates through time. The idea is appealing because it might be able to explain climatic changes of any length or intensity. However, it rests on no basis of fact, because measurements of solar heat output, made throughout the last half century or more, have not attained the extreme precision necessary to establish that fluctuation actually occurs. When and if fluctuation is established, the theory will be in business.

**Theory of volcanic ash.** A theory proposed long ago and still argued is that fine-grained volcanic ash, thrown high into the air by many explosive volcanoes, would baffle out part of the incoming solar radiation and thus lower air temperatures, perhaps enough to start a glacial age. Estimates of the volume of ash ejected in recent major eruptions and compilations of temperatures at many weather stations through periods of years have thrown light on the theory. A major explosive eruption can indeed create a veil of dust that spreads worldwide, and can reduce air temperatures by as much as 1°C. But the dust falls out within a few years at most and the temperature effect then disappears. In consequence, volcanic ash seems inadequate to explain a glacial age that lasts through thousands of years.

# CLIMATE IN THE FUTURE

Reliable prediction of any future climatic changes must be based on extrapolation—that is, estimation of the probable future continuation of a curve that shows past events of a similar kind. If, in 1940, we had extrapolated the curve seen in Figure 20.1, we would probably have extended the 1880–1940 curve upward. Later events would have proved us wrong, but therein lies the danger of extrapolation. However, if a long curve goes up and down in a regular manner, if its peaks or troughs are spaced in a repeating pattern, then extrapolation may be justified. Some scientists believe the curve in Figure 20.10 meets this requirement; others are skeptical. If one tried to extend that curve toward the left from Year 0 to say, the 50th year in the future, he would probably not feel confident of the result.

The curve in Figure 20.13 is derived from annual variations in the ratio of $^{18}O$ to $^{16}O$ in the composition of snow that has fallen at a place on the Greenland Ice Sheet through the last 700 years. The

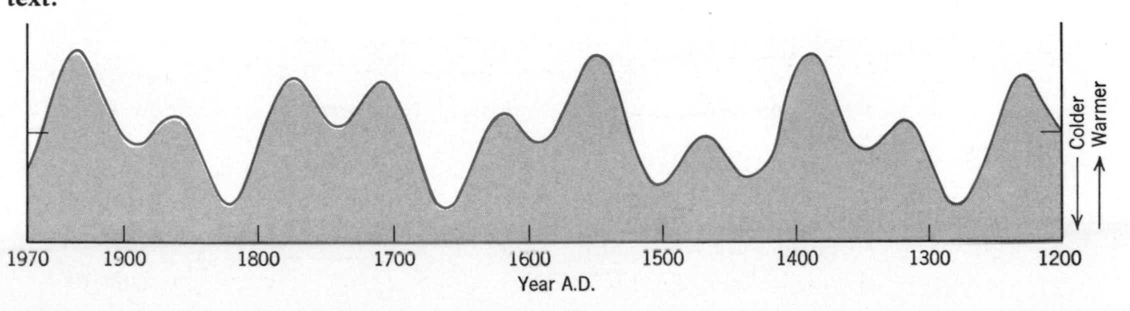


Figure 20.13
**Curve developed in 1970 from measured $^{18}O$ values in the Greenland Ice Sheet as explained in the text.**

ratio is closely related to air temperatures at the times of snowfall. The annual layers of fallen snow are recognizable in the upper part of a long core taken from the ice sheet. The curve possesses two periodicities, with wave-lengths of 80 years and 120 years respectively. This makes possible a try at extrapolation, which implies crude prediction.

The extrapolation predicts that temperature (in Greenland at least) will continue to get colder until near the end of the present century, when it will begin to rise again, reaching a warm peak around the year 2015. However, if account is taken of the quantities of $CO_2$ that are annually injected into Earth's atmosphere through man's industrial activities, the prediction may be greatly modified. In discussing man as a geologic agent (Chapter 23) we will refer to this again.

# SUMMARY

1. Former climates are determined mainly from relict fossils, relict physical features, and measurement of oxygen isotopes.

2. Changes in climates within the last 100 years are established by official instrumental records. Since the 1940s, temperatures have fallen.

3. Climatic changes during the last 1,000 years are established (less accurately) by historic records. In northern middle latitudes that period was characterized by a warm "Viking" climate followed by a cold "little ice-age."

4. Changes during the last 10,000 years are established mainly by the stratigraphy of pollen and by measured fluctuations. Glacial-age temperatures rose to a peak about 5,000 years ago, then fell during the Neoglaciations.

5. In the latest glacial age, temperatures were 5° to 15° lower than today's. They are determined from fossils on land and in deep-sea cores, and from evidence of lowered snowline.

6. The latest glacial age lasted 10,000 years or more. Glacial ages alternated with interglacial ages in which temperatures approximated those of today. Apparently the changes affected the whole Earth.

7. Glacial ages are discerned in many parts of the geologic column; they extend back more than two billion years.

8. Possible causes of glacial ages include (a) drifting of continents, and other physical and geographic changes, (b) variations in the geometry of Earth's orbit and axis, and (c) fluctuations of solar heat.

9. Prediction of future climatic change still lacks an adequate basis.

# SELECTED REFERENCES

Brooks, C. E. P., 1949, Climate through the ages: 2d. ed., London, Ernest Benn.

Calder, Nigel, 1974, The weather machine: New York, Viking Press.

Crowell, J. C., and Frakes, L. A., 1970, Phanerozoic glaciation and the causes of ice ages: American Jour. Sci., v. 268, p. 193–224.

Denton, G. H., and Porter, S. C., 1970, Neoglaciation: Sci. American, v. 222, p. 101–110.

Flint, R. F., 1971, Glacial and Quaternary geology: New York, Wiley.

Lamb, H. H., 1972, Climate: Present, past, and future: v. 1., Fundamentals and climate now: London, Methuen.

Lamb, H. H., and others, 1966, p. 174–217 in World climate 8000–0 B.C. Sawyer, J. S., ed., 1966, World climate from 8000 to 0 B.C.: Proc. Internat. Symposium on World Climate, Royal Meteorol. Soc., London.

Schwarzbach, Martin, 1963, Climates of the past: London, Van Nostrand.

# 21 SOURCES OF ENERGY



## MATERIALS AND ENERGY

When we speak of the present epoch in western civilization as "the industrial age," we mean the age of enormously increased use of fuels and metals. Another term would be "the age of intensive use of energy in industry." Stone-Age people had an industry too: they chipped and flaked mineral substances (mostly pieces of quartz) to make tools and weapons. But theirs was an industry with a low input of energy because the energy was supplied by human muscle. Although they were limited by this fact and by a very narrow choice of materials to work with, we must not underestimate them. They were as intelligent as we are, lacking only our accumulated experience and the skill that comes from experience.

This lack was gradually overcome by Stone-Age people and by their descendants. Here are a few of their accomplishments, discovered and dated by archeologists:

| | |
|---|---|
| 4000 B.C. | Chaldeans had become skilled workers in metals such as gold, silver, copper, lead, tin, and iron. |
| 3000 B.C. | Eastern Mediterranean peoples were making glass, glazed pottery, and porcelain. |
| 2500 B.C. | Babylonians were using petroleum instead of wood for fuel. |
| 1100 B.C. | Chinese were mining coal and were drilling wells hundreds of feet deep for natural gas. |

These accomplishments were arts learned by experience, and they implied the substitution of metals, glass, and other substances for stone. But the making of a copper implement used more energy than the manufacture of a stone tool. Still more was needed to make objects of glass. So with each advance, more energy and therefore more fuel were needed. Most of the energy for working iron, copper, lead, and other materials came from wood fuel and from the muscles of men and animals. But as time passed, other fuels came to be used: coal, oil, and natural gas. The new fuels were more efficient than the old. By using them, each man could produce many more implements than he could have produced with wood fuel or with muscle. The new fuels therefore sparked rapid growth of the amounts of metals used.

**Use of energy**

Metals to build machines and the energy needed to power machines are complementary. Energy is also complementary to today's intensive agriculture. In order to increase food yields and to feed the world's growing billions of people, more energy must be used (Fig. 21.1). Indeed, muscle power is now aided by supplementary energy in almost every part of our lives. It is humbling to stop and consider the inadequacies of muscle energy by comparison with present-day needs. A healthy, hard-working person can produce just enough muscle energy to keep a single 100-watt light bulb burning during an eight-hour working day. When we purchase the same amount of energy from our local electrical source, we pay less than five cents. Viewed strictly as machines, we aren't worth much. By comparison, the amount of supplementary energy used each eight-hour working day in North America could keep 300 of those bulbs burning for every person living there. Think of all this supplementary energy, if you will, as 300 silent "energy slaves" working eight hours a day to keep us well-fed and enjoying a high standard of living. It is the "energy slaves" that produce our high crop yields, keep industry productive, power our automobiles, and heat our homes. Every person on Earth now relies on at least some supplementary energy, which means everyone has some "energy slaves" working for them. In India the use of supplementary energy is now equivalent to 15 "energy slaves" for every man, woman, and child. In South America the figure is 30, in Japan 75, in Russia 120, and in Europe 150.

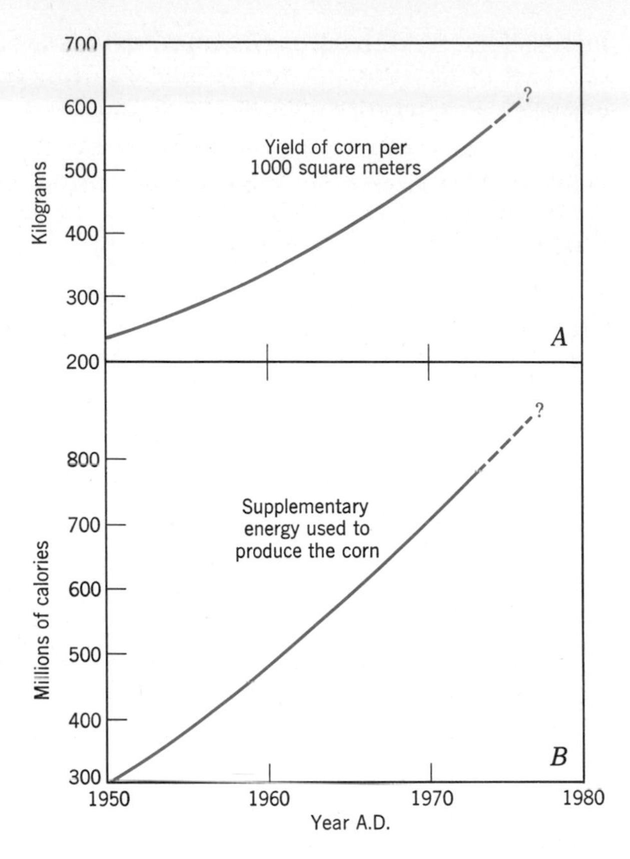


**Figure 21.1 Relation between increased crop yield and supplementary energy.**
**A. Between 1950 and 1970 the average yield of corn per 1,000 square meters in the United States doubled. The increase resulted mainly from the increased use of fertilizers and pesticides, and more intensive cultivation.**
**B. Supplementary energy required to produce the corn more than doubled. The graph was prepared by adding up the energy needed to manufacture the farm machinery, fertilizers, pesticides, and other farming aids, as well as the energy needed to drive the tractors and other farm machines. (*After Pimental and others, 1973, Science, v. 182, p. 443–449.*)**

To see where all the energy is used we have to look at society as a whole and sum up all the energy that is employed to transport food, to make clothes, to cut lumber for new homes, to light streets, to heat and cool office buildings, and to do myriad other things. All the uses can be grouped into three categories: transportation, industry (meaning all manufacturing and raw material processing), and home plus commercial uses. The present-day uses of energy in the United States are summarized in Figure 21.2.

### Supplies of energy

The chief sources of energy consumed in highly industrialized nations are few: the fossil fuels (coal, oil, natural gas), hydroelectric power, nuclear energy, wood, wind, and a very small amount of muscle energy. As recently as a century ago wood was an important fuel, but now it is used mainly for space heating in some dwellings. Wood and nuclear energy together supply only one percent of the energy consumed in the United States today. Excluding muscle (a resource now getting very little exercise in industrial countries), the sources of energy consumed in the United States are shown in Figure 21.2.

For Europe the breakdown of energy supply and use is different: Coal accounts for about half the energy consumed, and industry for more than 40 percent of the use. For the world as a whole, with its hundreds of millions of agricultural workers, the breakdown is different again; wood is a more substantial source of energy, and fossil fuels, especially oil and gas, bulk less.

Because our ability to produce and use Earth's mineral supplies depends on our having energy available to do the necessary work, we first discuss the principal sources of energy, and then, in Chapter 22, the sources of mineral supplies.

# COAL

The black, combustible sedimentary rock we call coal is the most abundant of the *fossil fuels.* These fuels are so called because they contain solar energy, locked up securely in chemical compounds by the plants or animals of former ages. Most of the coal that is mined is eventually burned under boilers to make steam for electrical generators, or is converted into coke, an essential ingredient in making steel. Varying amounts of liquid fuel can be derived from coal, and because internal-combustion engines are generally more efficient than steam engines, the future will probably see more and more coal used in this way. In addition to its uses as energy, coal is the chief raw material from which nylon and most of the other plastics, plus a multitude of chemicals are made.

### Origin of coal

Coal occurs in strata (miners call them *seams*) along with other sedimentary rocks, mostly shale and sandstone. A look through a magnifying glass at a piece of coal reveals bits of fossil wood, bark,

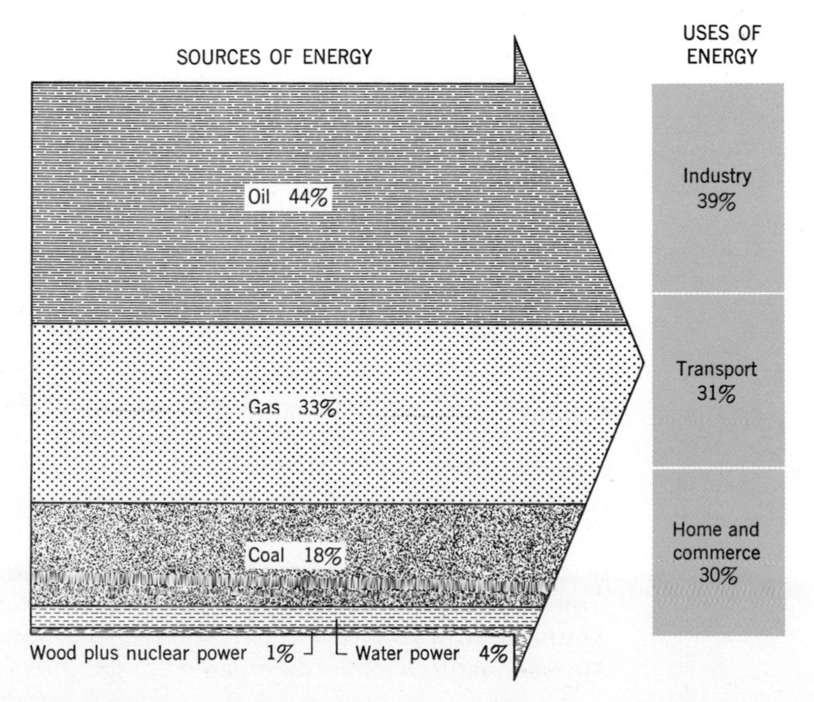


**Figure 21.2**
**Uses and sources of supplementary energy in the United States. Percentages remained about the same through the first five years of the 1970s.**

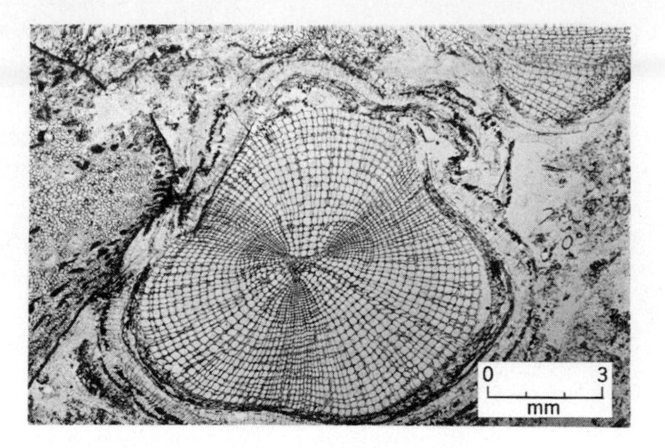


**Figure 21.3**
**Coal of Pennsylvanian age from Illinois, showing cellular structure in fossilized fragment of wood.** **(*General Biological Supply House; Julius Weber.*)**

leaves, roots, and other parts of land plants, chemically altered but still identifiable. This observation leads at once to the conclusion that coal is fossil plant matter (Fig. 21.3), a conclusion that we incorporate in our definition of ***coal*** as *a black sedimentary rock consisting chiefly of decomposed plant matter and containing less than 40 percent inorganic matter.*

Many coal seams include fossil tree stumps rooted in place in the underlying shale, evidently a former clay-rich soil. Unlike the material in most sedimentary rock, the sediment we now find in coal seams was not eroded, transported, and deposited; instead it accumulated right where the plants grew. It was recognized as long ago as 1778 that the places where coal accumulated were ancient swamps, because (1) a complete physical and chemical gradation exists from coal to peat, which today accumulates only in swamps, and (2) only under swamp conditions is the conversion of plant matter to coal chemically probable. On dry land, dead plant matter (composed chiefly of carbon, hydrogen, and oxygen) combines with atmospheric oxygen to form carbon dioxide and water; it rots away. Under stagnant or nearly stagnant swamp water, however, oxygen is excluded from dead plant matter and oxidation is prevented. Instead, the plant matter is attacked by anaerobic bacteria, which partly decompose it by splitting off oxygen and hydrogen. These two elements escape, combined in various gases, and the carbon gradually becomes concentrated in the residue. Although they work to destroy the vegetal matter, the bacteria themselves are destroyed before they can finish the job, because poisonous acid compounds they liberate from the dead plants kill them. This could not happen in a stream because the flowing water would bring in new oxygen to decompose the plants and would also dilute the poisons and permit the bacteria to complete their destructive process.

With the destruction of bacteria, the plant matter has been converted to peat. But as the peat is buried beneath more plant matter and beneath accumulating sand, silt, or clay, both temperature and pressure increase. These bring about a series of continuing changes (Fig. 21.4). The peat is compressed, water is squeezed out, and volatile organic compounds such as methane

($CH_4$) escape, leaving an ever-increasing proportion of carbon. The peat is converted successively into lignite, subbituminous coal, and bituminous coal. These coals are sedimentary rocks, but a still-later phase, anthracite, is a metamorphic rock. Since anthracite generally occurs in folded strata, we infer that it has undergone a further loss of volatiles and a concentration of carbon, as a result of the pressure and heat that accompany folding. Because of its low content of volatiles, anthracite is hard to ignite, but once alight, it burns with almost no smoke. In contrast, lignite, rich in volatiles, ignites so easily that it is dangerously subject to spontaneous ignition (in chemical terms, rapid oxidation), and burns smokily. In certain regions where folding has been intense, coal has been metamorphosed so thoroughly that it has been converted to graphite, in which all volatiles have been lost, leaving nothing but carbon. Graphite, therefore, will not burn.

### Occurrence

We have said that coal occurs in layers or seams, which are merely strata. Each seam is a flat, lens-shaped body corresponding in area to the area of the swamp in which it accumulated originally. Most coal seams are 0.5m to 3m thick, although some reach more than 30m. They tend to occur in groups: in western Pennsylvania, for example, there are about 60 beds of bituminous coal. This indicates that coal formed in slowly subsiding sedimentary basins. The occurrence of coal beds in the rock layers of every period later than Devonian indicates that during the last 300 million years or so, swamps rich in vegetal growth have been recurrent features of the land. Peat is accumulating today, at an average rate of about 1m every 100 years, in swamps on the Atlantic and Gulf Coastal Plains of the United States. The swamps now represented by coal beds must have been much the same as these.

Coal swamps seem to have formed in many environments, of which

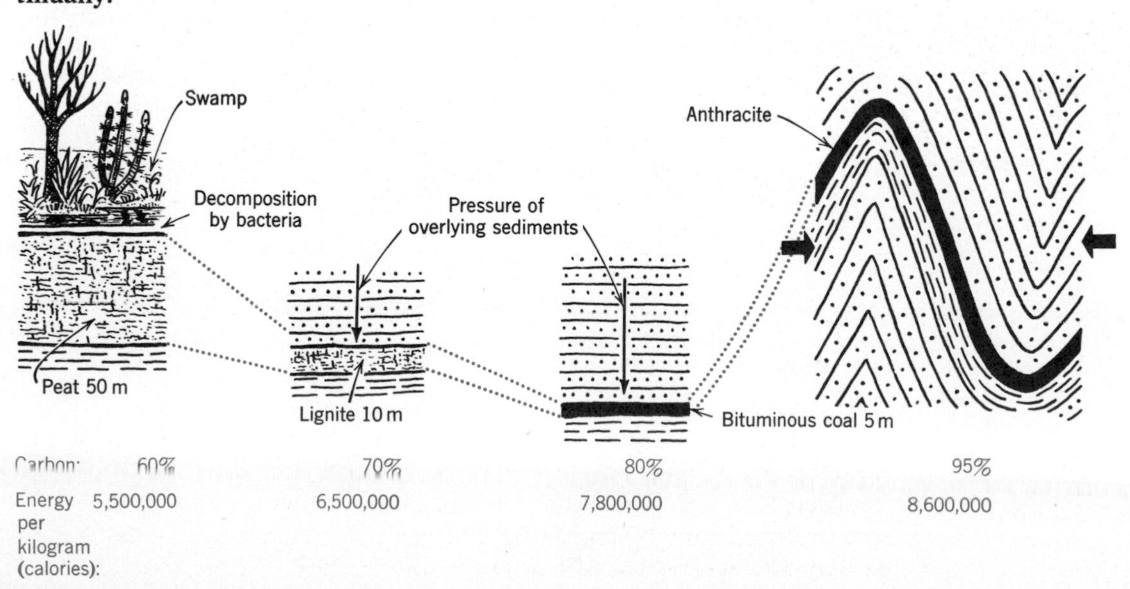


**Figure 21.4** **Accumulating plant matter is converted into coal by decomposition, pressure and heat. By the time it has become bituminous coal, a layer of peat has decreased to one-tenth of its original thickness. During the same time the proportion of carbon has increased and the calorific value (the amount of heat energy obtained when a kilogram of coal is burned) has risen continually.**

two predominated. One consists of slowly subsiding basins in continental interiors and the swampy margins of shallow inland seas. This is the home environment of the lignite seams in Montana and the Dakotas. The other consists of continental margins with wide continental shelves. This is the environment of the subbituminous coals of the Appalachian region. That same environment exists along the east coast of North America today, and it is there that one of the largest modern coal swamps is to be found. Dismal Swamp, in Virginia and North Carolina, with an area of 5,700km² contains an average thickness of two meters of peat.

Although peat can form under even subarctic conditions, it is clear that the luxuriant plant growth needed to form thick and extensive coal seams can occur most readily under a tropical or semitropical climate. Even the dense growth in Dismal Swamp is probably insufficient to produce, ultimately, a coal seam as thick as some of the seams in Pennsylvania. This means that probably the swamps in which most of the world's coal seams formed were within 30 degrees of the equator when the plant matter accumulated. The fact that we now find coal seams in high latitudes—even in Antarctica—can only be explained by continental drifting. The coal formed in equatorial regions but has been rafted by moving lithosphere to its present positions.

### Distribution

Coal is not only abundant; it is also rather widely distributed. Experts believe that most of the world's coal seams have been discovered; in that case a good estimate of the amount of mineable coal can be made. Coal seams thinner than 0.3m are too thin to be mined, and seams deeper than 2,000m make mining dangerous as well as expensive. Taking 0.3m as the lower limit of thickness and 2,000m as the depth limit for mining, experts estimate that 16,620 billion tons of coal are available and that of this, at least 50 percent can eventually be mined. Most of this reserve is in Asia

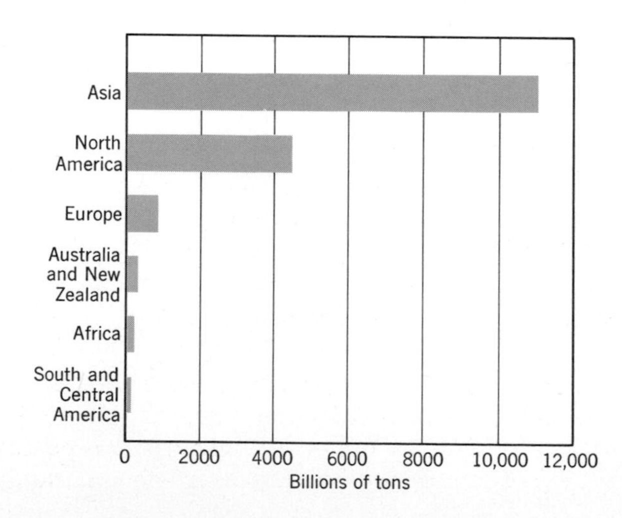


**Figure 21.5**
**Geographic distribution of the world's coal resources. With present-day mining methods, it is unlikely that more than 50 percent of this coal could be recovered. Future technological advances may raise the recovery percentage. (*Data from U.S. Geol. Survey, 1974.*)**

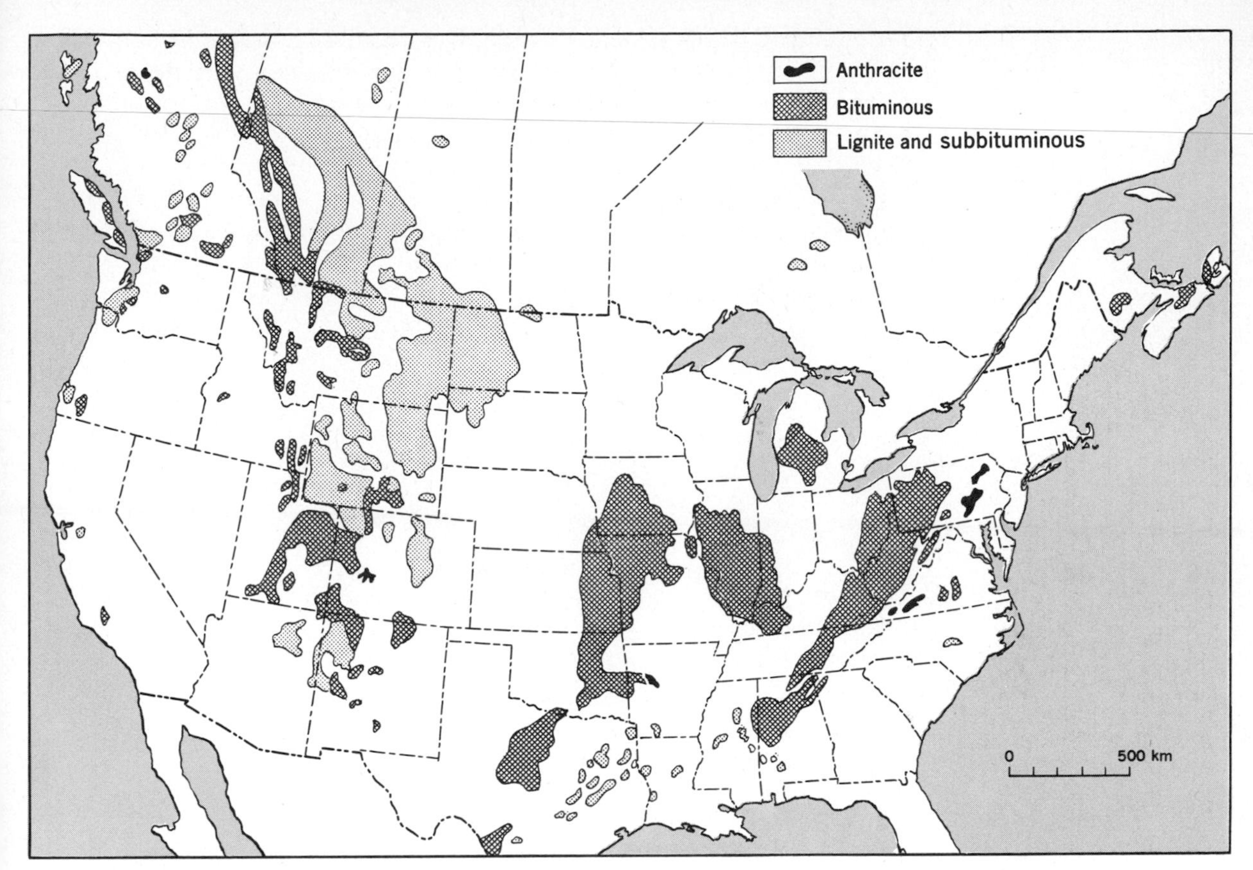


Figure 21.6
**Coal fields of United States and southern Canada.** (*Adapted from maps by U.S. Geol. Survey and Geol. Survey of Canada.*)

(Fig. 21.5) (principally in Russia) but North America is well-endowed too (Fig. 21.6).

Most of the world's coal resources are in the northern hemisphere. Land plants first evolved during the Silurian period, so no coal is older than Silurian. Evidently the southern continents have not spent much time close to the equator since the Silurian.

### Mining

The average thickness of all seams now mined in the United States is about 1.6m. In the past most coal has been obtained by mining underground, and historically, miners have recovered no more than about 50 percent of the coal present. Underground mining has always been a dangerous and unpleasant occupation. Increasingly, underground mining is being modernized by automatic coal-cutting and coal-loading machines, which have increased miners' safety, increased production of coal per man-hour by factors of 10 to 20 times, and increased the percentage of coal recovered from some seams to about 60 percent.

Much of the coal mined today is recovered by surface mining. When coal lies less than 60m beneath the surface, it can be mined by stripping away the overlying surface rock, and then digging out the exposed coal with huge power shovels that take 10 to 100 cubic

**Figure 21.7**
**This coal strip-mine at Wyodak, Wyoming, produces, each year, 300,000 tons of subbituminous coal from a seam 20m to 30m thick. (*U.S. Bur. Mines.*)**

meters at a single bite (Fig. 21.7). Referred to as *strip mining,* the method greatly reduces the danger to miners and permits recoveries of 90 percent or more of the coal in the ground. One drawback is, of course, that the surface is disturbed. Because much of the shallow, strippable coal occurs in rich farming areas, the farming is disturbed. The land can be restored after mining is completed, but reclamation is slow and sometimes expensive. The big complaint is that it is done in a shoddy manner! It is estimated that less than 10 percent of all coal in the United States is close enough to the surface to be reached by strip mining.

Another method of mining coal from the surface, without even going underground, employs huge augers. This method is used mainly in hilly areas of the Appalachian states. First a bench is cut with bulldozers to expose the coal seam; then two or more augers, about two meters in diameter, are installed. They bore horizontally into the seam and move the coal outward between their spiral blades to loading belts (Fig. 21.8). Although the method is cheap and efficient, only a very small percentage of all coal can be recovered by augers. Unfortunately, however, the cutting of benches in hilly country creates serious environmental problems (Fig. 23.3).

Experiments are now being made with burning coal while it is still in the ground, to produce gas for industrial use. This technique would eliminate mining altogether, and if successfully developed, would greatly reduce environmental damage.

# PETROLEUM: OIL AND NATURAL GAS

Oil was first produced in 1857, in Romania. Soon afterward, in 1859, the first oil well in North America was drilled at Titusville, Pennsylvania. That event was epoch making. The well, 23m deep, produced 25 barrels* per day of a substance that spelled the doom of candles and whale-oil lamps. More than a hundred years later, in 1976, the United States was producing nearly 10 million barrels of oil per day from about half a million wells, the deepest of which extended nearly 9km down into the crust. Of this huge production, 90 percent is used as fuel; the remainder goes into lubricants, without which the fuel would turn very few wheels, and into thousands of products manufactured by the petrochemical industry.

Oil and gas are the two chief kinds of petroleum. We define ***petroleum*** as *gaseous, liquid, and solid substances, occurring naturally and consisting chiefly of chemical compounds of carbon and hydrogen.* Oil and gas occur together and are searched for in the same way. We can therefore follow general practice and talk about oil pools, oil exploration, and the origin of oil with the understanding that we mean not only oil but gas as well. In discussing oil, we shall deal with its occurrence, its origin, the means used to find new occurrences, and distribution, in that order.

### Occurrence: oil pools

The accumulated experience of more than a century of exploration, drilling, and producing has taught us much about where and how

Figure 21.8
**A huge multiple auger in eastern Kentucky bores into a coal seam from an artificially excavated bench. Coal is cut, moved out, and loaded in a single operation. (*Courier-Journal and Louisville Times.*)**

* A barrel contains 42 gallons, and is the unit of volume generally used when oil is discussed.

oil occurs. Oil possesses two important properties that affect its occurrence. It is fluid, and it is generally lighter than water. Oil is produced from pools (an ***oil pool*** is *an underground accumulation of oil or gas in a reservoir limited by geologic barriers*). The word pool sometimes gives a wrong impression because an oil pool is not a lake of oil. It is a body of rock in which oil occupies the pore spaces. *A group of pools, usually of similar type, or a single pool in an isolated position* constitutes an ***oil field.*** The pools in a field can be side by side or one on top of the other.

For oil or gas to accumulate in a pool, five essential requirements must be met. (1) There must be a *reservoir rock* to hold the oil, and this rock must be permeable so that the oil can percolate through it. (2) The reservoir rock must be overlain by a layer of impermeable *roof rock,* such as shale, to prevent upward escape of the oil, which is floating on ground water. (3) The reservoir rock and roof rock must form a *trap* that holds the oil and prevents it from moving any farther under the pressure of the water beneath it (Fig. 21.9). These requirements are much like those of an artesian-water system but with the essential difference that the artesian aquifer connects with the surface, whereas the oil pool does not. Although these three features—reservoir, roof, and trap—are essential, they do not guarantee a pool. In many places where they occur together, drilling has shown that no pool exists, generally because of lack of a source from which oil could enter the trap. So, to the foregoing requirements for a pool we must add two others: (4) there must be *source rock* to provide oil and, (5) the *deformation* that forms the trap must occur *before all the oil has escaped* from the reservoir rock.

### Origin

Oil is believed to be a product of the decomposition of organic matter, both plant and animal. This belief arises principally because of two observations: (1) oil possesses optical properties known only in hydrocarbon compounds derived from organic matter, and (2) oil contains nitrogen and certain compounds (porphyrins) that scientists believe can only originate in living matter.

Oil is nearly always found in marine sedimentary strata. Indeed, in places on the sea floor, particularly on the continental shelves and at the bases of the continental slopes, sampling has shown that fine-grained sediment now accumulating contains up to eight percent organic matter—chemically good potential oil substance. Thus, on the basis of the Principle of Uniformity, we can make the general suggestion that the oil we now find in rock originated as organic matter deposited with marine sediment.

This suggestion is only a general one. Analyzing the organic matter in modern sediments, chemists have found significant differences between its composition and that of petroleum. Therefore, in order to effect the change from organic matter to petroleum, a complex and little-understood series of changes must occur. The following

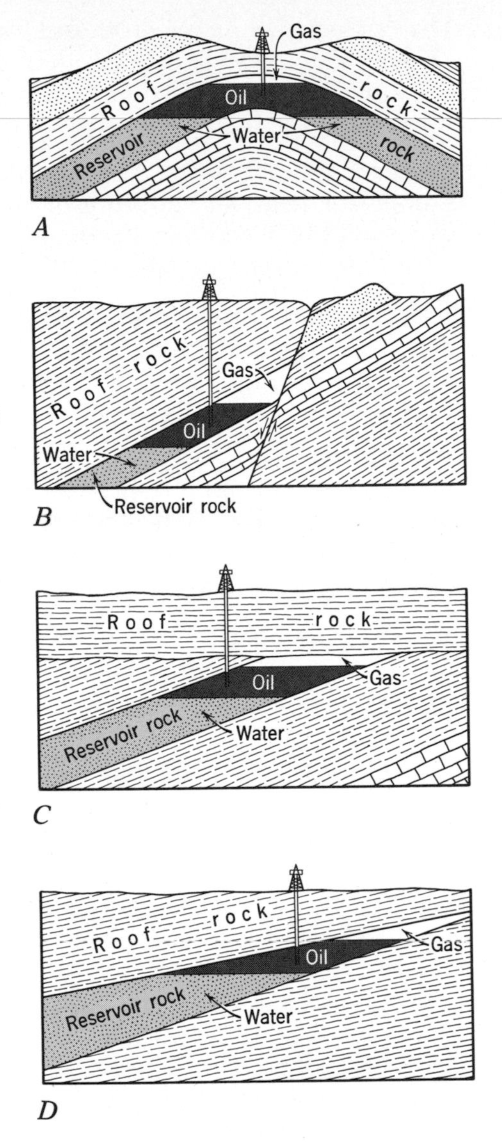


**Figure 21.9**
**Four of the many kinds of oil traps. *A* and *B* are structural traps; *C* and *D* are stratigraphic traps. Gas (white) overlies oil (black), which floats on ground water (blue), saturates reservoir rock, and is held down by roof of claystone. Oil fills only the pore spaces in the rock.**

simplified theory about how the steps occur is widely held and is supported by enough facts to be at least somewhere near the truth.

1. The raw material consists mainly of microscopic marine organisms, mostly plants, living in multitudes at and near the sea surface. Measurements show that the sea grows at least 31,500kg of organic matter per square kilometer per year, and the most productive inshore waters sometimes grow as much as six times this amount. The latter value represents more organic matter than could be harvested in a year from the richest farmland.

2. When the microscopic plants die, they fall to the bottom. Some of the dead plant matter is devoured by scavengers, some is oxidized by oxygen in the water, and the rest becomes buried in the accumulating pile of sediment. There it is attacked and

decomposed by bacteria, which split off and remove oxygen, nitrogen, and other elements, leaving residual carbon and hydrogen. Accumulating sediment that is rich in organic matter teems with bacteria.

3. Deep burial beneath further fine sediment destroys the bacteria and provides pressure, heat, and millions of years for further chemical changes that convert the substance into droplets of liquid oil and minute bubbles of gas.

4. Gradual compaction of the enclosing sediments, under the pressure of their own increasing weight, and any deformation that might occur by folding, reduces the space between the rock particles and squeezes out oil and gas substance into nearby layers of sand or sandstone, where open spaces are larger.

5. Aided by their buoyancy and by the circulation of artesian water, oil and gas migrate (generally upward) through the sand until they reach the surface and are lost, or until they are caught in a trap and form a pool.

We repeat that this statement is greatly oversimplified. A long and complex chain of chemical reactions is involved in the conversion of the original organic constituents to crude petroleum. Also, chemical changes may occur in oil and gas even after they have migrated into their reservoirs. This may help explain why chemical differences exist between the oil in one pool and that in another.

The migration of oil needs more explanation. The sediment in which oil substance is accumulating today is rich in clay minerals, whereas most of the strata that constitute oil pools are sandstones consisting of quartz grains, limestones and dolostones consisting of carbonate minerals, and much-fractured rock of other kinds (Fig. 21.10). It seems obvious, therefore, that oil forms in one kind of material and at some later time migrates to another. The migration process involves essentially the principles of movement of ground water. When, as we mentioned above, oil is squeezed out of the clay-rich sediment in which it originated and enters a body of sandstone or limestone somewhere above, it can migrate more easily than before. One reason is that sandstone, as explained in Chapter 8, is far more permeable than any clay-rich rock. Another reason is that the force of molecular attraction between oil and quartz or carbonate minerals is less strong than that between water and quartz or water and carbonate minerals. Hence, because

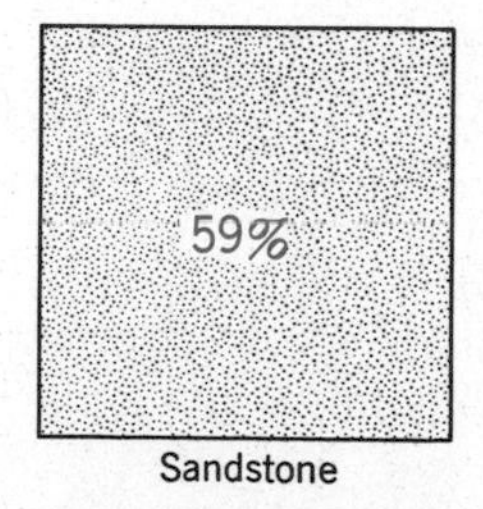


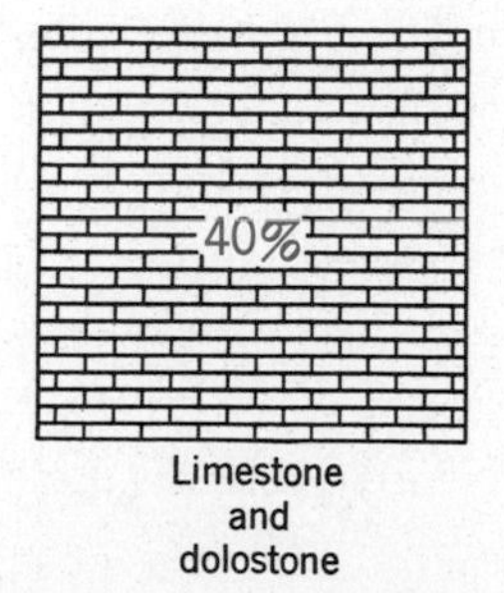


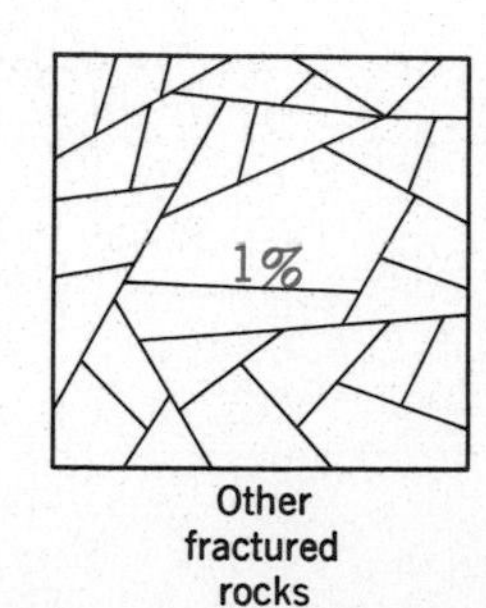


**Figure 21.10**
**Percentage of world's oil found in principal kinds of reservoir rocks.**

oil and water do not mix, water remains fastened to the quartz or carbonate grains while oil occupies the central parts of the larger openings (like those shown in Fig. 8.2). Because it is lighter than water, the oil tends to glide upward past the carbonate- and quartz-held water. In this way it becomes segregated from the water; and when it encounters a trap, it can form a pool.

Most of the oil that forms in sediments does not find a suitable trap and eventually makes its way, along with the artesian water, to the surface. It is estimated that no more than 0.1 percent of all the organic matter originally buried in a sediment is eventually trapped in an oil pool. Most of it escapes to the surface. It is not surprising, therefore, that the highest ratio of oil pools to volume of sediment is found in rock no older than 2.5 million years, and that nearly 60 percent of all the oil so far discovered has been found in strata of Cenozoic age (Fig. 21.11). This does not mean older rocks produced less oil. It simply means that oil in older rocks has had a longer time in which to escape. Another interesting discovery by oil drillers is that the amount of oil decreases as drilling proceeds deeper and deeper (Fig. 21.12). This too might be expected because deeper rocks are more highly compacted and the pressure that drives the oil and water upward is greater.

## Exploration

When U.S. commercial production began in western Pennsylvania in 1860, the first pools were discovered at places where oil and gas advertised their presence by seeping out at the surface. It soon became apparent that at all the producing wells, the strata were sedimentary and that at most wells the structure consisted of anticlines (Fig. 21.9). In 1861 the theory was advanced that oil had migrated upward into anticlines and lay trapped beneath their

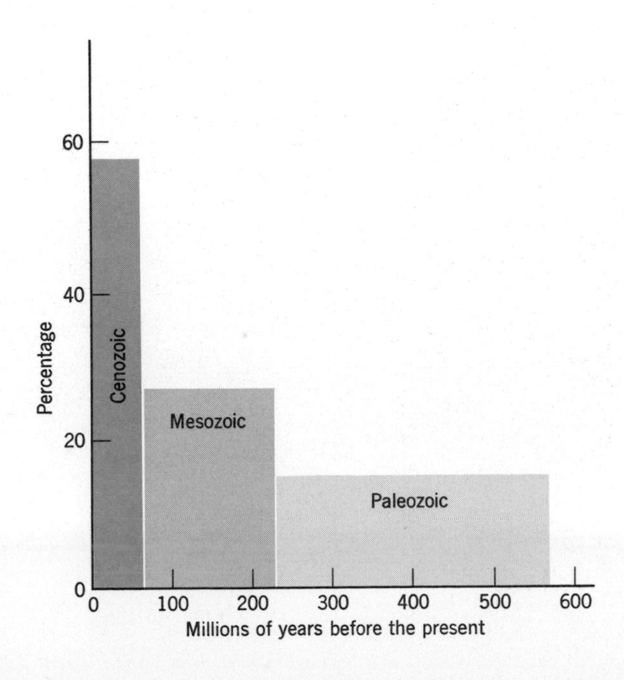


**Figure 21.11**
**Percentage of world's total oil production from strata of different ages.**

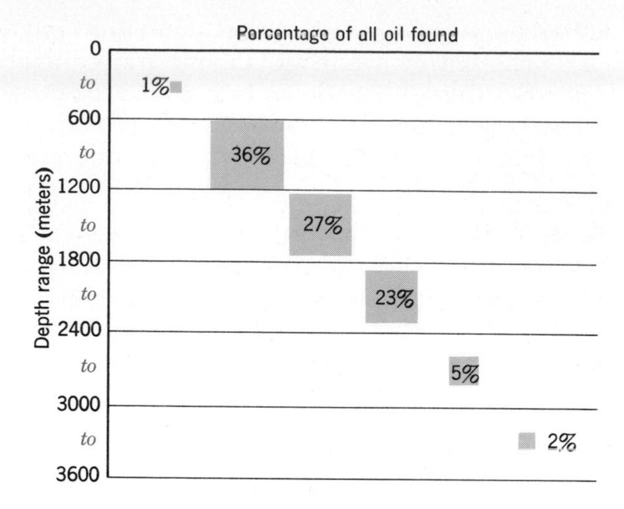


**Figure 21.12**
**Relation between depths of oil pools and percentage of world's oil production.**

crests. In the 1880s this theory was tested and confirmed by drilling holes in certain anticlines where no oil had been found previously; the holes were successful wells.

**Geologic exploration.** Proof of the anticline theory led to a period (1890 to 1925) in which geologists searched for and mapped anticlines from surface exposures. During this period traps other than anticlines were discovered (Fig. 21.9*B–D*), and the search for pools widened rapidly. Even today, however, 80 percent of all the oil found has been discovered in anticlines.

The occurrence of buried traps, such as the one in Fig. 21.9*C*, showed that mapping of exposures would have to be supplemented by methods that would reveal structural and stratigraphic traps not apparent at the surface (Fig. 21.9*D*). This was done first, and is still being done, by core drilling in a pattern beneath a suspected area. Study of the cores makes it possible to construct a graphic log from each core and to correlate the strata by physical character and contained fossils almost as well as though the rocks were exposed at the surface. The depths at which a recognizable bed is encountered by a series of drill holes gives a strong clue to structure (Fig. 21.13).

Core drilling, however, is expensive. This fact stimulated the development of electric logging, which can be done by means of holes drilled by cheaper methods. Common salt and other dissolved substances are present in the ground water that saturates

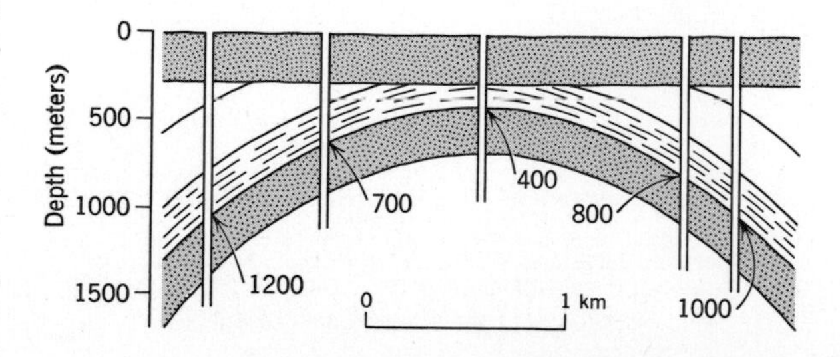


**Figure 21.13**
**Five core-drill holes intersect the same stratum at different depths, showing that flat-lying layers at the surface conceal an anticline lying unconformably beneath it. Numbers show depth (in meters) at which the top of the sandstone stratum was encountered.**

all deep-lying sedimentary rocks. Because salt solution is a good conductor, the conductivity at any desired depth is measured by devices lowered into the hole by a cable. The resulting electric logs are widely used in the correlation of strata.

**Geophysical exploration.** In the 1920s geophysical exploration began to be used widely. Twenty years later this kind of exploration was uncovering many of the new pools, simply because the majority of those discoverable by geologic methods alone had been found already. Geophysical exploration is designed to detect variations, within the rocks, of some physical property such as density, magnetism, or ability to transmit or reflect seismic waves. Because most of these variations are related to structures, they help to indicate the presence of possible pools. If, for instance, a stratum of high density has been lifted up during the making of an anticline, the pull of gravity will be increased very slightly compared with that in the surrounding area. The increase is not likely to be more than about one ten-millionth, so the measuring instrument (a gravity meter or *gravimeter*) has to be extremely sensitive.

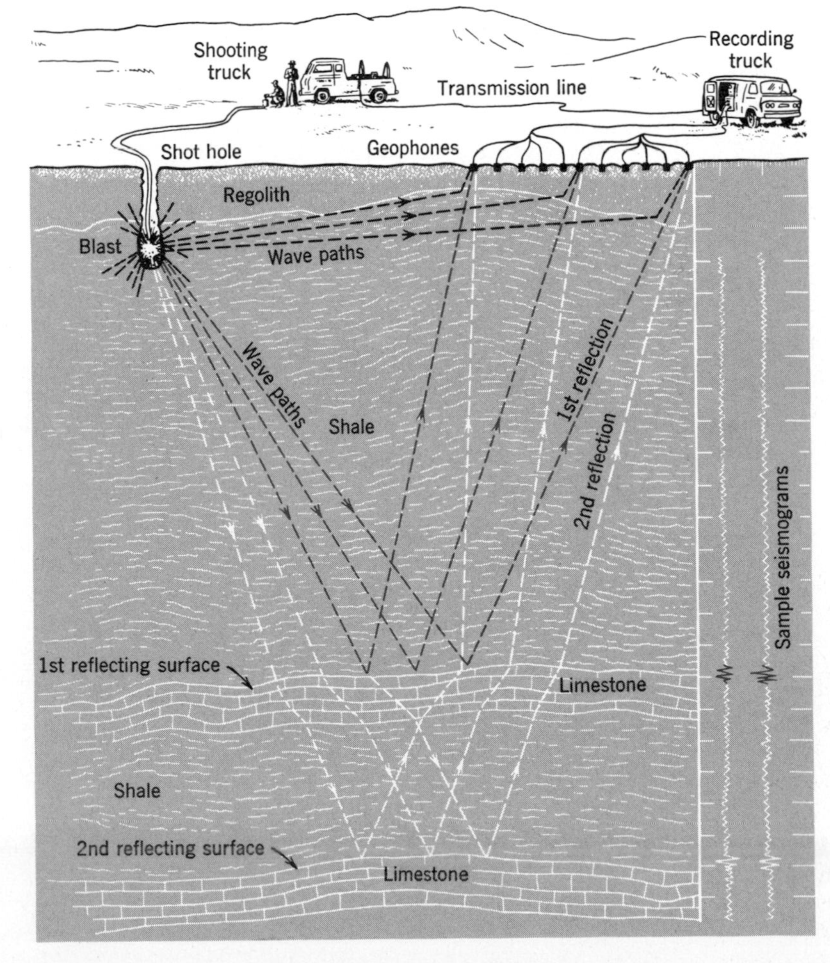


Figure 21.14
**Determination of structure of rock layers by seismic exploration.**
**Blast creates seismic waves that are reflected by interfaces between layers of limestone and shale. Analyses of reflections from interfaces reveal attitude of strata. (*After C. A. Heiland, Geophysical exploration. By permission of Prentice-Hall, Inc., Englewood Cliffs, N.J.*)**

The most useful geophysical exploration method is seismic surveying. An artificial earthquake is created by exploding buried dynamite, and, in the common method, the waves reflected from the upper surfaces of strata having high wave-reflection potential are picked up at ground level by portable seismographs located a kilometer or so from the site of the explosion. With velocity of wave transmission known, depths of the reflecting layers can be calculated from the travel time of the waves (Fig. 21.14). Special techniques have been developed for using seismic methods to locate pools beneath areas that are covered by water. Fig. 21.15 illustrates one of the techniques.

### Distribution

Petroleum, like coal, is widespread but distributed unevenly. The reasons for the uneven distribution are not so obvious as they are with coal. Suitable source sediments for oil are very widespread and seem as likely to form in subarctic waters as in tropical regions. The critical controls seem to be a supply of heat and pressure to effect the conversion of solid organic matter to liquid and gaseous forms, and the formation of a suitable trap before the petroleum has leaked away.

How much oil is there in the world? This is an extremely controversial question. Approximately 600 billion barrels of oil have been discovered, but a great deal remains to be found by drilling. Unlike coal, for which the volume of strata in a basin of sediment can be accurately estimated, the volume of undiscovered oil can only be guessed at. The way guesses are made is to use the accumulated experience of a century of drilling. Knowing how much oil has been found in an intensively drilled area, such as eastern Texas, experts make estimates of probable discoveries in

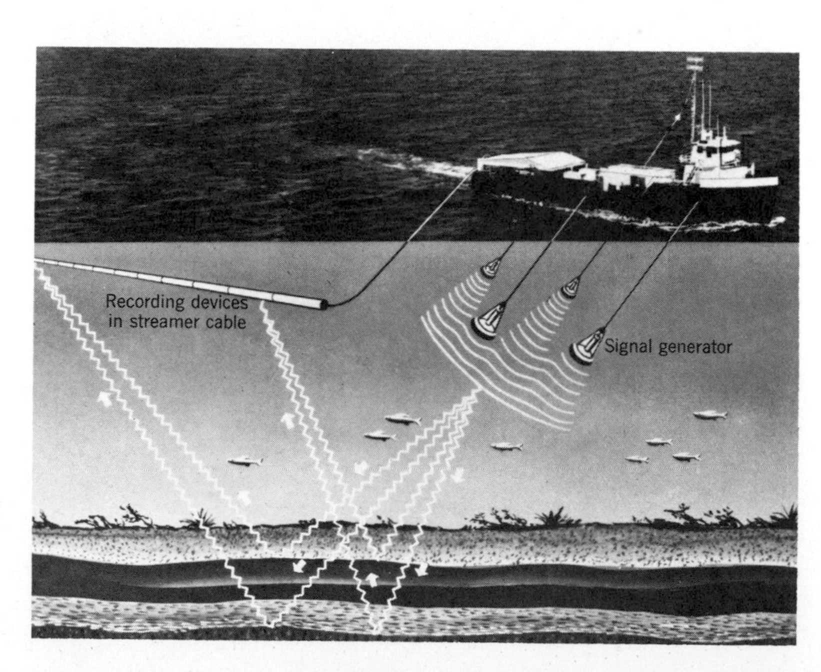


**Figure 21.15 Seismic exploration at sea.**
**Signal generators send out waves that travel through seawater but are reflected by layers of sediment beneath the sea floor. Reflected waves are recorded by detectors housed in the streamer cable, which may be as much as 2.5km in length.** ***(Courtesy Continental Oil Company and American Petroleum Institute.)***

other regions where rock types and structures are similar. Using this approach, and considering all the sedimentary basins of the world (Fig. 21.16), experts estimate that somewhere between 1,500 and 3,000 billion barrels of oil have already been, or eventually will be, discovered and produced. The estimate includes gas, which is added in on the basis of its calorific value, 1,470 cubic meters of gas being taken as equal in heating capacity to one barrel of oil. Considering that the world used 24 billion barrels of oil in 1975, and that consumption continues to grow despite embargoes, higher prices, and political difficulties, the total amount of oil left seems highly inadequate for probable future needs.

Not only does the supply seem inadequate, but also it is distributed unevenly on a geographic basis. The estimate takes into account all the sediments on the continental shelves around the world and the great piles of sediment at the bases of continental slopes. Most of the promising sediments occur along continental shelves that do not coincide with the margins of plates of lithosphere. Where continental margins coincide with convergent margins of plates, the sediment piles and potential for oil seems to be less. Because they are just being explored, the offshore regions offer the greatest potential for the future. These are difficult and expensive places to explore and drill, and many problems of ownership remain to be solved. Wells are already being drilled as far as 150km offshore in water depths as much as 200m. Eventually, much more distant and deeper-water sites will be drilled. The geographic regions, both onshore and offshore, where petroleum is known or estimated to occur are shown in Fig. 21.17. The same figure also shows where the world's oil presently comes from. It is significant, and to many people disturbing, that North America and Africa produce more petroleum by comparison with their total resources than other continents do.

When petroleum flows or is pumped from an oil pool, as much as 60 percent of the oil originally present remains trapped as coatings on mineral grains, and in innumerable tiny holes and fractures in the reservoir rock. Some of this trapped oil can be recovered by secondary processes such as blasting and underground heating, but as much as half of the original total still remains trapped and is not recoverable by any presently known methods. Trapped oil thus constitutes a potential resource, as large as flowing oil, but a great technological advance is needed before it can be recovered.

# TAR SANDS AND VISCOUS OIL

Oil that is exceedingly viscous and thick will not flow and cannot be pumped. Colloquially called *tar* or *asphalt,* heavy, viscous oil acts as a cementing agent between mineral grains. The tar can be recovered from the tar sands only if the sand is heated to make the tar flow. The resulting tar must then be processed to recover the valuable gasoline fraction. The heating can be done underground,

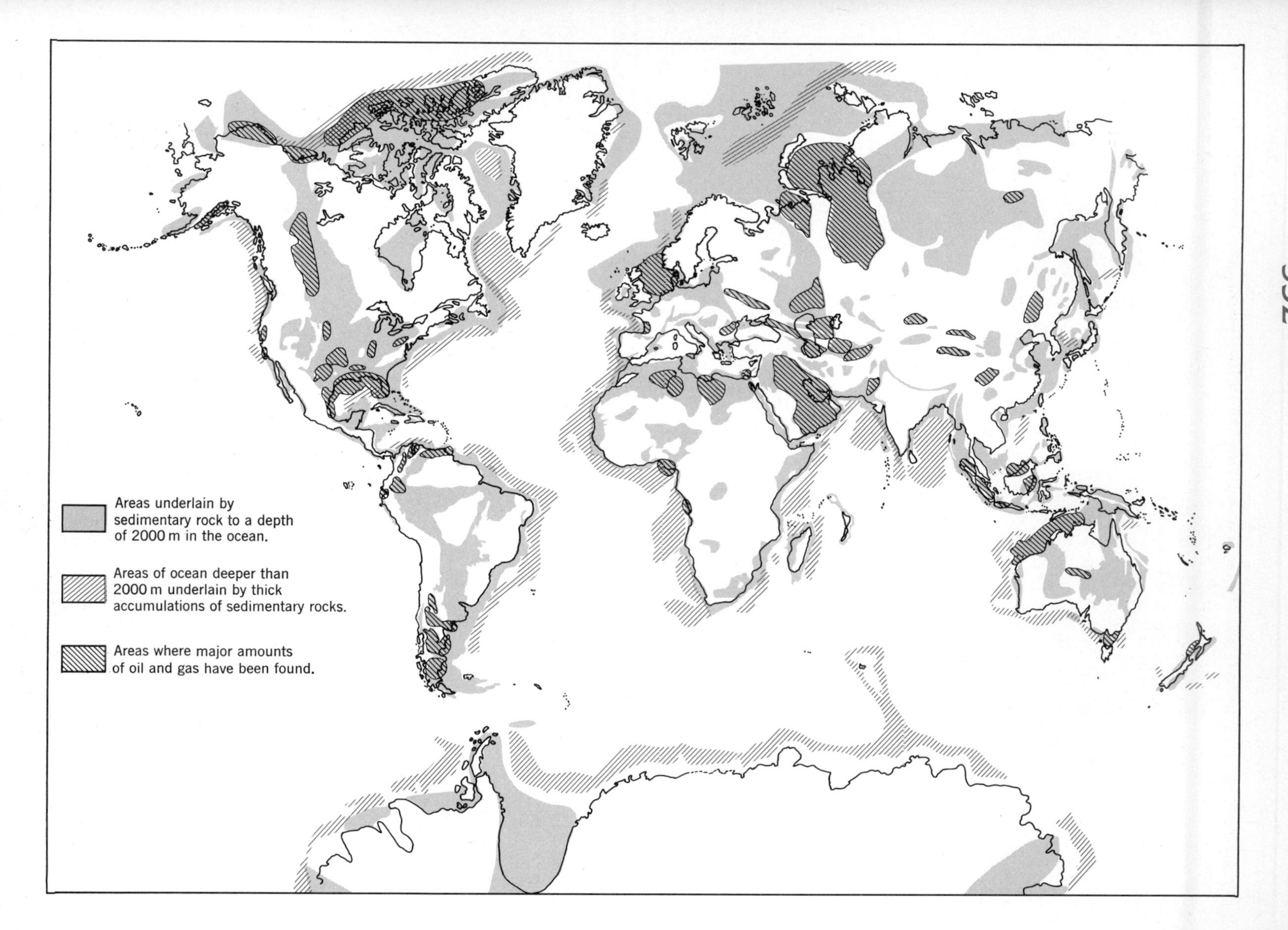
Areas underlain by sedimentary rock to a depth of 2000 m in the ocean.
Areas of ocean deeper than 2000 m underlain by thick accumulations of sedimentary rocks.
Areas where major amounts of oil and gas have been found.

**Figure 21.16**
**(see opposite page) World map showing areas underlain by sedimentary rock and regions where large accumulations of oil and gas have been located. Where the ocean is deeper than 2,000m, sedimentary rock has yet to be tested for its oil and gas potential.**

at least in theory, somewhat as in the secondary recovery of trapped oil. Present practice, however, relies on mining and surface processing. This means that deposits must be close enough to the surface to be worked by inexpensive surface-mining methods. By far the largest occurrence of tar sands that meet these requirements in North America is in Alberta, Canada, where the *Athabasca Tar Sand* covers an area of 50,000km$^2$ and reaches thicknesses of 60m (Fig. 21.18). Mining of the Athabasca Tar Sand has already commenced at the rate of 100,000 tons a day. Assuming 50 percent recovery, as much as 500 billion barrels of viscous tar might eventually be recovered. Deposits of tar sand almost as large as the Athabasca deposit are know in Venezuela and in the U.S.S.R. A great many smaller deposits of tar sand are also known around the world. Some experts suggest the total resource of tar might be as large as the sum of all the trapped and flowing oil. If just half this amount were recovered, tar sands would ultimately yield as much oil as would flowing oil and gas from oil pools.

# OIL SHALE

Another source of petroleum consists of solid organic matter enclosed in fine-grained sedimentary rock. If the organic matter is heated, the solid breaks down and liquid and gaseous hydrocarbons, similar to those in oil, can be distilled from it. All

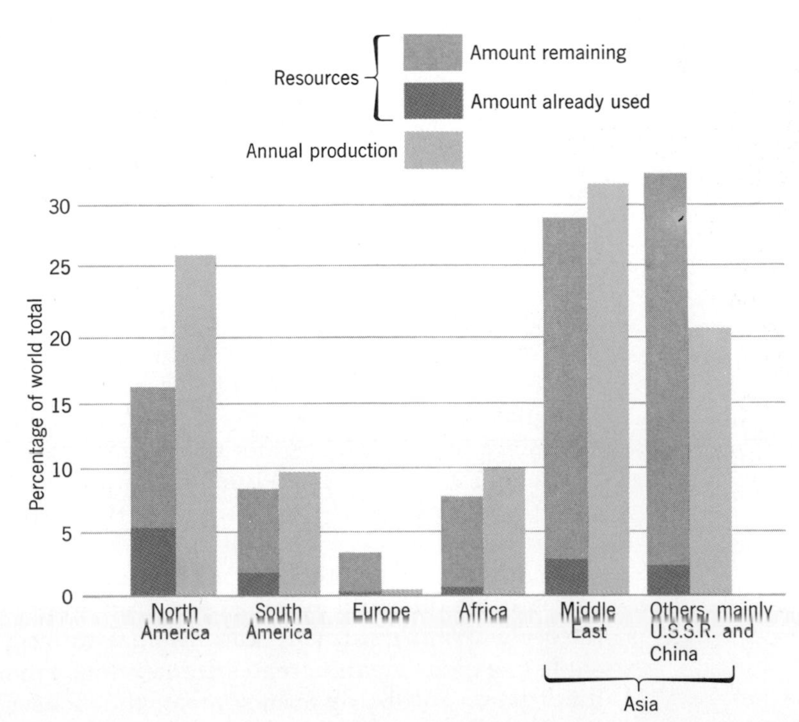


**Figure 21.17**
**World resources and present production of petroleum.**
**Percentage of total oil and gas resources, cumulative amount produced by 1975, and percentage of world's total annual production of oil and gas. North America produces more than its share, compared to its resources. Russia and China produce less than their relative share.**

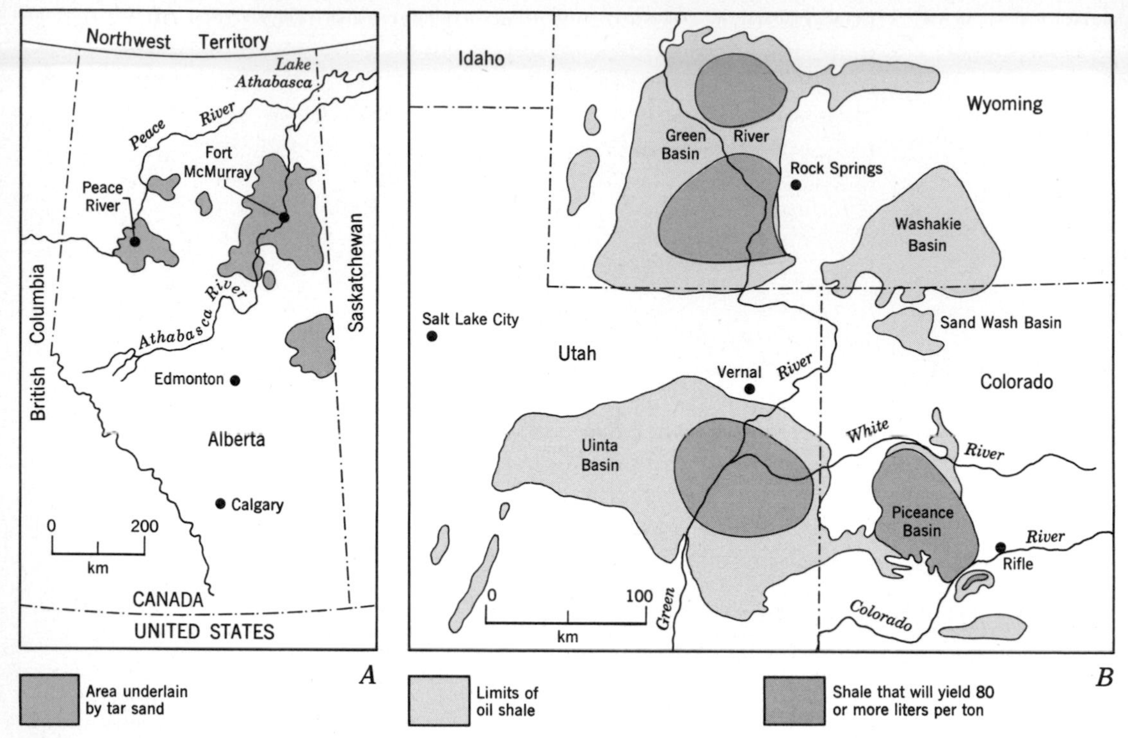


Figure 21.18
**Areas in North America underlain by rich tar sand and rich oil shale.**
***A.* Athabasca Tar Sand in Alberta.**
***B.* Green River Oil Shale, originally deposited as sediment rich in organic matter. The sediment accumulated in freshwater lakes that formed in shallow basins in Colorado, Wyoming, and Utah. Compaction and cementation of the sediment formed the shale we see today. (*U.S. Geol. Survey, 1968.*)**

sedimentary rock contains some organic matter, but to be considered an energy resource the organic matter must yield more energy than that required for the processes of mining and distillation. The only kind of sedimentary rock that contains sufficient solid organic matter to be given any attention is shale, and only those shales that yield 40 or more liters of distillate per ton can be considered, because the energy needed to mine and process a ton of shale is equivalent to that created by burning 40 liters of oil.

Unfortunately, most shales will not yield 40 liters of oil per ton, but a few particularly rich oil-shale deposits are known. In three places in the world, in Estonia, U.S.S.R., and China, shales that yield as much as 320 liters of oil per ton are already being worked. The world's largest deposit of rich oil shale, however, is in the United States. During the Eocene Epoch many large, shallow lakes existed in basins in Colorado, Wyoming, and Utah; in three of them was deposited a series of rich organic sediments that are now the *Green River Oil Shales* (Fig. 21.18). The richest shales were deposited in the lake of Colorado. These shales are capable of producing as much as 240 liters of oil per ton. Scientists of the U.S. Geological Survey estimate that oil-shale resources capable of producing 40 liters or more oil a ton, in the Green River Oil Shales alone, total about 2,000 billion barrels of oil. It is most unlikely that all this oil could ever be recovered. Many low-grade areas hardly warrant treatment, and because mining and processing the shale is expensive and creates tremendous environment disturbance, probably even some of the richest areas will not be touched. Nevertheless, experts suggest that 50 percent of the oil from the Green River Oil Shales may someday be recovered. Many of the same experts believe that large-scale mining and processing of the shale will be under way before the end of the 1980s.

Rich resources of oil shale in other parts of the world have not been adequately explored, but another huge deposit occurs in Brazil, in the Irati Shale. At this stage it is not possible to make an accurate worldwide estimate of the total amount of distillable oil in oil shale. What is clear, however, is that oil shale is by far the largest of the fossil fuel resources and is possibly a hundred or more times larger than coal. The trouble is that most of the oil shale is too lean ever to be worked, and as a consequence the amount of economically recoverable shale oil around the world may not even be as large as recoverable resources of flowing oil.

# HOW MUCH FOSSIL FUEL?

Because of their "one-crop" availability, are supplies of fossil fuel adequate to meet future demands? If we use a barrel of oil as our unit of measurement, we can compare quantities of all fossil fuels directly. Thus approximately 0.22 ton of coal produces the same amount of heat energy as one barrel of oil; so the $8310 \times 10^9$ tons of coal stated earlier to be the world's recoverable coal reserves are equivalent to about 37,800 billion barrels of oil.

Considering the 1975 use rate of 24 billion barrels of oil a year, and comparing the estimated recoverable amounts of fossil fuels (Table 21.1), it is apparent that only coal seems to have the capacity to meet long-continued demands. The two fuels on which we now rely most heavily, oil and gas, are the two least abundant. But production of oil and gas is very efficient and the least expensive of all the fossil fuels in terms of manpower and disruption of environments. As we shift, inevitably, to increased use of other

**Table 21.1**
**Amounts of Fossil Fuels Possibly Recoverable Worldwide. Unit of Comparison is a Barrel of Oil.**

| Fossil Fuel | Total Amount in Ground (Billion of Barrels) | Amount Possibly Recoverable (Billions of Barrels) |
|---|---|---|
| Coal | 75,600[a] | 37,800[a] |
| Oil and gas (flowing) | 1,500 to 3,000 | 1,500 to 3,000 |
| Trapped oil in pumped-out pools | 1,500 to 3,000 | 0 to ? |
| Viscous oil (tar sands) | 3,000 to 6,000 | 500 to ? |
| Oil shale | Total unknown. Much greater than coal. | 1,000 to ? |

[a] 0.22 tons of coal = 1 barrel of oil

fuels, a great many problems can be anticipated. It may even be necessary to decide that we should not mine and use fossil fuels on a large scale. We shall therefore consider other possible sources of energy on Planet Earth.

# OTHER SOURCES OF ENERGY

Two sources of energy other than fossil fuels have already been developed to some extent: hydroelectric energy and nuclear energy. Two others—tides and Earth's internal heat—have been tested and developed on a limited basis. Neither has yet been developed on a large scale in the United States, but the day may not be far in the future when both will become locally important.

### Hydroelectric energy

Hydroelectric energy is recovered from the potential energy of water in streams and rivers as they flow downward to the sea. Water ("hydro") power is an expression of solar power because it is the Sun's heat energy that drives the water cycle. That cycle is continuous; so energy obtained from flowing water is also continuous: Unlike coal and oil, it cannot be used up. However, to convert the power of flowing water into electricity, it is necessary to dam streams, and because a stream carries a suspended load of sediment, the reservoir behind the dam will eventually become filled with deposited sediment. Depending on the sediment load, a reservoir could be completely filled up within 50 to 200 years. Lake Nasser, the reservoir held behind the great Aswan Dam on the River Nile, will be almost half filled with silt by the year 2025. Thus, although water power is continuous, the reservoirs needed for the conversion of water power to electricity have limited lifetimes.

Water power has been used in small ways for thousands of years, but only in the 20th Century has widespread use been made for generating electricity. All the water flowing in the streams of the world has a total amount of recoverable energy that has been estimated as $2.2 \times 10^{19}$ calories per year. This is an amount of energy equivalent to burning 15 billion barrels of oil.

The distribution of water power around the world, and the extent to which it is already developed, is shown in Figure 21.19. There we see that the greatest undeveloped potential lies in two continents of the southern hemisphere, Africa and South America, both of which lack large coal reserves.

### Nuclear energy

Nuclear energy, unlike hydroelectric energy, is not limited in its potential. The use of nuclear energy in industry has already begun, and in time will probably play an important part in the world's economy. Nuclear energy is the heat energy produced during radioactive decay. Three of the same radioactive atoms that keep the Earth hot—$^{238}U$, $^{235}U$, and $^{232}Th$—can be mined and used to

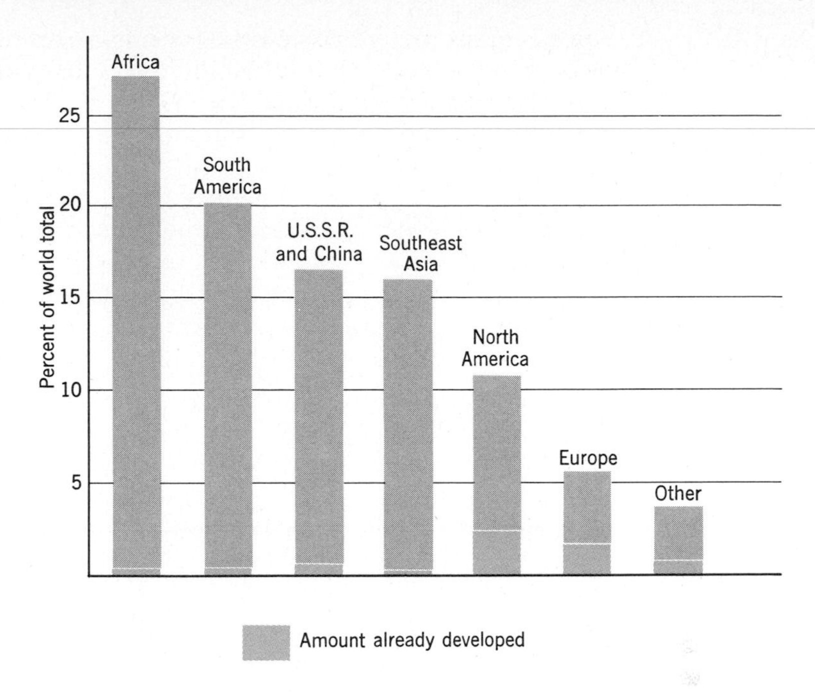


**Figure 21.19**
**Percentage of world's total water-power resources, and extent to which they have already been developed in hydroelectric power schemes.** ***(From F. L. Adams, after M. K. Hubbert, 1969.)***

produce energy by speeding up the decay rate under controlled conditions. The speedup is accomplished by bombarding the radioactive atoms with neutrons, thus accelerating the rate of disintegration and release of heat energy. The device in which this tricky operation is carried out is called a *pile.*

When $^{235}U$ disintegrates, it not only releases heat and forms new elements but also ejects some neutrons from its nucleus. These neutrons can then be used to cause more $^{235}U$ atoms to disintegrate and a continuous chain reaction occurs. The function of a pile is to control the flux of neutrons so that the rate of disintegration can be controlled. When a chain reaction proceeds without control, an atomic explosion occurs. Controlled nuclear decay, then, is the method used by nuclear power plants, and a tremendous amount of energy can be obtained in the process. The disintegration, or *fission,* of one gram of $^{235}U$ produces as much heat as the burning of 13.7 barrels of oil. Unfortunately, however, $^{235}U$ is the only natural radioactive atom that will maintain a chain reaction, and is the least abundant of the three useful atoms. Only one atom of each 141 atoms of uranium in nature is $^{235}U$. The remaining atoms are $^{238}U$, which will not sustain a chain reaction. All the present nuclear power plants use $^{235}U$. However, if $^{238}U$ is placed in a pile, with $^{235}U$ that is undergoing a chain reaction, some of the neutrons will bombard the $^{238}U$ and convert it to a new element, plutonium-239 ($^{239}Pu$). This new element can, under suitable conditions, sustain a chain reaction of its own. The pile in which the conversion of $^{238}U$ takes place is called a *breeder reactor* and the same kind of device can be used to convert $^{232}Th$ into a new atom, $^{233}U$, that will also sustain a chain reaction. But a number of technological problems remain before breeder reactors move from the experimental stage into widespread practical use.

Already there are nearly a hundred nuclear power plants operating around the world. They utilize the heat energy from radioactive disintegration to produce steam, which in turn drives turbines and generates electricity. In the United States alone, 24 nuclear power plants were operating at the beginning of 1975 and some were operating several piles. Some serious questions about nuclear energy remain to be answered. One concerns the amount and availability of uranium and thorium to serve as fuel.

Uranium is present in the continental crust in very small amounts, constituting only 0.00016 percent of average crustal rock. Thorium is slightly more abundant, but neither uranium nor thorium is a very common element. Fortunately, both uranium and thorium do occur in local enrichments—ore deposits—that can be mined. Uranium is much more susceptible to the processes that produce ore deposits; consequently, most interest has centered on it. If we were to use only $^{235}U$ atoms, supplies of nuclear fuel would be very limited because only the richest deposits of uranium ores could be worked. This is so because the cost of separating $^{235}U$ atoms from $^{238}U$ is so high that it is not feasible also to pay high costs for mining and recovering the raw uranium ore. At present the $^{238}U$ is of no use, and so is simply a by-product. If, however, breeder reactors can be perfected and if $^{238}U$ can be used also, 140 times more uranium would be available. By avoiding the expensive atomic separation, it would be possible to spend much more money on mining raw uranium. That is, we could afford to pay much more for uranium ore to be used in breeder reactors than we can as long as we are limited to $^{235}U$ alone. Higher prices mean that less-concentrated ores can be worked, and this in turn means a vast increase in the amount of uranium fuel that is available.

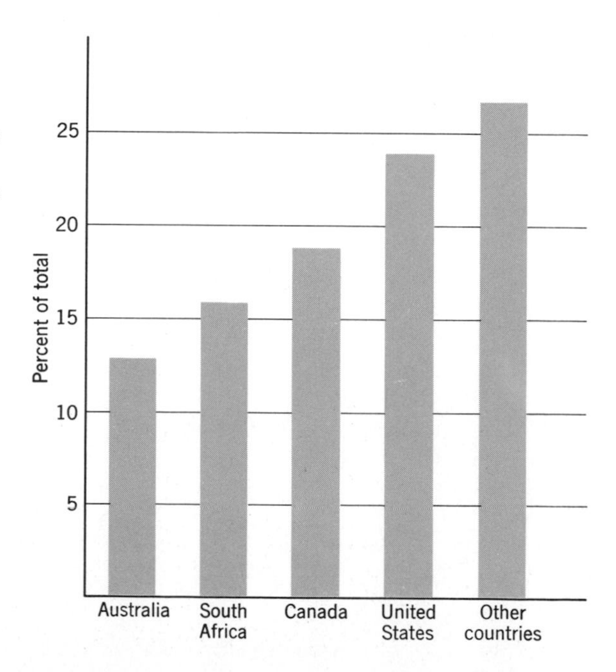


**Figure 21.20**
**Uranium ore from which uranium can be recovered for three cents per gram or less. Figures available for western world only. Data not available from other countries because of strategic importance of uranium for atomic weapons. (*Data from World Mining, July 1974.*)**

Uranium ores rich enough to be worked and used solely for $^{235}U$ plants are very limited. The cost of uranium before separation should be no more than about three cents per gram. Estimates of the total ore in Western countries amounts to only 1,665,000 tons (Fig. 21.20). If all this uranium were used in $^{235}U$ plants, corresponding heat energy would only be equivalent to the burning of 155 billion barrels of oil. But if all the $^{238}U$ could be used in a breeder reactor, the equivalent figure would be 21,700 billion barrels of oil.

Apart from the rich uranium deposits, it is difficult to make an assessment because the needed work has not been done. Within the United States alone, a general evaluation of low-grade source materials indicates that, at a cost of up to $1.00 per gram, as much as two billion tons of uranium might be recovered. If used in breeder reactors, this amount would generate heat equivalent to burning 27,000,000 billion barrels of oil. It is a small wonder, therefore, that many people view nuclear power as a principal answer to the decline of our resources of oil and gas. But many problems remain to be solved. One involves the safety of atomic piles, a point on which experts are still divided. Another involves the disposal of radioactive wastes. After nuclear fuel is spent, it is still highly radioactive. Some of the daughter products are themselves radioactive, and many, such as strontium-90 ($^{90}Sr$), are dangerous to people. Also they have long enough half-lives so that the wastes must be disposed of in places that will be safe for a thousand years or more. No generally acceptable solution to the problem has yet been found.

### Tidal energy

Tidal energy resembles hydroelectric energy in that it is recovered from flowing water and cannot be used up. People have harnessed tides to drive water wheels for many centuries; much of the flour used in Boston in the 17th and 18th Centuries was ground in a tidal mill. But today's interest in tides lies in their use as a source of water to drive electrical generators. Water in a restricted bay, retained behind a dam at high tide and allowed to flow out again at low tide, can drive a generator in the same way that river water can. One important difference between hydroelectric power and tidal power is that rivers flow continually, while tides can be impounded only twice a day, so that electricity is produced only periodically.

Along most of the world's coasts the differences in height between high tide and low tide are too small—about 2m—to drive generators. In some restricted bays, however, tidal ranges exceed 15m, and here it is possible to recover tidal energy. In France, at the mouth of the River Rance, a tidal-power plant has been operating successfully for several years. Another plant has been built in the U.S.S.R. near Murmansk on the White Sea. But no tidal-power plants have been built in the United States. When they are, the most likely sites are in Passamaquoddy Bay in Maine, and in some of the bays and inlets of Alaska.

Experts estimate that if all sites with suitable tidal ranges were developed to produce power, the total potential would only be equivalent to the burning of 170 million barrels of oil a year. Furthermore, about half the potential is along the sparsely inhabited Kimberley coast in northwestern Australia. Thus, tidal power could have significant local uses but could never be important on a worldwide scale.

### Geothermal energy

Geothermal energy, as Earth's internal-heat energy is called, has been used for more than fifty years in Italy and Iceland and more recently in other parts of the world, including the United States. The steam produced in hot-spring areas (Chapter 15) can be used to power generators in the same way as steam produced in coal- and oil-fired boilers. To capture steam in hot-spring areas, it is necessary to drill into the hot underground reservoirs that feed the springs, bring the steam or hot water to the surface in pipes, and feed it into a power plant (Fig. 21.21). The places where sources of heat are close enough to the surface to produce steam, and therefore where geothermal energy can be developed, are mostly in areas of current or recent volcanic activity, where magma or hot intrusive igneous rock is close to the surface and can serve as a source of heat. In the United States these areas include California, Nevada, Montana, Wyoming, New Mexico, Utah, and Alaska. In other countries, in addition to Italy and Iceland, large resources of geothermal energy exist in Japan, U.S.S.R., New Zealand, several countries in Central America, Ethiopia, and Kenya. Not surprisingly, most of the world's geothermal steam reservoirs are around the margins of plates, because plate margins are where most of the recent volcanic activity has occurred.

A depth of 3km seems to be a rough lower limit for big geothermal steam and hot-water pools. Experts from the U.S. Geological

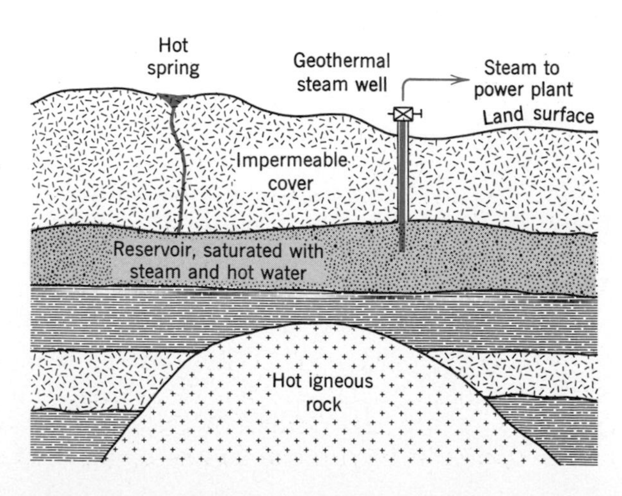


**Figure 21.21**
**A typical geothermal steam reservoir. Water in a permeable aquifier, such as sandstone, is heated by magma or hot igneous rock. As steam and hot water are withdrawn through the well, cold water flows into the reservoir through the aquifier.**

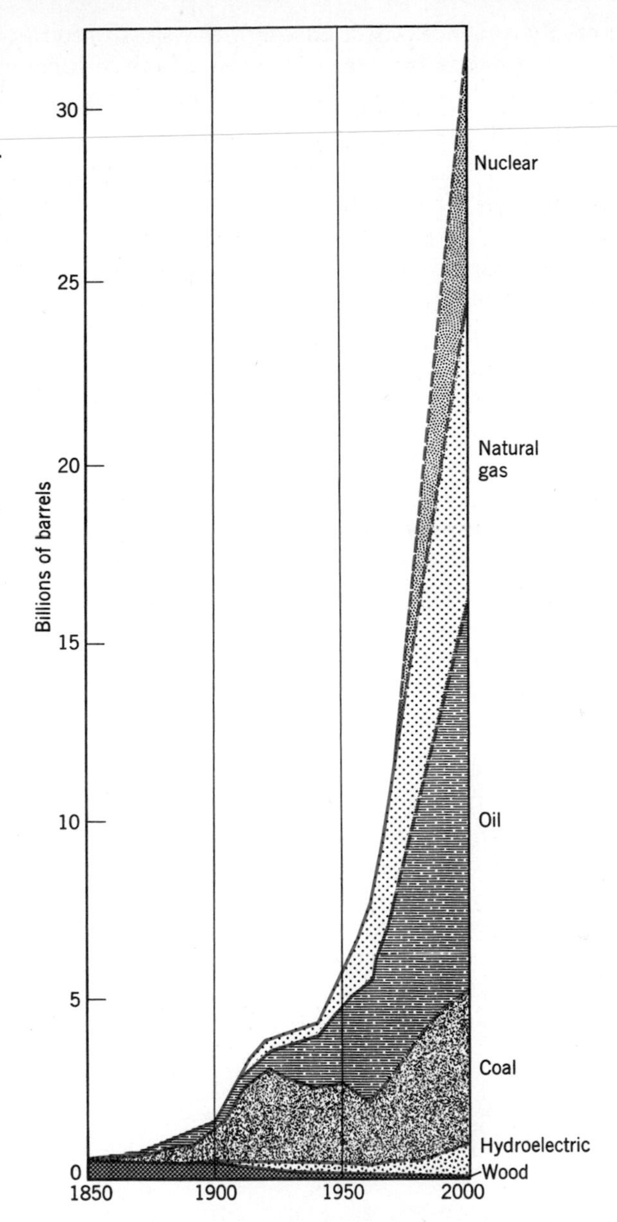


**Figure 21.22**
**Changing pattern of energy sources in the United States.**
**In 1850, firewood supplied most needs, but as energy demands increased, coal, oil, and natural gas became increasingly important. In the years ahead, the importance of nuclear energy is expected to grow significantly. Dashed lines are projections. Energy is expressed in barrels of oil.**

Survey have reported that down to 3km worldwide reservoirs of geothermal energy contain about $1.9 \times 10^{19}$ calories that can be recovered—equivalent to burning 13 billion barrels of oil. The experts' estimate takes note of the fact that experiences in New Zealand and Italy indicate that only about one percent of the energy in a geothermal reservoir is recoverable. If the recovery efficiency were to rise, the estimate of recoverable geothermal resources would also rise.

Interesting geothermal experiments are now being conducted in New Mexico. In the Jemez Mountains there is a recently extinct (but still hot) volcano. Scientists are planning to drill deep into the hot rock, shatter it with explosives to create an artificial reservoir, and then pump water through the shattered rock to produce steam. The success of the test will not be known for some years, but if it is successful, it will open a way to vastly increased use of Earth's geothermal energy.

# THE FUTURE

We face neither an energy crisis nor an energy shortage. Vast amounts of energy are available—more than we can ever use. But first we must learn how to use energy in different forms. The only crises we face are of our own making because we have, for too long, relied on limited supplies of oil and gas. But as we have just shown, many alternatives to oil and gas are already in use, at least in a small way. Still others are available and are hardly used at all. The Sun's heat rays, the energy of the wind to turn windmills, the waves that pound all seashores, the giant ocean currents, and even the temperature difference between warm water at the ocean surface and cold water at depth, are all potential energy resources. Still other potential sources are to be found in the possibility of duplicating the process of nuclear fusion that creates the Sun's heat energy. We can release the energy in a hydrogen bomb; why not in a slower controlled way?

As they have done in the past, energy sources will surely change in the future (Fig. 21.22). Changing energy sources will inevitably bring changes in life styles. As fossil fuels are depleted, it seems inevitable that present systems of transportation will change too. As energy sources alter, so will the machines that use the energy. Demand patterns for metals to build the machines must also change. How the demand patterns change will be conditioned, in part, by the metals available and how much it costs to mine and process them. We turn next, therefore, to a discussion of mineral resources.

# SUMMARY

1. Ancient industry based largely on wood and muscle has given way to industry that uses energy intensively and is based mainly on fossil fuels.

2. Coal originated as plant matter in ancient swamps, and is both abundant and widely distributed.

3. Oil and gas probably originated as organic matter sedimented on sea floors and decomposed chemically. Later these fluids moved through reservoir rocks and were caught in geologic traps to form pools.

4. Oil pools are found by both geologic and geophysical methods. Of the latter, seismic surveying is the most common.

5. Oil and gas are limited in abundance; the world's supplies cannot long sustain our present rate of use.

6. As much as half the oil in a reservoir remains trapped, as coatings on mineral grains and in pockets between grains, after flowing oil has been pumped out.

7. Heavy, nonflowing oil (tar) can be recovered by mining techniques. The amount of tar in the world probably equals the amount of flowing oil.

8. When heated, part of the solid organic matter found in shale will convert to oil and gas. Oil from shales is the world's largest resource of fossil fuel. Unfortunately, most shales contain so little solid organic matter that more oil must be burned to heat the shale than is produced by the conversion process.

9. Nuclear energy is derived from atomic nuclei of unstable elements, chiefly uranium. The amount of nuclear energy available from naturally occurring radioactive elements is the single largest energy resource available.

10. Other sources of energy currently used to some extent are geothermal heat, the tides, and energy from flowing streams.

# SELECTED REFERENCES

Averitt, Paul, 1975, Coal resources of the United States, January 1, 1974: U.S. Geol. Survey Bull. 1412.

Bateman, A. M., 1951, The formation of mineral deposits: New York, John Wiley.

Cloud, Preston (ed.), 1969, Resources and man: San Francisco, W. H. Freeman.

Levorsen, A. I., 1967, Geology of petroleum: 2nd ed., San Francisco, W. H. Freeman.

Skinner, B. J., 1976, Earth resources: 2nd ed., Englewood Cliffs, N.J., Prentice-Hall.

Skinner, B. J., and Turekian, K. K., 1973, Man and the ocean: Englewood Cliffs, N.J., Prentice-Hall.

U.S. Bureau of Mines Staff, 1970, Mineral facts and problems: U.S. Bureau of Mines Bull. 650.

Williamson, I. A., 1967, Coal mining geology: New York, Oxford Univ. Press.

# 22 SOURCES OF MATERIALS

Use of materials
Origin of mineral deposits
Useful mineral substances
Future supplies of minerals
Summary
Selected references

## USE OF MATERIALS

Having discussed the principal sources of *energy*, we now turn to the mineral substances that provide *materials* from which things can be made. The number and diversity of such substances is so great that to make a simple classification covering all of them is almost impossible. Nearly every mineral substance known can be used for something. A society such as ours, that possesses an energy-intensive industry, not only requires a diverse group of metals for machines, but also demands a host of nonmetallic mineral products, such as claystone and limestone for making cement, gypsum for making plaster, salt for making chemical compounds, and calcium phosphate (apatite) for making fertilizer. To supply all their needs, the people of the world now mine several hundred kinds of mineral products. The amounts used are truly enormous, and are still increasing. In the United States in 1975, the quantity of iron and steel used was equivalent to 600kg for every man, woman, and child in the country. In the same year the use of sand and gravel amounted to 4,100kg per person, of crushed stone 3,900kg, of aluminum 20kg, and of copper 10kg. When the mineral substances mined for energy (3,800kg of petroleum and 2,400kg of coal) are added to all the rest, we get a total of 15,000kg, or about 15 tons, of mineral substances mined in 1975 for every person living in the United States. For the world as a whole, the total is a staggering 15 billion tons, or 3.75 tons per person. It is difficult to imagine where all these materials go, but they are used in every corner of our lives. They build roads and skyscrapers, generate electricity, make automobiles and ashtrays, grow and transport food, and make clothes and indeed all the familiar objects around us. Even the paper on which these words are printed contains mineral substances in addition to woody vegetable matter.

### Supplies of minerals

The majority of industrialized nations possess a strong mineral base; that is, they are rich in many kinds of ***mineral deposits*** (*any volume of rock containing an enrichment of one or more minerals*) which they are exploiting vigorously. But because no nation is entirely self-sufficient in mineral supplies, each must rely to some extent on other nations to fulfill its needs (Fig. 22.1).

Mineral resources have certain peculiar aspects that differ from agriculture, grazing, forestry, or fisheries. First, occurrences of usable minerals are limited in abundance and distinctly localized at places within Earth's crust. This is the main reason why no nation is self-sufficient where mineral supplies are concerned. Because usable minerals are localized, they must be searched out, and the search ranges over the entire globe.

Second, the quantity or reserve of a given mineral available in any one country is rarely known with accuracy, and the likelihood that new deposits will be discovered is difficult to assess. As a result, production over a period of years may be difficult to predict. Thus

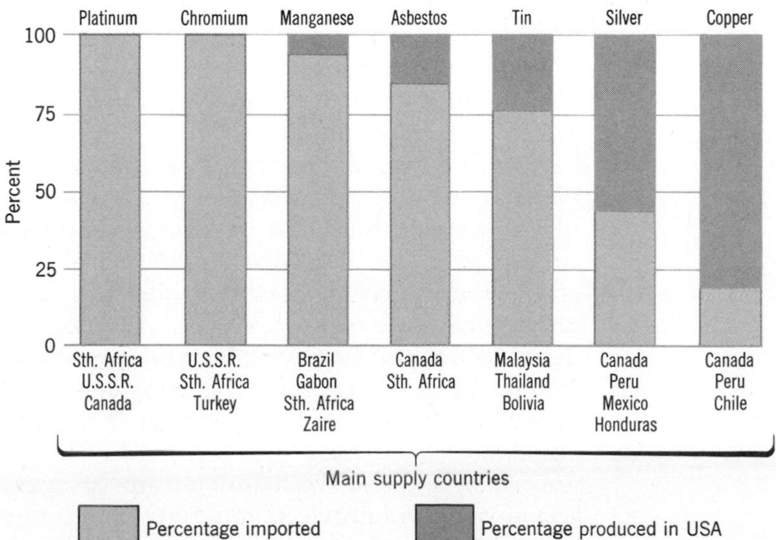


**Figure 22.1**
**Selected mineral substances for which United States consumption exceeds production. The difference must be supplied by imports.** ***(From a report by the Secretary of the Interior, to the U.S. Congress, 1973.)***

a country that today can supply its need for a given mineral substance may face a future in which it will have become an importing nation. A good example is England, once self-sufficient in most minerals, but today almost entirely an importing country. Third, unlike plants and animals that are cropped yearly or seasonally, deposits of minerals are depleted by mining, which eventually exhausts them. Minerals therefore have only a "one-crop" availability per occurrence; this disadvantage can be offset only by finding new occurrences or by making use of scrap—that is, by re-using the same material repeatedly.

These peculiarities of the mineral industry place a premium on the skills of geologists, prospectors, and engineers who play essential parts in finding and bringing to the surface mineral substances for use in industry. The task of finding and mining is accomplished through the application of the basic principles that have been set forth in this book, and by the use of additional specialized knowledge concerning the origin and distribution of mineral deposits. Much ingenuity has been expended in bringing the production of minerals to its present state. Because known deposits are being rapidly used up, while demands for minerals are increasing, we can be sure that even more ingenuity will be needed in the future.

## Ore

The form in which a mineral substance is used is rarely the form in which it is mined. This means that the mined product must be processed before use, and because processing costs are high and vary from mineral to mineral, some minerals are more desirable than others. Iron, which is used mainly as a metal, occurs in hundreds of different minerals. But only the oxide minerals hematite, magnetite, and limonite, and the carbonate mineral siderite ($FeCO_3$), are sought and mined. From each of these minerals metallic iron can be separated by relatively inexpensive smelting processes. By comparison, the cost of smelting iron from common silicate minerals such as garnet, biotite, or olivine, is many times greater.

Mineral substances for industry are therefore sought in deposits of certain preferred minerals from which the desired substances can be recovered easily. The more concentrated the preferred mineral, the more valuable the deposit. In some deposits the desired minerals are so highly concentrated that even very rare substances such as gold and platinum can be seen with the naked eye. But for all desired mineral substances there is a level of concentration below which the deposit cannot be worked economically. To distinguish between profitable and unprofitable mineral deposits, we use the word ***ore***, meaning *an aggregate of minerals from which one or more materials can be extracted profitably*. It is not always possible to say exactly how much of a given mineral must be present in order to constitute an ore. For example, two deposits containing the same iron minerals may contain 60 percent iron, but one is not ore because it is deeply buried, or is so remote that costs of mining and transport are too high for the final product—metallic iron—to be competitive with iron from other deposits. Furthermore, as both costs and market prices fluctuate, a particular aggregate of minerals may be an ore at one time but not at another.

Along with ore minerals, from which the desired substances are extracted, are other minerals collectively termed ***gangue*** (pronounced *gang*). These are *the nonvaluable minerals of an ore.* Familiar minerals that commonly occur as gangue are quartz, feldspar, mica, calcite, and dolomite.

The ore problem has always been twofold: first, to find the ores (which altogether underlie an infinitesimally small proportion of Earth's land area), and second, to mine the ore and get rid of the gangue as cheaply as possible. Getting rid of gangue and mining are both technical problems; engineers have been so successful in solving them that some deposits now considered ore are only one-sixth as rich as were the lowest-grade ores 100 years ago (Fig. 22.2).

# ORIGIN OF MINERAL DEPOSITS

Ores are mineral deposits, but not all mineral deposits are ores. Ore is an economic term and the processes that form mineral deposits cannot be classified on economic grounds, or even on the basis of the way in which mineral products are used. So we discuss the *origin* of mineral deposits without regard to questions of use and economics. Minerals become concentrated in five principal ways:

1. By concentration within magma

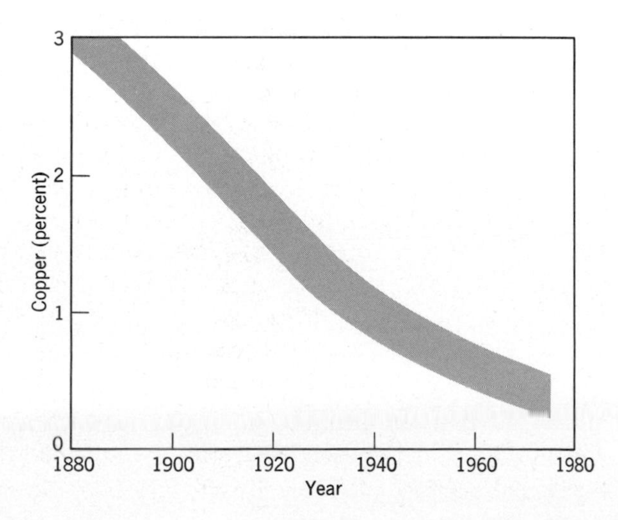


**Figure 22.2**
**Declining percentage of copper needed for a mineral deposit to be ore. The reduction occurred because large-volume mining with efficient machines led to steady reduction of mining costs. (*After U.S. Bur. Mines.*)**

2. By deposition from hot, watery solutions
3. By deposition from lake water or seawater
4. By concentration through weathering and through chemical activity of ground water
5. By mechanical concentration in flowing water.

### Concentration within magma

The processes of partial melting and fractional crystallization (Chapter 15) are, by their very definitions, ways of separating some substances from others. These special circumstances, however, lead to the creation of large and potentially valuable mineral deposits.

**Pegmatites.** When a magma undergoes differentiation by fractional crystallization, the residual liquid becomes progressively enriched in chemical elements that are not present in the early-crystallizing minerals. Separation and crystallization of the remaining liquid produces an igneous rock that contains all the concentrated elements. The most important example of this kind of concentration process is found in pegmatites (Appendix C). Apparently these unusual rocks form by extreme differentiation of deep-seated granitic magma. Commonly they contain significant enrichments of elements such as beryllium, tantalum, niobium, and lithium.

**Magmatic segregation.** We noted in Chapter 15 that another form of magmatic differentiation occurs when dense minerals sink through a magma and accumulate on the floor of the magma chamber. This process is magmatic segregation, and in some cases the segregated minerals make desirable ores. Most of the world's chromium ores were formed in this manner by accumulation of the mineral chromite ($FeCr_2O_4$). The largest chromite deposits are in South Africa, Rhodesia, and the U.S.S.R. Similarly, vast deposits of ilmenite ($FeTiO_3$), a source of titanium, were formed by magmatic differentiation. Large deposits occur in the Adirondack Mountains.

**Immiscible liquids.** Finally, concentration occurs by a kind of magmatic differentiation that we have not mentioned previously.

When ore containing sulfide minerals such as chalcopyrite and pyrrhotite is melted in a smelter, the molten sulfides, being heavy, sink to the bottom, whereas the molten silicates, being light, rise to the top as a sort of scum. Thus the ore melts into two immiscible liquids, just as oil and water, instead of mixing, separate into two layers. The property of immiscibility forms the basis of processing many ores. A similar form of concentration also occurs in nature when, for reasons not clearly understood, certain magmas separate into two liquids, of which one, a sulfide liquid, rich in copper and nickel, sinks to the floor of the magma chamber because it is dense. The world's greatest concentration of nickel ore, at Sudbury, Ontario, is believed to have formed in this fashion. Other great nickel deposits in Canada, Australia, and Rhodesia are believed to have formed in the same manner.

**Concentration by deposition from hot, watery solutions**
Many of the most famous mines in the world contain ores that were formed when their essential minerals were deposited from hot-water solutions. Probably, indeed, more mineral deposits are formed by deposition from hot-water solutions than by any other mechanism. But despite the importance of such ores, the origins and compositions of the solutions remain problematical. Deposition occurs deep underground where we cannot see it happen, and by the time the deposit is finally uncovered by erosion, the solution is no longer present. Nevertheless, enough clues have been found so that the process of deposition is understood at least in a general way.

**Composition of the solutions.** The principal ingredient of ore-forming solutions is water, although small amounts of carbon dioxide, methane, and other fluids are present also. This

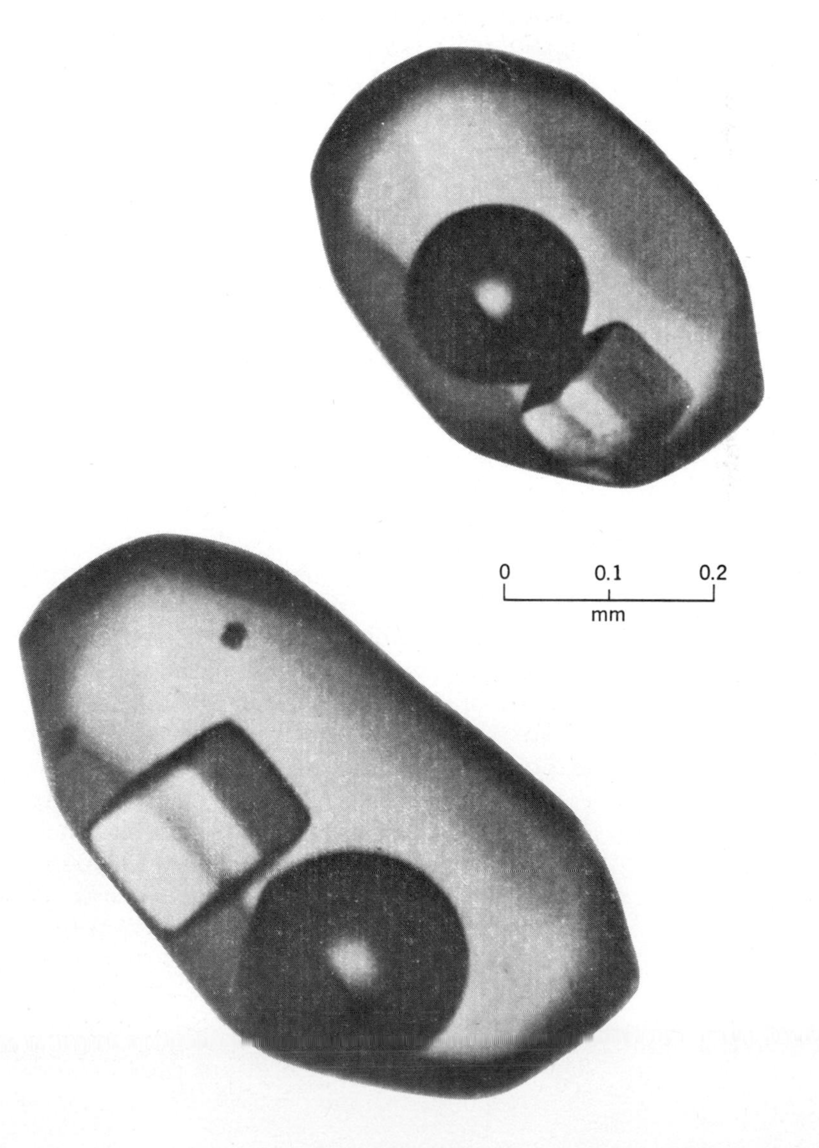


Figure 22.3
**Tiny inclusions of the solution that formed a small mineral deposit in the Swiss Alps were trapped inside a quartz crystal. Similar inclusions, visible only through a microscope, can be found in minerals from many deposits. Dark, circular objects are gas bubbles. When first formed, the inclusions were full of solution and no gas bubbles were present. Later, as temperatures dropped, the solution contracted and the gas bubbles formed. The small cubes are crystals of sodium chloride. They also formed as temperatures dropped and the originally high temperature solution became saturated in salt. If the quartz crystal is heated back to its original temperature of formation, the solution again becomes homogeneous and both gas bubbles and salt crystals disappear. (*Jacques Touret.*)**

information comes from analyses of tiny bubbles of ore fluid trapped inside crystals of quartz and other minerals of the ore (Fig. 22.3). The water is never pure and always contains dissolved within it, salts such as sodium chloride, potassium chloride, calcium sulfate, and calcium chloride. The amounts of such solutes vary, but most solutions range from about the saltiness of seawater (3.5 percent dissolved solids by weight) to about ten times as salty as seawater. The ore-forming solution is therefore a brine, and brines, unlike pure water, are capable of dissolving minute amounts of seemingly insoluble sulfide minerals such as chalcopyrite, galena, and sphalerite.

**Origins of the solutions.** Brines have many sources. The pores in all marine sediments and igneous rocks in the oceanic crust are filled with brines. But even fresh water can become a brine because traces of soluble salts exist in all rock. Water percolating through a sufficiently large volume of rock, therefore, can eventually become a brine. Likewise the percolating brine can dissolve some of the widely dispersed copper, lead, zinc, and other minerals contained in the rock through which it passes.

Reactions between circulating brines and enclosing rock are speeded greatly by high temperature. It is not surprising, therefore, that many mineral deposits seem to have derived their source materials from hot volcanic rock invaded by circulating fluids.

Ore-forming fluids can form also when seawater, trapped in the pores of buried sediments, is squeezed out by compaction. Still others form when granitic magma, or any other magma generated by wet partial melting (Chapter 15), gives off water as it cools. Ore-forming solutions having very similar compositions can apparently form in many ways. What seems to be important is not *where* the solution came from, but *what made it precipitate the valuable minerals it carried.*

**Causes of precipitation.** If a deposit-forming solution moves slowly upward, like ground water percolating through a confined aquifer, changes happen very slowly. If dissolved minerals are precipitated, they are spread out over great distances and will not be sufficiently concentrated to form a useful mineral deposit. But if a solution finds an open fracture, a body of shattered rock, or any other place where it can move rapidly, it can experience sudden changes such as boiling, rapid decrease of temperature and pressure, composition changes caused by reaction with adjacent rock, and dilution by mixing with ground water. Each change can cause dissolved minerals to be precipitated, and if precipitation is confined to a small space, a rich mineral deposit can form. If valuable minerals are involved, an ore results.

**Examples.** As noted earlier, deposits formed by precipitation from hot-water solutions are common. In most areas of metamorphic rock, quartz veins occur (Fig. 22.4). Probably such veins were deposited by hot salty solutions that slowly escaped upward as they responded to the pressure and heat involved in

Figure 22.4
**Quartz vein cutting across fine-grained metamorphic rock. Such veins are formed when dissolved minerals are deposited from hot, watery solutions. Ruler is scaled in inches. West Cummington, Massachusetts. (*B. M. Shaub.*)**

metamorphism. Most veins which are products of metamorphism are not, unfortunately, ore deposits.

In regions of volcanic activity, particularly where rhyolite and andesite occur, many veins containing valuable minerals are found. Probably this is because volcanism heats solutions very near the surface, making them effective ore-formers. The famous gold deposit at Cripple Creek, Colorado and the huge tin deposits in Bolivia were formed in this fashion. Formation of hot solutions by volcanism seems to occur even when volcanism takes place beneath the sea. The gold deposits at Kirkland Lake in Ontario and Kalgoorlie in Western Australia were formed apparently during submarine volcanic activity. So too were the famous copper deposits of Cyprus, and those in New Brunswick and many other areas.

When a body of granitic magma cools near the surface, it is a source of heat just as a volcano is, and so can become a source of hot solutions. Such a solution moves outward from the cooling intrusive body, flowing through any fracture or channel, altering the surrounding rock, and commonly depositing valuable minerals (Fig. 22.5). Many famous ore bodies are associated with shallow

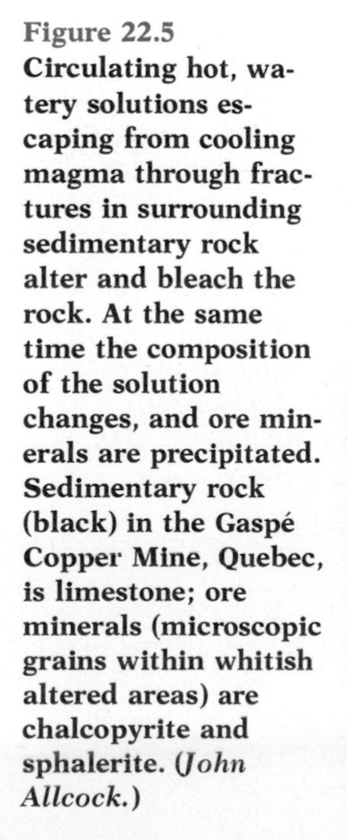

Figure 22.5
**Circulating hot, watery solutions escaping from cooling magma through fractures in surrounding sedimentary rock alter and bleach the rock. At the same time the composition of the solution changes, and ore minerals are precipitated. Sedimentary rock (black) in the Gaspé Copper Mine, Quebec, is limestone; ore minerals (microscopic grains within whitish altered areas) are chalcopyrite and sphalerite. (*John Allcock.*)**

intrusive rocks. The copper deposits at Butte, Montana, Bingham, Utah, and Bisbee, Arizona, are examples.

Some deposits form from solutions that apparently start as ordinary ground water. As such water circulates through deep aquifers, it becomes transformed into a hot, salty solution. This reacts with adjacent rock, preferentially leaching certain metals; then, when circulation brings the solutions close to the surface once again, deposits the metals in the form of valuable minerals. Examples are the zinc deposits near Picher, Oklahoma, the lead deposits in southeastern Missouri, and the uranium deposits of the Colorado Plateau region.

Many of the ore bodies formed by deposition from hot brines are alike in that they are related in one way or another to igneous activity. We have seen already that igneous activity tends to be concentrated along the boundaries of plates. It is not surprising, therefore, that certain kinds of ore deposits are also found to be concentrated near plate boundaries. This realization is now being studied in detail, because it may help in the discovery of new deposits—especially if former plate boundaries can be identified. The relation between plate boundaries and ore deposits provides a possible explanation for the formation of a ***metallogenic province.*** This is *a limited region within which ore deposits occur in unusually large numbers.* A striking example is the metallogenic province that runs along the western margin of the Americas. Within the province is the world's greatest concentration of large copper deposits (Fig. 22.6). These deposits are associated with porphyritic igneous rock and hence are called *porphyry coppers.* The igneous bodies, and therefore the deposits themselves, are believed to have formed as a consequence of subduction of plates. Although many details of the process are obscure, the close parallelism between the convergent plate margins shown in Figure 4.4 and the metallogenic province shown in Figure 22.6 strongly supports the notion that porphyry copper deposits must somehow be connected with subduction.

### Deposition from lake water and seawater

In many saline lakes such as Great Salt Lake, sodium chloride and other nonmetallic compounds are precipitated. These compounds are derived, through tributary streams, from the products of chemical weathering of the rocks exposed within the lake's drainage area. The sequence: weathering → stream transport → precipitation in a water body is a prime example of natural concentration. Many salts of sodium (including common salt), potassium, and boron are recovered from saline lakes and playas (Chapter 9; Fig. 22.7). Similarly, enormous quantities of common salt and of the calcium sulfates (gypsum and anhydrite) used in making plaster are recovered from sedimentary layers precipitated in cut-off arms of ancient shallow seas. Much of the production of salt in the United States comes from such layers in rocks of Silurian age, more than 400 million years old.

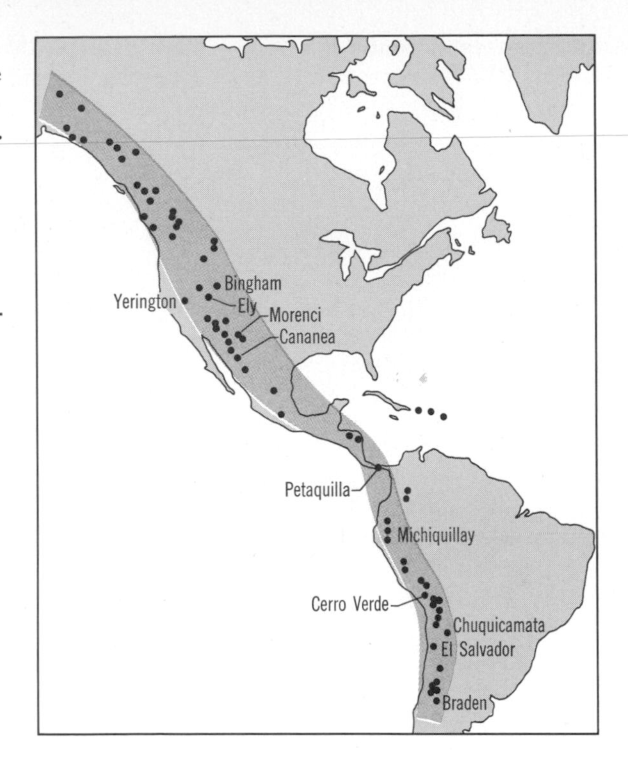


**Figure 22.6**
**Metallogenic province along the western edge of the Americas. The province parallels the edge of the American Plate, and is believed to have originated in igneous activity caused by subduction of the plate. Many of the world's largest copper deposits occur in the province. Some important mines are named.**

Salt, gypsum, and other nonmetallic substances have been concentrated by evaporation. Iron and the important fertilizer-mineral apatite have also been concentrated from solution in the sea, but their abundance in average seawater is so small that thick beds of iron minerals could not have been concentrated solely by evaporation; some other process must have been responsible.

**Figure 22.7**
**Salts, deposited by evaporation of ancient lake water, are mined near Boron, California. Boron salts (principally the mineral borax) were deposited about 15 million years ago. Evaporation of Rodgers Lake (white playa in background) is presently depositing sodium chloride. View looking southwest; pit is about 100m deep and 1.5km across. (*U.S. Borax Company.*)**

An example of such beds is the Clinton iron formation, occurring as lens-like bodies in several sedimentary units, one of them locally more than 10m thick, that extend from New York State southwestward for 1,100km into Alabama. The steel industry of Birmingham, Alabama, is based on the convenient proximity of iron-rich sediments to coal, and to limestone needed for smelting. Fossils show that the ore strata were deposited in a shallow sea during the Silurian Period. Ripple marks and mud cracks further indicate shallow water. Apparently these beds were parts of the thick pile of sediment that accumulated in the Appalachian geosyncline. The iron mineral is hematite, and the proportion of iron is 35 to 40 percent.

It is believed that the iron was dissolved from iron-bearing minerals in mafic igneous rock, carried (perhaps as a bicarbonate) by ground water to the shallow sea, and there precipitated as oxides when the ground water came into contact with seawater containing dissolved oxygen. The precipitate is a chemical sediment and the iron concentrated in this way amounts to billions of tons. The enormous iron-rich strata in the Alsace region in eastern France and similar strata in England, both originated in this way.

Large and impressive as the Clinton deposits and others like them are, they are all tiny by comparison with the class of deposits characterized by the Lake Superior iron deposits (Fig. 22.8). The deposits are the mainstay of the United States steel industry. Similar ores are found in Labrador, Venezuela, Brazil, U.S.S.R., India, and Australia, and are one of the most unusual kinds of chemical sediment that are known. They differ from the Clinton deposits in two important respects. First, whereas the Clinton deposits contain fossils and other detritus, the Lake Superior deposits are sediments of wholly chemical origin and are free of all detritus. Second, whereas the Clinton deposits contain calcite as a gangue, the Lake Superior deposits contain very fine bands of cherty quartz. Every aspect of the Lake Superior deposits indicates they are chemical precipitates. Because the deposits are so large, it is difficult to envision how so much iron could be transported in ground water. We therefore infer that transportation must have taken place in surface water, presumably by streams. This inference forces us to a further conclusion. Because the Lake Superior deposits, and others like them, formed more than two billion years ago, the composition and properties of the atmosphere and seawater may have been different from those of today. One important difference involved the atmosphere: It must have contained less oxygen, so the surface waters were less oxygenated than those of today and could transport large amounts of dissolved iron. The differences in atmosphere and seawater would lead to different styles of weathering, transport, and precipitation from those we see today and could account for the unusual composition of the iron-rich sediments. However the unusual Lake Superior deposits formed, they will be sources of iron for centuries ahead.

Figure 22.8
**A deep open-pit iron mine in Minnesota, where Lake Superior-type sedimentary iron ore is mined. Light-colored strata overlying the ore are visible on both sides of the pit. The iron ore is the dark rock at the bottom of the pit.** (*Minnesota Dept. Economic Development.*)

Precipitation from seawater is also the origin of most of the world's manganese deposits. The two largest deposits occur in the U.S.S.R. But manganese minerals are unusual in that they are precipitating, even today, on the deep-sea floor. Nodular growths of manganese oxides occur on the floors of all the oceans (Fig. 22.9). Ranging in diameter from 1cm to as much as 30cm, the nodules form by very slow precipitation from seawater. Individual nodules may take millions of years to grow. But the sea floor is so extensive, and in places the frequency of nodules is so high, that literally trillions of nodules are present. Recovery tests to dredge the nodules from the floor of the Pacific Ocean are already under way. The nodules contain, in addition to the manganese minerals, one percent or more of copper and nickel. The copper and nickel may prove to be even more valuable than the manganese.

### Concentration caused by weathering and ground water

Deposits are concentrated in two general ways: secondary enrichment and residual concentration.

**Secondary enrichment.** The related processes of sorting and concentration by ground water are chemical. They depend on the principle that change in the environment of a mineral may make it

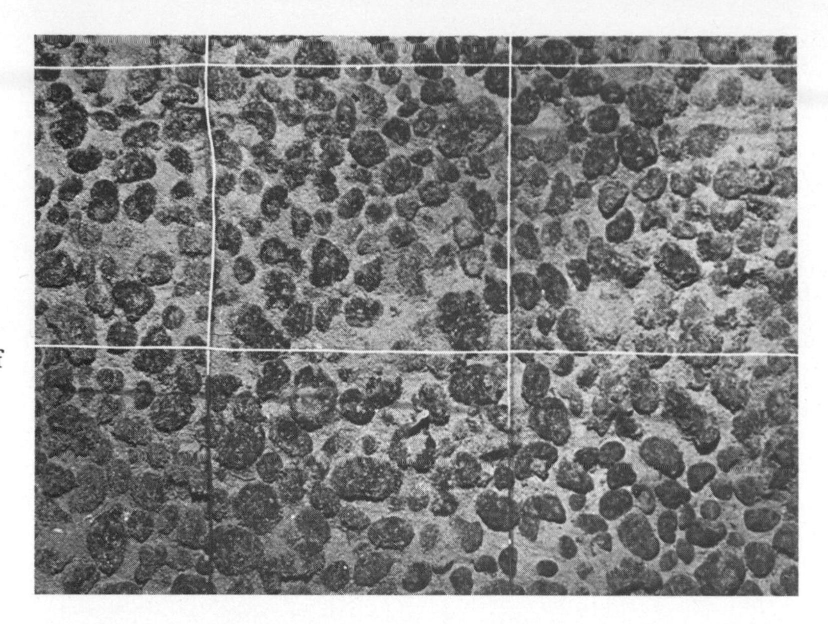

**Figure 22.9**
**Nodular deposits of manganese- and iron-oxide minerals are widely distributed on the deep-sea floor. This photograph, made on the floor of the Pacific Ocean, shows a device used in estimating the abundance and size of nodules for the purpose of mining them. Size of grid, 1 foot. (*Global Marine, Inc.*)**

vulnerable to chemical attack. Although stable at the high temperatures at which they crystallized from solution, many minerals found in rich deposits become unstable in the zone of aeration at Earth's surface. In areas where erosion has worn the surface down, even formerly deep-lying minerals are brought within the shallow zone in which atmospheric oxygen and ground water can attack them. Thus, in the new environment of low temperature and low pressure, these minerals decay. Ground water dissolves pyrite ($FeS_2$) creating sulfuric acid ($H_2SO_4$) and residual iron oxides. The acid dissolves copper minerals, forming copper sulfate. Entering the zone of saturation, the solution loses oxygen and reacts with minerals there, depositing copper as sulfides. In a vein, the added sulfides enrich the minerals already present and may convert a low-grade deposit into ore (Fig. 22.10). This process of addition is known as ***secondary enrichment***, defined as *natural enrichment of an ore body by addition of mineral matter, generally from percolating solutions.*

Secondary enrichment has been important at the Utah Copper Mine (Fig. 23.1) at Bingham Canyon, Utah, the largest copper producer in the United States. The ore lies in the upper part of a body of intensely fractured intrusive igneous rock. The minerals

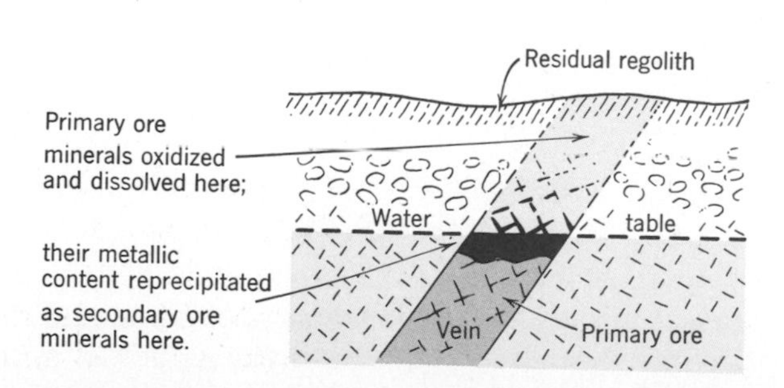


**Figure 22.10**
**Descending ground water impoverishes ore above water table by removing soluble ore minerals; it produces secondary enrichment below.**

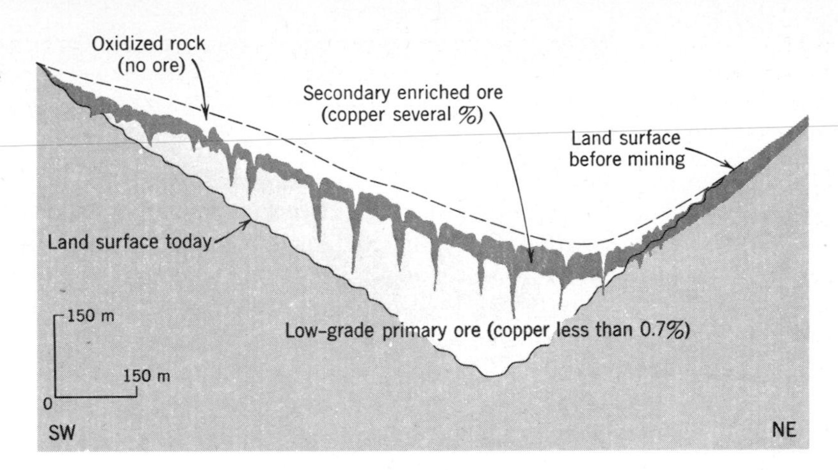


**Figure 22.11 Diagrammatic section across Utah Copper Mine (Fig. 23.1) at Bingham Canyon, Utah, shows how weathering concentrated copper in an enriched zone related to a land surface.**

containing the copper, gold, silver, and molybdenum now found in the deposit were carried upward by hot solutions from the magma underneath, and were deposited in the innumerable tiny fractures of the igneous rock. The upper part of the primary ore was secondarily enriched through a zone more than 30m thick (Fig. 22.11). This enormous low-grade deposit is ore only because it does not demand underground mining and can be worked by low-cost surface methods. The sides of the canyon have been cut into giant steps 15m to 25m high, making a huge pit resembling a Greek amphitheater. The ore is blasted out and is loaded by power shovels into railroad cars. Waste is dumped into side canyons. The ore already mined exceeds 1.25 billion tons, and some ore containing as little as 0.4 percent copper is being worked profitably.

**Residual concentration.** In secondary enrichment, ground water adds new material to an existing body. Some ores, however, are developed by subtraction of old material. This is ***residual concentration,*** defined as *the natural concentration of a mineral substance by removal of a different substance with which it was associated.* An example is bauxite, a mixture of various hydrous aluminum oxide minerals and the preferred source for aluminum.

Bauxite is a product of special chemical weathering. Aluminum is present as a constituent of original, primary silicate minerals in igneous rocks, schist, or clay. During weathering, silica is carried away in solution, gradually increasing the concentration of aluminum. We know this has happened because we can see bauxite grading downward into the underlying rock just as does any weathered regolith. From the locations of bauxite ores we know that the weathering has taken place beneath peneplains and other surfaces of low relief, with the water table close to the surface, with slow circulation of ground water, and always in tropical or subtropical climates. But the chemistry of the process is complex, and we still have much to learn about just what happens. Apparently ground water sometimes has just the acidity necessary to decompose the original silicate minerals and to carry silica away in solution, but not to remove the aluminum (and iron) oxides. Even though the rock originally contained no more than one percent aluminum, the resulting bauxite may reach a concentration of 40 percent.

Probably the richer iron ores of the districts in which ores similar to those around Lake Superior are found have originated in a similar way. The primary deposits contain about 25 percent iron, but near the surface pockets as rich as 66 percent are found. Apparently long-continued weathering has dissolved and removed the silica, leaving a residual concentration of almost pure hematite.

### Mechanical concentration in flowing water

The famous gold rush to California in 1849 resulted from the discovery that the sand and gravel in the bed of a small stream contained bits of gold. Indeed, many mining districts have been discovered by following trails of gold and other minerals upstream to their sources in veins in bedrock. Because pure gold is heavy

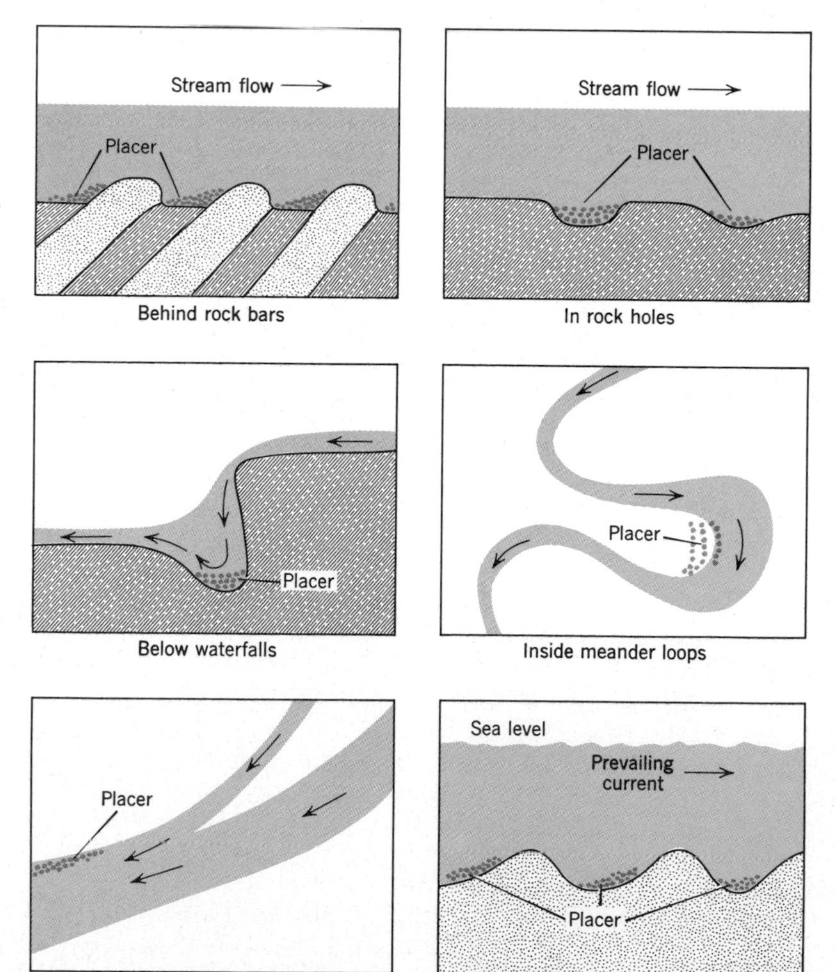


**Figure 22.12**
**Placers occur where barriers allow stream water to carry away suspended load of lightweight particles while trapping denser particles in bed load. Placers can form wherever water moves, but are most commonly associated with streams.**

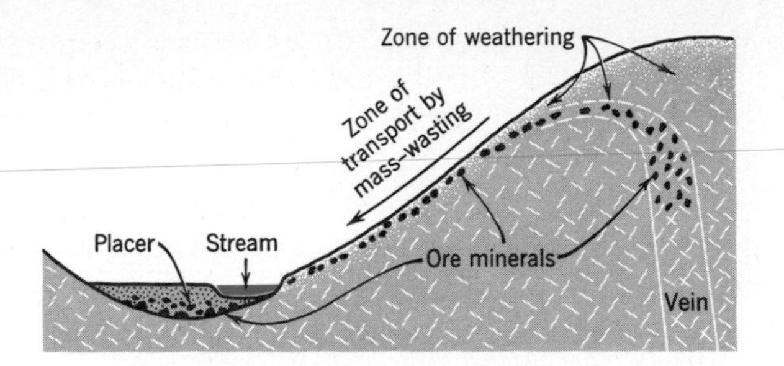


**Figure 22.13 Chemically resistant minerals from weathered fissure veins creep down slopes by mass-wasting and are redeposited by streams as placers.**

(density 19gm per cm$^3$), it is deposited from the bed load of a stream very quickly, while quartz, with a density of only 2.65gm per cm$^3$, is washed away. As most silicate minerals are light by comparison with gold, the gold becomes mechanically concentrated in places where the velocity of stream flow is least (Fig. 22.12). *A deposit of heavy minerals concentrated mechanically* is a ***placer.*** Besides gold, other heavy, durable metallic minerals form placers. These include minerals that occur as pure metals, such as platinum and copper, as well as tinstone (cassiterite, $SnO_2$), and the nonmetallic minerals diamond, ruby, sapphire, and other gemstones. Even if a vein contains a low percentage of gold or tin, the placer it yields may be quite rich.

Every phase of the conversion of gold in a fissure vein into placer gold has been traced. Chemical weathering of the exposed vein releases the gold, which then moves slowly downslope by mass-wasting (Fig. 22.13). In some places mass-wasting alone has concentrated gold or tinstone sufficiently to justify mining these metals.

More commonly, however, the mineral particles get into a stream, which concentrates them more effectively than mass-wasting can do. Most placer gold occurs in grains the size of silt particles, the "gold dust" of placer miners. Some of it is coarser; pebble-sized fragments are *nuggets* (Fig. 22.14), of which the largest ever recorded weighed 70.9kg and at the price of gold in December, 1975 (US $4.50 per gram) is worth $319,100. In following placers upstream, prospectors have learned that rounding and flattening

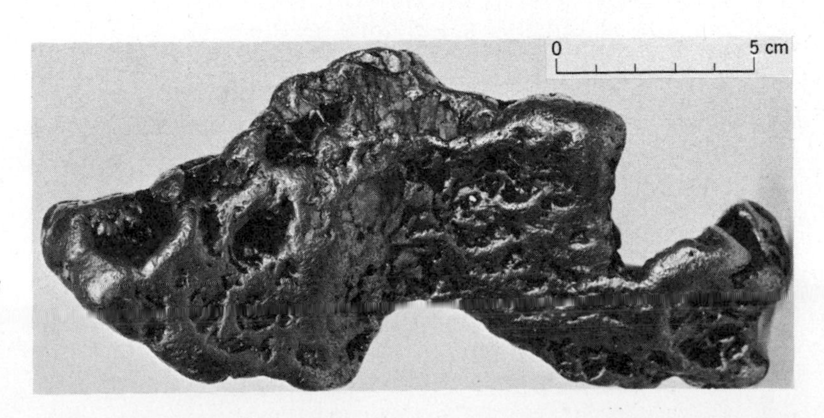


**Figure 22.14 Gold nugget, found near Greenville, California, weighs 2,550gm. Its surface shows clearly the pounding it received during stream transport. (*George Switzer; collection U.S. National Museum.*)**

(by pounding) increase downstream, just as does rounding of ordinary pebbles; when they find angular nuggets, prospectors know bedrock source is close.

Mining was done first by hand, simply by swirling stream sediment around in a small pan of water. Later it was done by jetting water under high pressure against gravel banks and washing the sediment through troughs that caught the heavy grains of gold behind cleats. Nowadays the more efficient method of dredging is used. The platinum placers of the Ural Mountains in the U.S.S.R., the rich diamond placers in Zaire, and the hundreds of tin placers in Malaya and Indonesia are examples of other mechanical concentrations by streams.

Gold, diamonds, and several other minerals have been concentrated in beaches by surf. Diamonds are being obtained in large quantities from gravelly beach placers, both above and below present sea level, along a 350-kilometer strip of the coast of South-West Africa. Weathered from deposits in the interior, the diamonds were transported by the Orange River to the coast and were spread southward by longshore drift. Later some beaches were lifted up, and others were submerged by rise of sea level.

Finally, gold is recovered in Mexico and in Australia from placers concentrated by wind. Quartz and other lightweight minerals have been deflated, leaving the heavier gold particles behind. Therefore, because mechanical concentration results from a variety of natural processes, the principles of erosion and deposition must be thoroughly understood by those who would discover new placers.

## USEFUL MINERAL SUBSTANCES

Mineral substances are so important in our daily lives that the question of adequacy of supply is important. Minerals and mineral products can be grouped for discussion on the basis of the way we use them. Excluding substances used for energy, there are two broad groups: (1) minerals from which metals such as iron, copper, and gold can be recovered, and (2) mineral substances such as salt, gypsum, and clay used not for the metals they contain but for their properties as chemical compounds. The nonmetallic substances can be further subdivided on the basis of more specialized uses (Table 22.1).

### Metals

Without exception, the useful metals are present in the crust in such low amounts that we can only mine and recover them when rich deposits can be located. Just how much richer than ordinary rock the mineral deposit must be before it is an ore deposit depends on the metals present. For example, iron must be enriched five times above its average value in the crust, while copper requires an enrichment of 75 times, platinum 600, and mercury an enormous 100,000 times (Fig. 22.15).

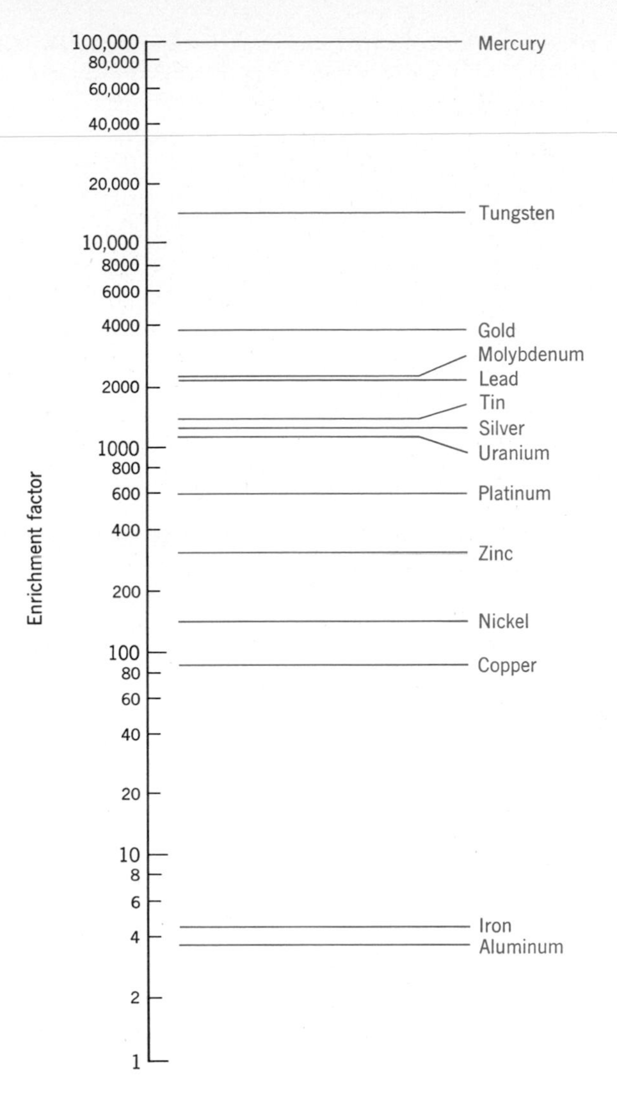


**Figure 22.15**
**Before a mineral deposit can be called an ore, the percentage of valuable metal in the deposit must be greatly enriched above its average percentage in Earth's crust. The enrichment is greatest for metals that are least abundant in the crust, such as gold and mercury. As mining and mineral processing become more efficient and less expensive, it is possible to work leaner ore and enrichment factors decline. Note that the scale is a magnitude (logarithmic) scale, in which the major divisions increase by multiples of ten.**

**Table 22.1**
**Principal Mineral Substances, Grouped According to Use.**

| |
|---|
| (1) *Metals*<br>Iron, aluminum, copper, lead, zinc, nickel, silver, gold, tin, mercury, silver, chromium, manganese, molybdenum, uranium, and many others |
| (2) *Nonmetallic Substances*<br>(a) Used for chemicals: salt, sodium carbonate, sulfur, borax, fluorite.<br>(b) Used for fertilizers: phosphate, potash, sulfur, calcium carbonate, sodium nitrate<br>(c) Used for building: gypsum, calcite, clay, asbestos, sand, gravel, shale (for cement)<br>(d) Used for ceramics and abrasives: clay, feldspar, quartz, diamond, garnet, corundum |

Metals can be usefully subdivided on the basis of their average percentage in the crust. Those present in such abundance that they make up 0.1 percent or more of the crust—iron, aluminum, manganese, magnesium, and titanium—are considered to be *geochemically abundant.* Geochemically abundant metals require comparatively small enrichment factors to form extremely large deposits. The size, distribution, and abundance of ore deposits of geochemically abundant metals is so great that we can consider the supplies to be virtually inexhaustible. Even if preferred ores, such as the bauxites now used as sources of aluminum, should be used up, there are many other kinds of deposits to which we could turn without a great increase in cost of the final product. For example, there are huge deposits of minerals such as kaolinite and anorthite that are almost as much enriched in aluminum as bauxite is, and the deposits are far more abundant.

Metals that make up less than 0.1 percent of the crust (and are said to be *geochemically scarce*) behave very differently from the geochemically abundant metals. Every common rock contains one or more minerals of iron and aluminum, and a great many contain the minerals of the other abundant metals, magnesium, titanium, and manganese. By contrast, with the exception of copper and chromium, minerals of the scarce metals are never present in common rocks. But chemical analysis of any rock reveals that even though the minerals are absent, geochemically scarce metals are certainly present. Further research reveals that the scarce metals are present exclusively in solid solution (Chapter 3) in the common rock-forming minerals. Atoms of the scarce metals substitute for common atoms such as magnesium and calcium. In order for a mineral deposit to form, therefore, some gathering and concentrating agent such as a hot brine must react with the rock-forming minerals, leach the scarce metals from them, transport the metals in solution and deposit them as sulfide and oxide minerals (or even, as in gold, as the metal itself) in a localized place. With such a complicated chain of events it is not surprising that deposits of geochemically scarce metals are rare, areally restricted, and very much smaller than deposits of geochemically abundant metals.

The number and size of ore deposits containing scarce metals is proportional to the percentage of the metals in the crust. In Figure 22.16 the sum of all the ore discovered in the United States is plotted against the crustal abundance of selected metals. Clearly, the amount of economically recoverable scarce metal is related to its average content in the crust. Even though iron is concentrated by many different processes, it is apparent in that figure that it too seems to fit the relationship. Perhaps, then, if all mineral deposits, whether ore or not, could be found, the amount of metal concentrated would be proportional to the abundance of metal in the crust. From this we might conclude that a prudent society should use metals at rates proportional to the supplies available and hence proportional to their geochemical abundances. In actual fact our worldwide rates of use are far from this ideal balance. By

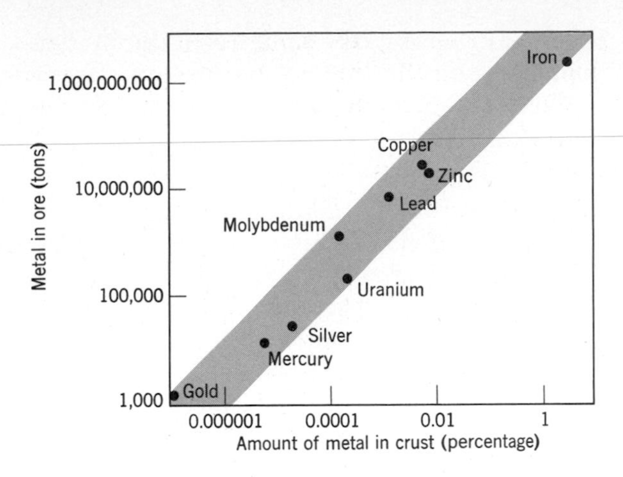


**Figure 22.16 Relation between the average content of a metal in the crust (geochemical abundance) and the metal present in all the ore deposits discovered in the United States.** ***(After V. E. McKelvey, 1973.)***

comparison with iron, which we have seen is very abundant, we use copper at a rate 13 times faster than we should, lead 41 times faster, and mercury 91 times faster. The scarcer the metal, it seems, the faster we are using it up. This leads to the prediction that within the lifetimes of people living today, demands for some of the geochemically scarcer metals such as mercury, silver, and gold will cease to be met. We will either have to develop ways of near-perfect recycling of scarce metals or have to learn eventually to do without them and to substitute more abundant materials instead.

### Nonmetallic substances

The great diversity of nonmetallic substances makes it difficult to devise an unambiguous classification scheme. Some substances have several different uses. Although far from perfect, the list in Table 22.1 includes the major uses of nonmetallic substances.

**Chemical materials.** The most important sources of materials for the chemical industry are organic materials—coal, oil, and gas—that are the raw materials needed for petrochemicals such as plastics, many drugs, pesticides, synthetic fibers, and countless other products. But many inorganic substances are needed also. The most important are sodium chloride, sodium carbonate, sodium sulfate, and minerals containing boron (such as borax, $Na_2B_4O_7 \cdot 10H_2O$). Each of these is recovered from evaporites. Fortunately the supplies of all of the chemical substances are very large.

**Fertilizer materials.** Growing plants draw the many chemical elements they need from the soil. Nature has its own way of recycling the necessary elements. When a plant dies and decays, rainwater washes the chemicals back into the soil. There they can be used again during the next growing season. Some are removed by ground water, but the loss is made up by chemical weathering of minerals in the bedrock. When we interfere with this process by growing crops and removing plant matter for food, the soil becomes progressively depleted in the needed elements. They must then be replaced in the form of soluble fertilizers added to the soil.

The three essential fertilizer elements are: (1) nitrogen, recovered by chemical means from the atmosphere, (2) potassium, recovered as the soluble salt potassium chloride from marine evaporite deposits, and (3) phosphorus, recovered as apatite ($Ca_5(PO_4)_3OH$) from a special class of marine chemical sediments known as *phosphorites*. For each essential fertilizer element the world's supplies are very large, and probably sufficient for all foreseeable needs. But there are difficulties associated with the distribution of fertilizers. They must be used in large amounts and costs of treatment are high. Also, deposits of phosphorus and potassium are geographically restricted, so that costs of transport become important. This makes it difficult for poor countries such as India to purchase all the fertilizers they need.

In addition, calcium carbonate and sulfur must be added to soils to keep a balance between acidity and alkalinity favorable for maximum plant growth. Calcium carbonate is produced from the abundant limestone strata around the world, but sulfur is more restricted, forming mainly in fumaroles (Chapter 15), in certain gypsum deposits in marine evaporites, and in natural gas. Supplies of sulfur are very large, but unfortunately for poor countries that need the sulfur, they are geographically restricted.

**Building materials.** Besides cut stone, crushed stone, and sand and gravel, of which the world has enormous resources, the main building materials are cement (manufactured from claystone and limestone), gypsum (used for plaster), clay (used for tile and brick), and asbestos (used for wallboards, sidings, and insulations). With the exception of asbestos, which is formed principally as an alteration product of peridotites by hot ground water, supplies of building materials are enormous. Problems associated with their use tend to be problems of transportation, siting of quarries in environmentally acceptable areas, and costs of treatment and preparation. It has been said that there will be a shortage of standing room on Earth before a shortage of building materials develops.

**Ceramic and abrasive materials.** We sometimes forget how extensively we use ceramic materials (including glass) in our everyday lives. Some ceramics have properties that make them desirable substitutes for certain metals; so we can anticipate that uses of ceramics will grow. Supplies of the essential raw materials—clays, feldspars, and quartz—are abundant and widespread.

Abrasives also play a widespread and vital part in our lives. Their importance is commonly not appreciated. Modern industry requires machines that work accurately and efficiently at high speeds. This requires precision grinding, shaping, and polishing of the machine parts, and for this a wide variety of abrasives is needed. The abrasives must be hard and tough enough not to fracture easily. The principal abrasive minerals are quartz and garnet (both common rock-forming minerals), corundum (rare as a mineral but easily synthesized), and diamond. It is not commonly appreciated that about 80 percent of all the diamonds produced are used as abrasives, the rest being cut as gems. Nor is it widely realized that although the primary source of diamonds is in a rare kind of igneous rock erupted from the mantle, most diamonds are recovered from placers. This is so because when diamonds do reach the surface, they are so hard and tough that even pounding in streams does not quickly wear them down. Because they are more dense than the silicate minerals with which they occur, they are readily concentrated in placers. Most of the world's production comes from placers in Zaire and from Siberia.

## FUTURE SUPPLIES OF MINERALS

Mineral deposits are exploited in the least expensive way possible. For some deposits this means underground mines in which miners drill and blast away at narrow veins (Fig. 22.17); the quantities of material dug from the ground in this fashion are truly enormous (Fig. 22.18). For other deposits, exploitation means huge open pits from which the ore is removed by the truck or trainload (Figs. 22.7, 23.1); for still others such as some salt bodies, exploitation may mean solution in hot water pumped down a drill hole and recovery of salt by evaporation from the resulting brine. All phases and kinds of mining have become increasingly efficient, so that once a deposit is discovered, it can be worked quickly. But how can we discover enough deposits? Minerals for industry, like fossil fuels,

Figure 22.17
**Drilling uranium ore in the New Quirke Mine, Ontario. The miner on the left is operating a drill. The holes he drills will be filled with explosives to break the ore free. (*Rio Algom Ltd.*)**

**Figure 22.18 Man-made mountains dot the landscape around Johannesburg, South Africa, adjacent to the world's largest gold mines. Underground mining to depths as great as 3.5km has removed vast quantities of gold ore. After the ore has been crushed and the gold separated, the residue, or tailing, is heaped on the surface. Eventually, trees and other vegetation will cover the heaps. (*Georg Gerster, Rapho Photo Researchers.*)**

are being used at ever-increasing rates, and mineral deposits, as we have said, are a one-crop-only resource.

At present, new mineral deposits are being found as fast as old ones are being exhausted. But the new finds are mostly in places such as South Africa, Australia, and Siberia, where intensive prospecting has begun rather recently. In Europe, which has been intensively prospected for many years, new deposits are found rarely; indeed, in the areas conquered and occupied by the Romans, no new deposits of metals used by them have been discovered since they departed more than 1,500 years ago. Perhaps, therefore, when all parts of Earth's surface have been prospected intensively, people will have to look to kinds of deposits not presently exploited. For geochemically scarce metals such as mercury, silver, gold, and molybdenum, that time may soon come. For these scarcer materials, careful conservation of supplies will some day have to be practiced, substitutes will have to be found for some of their applications, and efficient recycling will become essential. For materials that are abundant in the crust, such as iron, aluminum, common salt, and sulfur, supplies will probably always be sufficient to meet demand.

Probably the future of mineral supplies will be similar to the future of energy supplies. A sufficiency will be found, but present patterns of use will change; some materials will become more important, others less important, and yet other substances, not used today, will assume major importance. But one prediction can safely be made: Mining operations will get bigger, more materials will be used, and pollution will therefore be intensified. Environments will come under even heavier pressure than that which endangers them today. In the next chapter we will examine the environment further, and see how the material activities of civilization have become major factors in shaping the face of the Earth.

## SUMMARY

1. The mineral industry is based mainly on local concentrations of useful minerals in mineral deposits.

2. When a mineral deposit can be worked profitably it is called an ore deposit.

3. Mineral deposits form in five different ways: by concentration within magma; by deposition from hot, watery solutions; by deposition from lake water and seawater; by concentration caused by weathering and ground water; and by mechanical concentration in flowing water.

4. Deposits of geochemically abundant metals, which make up 0.1 percent or more of the crust, form in several ways. The amounts available for exploitation are enormous.

5. Geochemically scarce metals, present in the crust in amounts less than 0.1 percent, form deposits mainly by deposition of minerals from hot, watery solutions. Amounts of scarce metals available for exploitation are limited and geographically restricted.

6. Secondary enrichment by weathering and ground water has been important in further concentrating many copper deposits.

7. Gold, platinum, tinstone, diamonds, and other minerals are commonly found mechanically concentrated in placers.

8. Nonmetallic substances are used mainly in the chemical industry, for fertilizers, for building materials, and for ceramics and abrasives.

9. The most critical mineral substances, as far as future supplies are concerned, seem to be the geochemically scarce metals.

## SELECTED REFERENCES

Bateman, A. M., 1951, The formation of mineral deposits: New York, John Wiley.

Bates, R. L., 1960, Geology of the industrial minerals and rocks: New York, Harper and Row.

Brobst, D. A., and Pratt, W. P., eds., 1973, United States mineral resources: U.S. Geol. Survey Prof. Paper 820.

McDivitt, J. F., and Manners, Gerald, 1974, Minerals and men: Baltimore, Md., Johns Hopkins Univ. Press.

Skinner, B. J., 1976, Earth resources, 2nd ed.: Englewood Cliffs, N.J., Prentice-Hall.

Skinner, B. J., and Turekian, K. K., 1973, Man and the [illegible] Englewood Cliffs, N.J., Prentice-Hall.

U.S. Bureau of Mines Staff, 1970, Mineral facts and problems: U.S. Bureau of Mines Bull. 630

Warren, Kenneth, 1973, Mineral resources: New York, John Wiley.

# 23 MAN AS A GEOLOGIC AGENT



Across a span of perhaps 15,000 years, the human race has radically expanded the part it plays in Earth's ongoing natural processes. The succession of man-made changes set forth in Chapter 1 has involved four long-drawn-out events:

1. Extensive deforestation of Earth's surface.
2. An enormous increase in human population.
3. An accelerating increase in technical capability that has caused us to use ever-increasing amounts of natural resources.
4. Rapid urbanization (in industrialized countries), whereby the majority of people live crowded into comparatively restricted areas, under generally artificial conditions. Such conditions necessitate the importation of food and water and also demand the transport of people, over an average of about 16 kilometers per individual per day, between home and work.

These four have now grown so overwhelming that we must look closely at human interference with the steady state in lithosphere, hydrosphere, atmosphere, and biosphere. The rock cycle, water cycle, and other cycles have been distorted in a variety of ways. The present chapter describes examples of some of the principal kinds of interference for which mankind alone is responsible. *The application of geology to the problems involved in the interaction between man and the rest of the Earth* is the domain of ***environmental geology.***

# MAN AND THE LITHOSPHERE

**Direct shifting of rock and regolith**

The extent to which mankind has increased the old natural rates of erosion, merely by moving materials (Fig. 23.1), seems hard to believe. Nonetheless, the changes have been established by careful sampling. In a published report to a Maryland state commission, a geologist* remarked that the tonnage of sediment eroded from an acre of ground under construction in housing developments and highways can exceed 20,000 to 40,000 times the amount eroded naturally from an acre of woodland during the same length of time.

Rates of erosion of areas torn up by strip mining of coal probably are even greater. Although formerly most coal mining in the United States was done underground, leaving the surface almost undisturbed, surface stripping, which began on a large scale in 1915, now accounts for more than half of U.S. coal production. Stripping is done with gigantic power shovels that cut horizontal shelves following the contours of hills. The shelves, as much as 50m wide, are floored with a jumble of broken rock waste, and are backed by walls of rock as much as 30m high (Figures 23.2 and 23.3). By 1972 the Appalachian region had become scarred with more than 32,000km of such cuts.† Thus far, little has been done to remedy the instability the cuts have created. Stripping destroys both vegetation and soil, lowers the water table, acidifies the ground water and pollutes streams, creates landslides, and promotes erosion, which leads to the deposition of sediment on valley floors.

The examples cited are extreme, but the broad effect of human activities on erosion rates is reflected clearly in thoughtful estimates of man-induced erosion on nationwide and worldwide scales. In Table 23.1 (see also Fig. 1.1) we see the huge amounts of

* M. G. Wolman (1964).
† Open-pit mining of this and other kinds has scarred more than 45,000km$^2$—0.5 percent of the area of the United States.

**Table 23.1**
**Regolith and Rock Moved Artificially, According to an Estimate Made in 1971. (Metric tons per year.) (S. Judson, Geol. Soc. America Abstrs., 1971, p. 615)**

| Activity | USA | World |
|---|---|---|
| Mining and quarrying | 5 billion ($5 \times 10^9$) | 30 billion ($3 \times 10^{10}$) |
| Construction (roads, buildings, dams, etc.) | 10 billion ($10^{10}$) | 50 billion ($5 \times 10^{10}$) |
| Farming | 1 trillion ($10^{12}$) | 6.5 trillion $6.5 \times 10^{12}$) |

**Figure 23.1**
**Long-continued removal of ore has made this hole, nearly 600m deep, in a mountain range in Utah, and is steadily enlarging it. The Bingham Canyon Copper Mine is one of Earth's biggest man-made excavations.** (*Courtesy Kennecott Copper Corporation.*)

regolith and rock that are believed to be moved artificially each year. Farming accounts for the lion's share of the material moved, because the combined areas under cultivation far exceed the combined areas of all structures and all mines.

A large-scale example of shifting of regolith to convert sea into land in the Netherlands is described in Chapter 1.

### "Landslides" caused by man as an earth mover

The building of a dam to hold back a stream alters the natural steady state by steepening the stream's gradient, thereby increasing the potential for destructive erosion if the dam should fail. The reasons for failure are many, but all too commonly the failures lead to disaster. The earliest dam known to history, built across a tributary to the Nile in Egypt about 4,000 years ago, failed soon after it was built. The destructive failure of the St. Francis Dam in California is described in Chapter 7.

**Vaiont event.** A colossal "landslide" that destroyed a dam in Italy attained disaster proportions, because in it 2,600 people lost their lives. This occurrence was in the deep Vaiont Canyon, which traverses mountainous country in the Italian Alps (Fig. 23.4). A

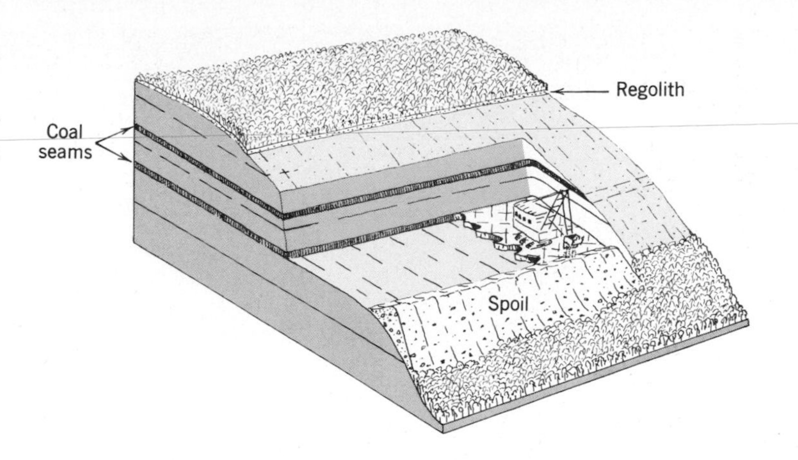


**Figure 23.2**
**Shelf, wall, and spoil pile made by strip mining of coal.** (*After a sketch published by R. D. Hill.*)

great dam 265m high had been built in the canyon in 1960, impounding a deep reservoir. On October 9, 1963, at 10:41 P.M., a tremendous body of rubble 1.8km long and 1.6km wide, consisting of an estimated 240 million $m^3$ of rock debris, thundered down the slopes on the south side of the canyon, setting up vibrations in the ground that were recorded throughout much of Europe. Within about one minute, debris had filled the reservoir and had piled up to more than 150m above the water surface. Air compressed by the rapidly flowing debris moved water and rock material more than 260m up the opposite (northern) side of the canyon.

Had the reservoir not been present, probably the disaster would have been far less serious. But the sudden filling with debris displaced a huge volume of water, which struck the dam in waves so large that they overtopped it by 100m. Although the dam itself withstood the impact, the water swept on as a flood. It destroyed every structure down the valley through 20km or more, all within a period of seven minutes.

Thorough investigation showed that two conditions had contributed greatly to the movement. One was an unfavorable geologic situation; the other was the presence of the reservoir. In other words, both natural conditions and human activity were involved. The bedrock consists of layers of limestone and claystone, weakened by much ancient deformation that has bent the layers and created many joints. Back in prehistoric times these conditions had caused earlier mass-wasting on a large scale, as inferred from debris. Then, in 1960, when the reservoir was first filled, water seeped from it into the rock of the valley sides, saturating the rock and changing the environment below ground. Moistening of rock layers that were previously dry caused the clay to increase in volume and become plastic, weakening the rock. In 1960, rockslide occurred on a small scale, showing that conditions were becoming worse.

Surveys made during the period between the 1960 and 1963 events showed that the ground was moving downslope at a rate of around

Figure 23.3
**Strip mining of coal in West Virginia. The three shelves, one above the other like a flight of steps, reveal the positions of three seams of coal. Waste from the lowest and largest step is creeping and sliding down the hillsides, among the trees.** (*U.S. Department of Agriculture—Soil Conservation Service.*)

1cm per week. In September 1963 heavy rains lasting two weeks intensified the movement, which increased to about 1cm per day and finally ended in catastrophe.

**Slump and debris flow on a small scale.** Artificial steepening of slopes often accompanies the construction of numbers of houses on hillsides. On some slopes the steepening creates instability that can lead to slump and debris flow (Chapter 6). Hundreds of such events (Fig. 23.5) in California alone have forced the adoption of ordinances to control cutting into stable natural slopes.

### Artificially induced subsidence

**Removal of support.** Human activities have long involved the moving of huge quantities of solids and fluids from one place to another, both on and below the ground. One of the results has been subsidence of the ground (Fig. 23.6). Sinking of the ground can occur either if support from below is withdrawn or if extra mass is added to it above. Examples of both are not hard to find.

Underground mining, especially of coal, comes to mind at once as an obvious way of creating big openings that remove support from the rock above. In some mines, active or abandoned, roofs have simply fallen in. In others, collapse of the overlying rock occurs slowly, causing the ground to subside gradually. Old coal mines beneath the city of Scranton, Pennsylvania, represent the space left by the removal of more than 150,000,000 $m^3$ of coal. Some of this space has been refilled with solid waste of various kinds, flushed down into it in places where heavy structures were to be placed above. Elsewhere the old tunnels and passages remain beneath the city, like the catacombs beneath the city of Rome.

Underlying western England, south of the city of Liverpool, the strata contain layers of salt, evaporated from ancient seawater. The thickest layers are as much as 180m thick and in places lie less than 60m below the surface. The salt has long been recovered by an extensive mining industry. Formerly it was worked in big underground mines, but nowadays the salt, dissolved in ground

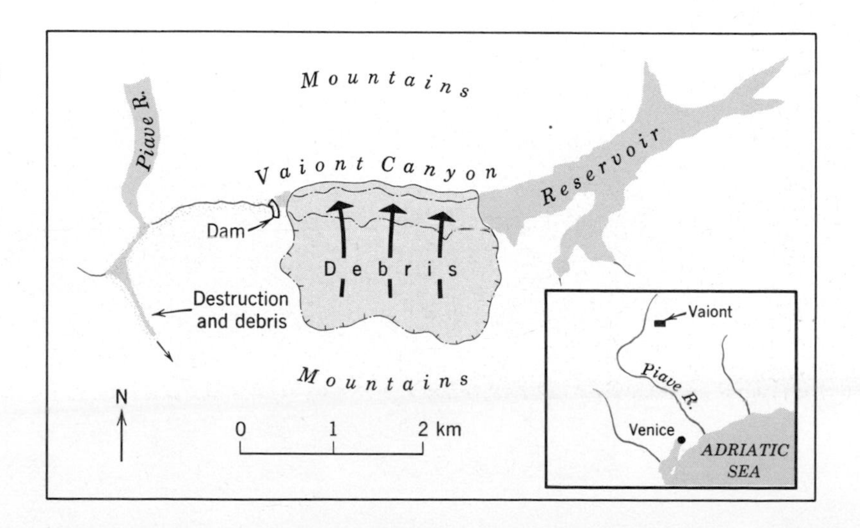


**Figure 23.4**
**Sketch map showing setting of the Vaiont event in 1963.**

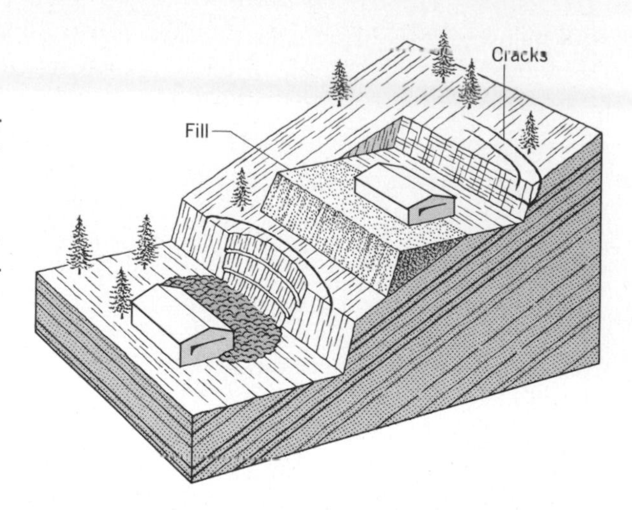


**Figure 23.5**
**Slump and debris flow caused by artificial alteration of a natural slope during construction of houses. The natural slope, conforming to the dip of strata, becomes unstable when cutting leaves strata unsupported. Uphill from the lower house, slump has occurred, grading outward into a debris flow that has partly enveloped the house. Cracks around the upper cut foretell future slump. (*After a sketch by R. H. Jahns.*)**

water to form a brine, is pumped out through wells and is recovered by evaporation. Many tens of millions of tons of salt have been recovered in this way. Removal of the salt by both mining and dissolution have been causing the ground to subside, locally at rates of more than 1m per year, destroying streets and buildings and forming basins that fill with rainwater and create lakes, some of them deep.

Even where dissolution is not involved, the removal of fluids from below ground can cause subsidence of the ground itself. In an aquifer consisting of sediment or sedimentary rock, the weight of the rock material that overlies it is supported not only by the particles of sediment, but also partly by the water itself. In addition, the water helps keep the rock particles loosely packed. When more water is withdrawn from the aquifer than is recharged into it, part of the support that had been borne by water is removed. The weight of the overburden begins to pack the rock

**Figure 23.6**
**This house, in southeastern Oklahoma, slid into the hole created by the collapsing roof of a zinc mine beneath it. (*U.S. Bureau of Mines.*)**

particles together more closely. Pore space in the aquifer is reduced permanently, and in some cases the ground above begins to subside.

In the central part of the San Joaquin Valley in California, ground water has been pumped for irrigation use since about 1920. As water was withdrawn, the sediment of the aquifer became compacted by as much as 50 percent and the ground above it subsided. Accurate surveys showed that between 1943 and 1960, within an area of more than 5000km$^2$, the ground subsided by amounts as great as 8m. If pumping were to cease and water were allowed to recharge the aquifer, the sediment in its compacted state would not hold as much water as formerly. Part of the capacity of the aquifer has been lost.

The Italian city of Venice, built on a low island of sediment, has been slowly subsiding over a long period. The causes are complex, but one of them is the pumping (mainly industrial) of ground water from beneath the city and its vicinity. In 1969 all pumping was halted by law. By 1975 a very slight rise of the land was measured. If the rise is confirmed by later observation, it can be concluded that ground water is increasingly sharing, with sediment, in the physical support of the city.

Near Las Vegas, Nevada, which derives much of its water from pumped wells, the ground within an area 8km in diameter sank to form a conical depression (Fig. 23.7, *left*) that reflected the form of the cone of depression created in the water table underneath. Between 1935 and 1950 maximum subsidence was 36cm. Withdrawal of ground water through more than 350 artesian wells resulted in sinking of the area of Mexico City, between 1891 and 1959, by as much as 7.5m. During 1953 the rate of subsidence reached 50cm/year. Subsequently the city reduced the trouble, by reducing the rate of withdrawal of water and by bringing in additional water from more distant sources.

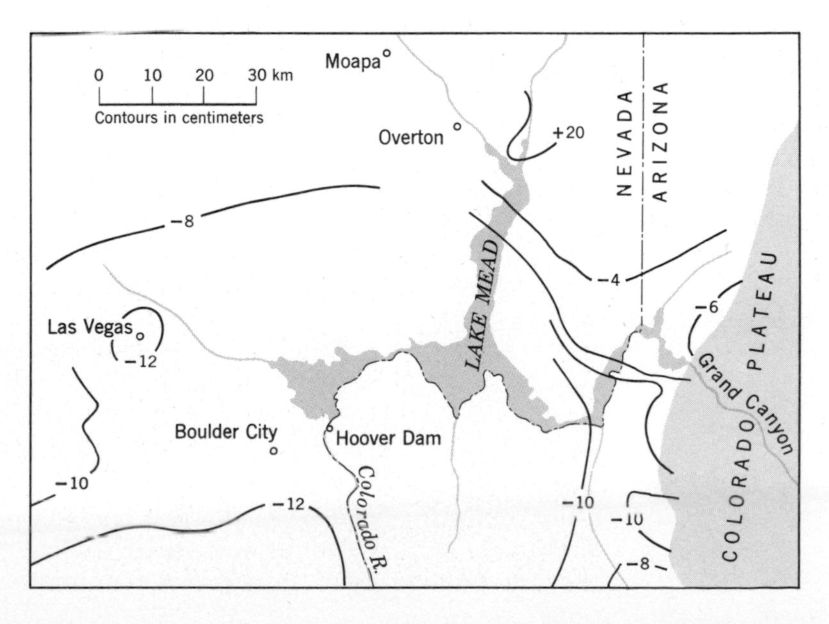


**Figure 23.7**
**Subsidence of the surface beneath and around Lake Mead, Nevada-Arizona, during the period 1935–1950, shown by contours scaled in centimeters. Greatest subsidence was 17cm.**
**The small circular contour at Las Vegas, Nevada shows more localized subsidence caused by pumping of ground water from beneath the city. (*Contours from U.S. Geol. Survey Circ. 346.*)**

Water is not the only fluid whose removal can cause subsidence. Between 1937 and 1962 the oilfield area around Long Beach Harbor, south of Los Angeles, subsided to form a basin with a maximum depth of nearly 8m, as a direct result of pumping of oil, gas, and water from the underlying strata. Inasmuch as a part of the area of subsidence had lain only 2 to 3m above sea level before pumping began in 1936, and furthermore was intensively industrialized, this result was serious. Because large economic benefits were expected, $30 million were spent between 1958 and 1962 on an elaborate scheme in which nearly 600 million barrels of water were injected into the ground through more than 200 wells. As a result, by 1963 subsidence near the shoreline had been locally reversed, and the ground had risen by as much as 15 percent of the earlier subsidence. It is unlikely, however, that full recovery in any part of the subsided area will ever be attained.

**Artificial loading.** In Figure 14.6 we saw the obvious effect of a vast, thick ice sheet that threw the crust beneath it out of isostatic equilibrium and caused it to subside, becoming hundreds of meters lower. A much smaller load likewise causes subsidence, although on a smaller scale. A neat example is Lake Mead, the big reservoir held in by Hoover Dam (Fig. 7.14). Created in 1936, the lake is nearly 200km long, and at the dam is more than 150m deep. The weight of the enormous volume of new lake water (more than 40 billion tons) plus the additional weight of the newly deposited sediment caused the crust beneath it to sag down. The sagging was measured by a series of very precise measurements beginning in 1935 and repeated at intervals. The measurements soon showed that in subsiding, the crust was forming a roughly circular basin having a radius of more than 40km and with its center at the lake itself. Figure 23.7 shows the amount of subsidence that had occurred by the end of 15 years. If, tomorrow, the dam were opened and the lake drained, we could expect the crust to rise toward the form it had before 1936, much as it is still rising today, in the region formerly covered with ice, toward the form it possessed before the latest of the glacial ages.

**Man-made earthquakes.** Other artificially induced movements of Earth's crust have been caused by the forcible injection into it of liquid industrial wastes. On the outskirts of Denver, Colorado, a hole was drilled to a depth of more than 3,600m. Between 1962 and 1967 large amounts of liquid waste were injected into it intermittently. About 40 days after injection commenced, local shallow earthquakes began to be recorded. Through the entire five-year period more than 1,500 earthquakes occurred. During two time intervals when injection ceased, the number of recorded earthquakes decreased sharply. Apparently the injections triggered the earthquakes, probably by building up stresses that were relieved by movement along faults.

Tampering with Earth's crust in this way not only threatens earthquake hazards, but also, especially where noxious wastes are forced into the ground, threatens contamination of otherwise pure ground water. Where extensive artesian-water systems exist, like those mentioned in Chapter 8, contamination could occur at distances as great as hundreds of kilometers from the points of injection of waste. Deep injection of waste is no more than storage underground. It amounts, in essence, to getting the waste out of sight—in other words, sweeping it under the carpet. With few exceptions such waste remains with little or no change, a potential threat to the quality of ground-water supplies.

### Creation and disposal of wastes

The people of all the cities in the United States together produce, each year, enough solid refuse to form a layer 3m thick overlying 600km$^2$ of land. But around the New York metropolitan area (Fig. 23.8) there isn't enough land to receive its share of so great a covering of waste. Some of the refuse goes to sea. Each year nearly 900,000,000 m$^3$ of waste, solid and liquid, are dumped from barges into the sea off New York. But when large-scale mining goes on beneath a sizable city far from the sea, waste disposal becomes a nightmare (Fig. 22.18).

Cities have always created waste. Archeologists have found evidence that through thousands of years, ancient cities gradually rose on their own debris, layer after layer, at rates averaging about 30cm per century. In later times many cities grew by spreading waste in low places, thus filling up swamps and lakes, narrowing river channels, and building seashores outward. Large parts of some cities consist of "made land" of this kind.

As city waste grows more bulky from year to year, there has ceased

Figure 23.8
**One day's normal production of trash by the people of New York City. (*New York Times photo.*)**

to be room for it in the cities themselves. It is deposited farther and farther away from urban centers and from smaller towns as well. Furthermore the waste includes increasing proportions of synthetic substances, such as certain plastics that are either very resistant to chemical weathering or impervious to destruction by bacterial decay; the latter have no natural enemies and are not biodegradable. Quantities of plastic packing material, an industrial waste, have been found floating in the Sargasso Sea, a part of the Atlantic Ocean (Fig. 13.2) that lies between Florida and northwestern Africa. The plastic is a long way from its probable points of origin, but it is neither consumed by organisms nor broken down in inorganic ways.

The bulk and character of city wastes are the joint result of increasing population, industrialization, and urbanization, and they pose serious problems. If burned, waste pollutes the air. If discharged into streams, lakes, or the ocean, it can pollute the water and also smother bottom-living life with blankets of slime. If buried, it can pollute ground water. In fact, waste in great quantity is a threat to public health and to the well-being of vast numbers of other organisms.

We say "in great quantity" because waste both solid and liquid plays a part in each of the natural cycles. All natural sediment is waste. As long as the steady state is not disturbed, as long as the input of waste into a cycle is moderate, the cycle absorbs it. But when the input of a substance is too great for the system to absorb, the steady state is disturbed and the substance becomes a pollutant. Indeed we define a ***pollutant*** as *any waste substance introduced into a natural system in greater amount than can be disposed of by the system.* Table 23.2 lists some of the principal sources of wastes that become pollutants of air and water. Table 23.3 gives estimates of rates of production of solid wastes in a highly industrialized country. From it we learn that within the United States, every year, nearly four billion metric tons (nearly 20 tons per person) of rock, metal, paper, wood, and animal wastes are being poured onto land, into water, and into the air.

**Radioactive wastes.** These pose a special problem, not because of their bulk, which is small, but because they include matter that is destructive to the biosphere. The generation of nuclear power and the manufacture of nuclear weapons involves atomic reactions. These create by-products—wastes—that are themselves radioactive. Their activity cannot be destroyed or neutralized artificially. Furthermore, the half-lives of some of the radioactive isotopes in these wastes are so long that activity will continue hundreds or even thousands of years into the future. Accordingly, the wastes must be stored as far away as possible from human contact, and they must be kept isolated for very long times. That means storage deep underground, apart from contact with ground water.

The problem of storage thus becomes a problem of geology. Present practice is to store the wastes as liquids, in tanks at ground level.

**Table 23.2**
**Some of the Wastes That Can Become Pollutants**

| Principal Sources | State S: solid L: liquid G: gas |
|---|---|
| Agriculture | |
| Residues from harvesting of crops | S |
| Residues from logging and pruning | S |
| Mineral fertilizers leached from fields | L |
| Animal wastes from feedlots and slaughterhouses | S,L |
| Salt dissolved by irrigation water and carried into streams | L |
| Mining and mineral processing | |
| Mines, blast furnaces, smelters, petrochemical plants | L,S |
| Radioactive wastes | L |
| Manufacturing and transportation | |
| Scrap metal, wood, paper, etc. | S |
| Cinders and ash; gases | S,G |
| Emissions from motors to atmosphere | G |
| Spills of oil, gasoline, etc., at wells, refineries, storage tanks, pipelines, ships, and filling stations | L |
| Wastes from burning of coal and petroleum | G,S |
| Salt placed on icy highways | L |
| Domestic | |
| Trash | S |
| Sewage | L |
| Waste from incinerators | G,S |

Plans have been made to place them, on an experimental basis, in an abandoned salt mine that can be expected to stay dry and undisturbed through "thousands of years." This practice, it is contended, would minimize the risk of escape of radioisotopes into the human environment. The wisdom of that contention has been widely debated, and salt-mine storage has not been put into practice.

**Table 23.3**
**Estimated Amounts of Solid Wastes Created Annually in the United States**

| Source of Waste | Estimated Amount (tons) |
|---|---|
| Agriculture | 2,000,000,000 |
| Mining and processing of minerals | 1,700,000,000 |
| Domestic | 250,000,000 |
| Junked cars | 12,000,000 |
| Total | 3,962,000,000 |

# MAN AND THE HYDROSPHERE

## Ground water

**Salt-water intrusion.** Earlier in the present chapter we mentioned that excessive pumping of ground water can deplete an aquifer and destroy the steady state of withdrawal and supply. This relationship was mentioned also in Chapter 8. When industrial or agricultural withdrawal does exceed recharge, not only will the future supply become uncertain, but ground water near a coast can become contaminated with salt water.

The ground water near a seacoast consists of salt water derived from the sea and fresh water derived from rainfall (Fig. 23.9*A*). The fresh water, having lower density, floats on the salt water, and normal recharge from above gives it enough volume and weight to hold in place the salt water beneath it. The interface between the two water bodies is distinct, and normally its position changes little or not at all. Wells that penetrate the fresh but not the salt layer can safely withdraw water as long as withdrawal does not exceed recharge.

However, in some coastal regions concentrated industry or intensive agriculture that demands irrigation has led to the drilling of many wells, which as a group pump fresh water in excess of recharge. What happens then is shown in Figure 23.9*B*. The water table is gradually lowered. As the layer of fresh water becomes thinner and exerts less pressure on the layer beneath, the interface between the layers rises. When the rising interface—the dividing surface—reaches the bottom of a well, that well begins to deliver salt water.

Intrusion of salt water into coastal wells has occurred commonly, especially in areas where population is dense and where sources of fresh water are limited. Areas of occurrence include Long Island and coastal New Jersey, southern Florida, coastal Texas, and Los

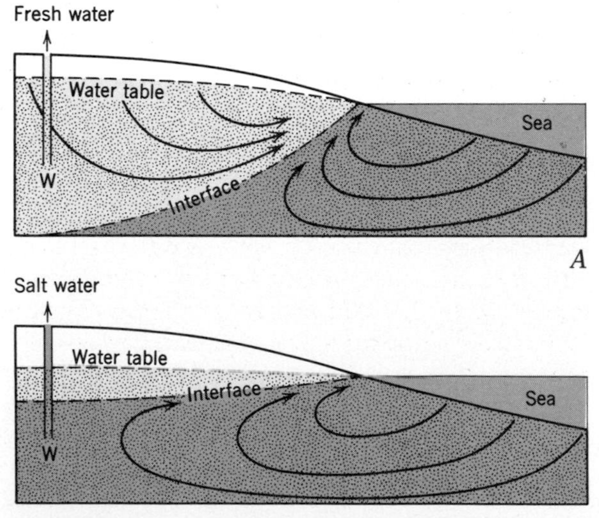


**Figure 23.9**
**Salt-water intrusion beneath a coastal area.**
***A*. Steady state in the ground-water system. Well (W) delivers fresh water.**
***B*. Result of over-pumping. Water table declines; salt water intrudes landward. Well delivers salt water.**

Angeles, as well as areas on coasts of other continents. In the case of the municipal water supply for Miami, Florida, begun in 1896, salt-water intrusion twice forced abandonment of a field of wells and their replacement with new fields in different areas. It has been estimated that by the year 1995 the projected population of Miami's Dade County would require an amount of water that would have to be supplied by recharge over an area of nearly 7,300 $km^2$.

### Streams

The locally increased rates of erosion caused by man's many earth-moving activities should be reflected in increased loads of sediment carried by streams. And indeed it is. The rivers of the United States are estimated to be carrying to the ocean, now, 1 billion ($10^9$) metric tons of sediment and dissolved load annually. This is twice the amount they are believed to have been carrying before European man began to cultivate American soil.

This 100 percent increase has occurred despite the fact that much has been done since about 1930 to reduce erosion on American farmland. At around that time a long succession of dry years coincided with the recent plowing of dry grassland to create farms. This combination disrupted the steady-state condition, mainly by wind erosion as described in Chapter 9. Many farms had to be abandoned, and distress was widespread. An educational campaign was mounted, and farmers not only in dry lands but throughout the nation were shown how to prevent and how to combat erosion by wind and by water. Undoubtedly the result has been a decrease in the sediment delivered from farmlands to streams, but undoubtedly also the decrease from that source has been offset, to an unknown extent, by the increasing pace of construction of roads, streets, and buildings, which likewise alters the steady state among natural processes.

Why is disruption of the steady state in this case a serious matter? The answer requires some explanation. Most of Earth's external processes involve whole chains of cause-and-effect activities. So we turn here from a cause (man's earth-moving capability) to one of its effects (on stream flow and stream transport of sediment), and under the latter heading continue the story, beginning with an answer to the question asked above.

Disruption of the steady state is serious here for this reason: In a climate sufficiently moist to permit the growth of forests, the leaves of trees shield the ground from the splashing, eroding impacts of raindrops. Beneath the surface, as we noted in Chapter 8, the regolith acts as a broad, thin sponge that absorbs water. This ground water percolates through the sponge at rates that are very slow compared with runoff over the surface, and little by little emerges in streams without having picked up any sediment at all. Fed with water at this slow rate, streams tend to fluctuate only moderately; they tend not to rise in high floods.

**Figure 23.10**
**On this steep hillside in western Oregon, cutting of timber has resulted in rapid erosion, aggravated by the presence of caterpillar gouges and a temporary road.** (*U.S. Department of Agriculture—Soil Conservation Service.*)

But when there are no trees and when regolith is covered only partly by vegetation, much of the fallen rain fails to enter the sponge below ground. Instead, it runs off along the surface, eroding regolith and picking up sediment as it runs (Fig. 23.10). Thus the water arrives at the nearest stream very quickly, swelling the stream and so creating a flood. The floodwater is turbid with sediment picked up on the hillslopes above. As the flood subsides, this load of mud is deposited somewhere downstream, often covering a meadowland with an unwanted layer of infertile silt.

So, an unbroken cover of vegetation on a watershed promotes the smooth operation of one part of the water cycle with minimum fluctuation, and is equally important for one part of the rock cycle because it promotes erosion and deposition of sediment at slow, steady rates.

It is interesting to watch the progress of a flood down a sizable river. A bulge, a sort of broad wave, travels down a river at a rate of a few kilometers per hour, and in some large rivers reaches 10m or more in height. This bulge is measured by gages fixed to the

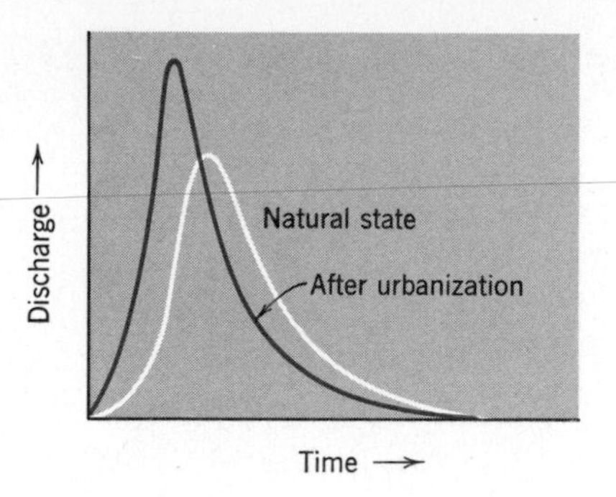


**Figure 23.11**
**Graph showing how the variation of flood discharge of a stream, through time, is distorted by urbanization of the stream's watershed. (*After L. B. Leopold, U.S. Geol. Survey Circ. 554.*)**

bank. They float, moving up or down in vertical slots that are graduated like a thermometer. Each gage marks the height of the flood as it passes by, so that from all the gages along a river the history of a flood can be reconstructed.

Figure 23.11 is not a picture of the bulge. It is a curve showing how discharge varies with time, increasing to a peak and then gradually tapering off. The white curve represents the flood when stream and watershed are in their natural states. The blue curve shows the flood after the watershed has been urbanized. For a given rainfall, the runoff from streets and roofs and through sewers is faster, and puts water into the stream more quickly; so the blue curve has a higher peak and steeper sides, and occurs farther to the left (that is, earlier in time) than the white curve. The difference between the two curves illustrates the special importance of advance planning for land use in city areas that occupy stream valleys. In this connection Figure 7.6 is worth looking at again.

**Floods: Mississippi River system.** What has just been said explains why new suburban communities sometimes experience flooding of streets and waterlogging of lawns and gardens despite the fact that in the years before construction began, flooding has been unknown. On a larger scale, it also affords one reason why overbank floods along the Mississippi River and its principal tributary streams became both higher and more frequent during the 19th Century and the first half of the 20th. As we noted in Chapter 7, the Mississippi River watershed stretches from Pennsylvania to Montana and includes about 41 percent of the area of the conterminous United States.

When spring rains add to the water from melting snow and frozen regolith, a large volume of water runs off into small streams, and so through larger streams into the channel of the Mississippi itself. The results—part of the steady-state condition among all the factors in the huge river system and not an event caused by man—are annual floods that occasionally go overbank.

This state of affairs was well known to the early French settlers in Louisiana, who saw the Mississippi spilling over its natural levees (Chapter 7), to convert the lower lands beyond into vast, shallow lakes (Fig. 23.12). To protect buildings and farmland the settlers,

Figure 23.12
**Mississippi River in flood near Louisiana, Missouri. The river's channel is in the foreground, bounded by a levee crowned with trees. Beyond are embanked roads with rows of trees showing above flooded land. (*Corps of Engineers, U.S. Army.*)**

**The inset shows Mississippi River floodwater pouring through a break in a levee and over a railroad near Alexandria, Missouri in June 1947, adding to the water over the floodplain. (*Wide World Photos.*)**

early in the 18th Century, began to heighten the natural levees by erecting artificial levees (dikes of earth with surface armor) on top of them. Heightening was continued and expanded, so that today the lower Mississippi is confined within a system of artificial levees. Together these have a combined length of more than 3,200km. Each levee is 90m wide from toe to toe and about 9m high.

Building the levees narrowed the natural channel by as much as one-third. Confined within a channel too narrow for it, the river when in flood eroded its bed and rose much higher than before; in general, the more recent floods rise higher than those of former years. The levee-heightening program was based on the assumption that it was principally the lower river that must be controlled if destructive floods were to be prevented. Gradually, however, it became evident that all the rest of the vast watershed must be considered also. The floods downstream began to be understood as partly the result of the hurrying of water across a field in Ohio or along a roadside ditch in Dakota, instead of letting it soak into the regolith and reaching the nearest stream by delayed action. Apart from the effect of the levees, the hurrying of little waters on the watershed made flood peaks both higher and narrower, as Figure 23.11 shows.

While the levees along the Mississippi were being built ever higher, these other things were happening over a large part of the watershed.

1. Deforestation.

2. Conversion of former woodland and grassland into cropland.

3. Draining of swamps and other wet areas on farms, the water flowing through tile drains instead of passing through the ground by percolation. (Between 1950 and 1970, two percent of all the marshland in the United States was filled or drained.)

4. Overgrazing, which impairs grassland sod and thereby partly destroys the sponge beneath the surface.

5. Construction of all kinds: buildings, paved streets, highways, and other structures, each with ditches and other conduits designed to lead rainwater off over the surface.

All these changes heightened and narrowed the peaks of floods and so made matters worse in downstream areas. Since the 1930s, the "hurrying of runoff" across farmland has been reduced somewhat, by the use of measures such as (a) plowing along contours instead of up and down the slopes, (b) alternating strips of grass with strips of crops along the contours of slopes (Fig. 23.13), and (c) leaving a litter of straw, stalks, and other organic trash scattered over bare fields to impede surface water.

**Pollution of streams.** Because streams are the avenues through which rainfall on the land returns to the ocean, and because many naturally and artificially formed substances are soluble in water, it follows that streams become polluted with the by-products of human life and human industry. Since long before the industrial

**Figure 23.13 Crops, planted in horizontal strips, wind around the contours of hills in southern Wisconsin farmland.** *(U.S. Dept. of Agriculture—Soil Conservation Service.)*

revolution, the easiest and most obvious way to get rid of an unwanted object or substance has been to throw it or pour it into a stream. Out of sight, out of mind. When the Industrial Revolution arrived, such wastes soon became pollutants.

Let us begin with salt. The fresh water of rivers becomes contaminated with salt from two principal sources, industrial and agricultural. Water is taken from a river or from the ground, used in various industrial processes that add salt to it, and the salty water is then poured into the river. In this way the salt content of some rivers has been increased by as much as 50 percent.

In arid parts of some western states, fresh water for irrigating cropland is withdrawn from rivers that originate in high mountains. In dry land the regolith is apt to be salty, because salt is precipitated by evaporation of ground water derived from infrequent rains. Irrigation water sinks into the regolith, dissolves salt, and returns to the river by seepage. The added salt impairs the quality of the water for domestic and other uses. Since large-scale irrigation along the Colorado River began, the concentration of salt in the river water has almost doubled.

Apart from salt, and much more pervasively distributed, are a host of pollutants in the waters of streams, and of lakes and the ground as well. Some, like salt, are dissolved; others are carried mechanically. They include agricultural fertilizers and pesticides leached from farmland, detergents from domestic and industrial sources, food wastes and sewage that include microorganisms, some of which carry disease, industrial wastes of many kinds, including poisonous metals, metallic compounds, acids, and plastics that cannot be broken down by organisms. The list is long and extremely varied because of man's great numbers and his ingenuity in industry.

### Lakes and oceans

A lake is a trap that catches and holds the water brought to it by streams, and also the sediment and substances dissolved in the water. Because it is a trap, a lake is specially vulnerable to contamination. Within limits, a stream can clean itself by flushing, but a lake cannot. An example is Lake Baikal in southwestern Siberia, with twice the volume of Lake Superior and a depth of nearly two kilometers. Fed by 336 streams, surrounded by forested mountains, and populated by a unique assemblage of animals and plants, it holds great scientific interest.

Decades ago, large-scale lumbering began along some of the tributary streams. As one result, much sediment was discharged into the lake, along with sewage and great numbers of cut logs, many of which sank to the lake bottom. By 1960 industrial development had begun on the lake shore itself, and the lake began to receive effluent from pulp and paper plants in amounts greater than could be absorbed. In consequence the lake is becoming rapidly polluted.

Lake Erie, a smaller, much shallower, and less spectacular water body than Lake Baikal, records a longer history of pollution and is in a condition of greater deterioration. From 1910 to 1960 the human population of the Lake Erie basin increased from 3 million to 10 million, and industrialization proceeded rapidly. Detergents, as well as fertilizer leached from agricultural areas poured, and are still pouring, into the lake. Their content of nitrogen and phosphorus has stimulated great growths of green algae in surface water where sunlight is greatest, thus screening sunlight from the bottom and inhibiting plant growth there. Further, dead algae drift down from above, decompose by oxidation, and so use up the oxygen in the bottom water. A result is suffocation of fish and other animals. The lake bottom is widely covered with slime, and in shallow places organic matter extends from the bottom to the surface. Fishing has been greatly reduced, as has recreational use of the lake. If these changes were to continue, the lake would become filled with organic matter and so be converted into a bog.

Others of the Great Lakes are in earlier stages of the same process. Pollutants are pouring into them also, disturbing the natural, steady-state condition. An ambitious and costly plan to halt and reverse this deterioration, a pilot plan financed jointly by public agencies at federal, state, and municipal levels, began to be implemented in 1971. Within a pilot region it diverts all sewage and industrial effluents away from Lake Michigan, and instead discharges them into specially created lagoons for thorough biochemical treatment. The treated liquid will then be distributed through irrigation systems to fertilize soil that although presently unproductive, can be made productive with the fertilizer. The plan, if adopted widely, is believed to hold promise of cleaning up the Great Lakes.

The influence of mankind on the world ocean is great in the aggregate, and consists of both physical changes and the dumping of pollutants. The placing of seawalls, groins, jetties, and other structures that interfere with normal processes along coasts (Chapter 11) have distorted the erosion/deposition steady state. The dredging of ship channels in rivers and harbors has produced an amount of spoil dumped in the ocean that far exceeds dumped material from all other sources.

Harbor spoil dumped into the sea is often a pollutant, but many other dumped materials are always pollutants. Garbage, sewage, sewage sludge, industrial wastes, construction debris, explosives, and radioactive materials are transferred to the ocean in large quantities, indirectly through streams and directly by transport in floating vessels. In addition there are oil spills from tankers, at times enormous, as well as seawater mixed with oil discharged during cleaning of bunkers. The impact of these pollutants on marine plants and animals, diving sea birds, and commercial fisheries is great.

# MAN AND THE ATMOSPHERE

### Pollution

Two effects are exerted by mankind on the atmosphere: pollution and change of climate. The first is a matter of here and now, the second mainly a matter of the near future. The here-and-now

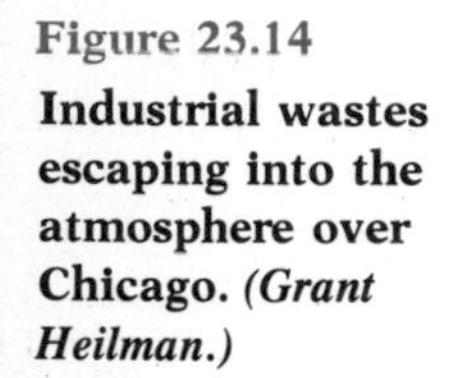

Figure 23.14
**Industrial wastes escaping into the atmosphere over Chicago.** ***(Grant Heilman.)***

pollution is evident in Figure 23.14 and can be seen frequently in major cities and intensely industrialized areas. The atmosphere is loaded with *particulates* (minute solid particles) that constitute smoke and soot, and so are visible pollution. In addition there is much invisible pollution, including (1) particulates much too small to be visible to the unaided eye, (2) carbon monoxide, (3) oxides of sulfur, (4) hydrocarbons, and (5) oxides of nitrogen. This collection of invisible pollutants comes from several sources, chief of which are exhausts of internal-combustion engines (chiefly vehicles) and by-products of combustion under steam boilers (mainly power plants and space heaters). The fuels in all cases are fossil fuels. Many of the polluting substances come from unburned and wasted fuel. Many are toxic, and constitute a serious threat to public health. Despite decrease of such pollutants in many areas, resulting from governmental measures designed to control them, the overall use of fossil fuels is increasing by several percent each year, thereby offsetting, or more than offsetting, the overall decrease of specific pollutants.

**Man's influence on climate**

Man's activities influence climate by adding three things to the atmosphere: (1) *carbon dioxide* (from burning fossil fuels), (2) small *solid particles* (smoke, soot, and ash from combustion, dust from earth-moving equipment, and agriculture), and (3) *heat* from combustion of all kinds.

Although by definition not a pollutant, carbon dioxide is an important product of combustion, and to a smaller extent, of the making of cement from limestone and the clearing of forests for dwellings (thereby reducing absorption of $CO_2$ by photosynthesis). Although it constitutes only about 0.03 percent of the atmosphere by volume, $CO_2$ influences temperature importantly. With increasing $CO_2$ content, the atmosphere becomes warmer because $CO_2$ absorbs and holds some of the heat of solar radiation that is re-radiated outward from lands and ocean, much as the glass roof of a greenhouse retains re-radiated solar heat and so warms the air beneath it. In addition, mankind-promoted combustion pours into the atmosphere direct heat as well as $CO_2$, and so helps further to warm it up.

Burning of fossil fuels on a large scale began about 1860, and each year since then has been increasing. Even though much of the $CO_2$ thus added to the atmosphere is soon absorbed by the ocean, the amount remaining in the atmosphere gradually increases, causing the temperature to rise. Temperature is increased further by the heat added directly by combustion. As for the particulates, their effect on temperature is complicated, but it is thought likely that as with volcanic "dust," their presence would on the whole tend to cool the atmosphere although probably not enough to offset the warming caused by $CO_2$. Evidently not enough is yet known about these effects to enable us to predict future temperatures with even a broad degree of accuracy.

# MAN AND THE BIOSPHERE

The circuitry that is built into the largest computer is simple compared with the circuitry within the biosphere and between it and the other spheres. It represents systems in which "everything is connected with everything else."

No wonder the circuitry is complex; the biosphere owes its existence to the presence of the other spheres. As we have noted already, short-wave solar energy passes through the atmosphere, and is absorbed by land and by surface water, which are warmed by it. The heat, as long-wave radiation, is re-radiated into the atmosphere where some of it is absorbed by water vapor and carbon dioxide. The atmosphere, thus warmed, acts as a thermal blanket that keeps the surfaces of lands and oceans warm. Because of this warmth, water substance can exist in most places as a liquid rather than as ice. This makes possible the existence of a biosphere, which originated under the protecting shield of the atmosphere.

Of course some of the re-radiated heat passes through the atmosphere and back into space, but in the system as a whole, income is balanced by outgo, with the atmosphere acting as a regulator. If the balance did not exist, lands and oceans would either cool off or heat up and would eventually reach a temperature at which all living things would be killed. Indeed, the biosphere developed and evolved to its present complex state precisely because there is an atmosphere, to screen and regulate the incoming solar energy that is the sole source of energy for living things. Through their ability to perform photosynthesis, by which they split molecules of carbon dioxide so as to separate oxygen from carbon, organisms use solar energy and water to build the carbon into the organic compounds that are the substance of plants. Animals, by respiring, extract free oxygen from atmosphere or hydrosphere, and with it oxidize the plant substance that constitutes their food. In respiring, they exhale $CO_2$, ready to be recycled once more by plants. Thus oxygen and carbon continually change hands, forming different chemical combinations as they do so. This means that Earth's spheres are mutually dependent.

It would require a large book to set forth even the little that is known about the effect of human activities on the rest of the biosphere. We must limit ourselves here to a single broad relationship that can be expressed in numbers: the time relationship between numbers of people and extinctions of species of other mammals and of birds during the last 300 years (Fig. 23.15). It was said of this chart: "Although the close parallelism of these two curves does not in itself prove cause and effect, it suggests that as man's numbers increase, more and more living species become exterminated."

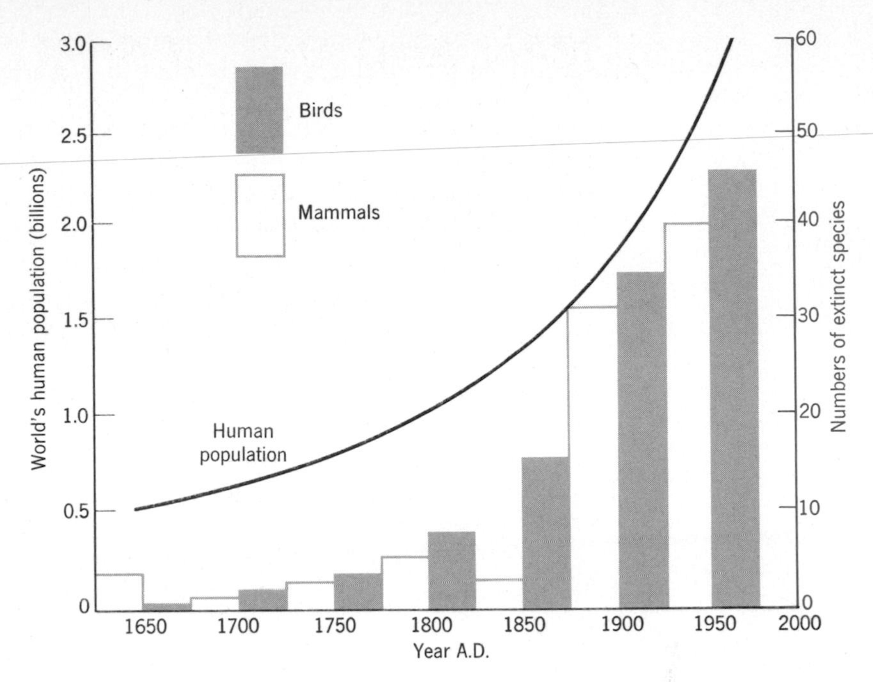


**Figure 23.15 Comparison of number of people with numbers of extinct species of other mammals and of birds, since A.D. 1650.** *(After V. Ziswiler, 1967, Extinct and vanishing species: New York, Springer-Verlag.)*

# POSTSCRIPT

We have touched on only a few of the ways in which man has interfered with natural processes, but we have chosen examples obviously pertinent to physical geology. Others, at least equally important, especially concern the biosphere. They include such matters as pollution of the atmosphere, pollution of the ocean, and the artificial synthesis of substances that can be decomposed with difficulty or not at all by chemical weathering.

The examples we have given demonstrate that man *is* a geologic agent, and a considerable one. As human beings we are together responsible for the fact that today various natural systems are operating at rates different from those which prevailed before our early ancestors began to bend natural processes. We (that is, all mankind) have been taking substances from the lithosphere, biosphere, hydrosphere, and atmosphere. All such substances participate in one or more of Earth's external processes. We have changed, combined, or fabricated many such substances, physically or chemically. Then we have put residues or used-up fabrications back into one cycle or another, mostly by throwing them away, to become pollutants. We have left the cycles to cope with this as best they could. Many of the things we take involve destruction. Many of those we throw away cause pollution. In both cases the steady state is usually disrupted, often with very disagreeable consequences.

Both the destruction and the pollution were begun unwittingly. They have continued as deeply rooted habits, exercised thoughtlessly or carelessly. For centuries, each new generation of people accepted the Earth as they found it, without knowledge of steady states and assuming that things were normal. With few exceptions they did not realize what they were doing. Then, just after the middle of the 20th Century, particularly in industrialized and urbanized countries, realization appeared and spread. It began to be evident that each of the natural cycles must be analyzed, not only qualitatively but also quantitatively, and in precise terms as well. The rapid growth of science has made such analysis possible, as a basis for change in existing practices.

Through our study of physical geology we fully realize that destructive human influence on Earth's ongoing cycles has been confined entirely to the very latest moment of geologic time—to an almost incredibly small part of Earth's history. Through all of the more than three billion years that have elapsed since the earliest fossils we know of today were living organisms, Earth's flow of energy and cycling of materials have been operating in mutual harmony. Significant human influence has existed during barely the last 15,000 years—only one two-hundred-thousandth part of that long time. If you should chisel about 4cm of rock off the top of Mt. Everest, you would be lowering the altitude of that peak by a proportional amount, one two-hundred-thousandth.

Man's unique skill in technology derives from his great intelligence compared with that of other organisms. That same intelligence is also the basis of man's ability to perceive disturbances in Earth's cycles, measure them scientifically, and plan their elimination. In other words, the basic human factors that led to human tampering with Earth's environments should be capable of intervening to restore steady states that have been thrown out of balance. Man is moving toward a future that is unknown, hoping for continuance of his species through long ages of time. It is this hope that may lead him to use his basic capacity to become, not an unsuccessful master, but a true son of the Earth from which he has risen.

## SUMMARY

1. Man is a geologic agent because he moves rock and regolith and changes the rates of many natural processes. By thus changing the steady state he distorts the processes.

2. Distortion of natural processes is the result chiefly of increased population and technical capability, rapid urbanization, and extensive deforestation.

3. Man moves more Earth materials than are moved by any other single external process. Farming, construction, and mining are the chief activities by which the materials are moved.

4. Avalanches, slump, debris flows, and other "landslides" can be caused or triggered by human agency.

5. When water, petroleum, or ores are removed from beneath the surface, or when a large artificial lake is created, the ground tends to subside.

6. When waste is introduced into a natural system in greater amount than can be disposed of by the system, it becomes a pollutant.

7. Disposal of solid waste by cities and towns involves pollution dangers of several kinds.

8. Industrial wastes, salt from streets and highways, and irrigation water increase the salt content of ground water and water in reservoirs.

9. Wastes that are radioactive pose a special storage problem that has not yet been solved.

10. Injection of liquid wastes into the ground can cause pollution of ground water and can even cause earthquakes.

11. Excessive withdrawal of ground water in a coastal area can cause intrusion of salt water from the sea, and in any area can cause the ground to subside.

12. Deforestation, swamp drainage, careless farming, and most kinds of construction tend to send runoff along the surface instead of through the ground. Among the results are increased erosion, increased flooding by streams, and lowered water tables.

13. Building artificial levees to contain river floods narrows the river channel and increases flood heights.

14. Wastes that contain nitrogen and phosphorus, poured to excess into a lake (a natural trap), act to suppress oxidation and to inhibit many kinds of animal life in the lake.

15. Man-caused pollutants in the atmosphere include particulates and various gases.

16. Carbon dioxide and particulates, derived from combustion, are added to the atmosphere in quantities sufficient to influence climates, but their precise effects are not fully known.

17. Human activities affect the rest of the biosphere in many ways, most of them destructive.

## SELECTED REFERENCES

Belt, C. B., 1975, The 1973 flood and man's constriction of the Mississippi River: Science, v. 189, p. 681–684.

Committee on Geological Sciences (National Research Council), 1972, The Earth and human affairs: New York, Harper & Row.

Commoner, Barry, 1971, The closing circle; nature, man, and technology: New York, Knopf. (Also 1971: New York, Bantam Books paperback.)

Detwyler, T. R. (ed.), 1971, Man's impact on environment: New York, McGraw-Hill.

Dolan, R. B., and others, 1974, Man's impact on the Colorado River in the Grand Canyon; American Scientist, v. 62, p. 392–401.

Flawn, Peter, 1970, Environmental geology: New York, Harper & Row.

Gates, D. M., 1972, Man and his environment: climate: New York, Harper & Row, Chap. 6.

Goldman, M. I., 1972, The spoils of progress. Environmental pollution in the Soviet Union: Cambridge, Mass., M.I.T. Press.

Legget, R. F., 1962, Geology and engineering, 2d ed.: New York, McGraw-Hill.

Legget, R. F., 1973, Cities and geology: New York, McGraw-Hill.

Leopold, L. B., 1968, Hydrology for urban land planning—a guidebook on the hydrologic effects of urban land use: U.S. Geol. Survey Circ. 554.

McKenzie, G. D., and Utgard, R. O., 1972, Man and his physical environment. Readings in environmental geology: Minneapolis, Burgess.

Piper, A. M., 1969, Disposal of liquid wastes by injection underground—neither myth nor millennium: U.S. Geol. Survey Circ. 631.

Russell, W. M. S., 1969, Man, nature, and history: New York, Natural History Press.

Schneider, W. J., 1970, Hydrologic implications of solid-waste disposal: U.S. Geol. Survey Circ. 601-F.

Singer, S. F., ed., 1970, Global effects of environmental pollution: New York, Springer-Verlag.

Thomas, H. E., 1954, First fourteen years of Lake Mead: U.S. Geol. Survey Circ. 346.

Thomas, William, and others, 1956, Man's role in changing the face of the Earth: University of Chicago Press.

Turk, Amos, Turk, Jonathan, and Wittes, J. T., 1972, Ecology, pollution, environment: Philadelphia, W. B. Saunders.

U.S. Department of the Interior, 1967, Surface mining and our environment: Washington, Government Printing Office.

U.S. Department of the Interior, 1968, Lake Erie report: Federal Water Pollution Control Administration, Great Lakes Region.

3

# APPENDIX A ORGANIZATION OF MATTER

**What is matter?**

Every object in the Universe is composed of matter. But can we say precisely what *matter* is? Ancient Greek philosophers first attempted the question and came close to a correct answer. They saw that matter occurs in three different states—solid, liquid, and gaseous—and that different forms of matter combine and react with each other, giving rise to new forms. Seeking some unifying principle, they examined properties such as color, hardness, taste, and odor of common materials, including rocks and minerals of many kinds. In a sense, some of those Greek philosophers were the first geologists, and the names they gave to some forms of matter are with us to this day. The modern word *copper* comes directly from the Greek word *cyprus,* and words such as *sapphire* and *magnetite* are little changed from the Greek originals.

Although the ancient philosophers learned much about matter, they were unable to resolve the question of what matter is. That question remained unanswered until the present era because no one was able to decide between two equally plausible possibilities, which can be stated in terms of questions that they themselves occasioned: can matter be endlessly subdivided into tiny pieces, all of which retain the properties of the whole; or is matter built up from submicroscopic particles of only a few kinds, so that the different properties of matter reflect different arrangements of the particles? There is some truth in both lines of thinking, but the second is

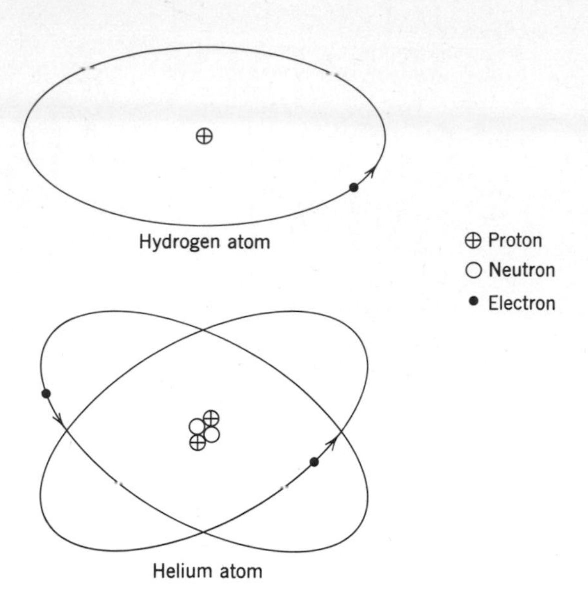


**Figure A.1**
**Hydrogen and helium atoms, shown schematically. Hydrogen is the simplest atom with one proton (positive) electrically balanced with one electron (negative). Helium has two electrons in orbit around a nucleus with two protons and two neutrons. The *atomic number* of helium is 2 (the number of protons); its *mass number* is 4 (the sum of protons and neutrons).**

essentially the correct one. Matter can be subdivided into atoms, and all atoms of a given chemical element have identical chemical properties. But atoms themselves are composed of still smaller subatomic particles, and these are the fundamental particles the Greek philosophers were seeking.

**Structure of Atoms.** Atoms are held together by electrical forces; we can therefore say that matter is basically electrical in character. Physicists have found that all atoms have a structure possessing a central portion called the *nucleus* which has a *positive electrical charge.* Spinning around the nucleus in defined paths, or orbits, are *negatively charged particles* called ***electrons.*** The positive charge of the nucleus exactly balances the negative charge of the orbiting electrons. Of all atoms, the simplest is hydrogen, its nucleus having a single positive charge. Only one orbiting electron, therefore, is required to provide electrical balance. The nucleus of the hydrogen atom gets its positive charge from a single particle called a proton. A ***proton*** is a *positively charged particle with a mass 1,832 times greater than the mass of an electron.*

The second simplest atom is helium. Its nucleus has two positive charges, indicating that it contains two protons and therefore requires two orbiting electrons to maintain electrical neutrality (Fig. A.1). Sensitive measurements show, however, that the mass of a helium nucleus is equal to four protons. The increase in mass results from the presence of yet another type of particle in the nucleus. The newcomer is *an electrically neutral particle with a mass 1,833 times greater than that of the electron,* called a ***neutron.***

Protons, neutrons and electrons are the three principal sub-atomic

**Table A.1**
**The Fundamental Particles of Matter**

| Name | Electric Charge | Relative Mass |
|---|---|---|
| Proton | +1 | 1832 |
| Neutron | 0 | 1833 |
| Electron | −1 | 1 |

**Table A.2**
**Alphabetical List of the Elements (After K. K. Turekian, 1969)**

| Element | Symbol | Atomic Number | Crustal Abundance, Weight Per Cent |
|---|---|---|---|
| Actinium | Ac | 89 | Man-made |
| Aluminum | Al | 13 | 8.00 |
| Americium | Am | 95 | Man-made |
| Antimony | Sb | 51 | 0.00002 |
| Argon | Ar | 18 | Not known |
| Arsenic | As | 33 | 0.00020 |
| Astatine | At | 85 | Man-made |
| Barium | Ba | 56 | 0.0380 |
| Berkelium | Bk | 97 | Man-made |
| Beryllium | Be | 4 | 0.00020 |
| Bismuth | Bi | 83 | 0.0000004 |
| Boron | B | 5 | 0.0007 |
| Bromine | Br | 35 | 0.00040 |
| Cadmium | Cd | 48 | 0.000018 |
| Calcium | Ca | 20 | 5.06 |
| Californium | Cf | 98 | Man-made |
| Carbon[a] | C | 6 | 0.02 |
| Cerium | Ce | 58 | 0.0083 |
| Cesium | Cs | 55 | 0.00016 |
| Chlorine | Cl | 17 | 0.0190 |
| Chromium | Cr | 24 | 0.0096 |
| Cobalt | Co | 27 | 0.0028 |
| Copper | Cu | 29 | 0.0058 |
| Curium | Cm | 96 | Man-made |
| Dysprosium | Dy | 66 | 0.00085 |
| Einsteinium | Es | 99 | Man-made |
| Erbium | Er | 68 | 0.00036 |
| Europium | Eu | 63 | 0.00022 |
| Fermium | Fm | 100 | Man-made |
| Fluorine | F | 9 | 0.0460 |
| Francium | Fr | 87 | Man-made |
| Gadolinium | Gd | 64 | 0.00063 |
| Gallium | Ga | 31 | 0.0017 |
| Germanium | Ge | 32 | 0.00013 |
| Gold | Au | 79 | 0.0000002 |
| Hafnium | Hf | 72 | 0.0004 |
| Helium | He | 2 | Not known |
| Holmium | Ho | 67 | 0.00016 |
| Hydrogen[b] | H | 1 | 0.14 |
| Indium | In | 49 | 0.00002 |
| Iodine | I | 53 | 0.00005 |
| Iridium | Ir | 77 | 0.00000002 |
| Iron | Fe | 26 | 5.80 |
| Krypton | Kr | 36 | Not known |
| Lanthanum | La | 57 | 0.0050 |
| Lawrencium | Lw | 103 | Man-made |
| Lead | Pb | 82 | 0.0010 |
| Lithium | Li | 3 | 0.0020 |
| Lutetium | Lu | 71 | 0.000080 |
| Magnesium | Mg | 12 | 2.77 |
| Manganese | Mn | 25 | 0.100 |
| Mendelevium | Md | 101 | Man-made |
| Mercury | Hg | 80 | 0.000002 |
| Molybdenum | Mo | 42 | 0.00012 |
| Neodymium | Nd | 60 | 0.0044 |
| Neon | Ne | 10 | Not known |
| Neptunium | Np | 93 | Man-made |
| Nickel | Ni | 28 | 0.0072 |
| Niobium | Nb | 41 | 0.0020 |
| Nitrogen | N | 7 | 0.0020 |
| Nobelium | No | 102 | Man-made |
| Osmium | Os | 76 | 0.00000002 |
| Oxygen[b] | O | 8 | 45.2 |
| Palladium | Pd | 46 | 0.0000003 |
| Phosphorus | P | 15 | 0.1010 |
| Platinum | Pt | 78 | 0.0000005 |
| Plutonium | Pu | 94 | Man-made |
| Polonium | Po | 84 | Footnote[d] |
| Potassium | K | 19 | 1.68 |
| Praseodymium | Pr | 59 | 0.0013 |
| Promethium | Pm | 61 | Man-made |
| Protactinium | Pa | 91 | Footnote[d] |
| Radium | Ra | 88 | Footnote[d] |
| Radon | Rn | 86 | Footnote[d] |
| Rhenium | Re | 75 | 0.00000004 |
| Rhodium[c] | Rh | 45 | 0.00000001 |
| Rubidium | Rb | 37 | 0.0070 |
| Ruthenium[c] | Ru | 44 | 0.00000001 |
| Samarium | Sm | 62 | 0.00077 |
| Scandium | Sc | 21 | 0.0022 |
| Selenium | Se | 34 | 0.000005 |
| Silicon | Si | 14 | 27.20 |
| Silver | Ag | 47 | 0.000008 |
| Sodium | Na | 11 | 2.32 |
| Strontium | Sr | 38 | 0.0450 |
| Sulfur | S | 16 | 0.030 |
| Tantalum | Ta | 73 | 0.00024 |
| Technetium | Tc | 43 | Man-made |
| Tellurium[c] | Te | 52 | 0.000001 |
| Terbium | Tb | 65 | 0.00010 |
| Thallium | Tl | 81 | 0.000047 |
| Thorium | Th | 90 | 0.00058 |
| Thulium | Tm | 69 | 0.000052 |
| Tin | Sn | 50 | 0.00015 |
| Titanium | Ti | 22 | 0.86 |
| Tungsten | W | 74 | 0.00010 |
| Uranium | U | 92 | 0.00016 |
| Vanadium | V | 23 | 0.0170 |
| Xenon | Xe | 54 | Not known |
| Ytterbium | Yb | 70 | 0.00034 |
| Yttrium | Y | 39 | 0.0035 |
| Zinc | Zn | 30 | 0.0082 |
| Zirconium | Zr | 40 | 0.0140 |

[a] Estimate from S. R. Taylor (1964). [b] Analyses of crustal rocks do not usually include separate determinations for hydrogen and oxygen. Both combine in essentially constant proportions with other elements, so abundances can be calculated. [c] Estimates are uncertain and have a very low reliability. [d] Elements formed by decay of uranium and thorium. The daughter products are radioactive with such short half-lives that crustal accumulations are too low to be measured accurately.

**Table A.3**
**Naturally Occurring Elements Listed in Order of Atomic Numbers, Together with the Naturally Occurring Isotopes of Each Element, Listed in Order of Mass Numbers**

| Atomic number[a] | Name | Symbol | Mass Numbers[b] of Natural Isotopes |
|---|---|---|---|
| 1 | Hydrogen | H | 1, (2, $\boxed{3}$[c]) - unstable |
| 2 | Helium | He | 3, 4 |
| 3 | Lithium | Li | 6, 7 |
| 4 | Beryllium | Be | 9 |
| 5 | Boron | B | 10, 11 |
| 6 | Carbon | C | 12, 13, $\boxed{14}$ |
| 7 | Nitrogen | N | 14, 15 |
| 8 | Oxygen | O | 16, 17, 18 |
| 9 | Fluorine | F | 19 |
| 10 | Neon | Ne | 20, 21, 22 |
| 11 | Sodium | Na | 23 |
| 12 | Magnesium | Mg | 24, 25, 26 |
| 13 | Aluminum | Al | 27 |
| 14 | Silicon | Si | 28, 29, 30 |
| 15 | Phosphorus | P | 31 |
| 16 | Sulfur | S | 32, 33, 34, 36 |
| 17 | Chlorine | Cl | 35, 37 |
| 18 | Argon | A | 36, 38, 40 |
| 19 | Potassium | K | 39, $\boxed{40}$, 41 |
| 20 | Calcium | Ca | 40, 42, 43, 44, 46, $\boxed{48}$ |
| 21 | Scandium | Sc | 45 |
| 22 | Titanium | Ti | 46, 47, 48, 49, 50 |
| 23 | Vanadium | V | $\boxed{50}$, 51 |
| 24 | Chromium | Cr | 50, 52, 53, 54 |
| 25 | Manganese | Mn | 55 |
| 26 | Iron | Fe | 54, 56, 57, 58 |
| 27 | Cobalt | Co | 59 |
| 28 | Nickel | Ni | 58, 60, 61, 62, 64 |
| 29 | Copper | Cu | 63, 65 |
| 30 | Zinc | Zn | 64, 66, 67, 68, 70 |
| 31 | Gallium | Ga | 69, 71 |
| 32 | Germanium | Ge | 70, 72, 73, 74, 76 |
| 33 | Arsenic | As | 75 |
| 34 | Selenium | Se | 74, 76, 77, 80, 82 |
| 35 | Bromine | Br | 79, 81 |
| 36 | Krypton | Kr | 78, 80, 82, 83, 84, 86 |
| 37 | Rubidium | Rb | 85, $\boxed{87}$ |
| 38 | Strontium | Sr | 84, 86, 87, 88 |
| 39 | Yttrium | Y | 89 |
| 40 | Zirconium | Zr | 90, 91, 92, 94, 96 |
| 41 | Niobium | Nb | 93 |
| 42 | Molybdenum | Mo | 92, 94, 95, 96, 97, 98, 100 |
| 44 | Ruthenium | Ru | 96, 98, 99, 100, 101, 102, 104 |
| 45 | Rhodium | Rh | 103 |

**Table A.3** (*Continued*)

| Atomic number[a] | Name | Symbol | Mass Numbers[b] of Natural Isotopes |
|---|---|---|---|
| 46 | Palladium | Pd | 102, 104, 105, 106, 108, 110 |
| 47 | Silver | Ag | 107, 109 |
| 48 | Cadmium | Cd | 106, 108, 110, 111, 112, 113, 114, 116 |
| 49 | Indium | In | 113, $\boxed{115}$ |
| 50 | Tin | Sn | 112, 114, 115, 116, 117, 118, 119, 120, 122, 124 |
| 51 | Antimony | Sb | 121, 123 |
| 52 | Tellurium | Te | 120, 122, 123, 124, 125, 126, 128, 130 |
| 53 | Iodine | I | 127 |
| 54 | Xenon | Xe | 124, 126, 128, 129, 130, 131, 132, 134, 136 |
| 55 | Cesium | Cs | 133 |
| 56 | Barium | Ba | 130, 132, 134, 135, 136, 137, 138 |
| 57 | Lanthanum | La | $\boxed{138}$, 139 |
| 58 | Cerium | Ce | 136, 138, 140, $\boxed{142}$ |
| 59 | Praseodymium | Pr | 141 |
| 60 | Neodymium | Nd | 142, 143, $\boxed{144}$, 145, 146, 148, 150 |
| 62 | Samarium | Sm | 144, $\boxed{147}$, $\boxed{148}$, $\boxed{149}$, 150, 152, 154 |
| 63 | Europium | Eu | 151, 153 |
| 64 | Gadolinium | Gd | $\boxed{152}$, 154, 155, 156, 157, 158, 160 |
| 65 | Terbium | Tb | 159 |
| 66 | Dysprosium | Dy | 156, 158, 160, 161, 162, 163, 164 |
| 67 | Holmium | Ho | 165 |
| 68 | Erbium | Er | 162, 166, 167, 168, 170 |
| 69 | Thulium | Tm | 169 |
| 70 | Ytterbium | Yb | 168, 170, 171, 172, 173, 174, 176 |
| 71 | Lutetium | Lu | 175, $\boxed{176}$ |
| 72 | Hafnium | Hf | 174, 176, 177, 178, 179, 180 |
| 73 | Tantalum | Ta | 180, 181 |
| 74 | Tungsten | W | 180, 182, 183, 184, 186 |
| 75 | Rhenium | Re | 185, $\boxed{187}$ |
| 76 | Osmium | Os | 184, 186, 187, 188, 189, 190, 192 |
| 77 | Iridium | Ir | 191, 193 |
| 78 | Platinum | Pt | 190, $\boxed{192}$, 195, 196, 198 |
| 79 | Gold | Au | 197 |
| 80 | Mercury | Hg | 196, 198, 199, 200, 201, 202, 204 |
| 81 | Thallium | Tl | 203, 205 |
| 82 | Lead | Pb | 204, 206, 207, 208 |
| 83 | Bismuth | Bi | 209 |
| 84 | Polonium | Po | $\boxed{210}$ |
| 86 | Radon | Rn | $\boxed{222}$ |
| 88 | Radium | Ra | $\boxed{226}$ |
| 90 | Thorium | Th | $\boxed{232}$ |
| 91 | Protactinium | Pa | $\boxed{231}$ |
| 92 | Uranium | U | $\boxed{234}$, $\boxed{235}$, $\boxed{238}$ |

[a] Atomic number = number of protons.
[b] Mass number = protons + neutrons.
[c] $\square$ Indicates isotope is radioactive.

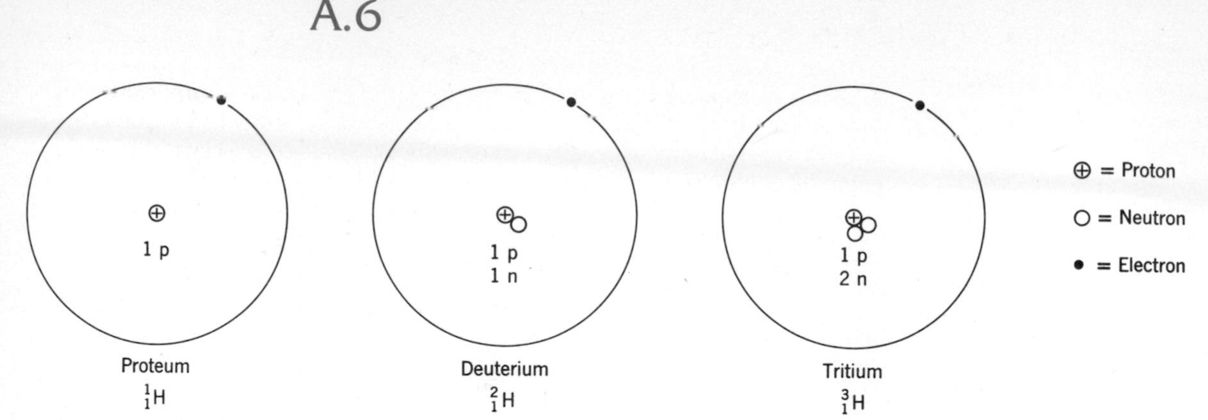


**Figure A.2**
**Three isotopes of hydrogen, each of which has one proton in the nucleus and one electron in orbit. The isotopes differ in the number of neutrons each nucleus contains. All isotopes of an element have the same atomic number, but differ in their mass numbers.**

particles that combine to form all atoms, and they are always the same, no matter in what atoms they occur. The way they fit together in a given atom determines the chemical properties of the atom. Protons, neutrons and electrons are therefore the fundamental particles of matter and, as the Greeks speculated, their different arrangements determine the different properties of matter.

Because protons and neutrons are so much heavier than electrons (Table A.1), 99.9 percent of the mass of an atom resides in the nucleus. We are already aware that atoms are tiny particles: the radius of an atom is about $10^{-8}$ centimeters. The radius of the atomic nucleus, however, is very much smaller—about $10^{-12}$ centimeters. Therefore, as Figure A.1 reveals, an atom is mostly open space. It has a tiny but heavy nucleus, surrounded by a diffuse cloud of electrons that move in distant orbits.

Elements are built systematically, starting with hydrogen, which has one proton and one electron, then helium, with two protons and two electrons, and so on. *The number of protons in the nucleus of an atom* is the ***atomic number***. All the known elements, with their atomic numbers and their abundance in Earth's crust, are listed in Table A.2.

### Isotopes and radioactivity

Although the number of neutrons that fit into the nucleus of an atom does not follow a simple pattern, nevertheless it is an important property of the atom because neutrons and protons together determine its mass. *The sum of the protons and neutrons in the nucleus of an atom* is called the ***mass***

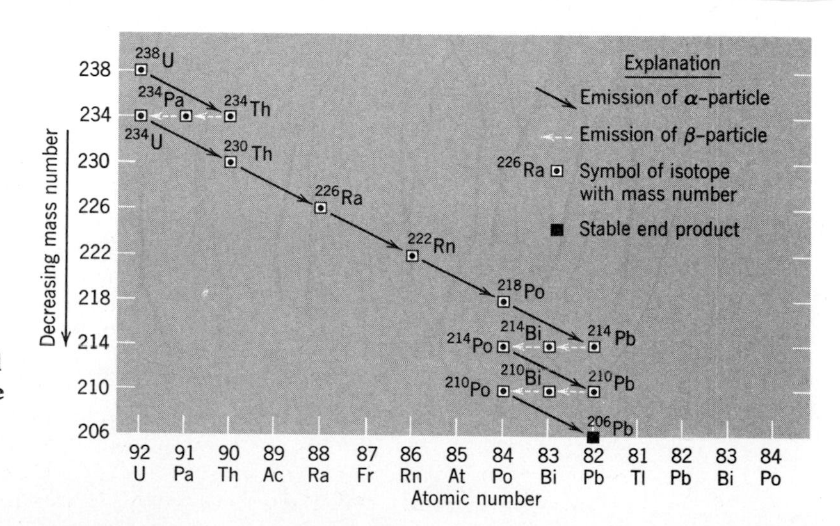


**Figure A.3**
**Example of a radioactive-decay series. It commences with the most common isotope of uranium, $^{238}_{92}U$. The stable end-product is an isotope of lead, $^{206}_{82}Pb$, but in order to reach it a number of different isotopes and different elements are produced as intermediate steps.**

***number.*** Both atomic number and mass number are such important properties of atoms that they are recorded as subscripts and superscripts respectively, usually on the left-hand side of the chemical symbol for the element. For example, $^{40}_{19}K$ means an atom of potassium with mass number 40 and atomic number 19. The atom therefore contains 21 neutrons and 19 protons.

We learned in Chapter 2 that *atoms having the same atomic number but differing numbers of neutrons in the nucleus* are called ***isotopes.*** Isotopes of an element therefore have identical atomic numbers but different mass numbers. All isotopes that are known to occur in the Earth are listed in Table A.3.

Hydrogen has three isotopes. Each has one proton and one electron. As shown in Figure A.2, the three isotopes of hydrogen are $^{1}_{1}H$, $^{2}_{1}H$, and $^{3}_{1}H$, known in scientific language as proteum, deuterium, and tritium, respectively. Every element has two or more isotopes, and for some elements as many as ten isotopes have been discovered. The number of isotopes that are possible for any given element is limited, however, because the mass numbers of elements can vary only within certain limits. Many combinations of protons and neutrons are unstable and break up spontaneously. *The decay process by which an unstable atomic nucleus spontaneously disintegrates,* emitting energy in the process, is called ***radioactivity.***

The rates at which radioactive isotopes disintegrate are inherent properties of each isotope and are unaffected by changes in temperature or pressure. As we noted in Chapter 5, the constancy of decay rates is very important because decay rates of certain isotopes such as those of $^{14}_{6}C$ and $^{40}_{19}K$ are so slow that we can use them as accurate clocks for timing events in Earth's history that happened very long ago.

When a radioactive isotope decays, it may form either a new isotope of the same element or an isotope of a different element. The decay products are called *daughter products,* which may, in turn, be either stable or radioactive. If they are radioactive, then further disintegrations will eventually occur, and still newer daughter products will form.

When an atom disintegrates radioactively, some of the energy that holds the nucleus together is released. The disintegrating atom may also lose some of its sub-atomic particles. These may be either *alpha-particles* ($\alpha$-particles) or *beta-particles* ($\beta$-particles). Alpha-particles are $^{4}_{2}He$ nuclei, stripped of their electrons so that their loss reduces the mass number by 4 and the atomic number by 2. Beta-particles are electrons expelled from the nucleus; their loss converts a neutron to a proton and thus increases the atomic number by 1 but leaves the mass number unchanged.

The results of radioactive emissions of $\alpha$- and $\beta$-particles are illustrated in Figure A.3, which depicts a sequence of decays beginning with the most common isotope of uranium, $^{238}_{92}U$. The uranium atom emits an $\alpha$-particle to form a daughter atom that is an isotope of thorium, $^{234}_{90}Th$. The thorium isotope then emits a $\beta$-particle to become $^{234}_{91}Pa$, and the new daughter product in turn emits a second $\beta$-particle to become $^{234}_{92}U$. The decay process continues through a total of fourteen radioactive daughter products until a stable atom, $^{206}_{82}Pb$, is formed. We infer from this decay scheme that the uranium content of the Earth must be slowly decreasing and, conversely, that the total lead content must be slowly increasing. Fortunately, the rate of decay of uranium is quite slow, and there is plenty left for us to mine. The rate of decay is also slow enough so that it has been possible to measure the way in which Earth's lead content has changed through its history, and as we saw in Chapter 5, the measurements allow us to infer a great deal about the age of the Earth.

When an atom decays radioactively, the principal way energy is released is as electromagnetic rays of very short wavelength, called *gamma-rays* ($\gamma$-rays), and as heat, which is generated whenever a nuclear disintegration occurs. As we noted in Chapter 2, the generation of heat by spontaneous nuclear decay has played a vitally important part in the history of Planet Earth. Radioactive decay of natural elements is believed to be the fuel that drives the planet's great internal processes.

**Matter and energy**

We stated earlier that the nature of matter is essentially electrical. We also know that electricity is a form of energy so we can modify the earlier statement to say that matter is essentially energy. The first person to understand this important concept was Albert Einstein who, in 1905, showed that matter and energy are connected by the equation

$$E = mc^2$$

where E is energy, $m$ is mass and $c$ is the velocity of electromagnetic waves.

When we consider the structure of atoms, the equivalence of matter and energy become vitally important. For example, the nucleus of an atom of $^{4}_{2}He$ contains two protons and two neutrons. Expressed in atomic mass units (AMU), a proton weighs 1.00758 AMU and a neutron 1.00893 AMU. The mass of $^{4}_{2}He$ should therefore be $(2 \times 1.00758) + (2 \times 1.00893) =$ 4.03303 AMU. When the helium nucleus is weighed, however, it is found to be only 4.00260 AMU—some of the mass has been lost. Explaining the mystery of the lost mass was one of the greatest triumphs of atomic physics. When the particles join together to form a helium nucleus, some of their mass is converted to energy, which appears partly as electromagnetic radiation and, because the new atomic nucleus has a high speed, partly as heat energy. The fusion of light particles to form heavier ones proceeds continuously in the Sun. Since nuclei of hydrogen, which are simply protons, are abundant in the Sun, astronomers believe that hydrogen fusion is the main source of solar energy. As a result of fusion, the Sun radiates a continuous stream of electromagnetic waves into space, and is slowly converting more of its matter into energy. *Fusion* of light particles to form heavy ones does not happen naturally on Earth, because the high temperatures needed to trigger it do not occur, but it is the process that goes on in the hydrogen bomb.

A nucleus can be broken down into its constituent particles only by the addition of sufficient energy to replace that lost during the joining process. The missing mass of the particles in the nucleus acts, in a sense, as a sort of "glue" that binds the nucleus together. The energy given off when protons and neutrons combine to form an atomic nucleus is, therefore, called the *binding energy* of the nucleus. Binding energy increases as the mass number of an element increases, but as we can see in Figure A.4, the relation is not a simple one. The curve dips down for elements with very high mass numbers. This means that if we split a nucleus such as uranium into one or more elements with low mass numbers, we move up the binding-energy curve, and energy is released in the same way as it is when light nuclei fuse to form heavier ones. Splitting a heavy atom, $^{235}_{92}U$, into lighter atoms of barium and krypton produces the energy of the atom bomb (Fig. A.5). We speak of heavy-atom splitting as *fission*.

Energy released by atomic fission appears both as electromagnetic waves

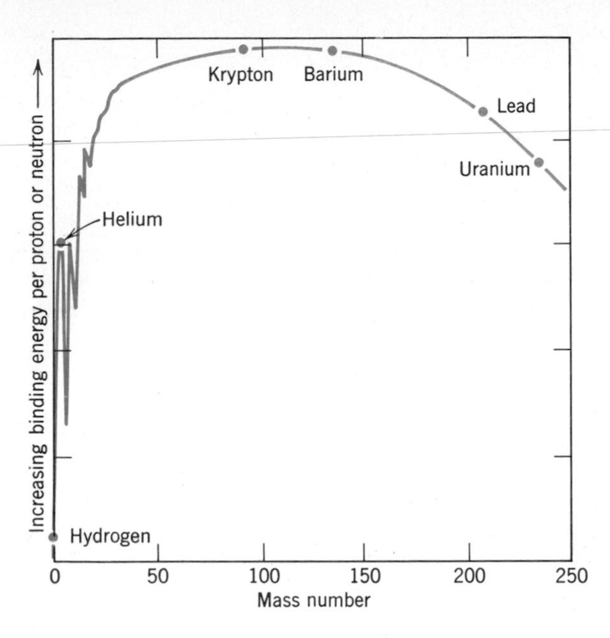


**Figure A.4**
**Binding energy, the amount of energy produced by conversion of some of the mass of each proton and neutron combined in a nucleus, varies with the mass number of the nucleus (see text for further explanation). The higher an element is on the curve, the more energy will be given off when protons and neutrons combine to form its nucleus. The fusion of four hydrogen nucleii to form helium releases a vast amount of energy. This is the process that produces the Sun's energy and that indirectly drives most of Earth's external activities. The splitting of a heavy uranium atom into lighter krypton and barium atoms also releases energy. This is the process that occurs in the atom bomb. Lead sits higher on the curve than uranium. Natural radioactive decay of uranium to lead produces energy, and is one of the principal sources of energy for Earth's internal activities.**

and as heat. If we refer to Figure A.3, in which the natural radioactive decay of $^{238}_{92}U$ is depicted, we see that a stable daughter, $^{206}_{82}Pb$, results. Lead sits higher on the binding-energy curve than uranium; so energy is released during the process of natural decay of uranium. The fraction that is heat energy is an important source of Earth's internal heat energy.

### Compounds

The chemical properties of an element (that is, the way elements combine together to form compounds) are determined by the orbiting electrons. Electrons are confined to specific orbits which are arranged at predetermined distances from the nucleus. Because the electrons in each orbit have a specific amount of energy characteristic for that orbit, the orbit distances are commonly called *energy-level shells.* The maximum number of electrons that can occupy a given energy-level shell is fixed. Shell 1, closest to the nucleus, is small and can accommodate only 2 electrons; shell 2, however, can accommodate 8 electrons; shell 3, 18; and shell 4, 32.

When an energy-level shell is filled with electrons it is very stable, like an evenly loaded boat. To fill their energy-level shells and so reach a stable configuration, atoms share or transfer electrons among themselves. The movement of electrons naturally upsets the balance of electrical forces, for an atom that loses an electron has lost a negative electrical charge and therefore has a net positive charge, while one that gains an electron has a net negative charge. The sharing and transfer of electrons gives rise to electronic forces, or *bonds,* that bind atoms together into *compounds.* For example, as can be seen in Figure A.6, lithium has shell 1 filled, but has only one electron in shell 2. The lone outer electron is loosely held and easily transferred to an element such as fluorine, which has 7 electrons in shell 2, needing only one more to be completely filled. In this fashion both the lithium and fluorine finish with filled shells, and the resulting positive charge on the lithium (written $Li^{+1}$) and the negative charge on the fluorine ($F^{-1}$) bind the two atoms together. Lithium and fluorine form the compound lithium fluoride, which is written LiF to indicate that for every Li atom there is a counterbalancing F atom. Properties of compounds are quite different from the properties of their constituent elements. A single

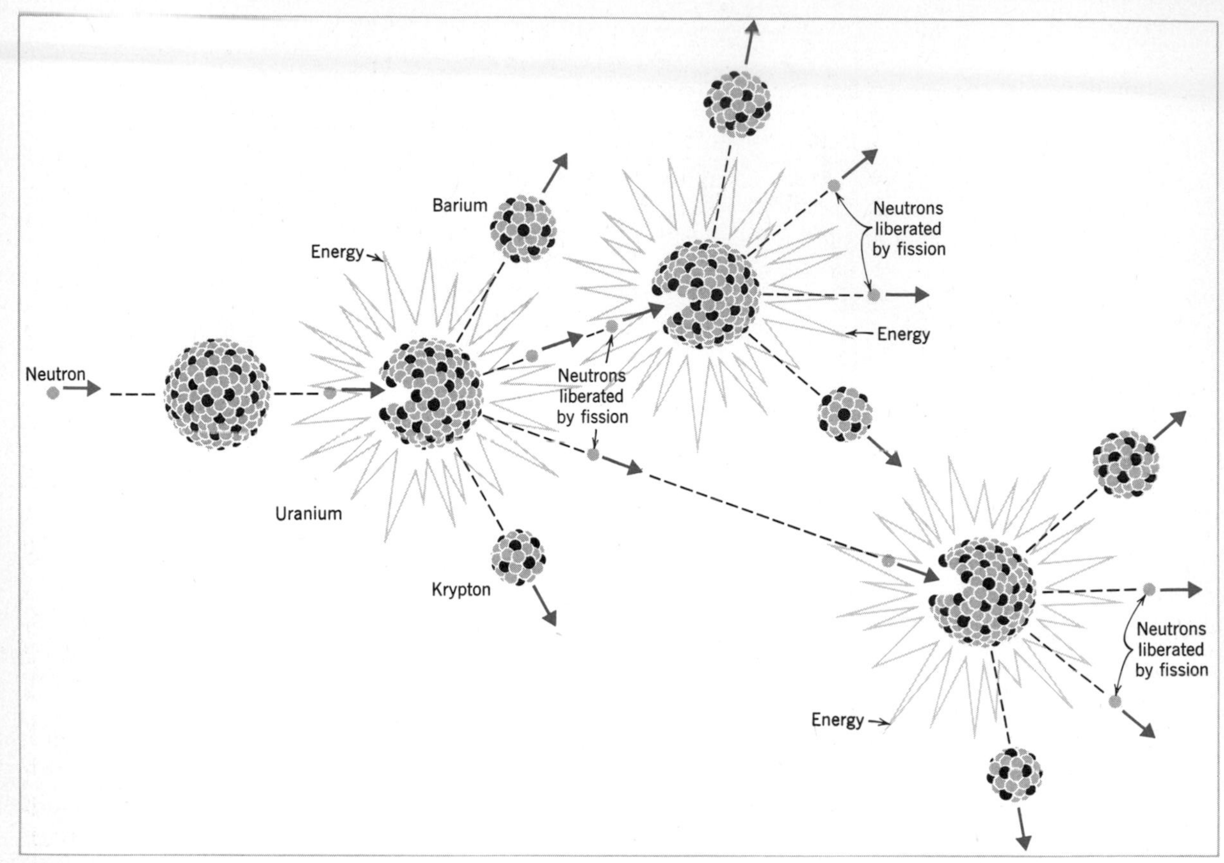


**Figure A.5**

**Release of energy by a chain reaction in uranium. A neutron strikes the nucleus of a $^{235}_{92}U$ atom, causing it to split into lighter elements such as barium and krypton, releasing energy and more neutrons. The new neutrons cause more $^{235}_{92}U$ atoms to disintegrate, producing a continuous chain reaction.**

LiF pair is called a molecule of lithium fluoride. A ***molecule*** is *the smallest unit that retains all the properties of a compound.*

Some elements occur naturally with their energy-level shells completely filled. They occur as individual, electrically neutral atoms and show little or no tendency to react with other elements and form compounds. Because they are so unreactive, elements with normally filled outer shells are called *noble gases.* The noble-gas elements are helium, neon, argon, krypton, xenon, and radon. All elements other than the noble gases readily bond with other atoms and form compounds. Atoms form bonds with like- or unlike atoms, but regardless of the pairing, when they transfer electrons they finish with a net positive or negative charge, depending on whether they give up or receive the transferred electrons.

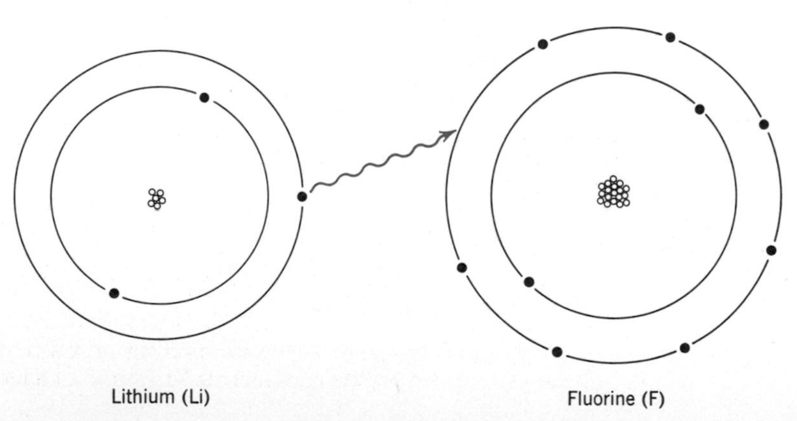


**Figure A.6**

**To form lithium fluoride, an atom of lithium combines with an atom of fluorine. The lithium atom transfers its lone outer-shell electron to fill the fluorine atom's outer shell.**

# APPENDIX B IDENTIFICATION OF COMMON MINERALS

Minerals that are abundant in rocks or common as ores number only a few dozens, and many can be identified without special equipment if sizable pieces are available. The techniques to be described consist of direct observations and simple tests. With these it is possible to recognize groups of rock-forming minerals and the common ore minerals. The most helpful accessory equipment to aid in making the tests is (1) a pocket knife, (2) a ten-power magnifying lens, (3) a piece of white, broken porcelain with a rough, unglazed surface, (4) a small hand magnet (the same tests can be performed by magnetizing the blade of your pocket knife).

### Physical properties

The properties of minerals are determined by their compositions and their crystal structures. Once we know which properties are characteristic of which minerals, we can use those properties to identify the minerals. It is not necessary, therefore, to analyze a mineral chemically or to determine its crystal structure in order to identify most common ones. The characteristics most often used in identifying minerals are the obvious physical properties, such as color, shape of crystal, and hardness, plus some less obvious properties, such as cleavage and specific gravity. Each property is discussed below.

**Crystal Form and Habit.** When a mineral grows freely, without obstruction from adjacent minerals, it forms a characteristic *geometric*

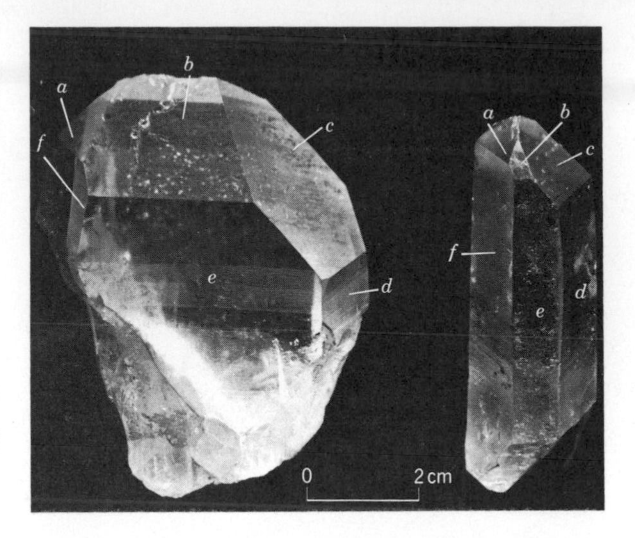


**Figure B.1**
**Two quartz crystals with the same crystal forms. Although the size of the individual faces differ markedly between the two crystals, it is clear that each face on one crystal is parallel to an equivalent face on the other crystal. The angles between adjacent faces are identical for all crystals of the same mineral.** (***Yale Peabody Museum.***)

*solid that is bounded by symmetrically arranged plane surfaces.* The characteristic solid is called the ***crystal form*** of a mineral. The plane faces of a crystal are an external expression of the strict, internal geometric arrangement of the constituent atoms. Each plane surface corresponds to a plane of atoms in the crystal structure. Unfortunately, crystals do not commonly grow in open, unobstructed spaces. Well-formed crystals are therefore rare. When nice crystals are found, however, an examination of the crystal form tells much about the crystal structure and immediately aids us in identification.

The sizes of individual crystal faces differ. Under some circumstances a mineral may grow a long, thin crystal; under others, a short, fat one. Superficially the two crystals may look very different; however, the unique characteristic of crystals is not the relative sizes of the individual crystal faces, but the angles between the faces. The angle between any two adjacent crystal faces in a mineral is a constant and is the same for all specimens of the mineral. Two crystals of quartz ($SiO_2$) are shown in Figure B-1. One is flattened, the other elongate, but it is clear that the same sets of crystal faces occur on both minerals. It is also clear, however, that on the two crystals the sets of faces are parallel and therefore the angle between any two equivalent faces must be the same on each crystal.

Every mineral has a unique crystal form. Some minerals also form crystals with distinctive shapes (Fig. B-2). For example, the mineral pyrite ($FeS_2$) is commonly found as intergrown cubes (Fig. B-3) with markedly striated faces, while the mineral stibnite ($Sb_2S_3$) almost invariably forms long, needle-like crystals (Fig. B-4).

As noted earlier, most minerals do not grow freely into open spaces and therefore do not develop well-shaped crystals. Instead, the growing minerals usually encounter other minerals and other obstructions that prevent the development of crystal faces. Usually, then, we cannot use crystal form to identify minerals, but we can sometimes make use of distinctive growth habits to aid identification. For example, Figure B-5 shows asbestos, a variety of the mineral serpentine that characteristically grows as fine, elongate threads. Another example of a distinctive growth habit is shown in Figure B-6. This is psilomelane, a common manganese oxide that possesses *botryoidal* structure, a collection of smooth, rounded

*A*

*B*

**Figure B.2**
***A.* Crystals of garnet have a characteristic crystal form characterized by four-sided crystal faces. This striking group of crystals was found near Russell, Massachusetts. (*Yale Peabody Museum.*)**
***B.* Small hexagonal quartz crystals showing three of the six prismatic sides. The characteristic crystal form of quartz is one of its most diagnostic characteristics. (*B. M. Shaub.*)**

surfaces together resembling grapes closely bunched. Other common habits are the *earthy* form of many iron oxides, which are crumbly like soil, and *micaceous,* which is a habit of many silicate minerals which cleave into thin plate-like fragments.

**Cleavage and Fracture.** If we break a mineral specimen with a hammer, or drop the specimen on the floor so that it shatters, the broken fragments are seen to be bounded by smooth, plane faces, so that the fragments resemble small crystals. A closer look shows that all fragments break along similar planes. *The tendency of a mineral to break in preferred directions along plane surfaces is called* ***cleavage.*** The plane surfaces along

Figure B.3
**Crystals of pyrite ($FeS_2$) showing a distinctive habit—intergrown cubes with striated faces.** (*B. M. Shaub.*)

Figure B.4
**Crystals of stibnite ($Sb_2S_3$) almost invariably occur as long, thin, needle-like crystals. The white, tabular-shaped crystals intergrown with the needles are calcite ($CaCO_3$).** (*Yale Peabody Museum.*)

Figure B.5
**Some minerals have distinctive growth habits even though they do not develop well-formed crystal faces. Asbestos, a variety of the mineral serbentine, grows as fine, cotton-like threads that can be separated and woven into fireproof fabric.** (*B. M. Shaub.*)

Figure B.6
**Psilomelane, a manganese oxide, commonly displays botryoidal surfaces. This habit is also common in various iron oxide minerals.** (*B. M. Shaub.*)

*A*

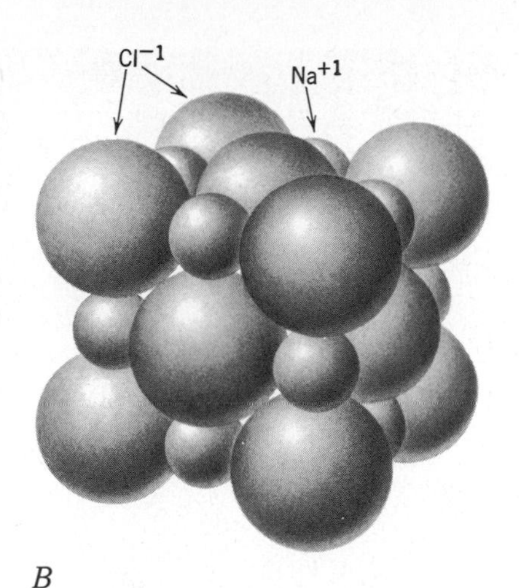


*B*

**Figure B.7**
**Relation between crystal structure and cleavage.**
***A.* Halite (NaCl) has well-defined cleavage planes; it breaks into fragments bounded by perpendicular faces.**
***B.* The crystal structure, in the same orientation as the cleavage fragments, shows that the plane of breakage is a plane in the crystal in which sodium and chlorine atoms occur in equal numbers. (*A, B. M. Shaub.*)**

which cleavage occurs are governed by the crystal structure (Fig. B-7). They are planes along which the bonds between atoms are relatively weak. Because the cleavage planes are direct expressions of the crystal structure, they are valuable guides for the identification of minerals.

Many minerals have distinctive cleavage planes. One of the most distinctive is found in the mineral *muscovite* (Fig. B-8). Clay minerals also have distinctive cleavage, and it is this easy cleavage direction that makes them feel smooth and slippery when rubbed between the fingers. Another mineral with a highly distinctive cleavage is calcite, which breaks into perfect rhombs (Fig. B-9). Besides micas, clays, and calcite, a number of other common minerals such as feldspar, amphibole, pyroxene, and galena have distinctive cleavages. Indeed, it is cleavage that allows us readily to distinguish amphibole from pyroxene—two groups of minerals that, in most properties, are nearly identical. The reason why cleavage is distinctive is because cleavages accurately reflect the ways in which silica tetrahedra polymerize in the structures (Fig. B-10).

Not all minerals have distinctive cleavages, and a few lack cleavage planes altogether and thus are distinctive in the opposite sense. Minerals lacking perceptible cleavage include garnet, which breaks along irregular fractures, quartz (Fig. B-11), and olivine, which fracture irregularly or display ***conchoidal fracture,*** *breakage resulting in smooth curved surfaces.* Some minerals break along splintery surfaces resembling those of wood.

**Figure B.8**
**Perfect cleavage of the mica mineral, muscovite shown by very thin, plane flakes into which this six-sided crystal has been split. The cleavage flakes suggest leaves of a book, a resemblance embodied in the name "books of mica" for crystals elongated in a direction perpendicular to the cleavage flakes. (*Ward's Natural Science Establishment.*)**

**Figure B.9**
**Perfect rhombs of *calcite* (calcium carbonate) formed by cleavage planes in three directions.** ***(Ward's Natural Science Establishment.)***

Because of a distinctive habit of growth within the crystals the cleavage surfaces of plagioclase nearly always appear to be *striated.* Striations are a reliable means of distinguishing plagioclase from potassium feldspar. They are seen to best advantage with a hand lens when the cleavage surface reflects a bright light.

**Color.** The color of a mineral is one of its striking properties, but unfortunately is not a very reliable means of identification. Color commonly results from impurities, which are present in only small amounts. Some minerals display various colors. Quartz, for example, can be clear and colorless, milky white, rose-colored, violet, and dark gray to black. Calcite, likewise, can be clear, milky white, pink, green, and gray. Among feldspars flesh-colored, cream-colored, pink, and light green characterize potassium feldspar and its relatives, whereas dead white and light blue typify plagioclases.

Color in minerals is determined by their composition, and, as we have seen, solid solution causes the compositions of minerals to vary within small ranges. Some elements can create strong color effects even when they are present in very small amounts. For example (Fig. B-12), the mineral corundum ($Al_2O_3$) is commonly white or grayish in color, but when small amounts of Cr have replaced Al by solid solution, this mineral is blood red, forming *ruby,* a prized gem variety of corundum. Similarly, when small amounts of Fe and Ti are present, the corundum is deep blue and another prized gemstone, *sapphire,* is the result.

**Luster.** *The quality and intensity of light reflected from a mineral* produce an effect known as ***luster.*** Two minerals with almost the same color can have totally different lusters. The more important are described as *metallic,* like that on a polished metal surface; *vitreous,* like that on glass; *resinous,* like that of yellow resin; *pearly,* like that of pearl; *greasy,* as if the surface were covered with a film of oil; and *adamantine,* having the brilliance of a diamond.

**Figure B.10**
**Sketches showing cleavage in *pyroxene*, and *amphibole*. Pyroxenes contain polymerized chains of silica tetrahedra, amphiboles contain double chains. Cleavage occurs by breakage parallel to the silica chains. The angle between cleavage surfaces is approximately 90° in pyroxenes, but only 56° in amphiboles.**

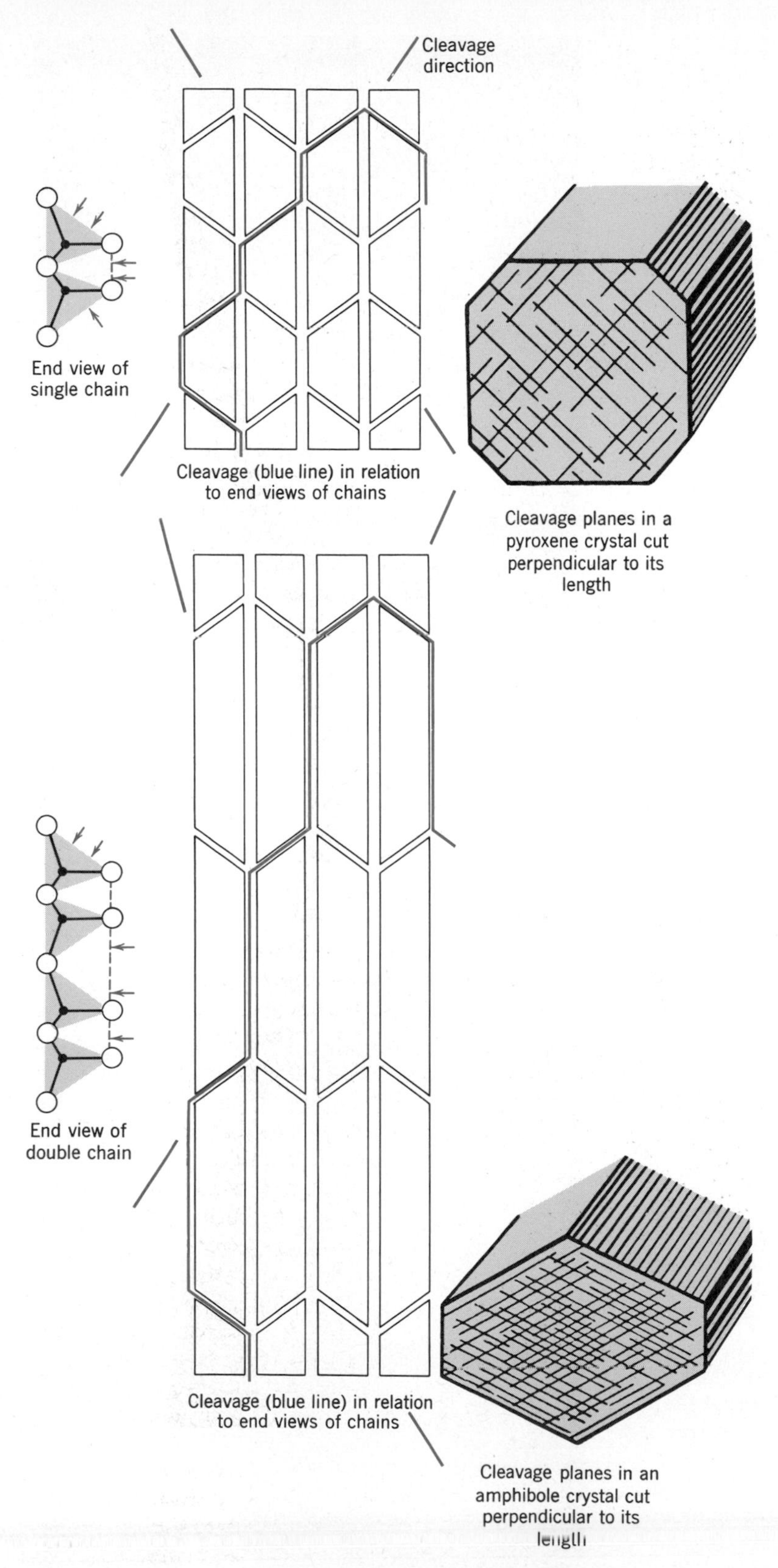

**Figure B.11**
**Irregular fracture of quartz. Curved fracture surfaces at end of quartz crystal from Arkansas. (*B. M. Shaub.*)**

Mineral identification will soon become familiar procedure to those who approach it systematically. Minerals with metallic luster should be set aside for tests of streak and magnetic susceptibility. Most ore minerals have a metallic luster. Minerals with vitreous luster should be checked for hardness, inspected for kinds of breakage surfaces, and given chemical tests.

**Streak.** The ***streak*** is *a thin layer of powdered mineral made by rubbing a specimen on a nonglazed porcelain plate.* The powder diffuses light and

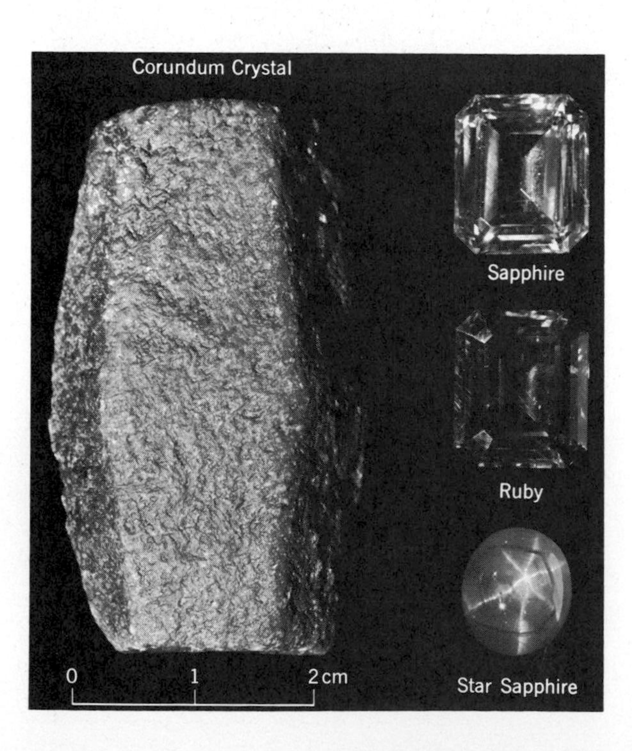


**Figure B.12**
**Corundum ($Al_2O_3$) commonly occurs as barrel-shaped, six-sided crystals with a variety of colors. Transparent crystals of corundum are the familiar gemstones ruby (red) and sapphire (blue). Sometimes tiny inclusions of other minerals are trapped within a corundum crystal. In star sapphires the inclusions are arranged parallel to the six crystal faces and disperse light so as to form a striking, six-pointed star. (*Yale Peabody Museum.*)**

gives a reliable color effect that is independent of the form and luster of the mineral specimen. Red streak characterizes hematite whether the specimen itself is red and earthy, like the streak, or black and metallic, like magnetite. Limonite streaks brown, and magnetite streaks black. Many minerals, particularly those with vitreous luster, streak an undiagnostic white.

**Hardness.** *Relative resistance of a mineral to scratching* is ***hardness,*** another distinctive property of minerals. Hardness, like crystal form and cleavage, is governed by crystal structure and by the strength of the bonds between atoms. The stronger the bonds, the harder the mineral. Degree of hardness can be decided in a relative fashion by determining the ease or difficulty with which one mineral will scratch another. Talc, the basic ingredient of most body ("talcum") powders, is the softest mineral known, and diamond the hardest. A relative hardness scale between talc (number 1) and diamond (number 10) is divided into ten steps, each marked by one of ten common minerals (Table B.1). These steps do not represent equal intervals of hardness, but the important feature of the hardness scale is that any mineral on the scale will scratch all minerals below it. Minerals on the same step of the scale are just capable of scratching each other. For convenience we often test relative hardness by using a common object such as a penny, or a penknife, as the scratching instrument.

In tests of hardness several precautions are necessary. A mineral softer than another may leave a mark that looks like a scratch, just as a soft pencil leaves its mark. A real scratch does not rub off. The physical structure of some minerals may make the hardness test difficult; if a specimen is powdery or in fine grains or if it breaks easily into splinters, an apparent scratch may be deceptive.

**Density and specific gravity.** The final obvious physical property of a mineral is its density, which in practical terms means how heavy it feels. We know that equal-sized baskets of feathers and of rocks have different weights: feathers are light, rocks are heavy. The property that causes this difference is ***density,*** or *the average weight per unit volume.* The units of density are numbers of grams per cubic centimeter. Minerals with a high density, such as gold, have their atoms closely packed. Minerals with low density, such as ice, have loosely packed atoms.

Minerals are divided into a heaviness or density scale. Gold has the highest density of all minerals, 19.3gms per cubic centimeter, but many others

**Table B.1**
**Scale of Hardness**

| | Relative Number in the Scale | Mineral | Hardness of Common Objects |
|---|---|---|---|
| Decreasing ↓ | 10. | Diamond | |
| | 9. | Corundum | |
| | 8. | Topaz | |
| | 7. | Quartz | |
| | 6. | Orthoclase | Pocket knife; glass |
| | 5. | Apatite | |
| | 4. | Fluorite | |
| | 3. | Calcite | Copper penny |
| | 2. | Gypsum | Fingernail |
| | 1. | Talc | |

## Table B.2

## Properties of the Rock-Forming Minerals (Ore Minerals, Some of Which Are Also Rock-Forming Minerals, Are Listed in Table B.3)

| Mineral | Chemical composition | Form and habit | Cleavage | Hardness | Specific gravity | Other properties | Most distinctive properties |
|---|---|---|---|---|---|---|---|
| Amphiboles. (A complex family of minerals *Hornblende* is most common.) | $X_2Y_5Si_8O_{22}(OH)_2$ where X = Ca, Na, Y = Mg, Fe, Al | Long, six-sided crystals: also fibers and irregular grains | Two; intersecting at 56° and 124°. | 5–6 | 2.9–3.8 | Common in metamorphic and igneous rocks. *Hornblende* is dark green to black; *actinolite*, green; *tremolite*, white. | Cleavage, habit. |
| Andalusite | $Al_2SiO_5$ | Long crystals, often square in cross-section. | Weak, parallel to length of crystal. | 7.5 | 3.2 | Found in metamorphic rocks. Often flesh colored. | Hardness, form. |
| Anhydrite | $CaSO_4$ | Crystals are rare. Irregular grains or fibers. | Three, at right angles. | 3 | 2.9 | Alters to gypsum. Pearly luster, white or colorless. | Cleavage, hardness. |
| Apatite | $Ca_5(PO_4)_3(F, OH, Cl)$ | Granular masses. Perfect six-sided crystals. | Poor. One direction. | 5 | 3.2 | Green, brown, blue or white. Common in many kinds of rocks in small amounts. | Hardness, form . |
| Aragonite | $CaCO_3$ | Massive, or slender, needle-like crystals. | Poor. Two directions. | 3.5 | 2.9 | Colorless or white. Effervesces with dilute HCl. | Effervescence with acid. Poor cleavage distinguishes from calcite. |
| Asbestos | | | See Serpentine | | | | |
| Augite | | | See Pyroxene | | | | |
| Biotite | | | See Mica | | | | |
| Calcite | $CaCO_3$ | Tapering crystals and granular masses. | Three perfect; at oblique angles to give a rhomb-shaped fragment. | 3 | 2.7 | Colorless or white. Effervesces with dilute HCL. | Cleavage, effervescence with acid. |
| Chlorite | $(Mg, Fe)_5(Al,Fe)_2Si_3O_{10}(OH)_8$ | Flaky masses of minute scales. | One perfect; parallel to flakes. | 2–2.5 | 2.6–2.9 | Common in metamorphic rocks. Light to dark green. Greasy luster. | Cleavage—flakes not elastic, distinguishes from mica. Color. |
| Dolomite | $CaMg(CO_3)_2$ | Crystals with rhomb-shaped faces. Granular masses. | Perfect in three directions as in calcite. | 3.5 | 2.8 | White or gray. Does not effervesce in cold, dilute HCl unless powdered. Pearly luster. | Clevage. Lack of effervescence with acid. |
| Epidote | Complex silicate of Ca, Fe and Al. | Small elongate crystals. Fibrous. | One perfect, one poor. | 6–7 | 3.4 | Yellow-green to dark green. Common in metamorphic rocks. | Habit, color. Hardness distinguishes from chlorite. |
| Feldspars: Potassium feldspar (*orthoclase* is a common variety) | $KAlSi_3O_8$ | Prism-shaped crystals, granular masses. | Two perfect, at right angles. | 6 | 2.6 | Common mineral. Flesh colored, pink, white or gray. | Color, cleavage. |
| Plagioclase | $NaAlSi_3O_8$ (albite) and $CaAl_2Si_2O_8$ (anorthite) and all compositions between. | Irregular grains, cleavable masses. Rarely as tabular crystals. | Two perfect, not quite at right angles. | 6–6.5 | 2.6–2.7 | White to dark gray. Cleavage planes may show fine parallel striations. | Cleavage. Striations on cleavage planes will distinguish from orthoclase. |
| Fluorite | $CaF_2$ | Cubic crystals, granular masses. | Perfect in four directions. | 4 | 3.2 | Colorless, blue green. Always an accessory mineral. | Hardness, cleavage, does not effervesce with acid. |
| Garnets | $X_3Y_2(SiO_4)_3$; X = Ca, Mg, Fe, Mn, Y = Al, Fe, Ti, Cr. | Perfect crystals with 12 or 24 sides. Granular masses. | None. Uneven fracture. | 6.5–7.5 | 3.5–4.3 | Common in metamorphic rocks. Red, brown, yellow green, black. | Crystals, hardness, no cleavage. |

## Table B.2 (*Continued*)

| Mineral | Chemical composition | Form and habit | Cleavage | Hardness | Specific gravity | Other properties | Most distinctive properties |
|---|---|---|---|---|---|---|---|
| Graphite | C | Scaly masses. | One, perfect. Forms slippery flakes. | 1–2 | 2.2 | Metamorphic rocks. Black with metallic to dull luster. | Cleavage, color. Marks paper. |
| Gypsum | $CaSO_4 \cdot 2H_2O$ | Elongate or tabular crystals. Fibrous and earthy masses. | One, perfect. Flakes bend but are not elastic. | 2 | 2.3 | Vitreous to pearly luster. Colorless. | Hardness, cleavage. |
| Halite | NaCl | Cubic crystals. | Perfect to give cubes. | 2.5 | 2.2 | Tastes salty. Colorless, blue. | Taste, cleavage. |
| Hornblende | | | See Amphibole | | | | |
| Kaolinite | $Al_2Si_2O_5(OH)_4$ | Soft, earthy masses. Submicroscopic crystals. | One, perfect. | 2–2.5 | 2.6 | White, yellowish. Plastic when wet; emits clay odor. Dull luster. | Feel, plasticity, odor. |
| Kyanite | $Al_2SiO_5$ | Bladed crystals. | One perfect. One imperfect. | 4.5 parallel to blade, 7 across blade | 3.6 | Blue, white, gray. Common in metamorphic rocks. | Variable hardness, distinguishes from sillimanite. Color. |
| Mica: Biotite | $K(Mg, Fe)_3AlSi_3O_{10}(OH)_2$ | Irregular masses of flakes. | One, perfect. | 2.5–3 | 2.8–3.2 | Common in igenous and metamorphic rocks. Black, brown, dark green. | Cleavage color. Flakes are elastic. |
| Muscovite | $KAl_3Si_3O_{10}(OH)_2$ | Thin flakes. | One perfect. | 2–2.5 | 2.7– | Common in igneous and metamorphic rocks. Colorless, pale green or brown. | Cleavage, color. Flakes are elastic. |
| Olivine | $(Mg, Fe)_2SiO_4$ | Small grains, granular masses. | None. Conchoidal fracture. | 6.5–7 | 3.2–4.3 | Igneous rocks. Olive green to yellow green. | Color, fracture, habit. |
| Orthoclase | | | See Feldspar | | | | |
| Plagioclase | | | See Feldspar | | | | |
| Pyroxene (A complex family of minerals. *Augite* is most common.) | $XY(SiO_3)_2$ $X = Y = Ca, Mg, Fe$ | 8-sided stubby crystals. Granular masses. | Two perfect, nearly at right angles. | 5–6 | 3.2–3.9 | Igneous and metamorphic rocks. *Augite*, dark green to black; other varieties, white to green. | Cleavage. |
| Quartz | $SiO_2$ | 6-sided crystals, granular masses. | None. Conchoidal fracture. | 7 | 2.6 | Colorless, white, gray, but may have any color, depending on impurities. Vitreous to greasy luster. | Form, fracture, striations across crystal faces at right angles to long dimension. |
| Serpentine (Fibrous variety is *asbestos*) | $Mg_3Si_2O_5(OH)_4$ | Platy or fibrous. | One, perfect. | 2.5–5 | 2.2–2.6 | Light to dark green. Smooth, greasy feel. | Habit, hardness. |
| Sillimanite | $Al_2SiO_5$ | Long needle crystals, fibers. | Breaks irregularly, except in fibrous variety. | 6–7 | 3.2 | White, gray. Metamorphic rocks. | Hardness distinguishes from kyanite. Habit. |
| Talc | $Mg_3Si_4O_{10}(OH)_2$ | Small scales, compact masses. | One, perfect. | 1 | 2.6–2.8 | Feels slippery. Pearly luster. White to greenish. | Hardness, luster, feel, cleavage. |
| Tourmaline | Complex silicate of B, Al, Na, Ca, Fe, Li and Mg. | Elongate crystals, commonly with triangular cross section. | None. | 7–7.5 | 3–3.3 | Black, brown, red, pink, green, blue and yellow. An accessory mineral in many rocks. | Habit. |
| Wollastonite | $CaSiO_3$ | Fibrous or bladed aggregates of crystals. | Two, perfect. | 4.5–5 | 2.8–2.9 | Colorless, white, yellowish. Metamorphic rocks. Soluble in HCl. | Habit. Solubility in HCl and hardness distinguish amphiboles, kyanite, sillimanite. |

**Table B.3**
**Properties of the Common Ore Minerals (Some Ore Minerals Are Also Rock-Forming Minerals)**

| Mineral | Chemical composition | Form and habit | Cleavage | Hardness | Specific gravity | Other properties | Most distinctive properties |
|---|---|---|---|---|---|---|---|
| Bornite | $Cu_5FeS_4$ | Massive. Crystals very rare. | None. Uneven fracture. | 3 | 5 | Brownish bronze on fresh surface. Tarnishes purple, blue and black. Gray black streak. | Color, streak. |
| Chalcocite | $Cu_2S$ | Massive. Crystals very rare. | None. Conchoidal fracture. | 2.5 | 5.7 | Steel-gray to black. Dark gray streak. | Streak. |
| Chalcopyrite | $CuFeS_2$ | Massive or granular. | None. Uneven fracture. | 3.5–4 | 4.2 | Golden yellow to brassy yellow. Dark green to black streak. | Streak. Hardness distinguishes from pyrite. |
| Chromite | $FeCr_2O_4$ | Massive or granular. | None. Uneven fracture. | 5.5 | 4.6 | Iron black to brownish black. Dark brown streak | Streak and lack of magnetism distinguishes from ilmenite and magnetite. |
| Copper | Cu | Massive, twisted leaves and wires. | None. Can be cut with a knife. | 2.5–3 | 9 | Copper color but commonly stained green. | Color, specific gravity, malleable. |
| Galena | PbS | Cubic crystals, coarse or fine grained granular masses. | Perfect in three directions at right angles. | 2.5 | 7.6 | Lead gray color. Gray to gray black streak. | Cleavage and streak. |
| Gold | Au | Small irregular grains. | None. Malleable. | 2.5 | 19.3 | Gold color. Can be flattened without breakage. | Color, specific gravity, malleability. |
| Hematite | $Fe_2O_3$ | Massive, granular, micaceous | Uneven fracture. | 5–6 | 5 | Red-brown, gray to black. Red-brown streak. | Streak, hardness. |
| Ilmenite | $FeTiO_3$ | Massive or irregular grains. | Uneven fracture. | 5.5–6 | 4.7 | Iron-black. Brown-red streak differing from hematite. | Streak distinguishes hematite. Lack of magnetism distinguishes magnetite. |
| Limonite (*Goethite* is most common.) | A complex mixture of minerals, mainly hydrous iron oxides. | Massive, coatings, botryoidal crusts, earthy masses. | None. | 1–5.5 | 3.5–4 | Yellow, brown, black. Yellow-brown streak. | Streak. |
| Magnetite | $Fe_3O_4$ | Massive, granular. Crystals have octahedral shape. | None. Uneven fracture. | 5.5–6.5 | 5 | Black, Black streak. Strongly attracted to a magnet. | Streak, magnetism. |
| Pyrite ("Fool's gold") | $FeS_2$ | Cubic crystals with striated faces. Massive. | None. Uneven fracture. | 6–6.5 | 5.2 | Pale brass-yellow, darker if tarnished. Greenish black streak. | Streak. Hardness distinguishes from chalcopyrite. Not malleable, which distinguishes from gold. |
| Pyrolusite | $MnO_2$ | Crystals rare. Massive, coatings on fracture surfaces. | Crystals have a perfect cleavage. Massive breaks unevenly. | 2–6.5 | 5 | Dark gray, black on bluish black. Black streak. | Color, streak. |

**Table B.3** *(Continued)*

| Mineral | Chemical composition | Form and habit | Cleavage | Hardness | Specific gravity | Other properties | Most distinctive properties |
|---|---|---|---|---|---|---|---|
| Pyrrhotite | FeS | Crystals rare. Massive or granular. | None. Conchoidal fracture. | 4 | 4.6 | Brownish-bronze. Black streak. Magnetic. | Color and hardness distinguish from pyrite, magnetism from chalcopyrite. |
| Rutile | $TiO_2$ | Slender, prismatic crystals or granular masses. | Good in one direction. Conchoidal fracture in others. | 6–6.5 | 4.2 | Red-brown (common), black (rare). Brownish streak. Adamantine luster. | Luster, habit, hardness. |
| * Sphalerite (zinc blende) | ZnS | Fine to coarse granular masses. Tetrahedron shaped crystals. | Perfect in six directions. | 3.5–4 | 4 | Yellow brown to black. White to yellow-brown streak. Resinous luster. | Cleavage, hardness, luster. |
| Uraninite | $UO_2$ to $U_3O_8$ | Massive, with botryoidal forms. Rare crystals with cube shapes. | None. Uneven fracture. | 5–6 | 6.5–10 | Black to dark brown. Streak black to dark brown. Dull luster. | Luster and specific gravity distinguish from magnetite. Streak distinguishes from ilmenite and hematite. |

* Only one with 6 planes of cleavage

such as galena (7.5), magnetite (5.2), and hematite (5.3) feel heavy by comparison with most silicate minerals, which have densities between 2.5 and 3.0.

Density is not commonly measured directly. Instead, unit weights of different minerals are compared against the unit weight of a standard substance. In comparing unit weights, we commonly use water as a standard. The ***specific gravity*** of any substance is expressed as *a number stating the ratio of the weight of the substance to the weight of an equal volume of pure water.* Specific gravity can be approximated by comparing different minerals held in the hand. Metallic minerals such as galena feel "heavy" whereas nearly all others feel "light."

**Magnetism.** Of the common minerals only magnetite and pyrrhotite are strongly magnetic. They can be singled out at once by their strong attraction to a small magnet.

### Chemical Properties

Only two chemical tests are commonly used in the beginning study of minerals: (1) taste test for halite and (2) acid test for calcite and dolomite. The salty taste of halite is distinctive. Carbonate minerals effervesce (make bubbles) in dilute hydrochloric acid. Calcite effervesces freely no matter what the size of the particles. Dolomite may not effervesce at all unless the specimen is powdered or the acid is heated. Dolomite powder effervesces slowly in cool dilute acid.

*Caution.* Hydrochloric acid is hard on teeth and has an unpleasant taste. Where many students share mineral specimens caution in use of acid is necessary. Acid should be applied in small drops and when the test is finished the specimen should be blotted dry. The next user of the specimen may decide to try the taste test.

**Table B.4**
**Properties of Some Common Gemstones**

| Mineral and variety | Composition | Form and habit | Cleavage | Hardness / Specific gravity | Other properties | Most distinctive properties |
|---|---|---|---|---|---|---|
| Beryl: *Aquamarine* (blue) *Emerald* (green) *Golden beryl* (golden-yellow) | $Be_3Al_2Si_6O_{18}$ | Six-sided, elongate cyrstals common. | Weak. | 7.5–8 / 2.75 | Bluish green, green, yellow, white colorless. Common in Pegmatites. | Form. Distinguished from apatite by its hardness. |
| Corundum: *Ruby* (red) *Sapphire* (blue) | $Al_2O_3$ | Six-sided, barrel-shaped crystals. | None, but breaks easily across its crystal. | 9 / 4 | Brown, pink, red, blue, colorless, Common in metamorphic rocks. Star sapphire is opalescent with a six-sided light spot showing. | Hardness. |
| Diamond | C | Octahedron-shaped crystals. | Perfect, parallel to faces of octahedron. | 10 / 3.5 | Colorless, yellow, rarely red, orange, green, blue or black. | Hardness, cleavage |
| Garnet: *Almandite* (red) *Grossularite* (green, cinnamon-brown) *Demantoid* (green) | A rock-forming mineral—See Table B.2 | | | | | |
| Opal (A mineraloid) | $SiO_2 \cdot nH_2O$ | Massive, thin coating. Amorphous. | None. Conchoidal fracture. | 5–6 / 2–2.2 | Colorless, white, yellow, red, brown, green, gray, opalescent. | Hardness, color, form. |
| Quartz: (1) Coarse crystals *Amethyst* (violet) *Cairngorm* (brown) *Citrine* (yellow) *Rock crystal* (colorless) *Rose quartz* (pink) (2) Fine-grained *Agate* (banded, many colors) *Chalcedony* (brown, gray) *Heliotrope* (green) *Jasper* (red) | A rock-forming mineral—See Table B.2 | | | | | |
| Topaz | $Al_2SiO_4(OH, F)_2$ | Prism shaped crystals, granular masses. | One, perfect. | 8 / 3.5 | Colorless, yellow, blue, brown. | Hardness, form, color. |
| Tourmaline | A rock-forming mineral—See Table B.2 | | | | | |
| Zircon | $ZrSiO_4$ | Four-sided elongate crystals, square in cross-section. | None. | 7.5 / 4.7 | Brown, red, green, blue, black. | Habit, hardness. |

**Identification of Minerals**

For convenience, the common rock-forming minerals are arranged alphabetically in Table B.2, and the common ore minerals in Table B.3. The chemical formulas in the second column of both tables are there for reference only, not to aid in identifications. The final column in Tables B.2 and B.3 list the physical properties most characteristic for each mineral, together with suggestions for distinguishing between similar-looking minerals. Most rock-forming minerals have white streak and vitreous luster and are either transparent or translucent. Ore minerals are almost all opaque, have metallic luster and high specific gravities. Most ore minerals will be encountered only in mineral deposits. A few, notably chalcopyrite, hematite, ilmenite, limonite, magnetite, pyrite, pyrrhotite, and rutile, may also be encountered as accessory minerals in common igneous, sedimentary or metamorphic rocks.

Gemstones are minerals with desirable properties of color and wearing ability. A list of the commonly observed gem minerals, together with the names of the gem varieties, is given in Table B.4.

# SELECTED REFERENCES

Dietrich, R. V., 1966, Mineral tables—hand specimen properties of 1500 minerals: Blacksburg, Va., Virginia Polytechnic Inst. Bull. 160.

Pearl, R. M., 1962, Successful mineral collecting and prospecting: New York, New American Library.

Tennissen, A. C., 1974, Nature of Earth materials: Englewood Cliffs, N.J., Prentice-Hall.

Vanders, Iris, and Kerr, P. F., 1967, Mineral recognition: New York, John Wiley.

# APPENDIX C IDENTIFICATION OF COMMON ROCKS

**Diagnostic features**

The three major classes of rocks (igneous, sedimentary, metamorphic) are defined and discussed in Chap. 3. In each class there are many distinct kinds, and each has a specific name. Fortunately, fewer than 30 in all make up the great bulk of the visible part of the Earth's crust. These common kinds must be learned well if we are to read correctly the history recorded in the crust. By studying representative specimens of all the important kinds, we can learn the properties some have in common, and the distinctive features by which each kind is identified.

Ideally, we should see each specimen in the field as part of an exposure in which the larger features and relations are clearly shown. The sedimentary layering of sandstone or limestone, and the intrusive relation of a dike tell us at once that the rock is sedimentary or igneous. Some hand specimens in a laboratory may not have clear indications of their general classification. But any systematic description of the common rocks lists them according to class, and we welcome any clue that may tell us at the start whether an unknown specimen is igneous, sedimentary, or metamorphic. A number of such clues are found with practice, and even without them a specimen can ordinarily be traced quickly to its class by a process of elimination.

**Texture.** We examine a rock specimen closely for the pattern of visible constituents, just as we inspect the weave (texture) in cloth. Many rocks have visible mineral grains, and if these can be made out with the unaided eye over the entire surface, the rock has *granular texture, coarse* if the average grains are 5mm or more across, *medium* if the average is 1 to 5mm across and, *fine* if less than 1mm across. Some igneous rocks have *glassy* texture; the cooling from magma was too rapid for any grains to form.

Many igneous rocks are porphyritic, with distinct phenocrysts (Fig. 3.15); this proves igneous origin. Certain textures, then, tell us the general classification of some rock specimens as a first step toward their identification.

In examining texture, we are interested not only in the size of grains but also in their shapes and the way they fit together. If the grains are angular and dovetail one into another to fill all the space, they must have been formed by crystallization, and the rock is probably either igneous or metamorphic. If the grains are separated by irregular spaces filled with fine cementing material, probably they are fragments and the rock may be of either sedimentary or volcanic origin.

**Mineral Assemblage.** Many rocks are identified by their component minerals. The critical minerals are determined by their physical properties, as explained in Appendix B. Grains large enough for clear visibility are required for study without a microscope. A hand lens that magnifies about 10 times is a useful aid in studying the mineral grains, even in coarse-textured rocks.

**Other Properties.** Some limestones look superficially like fine-grained igneous rocks but are much softer. Every rock specimen under study should be tested for hardness.

Some kinds of rocks show characteristic forms on fracture surfaces. Other tests are mentioned in the descriptions of specific rocks.

The principal group of rocks discussed in this Appendix is the igneous group. They offer more trouble to a beginning student than sedimentary or metamorphic rocks. The following discussion, and Table C.1, therefore present the necessary data for identifying common igneous rocks. Identification of sedimentary and metamorphic rocks can be effected by studying Chap. 3 and using Tables C.2 and C.3 respectively.

### Kinds of igneous rock

In an introductory study igneous rocks can be classified more systematically than others, and details required for satisfactory analysis of laboratory specimens are here separated from the general treatment in the body of the book. Table C.1 supplements Figure 3.16 and the two can be used together with profit. Igneous rocks are grouped in the table according to (1) texture and (2) composition. Any classification for use with hand specimens must be general, and the number of names in the table is reduced to a minimum.

Nature does not draw sharp boundaries, there are all conceivable gradations in texture and composition. The separating lines in the table, therefore, are somewhat fictitious. We can find a series of specimens that will bridge the gaps in composition between granite and granodiorite, granodiorite and diorite.

Minerals of igneous rocks are divided generally into a light-colored group (including the feldspars, quartz, and muscovite) and a dark-colored group (including biotite, pyroxene, hornblende, and olivine). Quartz grains in an

**Table C.1**
**Aids in Identification of Common Igneous Rock**
To use the table, first determine the texture, and find which entries in Column 1 best describe it. Then determine if quartz is present, and if it is sparse or abundant. These options are listed in Column 2. Finally, determine which feldspars are present (Column 3), and what the remaining minerals are (Column 4). These four sets of observations uniquely determine the rock type, listed in Column 5. Column 6 is a review column in which the key identification features are mentioned together with suggestions on general rock features that may be helpful.

| | Minerals | | | | |
|---|---|---|---|---|---|
| Texture | Quartz | Feldspar | Other | Rock Name | Helpful Distinguishing Features |
| Coarse-grained. Grains uniform in size. | Abundant. | Abundant. Orthoclase exceeds plagioclase. | Muscovite and/or biotite common. Hornblende sometimes present. | **Granite** | Quartz and feldspar predominant. Light-colored rock, commonly pink, white, shades of gray. Make sure orthoclase exceeds plagioclase. Easily confused with granodiorite. |
| Coarse-grained. Grains uniform in size. | Abundant. | Abundant. Plagioclase exceeds orthoclase. | Muscovite and/or biotite common. Hornblende sometimes present. | **Grandiorite** | Quartz and feldspar predominant. Shades of gray. |
| Coarse-grained. Grains uniform in size. | Sparse or absent. | Abundant plagioclase. Orthoclase rare or absent. | Biotite and/or hornblende common. Pyroxene sometimes present. | **Diorite** | About equal amounts of light- and dark-colored minerals. A darker rock than grandiorite. Absence or sparsity of quartz is diagnostic. |
| Coarse-grained. Grains uniform in size. | Sparse or absent. | Abundant. Orthoclase exceeds plagioclase. | Biotite, hornblende, nepheline may be present. | **Syenite** | Commonly pink or red. Distinguish from granite by quartz content. |
| Coarse-grained. Grains uniform in size. | Absent. | Common. Plagioclase only. | Pyroxene abundant. Olivine may be present. | **Gabbro** | Dark minerals exceed light. A dark-colored rock. Distinguish from peridotite and pyroxenite by common plagioclase. |
| Coarse-grained. Grains uniform in size. | Absent. | Rare or absent. | Pyroxene abundant. Olivine may be present. | **Pyroxenite** | A dark-colored rock consisting very largely of pyroxenes. |
| Coarse-grained. Grains uniform in size. | Absent. | Rare or absent. | Olivine abundant. Pyroxene common to abundant. | **Peridotite** | Dark-colored rock. Olivine is commonly a clear green and grains rounded. |

igneous rock indicate a surplus of silica in the parent magma; therefore the presence or absence of quartz grains is a logical basis for drawing a line between granitic rocks and diorite. In the rocks that have no quartz, the feldspars divide honors with the dark-colored minerals, and the boundary between diorite and gabbro is drawn where the dark minerals exceed 50 percent of the total. This boundary is carried through between andesite and basalt on the basis of color. Inspecting thin edges of specimens in strong light is the best test; andesite transmits some of the light, basalt is opaque.

**Granite.** Feldspar and quartz are the chief minerals in granite. Some biotite usually is present, and many granites have scattered grains of hornblende. The dark minerals commonly are in nearly perfect crystals; this suggests that they formed first, while most of the mass was molten. The feldspar formed next, and the grains crowded against and hampered

IDENTIFICATION OF COMMON ROCKS

## A.29

| | Minerals | | | | |
|---|---|---|---|---|---|
| Texture | Quartz | Feldspar | Other | Rock Name | Helpful Distinguishing Features |
| Medium-grained. Grains uniform in size. | Rare or absent. | Abundant. Plagioclase only. | Pyroxene common. Olivine may be present. | **Diabase** | A common medium-grained, dark gray-colored rock. Look for pyroxene and plagioclase. Distinguish from basalt by grain size and lack of extrusive volcanic features. Often called trap rock. |
| Fine grained. Grains uniform in size. | Abundant. Hard to see because of grain size. | Abundant. Orthoclass exceeds plagioclase. | Hornblende, biotite may be present. | **Rhyolite** | A light-colored volcanic rock. White, gray, red, purple. May contain some glass. Often shows signs of flowage. |
| Fine-grained. Grains uniform in size. | Sparse or absent. | Abundant. Plagioclase exceeds orthoclase. | Pyroxene, hornblende, biotite may be present. | **Andesite** | A dark-colored volcanic rock. Shades of gray, brown, green. Glass is not common. |
| Fine-grained. Grains uniform in size. | Absent. | Abundant. Plagioclase only. | Pyroxene common. Olivine often present. | **Basalt** | A common dark-colored volcanic rock. No quartz present. Often rings like a bell when struck with a hammer. |
| Glassy | — | — | — | **Obsidian** | A dense glass. May contain some vesicles. |
| Glassy | — | A few feldspar crystals may be present. | — | **Pumice** | A glassy froth. |
| Phenocrysts in a coarse- or medium-grained ground mass. | Determine overall composition of rock. If a granite composition, rock is a **granite porphyry,** if a gabbro composition, rock is a **gabbro porphyry.** Common varieties are granite, granodiorite and diorite porphyries. Phenocrysts commonly quartz, feldspar, hornblende. | | | | |
| Phenocrysts in a fine-grained ground mass. | Determine overall composition of rock. Texture indicates a prophyritic volcanic rock. Most common varieties are **rhyolite prophyry** and **andesite porphyry.** Phenocrysts commonly quartz, feldspar, biotite. | | | | |

each other in growth. Quartz, the surplus silica, crystallized last and so is molded around the angular grains of the earlier minerals. This *interlocking arrangement of visible mineral grains characteristic of granite* is called ***granular texture.***

Technically, the term granite is applied only to quartz-bearing rocks in which orthoclase is predominant, and the name granodiorite applies to similar rocks in which plagioclase is the chief feldspar. Without special equipment, the differences in feldspars are not always easily recognized, and in a general study the term granite sometimes is extended to this whole group of rocks. We sometimes recognize the variation in mineral composition by speaking of the *granitic rocks.* They are widespread in all the continents.

***Pegmatite,*** or "giant granite," is a special kind of *granite which has abnormally large grains.* Quarries in pegmatite produce in commercial quantities large sheets of mica and minerals that yield the valuable elements lithium and beryllium.

**Diorite.** The chief mineral in diorite is feldspar, mainly plagioclase, though this may not be evident to the unaided eye. Generally the dark minerals are more abundant than in granite. Diorite forms many large masses, but it is not nearly so abundant as granitic rocks.

**Gabbro.** Dark diorite grades into *gabbro* as the dark minerals exceed 50 percent of the rock and plagioclase becomes subordinate. The chief dark mineral in gabbro is pyroxene, commonly with some olivine. These minerals are heavier than feldspar, and gabbro is distinctly heavier than granite and average diorite.

*Diabase* is fine-grained, intermediate in texture between gabbro and basalt.

**Pyroxenite and Peridotite.** As the dark minerals displace plagioclase entirely, we reach the extreme in composition from that of granite. A granular rock composed almost entirely of pyroxene is *pyroxenite.* If considerable olivine is present with the pyroxene, the rock is *peridotite.* Both these rocks are very dark and heavy and both are commonly associated with ores containing the metals nickel, platinum, and iron. In many masses the pyroxene and olivine have been partly or completely altered to serpentine.

**Porphyritic Rock.** Both coarse- and fine-grained igneous rocks commonly have prominent phenocrysts. If the larger grains make up less than about 25 percent of the rock mass, we say that the rock is *porphyritic* and give it the name suited to its groundmass (*e.g.,* porphyritic granite, porphyritic diorite, porphyritic andesite). If the proportion of phenocrysts is more than 25 percent, we call the rock a *porphyry* and combine this term with the name that is proper for the groundmass (*e.g.,* granite porphyry, diorite porphyry, rhyolite porphyry).

**Rhyolite.** A fine-grained rock with phenocrysts of quartz is *rhyolite.* The quartz indicates an excess of silica and therefore a close chemical kinship to granite. Rhyolites usually have phenocrysts of feldspar and biotite as well. Colors of the groundmass range from nearly white through shades of gray, yellow, red, or purple. Rhyolite commonly has irregular bands made by flowage of stiff magma shortly before it become solid.

**Andesite.** A fine-grained rock generally similar to rhyolite but lacking the quartz phenocrysts is *andesite.* Usually it has phenocrysts of feldspar and dark minerals. Common colors are shades of gray and green, but some andesites are very dark, even black. Freshly broken, thin edges of dark andesites transmit some light and appear almost white when held before a bright source of light. In this way they are distinguished from basalt, which is opaque even on thin edges. The lighter-colored andesites commonly have irregular banding similar to that of rhyolite.

Andesite is extremely abundant as a volcanic rock, especially around the margins of the Pacific Ocean. The name comes from the Andes of South America.

**Basalt.** *Basalt* is a fine-grained rock that appears dark even on freshly broken thin edges. Common colors are black, dark brown or green, and very dark gray. In the upper parts of lava flows the rock generally is *vesicular*—filled with small openings or *vesicles* made by escaping gases. In many flows these openings have been filled with calcite, quartz, or some other mineral deposited from solution.

**Glassy Rock.** Quick chilling of magma forms natural glass. *Obsidian* is a highly lustrous glassy rock. Obsidian displays a conchoidal pattern when broken.

Clear natural glass is not unknown, but most obsidians appear dark, even black. Because many of them correspond in chemical composition to rhyolite and granite, they seem to contradict the rule that rocks with high content of silica are light-colored. But obsidian chipped to a thin edge appears white, even transparent. The dark coloring results from a small content of dark mineral matter distributed evenly in the glass.

*Pumice* is glass froth, full of cavities made by gases escaping through stiff, rapidly cooling magma. Because many of these cavities are small winding tubes, some of them sealed, pumice will float on water for a long time. As the thin walls of the cavities transmit light, pumice is almost white, though it may form the cap of a black sheet of obsidian.

*Basalt glass* forms in the outer parts of some basaltic flows. It is opaque, like ordinary basalt, but has glassy luster.

**Extrusive Fragmental Rock.** *Volcanic ash,* the fine debris from explosive eruptions, becomes consolidated to form *tuff.* Commonly, the particles of magma are made into froth by the expanding gas, and therefore the flakes making up an ash deposit are in large part pumice, mixed with pieces of older bedrock blasted from the walls of the vent.

The particles of ash in some falls are very hot, and when they land, the temperature is high enough so the particles weld together. *A glassy or fine-grained rock formed by fusion of volcanic ash during deposition is* ***welded tuff.*** Some natural glasses are welded tuffs rather than chilled lava flows. Volcanic rocks in southeastern Arizona, in Utah, and Nevada that strongly resemble flows of rhyolite are really welded tuffs.

With increase in size of particles, tuff grades into *volcanic breccia.*

Many volcanic tuffs and breccias are stratified and look much like sedimentary rocks. Successive layers of ash are spread by air currents and become solidified as distinct beds. Furthermore, many explosive eruptions are accompanied by heavy rains, and the ash is swept out in thin uniform beds. The loose volcanic debris on steep slopes becomes saturated, and masses of it move down as mudflows and debris. Some pyroclastic rocks, therefore, are hybrid in classification—partly igneous and partly sedimentary.

### Kinds of sedimentary rock

A study of the common sedimentary rocks, with good specimens in a laboratory, is a convenient way to become familiar with the different kinds of sedimentary rock. To aid this study, a brief description of common sedimentary rock types is given below, and is summarized in Table C.2.

**Clastic Sedimentary Rock.** All clastic sedimentary rock is composed of particles of broken rock transported, deposited, compacted, and then cemented to form a coherent mass. *A clastic sedimentary rock that contains numerous rounded pebbles or larger particles is* ***conglomerate.*** The pebbles, cobbles, and boulders have been more or less rounded during transport by streams or glacier ice or in buffeting by waves along a shore. They can consist of any kind of rock but most commonly of the kinds rich in the durable mineral quartz. Usually the spaces between pebbles contain sand cemented with silica, clay, limonite, or calcite.

**Sedimentary breccia** is *a clastic sedimentary rock that resembles conglomerate, but most of whose fragments are angular instead of rounded.* We find all gradations between conglomerate and breccia.

The term *breccia* should always be used with a modifier to indicate origin; for example, sedimentary breccia, or volcanic breccia.

*A clastic sedimentary rock consisting of cemented sand grains is* ***sandstone.***

With progressive change in size of grain, coarse sandstone grades into *siltstone*. In many rocks, grain sizes are mixed; so we speak of conglomeratic sandstone or sandy siltstone.

In sandstone the grains consist almost entirely of quartz. The cementing material varies, as in conglomerate; calcium carbonate is common, but silica makes a more durable rock. Color in sandstone, produced partly by the color of the grains, partly by that of the cementing material, varies within a wide range.

***Arkose*** *is a variety of sandstone with a large proportion of feldspar grains.* A composition consisting of feldspar and quartz suggests granite, so that arkose might be mistaken for it. In arkose the grains do not interlock; they are rounded and separated by fine-grained cementing material.

***Claystone*** (also called *shale*) is *a clastic sedimentary rock made of compacted clay and silt.* It is so fine grained that to the unaided eye it seems homogeneous. Claystone is soft and generally feels smooth and greasy, but some fine sand or coarse silt may make it feel gritty. Claystones generally split into thin layers or flakes parallel to the sedimentary layering. Rocks of similar composition but with thick, blocky layers are termed *mudstone.*

The color of claystones and mudstones ranges through shades of gray, green, red, and brown. Some layers that contain considerable carbon are black.

**Chemical Sedimentary Rock.** The difference between chemical sedimentary rock and clastic sedimentary rock is principally one of transportation. In chemical sedimentary rock the constituents were transported in solution and precipitate by inorganic or organic chemical reactions.

***Limestone,*** which may be either *a clastic or chemical sedimentary rock, consisting chiefly of calcite,* has many impurities and varies greatly in appearance. Limestone belongs as a chemical sedimentary rock because even the clastic particles, such as shell fragments and corals, were formed first by chemical and biological activity. Some limestones that are uniformly fine-grained were formed purely as chemical precipitates, aided more or less by tiny organisms. Some of the sediment on today's sea floors probably represents an early stage in the formation of fine-grained limestone. By contrast, many limestones are coarse-grained, either from crystallization of the calcium carbonate or because they are made largely of detrital shell fragments.

***Dolostone,*** like limestone, is either *a clastic or a chemical sedimentary rock consisting chiefly of* the mineral *dolomite,* $CaMg(CO_3)_2$. Dolomite looks like calcite, which is why dolostone looks like limestone. But dolomite is slightly harder than calcite and only effervesces with acid on a scratched surface or in powdered form.

### Kinds of metamorphic rock

Metamorphic rocks are classified on the basis of texture (Chapter 3). Where a particular mineral is very obvious, its name may be used as a prefix. Examples of the seven common metamorphic rocks are given below and identifying features are summarized in Table C.3.

***Slate*** is *fine-grained metamorphic rock with a pronounced cleavage.* Cleavage planes separate slates into thin, flat plates, commonly cutting across original sedimentary layering. Although surfaces of the cleavage slabs have considerable luster, mineral grains can be seen only with very high

**Table C.2**
**Aids in Identification of Sedimentary Rock**

| Rock Name | Composition | Critical Tests |
|---|---|---|
| 1. Clastic sedimentary rock | | |
| **Conglomerate** | Cemented particles, somewhat rounded, considerable percentage of pebble size | Larger particles more than 2mm in diameter; smaller particles and binding cement in interstices |
| **Breccia** | Fragments conspicuously angular, with binding cement | Large particles of pebble size or larger |
| **Sandstone** | Rounded fragments of sand size, 0.02 to 2mm; binding cement | Grains commonly quartz, but other rock materials qualify in general classification |
| **Arkose** | Important percentage of feldspar grains, sand size or larger | Essential that feldspar grains make 25 per cent or more of rock; some may be larger than sand size |
| **Graywacke** | Fragments of quartz, feldspar, rock fragments of any kind, with considerable clay | Poor assortment of several kinds of ingredients, with considerable clay in matrix |
| **Siltstone** | Chiefly silt particles, some clay | Surface is slightly gritty to feel |
| **Clay** and **Shale** | Chiefly clay minerals | Surface has smooth feel, no grit apparent |
| 2. Chemical sedimentary rock | | |
| **Limestone** | Calcite; may be even grained and crystalline | Easily scratched with knife; effervesces in cold dilute hydrochloric acid |
| **Dolostone** | Dolomite; may be even grained and crystalline | Harder than limestone, softer than steel; requires scratching or powdering for effervescence in cold dilute hydrochloric acid |

magnification (Fig. 3.20). A common color is dark bluish gray, generally known as "slate color," but many slates are red, green, or black.

***Phyllite*** is *exceptionally lustrous rock representing a higher stage of metamorphism than slate.* The mica flakes responsible for the luster can be seen only with magnification, but the mineral grains are coarser than in slates and some phyllites have visible grains of garnet and other minerals (Fig. 3.18). The cleavage plates commonly are wrinkled or even sharply bent.

***Schist****, a well-foliated metamorphic rock in which the component platy minerals are clearly visible,* represents a higher stage in metamorphism than phyllite. Mica schist is rich in mica (either biotite, muscovite, or both together). Chlorite schist and hornblende schist also are common. Quartz is abundant in all kinds of schists and many are studded with garnets and other metamorphic minerals.

***Gneiss*** is *coarse-grained, foliated metamorphic rock, commonly with marked layering but with imperfect cleavage.* Many gneisses have a streaky, roughly

**Table C.3**
**Aids in Identification of Metamorphic Rock**

| Rock Name | Distinguishing Characteristics |
|---|---|
| 1. Foliated metamorphic rock | |
| **Slate** | Cleaves into thin, plane plates that have considerable luster; commonly the sedimentary layers of parent rock make lines on plates; thin slabs ring when they are tapped sharply |
| **Phyllite** | Surfaces of plates highly lustrous; plates commonly wrinkled or sharply bent; grains of garnets and other minerals on some plates |
| **Schist** | Well foliated, with visible flaky or elongate minerals (mica, chlorite, hornblende); quartz a prominent ingredient; grains of garnet and other accessory minerals common; foliae may be wrinkled |
| **Gneiss** | Generally coarse-grained, with imperfect but conspicuous foliation; lenses and layers differ in mineral composition; feldspar, quartz, and mica are common ingredients |
| 2. Nonfoliated metamorphic rock | |
| **Quartzite** | Consists wholly of quartz sand cemented with quartz; outlines of sand grains show on broken surfaces; the breaks passing through the grains; wide range in shades of color |
| **Marble** | Wholly crystallized limestone or dolostone; grain varies from coarse to fine; responds to hydrochloric acid test, as do calcite and dolomite; accessory minerals have developed from impurities in original rock |
| **Hornfels** | Hard, massive, fine-grained rock, commonly with scattered grains or crystals of garnet, andalusite, staurolite, or other minerals that are common in zones of contact metamorphism |

banded appearance, caused by alternating layers that differ in mineral composition (Fig. 3.19). Feldspars, quartz, micas, amphiboles, and garnets are common minerals in gneisses. ***Granite gneiss*** is *distinctly banded rock with the mineral composition of granite.*

Marble is created by metamorphism of limestone, during which the calcite grains grow until the entire rock mass has become coarse grained. Dolomite marble is formed in the same way from dolostone. ***Marble,*** then, is merely *coarsely crystalline limestone or dolostone.* Impurities in the original rock may cause the growth of small amounts of pyroxene, amphibole, and other minerals, which in many marbles make striking patterns.

Some commercial "marbles" are actually nonmetamorphosed limestones and dolostones that "take a good polish."

***Quartzite*** is *a metamorphic rock developed from sandstone by introduction of silica into all spaces between the original grains of quartz.* When quartzite is broken, the fracture passes through the original quartz grains, not around them as in ordinary sandstone. Quartzites usually have no patterns of foliation. Quartzite and marble are examples of nonfoliated metamorphic rocks, in contrast to gneiss, schist, phyllite, and slate, which are all foliated.

**Hornfels** is a rock formed near intrusive igneous bodies where the invaded rock is greatly altered by high temperature. Shale and some other fine-grained rocks are thus changed to **hornfels,** *a very hard, nonfoliated metamorphic rock, commonly studded with small crystals of mica and garnet.*

## SELECTED REFERENCES

Pirsson, L. V. and Knopf, A., 1946, Rocks and rock minerals: New York, John Wiley.

Tennissen, A. C., 1974, Nature of Earth materials: Englewood Cliffs, N.J., Prentice-Hall.

# APPENDIX D
# MAPS, CROSS SECTIONS, FIELD MEASUREMENTS, INTERPRETATIONS

### Uses of maps

An important part of the accumulated information about the geology and morphology of the Earth's crust exists in the form of maps. Nearly everyone has used automobile road maps in planning a trip or in following an unmarked road. A road map of a state, province, or county does what all maps have done since their invention at some unknown time more than 5,000 years ago; it reduces the pattern of part of the Earth's surface to a size small enough to be seen as a whole. Maps are especially important for an understanding of geologic relations because a continent, a mountain chain, and a major river valley are of such large size that they cannot be viewed as a whole unless represented on a map.

Furthermore, although most geologic maps are supplemented by text material, they contain information that cannot be stated as fully and accurately in words or by any other means. Maps are the core of most geologic reports based on field study and are supplementary to many other kinds of geologic papers. This appendix is designed to outline basic information about maps and about common field measurements used in the construction of geologic maps.

A map can be made to express much information within a small space by the use of various kinds of symbols. Just as some aspects of physics and chemistry use the symbolic language of mathematics to express significant relationships, so many aspects of geology use the simple symbolic language of maps to depict relationships too large to be observed within a single view. Maps made or used by geologists generally depict either one or two of three sorts of things:

1. Hills, valleys, and other surface forms. Most maps of this kind are *geographic maps.*
2. Distribution and attitudes of bodies of rock or regolith. Maps showing such things are *geologic maps.*
3. Geographic features such as mountains, rivers, and seas, not as they are today but reconstructed as they are inferred to have been at some time in the past (Fig. 12.17). Maps of this kind are *paleogeographic maps.*

The first two express the results of direct observation and measurement and are frequently included on a single map. The third expresses concepts built up from whole groups of observations and necessarily is shown on special maps separate from those representing present-day features.

### Base Maps

Every map is made for some special purpose. Road maps, charts for sea or air navigation, and geologic maps are examples of three special purposes. But whatever the purpose, all maps have two classes of data: *base* data and *special-purpose* data. As base data, most geologic maps show a latitude–longitude grid, streams, and inhabited places; many also show roads and railroads. A geologist may take an existing *base map* containing such data and plot geologic information on it, or he may start with blank paper and plot on it both base and geologic data, a much slower process if the map is made accurately.

**Two-Dimensional Base Maps.** Many base maps used for plotting geologic data are two-dimensional; that is, they represent length and breadth but not height. A point can be located only in terms of its horizontal distance, in a particular direction, from some other point. Hence a base map always embodies the basic concepts of direction and distance. Two natural reference points on Earth are the North and South Poles. Using these two, the ancient Greeks established a grid by means of which any other point

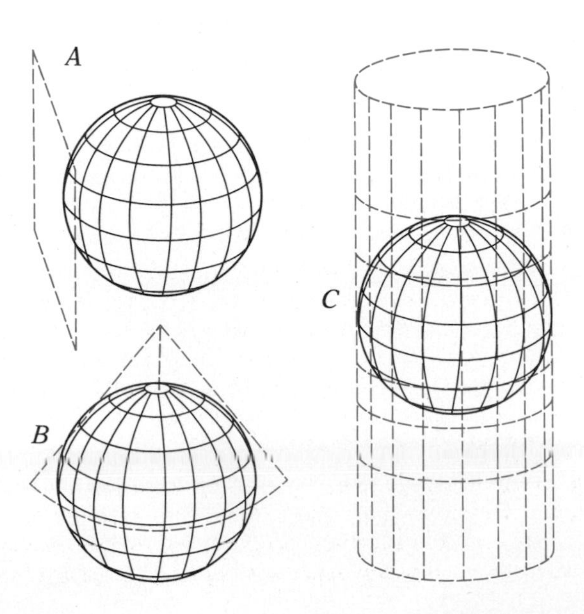


**Figure D.1** **The Earth's latitude-longitude grid can be projected onto a plane *A*, a cylinder *C*, or a cone *B* that theoretically can be cut and flattened out.**

could be located. The grid we use now consists of lines of *longitude* (half circles joining the poles) and *latitude* (parallel circles concentric to the poles) (Fig. D.1). The longitude lines (*meridians*) run exactly north–south, crossing the east–west *parallels* of latitude at right angles. Since the circumference of the Earth at its Equator and the somewhat smaller circumference through its two poles is known with fair accuracy, it is possible to define any point on the Earth in terms of direction and distance from either pole or from the point of intersection of any parallel with any meridian.

For convenience in reading, most maps are drawn so that the north direction is at the top or upper edge of the map. This is an arbitrary convention adopted mainly to save time. The north direction could just as well be placed elsewhere, provided its position is clearly indicated.

The accuracy with which distance is represented determines the accuracy of the map. *The proportion between a unit of distance on a map and the unit it represents on the Earth's surface* is the **scale** of the map. It is expressed as a simple proportion, such as 1:1,000,000. This ratio means that 1 meter, foot, or other unit on the map represents exactly 1,000,000 meters, feet, or other units on the Earth's surface, and works out to 1 cm = 10 km and 1 inch = about 16 miles. It is approximately the scale of many of the road maps widely used by motorists in North America. Scale is also expressed graphically by means of a numbered bar, as is done on most of the maps in this book. A map with a latitude–longitude grid needs no other indication of scale (except for convenience) because the lengths of a degree of longitude (varying from 110.7 km at the Equator to 0 at the poles) and of latitude (varying from 109.9 km at the Equator to 110.9 km at the poles) are known.

**Map Projections.** The Earth's surface is nearly spherical, whereas nearly all maps other than globes are planes, usually sheets of paper. It is geometrically impossible to represent any part of a spherical surface on a plane surface without distortion (Fig. D.2). The latitude–longitude grid has to be *projected* from the curved surface to the flat one. This can be done in various ways, each of which has advantages, but all of which represent a sacrifice of accuracy in that the resulting scale on the flat map will vary from one part of the map to another. Examples of cylindrical projections are Figs. 13.2 and 12.4. The most famous of these is the Mercator projection; although it distorts the Polar Regions very greatly, compass directions drawn on it are straight lines. Because this is of enormous value in navigation, the Mercator projection is widely used in navigators' charts.

Figure 15.13*A* is an example of conic projection. Some commonly used varieties are polyconic, in which not one cone, as in Fig. D.1*B*, but several

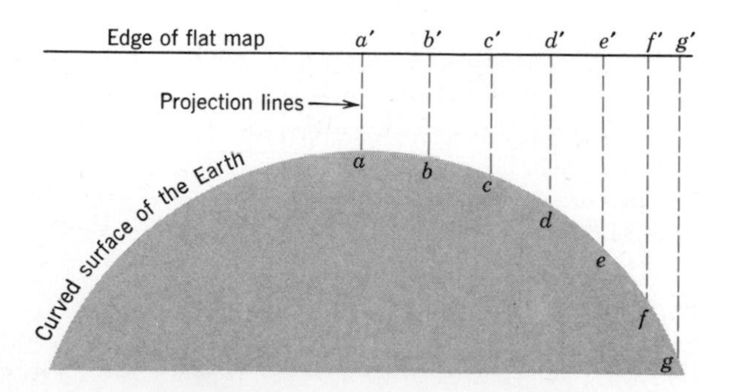


**Figure D.2**
**Equally spaced points (*a*, *b*, *c*, etc.) along a line in any direction on the Earth's surface become unequally spaced when projected onto a plane. This is why all flat maps are distorted.**

cones are employed, each one tangent to the globe at a different latitude. This device reduces distortion.

Figure 21.16 represents a plane projection, and Fig. 9.1 is a projection with certain areas, unimportant for the purpose of this map, eliminated. In a map of a very small area, such as Fig. 7.29, the distortion is of course slight, but it is there nevertheless.

**Topographic Maps.** A more complete kind of base map is three-dimensional; it represents not only length and breadth but also height. Therefore it shows ***relief*** (*the difference in altitude between the high and low parts of a land surface*) and also ***topography,*** defined as *the relief and form of the land. A map that shows topography* is a ***topographic map.*** Topographic maps can give the form of the land in various ways. The maps most commonly used by geologists show it by contour lines.

*Contours.* A **contour line** (often called simply a *contour*) is *a line passing through points having the same altitude above sea level.* If we start at a certain altitude on an irregular surface and walk in such a way as to go neither uphill nor downhill, we will trace out a path that corresponds to a contour line. Such a path will curve around hills, bend upstream in valleys, and swing outward around spurs. Viewed broadly, every contour must be a closed line, just as the shoreline of an island or of a continent returns upon itself, however long it may be. Even on maps of small areas, many contours are closed lines, such as those at or near the tops of hills. Many, however, do not close within a given map area; they extend to the edges of the map and join the contours on adjacent maps.

Imagine an island in the sea crowned by two prominent isolated hills, with much steeper slopes on one side than on the other and with an irregular shoreline. The shoreline is a contour line (the zero contour) because the surface of the water is horizontal. If the island is pictured as submerged until only the two isolated peaks project above the sea, and then raised above the sea 20 feet at a time, the successive new shorelines will form a series of contour lines separated by 20-foot contour intervals. (A ***contour interval*** is *the vertical distance between two successive contour lines,* and is commonly the same throughout any one map.) At first, two small islands will appear, each with its own shoreline, and the contours marking their shorelines will have the form of two closed lines. When the main mass of the island rises above the water, the remaining shorelines or contours will pass completely around the land mass. The final shoreline is represented by the zero contour, which now forms the lowest of a series of contours separated by vertical distances of 20 feet.

As the island is raised, the successive new shorelines are not displaced through so great a horizontal distance where the slope is steep as where it is more gradual. In other words, the water's edge retreats through a shorter horizontal distance in falling from one level to the next along the steep slope than along the gentle slope. Therefore, when these successive shorelines are projected upon the flat surface of a map, they will be crowded where the slope is steep and farther apart where it is moderate. In order to facilitate reading the contours on a map, certain contours (usually every fifth line) are drawn with a wider line. Contours are numbered at convenient intervals for ready identification. The numbers are always multiples of the contour interval and are placed between broken ends of the contour they designate.

Because the contours that represent a depression without an outlet resemble those of an isolated hill, it is necessary to give them a distinctive

**Figure D.3**
**Perspective sketch of a landscape.** *(Modified from U.S. Geol. Survey.)*

appearance. Depression contours therefore are *hatched;* that is, they are marked on the downslope side with short transverse lines called *hachures.* An example is shown on one contour in Figure D.4. The contour interval employed is the same as in other contours on the same map.

*Idealized Example.* Figures D.3 and D.4 show the relation between the surface of the land and the contour map representing it. Figure D.3, a perspective sketch, shows a stream valley between two hills, viewed from the south. In the foreground is the sea, with a bay sheltered by a curving spit. Terraces in which small streams have excavated gullies border the valley. The hill on the east has a rounded summit and sloping spurs. Some of the spurs are truncated at their lower ends by a wave-cut cliff, at the base of which is a beach. The hill on the west stands abruptly above the valley with a steep scarp and slopes gently westward, trenched by a few shallow gullies.

Each of the features on the map (Fig. D.4) is represented by contours directly beneath its position in the sketch.

### Air Photographs

Figures 9.24 and 10.4 are photographs made from airplanes, with cameras pointing obliquely at the Earth's surface. Oblique air photographs enable us to see over a much larger area than could be seen from a single point on the ground and are therefore useful in making clear the broad rather than the detailed relations of various geologic features. The figures referred to above were selected for this very purpose. All oblique photographs are "pictorial" in that they show hills and valleys in perspective.

In contrast, many air photographs are made with cameras pointing down vertically at the Earth's surface. Unlike oblique views, vertical photographs do not show perspective; they look "flat." But although distortion is always present (Fig. D.2), they show the pattern of the ground with distortion at a minimum. The vertical air photograph, therefore, is one kind of map. It is used widely as a base map on which geologic data are plotted in the field. Generally the scale of such a photograph must be 1:20,000 or larger, for it shows so much detail that on a smaller scale the various features can become blurred.

Used by itself, a vertical air photograph is a two-dimensional map. However, two photographs taken in sequence from a flying airplane, so that they overlap by a mile or two, can form a three-dimensional map

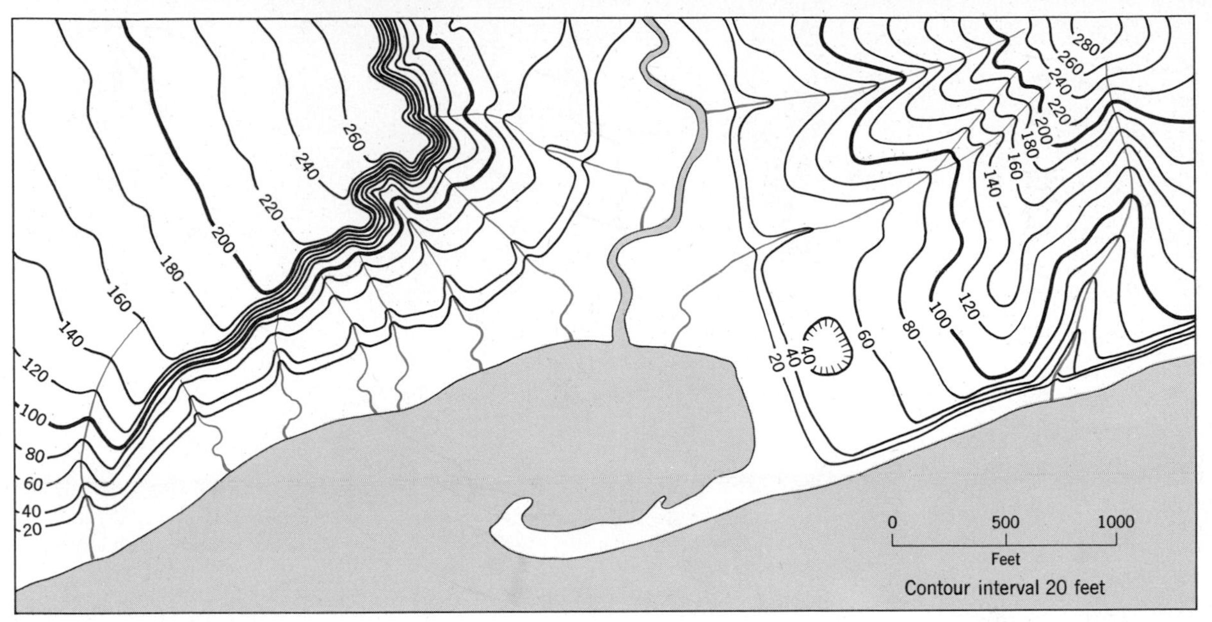


**Figure D.4**
**Contour map of the area shown in Fig. D.3. Note that this map is scaled in feet. (*Modified from U.S. Geol. Survey.*)**

with the vertical dimension exaggerated. The camera lens acts as a series of "eyes" with an interpupillary separation equal to the distance flown between successive pictures. This separation is so much greater than the distance between the two eyes of a human being that when overlapping photographs are viewed through a stereoscope, the relief of the ground surface becomes startlingly apparent. In fact we can see much more than we could by looking down from above at the same field of view. Not only is this a direct aid in plotting geologic data on the photograph, but also it forms the basis of a method of drawing an accurate contour map of the area photographed without the necessity of making a slow, laborious, and costly ground survey. Topographic contour maps are made largely by methods based on this principle. Finally, an individual air photograph can be used as a map, on which geologic features are then plotted.

### Measurements made in the field

The most common kind of ***geologic map*** is *a map that shows the distribution, at the surface, of rocks of various kinds or of various ages.* Examples are Figure 7.13, a sketch map of bodies of alluvium of successive ages, Figure 14.21*B*, a sketch of eroded and folded strata, and Figure 15.13, sketch maps showing the distribution of basalt compared with that of other rocks.

**Field Equipment.** A first essential for making a geologic map in the field is a base map, preferably a topographic map with a contour interval no greater than 3 to 5 meters. Other probable necessities are a hammer, a steel tape for measuring thicknesses of strata, and a pocket transit (geologists' compass) known almost universally in the United States and Canada as a Brunton compass, or simply a brunton, after the name of its designer. This compact instrument (Fig. D.5) is a compass, a clinometer, and a sighting device used in reading compass directions and in hand leveling.

**Strike and Dip.** One of the commonest kinds of field measurement is

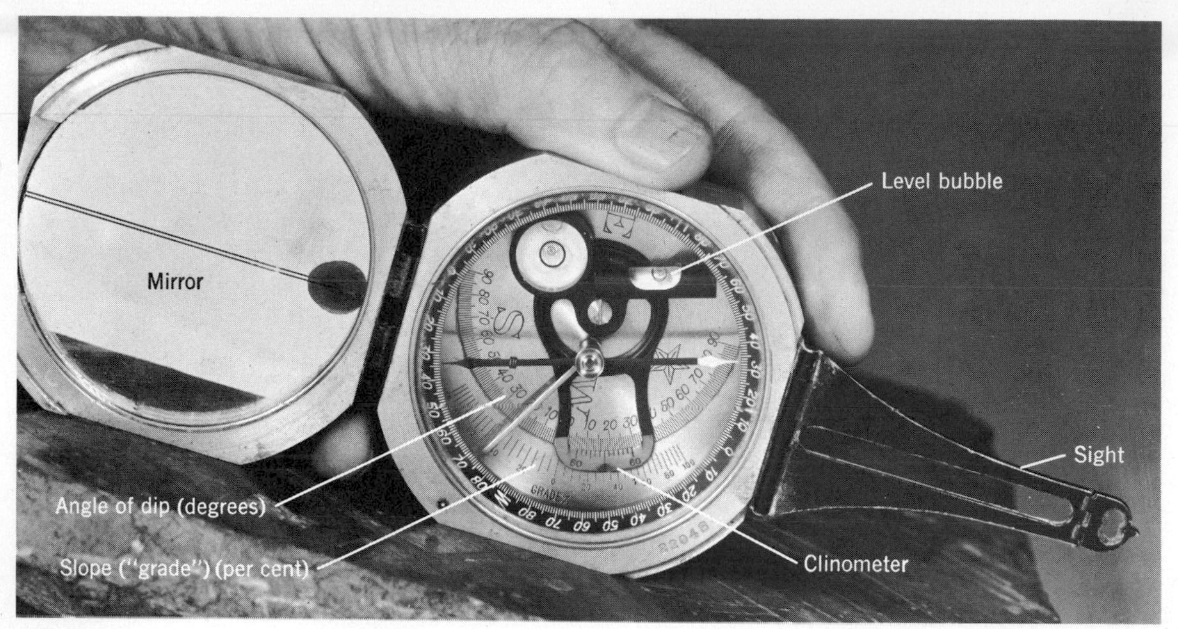


Figure D.5
**Use a brunton (pocket transit) to measure angle of dip of a stratum. The sides of the brunton are plane, parallel surfaces. With one side placed on a sloping plane surface in the direction of dip, the level bubble of the clinometer is centered by means of a lever on the back of the case (not visible). The angle is read on one of the two arcs below. The inner semicircular arc, calibrated in degrees from 0° to 90°, shows a reading of 17.5°; the outer, shorter scale, calibrated in percent, reads 31 percent. Compass directions are read on the dark outer scale when the instrument is held face up and is leveled by use of the circular level bubble. The hinged mirror and sight aid in taking bearings on selected points, and in using the brunton as a hand level (Fig. D.8), with the clinometer set to read zero degrees.**

determination of the attitude of a stratum. To represent the orientation of an inclined plane, we need to remember two principles of geometry: (1) the intersection of two planes defines a line; (2) in an inclined plane only one horizontal direction exists. The horizontal line formed by the intersection of the inclined plane with the horizontal plane can be visualized as the water line on a boat-launching ramp or by placing the edge of a carpenter's level in a horizontal position on a sloping plane (Fig. D.6). Instead of the level we can place one edge of a brunton, in the level position, against the inclined surface of a stratum.

*The compass direction of the horizontal line in an inclined plane* is the ***strike*** of the plane. In the United States and Canada, strike and other compass directions commonly are expressed as angles between 0° and 90° east or west of true north. A strike trending 20° east of north would be written N20E (Fig. D.7*B*); one trending 72° west of north would be written N72W. In Europe and elsewhere strikes are measured as angles clockwise from true north (0°), through a full circle of 360°.

Once we know the strike we need only one more measurement to fix the orientation of the plane. That is the ***dip,*** *the angle in degrees between a horizontal plane and the inclined plane, measured down from horizontal in a plane perpendicular to the strike.* Dip is measured with a *clinometer,* usually a brunton in the position seen in Figure D.5. Not only the angle but the

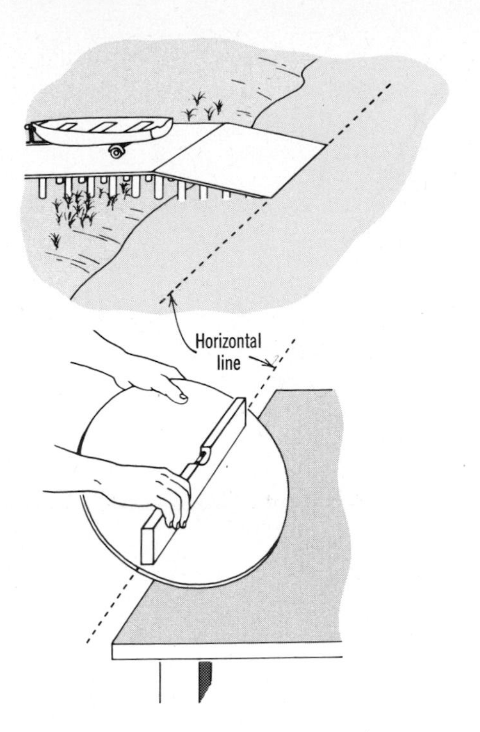


**Figure D.6**
**The one horizontal direction in an inclined plane is illustrated by the water line against a boat-launching ramp, and by a carpenter's level held against an inclined board. The two horizontal lines shown are the directions of strike of the two inclined planes (see text).**

direction of dip (always at right angles to the strike) must be noted.

When plotted on a map as symbols (Fig. D.7*A*), orientation measurements made at each locality graphically convey the significant structural features of an area (Fig. D.7*C*). In a similar manner dip directions of cross strata can be plotted on a map to indicate directions of flow of ancient currents (Fig. 12.17).

**Figure D.7**
**Plotting strike and dip.**

***A*. Strike and dip are plotted on maps with the T-shaped symbol shown in blue.**

***B*. Strike and direction of dip shown in *A* are indicated by blue lines on the face of a compass. Strike is N20E, direction of dip S70E. Angle of dip, being a vertical angle, can not be shown.**

***C*. Strike and dip symbols, from measurements at several localities on the same stratum, plotted on a geologic map. Evidently the structure is a syncline plunging southeast, the northeast limb dipping more steeply than the southwest (Fig. 14.20).**

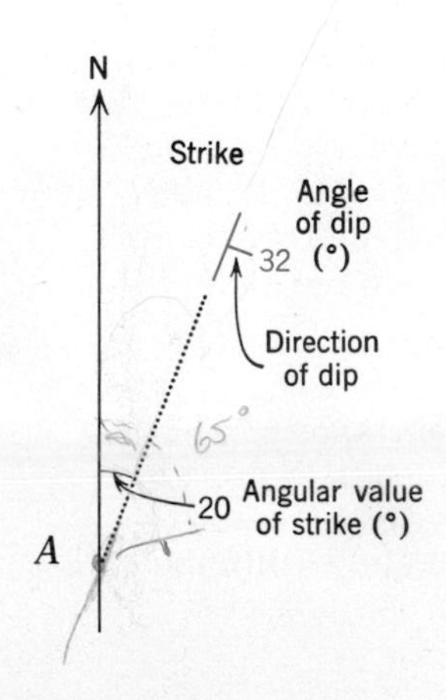


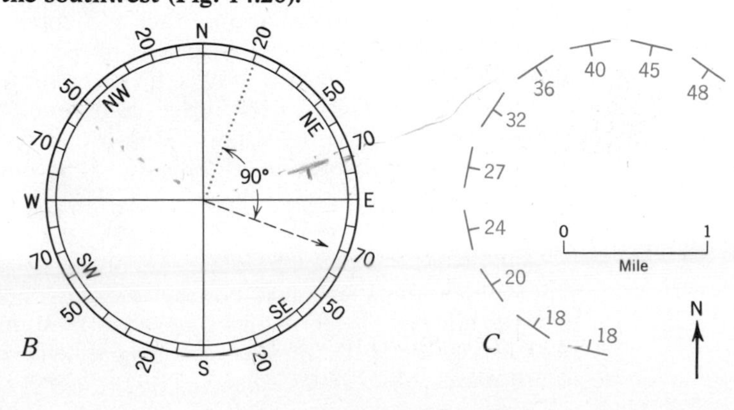

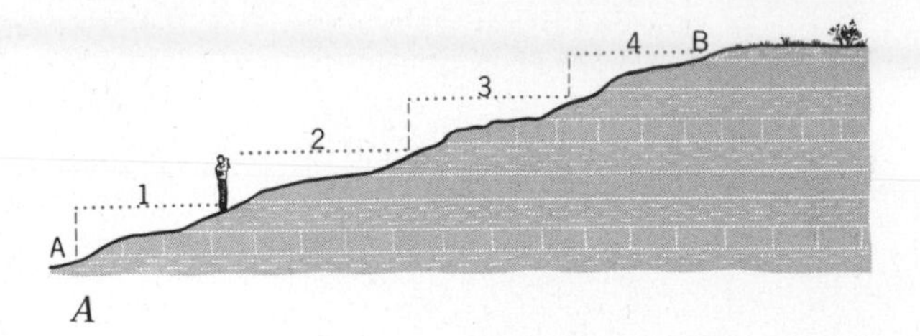


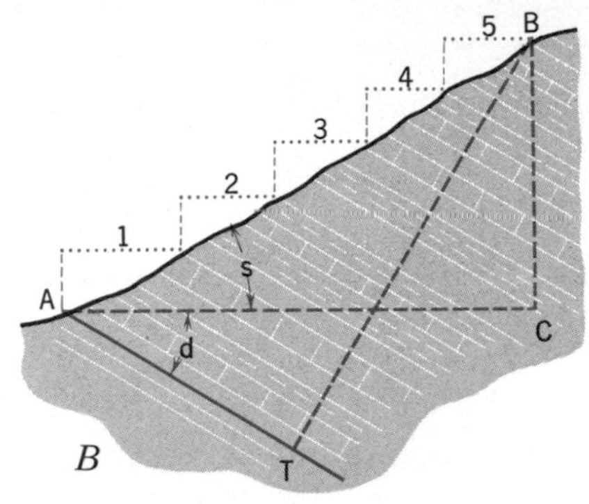


Problem: to determine thickness, BT, of strata exposed on a hillslope between A and B:

1 to 5 = Sights with hand level
BC = Height of B above A
AC = Horizontal line
s = Angle of slope
d = Angle of dip

Then $AB = \frac{BC}{\sin s}$

$BT = AB \sin (s + d)$

**Figure D.8**
**Use of hand level in measuring heights, and thicknesses of strata exposed in slopes.**
***A*. Thicknesses of horizontal strata are measured directly by hand level. The geologist determines accurately the height of his eyes above the ground, and multiplies this figure by the number of "sights" (1, 2, 3, etc.) he makes. If, instead of a hand level, he uses an altimeter, he computes differences between readings at critical points.**
***B*. Where strata are inclined, vertical distance between base and top is measured, up the slope, by hand level. Angle of dip is measured with a clinometer. Thickness is determined by the trigonometric computation given above.**

*Thicknesses of Strata.* The perpendicular distance between the upper and lower surfaces of a stratum, the *thickness* of the stratum, can be measured directly with a steel tape or other scale or with a hand level (Fig. D.8*A*) or altimeter. Thickness of an inclined stratum is usually calculated from simple trigonometric data (Fig. D.8*B*).

### Geologic maps

**Patterns Made by Strata.** *Horizontal Strata.* Figure D.9*A* is a geologic sketch map showing three units: two shale units with a unit of limestone between them. Each is a *formation* as defined in Chapter 12. The sketch map has been constructed on a topographic base representing two rounded hills with a stream between them. The two black lines represent the traces, along the land surface, of the interface of contact between the base of the limestone and underlying shale and of that between the top of the limestone and overlying shale. Symbols (Fig. D.10*A*) at five places indicate that the strata exposed at those places were found to be horizontal; that is, their dip is zero.

The geologist who drew the black lines, or "contacts" determined the altitude of the lower one by hand level from a known point farther down the slope and then walked around each hill, following by eye, and then plotting on his map, the change in type of rock from shale below to limestone above. His circuit of each hill brought him back to his starting point, and the contact he had plotted maintained a constant altitude around both hills. The line representing it therefore extends between the same two contours around both hills and is parallel with the contours.

The geologist then repeated the process with the higher contact, thereby completing the map. By hand leveling (Fig. D.8*A*) he could measure the vertical distance between base and top of the limestone and thereby find that the stratum is 8m thick. For the thicknesses of each of the two layers of shale, however, he could measure only minimum values, because neither the base of one nor the top of the other is exposed within the area of the map.

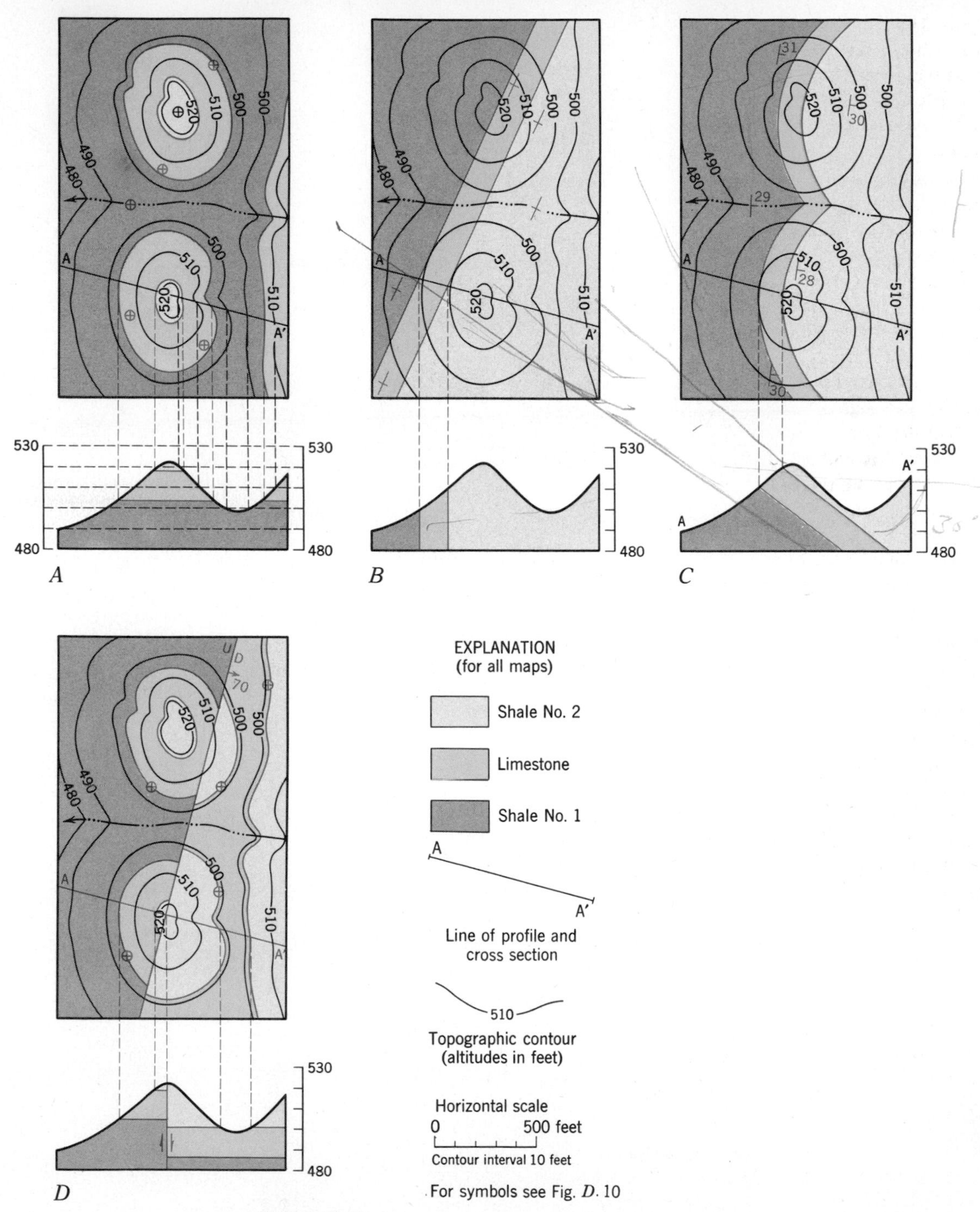


Figure D.9

**Simple geologic sketch maps, each with topographic profile and geologic cross section below it, showing relation between pattern on map and geometry of section in four situations. Scale is in feet.**

*A.* **Strata horizontal.**

*B.* **Strata vertical.**

*C.* **Strata dipping east at 30° angle.**

*D.* **Horizontal strata cut by a fault.**

**Figure D.10**
*A.* **Symbols commonly used to show structure on geologic maps.**
*B.* **Representative patterns commonly, but not universally, used to show kinds of rock in geologic cross sections.**

| Symbol | Explanation |
|---|---|
| 18 | Strike and dip of strata |
| 90 | Strike of vertical strata; tops of strata are on side marked with angle of dip |
| | Structure of horizontal strata; no strike, dip = 0 |
| 43 | Strike and dip of foliation in metamorphic rocks |
| | Strike of vertical foliation |
| | Anticline; arrows show directions of dip away from axis |
| | Syncline; arrows show directions of dip toward axis |
| 21 | Anticline, showing direction and angle of plunge |
| 15 | Syncline, showing direction and angle of plunge |
| | Normal fault; hachures on downthrown side |
| | Reverse fault; arrow shows direction of dip, hachures on downthrown side |
| 50 U D | Dip of fault surface; D, downthrown side; U, upthrown side |
| | Directions of relative horizontal movement along a fault |
| | Low-angle thrust fault; barbs on upper block |

*A*

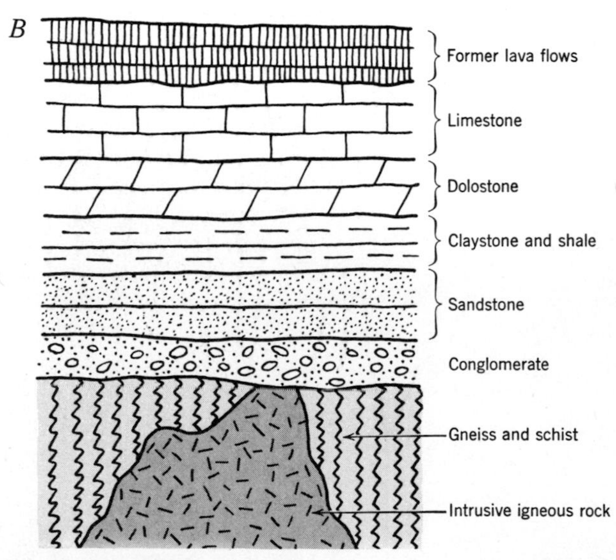

*The area, on a geologic map, shown as occupied by a particular rock unit* is the ***outcrop area*** of that unit. An outcrop area, therefore, is the area of the Earth's surface in which some particular rock unit constitutes the highest part of the underlying rock, whether exposed at the surface or covered by regolith derived from it. Generally the aggregate area of actual exposure, free from overlying regolith, is far less than the outcrop area.

*Vertical Strata.* The lines representing contacts between strata whose dip is vertical appear as straight lines on a geologic map (Fig. D.9*B*). They change direction only where the strike of an interface of contact changes, or where the contact is offset by a fault.

*Inclined Strata.* On a geologic map, the lines of contact between strata that are inclined cross the contours in directions that vary with angle of dip of the strata and with slope of the surface as represented by trend and spacing of contours. In Figure D.9*C* we note that the lines representing the top and base of the limestone layer swing west through broad arcs in each of the two hills and bend east in a chevronlike pattern as they cross the valley. Measurements of strike and dip at five localities show that all three strata are dipping east. If the dip were west instead of east, the pattern would be reversed, with arcs swinging east and the chevron pointing west.

**Faults.** On a geologic map, faults (which are, of course, one kind of surface of contact) are represented by lines thicker than those used for all other kinds of contact. Where faults displace rock units they also displace other contacts between the units; these appear on a map as offsets (Fig. D.9*D*). The details of faults and other structures are brought out on geologic maps by means of special symbols (Fig. D.10*A*).

### Geologic cross

Once a geologic map has been constructed, it can be used as the basis for constructing cross sections along planes that extend downward below the surface, at right angles to the plane of the map. A ***geologic cross section*** is *a diagram showing the arrangement of rocks in a vertical plane.* It represents what would be revealed, as outcrops, in the vertical wall of a deep trench. In a natural trench such as the Grand Canyon we could construct a geologic cross section by measuring the exposed rocks directly, with an accuracy as great as would be possible in making a geologic map of a comparable area. Most sections, however, are made by geometric projection of the contacts already drawn on a map. The accuracy of a section constructed in this way varies with the kinds of rock bodies present. Where strata of originally wide extent have been deformed, surfaces of contact are generally parallel and can be projected downward with fair accuracy. But where rock bodies having irregular surfaces of contact are present and where exposures of the rocks are few, sections are less accurate and some are little more than speculative. Sections of this kind can be checked and corrected only by comparing them with subsurface data obtained from the records of drilling or geophysical exploration.

The method of constructing geologic cross sections is illustrated in Figure D.9. A distinctive unit of limestone, shown in four geologic sketch maps, is horizontal in *A*, vertical in *B*, dipping at 30° in *C*, and cut by a fault in *D*. What would be the geometry of this unit as seen in a vertical plane?

Below the map we construct a grid of horizontal lines, spaced to represent the contours. Dashed vertical lines are drawn from the intersection of line

AA′ with each contour on the map to corresponding horizontal lines on the grid. By connecting the points of intersection, we construct an east–west profile of the hill, a topographic profile. Next we draw lines from borders of the outcropping unit to intersect the profile on the grid. By connecting points on the profile that represent, respectively, the top and bottom of the unit, the cross section of the layer is completed.

Because the topography on all four base maps is the same, construction lines for making the topographic profile are shown for *A* only. On the identical profiles for *B*, *C*, and *D*, the lines dropped from lower and upper boundaries of the outcropping limestone unit give the points required for completing the sections.

Patterns commonly used in geologic cross sections to represent rocks of various kinds are illustrated in Figure D.10*B*.

### Maps of igneous and metamorphic rocks

Because sedimentary rocks are commonly stratified, with rather distinct upper and lower surfaces of contact, and because many of them are widely extensive, they present fewer mapping problems than do most igneous and metamorphic rocks. An exception is volcanic rocks that occur in widespread sheets like sedimentary strata and that may be interbedded with layers of volcanic ash. In some districts it is possible to map these units separately just as sedimentary formations are mapped. But in many volcanic fields, bodies of extrusive igneous rock and bodies of volcanic ash, erupted from a number of centers at irregular intervals, are mixed in great confusion. Such complexes can be mapped only as general assemblages, without regard to the various kinds of volcanic rock that compose them.

Intrusive igneous bodies exposed at the surface are highly varied in form and in their relation to associated rock. Commonly they cut across older sedimentary units and their outcrop areas range widely in shape and in size. Those large enough to be shown clearly to the scale of the map are outlined and marked with distinctive colors or patterns.

Most metamorphic rocks have complex structure and are not easily divisible into distinctive units. Many such rocks therefore are represented on maps by a single color or pattern. In many mountain zones, metamorphic rocks intermixed with bodies of granite or other intrusive rocks are treated as a unit and are identified on many maps as "basement complex."

In some areas, on the other hand, metamorphic rocks are mapped in great detail. Various kinds of schists and gneisses are mapped individually, as formations; attitudes of foliation are recorded at many points by symbols (Fig. D.10*A*), and faults are mapped, many of them through long distances. In such areas the various intensities of metamorphism that have affected the rocks are identified through critical minerals. With such identification at a large number of points, it is possible to draw a series of isograds. In this case an ***isograd*** is *a line on a map, connecting points of first occurrence of a given mineral in metamorphic rocks.* The trends and patterns of isograds give clues to the directions in which the former pressures were applied, and also suggest former temperature gradients.

### Isopach maps

Somewhat analogous to an isograd is an ***isopach,*** *a line on a map, that connects points of equal thickness of a rock unit.* Isopachs are usually shown on special maps of limited areas where thickness is particularly significant. For example, an isopach map might show the thickness of a surface cover of loess (Fig. 9.23), volcanic ash (Fig. 15.3), alluvium, or glacial drift. Again it might represent the thinning, in some direction, of a buried unit similar to the one appearing in cross section in Fig. 21.9*D*. An isopach map of that unit probably would be useful in a local search for petroleum.

**Field study of a sequence of strata**

Here is an example of the way a sequence of strata is recognized in the field, developed, and recorded (Fig. D.11). We examine the rocks systematically by climbing up the slope, taking notes as we go. We identify the kinds of rock present and determine the thickness of each of the units, of which the figure shows five, lettered *A* to *E*. The surfaces of contact between any two units are not alike. The contacts between *D* and *E* and between *C* and *D* are sharp and distinct, whereas those between *A* and *B* and between *B* and *C* are transitional; that is, they represent a gradation between the layer beneath and the layer above. Looking for fossils to help determine the origins of the rocks, we find fossils of marine animals in units *A*, *B*, *C*, and *D* and fossil bones of land animals in unit *E*.

Examining sandstone *A*, we find it is made up of grains of quartz and other minerals. All the minerals in this group occur in igneous rock; so we infer that the sandstone was built of the sediment resulting from erosion of igneous rock. The sediment might have been derived from its parent rock either directly or remotely, for it could, for a long time, have formed part of one or more sandstones older than the one we are examining. In either case the sandstone shows sorting, for the proportion of quartz to other minerals is far greater than in most igneous rocks. Probably this

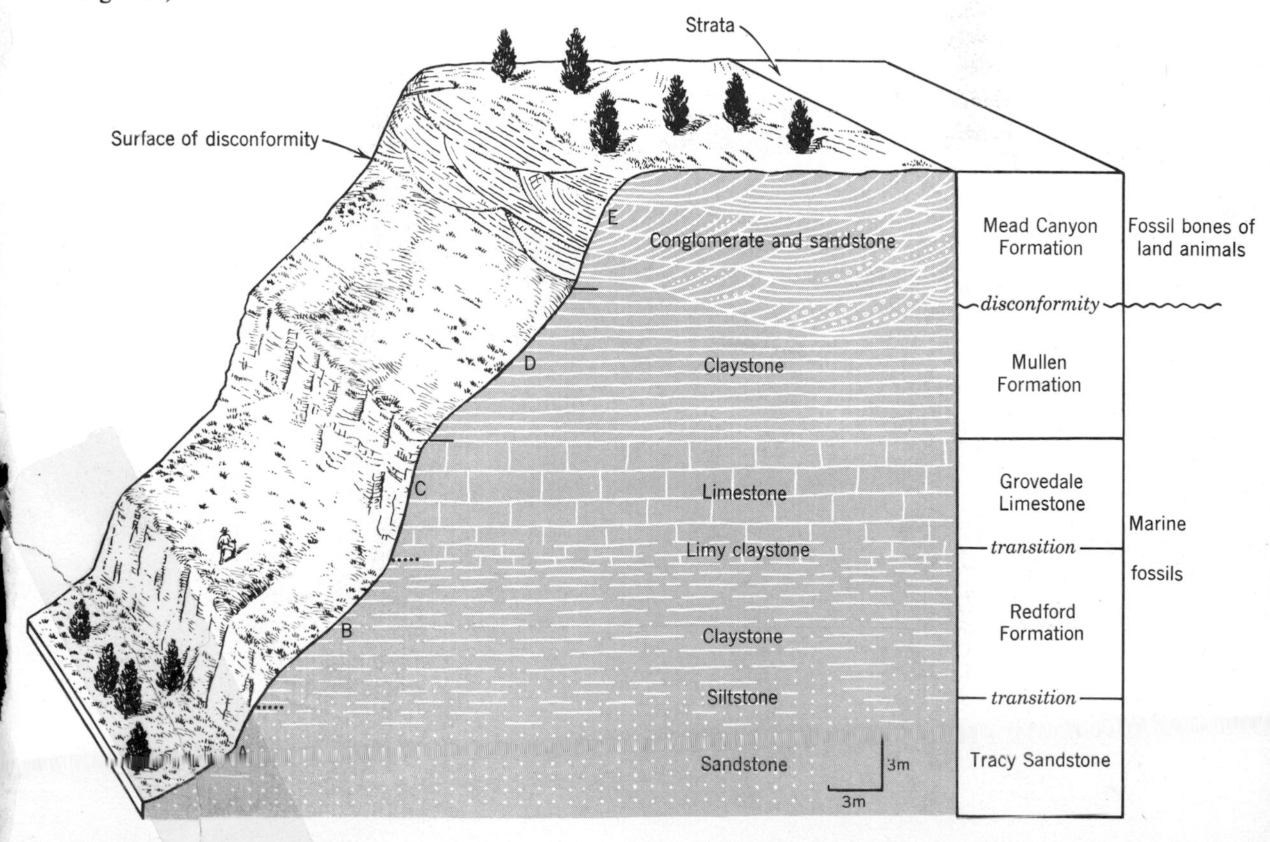


**Figure D.11**
**Idealized sequence of strata exposed on a steep slope (compare Fig. D.8).**

means that the durable quartz grains survived their trip from the region of erosion to the region of deposition in much greater numbers than did their less resistant associates. However, under a hand lens we can see that even the quartz grains have become moderately well rounded. From this we judge roughly that the journey involved considerable distances, long periods of time, or both.

Stratification of the sandstone is indistinct. The rock lacks well-marked layers because the diameters of all the sand grains are nearly the same; hence there is little distinction between one layer and the next. Yet the faint layers that are present lie flat and nearly parallel. From this we infer the sand was deposited in water deep enough to be beyond the reach of waves and strong currents. We can suppose the sand was either derived from the wave erosion of cliffs of an older sandstone or brought into the sea by rivers that carried some of the products of weathering of inland rocks. In the latter case the streams must have possessed enough energy to enable them to move sand grains; this in turn suggests that the slope was fairly steep and therefore the tributary region was one of hills or mountains. These are only suggestions, which can hardly be verified from a study of this one exposure. Verification could come only from a wide study of the sandstone and of the tributary region.

We noted that upward the sand becomes finer and is accompanied by silt. The silt then increases and is accompanied by clay, so that we pass upward from sandstone through siltstone into claystone (unit *B*). In the claystone we find the same near-parallelism of the laminae and much the same kinds of marine fossils. Hence the only properties that have changed, as compared with the sandstone beneath, are sizes and shapes of grains and also kinds of minerals present, because the fine particles that constitute claystone consist not of quartz but of clay minerals. We can explain the change in one of two ways. Either the water deepened, shifting the zone of sand deposition farther toward shore and causing deposition of mud here or the contributing streams lost energy and deposited their sand before reaching the sea. To determine the cause we would have to study a much wider region.

The upper part of the claystone becomes limy; that is, it begins to include calcite along with the clay minerals. Also it grades upward into nearly pure limestone (unit *C*) containing the same general kinds of fossils as those beneath. On close inspection the limestone is seen to be a mechanical deposit of tiny shell fragments. We can conclude with confidence that, in part of the sea, the water cleared and marine organisms became abundant. This could have been brought about by continued deepening.

Examining unit *D*, we find no transtion from *C*, but instead an abrupt change from limestone to claystone. Here we find the same kinds of fossils and the same flat-lying, parallel layers; only the composition of the rock has changed. We infer sudden shoaling of the water or sudden increase in stream energy rather than a gradual change.

A summary of our findings on units *A*, *B*, *C*, and *D* suggests a former sea, deepening gradually and perhaps somewhat irregularly or fed with sediment by rivers with changing characteristics. Hence we are left with uncertainties. However, these are not unexpected, for the nature of the evidence we must rely on often permits us to do no more than judge the relative probability of two possible explanations of the same feature.

However, from our observation that units $A$, $B$, $C$, and $D$ are parallel and show no sign of intervening breaks, we infer with little doubt that together they represent continuous, unbroken sedimentation on the sea floor. Such strata are said to be *conformable.*

Unit $E$ differs from underlying units. It is part sandstone and part conglomerate and is cross-stratified in wedge-shaped units. The repetition of wedge-shaped beds suggests currents that shifted in position and direction; the occurrence of both sand and gravel indicates changes in energy as well; and finally the common direction of inclination of the cross-strata suggests general flow from right to left.

Also unit $E$ contains the fossil bones of land animals. This fact makes a marine origin unlikely; the presence of conglomerate rules out the wind. Accordingly we conclude that probably $E$ was deposited by a stream or streams, and the dimensions of the wedge-shaped beds give a clue to the size of the stream channels. Streams today are seen to deposit similar beds.

Evidently $E$ overlies $D$ with a contact that is notably irregular and that cuts through some of the uppermost layers in $D$. This relationship could result only from erosion of $D$ before or during the deposition of $E$. When two parallel strata are separated by an irregular surface of erosion corresponding to a time gap in sedimentary deposition, the relationship between them is *disconformable.* The irregular interface is a *disconformity.* The implications of this relationship are set forth in Chapter 14 under the heading *Unconformity.*

From these facts we judge that the sea floor emerged for some reason not shown by the evidence at this place. Unit $D$ thereby became land, and a stream flowed over it, eroding an unknown amount of its upper part. To determine how much of $D$ was removed during this erosion and whether to attribute all of the erosion to the stream or part of it to some agency that antedated the stream, we should have to examine $D$ at many other places. At any rate, the stream possessed rather high energy because it carried a bed load of pebbles as well as sand and because the load was so abundant that some was deposited. Evidently the stream was gradually building up its bed because the wedge-shaped layers are piled one on another.

Like most sand grains, those of $E$ are mainly quartz, but the pebbles are samples of various kinds of rock. Here is a clue to the region from which the stream brought the pebbles. To make use of the clue we should have to identify the kinds of rock in the pebbles and to search (probably in the direction to the right of Fig. D.11) for areas in which those same kinds of rock occur as bedrock. A further clue to the distance the sediment has been transported is found in the degree of rounding of the particles.

This is the way in which sedimentary rocks are commonly studied in the field. From such a study it is possible to re-create a general sequence of events, but not all our inferences are firm; we cannot always rule out other possibilities; so the study leads to questions that can be answered, if at all, only by extending the inquiry into the surrounding region. If the sequence is confirmed by sequences seen in additional exposures, we are probably justified in naming the stratigraphic units according to accepted rules (Chapter 12). The names chosen are included in Figure D.11. They replace the temporary letters we used in the field study.

# SELECTED REFERENCES


## BASE MAPS

Raisz, Erwin, 1948, General cartography, 2nd ed.: New York, McGraw-Hill.

Robinson, A. H., 1960, Elements of cartography, 2nd ed.: New York, John Wiley.

## AIR PHOTOGRAPHS

American Society of Photogrammetry, 1952, Manual of photogrammetry, 2nd ed.: Washington, American Society of Photogrammetry.

American Society of Photogrammetry, 1960, Manual of photographic interpretation: Washington, American Society of Photogrammetry.

Lobeck, A. K., and Tellington, W. J., 1944, Military maps and air photographs: New York, McGraw-Hill, p. 199–250.

Miller, V. C., and Miller, C. F., 1961, Photogeology: New York, McGraw-Hill.

## GEOLOGIC MAPS AND CROSS SECTIONS

Blyth, F. G. H., 1965, Geological maps and their interpretation: London, Edward Arnold.

## FIELD TECHNIQUES

Compton, R. R., 1961, Manual of field geology: New York, John Wiley.

Lahee, F. H., 1961, Field geology: 6th ed., New York, McGraw-Hill.

# APPENDIX E UNITS AND THEIR CONVERSIONS

## Representation of numbers

### Multiples, Submultiples, and Prefixes

| | |
|---|---|
| $1{,}000{,}000 = 10^6$ | mega |
| $1{,}000 = 10^3$ | kilo |
| $100 = 10^2$ | hecto |
| $10 = 10$ | deka |
| $0.1 = 10^{-1}$ | deci |
| $0.01 = 10^{-2}$ | centi |
| $0.001 = 10^{-3}$ | milli |
| $0.000001 = 10^{-6}$ | micro |

## Units of measure

### Linear Measure

| | |
|---|---|
| 1 mile (mi.) | = 5,280 feet (ft) |
| 1 foot (ft) | = 12 inches (in.) |
| 1 kilometer (km) | = 1000 meters (m) = $10^3$m |
| 1 meter (m) | = 100 centimeters (cm) = $10^2$cm |
| 1 centimeter (cm) | = 0.01m = $10^{-2}$m |
| 1 millimeter (mm) | = 0.001m = $10^{-3}$m |
| 1 angstrom (Å) | = 0.0000000001m = $10^{-10}$m |
| 1 micron ($\mu$) | = 0.001mm |

**Area Measure**

| | |
|---|---|
| 1 square mile | = 640 acres |
| 1 acre | = 43,650 square feet |
| 1 acre | = 4,840 square yards |
| 1 square meter | = 10,000 square centimeters ($cm^2$) |
| 1 square kilometer | = 100 hectares |

**Volume and Cubic Measure**

| | |
|---|---|
| 1 cubic foot | = 1,728 cubic inches |
| 1 cubic yard | = 27 cubic feet |
| 1 barrel (oil) | = 42 gallons |
| 1 liter | = 0.001 cubic meter |
| | = 1 cubic decimeter |
| 1 liter | = 1,000 milliliters |
| 10 milliliters | = 1 centiliter |
| 1 milliliter | = approximately 1 cubic centimeter (cc) |
| 1 cubic meter ($m^3$) | = 1,000,000 cubic centimeters |

**Weights and Masses**

| | |
|---|---|
| 1 short ton | = 2,000 pounds |
| 1 long ton | = 2,240 pounds |
| 1 pound (avoirdupois) | = 7,000 grains |
| 1 ounce (avoirdupois) | = 437.5 grains |
| 1 gram | = 15.432 grains |
| 1,000 grams | = 1 kilogram |
| 1,000 kilograms | = 1 metric ton |

## Units of energy and power

**Energy**

1 erg = the work done by a force of 1 dyne when its point of application moves through a distance of 1 centimeter in the direction of the force.

1 calorie (cal) = the amount of heat that will raise the temperature of 1 gram of water 1 degree Celsius with the water at 4 degrees Celsius.

1 erg = $9.48 \times 10^{-11}$ British thermal unit (BTU)
1 erg = $7.367 \times 10^{-8}$ foot-pounds
1 erg = $2.778 \times 10^{-14}$ kilowatt-hours
1 kilowatt-hour = 3,413 BTU = $3.6 \times 10^{13}$ ergs
1 BTU = $2.928 \times 10^{-4}$ kilowatt-hours
= $1.0548 \times 10^{10}$ ergs
1 joule = $1 \times 10^{7}$ ergs
1 calorie = $3.9685 \times 10^{-3}$ BTU
= $4.186 \times 10^{7}$ ergs

**Power (= energy per unit time)**

1 watt = 3.4129 BTU/hour
1 watt = $1.341 \times 10^{-3}$ horsepower
1 watt = 1 joule per second
1 watt = 14.34 calories per minute

## Conversions

### English–Metric Conversions

| | |
|---|---|
| 1 inch | = 25.4 millimeters |
| 1 foot | = 0.3048 meter |
| 1 yard | = 0.9144 meter |
| 1 mile | = 1.609 kilometers |
| 1 square inch | = 6.4516 square centimeters |
| 1 square foot | = 0.0929 square meter |
| 1 square yard | = 0.836 square meter |
| 1 acre | = 0.4047 hectare |
| 1 square mile | = 2.590 square kilometers |
| 1 cubic inch | = 16.39 cubic centimeters |
| 1 cubic foot | = 0.0283 cubic meter |
| 1 cubic yard | = 0.7646 cubic meter |
| 1 acre-foot | = 1,233.46 cubic meters |
| 1 cubic mile | = 4.168 cubic kilometers |
| 1 gallon | = 3.784 liters |
| 1 ounce | = 28.33 grams |
| 1 pound | = 0.4536 kilograms |

### Metric–English Conversions

| | |
|---|---|
| 1 millimeter | = 0.0394 inch |
| 1 meter | = 3.281 feet |
| 1 meter | = 1.094 yards |
| 1 kilometer | = 0.6214 mile |
| 1 sq centimeter | = 0.155 sq inch |
| 1 sq meter | = 10.764 sq feet |
| 1 sq meter | = 1.196 sq yards |
| 1 hectare | = 2.471 acres |
| 1 sq kilometer | = 0.386 sq mile |
| 1 cu centimeter | = 0.061 cu inch |
| 1 cu meter | = 35.3 cu feet |
| 1 cu meter | = 1.308 cu yards |
| 1 liter | = 1.057 quarts |
| 1 cu meter | = 264.2 gallons (U.S.) |
| 1 cu kilometer | = 0.240 cu miles |
| 1 gram | = 0.0353 ounce |
| 1 kilogram | = 2.205 pounds |

### Temperature

To change from Fahrenheit (F) to Celsius (C)

$$°C = \frac{(°F - 32°)}{1.8}$$

To change from Celsius (C) to Fahrenheit (F)

$$°F = (°C \times 1.8) + 32°$$

# GLOSSARY

Some definitions are not included in the glossary: *mineral names* can be found in Tables B.1, B.2, and B.3; *rock names* are defined in Tables C.1, C.2, and C.3, and Figure 3.16; *units of measurement* are defined in Appendix E.

***Abrasion*** (caused by external processes). The mechanical wear of rock on rock.

***Abyssal plain.*** A large flat area of deep-sea floor having slopes less than about 1 m per km.

***Accretion.*** The process by which solid particles gather together to form a planet.

***Alluvial fill.*** A body of alluvium, occupying a stream valley, and conspicuously thicker than the depth of the stream.

***Alluvium.*** Sediment deposited by streams in land environments.

***Angle of repose.*** The steepest angle, measured from the horizontal, at which material remains stable.

***Angular unconformity.*** Unconformity marked by angular divergence between older and younger rocks.

***Anion.*** A negatively charged ion.

***Antecedent stream.*** A stream that has maintained its course across an area of the crust that was raised across its path by folding or faulting.

***Anticline.*** An upfold in the form of an arch.

***Aquifer.*** A body of permeable rock or regolith through which ground water moves.

***Arête.*** A jagged, knife-edge ridge created where two groups of cirque glaciers have eaten into the ridge from both sides.

***Artesian well.*** A well in which water rises above the aquifer.

***Asthenosphere.*** That zone of the solid Earth immediately below the lithosphere. The top of the asthenosphere is approximately 100km below the surface, the bottom approximately 350km.

***Atmosphere.*** The gaseous envelope that surrounds the Earth.

***Atom.*** The smallest individual particle that retains all the properties of a given chemical element.

***Atomic number.*** The number of protons in the nucleus of an atom.

***Axial plane.*** An imaginary plane that divides a fold as symmetrically as possible, and that passes through the axis.

*Axis.* The median line between the limbs, along the crest of an anticline or the trough of a syncline..

*Barrier island.* A long island built of sand, lying offshore and parallel to the coast.

*Base level.* The limiting level below which a stream cannot erode the land.

*Base level, local.* The level of a lake or any other base level that stands above sea level.

*Base level, ultimate.* Sea level, projected inland as an imaginary surface beneath all streams.

*Batholith.* A very large intrusive igneous body, of irregular shape, that cuts across the layering of the intruded rocks.

*Bay barrier.* A ridge of sand or gravel that completely blocks the mouth of a bay.

*Beach.* The wave-washed sediment along a coast, extending throughout the surf zone.

*Bed load.* The coarse solid particles, within a body of flowing fluid, moving along or close above the bed.

*Bedrock.* The continuous solid rock of the continental crust.

*Biosphere.* The totality of Earth's organisms.

*Blowout.* A deflation basin excavated in shifting sand.

*Bottomset layer.* The gently sloping, fine, thin part of each layer in a delta.

*Boulder train.* A group of erratics spread out fanwise.

*Caldera.* A roughly circular, steep-walled basin several kilometers or more in diameter.

*Caliche.* A whitish accumulation of calcium carbonate developed in a soil profile. Also known as *calcrete.*

*Calorie.* The amount of heat energy needed to raise the temperature of one gram of water by one degree Celsius.

*Cation.* A positively charged ion.

*Cavern.* A large, roofed-over cavity in any kind of rock.

*Chemical elements.* The most fundamental substances into which matter can be separated by chemical means.

*Chemical sediment.* Sediment formed by precipitation of minerals from solution in waters of the Earth.

*Cirque.* A steep-walled niche, shaped like a half bowl, in a mountainside, excavated mainly by frost wedging and glacial plucking.

*Cirque glacier.* A very small glacier that occupies a cirque.

*Clastic sediment.* See ***Detritus.***

*Cleavage* (in minerals). The tendency of a mineral to break in preferred directions along plane surfaces.

*Climate.* The average weather conditions of a place or area over a period of years.

*Col.* A gap or pass in a mountain crest where the headwalls of two cirques intersect each other.

*Colluvium.* Sediment deposited by any process of mass-wasting or by overland flow.

*Column.* (1) A stalactite connected with a stalagmite. (2) Sometimes used as an abbreviation for *geologic column.*

*Columnar joints.* Joints that split igneous rocks into long prisms or columns.

*Composite volcano.* A volcano that emits both fragmental material and viscous lava and that builds up a steep conical mound. Also called a *stratovolcano.*

*Compound.* A combination of atoms of different elements bonded together.

*Conchoidal fracture.* Breakage resulting in smooth curved surfaces.

*Concretion.* A localized body having distinct boundaries, enclosed in sedimentary rock, and consisting of a substance precipitated from solution, commonly around a nucleus.

*Cone of depression.* A conical depression in the water table immediately surrounding a well.

*Conglomerate.* A clastic sedimentary rock containing numerous rounded pebbles or larger particles.

*Consequent stream.* A stream whose pattern is determined solely by the direction of the slope of the land.

*Contact* (between rock units). A surface along which two rock units meet.

*Contact metamorphism.* Metamorphism in the vicinity of a pluton and resulting from the effects of the magma on its surrounding rocks.

*Continental shelf.* A submerged platform of variable width, that forms a fringe around a continent.

*Continental shield.* Mass of Precambrian rock around which a continent has accumulated.

*Continental slope.* A pronounced slope beyond the seaward margin of a continental shelf.

*Contour.* See ***Contour line.***

*Contour interval.* The vertical distance between two successive contour lines.

*Contour line.* A line passing through points having the same altitude above sea level; often simply called a contour.

*Convection current.* The movement of material, within a closed system, as a result of thermal convection arising from the unequal distribution of heat.

***Convergent margin.*** A line along which two plates move toward each other and where old lithosphere is subducted, that is, forced downward into the mantle. Also called an *edge of consumption* or a ***subduction zone.***

***Coral reef.*** A ridge of limestone built by colonial marine organisms.

***Core*** (Earth's). The innermost zone of the Earth. A spherical mass of metallic iron and nickel, 3488km in diameter, consisting of a molten outer part approximately 2128km thick and a solid inner sphere with a radius of 1360km.

***Correlation.*** Determination of equivalence, in geologic age and position, of the sequences of strata found in two or more different areas.

***Creep.*** The imperceptibly slow downslope movement of regolith.

***Crevasse.*** A deep, gaping crack in the upper surface of a glacier.

***Cross section.*** See ***Geologic cross section.***

***Cross-strata.*** Strata that are inclined with respect to a thicker stratum within which they occur.

***Crust*** (Earth's). The outer part of the lithosphere.

***Crust, continental.*** The Earth's crust beneath continents, 20 to 40km thick, consisting of an upper part having the same elastic properties as sialic rocks and a lower part having the same elastic properties as mafic rocks.

***Crust, oceanic.*** The Earth's crust beneath ocean basins, 10 to 15km thick, and consisting of material having the same elastic properties as mafic rocks.

***Crystal.*** A solid compound composed of ordered, three-dimensional arrays of atoms or ions chemically bonded together.

***Crystal form.*** Geometric solid that is bounded by symmetrically arranged plane surfaces.

***Crystal structure.*** The geometric pattern that atoms assume in a mineral.

***Curie point*** (of a mineral). A temperature above which all magnetism is destroyed.

***Debris flow.*** The rapid downslope plastic flow of a mass of debris.

***Declination.*** See ***Magnetic declination.***

***Decomposition*** (as a process of weathering). The chemical alteration of rock materials.

***Deflation.*** The picking up and removal of loose rock particles by wind.

***Deflation armor.*** A surface layer of coarse particles concentrated chiefly by deflation.

***Delta.*** A body of sediment deposited by a stream where it flows into standing water. The name comes from the similarity of the plan view to the shape of the Greek letter Δ.

***Dendritic pattern.*** A stream pattern characterized by irregular branching in many directions.

***Density.*** The average weight per unit volume.

***Density current.*** A localized current, within a body of water, caused by dense water sinking through less-dense water.

***Detritus.*** The accumulated particles of broken rock and of skeletal remains of dead organisms.

***Dike.*** A sheet of intrusive igneous rock cutting across the layering of pre-existing rock.

***Dip*** (of a magnetic needle). See ***Magnetic inclination.***
(of a stratum). The angle in degrees between a horizontal plane and an inclined plane, measured down from horizontal in a plane perpendicular to the strike.

***Discharge.*** The quantity of water that passes a given point in a given unit of time.

***Disconformity.*** Unconformity that is not marked by angular divergence between two groups of strata in contact.

***Dissolved load.*** Matter dissolved in the water of a stream.

***Divergent margin.*** A line along which two adjacent plates move apart from each other, and along which new lithosphere is made. Also called a ***spreading edge.***

***Divide.*** The line that separates adjacent drainage basins.

***Drainage basin.*** The total area that contributes water to a stream.

***Drift*** (glacial). Sediment deposited directly by glaciers (till, q. v.), or indirectly in glacial streams, lakes, and the sea (stratified drift, q. v.).

***Dripstone.*** Material chemically precipitated from dripping water in an air-filled cavity.

***Drumlin.*** A streamline hill consisting of drift, generally till, and elongated parallel with the direction of glacier movement.

***Dune.*** A mound or ridge of sand deposited by the wind.

***Dynamic equilibrium.*** See ***Steady state.***

***Earthquake focus.*** The point of first release of the energy that causes an earthquake.

***Economy*** (of a stream). The input and consumption of energy within a stream or other system and the changes that result.

***Edge of consumption.*** See ***Convergent margin.***

***Electron.*** A negatively charged particle.

***Elements.*** See ***Chemical elements.***

***End moraine.*** A ridgelike accumulation of drift, deposited by a glacier along its margin.

***Energy.*** The capacity to produce activity.

*Environmental geology.* The application of geology to the problems involved in the interaction between man and the rest of the Earth.

*Epicenter.* The point, on Earth's surface, that lies vertically above the focus of an earthquake.

*Erosion.* The complex group of related processes by which rock is broken down physically and chemically and its products removed.

*Erratic* (glacial). A glacially deposited fragment of rock that differs from the bedrock beneath it. The agent of transport was commonly glacier ice or floating ice.

*Esker.* A long narrow ridge, commonly sinuous, of stratified drift.

*Evaporite.* A nonclastic sedimentary rock whose constituent minerals were precipitated from water solution as a result of evaporation.

*Exfoliation.* The separation, during weathering, of successive shells from rock.

*Exposure.* A place where solid rock is exposed at Earth's surface.

*Extrusive igneous rock.* Rock formed by the cooling of magma poured out onto the Earth's surface.

*Facies.* A distinctive group of characteristics, within a rock unit, that differs as a group from those elsewhere in the same unit.

*Fan.* A fan-shaped body of alluvium built at the base of a steep slope. Known also as an *alluvial fan.*

*Fault.* A fracture along which the opposite sides have been displaced relative to each other. For kinds of faults see Figure 14.11.

*Faunal succession, law of.* The law, discovered by William Smith, that states that fossil faunas and floras succeed one another in a definite, recognizable order.

*Ferromagnesian minerals.* Silicate minerals in which iron and/or magnesium are the principal cations. Examples are the biotites, amphiboles, pyroxenes, and olivines.

*Fiord.* A glaciated trough partly submerged by the sea.

*Fissure eruption.* Extrusion of volcanic materials along an extensive fracture.

*Floodplain.* That part of any stream valley which is inundated during floods.

*Flowstone.* Material chemically precipitated from flowing water in the open air or in an air-filled cavity.

*Fold.* A pronounced bend in layers of rock. For kinds of fold see Figure 14.20.

*Foliation.* A parallel or nearly parallel structure in metamorphic rock, caused by a parallel arrangement of platy minerals.

*Footwall.* The surface of the block of rock below an inclined fault.

*Foreset layer.* The coarse, thick, steeply sloping part of each layer in a delta.

*Formation.* A succession of strata distinctive enough to constitute a basic unit for mapping and description.

*Fossil.* The naturally preserved remains or traces of an animal or a plant.

*Fossil fuel.* A fuel that contains solar energy, locked up securely in chemical compounds by the plants or animals of former ages.

*Frost heaving.* The lifting of regolith by freezing of contained water.

*Frost wedging.* The pushing apart of rock particles by water freezing to ice within rock or regolith.

*Fumaróle.* A volcano that emits only gases.

*Gangue.* The nonvaluable minerals of an ore.

*Geologic column.* A composite diagram combining in a single column the succession of all known strata, fitted together on the basis of their fossils or of other evidence of relative age.

*Geologic cross section.* A diagram showing the arrangement of rocks in a vertical plane.

*Geologic map.* A map that shows the distribution, at the surface, of rocks of various kinds or of various ages.

*Geology.* The science of the Earth.

*Geosyncline.* A great trough that has received thick deposits of sediment during its slow subsidence through long geologic periods.

*Geothermal gradient.* The rate of increase of temperature downward in the Earth.

*Geyser.* A hot spring equipped with a system of plumbing and heating that causes intermittent eruptions of water and steam.

*Glaciation.* The alteration of a land surface by the massive movement over it of glacier ice.

*Glacier.* A body of ice, consisting mainly of recrystallized snow, flowing on a land surface.

*Graben.* A trench-like structure bounded by parallel normal faults.

*Graded layer.* A layer in which the particles grade upward from coarse to finer.

*Gradient* (of a stream). The slope measured along the stream, on the water surface or on the bottom.

*Granular texture.* Interlocking arrangement of visible mineral grains characteristic of granite.

*Gravimeter* (or *gravity meter*). A sensitive device for measuring the force of gravity at any locality.

*Gravity* (of Earth). The inward-acting force with which Earth tends to pull all objects toward its center.

*Groin.* A low wall, built on a beach, that crosses the shoreline at a right angle.

*Ground moraine.* Widespread thin drift with a smooth surface consisting of gently sloping knolls and shallow closed depressions.

*Ground water.* All the water contained in spaces within bedrock and regolith.

*Guyot.* A seamount with a conspicuously flat top well below sea level.

*Half-life* (in radioactive decay). The time required to reduce the number of parent atoms by one-half.

*Hand specimen.* A piece of rock that can be conveniently held in the hand for close study.

*Hanging wall.* The surface of the block of rock above an inclined fault.

*Hardness.* Relative resistance of a mineral to scratching.

*Horn.* A bare, pyramid-shaped peak left standing where glacial action in cirques has eaten into it from three or more sides.

*Horst.* An elevated elongate block bounded by parallel normal faults.

*Humus.* The decomposed residue of plant and animal tissues.

*Hydrosphere.* The Earth's discontinuous water envelope, including the oceans, lakes, streams, ground water, snow, and ice.

*Ice cap.* A small ice sheet.

*Ice-contact stratified drift.* Stratified drift deposited in contact with its supporting ice.

*Ice sheet.* A broad glacier of irregular shape, generally blanketing an extensive land surface.

*Igneous rock.* Rock formed by the cooling and solidification of magma; an interlocking aggregate of silicate minerals. For varieties of igneous rocks, and rock names, see Figure 3.16 and Appendix C.

*Intrusive igneous rock.* Any igneous rock formed by cooling and solidification of magma below Earth's surface.

*Ion.* An atom that has excess positive or negative charges caused by electron transfers.

*Island arc.* Great arcuate belt of andesitic and basaltic volcanic islands.

*Isograd.* A line on a map, connecting points of first occurrence of a given mineral in metamorphic rocks.

*Isopach.* A line on a map, that connects points of equal thickness of a rock unit.

*Isostasy.* The ideal property of flotational balance among segments of the lithosphere.

*Isotopes.* Atoms having the same atomic number but differing numbers of neutrons in the nucleus.

*Joint.* Fracture on which movement has not occurred in a direction parallel to the plane of the fracture.

*Joint set.* A widespread group of parallel joints.

*Joint system.* A combination of two or more intersecting sets of joints.

*Kame.* A body of ice-contact stratified drift in the form of a knoll or hummock.

*Kame terrace.* A terracelike body of ice-contact stratified drift along the side of a valley.

*Karst topography.* An assemblage of topographic forms consisting primarily of closely spaced sinks; strikingly developed in the Karst region of Yugoslavia.

*Kettle* (glacial). A basin in glacial drift, created by melting-out of a mass of underlying ice.

*Kimberlite pipes.* Narrow, pipe-like masses of igneous rock, some of which contain diamonds, that intrude the crust but originate deep in the mantle.

*Laccolith.* A lenticular intrusive igneous body above which the layers of invaded bedrock have been bent upward to form a dome.

*Lagoon.* A bay inshore from a line of barrier islands or from a coral reef.

*Lava.* Magma that reaches Earth's surface through a volcanic vent, and flows out as hot streams or sheets.

*Leaching.* The continued removal, by water, of soluble matter from regolith or bedrock.

*Levee, natural.* A broad, low ridge of fine alluvium built along both sides of a stream channel by water that spreads out of the channel during floods.

*Limbs.* The sides of a fold.

*Lithosphere.* The outer zone of the solid Earth. The lithosphere includes the crust and the upper part of the mantle. The lower boundary of the lithosphere is the low-velocity zone.

*Load* (of a stream). The material the stream carries.

*Loess.* Wind-deposited silt, commonly accompanied by some clay and some fine sand.

*Long profile* (of a stream). A line connecting points on the surface of a stream.

*Longshore current.* A current, within the surf zone, that flows parallel to the shore.

*Luster.* The quality and intensity of light reflected from a mineral.

*M-discontinuity.* See *Mohorovičić discontinuity.*

*Magma.* Molten silicate materials beneath the Earth's surface, including crystals derived from them and gases dissolved in them.

*Magmatic differentiation by crystallization.* The compositional changes that occur in magma by the separation of early-formed minerals from residual liquids.

***Magmatic differentiation by partial melting.*** The process of forming magma with differing compositions by the incomplete melting of rocks.

***Magnetic declination.*** The clockwise angle from true north assumed by a magnetic needle.

***Magnetic field*** (Earth's). The magnetic lines of force surrounding the Earth.

***Magnetic inclination*** (or *dip*). The angle with the horizontal assumed by a magnetic needle.

***Mantle*** (Earth's). A zone of rocky matter, about 2800km thick, surrounding the Earth's core and covered by the thin crust. The mantle occupies about 80 percent of the total volume of the Earth.

***Marine sediment.*** Sediment deposited in the sea.

***Mascon*** (on the Moon). A region with an anomalously high gravity pull.

***Massive rock.*** Rock that is fairly uniform in appearance and lacks any breakage surfaces.

***Mass number.*** The sum of the protons and neutrons in the nucleus of an atom.

***Mass-wasting.*** The movement of regolith downslope by gravity without the aid of a stream, a glacier, or wind.

***Mature soil.*** A soil that has a fully developed profile.

***Meander.*** A looplike bend of a stream channel.

***Mélange.*** A chaotic mixture of rocks.

***Metallogenic province.*** A limited region within which ore deposits occur in unusually large numbers.

***Metamorphic facies.*** Contrasting assemblages of minerals that reach equilibrium during metamorphism within a specific range of physical conditions.

***Metamorphic rock.*** Rock formed within Earth's crust by transformation, in the solid state, of pre-existing igneous or sedimentary rocks as a result of high temperature, high pressure, or both. For varieties of high metamorphic rocks, and rock names, see Appendix C.

***Metamorphism.*** The changes in mineral assemblage and rock texture, or both, that take place in the solid state within the Earth's crust as a result of high temperatures and high pressures.

***Meteorite.*** A small solid body formed during condensation of the planetary nebula.

***Mid-ocean ridge.*** See ***Oceanic ridge.***

***Migmatite.*** A composite rock containing both igneous and metamorphic portions.

***Mineral assemblage.*** The variety and abundance of minerals present in a rock.

***Mineral deposit.*** Any volume of rock that contains an enrichment of one or more minerals.

***Minerals.*** All naturally occurring, crystalline, inorganic materials. For mineral names, compositions, and properties, see Appendix B.

***Moho.*** See ***Mohorovičić discontinuity.***

***Mohorovičić discontinuity.*** The base of the crust. Also known as the ***M-discontinuity,*** and as the ***moho.***

***Molecule.*** The smallest unit that retains all the properties of a compound.

***Monocline.*** A one-limbed flexure, on both sides of which the strata either are horizontal or dip uniformly at low angles.

***Mountain chain.*** An elongate unit consisting of numerous ranges or systems, regardless of similarity in form or equivalence in age.

***Mountain range.*** An elongate series of mountains belonging to a single geologic unit.

***Mountain system.*** A group of ranges similar in general form, structure, and alignment, and presumably owing their origin to the same general causes.

***Mud crack.*** Crack caused by shrinkage of wet mud as its surface becomes dry.

***Mudflow.*** A debris flow in which the consistency of the flowing substance is that of mud.

***Natural levee.*** See ***Levee, natural.***

***Neutron.*** An electrically neutral particle with a mass 1833 times greater than that of the electron.

***Oceanic ridge*** (also called ***mid-ocean ridge*** and ***oceanic rise***). A continuous rocky ridge on the ocean floor, many hundred to a few thousand kilometers wide, with a relief of more than 600m.

***Oceanic rise.*** See ***Oceanic ridge.***

***Oil field.*** A group of oil pools, usually of similar type, or a single pool in an isolated position.

***Oil pool.*** An underground accumulation of oil or gas in a reservoir limited by geologic barriers.

***Ophiolite complex.*** Fragment of ancient oceanic crust.

***Ore.*** An aggregate of minerals from which one or more materials can be extracted profitably.

***Outcrop area.*** The area, on a geologic map, shown as occupied by a particular rock unit.

***Outwash.*** Stratified drift deposited by streams of meltwater as they flow away from a glacier.

***Outwash plain.*** A body of outwash that forms a broad plain beyond the moraine.

***Overland flow.*** The movement of runoff in broad sheets or groups of small interconnecting rills.

***Paleomagnetism.*** Magnetism in ancient rock.

***Parallel strata.*** Strata whose individual layers are parallel.

***Pedalfer.*** A soil in which much clay and iron have been added to the B horizon.

***Pediment.*** A sloping surface, cut across bedrock, adjacent to the base of a highland in an arid climate.

***Pedocal.*** A soil with calcium-rich upper horizons.

***Peneplain.*** "Almost a plain." A land surface worn down to very low relief by streams and mass-wasting.

***Perched water body.*** A water body that occupies a basin in impermeable material, perched in a position higher than the main water table.

***Permafrost.*** Ground that is frozen perennially.

***Permeability.*** Capacity for transmitting fluids.

***Petroleum.*** Gaseous, liquid, and solid substances, occurring naturally and consisting chiefly of chemical compounds of carbon and hydrogen.

***Phase transition.*** Atomic repacking caused by changes in pressure or temperature.

***Phenocrysts.*** The isolated large crystals in porphyry.

***Piedmont glacier.*** A glacier on a lowland at the base of a mountain, fed by one or more valley glaciers.

***Placer.*** A deposit of heavy minerals concentrated mechanically.

***Planetary nebula.*** A flattened rotating disk of gas.

***Planetology.*** A comparative study of Earth with the Moon and with the other planets.

***Plateau basalt.*** Fissure basalt that forms widespread, nearly horizontal layers on continents.

***Playa.*** The dry bed of a playa lake (q. v.); a nearly level area on the floor of an intermontane basin. Occasional floods deposit silt and clay that mantle a playa.

***Playa lake.*** An ephemeral shallow lake in a desert basin.

***Plucking (glacial).*** The lifting out and removal of fragments of bedrock by a glacier.

***Plunge.*** The angle between a fold axis and the horizontal.

***Plunging fold.*** A fold with an inclined axis.

***Pluton.*** An intrusive igneous body, regardless of shape, size, or composition.

***Pluvial lake.*** A lake that existed under a former climate, when rainfall on the surrounding region was greater than today's.

***Pollutant.*** Any waste substance introduced into a natural system in greater amount than can be disposed of by the system.

***Polymerization*** (applied to silicate minerals). The process of linking silica tetrahedra into larger groups.

***Polymorph.*** A compound that occurs in more than one crystal form.

***Porosity.*** The proportion, in percent, of the total volume of a given body of bedrock or of regolith that consists of pore spaces.

***Potential energy.*** Stored energy waiting to be used.

***Principle of Uniformity.*** The external and internal processes we recognize today have been operating unchanged, and at the same set of rates, throughout most of Earth's history.

***Proton.*** A positively charged particle with a mass 1832 times greater than the mass of an electron.

***Radioactivity.*** The decay process by which an unstable atomic nucleus spontaneously disintegrates.

***Radiometric dating.*** Determination of absolute age of a mineral through measurement of the radioactivity of certain elements in the mineral.

***Recharge.*** The addition of water to the saturated zone.

***Rectangular pattern.*** A stream pattern characterized by right-angle bends in the streams.

***Refraction.*** See ***Wave refraction.***

***Regolith.*** The blanket, consisting of loose, noncemented rock particles, that commonly overlies bedrock.

***Rejuvenation.*** The development of youthful topographic features in a land mass further advanced in the cycle of erosion. The common cause is regional uplift.

***Relief.*** The difference in altitude between the high and low parts of a land surface.

***Replacement.*** The process by which a fluid dissolves matter already present and at the same time deposits from solution an equal volume of a different substance.

***Residual concentration.*** The natural concentration of a mineral substance by removal of a different substance with which it was associated.

***Rock.*** Any naturally formed, firm, and coherent aggregate or mass of mineral matter that constitutes part of Earth's crust. For varieties of rocks and rock names see Appendix C.

***Rock avalanche.*** The rapid downslope flow of a mass of dry rock particles.

***Rock cleavage.*** The property by which a rock breaks into plate-like fragments along flat planes.

***Rock cycle.*** The cyclic movement of rock material, in the course of which rock is created, destroyed, and altered through the operation of internal and external Earth processes.

***Rockfall*** (and ***debris fall***). The rapid descent of a mass of rock (debris), vertically from a cliff or by leaps down a slope.

***Rock flour.*** Fine sand and silt produced by crushing and grinding in a glacier.

***Rock glacier.*** A lobate, steep-fronted mass of coarse, angular regolith that extends out from the base of a cliff in a cold climate and that moves downslope with the aid of interstitial water and ice.

***Rockslide*** (and ***debris slide***). The rapid descent of a mass of rock (debris) by sliding down a slope.

***Runoff.*** Water that flows over the lands.

***Saturated zone.*** The subsurface zone in which all openings are filled with water.

***Scale*** (of a map). The proportion between a unit of distance on a map and the unit it represents on the Earth's surface.

***Sea-floor trench.*** A long, narrow, very deep basin in the sea floor.

***Seamount.*** An isolated volcanic hill standing more than 1000m above the sea floor.

***Secondary enrichment.*** Natural enrichment of an ore body by later addition of mineral matter, generally from percolating solutions.

***Sediment.*** Regolith that is being or has been transported by any of Earth's external processes.

***Sedimentary rock.*** Rock formed from sediment by cementation or by other processes acting at ordinary temperatures at or near Earth's surface. For varieties of sedimentary rocks, and rock names, see Appendix C.

***Seismic belt.*** Large tract subject to frequent earthquake shocks.

***Sheet erosion.*** The erosion performed by overland flow (q. v.).

***Shield volcano.*** A volcano that emits fluid lava and builds up a broad, dome-shaped edifice, convex upward, with a surface slope of only a few degrees.

***Sill.*** A sheet of intrusive igneous rock that is parallel to the layering of pre-existing rock.

***Sink.*** A large solution cavity open to the sky.

***Slickensides.*** Striated or highly polished surfaces on hard rocks, abraded by movement along a fault.

***Sliderock.*** Colluvium that constitutes a talus.

***Slip face.*** The straight, lee slope of a dune.

***Slump.*** The downward slipping of a coherent body of rock or regolith along a curved surface of rupture.

***Snowfield.*** A wide cover, bank, or patch of snow lying above the snowline that persists throughout the summer season.

***Snowline.*** The lower limit of perennial snow.

***Soil.*** That part of the regolith which can support rooted plants.

***Soil profile.*** The succession of distinctive horizons in a soil, from the surface down to the unchanged parent material beneath it.

***Solar System.*** The system comprised of the Sun and the group of planetary bodies that revolve around it.

***Solid solution.*** The substitution of one atom for another in a random fashion throughout a crystal structure.

***Solifluction.*** Imperceptibly slow, downslope flow of water-saturated regolith.

***Specific gravity.*** A number stating the ratio of the weight of the substance to the weight of an equal volume of pure water.

***Spit.*** An elongate ridge of sand or gravel that projects from land and ends in open water.

***Spreading edge.*** See ***Divergent margin.***

***Spring.*** A flow of ground water emerging naturally at the ground surface.

***Stalactite.*** An icicle-like form of dripstone and flowstone that hangs from the ceiling of a cavern.

***Stalagmite.*** A blunt, icicle-like form of flowstone that projects upward from the floor of a cavern.

***Steady state.*** A condition in which the rate of arrival of some materials equals the rate of escape of other materials.

***Stock.*** A small body with the same characteristics as a batholith.

***Stratification.*** Layered arrangement of the particles that constitute sediment or sedimentary rock.

***Stratified drift.*** Drift that is sorted and stratified.

***Stratigraphic superposition, principle of.*** The principle which states that in any sequence of strata, not later disturbed, the order in which they were deposited is from bottom to top.

***Stratovolcano.*** See ***Composite volcano.***

***Stratum*** (plural, ***strata***). A distinct layer of sedimentary or igneous rock consisting of material that has been spread out upon the Earth's surface.

***Streak.*** A thin layer of powdered mineral made by rubbing a specimen on a nonglazed porcelain plate.

***Stream.*** A body of water that carries rock particles and dissolved substances, and that flows down a slope along a definite path.

***Stream capture.*** (Known also as ***stream piracy.***) The diversion of a stream by the headward growth of another stream.

***Stream flow.*** The flow of surface water between well-defined banks.

***Stream system, adjusted.*** A stream system in which most of the streams occupy weak-rock positions.

***Stream terrace.*** A bench along the side of a valley, the upper surface of which was formerly the alluvial floor of the valley.

***Striations (glacial).*** Scratches and grooves on bedrock surfaces, caused by grinding of rock against rock during movement of glacier ice.

***Strike.*** The compass direction of the horizontal line in an inclined plane.

***Subduction zone.*** See ***Convergent margin.***

***Subsequent stream.*** A stream whose course has become adjusted so that it occupies belts of weak rock.

***Superposed stream.*** A stream that was let down, or superposed, from overlying strata onto buried bedrock having composition or structure unlike that of the covering strata.

***Surf.*** Wave activity between the line of breakers and the shore.

***Surface ocean currents.*** Broad, slow drifts of surface water.

***Suspended load.*** Fine particles suspended in the stream.

***Syncline.*** A downfold with a troughlike form.

***Talus.*** An apron of rock waste that slopes outward from a cliff from which it is derived.

***Tectonics.*** The study of Earth's broad structural features.

***Texture*** (of a rock). The sizes and shapes of the individual particles in a rock, and the mutual relationship between them.

***Thin section.*** A thin rock slice prepared for microscopic study.

***Thrust fault*** (also called a ***thrust***). Low-angle reverse fault with dip generally less than 45°.

***Till.*** Nonsorted glacial drift.

***Tillite.*** Till converted into solid rock.

***Tombolo.*** A ridge of sand or gravel that connects an island to the mainland or to another island.

***Topographic map.*** A map that delineates surface forms.

***Topography.*** The relief and form of the land.

***Topset layer.*** Stream sediment that overlies the foreset layers in a delta.

***Transform fault.*** A special class of strike-slip fault that links major structural features.

***Transpiration.*** The giving off of moisture through leaves and other parts of plants. Some authors include also perspiration by animals.

***Trellis pattern.*** A rectangular stream pattern in which tributary streams are parallel and very long.

***Turbidite.*** Sediment deposited by a turbidity current.

***Turbidity current.*** A density current whose excess density results from sediment suspended in them.

***Unconformity.*** A lack of continuity of sedimentary deposition between two units of rock in contact, corresponding to a gap in the geologic record.

***Valley glacier.*** A glacier that flows downward through a valley.

***Valley train.*** A body of outwash that partly fills a valley.

***Varve.*** A pair of sedimentary layers deposited during the cycle of the year with its seasons.

***Ventifact.*** A stone, the surface of which has been abraded by wind-blown sediment.

***Vesicle.*** Small opening, in extrusive igneous rock, made by escaping gas originally held in solution under high pressure while the parent magma was underground.

***Viscosity.*** The internal property of a substance that offers resistance to flow.

***Volcanic crater.*** A funnel-shaped depression from which gases, fragments of rock, and lava are ejected.

***Volcano.*** The vent from which molten igneous matter, solid rock debris, and gases are erupted.

***Water cycle.*** The cyclic movement of water substance through evaporation, wind transport, stream flow, percolation, and related processes. (Also called ***hydrologic cycle***.)

***Water gap.*** A pass, in a ridge or mountain, through which a stream flows.

***Water table.*** The upper surface of the saturated zone.

***Wave-built terrace.*** The body of wave-washed sediment that extends seaward from the breakers.

***Wave-cut bench.*** A bench or platform cut across bedrock by surf.

***Wave-cut cliff.*** A coastal cliff cut by surf.

***Wave refraction.*** The process by which the direction of a series of waves, moving in shallow water at an angle to the shoreline, is changed.

***Weathering.*** The chemical alteration and mechanical breakdown of rock materials during exposure to air, moisture, and organic matter.

***Zone of aeration.*** The zone in which the open spaces in regolith or bedrock are normally filled mainly with air.

***Zone of saturation.*** The subsurface zone in which all openings are filled with water.

# INDEX

Numbers of pages on which terms are defined are in **boldface.** Asterisks indicate illustrations.

A indicates page is in the Appendices.
G indicates page is in the Glossary.

PLATE B
MORPHOLOGY OF EARTH'S CRUST
(Based on diagrams by B. C. Heezen, M. C. Tharp and other sources)
Lands
Sea floor of the continental shelves
Major ice sheets
Sea floor beyond the continental shelves
Mercator projection
120°
150°
180°
150°
120°
90°
80°
75°
70°
60°
45°
30°
15°
0°
15°
30°
45°
60°
70°
ALEUTIAN TRENCH
KURILE TRENCH
EMPEROR SEAMOUNTS
JAPAN TR.
MENDOCINO FRACTURE ZONE
MARIANA TRENCH
MINDANAO TR.
CLARION FRACTURE ZONE
GALAPAGOS FRACTURE ZONE
TONGA-KERMADEC TRENCH
EAST PACIFIC RISE
PERU-CHILE TRENCH